D1665793

Edited by
Vladislav V. Kharton

**Solid State
Electrochemistry II**

Related Titles

Kharton, V. V. (ed.)

Solid State Electrochemistry I:

Fundamentals, Materials and their Applications

2009

ISBN: 978-3-527-32318-0

Vielstich, W., Gasteiger, H. A., Yokokawa, H. (eds.)

Handbook of Fuel Cells

Advances in Electrocatalysis, Materials, Diagnostics and Durability, Volumes 5 & 6

2009

ISBN: 978-0-470-72311-1

Endres, F., MacFarlane, D., Abbott, A. (eds.)

Electrodeposition from Ionic Liquids

2008

ISBN: 978-3-527-31565-9

Bard, A. J., Stratmann, M., Gileadi, E., Urbakh, M., Calvo, E. J., Unwin, P. R., Frankel, G. S., Macdonald, D., Licht, S., Schäfer, H. J., Wilson, G. S., Rubinstein, I., Fujihira, M., Schmuki, P., Scholz, F., Pickett, C. J., Rusling, J. F. (eds.)

Encyclopedia of Electrochemistry

11 Volume Set

2007

ISBN: 978-3-527-30250-5

Hamann, C. H., Hamnett, A., Vielstich, W.

Electrochemistry

2007

ISBN: 978-3-527-31069-2

Staikov, G. T. (ed.)

Electrocrystallization in Nanotechnology

2007

ISBN: 978-3-527-31515-4

Edited by
Vladislav V. Kharton

Solid State Electrochemistry II

Electrodes, Interfaces and Ceramic Membranes

WILEY-VCH

WILEY-VCH Verlag GmbH & Co. KGaA

The Editor

Vladislav V. Kharton
University of Aveiro
CICECO, Dept. of Ceramics and Glass Engin.
3810-193 Aveiro
Portugal

All books published by **Wiley-VCH** are carefully produced. Nevertheless, authors, editors, and publisher do not warrant the information contained in these books, including this book, to be free of errors. Readers are advised to keep in mind that statements, data, illustrations, procedural details or other items may inadvertently be inaccurate.

Library of Congress Card No.: applied for

British Library Cataloguing-in-Publication Data
A catalogue record for this book is available from the British Library.

Bibliographic information published by the Deutsche Nationalbibliothek
The Deutsche Nationalbibliothek lists this publication in the Deutsche Nationalbibliografie; detailed bibliographic data are available on the Internet at http://dnb.d-nb.de.

© 2011 Wiley-VCH Verlag & Co. KGaA, Boschstr. 12, 69469 Weinheim, Germany

All rights reserved (including those of translation into other languages). No part of this book may be reproduced in any form – by photoprinting, microfilm, or any other means – nor transmitted or translated into a machine language without written permission from the publishers. Registered names, trademarks, etc. used in this book, even when not specifically marked as such, are not to be considered unprotected by law.

Typesetting Thomson Digital, Noida, India
Printing and Binding betz-druck GmbH, Darmstadt
Cover Design Grafik-Design Schulz, Fußgönheim

Printed in the Federal Republic of Germany
Printed on acid-free paper

ISBN Print: 978-3-527-32638-9
ISBN oBook: 978-3-527-63556-6
ISBN ePDF: 978-3-527-63558-0
ISBN ePub: 978-3-527-63557-3
ISBN Mobi: 978-3-527-63559-7

Contents

Preface *XV*
List of Contributors *XIX*

1	**Ionic Memory Technology** *1*	
	An Chen	
1.1	Introduction *1*	
1.2	Ionic Memory Switching Mechanisms *4*	
1.2.1	Cation-Based Resistive Switching Mechanism *4*	
1.2.2	Anion-Based Resistive Switching Mechanism *6*	
1.3	Materials for Ionic Memories *8*	
1.3.1	Metal Sulfide Solid Electrolytes *9*	
1.3.2	Ge-Based Chalcogenide Solid Electrolytes *9*	
1.3.3	Oxide Solid Electrolytes *10*	
1.3.4	Miscellaneous Other Solid Electrolytes *11*	
1.4	Electrical Characteristics of Ionic Memories *12*	
1.4.1	Ionic Memory Device Characteristics *12*	
1.4.2	Ionic Memory Array Characteristics *18*	
1.4.3	Comparing Ionic Memory with Other Memories *19*	
1.5	Architectures for Ionic Memories *20*	
1.5.1	CMOS-Integrated Architecture *20*	
1.5.2	Crossbar Array Architecture *21*	
1.5.3	CMOS/Hybrid Architecture *21*	
1.6	Challenges of Ionic Memories *22*	
1.6.1	Overprogramming and Overerasing *22*	
1.6.2	Random Diffusion of Metal Ions and Atoms *22*	
1.6.3	Thermal Stability *23*	
1.6.4	Switching Speed *23*	
1.6.5	Degradation of Inert Electrodes *24*	
1.7	Applications of Ionic Memories *24*	
1.7.1	Stand-Alone Memory *25*	
1.7.2	Embedded Memory *25*	
1.7.3	Storage Class Memory *25*	

1.8	Summary 26
	References 26

2	**Composite Solid Electrolytes** 31
	Nikolai F. Uvarov
2.1	Introduction 31
2.2	Interface Interactions and Defect Equilibria in Composite Electrolytes 32
2.2.1	Defect Thermodynamics: Free Surface of an Ionic Crystal 33
2.2.2	Defect Thermodynamics in Composites: Interfaces 37
2.2.3	Thermodynamic Stability Criteria and Surface Spreading 39
2.3	Nanocomposite Solid Electrolytes: Grain Size Effects 42
2.4	Ionic Transport 50
2.5	Other Properties 55
2.6	Computer Simulations 56
2.7	Design of the Composite Solid Electrolytes: General Approaches and Perspectives 58
2.7.1	Varying Chemical Nature of the Ionic Salt–Oxide Pair 58
2.7.2	Changing the Physical State of Ionic Salt 60
2.7.3	Changing the Physical State of the "Inert" Component 61
2.7.4	Alteration of the Inert Component Morphology: The Geometric Aspects 61
2.7.5	Metacomposite Materials and Nanostructured Systems 62
2.7.6	Achieving the Nanoscale Level 62
2.8	Composite Materials Operating at Elevated Temperatures 64
2.9	Conclusions 65
	References 66

3	**Advances in the Theoretical Description of Solid–Electrolyte Solution Interfaces** 73
	Orest Pizio and Stefan Sokołowski
3.1	Introduction 73
3.2	Theoretical Approaches 74
3.2.1	Background 74
3.2.2	Inhomogeneous Electrolyte Solutions: Generalities 77
3.2.3	Integral Equations for Solid–Electrolyte Solution Problems 80
3.2.4	Density Functional Theories 88
3.3	Computer Simulations 100
3.3.1	Summation of Long-Range Forces 101
3.3.2	Simulations of Simple Models for Electric Double Layer 104
3.3.3	Solvent Effects on Electric Double Layer: Simple Models 109
3.3.4	Solvent Effects on Electric Double Layer: Models of Water 112
3.3.5	Summary 116
	References 118

4	**Dynamical Instabilities in Electrochemical Processes** 125	
	István Z. Kiss, Timea Nagy, and Vilmos Gáspár	
4.1	Introduction 125	
4.2	Origin and Classification of Dynamical Instabilities in Electrochemical Systems 126	
4.2.1	Classification Based on Essential Species 127	
4.2.1.1	"Truly" or "Strictly" Potentiostatic Systems 128	
4.2.1.2	Negative Differential Resistance Systems 128	
4.2.2	Classification Based on Nonlinear Dynamics 131	
4.3	Methodology 136	
4.3.1	Experimental Techniques 136	
4.3.1.1	Techniques for Temporal Dynamics 136	
4.3.1.2	Techniques for Spatial Dynamics 136	
4.3.2	Data Processing 138	
4.3.2.1	Digital Signal Processing 139	
4.3.2.2	Analysis of Time Series Data with Tools of Nonlinear Dynamics 140	
4.4	Dynamics 142	
4.4.1	Bistability 143	
4.4.2	Oscillations 144	
4.4.3	Chaos 150	
4.4.4	Bursting 151	
4.4.5	Dynamics of Coupled Electrodes 153	
4.4.6	Dynamics of Pattern Formation 158	
4.5	Control of Dynamics 159	
4.5.1	IR Compensation 160	
4.5.2	Periodic Forcing 160	
4.5.3	Chaos Control 162	
4.5.4	Delayed Feedback and Tracking 163	
4.5.5	Adaptive Control 163	
4.5.6	Synchronization Engineering 165	
4.5.7	Effect of Noise 167	
4.6	Toward Applications 167	
4.7	Summary and Outlook 171	
4.7.1	Electrochemistry of Small Systems 172	
4.7.2	Description of Large, Complex Systems 172	
4.7.3	Engineering Structures 172	
4.7.4	Integration of Electrochemical and Biological Systems 173	
	References 173	
5	**Fuel Cells: Advances and Challenges** 179	
	San Ping Jiang and Xin Wang	
5.1	Introduction 179	
5.1.1	Principle and Classification of Fuel Cells 180	
5.1.2	Fuels for Electrochemical Cells 183	

5.1.2.1	Hydrogen, Methanol, Formic Acid, and Liquid Fuels	183
5.1.2.2	Methane, Hydrocarbons, and Biomass-Derived Renewable Fuels	185
5.2	Alkaline and Alkaline Membrane Fuel Cells	187
5.2.1	Alkaline Fuel Cells	187
5.2.1.1	Historic Development and Prospects	188
5.2.1.2	Gas Diffusion Cathode, Electrolyte Carbonation, and Corrosion	188
5.2.2	Alkaline Membrane Fuel Cells	190
5.2.2.1	Alkaline Anion Exchange Membranes	190
5.2.2.2	Electrocatalysts in Alkaline Medium	192
5.3	Polymer Electrolyte Membrane Fuel Cells	194
5.3.1	Technology Development of PEMFCs	194
5.3.1.1	Nafion and Other Perfluorinated Membranes	195
5.3.1.2	Pt, PtRu, and Nonprecious Metal Composite Electrocatalysts	196
5.3.1.3	Self-Humidification and Electrode Structure Optimization	199
5.3.1.4	Bipolar Plates and Stack Design	200
5.3.1.5	Heat, Water Management, and Modeling	202
5.3.1.6	Durability and Degradation of Electrocatalysts, Membrane and Cell Performance	203
5.3.2	PEMFCs with Liquid Fuels	204
5.3.2.1	Direct Methanol Fuel Cells	205
5.3.2.2	Direct Formic Acid Fuel Cells	208
5.3.2.3	Direct Ethanol Fuel Cells	210
5.4	Phosphoric Acid Fuel Cells and Molten Carbonate Fuel Cells	211
5.4.1	Phosphoric Acid Fuel Cells	211
5.4.1.1	Electrocatalysts, Electrolyte, and Gas Diffusion Electrode	212
5.4.1.2	Sustainability and Challenges	213
5.4.2	Molten Carbonate Fuel Cells	213
5.4.2.1	NiO Cathode	214
5.4.2.2	Anode, Stability and Retention of Electrolyte, Corrosion of Separate Plate, and Component Stability	215
5.4.2.3	Direct Internal Reforming MCFCs	216
5.5	Solid Oxide Fuel Cells	217
5.5.1	Development of Key Engineering Materials	218
5.5.1.1	Anode	218
5.5.1.2	Cathode	222
5.5.1.3	Electrolytes	225
5.5.1.4	Interconnect, Sealing, and Balance of Plant	228
5.5.2	SOFC Structures and Configurations	230
5.5.2.1	Cell Structures	230
5.5.2.2	Stack Design and Configurations	231
5.6	Emerging Fuel Cells	232
5.6.1	Protic Ionic Liquid Electrolyte Fuel Cells	232
5.6.2	Microbial Fuel Cells	232

5.6.3	Biofuel Cells *235*	
5.6.4	Microfluidic Fuel Cells *237*	
5.6.5	High-Temperature Proton Exchange Membrane Fuel Cells *238*	
5.6.5.1	HT-PEMFCs Based on Phosphoric Acid-Doped PBI PEM *238*	
5.6.5.2	HT-PEMFCs Based on Inorganic and Ceramic PEM *239*	
5.6.6	Single-Chamber Solid Oxide Fuel Cells *241*	
5.6.7	Microsolid Oxide Fuel Cells *242*	
5.6.8	Direct Carbon Fuel Cells *246*	
5.7	Applications of Fuel Cells *247*	
5.8	Final Remarks *249*	
	References *252*	
6	**Electrodes for High-Temperature Electrochemical Cells: Novel Materials and Recent Trends** *265*	
	Ekaterina V. Tsipis and Vladislav V. Kharton	
6.1	Introduction *265*	
6.2	General Comments *266*	
6.3	Novel Cathode Materials for Solid Oxide Fuel Cells: Selected Trends and Compositions *267*	
6.4	Oxide and Cermet SOFC Anodes: Relevant Trends *286*	
6.5	Other Fuel Cell Concepts: Single-Chamber, Micro-, and Symmetrical SOFCs *301*	
6.6	Alternative Fuels: Direct Hydrocarbon and Direct Carbon SOFCs *309*	
6.7	Electrode Materials for High-Temperature Fuel Cells with Proton-Conducting Electrolytes *312*	
6.8	Electrolyzers, Reactors, and Other Applications Based on Oxygen Ion- and Proton-Conducting Solid Electrolytes *317*	
6.9	Concluding Remarks *321*	
	References *322*	
7	**Advances in Fabrication, Characterization, Testing, and Diagnosis of High-Performance Electrodes for PEM Fuel Cells** *331*	
	Jinfeng Wu, Wei Dai, Hui Li, and Haijiang Wang	
7.1	Introduction *331*	
7.2	Advanced Fabrication Methods for High-Performance Electrodes *333*	
7.2.1	Conventional Hydrophobic PTFE-Bonded Gas Diffusion Electrode *334*	
7.2.1.1	Catalyst Layer Coating onto Gas Diffusion Layer *334*	
7.2.1.2	Nafion Impregnation *335*	
7.2.1.3	MEA Assembly: Hot Pressing *336*	
7.2.2	Thin-Film Hydrophilic Catalyst Coated Membrane Method *336*	
7.2.2.1	Screen Printing Method *337*	
7.2.2.2	Modified Decal Method *338*	
7.2.2.3	Inkjet Printing Technology *338*	

7.2.2.4	Direct Spray Coating 339
7.2.3	Advanced Fabrication Methods for Low Pt Loading Electrodes 340
7.2.3.1	Electrophoretic Deposition 340
7.2.3.2	Pulse Electrodeposition 342
7.2.3.3	Electrospray 345
7.2.3.4	Plasma Sputtering 345
7.2.3.5	Sol-Gel Method 346
7.2.4	Fabrication of Gas Diffusion Media 347
7.2.4.1	Fabrication of Gas Diffusion Layer 347
7.2.4.2	Fabrication of Microporous Layer 347
7.3	Characterization of PEM Fuel Cell Electrodes 348
7.3.1	Surface Morphology Characteristics 348
7.3.1.1	Optical Microscopy 348
7.3.1.2	Scanning Electron Microscopy 349
7.3.1.3	Scanning Probe Microscopy 352
7.3.2	Microstructure Analysis 354
7.3.3	Physical Characteristics 356
7.3.3.1	Porosimetry 356
7.3.3.2	Permeability and Gas Diffusivity 357
7.3.3.3	Wetting 358
7.3.3.4	Conductivity 360
7.3.4	Composition Analysis 361
7.3.4.1	Energy Dispersive Spectrometry 361
7.3.4.2	Thermal Gravimetric Analysis 362
7.4	Testing and Diagnosis of PEM Fuel Cell Electrodes 364
7.4.1	Electrochemical Techniques 364
7.4.1.1	Polarization Curves 364
7.4.1.2	Current Interrupt 365
7.4.1.3	Electrochemical Impedance Spectroscopy 366
7.4.1.4	Other Electrochemical Methods 369
7.4.2	Physical and Chemical Methods 371
7.4.2.1	Species Distribution Mapping 371
7.4.2.2	Temperature Distribution Mapping 373
7.4.2.3	Current Distribution Mapping 374
7.5	Final Comments 377
	References 378
8	**Nanostructured Electrodes for Lithium Ion Batteries** 383
	Ricardo Alcántara, Pedro Lavela, Carlos Pérez-Vicente, and José L. Tirado
8.1	Introduction 383
8.2	Positive Electrodes: Nanoparticles, Nanoarchitectures, and Coatings 384
8.2.1	Layered Oxides: $LiMO_2$ 384
8.2.2	Spinel Compounds 385

8.2.2.1	Materials for 4 V Electrodes	385
8.2.2.2	Materials for Higher Voltages	386
8.2.3	Olivine Phosphates	387
8.2.4	Silicates	390
8.2.5	Transition Metal Fluorides	392
8.3	Negative Electrodes	393
8.3.1	Intercalation Nanomaterials	393
8.3.1.1	Graphene, Fullerene, and Carbon Nanotubes	393
8.3.1.2	Titanium Oxides	396
8.3.2	Intermetallic Compounds	397
8.3.2.1	Tin and Tin Composites	397
8.3.2.2	Silicon	400
8.3.3	Nanomaterials Obtained *In Situ*	401
8.3.3.1	Binary Oxides for Conversion Electrodes	401
8.3.3.2	Multinary Oxides	404
8.3.3.3	Transition Metal Oxysalts	404
8.4	Concluding Remarks	406
	References	407
9	**Materials Science Aspects Relevant for High-Temperature Electrochemistry**	**415**
	Annika Eriksson, Mari-Ann Einarsrud, and Tor Grande	
9.1	Introduction	415
9.2	Powder Preparation, Forming Processes, and Sintering Phenomena	416
9.2.1	Powder Processing and Forming Techniques	416
9.2.2	Densification, Grain Growth, and Pore Coalescence	420
9.2.3	Sintering of Oxide Electrolytes and Ceramic Membrane Materials	424
9.3	Cation Diffusion	426
9.3.1	Theoretical Aspects of Cation Diffusion	426
9.3.2	Grain Boundary and Bulk Diffusion	429
9.3.3	Experimental Methods for Determination of Cation Diffusion	430
9.3.4	Cation Diffusion in Perovskite and Fluorite Oxide Materials	433
9.3.5	Kinetic Demixing and Decomposition	434
9.4	Thermomechanical Stability	437
9.4.1	Thermal Expansion of Oxide Electrolyte and Mixed Conducting Ceramics	437
9.4.2	Chemical Expansion	439
9.4.3	Mechanical Properties	440
9.4.4	Degradation Due to Fracture	442
9.4.5	Chemical Compatibility of Materials	443
9.4.6	High-Temperature Creep	444

9.5	Thermodynamic Stability of Materials 447
9.5.1	Phase Decomposition and Solid-State Transformation 447
9.5.2	Reactions with Gaseous Species 450
9.5.3	Volatilization of Components 453
	References 454

10 Oxygen- and Hydrogen-Permeable Dense Ceramic Membranes 467
Jay Kniep and Jerry Y.S. Lin

10.1	Introduction 467
10.2	Structure of Membrane Materials 468
10.2.1	Fluorite Structure 468
10.2.2	Perovskite Structure 469
10.2.3	$Sr_4Fe_6O_{13\pm\delta}$ and Its Derivatives 470
10.2.4	Brownmillerite and Other Perovskite-Related Structures 470
10.2.5	Dual-Phase Membranes 471
10.3	Synthesis and Permeation Experimental Methods 471
10.4	Gas Permeation Models 473
10.5	Characteristics of Oxygen-Permeable Membranes 476
10.5.1	Electrical and Ionic Transport Properties 476
10.5.2	Oxygen Permeation under Air/Inert Gas Gradients 479
10.5.3	Oxygen Permeation in Reducing Gases 482
10.5.4	Membrane Stability and Mechanical Properties 483
10.6	Characteristics of Hydrogen-Permeable Membranes 484
10.6.1	Electrical and Ionic Transport Properties 484
10.6.2	Hydrogen Permeation under Oxidizing Conditions 485
10.6.3	Hydrogen Permeation in Inert Sweep Gases 487
10.6.4	Membrane Stability 489
10.6.5	Comparison with Oxygen-Permeable Membranes 489
10.7	Applications of Membranes 490
10.7.1	Gas Separation and Purification 490
10.7.2	Membrane Reactors 491
10.8	Summary and Conclusions 494
	References 495

11 Interfacial Phenomena in Mixed Conducting Membranes: Surface Oxygen Exchange- and Microstructure-Related Factors 501
Xuefeng Zhu and Weishen Yang

11.1	Introduction 501
11.2	Surface Exchange 503
11.2.1	Theoretical Analysis of Surface Effects: One Relevant Approach 504
11.2.2	Modeling Formulas 510
11.2.3	Selected Experimental Methods 514
11.2.3.1	Isotopic Exchange 514
11.2.3.2	Electrical Conductivity Relaxation 516

11.2.4	Reduction and Elimination of Surface Exchange Limitations *517*	
11.3	Microstructural Effects in Mixed Conducting Membranes *520*	
11.3.1	Selected Experimental Methods *520*	
11.3.2	Microstructural Phenomena in Perovskite-Type Membranes *521*	
11.3.3	Composite Membranes *524*	
11.3.3.1	Perovskite Membranes with Second Phase or Impurities *524*	
11.3.3.2	Dual-Phase Membranes *526*	
11.3.4	Asymmetric Membranes *529*	
11.4	Thermodynamic and Kinetic Stability *530*	
11.4.1	Surface Limitations *530*	
11.4.2	Microstructures and Kinetic Stability *531*	
	References *532*	

Index *541*

Preface

Aiming to combine the fundamental information and brief overview on recent advances in solid-state electrochemistry, this handbook primarily focuses on the most important methodological, theoretical, and technological aspects, novel materials for solid-state electrochemical devices, factors determining their performance and reliability, and their practical applications. Main priority has been given, therefore, to the information that may be of interest to researchers, engineers, and other specialists working in this and closely related scientific areas. At the same time, numerous definitions, basic equations and schemes, and reference data are also included in many chapters to provide necessary introductory information for newcomers to this intriguing field. In general, solid-state electrochemistry is an important, interdisciplinary, and rapidly developing science that integrates many aspects of the classical electrochemical science and engineering, materials science, solid-state chemistry and physics, heterogeneous catalysis, and other areas of physical chemistry. This field comprises, but is not limited to, electrochemistry of solid materials, thermodynamics and kinetics of electrochemical reactions involving at least one solid phase, transport of ions and electrons in solids, and interactions between solid, liquid, and/or gaseous phases whenever these processes are essentially determined by properties of solids and are relevant to the electrochemical reactions. The range of applications includes many types of batteries, fuel cells, and sensors, solid-state electrolyzers and electrocatalytic reactors, ceramic membranes with ionic or mixed ionic–electronic conductivity, accumulators and supercapacitors, electrochromic and memory devices, processing of new materials with improved properties, corrosion protection, electrochemical pumps and compressors, and a variety of other appliances. Although it has been impossible to cover the rich diversity of solid-state electrochemical devices, methods, and processes, the handbook is intended to reflect state-of-the-art in this scientific area, recent developments, and key research trends. The readers looking for more detailed information on specific aspects and applications may refer to the list of recommended literature [1–23] that includes several classical references and recent interdisciplinary and specialized books.

The first volume of the handbook [24], contributed by leading scientists from 11 countries, was centered on the general methodology of solid-state electrochemistry, major groups of solid electrolytes and mixed ionic–electronic conductors, and selected applications of the electrochemical cells. Attention was drawn to the general aspects and perspectives of solid-state electrochemical science and technology (Chapters 1–6), nanostructured solids and electrochemical reactions involving nano- and microparticles in a liquid electrolyte environment (Chapters 4 and 6), insertion electrodes (Chapter 5), superionics and mixed conductors (Chapters 2 and 7–9), polymer and hybrid materials (Chapters 10 and 11), principles of selected solid electrolyte devices such as fuel cells and electrochemical pumps (Chapter 12), and solid-state electrochemical sensors (Chapter 13). The fundamental principles of mixed conducting membrane operation and bulk transport properties of selected single-phase materials were briefly analyzed in Chapters 3, 9 and 12. This volume entitled *Solid State Electrochemistry II: Electrodes, Interfaces, and Ceramic Membranes* continues in these directions, with a major emphasis on the interface- and surface-related processes, electrode materials and reactions, and selected practical applications of ion-conducting solids.

Opening the second volume, Chapter 1 is dedicated to the ionic memory devices and related technologies, an emerging area with new horizons for solid-state electrochemistry. Chapter 2 presents an overview of composite solid electrolytes, a separate class of ion-conducting materials where the transport properties are essentially governed by interfacial phenomena. Chapters 3 and 4 deal with the key aspects of theoretical description and analysis of surface and interfacial processes, started in the first volume, again with a special attention on methodology and modeling. Chapter 5 provides an exhaustive review on the conventional and emerging fuel cell technologies, giving a brief summary on the relevant processes, materials, recent achievements, and future challenges. Continuing this survey, Chapters 6 and 7 are centered on the developments of novel materials and technologies for electrodes of the electrochemical cells with solid oxide electrolytes and polymer electrolyte membranes, while Chapter 8 briefly reviews the nanostructured electrodes for Li-ion batteries. The important aspects of materials science and processing technologies, with numerous examples on solid oxide fuel cells and ceramic membranes, are discussed in Chapter 9. Finally, Chapters 10 and 11 present reviews on the mixed conducting ceramic membranes for gas separation and catalytic reactors, membrane materials, selected models and experimental methods, and interfacial phenomena governing the membrane performance. All the chapters are written by leading international experts from 12 countries, namely, Australia, Canada, China, Hungary, Mexico, Norway, Poland, Portugal, Russia, Singapore, Spain, and the United States. After presenting a brief overview of the handbook, the authors and the editor trust that readers would find the contents useful, interesting, and stimulating.

References

1 Kröger, F.A. (1964) *The Chemistry of Imperfect Crystals*, North-Holland Publishing Company, Amsterdam.
2 Kofstad, P. (1972) *Nonstoichiometry, Diffusion, and Electrical Conductivity of Binary Metal Oxides*, Wiley-Interscience, New York.
3 Geller, S. (ed.) (1977) *Solid Electrolytes*, Springer, Berlin.

4. Takahashi, T. and Kozawa, A. (eds) (1980) *Applications of Solid Electrolytes*, JEC Press, Cleveland, OH.
5. Rickert, H. (1982) *Electrochemistry of Solids: An Introduction*, Springer, Berlin.
6. Chebotin, V.N. (1989) *Chemical Diffusion in Solids*, Nauka, Moscow.
7. Bruce, P.G. (ed.) (1995) *Solid State Electrochemistry*, Cambridge University Press, Cambridge.
8. Gellings, P.J. and Bouwmeester, H.J.M. (eds) (1997) *Handbook of Solid State Electrochemistry*, CRC Press, Boca Raton, FL.
9. Allnatt, A.R. and Lidiard, A.B. (2003) *Atomic Transport in Solids*, Cambridge University Press, Cambridge.
10. Bard, A.J., Inzelt, G., and Scholz, F. (eds) (2008) *Electrochemical Dictionary*, Springer, Berlin.
11. West, A.R. (1984) *Solid State Chemistry and Its Applications*, John Wiley & Sons, Ltd, Chichester.
12. Goto, K.S. (1988) *Solid State Electrochemistry and Its Applications to Sensors and Electronic Devices*, Elsevier, Amsterdam.
13. Schmalzried, H. (1995) *Chemical Kinetics of Solids*, Wiley-VCH Verlag GmbH, Weinheim.
14. Munshi, M.Z.A. (ed.) (1995) *Handbook of Solid State Batteries and Capacitors*, World Scientific, Singapore.
15. Vayenas, C.G., Bebelis, S., Pliangos, C., Brosda, S., and Tsiplakides, D. (2001) *Electrochemical Activation of Catalysis: Promotion, Electrochemical Promotion, and Metal-Support Interaction*, Kluwer/Plenum, New York.
16. Alkire, Richard C. and Kolb, Dieter M. (eds) (2002) *Advances in Electrochemical Science and Engineering*, vol. 8, Wiley-VCH Verlag GmbH, Weinheim.
17. Hoogers, G. (ed.) (2003) *Fuel Cell Technology Handbook*, CRC Press, Boca Raton, FL.
18. Wieckowski, A., Savinova, E.R., and Vayenas, C.G. (eds) (2003) *Catalysis and Electrocatalysis at Nanoparticle Surfaces*, Marcel Dekker, New York.
19. Balbuena, P.B. and Wang, Y. (eds) (2004) *Lithium-Ion Batteries: Solid-Electrolyte Interphase*, Imperial College Press, London.
20. Sammes, N. (ed.) (2006) *Fuel Cell Technology: Reaching Towards Commercialization*, Springer, London.
21. Monk, P.M.S., Mortimer, R.J., and Rosseinsky, D.R. (2007) *Electrochromism and Electrochromic Devices*, 2nd edn, Cambridge University Press, Cambridge.
22. Zhuiykov, S. (2007) *Electrochemistry of Zirconia Gas Sensors*, CRC Press, Boca Raton, FL.
23. Li, K. (2007) *Ceramic Membranes for Separation and Reaction*, John Wiley & Sons, Ltd, Chichester.
24. Kharton, V. (ed.) (2009) *Solid State Electrochemistry I: Fundamentals, Materials and Their Applications*, Wiley-VCH Verlag GmbH, Weinheim.

Vladislav V. Kharton
University of Aveiro, Portugal

List of Contributors

Ricardo Alcántara
Universidad de Córdoba
Laboratorio de Química Inorgánica
Edificio Marie Curie C3
Campus de Rabanales
14071 Córdoba
Spain

An Chen
GlobalFoundries
Apt. P6, 260 N. Mathilda Avenue
Sunnyvale, CA 94086
USA

Wei Dai
National Research Council Canada
Institute for Fuel Cell Innovation
4250 Wesbrook Mall
Vancouver, BC
Canada V6T 1W5

Mari-Ann Einarsrud
Norwegian University of Science and Technology
Department of Materials Science and Engineering
7491 Trondheim
Norway

Annika Eriksson
Norwegian University of Science and Technology
Department of Materials Science and Engineering
7491 Trondheim
Norway

Vilmos Gáspár
University of Debrecen
Institute of Physical Chemistry
P.O. Box 7
4010 Debrecen
Hungary

Tor Grande
Norwegian University of Science and Technology
Department of Materials Science and Engineering
7491 Trondheim
Norway

San Ping Jiang
Curtin University of Technology
Department of Chemical Engineering
Curtin Centre for Advanced Energy Science and Engineering
1 Turner Avenue
Perth, WA 6845
Australia

List of Contributors

Vladislav V. Kharton
University of Aveiro
CICECO, Department of Ceramics and
Glass Engineering
Campus de Santiago, 3810–193 Aveiro
Portugal

István Z. Kiss
Saint Louis University
Department of Chemistry
3501 Laclede Avenue
St. Louis, MO 63103
USA

Jay Kniep
Arizona State University
Department of Chemical Engineering
Engineering Center
G Wing 301
Tempe, AZ 85287–6006
USA

Pedro Lavela
Universidad de Córdoba
Laboratorio de Química Inorgánica
Edificio Marie Curie C3
Campus de Rabanales
14071 Córdoba
Spain

Hui Li
National Research Council Canada
Institute for Fuel Cell Innovation
4250 Wesbrook Mall
Vancouver, BC
Canada V6T 1W5

Jerry Y.S. Lin
Arizona State University
Department of Chemical Engineering
Engineering Center
G Wing 301
Tempe, AZ 85287-6006
USA

Timea Nagy
Saint Louis University
Department of Chemistry
3501 Laclede Avenue
St. Louis, MO 63103
USA

Carlos Pérez-Vicente
Universidad de Córdoba
Laboratorio de Química Inorgánica
Edificio Marie Curie C3
Campus de Rabanales
14071 Córdoba
Spain

Orest Pizio
Universidad Nacional Autónoma de
México
Instituto de Química
Coyoacán
04510 México, DF
Mexico

Stefan Sokołowski
Maria Curie-Sklodowska University
Department for Modelling of
Physicochemical Processes
Gliniana 33, 20031 Lublin
Poland

José L. Tirado
Universidad de Córdoba
Laboratorio de Química Inorgánica
Edificio Marie Curie C3
Campus de Rabanales
14071 Córdoba
Spain

Ekaterina V. Tsipis
Instituto Tecnológico e Nuclear
Estrada Nacional 10
2686-953 Sacavém
Portugal

Nikolai F. Uvarov
Institute of Solid State Chemistry and
Mechanochemistry
Siberian Branch of the Russian
Academy of Sciences
Kutateladze 18
630128 Novosibirsk
Russia

Haijiang Wang
National Research Council Canada
Institute for Fuel Cell Innovation
4250 Wesbrook Mall
Vancouver, BC
Canada V6T 1W5

Xin Wang
Nanyang Technological University
School of Chemical and Biomedical
Engineering
639798 Singapore
Singapore

Jinfeng Wu
National Research Council Canada
Institute for Fuel Cell Innovation
4250 Wesbrook Mall
Vancouver, BC
Canada V6T 1W5

Weishen Yang
Chinese Academy of Sciences
Dalian Institute of Physical Chemistry
State Key Laboratory of Catalysis
457 Zhongshan Road
116023 Dalian
China

Xuefeng Zhu
Chinese Academy of Sciences
Dalian Institute of Physical Chemistry
State Key Laboratory of Catalysis
457 Zhongshan Road
116023 Dalian
China

1
Ionic Memory Technology
An Chen

Ionic memory devices based on ion migration and electrochemical reactions have shown promising characteristics for next-generation memory technology. Both cations (e.g., Cu^+, Ag^+) and anions (e.g., O^{2-}) may contribute to a bipolar resistive switching phenomenon that can be utilized to make nonvolatile memory devices. With simple two-terminal structures, these devices can be integrated into CMOS (complementary metal–oxide–semiconductor) architecture or fabricated with novel architectures (e.g., crossbar arrays or 3D stackable memory). Large memory arrays made with standard CMOS process have been demonstrated in industry R&D. Although ionic memory technology has seen significant progress recently, some challenges still exist in device reliability and controllability. Ionic memories may present a promising candidate for stand-alone and storage class memory applications.

1.1
Introduction

With flash memories quickly approaching their scaling limit, numerous novel memory technologies have emerged as candidates for next-generation nonvolatile memories. Examples include phase change memory (PCM), magnetic random access memory (MRAM), ferroelectric RAM (FeRAM), resistive switching memory (also known as RRAM or resistive random access memory), polymer-based memory, molecular memory, and so on [1–3]. Figure 1.1 shows a classification of various memories presented by the International Technology Roadmap of Semiconductor (ITRS) [1]. Static random access memory (SRAM) and dynamic random access memory (DRAM) are called "volatile" memories because information stored in these memories cannot be retained when power is turned off. On the other hand, nonvolatile memories are able to retain information for a long period of time after power is turned off. A typical requirement of data retention is 10 years at room temperature. The mainstream nonvolatile memory in the market today is flash memory, which is divided into two categories, NAND and NOR, based on two

Solid State Electrochemistry II: Electrodes, Interfaces and Ceramic Membranes.
Edited by Vladislav V. Kharton.
© 2011 Wiley-VCH Verlag GmbH & Co. KGaA. Published 2011 by Wiley-VCH Verlag GmbH & Co. KGaA.

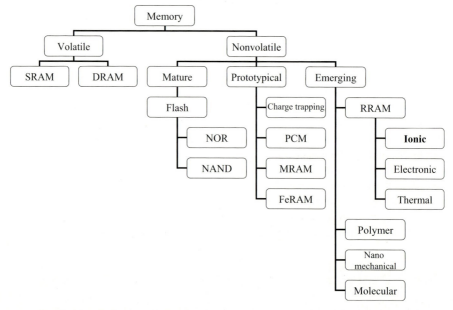

Figure 1.1 Classification of memory technologies based on information in ITRS roadmap. The abbreviations are explained in the text.

different memory architectures. NOR flash memories are preferred for code storage because of their random access capability, while NAND flash memories are more suitable for data storage due to their sequential access in a block of data. These three types of mature memory technologies – SRAM, DRAM, and flash memory – are all based on Si complementary metal–oxide–semiconductor (CMOS) technology. Their development has followed the so-called Moore's law; that is, the transistor density in the integrated circuits has doubled approximately every 2 years. This is achieved by shrinking the size of the Si CMOS transistors, a trend that has successfully continued for several decades. When transistor size is reduced, not only more bits of information can be stored on the same area but also better device/circuit performance can be achieved, for example, fast speed and lower power consumption.

However, with the transistor size being reduced to 22 nm and below, Si CMOS technology today is facing some fundamental challenges. Although power consumed by each transistor has decreased with scaling, the overall power density of the wafers has increased because of growing transistor density. Increasing power density induces more Joule heating and raises wafer temperature, which degrades transistor performance. The wafer temperature today is reaching the limit of practical cooling techniques, constraining further scaling of transistor size. Although it is believed that Si CMOS technology can be scaled down to 22 nm, it is not clear how much further it can go. These mainstream memory technologies based on Si CMOS are facing the same obstacles. Therefore, it has become increasingly important to explore alternative nonvolatile memory technologies that may potentially replace Si-based memories when they reach their limits.

1.1 Introduction

Among these emerging memory devices, resistive switching memory (i.e., RRAM) is a broad category involving a large variety of materials and switching characteristics. These memory devices are usually made in a two-terminal metal–insulator–metal (MIM) structure. They can be electrically switched between a high-resistance state (HRS) and a low-resistance state (LRS), and both states can be nonvolatile. Commonly used terminology refers LRS as the on state and HRS as the off state. Binary digital data can be recorded in these resistance states, for example, LRS for logic "0" and HRS for logic "1." The HRS-to-LRS switching is called "program" (or "write," "set") and the LRS-to-HRS switching "erase" (or "reset"). Promising characteristics have been reported on these devices. However, in many reports of resistive switching memories the switching mechanisms are not clearly understood. Some hints of the switching mechanism may be found in switching characteristics, for example, current–voltage ($I–V$) relationships, the voltage polarity dependence of switching, the presence or absence of "forming" processes, the effect of the electrodes on the switching properties, device size dependence, temperature effect, variation in transport properties, cycling stability, and so on. Unfortunately, systematic study on all these aspects is still lacking for many resistive switching materials, and controversial interpretations of the switching mechanism are widely presented in the literature.

A coarse-grained classification has been proposed to divide resistive switching memories into three types based on the nature of the dominating switching processes: *electronic effect*, *thermal effect*, and *ionic effect*. In electronic effect resistive switching memories, some electronic processes (e.g., charge trapping, Mott metal–insulator transition, or ferroelectric polarization reversal) alter the band structure and transport properties in the bulk or at the interface and trigger resistance changes. Thermal effect resistive switching memories are related to electric power-induced Joule heating and often involve the formation and rupture of some localized conduction paths in an insulating material.

The third type, ionic effect resistive switching memories, involves the transport and electrochemical reactions of cations (e.g., Ag^+, Cu^+) or anions (e.g., O^{2-}). The switching is usually bipolar; that is, programming and erasing are in opposite voltage polarities. This is because the switching between LRS and HRS is realized by driving charged ions in opposite directions to induce different electrochemical reactions. The switching process related to the migration and reaction of *cations* is well understood, and the switching process can be captured in microscopic observations. However, resistive switching process involving *anions* is less well understood, with many open questions regarding the details of the anion transport and electrochemical redox reactions. The discussion of ionic memories in this chapter will mainly focus on these memory devices based on cation migration and reactions. The resistive switching mechanisms and materials involving anions, mainly oxygen ions or vacancies, will also be briefly discussed.

This chapter is organized in the following sections. Section 1.2 discusses the ionic resistive switching mechanisms, followed by a review of materials used in these devices in Section 1.3. Electrical characteristics of ionic memories, including individual device properties and memory array statistics, are summarized in

Section 1.4. Section 1.5 addresses issues in the architecture design of ionic memories. In Section 1.6, challenges of ionic memories are discussed. Section 1.7 provides some information of potential applications of ionic memories. Finally, the chapter ends with a brief summary in Section 1.8.

1.2
Ionic Memory Switching Mechanisms

The switching mechanisms of cation-based devices and anion-based memories are different [3]. The resistance change in the *cation-based* devices is due to the electrochemical formation and dissolution of metallic filaments, which can be observed in sufficiently large devices in well-designed experiments [4–10]. For *anion-based* memories, the switching is generally believed to be triggered by the transport of oxygen ions/vacancies and some redox processes; however, the exact process is still not clear. In some cases, it is even unclear which ions are involved in the switching process and whether the device falls into the category of anion-based ionic memories.

1.2.1
Cation-Based Resistive Switching Mechanism

In the MIM structure of cation-based ionic memories, one of the two electrodes is made of electrochemically active materials and the other electrode is inert (e.g., Au or Pt). In most reported cation-based ionic memory devices, the active electrode is either Ag or Cu. The two electrodes are separated by a solid-state electrolyte "I" layer, in which cations can transport with the mobility much higher than that in regular solid-state materials. These solid electrolytes are sometimes called superionic materials (see Chapters 2 and 7 of the first volume). A typical switching process is illustrated in Figure 1.2. The solid-state electrolytes normally have high resistance initially and are considered to be insulators (Figure 1.2a). When positive voltage is applied to the active electrode, as the active electrode (acting as "anode" in this voltage configuration) is made of electrochemically active materials, metal atoms of the anode are oxidized and dissolved into the solid electrolyte. These metal cations migrate toward the cathode under electrical field and are reduced there. Therefore, under electrical field, oxidation and reduction reactions take places at the anode and cathode, respectively: $M^+ + e^- \underset{\text{oxidation}}{\overset{\text{reduction}}{\rightleftarrows}} M$. The reduced metal atoms form metal filaments that grow from the cathode toward the anode. When the anode and cathode are connected by complete metal filament(s), the MIM device switches from HRS to LRS (Figure 1.2b). When voltage polarity on the two electrodes is reversed, metal atoms dissolve at the edge of the metal filament(s) and eventually break the conductive filament(s) between the anode and the cathode. Current-induced Joule heating may also contribute to the rupture of the filament(s). Consequently, the MIM device is switched back to a high-resistance state (Figure 1.2c). Note that the metal filament(s)

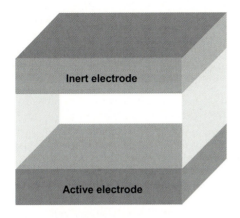
(a) Initial state

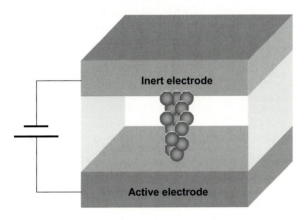
(b) ON state

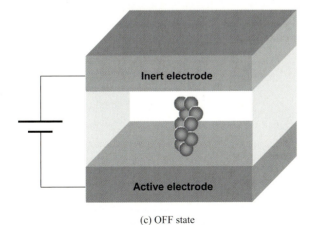
(c) OFF state

Figure 1.2 Schematic illustration of cation-based ionic resistive switching process.

need to be broken only partially to cause significant increase in resistance, and it is not necessary to completely annihilate the metal filament(s) from the "I" layer. As a result, the HRS off state during repeated switching processes may not be as insulating as the original state; however, sufficiently high on/off ratio between the LRS and the HRS can still be achieved. When the active electrode is positively biased again, the metal filament(s) can be "repaired" by the same cation migration and redox reactions. The HRS-to-LRS and LRS-to-HRS switching processes can be repeated continuously. Since the electrochemical reactions ideally do not cause significant damage to the MIM structure, the switching process may in principle work for many cycles.

From the switching process described above, it is clear that cation-based ionic memories have to be bipolar; that is, programming and erasing processes have to be done with opposite voltage polarities. This bipolar switching is considered one of the signatures of ionic memories. Another key feature of ionic memories is the localized conduction path of metal filament(s), which has been suggested as an evidence of excellent scalability of ionic memories. In principle, ionic memory devices can be made as small as one atomic chain of metal atoms. It is also easy to understand that the formation of these filaments is a self-limiting process. As soon as one conductive filament is formed, resistance of the "I" layer reduces dramatically, which results in significant decrease in electric field and the chance of forming additional filaments.

Different names have been given to the cation migration-based resistive switching devices, such as "atomic switch" [11], "programmable metallization cell" [7], "nanoBridge" [8], "solid-state electrolyte memory" [12], "conductive bridging RAM" [13], and so on.

1.2.2
Anion-Based Resistive Switching Mechanism

In many oxides, especially transition metal oxides, oxygen ions or vacancies are much more mobile than cations. The migration of oxygen ions and vacancies may introduce redox reaction at the electrode or doping effects in the metal oxides, which may alter the transport properties of the structure and cause resistance changes. Numerous models have been proposed to describe the details of the switching processes involving oxygen ions and vacancies; however, the exact microscopic processes are still not clear. The following are a few examples of resistive switching phenomena that have been suggested to be caused by oxygen ions or vacancies.

A single crystal of 0.2 mol% Cr-doped $SrTiO_3$ is found to switch from an initially insulating state to a conductive state after it is exposed to an electrical field of $10^5 \, V \, cm^{-1}$ for about 30 min, a process known as "conditioning" (or "forming"). High-temperature "hot spots" and high concentration of oxygen vacancies are both found close to the anode, by infrared thermal microscopy and laterally resolved micro-X-ray absorption spectroscopy, respectively. It is suggested that the conditioning process introduces a path of oxygen vacancies in the memory, which provides free carriers in the Ti 3d band and leads to metallic conduction. The Cr dopants play the role of a seed for oxygen vacancies. The resistive switching of the "conditioned" Cr–SrTiO$_3$ is believed to involve a drift of oxygen vacancies along the applied electrical field. When the anode is negatively biased, positively charged oxygen vacancies are attracted to the anode and a low-resistance state is obtained. On the other hand, when a positive bias is applied on the anode, oxygen vacancies retract from the anode and the memory device switches back to a high-resistance state [14]. Another study of resistive switching in SrTiO$_3$ suggests that the switching phenomenon originates from local modulations of the oxygen content and is related to the self-doping capability of the transition metal oxide [15].

The I–V characteristics of a Pt/TiO$_2$/Pt structure can be made rectifying (i.e., current can pass in one direction but not in the opposite direction) by applying a "programming" voltage to one of the Pt electrodes. The polarity of the rectification can be reversed by applying the "programming" voltage in the opposite direction. The proposed mechanism of this "field programmable rectification" involves field-induced motion of oxygen vacancies. Initially, the two Pt electrodes form Schottky barriers at the TiO$_2$ interface, giving the device an insulating state. When a programming voltage is applied, oxygen vacancies are pushed toward the anode and the accumulation of oxygen vacancies increases the doping of TiO$_2$ near the anode, which gradually eliminates the Schottky barrier at this electrode. Consequently, the I–V characteristics become rectifying, dominated by the Schottky barrier at the other electrode. The reversal of the programming voltage may push oxygen vacancies toward the other electrode and the direction of the rectification is eventually reversed. During their transit between electrodes, oxygen vacancies may not lie next to either electrode, which return the device to an insulating state. Therefore, resistive switching between insulating state and a rectifying conduction state can be achieved under well-controlled conditions. Since the motion of oxygen vacancies is thermally activated, it is also possible that the migration of oxygen vacancies is driven by current-induced Joule heating, in addition to the electrical field [16]. In another study of TiO$_2$-based resistive switching devices, a filament model is suggested for the switching mechanism. In this model, the switching between HRS and LRS is explained by the propagation of portions of the filaments close to the anode and cathode, accompanied by the transfer of O^{2-} along the filaments [17].

The resistive switching characteristics of NiO$_x$ are found to be highly sensitive to the oxygen flow ratio during the reactive ion beam sputtering deposition process used to prepare NiO$_x$. The window of resistance change (or on/off ratio) increases with the increase in oxygen flow ratio, which also correlates with the increase in barrier height change between on and off states. It is expected that oxygen ions or vacancies are contributing to the control of the switching process, although their exact functions are not yet clear [18]. Oxygen content has also been found to play crucial role in the conduction and resistive switching of La$_{0.7}$Ca$_{0.3}$MnO$_3$ (LCMO) [19] and Pr$_{0.7}$Ca$_{0.3}$MnO$_3$ (PCMO) [20]. X-ray photoelectron spectroscopy measurement shows that excess oxygen in PCMO introduced by oxygen annealing causes an increase in Mn^{4+} content and hence a change in the Mn^{4+}/Mn^{3+} ratio at the PCMO surface. As a result, the on/off resistance ratio is increased by oxygen annealing of the PCMO thin film [21]. Similarly, the resistive switching characteristics of LCMO are also improved by oxygen annealing [19].

As shown by the examples above, there are different ways that oxygen ions or vacancies may introduce resistance changes in metal oxides. A generic picture of the switching process can be summarized as the following. Driven by electrical field or Joule heating, mobile anions may migrate from the negatively biased electrode to the positively biased electrode. The migration of the anions may introduce electrochemical reactions at the electrode, create doping effects to change carrier density or band structure, or simply alter the properties of some localized conduction paths ("filaments"). Consequently, resistive switching may take place when sufficient change in the transport properties is reached. Further study is required to elucidate the microscopic nature of the processes that causes the change in the transport properties of the metal oxides.

1.3
Materials for Ionic Memories

Resistive switching materials involving anions (oxygen ions or vacancies) are usually metal oxides, especially transition metal oxides. Resistive switching phenomena explained by anion migration and reactions have been demonstrated in various oxides including NiO_x, TiO_2, $SrTiO_3$, LCMO, PCMO, and so on. Properties of these oxides are highly sensitive to composition; therefore, even slight change in oxygen content may trigger large change in conductivity. Since the exact resistive switching mechanisms of these transition metal oxides are not yet clear, it is possible that other physical processes may also contribute to the switching, for example, Joule heating, charge trapping, and so on. The discussion on ionic memory materials in this section will mainly focus on cation-based ionic memories that can be more clearly defined.

In the MIM structure for cation-based ionic memory devices, the key components are the electrochemically active electrode and the solid-state electrolyte. The inert electrode typically uses Pt, Au, or W. The active electrode contributes the mobile ions for the formation of metallic filaments, and the solid electrolyte provides an environment where cations migrate and metal filaments grow. Most ionic memories use Ag or Cu as the active electrode because of the high mobility of Ag^+ or Cu^+ ions. The most commonly used solid-state electrolytes are the sulfides of these elements (i.e., Ag_2S, Cu_2S) and Ge-based chalcogenides (i.e., GeSe, GeS). It is also found that many oxides when combined with Ag or Cu electrodes can work as ionic switching devices. SiO_x and Ta_2O_5 are the two oxides frequently used. Figure 1.3 plots a matrix showing the combinations of these two active electrodes (Ag or Cu) with various solid-state electrolytes [6–11, 22–60]. Each dot in the figure represents an ionic memory device reported in the literature with reference numbers. In these ionic memories with superionic metal sulfides as the solid electrolyte (e.g., Ag_2S), mobile ions (e.g., Ag^+) already exist in the solid electrolyte. However, for the other solid electrolytes, the mobile ions are external to the electrolytes and usually a "forming"

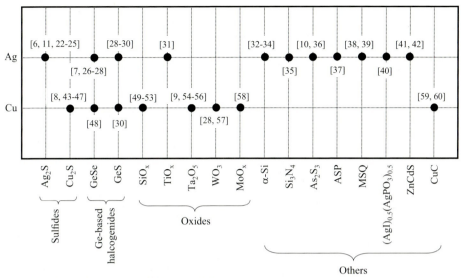

Figure 1.3 Materials for cation-based ionic resistive switching memories.

process is performed to drive these ions into these materials, before stable resistive switching can be achieved.

1.3.1
Metal Sulfide Solid Electrolytes

In a scanning tunneling microscope (STM) experiment using Ag_2S as the tip, it was found that Ag protrusion grew on the top of Ag_2S tip, which was explained by solid electrochemical reaction of Ag ions in Ag_2S [23]. A filamentary shape Ag cluster with the length of ~200 nm and the diameter of ~70 nm could be clearly observed in scanning electron microscope (SEM) [6]. This field-driven growth and rupture of Ag filament inside Ag_2S was then utilized to make a switching device [11, 24, 25]. In Ag/Ag_2S devices, the Ag_2S layer is formed by sulfidization of Ag in a thermal process [11, 24] or an electrochemical process [25]. Similarly, Cu_2S is a natural solid electrolyte for Cu^+ to form Cu/Cu_2S ionic memories. Cu_2S can be made by electrochemically sulfidizing Cu films in Na_2S solution [43] or pulsed laser deposition (PLD) [45, 47].

1.3.2
Ge-Based Chalcogenide Solid Electrolytes

Ag and Cu ions exhibit high mobility and good thermal stability in Ge-based chalcogenide glasses such as GeSe or GeS. For example, AgGeSe is shown to take the form of a continuous glassy Ge_2Se_3 host and a dispersed nanoscale Ag_2Se phase, which possesses superionic properties [7]. The metal-rich Ag_2Se phase is a mixed ionic–electronic conductor (MIEC), but the Ge_2Se_3 host material that separates these conductive regions is a good insulator. The overall resistance of the material prior to ionic switching is high. Ag or Cu can be introduced to the base glass (GeSe or GeS) using a photodriven process (exposure to ultraviolet light) with or without high temperature. After this "treatment," bipolar resistive switching can be achieved by applying appropriate voltages. A potential challenge with GeSe-based devices is their thermal stability. Ag-doped GeSe electrolytes cannot tolerate processing conditions beyond 200 °C. It is found that devices made of GeS are more stable than GeSe-based devices. For example, Ag/GeS devices become functional after annealing at 300 °C for 15 min in oxygen ambient [30]. Cu/GeS memory devices can be operated at 125 °C [30].

1.3.3
Oxide Solid Electrolytes

Various oxides can also be used as solid electrolytes in ionic memories. Since deposited SiO_x is a major component of back-end-of-line (BEOL) process in CMOS and Cu has also been widely used in CMOS process, the Cu/SiO_x combination is a particularly desirable structure for ionic switching memories. The SiO_x film can be prepared by physical vapor deposition [49], electron beam (e-beam) evaporation [50], or radio frequency (rf) sputtering [51]. Cu is then introduced into SiO_x film by thermal diffusion at 610 °C, and stable resistive switching can be achieved afterward. The photodiffusion "forming" technique used for Ge chalcogenide materials does not work for Cu/SiO_x devices, which may be explained by the high rigidity of SiO_x network that limits Cu photodiffusion [50]. Unlike e-beam evaporated SiO_x, sputtered SiO_x is found to require no intentional "forming" process and devices can be switched with very low current compliances (e.g., several tens of pA) [51].

Polycrystalline TiO_2 formed by thermal oxidation of Ti films can also act as a solid electrolyte for Ag^+ ions, and no "forming" is needed for stable switching [31]. If tungsten oxide (WO_3) is sufficiently porous, Cu can be introduced into this WO_3 base glass through diffusion by ultraviolet illumination [28]. WO_3 can be grown by oxygen plasma treatment of a tungsten metal electrode or by high-vacuum thermal evaporation from high-purity tungsten oxide source [57]. Devices made of plasma oxidation WO_3 have variable switching voltage and poor retention, whereas devices based on deposited WO_3 show stable switching voltage and good retention, even at high temperature (>125 °C). The difference may be explained by the observation that Cu electrode tends to oxidize in plasma oxidation WO_3 but is more stable in deposited WO_3. Cu/Ta_2O_5 devices also demonstrate bipolar resistive switching. However, because Cu^+ diffusion coefficient is relatively low in Ta_2O_5, it is found that the threshold voltage of Cu/Ta_2O_5 switching devices is higher than that of other switching devices and the data retention time is longer [9, 54]. By using transmission electron microscopy and energy dispersive X-ray spectrometry (EDX), it can be clearly identified that the conductive path consists of Cu metallic islands separated by tunneling barriers [9]. Ta$_2$O$_5$ can be prepared by PLD [9] or sputtering [56]. Another oxide electrolyte for Cu$^+$ is MoO$_x$, prepared by sputtering a 0.5% Cu-doped MoO$_3$ target on Cu substrate at 500 °C [58]. The Cu concentration in the deposited MoO$_x$ film is measured to be more than 5% (10 times higher than the target concentration) because of Cu diffusion from the bottom electrode during high-temperature sputtering. This diffusion process helps to form a Cu-rich MoO$_x$ solid electrolyte for ionic switching.

1.3.4
Miscellaneous Other Solid Electrolytes

Numerous other materials are shown to be good ionic conductors for Ag$^+$ or Cu$^+$, and they are also suitable for solid electrolytes in ionic memory devices. Some examples are discussed in this section.

Ag electrode can be combined with hydrogenated amorphous Si (α-Si) deposited by chemical vapor deposition (CVD) to form ionic memories [33]. In addition to planar device structures, α-Si nanowires (NWs) are also used with Ag metal lines (perpendicular to the direction of the NWs) to form a cross-point array of ionic memories [34]. An interesting observation is the rectifying I–V characteristics in LRS for both planar and NW Ag/α-Si devices, which is rarely found in other ionic memory devices. Si$_3$N$_4$ is another material fully compatible with CMOS processing, and Si$_3$N$_4$ films deposited by plasma enhanced CVD (PECVD) exhibit bipolar resistive switching with Ag electrode [35]. The growth of Ag filaments inside As$_2$S$_3$ films and the polarity-dependent switching were observed more than 30 years ago [10]. In a more recent study, As$_2$S$_3$ film was made by vacuum evaporation and Ag was dissolved into it through photoassisted and thermal-induced doping to make Ag/As$_2$S$_3$ ionic switching devices [36]. Ag$_2$S-doped AgPO$_3$ (ASP) has been studied for applications in solid-state batteries and electrochemical devices. A recent study reports bipolar resistive switching in Ag/ASP devices prepared by PLD, showing potential for memory applications [37]. A similar solid electrolyte is glassy (AgI)$_{0.5}$(AgPO$_3$)$_{0.5}$ that can also be deposited by PLD and works as an ionic memory device with Ag electrode [40]. Spin-on glass methyl silsesquioxane (MSQ) is not only proven to be a functional solid electrolyte for Ag$^+$ but also particularly useful in making nano-crossbar memory structures through UV nanoimprint lithography (UV NIL) [38, 39]. Devices with 64 × 64 bit crossbar memory array and 100 × 100 nm size Si/MSQ have been demonstrated. Resistive switching behavior of an Ag/Zn$_x$Cd$_{1-x}$S device was first attributed to ferroelectric polarization reversal, but later it was suggested that the formation and annihilation of Ag-rich deposits were more likely to be the switching mechanism [42]. Different metals (Ag, Cu, Zn, Au, Ta, and Pt) have been experimentally tested as electrodes on Zn$_x$Cd$_{1-x}$S, and only devices with electrochemical active metals (i.e., Ag, Cu, and Zn) demonstrated bipolar resistive switching behaviors consistent with the ionic switching model [41].

Cu-doped amorphous carbon (CuC) film is another potential candidate for solid electrolyte of Cu$^+$-based ionic memories because of its insulator properties and high Cu diffusivity. The CuC film can be deposited by rf magnetron reactive sputtering using a Cu target in Ar, CH_4, and O_2 at room temperature [59, 60]. Here, O_2 acts as a catalyst for the decomposition of CH_4 to provide a source of carbon. After a voltage stress "forming" process, Cu/CuC devices can be electrically switched repeatedly.

1.4
Electrical Characteristics of Ionic Memories

Many ionic memory devices have been characterized on individual device level. Device-level characterization is relatively easy to implement and helps to understand the switching mechanisms. Large-scale memory arrays have also been fabricated and tested by several companies. Array-level characterization provides important statistics for the evaluation of ionic memory technologies and more accurate parameters of memory performance. For example, parasitics inevitably exist in individual device structures and the measurement setup, which limits the accuracy of switching speed measurement. High-quality memory arrays, especially those fabricated in standard CMOS, are more suitable for high-speed measurement.

1.4.1
Ionic Memory Device Characteristics

The individual device characteristics of various Ag- and Cu-based ionic memories are summarized in Table 1.1. These switching parameters are collected from published papers, and reference numbers are given in the last column. Unless specified, all data in this table are measured at room temperature. It should be noted that many parameters are not accurately identified in these papers; therefore, their values are recorded as ranges or approximations rather than precise numbers. They may also vary in a large range depending on devices and measurement conditions. In addition, some values are not reported in these papers and are left as blank in Table 1.1.

The thickness t_{se} refers to the thickness of the solid-state electrolyte and the diameter D_d is the diameter of the device (usually the size of the smaller electrode or vias where the device is located). Both parameters are given in nanometers. Switching time τ_{sw} is usually reported as the minimum pulse width that can successfully switch devices in alternating current (AC) measurements. This minimum switching pulse width may be limited by equipment capability or external parasitics, so many values are given as the upper limit of the switching time. The intrinsic switching speed may be much faster than the measured speed. The "on/off ratio" is the resistance ratio between HRS and LRS, measured at typical reading voltages. "Retention" has been measured on many devices, to demonstrate their feasibility for nonvolatile memory applications. A typical retention criterion for nonvolatile memory is 10 years at room temperature. Most retention data summarized in Table 1.1 are direct measurements of the device resistance state over a prolonged period of time. Retention of 10^5 s (28 h) or longer has been

1.4 Electrical Characteristics of Ionic Memories

Table 1.1 Device characteristics of Ag-based and Cu-based ionic memories.

Solid electrolyte	t_{se} (nm)	D_d (nm)	HRS → LRS V_{sw} (V)	HRS → LRS I_{sw} (A)	LRS → HRS V_{sw} (V)	LRS → HRS I_{sw} (A)	τ_{sw} (s)	On/off ratio	Retention (s)	Cycle	References
Ag electrode											
Ag_2S	—	0.1–0.15	<0.4	—	<0.4	—	<10^{-6}	>10^3	—	>10^5	[11, 24, 25]
GeSe	~50	75	<0.5	<10^{-6}	<0.2	<10^{-6}	<10^{-7}	~10^5	—	>10^{11}	[7, 26–28]
GeS	60	240	~0.5	<10^{-5}	~0.3	<10^{-5}	<10^{-7}	~10^6	>10^5	—	[28–30]
TiO_x	20	10^5	~0.3	~10^{-4}	~0.2	~10^{-4}	—	~10^6	10^4	—	[31]
a-Si-nanowire	5	—	~3.0	~10^{-5}	~3.0	—	~10^{-7}	>10^4	>2 weeks	>10^4	[34]
Si_3N_4	50	10^5	0.35–3.5	~10^{-3}	0.3–0.5	~10^{-3}	—	>10^2	>10^4	—	[35]
As_2S_3	—	~10^7	~0.2	~10^{-2}	>1	~10^{-2}	~10^{-5}	~10^2	—	—	[36]
ASP	200	10^3	~2.1	~0.05	~1.25	~0.05	~10^{-4}	~10^4	—	>10^5	[37]
MSQ	130	100	1.6	5×10^{-5}	~0.2	~10^{-5}	~10^{-8}	~10^3	—	>250	[38]
$(AgI)_{0.5}(AgPO_3)_{0.5}$	200	10^3	~0.2	~10^{-2}	~0.1	~10^{-3}	~10^{-7}	~10^6	—	>10^5	[40]
ZnCdS	200	—	~0.5	~10^{-3}	~0.5	~10^{-3}	<10^{-7}	~10^7	—	—	[42]
Cu electrode											
Cu_2S	~40	~30	~0.3	~10^{-3}	~0.1	~10^{-3}	<10^{-4}	~10^6	~0.25 years	>10^3	[8, 43]
GeSe	40	200	~0.6	~10^{-4}	~1.2	~10^{-4}	—	>10^2	~10^5	>10^5	[48]
GeS	60	300	~0.3	~10^{-5}	~0.1	~2×10^{-6}	<10^{-7}	~10^5	—	—	[30]
SiO_x	12	10^3	~1.0	~10^{-6}	~0.2	~10^{-6}	<10^{-6}	~10^3	>10^5	~10^7	[50]
Ta_2O_5	15	200–10^4	~2.5	10^{-4}	~1.0	~10^{-3}	10^{-5}–10^{-4}	~10^4	10 years	10^4	[9, 54]

(Continued)

Table 1.1 (*Continued*)

Solid electrolyte	t_{se} (nm)	D_d (nm)	HRS → LRS		LRS → HRS		τ_{sw} (s)	On/off ratio	Retention (s)	Cycle	References
			V_{sw} (V)	I_{sw} (A)	V_{sw} (V)	I_{sw} (A)					
WO_3	50	240	~0.4	~10^{-6}	~0.2	~10^{-6}	—	>10^4	>10^4	>10^4	[28, 57]
MoO_x	400	—	~0.5	~10^{-1}	~1.5	~10^{-1}	<10^{-6}	~10	>10^5	>10^6	[58]
CuC	500	5000	<1	~10^{-4}	~0.5	~10^{-4}	<10^{-6}	~10^2	>10^4 (85 °C)	>10^3	[59]

Notes:
1) Symbols: t_{se}, thickness of the solid electrolyte; D_d, diameter of the devices; V_{sw}, switching voltage; I_{sw}, switching current; τ_{sw}, switching time. The "HRS → LRS" switching is known as "programming" or "set" and the "LRS → HRS" switching is known as "erasing" or "reset."
2) The switching voltage listed in the table refers to the DC switching voltage; AC switching voltage can be much higher and may vary depending on pulse width.
3) The switching is bipolar, therefore, V_{sw} values have opposite sign for "HRS → LRS" and "LRS → HRS" processes. V_{sw} values in the table are absolute values.
4) Solid electrolytes: ASP = $Ag_{30}S_2P_{14}O_{42}$; MSQ = methyl silsesquioxane.
5) The Cu/WO_3 device here is based on deposited WO_3. Devices based on WO_3 grown by plasma oxidation of W have different characteristics (less stable).

measured on many devices. To confirm 10-year retention in product-level measurements, tests under thermal or electrical stress are needed and 10-year retention can be extrapolated from measured data based on established retention models. Most ionic memory devices are reprogrammable and cycling endurance is an important parameter. Cycling endurance is a critical indicator of stability and reliability. Endurance of 10^4–10^5 cycles has been observed in many devices. Ag/GeSe device has demonstrated the endurance as long as 10^{11} cycles.

The switching voltage and current in Table 1.1 are measured in direct current (DC) operations. In a typical DC measurement, voltage applied to the electrodes is swept through a range, and current compliance is often applied to prevent device damage. Hysteretic loops exist in measured I–V characteristics, indicating memory effects. Voltages required for AC operation are usually much higher than those for DC operation. Some studies have shown that the switching voltage and pulse width strongly depend on each other. For example, the switching time of Ag_2S/Ag devices decreases exponentially with the increase in switching voltage [24]. Although devices can be switched at $\sim$0.2 V with 10 ms wide pulses, switching voltage $\sim$0.4 V is required to operate them with 1 μs. Since the ionic switching process involves the drift of ions and electrochemical reactions, it is reasonable to expect that higher electrical field is needed to accomplish the same ionic switching process in shorter time.

An important characteristic of ionic memories is the low operation voltage. Figure 1.4 plots the "set" switching voltage and current for Ag^+-based devices (a) and Cu^+-based devices (b), using the data in Table 1.1. The labels are the solid electrolytes of these devices. Most devices can be programmed with voltage below 0.5 V, which is significantly lower than the operation voltage of flash memory and also lower than many other emerging memories. Flash memory circuits require special design to provide high voltages, which not only creates an area/design overhead but also causes reliability concerns. Low-voltage operation is an advantage of ionic memories. On the other hand, devices that can be switched with very low voltage (e.g., $\leq$0.2 V) also need to be carefully designed and protected from unintentional electrical shock.

Switching current of ionic memories distributes in a wide range, as shown in Figure 1.4. Low-power operation requires both low voltage and low current. Moreover, high switching current requires large transistors to provide sufficient drive current, diminishing the scalability advantages of ionic memories. It is desirable to control the switching current below 10 μA. It is found that by introducing a thin oxide layer within GeSe chalcogenide film, the switching current for an Ag/GeSe/Pt device (diameter of 2.5 μm) can be reduced to as low as 1 nA [26]. Some studies even claim much lower switching current, for example, 10 pA for Cu/SiO_x devices and 5 pA for Cu/Ta_2O_5 devices [55, 56]. The feasibility of such low operation current needs to be further tested in large memory arrays and product-like operation conditions.

In many ionic memories, the first switching cycle is different from the following cycles, which is also known as the "electroforming" cycle. Switching voltage of Cu/SiO_x devices in the "electroforming" cycle is found to scale linearly with the oxide thickness [53]. This thickness dependence can be explained by the fact that the formation of conductive path(s) requires the drift of ions through the whole thickness

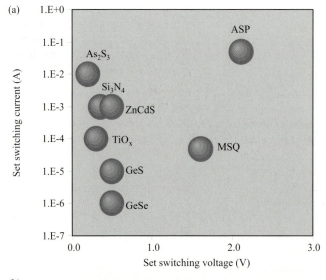

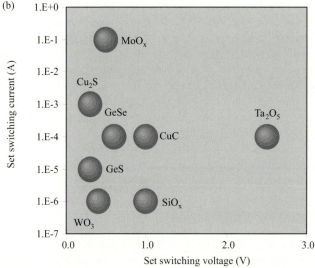

Figure 1.4 Distribution of set switching current and voltage for Cu^+-based (a) and Ag^+-based (b) ionic memory devices.

of the film. The switching voltages after the "forming" cycle are nearly independent of the film thickness. An intuitive explanation for this thickness independence is the incomplete dissolution of the Cu filament(s) after the "forming" process [45]. Another possible reason is that ion transport is no longer a rate-determining factor after the "electroforming" cycle [53].

As described in Section 1.2.1, the set switching process involves the following steps: (1) anodic dissolution of active electrode: $M \rightarrow M^{z+} + ze^-$; (2) drift of the

1.4 Electrical Characteristics of Ionic Memories

cations M^{z+} across the solid electrolyte under electrical field; and (3) cathodic reaction at the inert electrode by the nucleation and growth of the metal M electrodeposits: $M^{z+} + ze^- \rightarrow M$. It is still controversial on which step dominates the switching process. A study on Cu/Cu_2S devices shows that the migration of Cu^+ (step 2) is the dominating step in the set switching process because measurements show that the electrochemical reaction rate is much higher than the Cu^+ ion diffusion rate [45]. However, an exponential dependence of switching voltage on the switching rate is observed on Cu/SiO_x devices. An estimation of the ion hopping distance in the ion migration step suggests that this step cannot explain this exponential dependence. Therefore, it is proposed that the set switching kinetics of Cu/SiO_x devices is determined by the electrocrystallization process at the cathode (step 3) [53]. More studies are needed to further elucidate the dynamics of the ionic switching process, which may have a significant impact on the control of switching parameters and optimization of switching speed.

Since both the diffusivity of ions and the rate of electrochemical reactions decrease with cooling, it is expected that the switching voltage of ionic memories will increase at lower temperature because the reduction of ion activity needs to be compensated by higher electrical field. In Cu/Cu_2S devices, both set and reset switching voltages increase exponentially with the inverse of temperature ($1/T$) [45]. This temperature dependence of switching voltages correlates well with the temperature dependence of the measured Cu^+ diffusion coefficient [45].

Since the formation of one filament can significantly reduce the resistance of the solid electrolyte, both voltage drop on the device and the electrical field in the electrolyte may decrease abruptly. This makes it difficult for more filaments to form subsequently; that is, the filament formation in the ionic switching process is more or less self-limiting. Therefore, the on-state conduction of ionic memories is likely to be a few localized paths, rather than large amount of distributed paths. This prediction has been confirmed by the area dependence measurement of ionic memories. For example, the LRS resistance of Cu/CuC devices varies less than one order of magnitude over four orders of magnitude change of device size [60]. In other words, the LRS resistance is nearly area independent. The HRS resistance, on the other hand, shows much stronger area dependence. Since the HRS resistance increases with the decrease in device size and LRS resistance is nearly unchanged, the on/off ratio (= HRS resistance/LRS resistance) is expected to increase when device size is reduced. This enhanced signal strength at smaller device size is considered a unique scaling advantage of ionic memory devices.

The ionic switching model also clearly identifies these localized conduction paths as metallic filaments. The metallic nature of these filaments has been confirmed by the linear I–V characteristics and the negative temperature coefficient of LRS resistance. For example, measurements have shown that LRS conductance of Cu/CuC [60] and Cu/Cu_2S [43] decreases with the increase in temperature. On the other hand, HRS conductance shows positive temperature coefficient and the conduction mechanism can be explained by thermally active transport.

The on resistance of many RRAM devices has shown dependence on the programming current, which may have different explanations based on various switching mechanisms. Ionic memories have also exhibited a similar dependence; for example, the on resistance of Cu/Ta_2O_5 decreases with the increase in the programming current. An almost linear dependence exists between the on resistance and the programming current on a logarithm–logarithm scale [55]. This dependence of ionic memories can be generally explained by the enhancement of conductive filaments (i.e., lower resistance) under higher programming current.

The measured cycling endurance of ionic memories varies from several hundred cycles to 10^{11} cycles. The impressive endurance of 10^{11} cycles is reported on Ag/GeSe devices [28]. As discussed later, several factors may cause reliability issues in ionic memories and reduce the cycling endurance. If strong cycling endurance can be achieved uniformly on a large array of ionic memories, it will become a key advantage for ionic memories and may significantly expand their applications.

1.4.2
Ionic Memory Array Characteristics

Several companies have participated in ionic memory R&D and developed CMOS-compatible ionic memory processing and architectures. Some of them have reported array-level characteristics with more accurate device parameters and statistics. Table 1.2 summarizes the array characteristics measured on three ionic memories: Cu/Cu_2S (NEC/JSTA) [8], Ag/GeSe or Ag/GeS (Qimonda) [13, 61, 62], and Cu–Te/GdO$_x$ (Sony) [63]. They are all made with standard CMOS process flow and the

Table 1.2 Array characteristics of ionic memories.

Company	NEC/JSTA[a]	Qimonda	Sony
Tested array size	1 kbit	2 Mbit	4 kbit
Material system	Cu/Cu_2S	Ag/GeSe or GeS	Cu–Te/GdO$_x$
Technology node	250 nm CMOS	90 nm CMOS	180 nm CMOS
Minimum tested cell size	$D \sim 30$ nm	$D = 20$ nm	$D = 20$ nm
Memory structure	1T1R	1T1R	1T1R
Programming (HRS → LRS)			
Voltage	1.1 V	≥ 0.6 V	3 V
Current	—	10 μA	110 μA
Pulse width	5–32 μs	≤ 50 ns	5 ns
Erasing (LRS → HRS)			
Voltage	1.1 V	≤ 0.2 V	1.7 V
Current	—	20 μA	125 μA
Pulse width	5–32 μs	≤ 50 ns	1 ns
Retention			
Measured	10^3 s under 35 mV	10^5 s @ 70 °C	100 h @ 130 °C
Projected	3 months	10 years	10 years
R_{on}/R_{off}	$\leq 10^2 \Omega / \geq 10^3 \Omega$	$10^4 \Omega / 10^{11} \Omega$	$10^4 \Omega / 10^6$–$10^8 \Omega$
Endurance	10^3–10^4 cycles	10^6 cycles	10^7 cycles
Operating temperature	—	≥ 110 °C	≥ 130 °C
References	[8]	[13, 61, 62]	[63]

a) JSTA: Japan Science and Technology Agency.

memory cell uses a 1-transistor–1-resistor structure (1T1R) structure. The access transistor not only provides selection function but also controls the switching current. Array characterization provides more accurate results on some parameters; for example, switching speed as fast as several nanoseconds is measured on Cu–Te/GdO_x devices [63]. The switching voltage and current in Table 1.2 are measured in AC operation, and the switching voltages are generally higher than DC voltages in Table 1.1. Array statistics have shown reasonably good uniformity of device parameters (e.g., switching voltage and current, on and off resistance, etc.) across the array.

1.4.3
Comparing Ionic Memory with Other Memories

Figure 1.5 compares the performance of ionic memories with other mature and emerging memory technologies, based on switching speed, switching energy (per bit), and operation voltage. Data used in this figure are taken from ITRS Emerging Research Device (ERD) chapter [1]. The label of "ionic effect" resistive switching memory is highlighted in bold. In addition to the ionic effect, electronic effect and thermal effect are the other two types of resistive switching memories (i.e., RRAM). The *X*- and *Y*-axes in Figure 1.5 represent switching energy and switching speed, respectively. Size of the symbols is proportional to the typical operation voltage of each memory. Although SRAM and DRAM have the best switching characteristics represented by fast speed, low energy, and small operation voltage, they cannot retain information without power. Mainstream NOR and NAND flash memories require high operation voltage and switch slowly due to the tunneling-based programming/erasing mechanism. Among the emerging memory technologies, ionic memories have demonstrated some promising switching characteristics, including low operation voltage, low switching power, and fast switching speed.

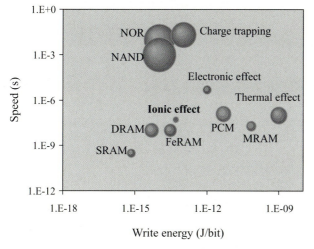

Figure 1.5 Comparison of ionic resistive switching devices with other memory technologies on switching speed (*Y*-axis), write energy (*X*-axis), and operation voltage (size of the symbols).

1.5
Architectures for Ionic Memories

Ionic memory devices can be integrated into CMOS in the BEOL process. However, the density of CMOS-integrated ionic memory architecture is usually constrained by the access transistors, and the scaling advantage of ionic memories is not utilized. With simple two-terminal structure, ionic memories are suitable for crossbar architecture that can achieve high memory density. Some hybrid/CMOS architectures are also proposed to take advantage of both the mature infrastructure of CMOS and the density advantages of crossbar arrays.

1.5.1
CMOS-Integrated Architecture

Figure 1.6 illustrates a cross-sectional view of ionic memories integrated into CMOS, where the ionic memory elements are built at via locations. This architecture is particularly suitable for MIM devices where the "I" layer can be formed by some treatment of the bottom electrode (e.g., oxidation) because the CMOS process flow needs to be only slightly modified to incorporate these memory elements. For ionic memory elements formed by deposition or sputtering, additional steps for patterning are required. Building memory devices at via locations incurs no area penalty in CMOS layout. The access transistors not only provide selection functions but also control the switching conditions (e.g., current limit). It has been found that the current control during switching could significantly impact the uniformity of device characteristics. Almost all the functional ionic memory arrays demonstrated so far have been built in CMOS architecture. The CMOS-integrated architecture is particularly useful as testing vehicles in R&D stage. However, since CMOS-integrated architecture does not truly utilize the scalability of ionic memories, it may not make ionic memories much more competitive than conventional memories.

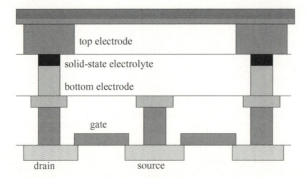

Figure 1.6 Cross-sectional view of CMOS-integrated architecture for ionic memories.

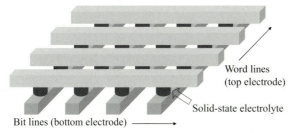

Figure 1.7 Crossbar architecture for ionic memories.

1.5.2
Crossbar Array Architecture

In crossbar architectures, the memory devices are built at the crossing points of orthogonally arranged horizontal and vertical access lines, as shown in Figure 1.7 [64, 65]. Obviously, higher memory density can be achieved in crossbar architectures than in CMOS architectures. If these crossbar memory arrays can be stacked in 3D structures, the memory density can be further increased. On the other hand, there are still severe challenges associated with the crossbar architecture, including the control of read/write disturbance, lack of signal gain, array size constraint determined by the asymmetry of device switching characteristics, and so on. Crossbar array also requires good uniformity of device switching parameters. In addition, asymmetric $I-V$ characteristics are desirable for crossbar architectures. In planar Ag/α-Si devices [33] and Ag/α-Si-nanowire devices [34], rectifying $I-V$ characteristics have been reported. However, most ionic memory devices have shown symmetric ohmic conduction in on state and may require diode-like selection devices.

Since the crossbar itself is a passive array, external CMOS circuitry is required for programming, erasing, and readout. In order to have distinguishable states, the switching devices need to have high on/off ratio; otherwise, an advanced sensing scheme is required. Two-terminal selection devices (e.g., diodes) are desirable for crossbar architectures. Although some two-terminal selection devices have been proposed, there is still no proven solution of selection devices for truly functional crossbar memory arrays [66].

1.5.3
CMOS/Hybrid Architecture

In the so-called "CMOL" architecture, crossbar of two-terminal switching devices can be built on top of CMOS circuits and is connected to CMOS circuits through interface pins [67]. Each switching device plays the role of one-bit memory cell, while the CMOS subsystem may be used for coding/decoding, sensing, input/output functions, and so on. This hybrid architecture utilizes the advantages of both CMOS and crossbar architectures. Although it is believed that CMOL provides various advantages over conventional CMOS and simple crossbar structure, functional CMOL circuits have not yet been demonstrated in experiments.

1.6
Challenges of Ionic Memories

As shown by the device and array characteristics, ionic memories have some unique advantages: low operation voltage, good scalability, nonvolatility, CMOS compatibility, and so on. The switching mechanism is clearly understood for cation-based ionic memories. These advantages make ionic memories attractive candidates for the next-generation memory beyond flash memories. However, ionic memory technology is still in research stage and there are numerous challenges. This section discusses some challenges and obstacles for ionic memory technologies that need to be addressed before ionic memories can enter mainstream market.

1.6.1
Overprogramming and Overerasing

Overprogramming/erasing happens when voltage is continuously applied even after the device has turned on or off, leading to an excessive thickening or dissolution of the atomic filament(s). As a result, the subsequent erasing or programming operations take longer, and in the worst case switching may fail [8]. Overprogramming/erasing may also decrease cycling endurance because it overstresses the devices. Since nonuniformity inevitably exists in large memory arrays, different devices may require different programming/erasing conditions. However, it is desirable from both design and product perspectives to apply uniform programming conditions on all devices in the array. The operation voltage suitable for majority of devices may become an overstress for devices that are easier to be programmed or erased. The improvement of device uniformity, mainly through material engineering and fabrication technology, will help to alleviate this problem.

Another solution uses an "adaptive" design to control the programming/erasing voltage and the biasing time by monitoring the states of the devices [8]. A sensing and control circuit could be added to ensure that voltage bias is turned off when the device switching is completed. In other words, different operation conditions are provided for devices with different characteristics. However, the control circuits for this "adaptive" operation complicate circuit/product design, induce memory density penalty by adding more peripheral circuits, and increase the total operation time due to additional "testing" steps.

1.6.2
Random Diffusion of Metal Ions and Atoms

The switching of ionic memories involves the drifting of charged metal ions and the growth of atomic filament(s) between two electrodes. These ions and atoms may diffuse randomly, driven by several factors, for example, thermal fluctuation. This random diffusion could change the size of the filament(s) and cause variation of on- and off-state resistance [30]. In severe cases, it may induce retention degradation or even device failure.

In retention measurement of an Ag/GeSe device, the on-state resistance (R_{on}) increases with time and exhibits an almost linear trend in logarithm–logarithm scale, that is, $R_{on} \propto t^n$ [28]. In this particular case, the exponent n is approximately equal to 0.25. It is suspected that this resistance change (i.e., loss of on-state retention) is caused by slow diffusion of metal atoms from the electrodeposited filament(s) into local defects in the electrolyte. In a study of on-state stability of Ag/GeS devices, it was found that negative bias stress (against programming voltage direction) tends to slowly increase on resistance whereas a positive bias stress results in a decrease in on resistance [29]. This observation can be explained by the migration of ions/atoms, which may effectively change the on or off states of ionic memories. This random diffusion also presents a trade-off between switching speed and device stability (or retention). High activity of ions, represented by high mobility, is desirable for fast switching. However, stability requirement would prefer low activity of these mobile ions, to minimize device state variation caused by electrical disturbance and thermal fluctuation that exists in memory arrays.

1.6.3
Thermal Stability

In addition to the unintentional thermal disturbance mentioned above, thermal processing steps are frequently used in semiconductor fabrication, including BEOL process. Ionic memories suitable for practical products need to be able to tolerate a certain range of temperature variation. Since the ion transport and electrochemical reactions are usually temperature sensitive, the switching parameters may shift significantly with the increase in temperature [57]. For example, when temperature is increased from 25 to 135 °C, the erasing voltage of Cu/WO$_3$ devices decreases from ∼0.3 to ∼0.1 V. Some ionic switching materials are found to be more stable; for example, Ag/GeS devices are found to tolerate higher annealing temperatures than Ag/GeSe devices [30]. Memory arrays made of Ag/Ge chalcogenide devices have demonstrated relatively stable on-/off-state resistance and programming/erasing voltages at temperatures up to 110 °C [61]. These results indicate that it is possible to build robust ionic memory products by carefully choosing materials and optimizing device structures.

1.6.4
Switching Speed

Two critical steps controlling the ionic switching process are ion migration in the solid electrolyte and the electrochemical reactions on the electrodes. The speed of both steps determines the switching speed. Although ion transport in superionic materials is much faster than that in regular solid materials, ion mobility is still significantly slower than electron mobility. As shown in Tables 1.1 and 1.2, the switching time of ionic memories varies in a wide range. First-order estimation shows that the speed of ion transport may be limited to the order of nanoseconds. Some ionic memory devices have demonstrated promising switching speed of ∼5 ns [63]. This speed is better than that of mature flash memory devices; however, it may not be competitive in comparison with some other emerging memory technologies.

1.6.5
Degradation of Inert Electrodes

As mobile ions migrate from the active electrode and deposit on the inert electrode, metal atoms start to accumulate on the inert electrode. Although metal filaments are broken during the erasing process, they may not be completely removed from the solid electrolyte or from the surface of the inert electrode. After certain number of switching cycles, more and more metal atom residuals may accumulate on the surface of the inert electrode. Eventually, sufficient amount of metal atoms from the active electrode may build up on the surface of the inert electrode and effectively turn it into another active electrode. Once this happens, programming may occur in both voltage bias directions because both electrodes can be a source of mobile ions. Both bipolar and unipolar resistive switching phenomena have been observed in a $Cu/SiO_x/W$ device, where W electrode is the inert electrode [50]. The unipolar switching can take place in both voltage bias directions after certain number of cycles of bipolar ionic switching. It can be explained by the accumulation of Cu atoms on the surface of W electrode, converting it into a "Cu electrode." Erasing in this unipolar switching process is probably due to Joule heating. Similar process is also found in Ag/TiO_2 devices [31]. The degradation of the inert electrode and the transition from bipolar to unipolar switching, if not controlled well, may raise reliability concerns on ionic memories.

1.7
Applications of Ionic Memories

Ionic memories provide unique properties and potential advantages in performance over conventional flash memories. Some major memory companies have started exploring this emerging technology for potential future products. These industry players include companies in the United States (e.g., Axon, Unity Semiconductor, and Micron), Japan (e.g., NEC, Sony, Fujitsu, and Toshiba), and Europe (e.g., Qimonda). Ionic memories can be used as stand-alone memories, embedded memories, and storage class memories. They have also been studied for logic applications, for example, reprogrammable logic.

1.7.1
Stand-Alone Memory

For stand-alone memory applications, ionic memories are usually compared with flash memories. Ionic memories have demonstrated some advantages over mainstream flash memories, including lower operation voltage and current, faster speed, better scalability, and so on. However, there also exist several other types of emerging memory devices that provide similar advantages over conventional memories; therefore, ionic memories also need to be benchmarked against these competitors. Density is an important figure of merit for stand-alone memories. Ionic memories have unique advantages on density due to the localized switching process and the highly scalable size of conductive filaments.

1.7.2
Embedded Memory

Embedded applications have different requirements on memory performance from those of stand-alone memory applications. The key parameters are speed, power, and endurance. Memory speed needs to be compatible with logic components, which switch at the speed faster than picoseconds. Very few emerging memory candidates meet the speed requirement, and ionic memories are facing the same challenge. Embedded memories need to have almost infinite cycling endurance because of the frequent access of logic components to embedded memories. Although relatively long endurance (e.g., 10^{11} cycles) has been demonstrated on ionic memories, most ionic memories are not yet proven to have such long endurance and even 10^{11} cycles may not be sufficient for some embedded applications. The nonvolatility of ionic memories is a unique advantage that may enable unprecedented functions in embedded applications.

1.7.3
Storage Class Memory

Storage class memories require not only high density and long retention but also low cost per bit to manufacture. Magnetic hard disk drives still dominate the data storage applications, owing to their low cost. Flash memory-based solid-state drives (SSDs) have gained increasing attention because of their unique advantages, for example, low power, fast speed, and so on. However, the high cost still limits them in niche applications. Ionic memories may be manufactured with the cost lower than that of flash memories. If 3D stackable ionic memory architectures can be achieved, memory density can be significantly improved and cost per bit may become even more competitive.

In addition to memory applications, ionic resistive switching devices may also be used for logic computation. For example, ionic switching devices have been experimented for applications in reprogrammable logic, which utilizes their nonvolatility and scalability to achieve better circuit performance than conventional field programmable gated array (FPGA) based on SRAM. Functional configurable logic cells based on lookup tables made of Cu/Cu_2S devices have been demonstrated in experiments [8]. It is expected that these novel reprogrammable logic circuits can be significantly smaller than conventional FPGA because of the small size of the ionic memory elements.

1.8
Summary

Ionic effect resistive switching phenomena are caused by field-driven ionic transport and electrochemical reactions. The hysteretic I–V characteristics induced by ionic resistive switching are promising for memory applications. The ionic resistive switching devices are highly scalable, owing to the nature of the atomic filaments formed between the electrodes to switch the devices to the on state. Ionic switching can take place at low voltage and current, which is attractive for low-power applications. The switching mechanism of cation migration-based ionic memories is well understood, although the switching process induced by anion migration and reactions is still unclear. A wide range of materials have demonstrated ionic resistive switching behaviors, including metal sulfides, chalcogenides, metal oxides, and so on. Promising device characteristics have been demonstrated on large memory arrays. Because of their simple two-terminal structures, ionic switching devices may enable novel memory architectures and reconfigurable circuits. Although significant progress has been made in the past 10 years, some important challenges still exist for ionic memories. Overprogramming/erasing, random diffusion of ions/atoms, and thermal stability may cause variation and degradation in device characteristics, or even operation failures. Although promising switching parameters have been reported on individual devices and even small memory arrays, uniformity and stability of large-scale ionic memory arrays still need to be studied. Similar to many other emerging memory devices and nanotechnologies, defects and random failures inevitably exist in ionic memories, even with improved processing technologies. It is expected that the post-flash memory devices may not be as robust as CMOS-based flash memories, although devices' switching characteristics may be improved. Therefore, defect-tolerant design may become important for these emerging memory technologies, including ionic memories.

References

1 International Technology Roadmap of Semiconductor (ITRS), Emerging Research Device (ERD) chapter, 2009.
2 Muller, G., Happ, T., Kund, M., Lee, G.Y., Nagel, N., and Sezi, R. (2004) Status and outlook of emerging nonvolatile memory technologies. *IEDM Technical Digest*, pp. 567–570.
3 Waser, R. and Aono, M. (2007) Nanoionics-based resistive

switching memories. *Nat. Mater.*, **6**, 833–840.
4 Kozicki, M.N., Ratnakumar, C., and Mitkova, M. (2006) Electrodeposit formation in solid electrolytes. IEEE Proceedings: Non-Volatile Memory Technology Symposium, pp. 111–115.
5 Guo, X., Schindler, C., Menzel, S., and Waser, R. (2007) Understanding the switching-off mechanism in Ag^+ migration based resistively switching model systems. *Appl. Phys. Lett.*, **91**, 133513.
6 Terabe, K., Nakayama, T., Hasegawa, T., and Aonob, M. (2002) Formation and disappearance of a nanoscale silver cluster realized by solid electrochemical reaction. *J. Appl. Phys.*, **91**, 10110–10114.
7 Kozicki, M.N., Park, M., and Mitkova, M. (2005) Nanoscale memory elements based on solid-state electrolytes. *IEEE Trans. Nanotechnol.*, **4**, 331–338.
8 Kaeriyama, S., Sakamoto, T., Sunamura, H., Mizuno, M., Kawaura, H., Hasegawa, T., Terabe, K., Nakayama, T., and Aono, M. (2005) A nonvolatile programmable solid-electrolyte nanometer switch. *IEEE J. Solid-State Circuits*, **40**, 168–176.
9 Sakamoto, T., Lister, K., Banno, N., Hasegawa, T., Terabe, K., and Aono, M. (2007) Electronic transport in Ta_2O_5 resistive switch. *Appl. Phys. Lett.*, **91**, 092110-1–092110-3.
10 Hirose, Y. and Hirose, H. (1976). Polarity-dependent memory switching and behavior of Ag dendrite in Ag-photodoped amorphous As_2S_3 films. *J. Appl. Phys.*, **47**, 2767–2772.
11 Terabe, K., Hasegawa, T., Nakayama, T., and Aono, M. (2005) Quantized conductance atomic switch. *Nature*, **433**, 47–50.
12 Aratani, K., Ohba, K., Mizuguchi, T., Yasuda, S., Shiimoto, T., Tsushima, T., Sone, T., Endo, K., Kouchiyama, A., Sasaki, S., Maesaka, A., Yamada, N., and Narisawa, H. (2007) A novel resistance memory with high scalability and nanosecond switching. IEDM Technical Digest, pp. 783–786.
13 Kund, M., Beitel, G., Pinnow, C.-U., Röhr, T., Schumann, J., Symanczyk, R., Ufert, K.-D., and Müller, G. (2005) Conductive bridging RAM (CBRAM): an emerging non-volatile memory technology scalable to sub 20nm. IEDM Technical Digest, pp. 754–757.
14 Janousch, M., Meijer, G.I., Staub, U., Delley, B., Karg, S.F., and Andreasson, B.P. (2007) Role of oxygen vacancies in Cr-doped $SrTiO_3$ for resistance-change memory. *Adv. Mater.*, **19**, 2232–2235.
15 Szot, K., Speier, W., Bihlmayer, G., and Waser, R. (2006) Switching the electrical resistance of individual dislocations in single-crystalline $SrTiO_3$. *Nat. Mater.*, **5**, 312–320.
16 Jameson, J.R., Fukuzumi, Y., Wang, Z., Griffin, P., Tsunoda, K., Meijer, G.I., and Nishi, Y. (2007) Field-programmable rectification in rutile TiO_2 crystals. *Appl. Phys. Lett.*, **91**, 112101- 1-3.
17 Kim, K.M., Choi, B.J., and Hwang, C.S. (2007) Localized switching mechanism in resistive switching of atomic-layer-deposited TiO_2 thin films. *Appl. Phys. Lett.*, **90**, 242906-1–242906-3.
18 Lee, M.D., Ho, C.H., Lo, C.K., Peng, T.Y., and Yao, Y.D. (2007) Effect of oxygen concentration on characteristics of NiO_x-based resistance random access memory. *IEEE Trans. Magn.*, **43**, 939–942.
19 Dong, R., Xiang, W.F., Lee, D.S., Oh, S.J., Seong, D.J., Heo, S.H., Choi, H.J., Kwon, M.J., Chang, M., Jo, M., Hasan, M., and Hwang, H. (2007) Improvement of reproducible hysteresis and resistive switching in metal–$La_{0.7}Ca_{0.3}MnO_3$–metal heterostructures by oxygen annealing. *Appl. Phys. Lett.*, **90**, 182118-1–182118-3.
20 Ignatiev, A., Wu, N.J., Chen, X., Liu, S.Q., Papagianni, C., and Strozier, J. (2006) Resistance switching in perovskite thin films. *Phys. Status Solidi (b)*, **243**, 2089–2097.
21 Kim, D.S., Lee, C.E., Kim, Y.H., and Kim, Y.T. (2006) Effect of oxygen annealing on $Pr_{0.7}Ca_{0.3}MnO_3$ thin film for colossal electroresistance at room temperature. *J. Appl. Phys.*, **100**, 093901-1–093901-4.

22 Terabe, K., Hasegawa, T., Nakayama, T., and Aono, M. (2001) Quantum point contact switch realized by solid electrochemical reaction. *RIKEN Rev.*, **37**, 7–8.

23 Terabe, K., Nakayama, T., Hasegawa, T., and Aono, M. (2002) Ionic/electronic mixed conductor tip of a scanning tunneling microscope as a metal atom source for nanostructuring. *Appl. Phys. Lett.*, **80**, 4009–4011.

24 Tamura, T., Hasegawa, T., Terabe, K., Nakayama, T., Sakamoto, T., Sunamura, H., Kawaura, H., Hosaka, S., and Aono, M. (2006) Switching property of atomic switch controlled by solid electrochemical reaction. *Jpn. J. Appl. Phys.*, **45**, L364–L366.

25 Liang, C., Terabe, K., Hasegawa, T., and Aono, M. (2007) Resistance switching of an individual Ag_2S/Ag nanowire heterostructure. *Nanotechnology*, **18**, 485202-1–485202-5.

26 Schindler, C., Meier, M., Waser, R., and Kozicki, M.N. (2007) Resistive switching in Ag–Ge–Se with extremely low write currents. IEEE Proceedings: Non-Volatile Memory Technology Symposium, pp. 82–85.

27 Chen, L., Liu, Z., Xia, Y., Yin, K., Gao, L., and Yin, J. (2009) Electrical field induced precipitation reaction and percolation in $Ag_{30}Ge_{17}Se_{53}$ amorphous electrolyte films. *Appl. Phys. Lett.*, **94**, 162112-1–162112-3.

28 Kozicki, M.N., Gopalan, C., Balakrishnan, M., Park, M., and Mitkova, M. (2004) Non-volatile memory based on solid electrolytes. IEEE Proceedings: Non-Volatile Memory Technology Symposium, pp. 10–17.

29 Kamalanathan, D., Baliga, S., Thermadam, S.C.P., and Kozicki, M. (2007) ON state stability of programmable metalization cell (PMC) memory. IEEE Proceedings: Non-Volatile Memory Technology Symposium, pp. 91–96.

30 Kozicki, M.N., Balakrishnan, M., Gopalan, C., Ratnakumar, C., and Mitkova, M. (2005) Programmable metalization cell memory based on Ag–Ge–S and Cu–Se–S solid electrolytes. IEEE Proceedings: Non-Volatile Memory Technology Symposium, pp. 83–89.

31 Tsunoda, K., Fukuzumi, Y., Jameson, J.R., Wang, Z., Griffin, P.B., and Nishi, Y. (2007) Bipolar resistive switching in polycrystalline TiO_2 films. *Appl. Phys. Lett.*, **90**, 113501-1–113501-3.

32 Hu, J., Branz, H.M., Crandall, R.S., Ward, S., and Wang, Q. (2003) Switching and filament formation in hot-wire CVD p-type a-Si:H devices. *Thin Solid Films*, **430**, 249–252.

33 Jo, S.H. and Lu, W. (2006) Ag/a-Si:H/c-Si resistive switching nonvolatile memory devices. IEEE Nanotechnology Materials and Devices Conference, pp. 116–117.

34 Dong, Y., Yu, G., McAlpine, M.C., Lu, W., and Lieber, C.M. (2008) Si/a-Si core/shell nanowires as nonvolatile crossbar switches. *Nano Lett.*, **2**, 386–391.

35 Sun, B., Liu, L.F., Wang, Y., Han, D.D., Liu, X.Y., Han, R.Q., and Kang, J.F. (2008) Bipolar resistive switching behaviors of Ag/Si_3N_4/Pt memory device. International Conference on Solid-State and Integrated Circuit Technology, pp. 925–927.

36 Stratan, I., Tsiulyanu, D., and Eisele, I. (2006) Polarity: a programmable metallization cell based on $Ag-As_2S_3$. *J. Optoelectron. Adv. Mater.*, **8**, 2117–2119.

37 Guo, H.X., Gao, L.G., Xia, Y.D., Jiang, K., Xu, B., Liu, Z.G., and Yin, J. (2009) The growth of metallic nanofilaments in resistive switching memory devices based on solid electrolytes. *Appl. Phys. Lett.*, **94**, 153504-1–153504-3.

38 Kügeler, C., Nauenheim, C., Meier, M., Rüdiger, A., and Waser, R. (2008) Fast resistance switching of TiO_2 and MSQ thin films for non-volatile memory applications (RRAM). IEEE Proceedings: Non-Volatile Memory Technology Symposium, pp. 1–6.

39 Meier, M., Rosezin, R., Gilles, S., Rüdiger, A., Kügeler, C., and Waser, R. (2009) A multilayer RRAM nanoarchitecture with resistively switching Ag-doped spin-on glass. International Conference on Ultimate Integration of Silicon, pp. 143–146.

40 Guo, H.X., Yang, B., Chen, L., Xia, Y.D., Yin, K.B., and Liu, Z.G. (2007) Resistive switching devices based on nanocrystalline solid electrolyte $(AgI)_{0.5}(AgPO_3)_{0.5}$. *Appl. Phys. Lett.*, **91**, 243513.

41 Wang, Z., Griffin, P.B., McVittie, J., Wong, S., McIntyre, P.C., and Nishi, Y. (2007) Resistive switching mechanism in $Zn_xCd_{1-x}S$ nonvolatile memory devices. *IEEE Trans. Electron Dev. Lett.*, **28**, 14–16.

42 van der Sluis, P. (2003) Non-volatile memory cells based on $Zn_xCd_{1-x}S$ ferroelectric Schottky diodes. *Appl. Phys. Lett.*, **82**, 4089; van der Sluis, P. (2004) Addendum: non-volatile memory cells based on $Zn_xCd_{1-x}S$ ferroelectric Schottky diodes. *Appl. Phys. Lett.* **84**, 2211.

43 Sakamoto, T., Sunamura, H., Kawaura, H., Hasegawa, T., Nakayama, T., and Aono, M. (2003) Nanometer-scale switches using copper sulfide. *Appl. Phys. Lett.*, **82**, 3032–3034.

44 Sakamoto, T., Kaeriyama, S., Sunamura, H., Mizuno, M., Kawaura, H., Hasegawa, T., Terabe, K., Nakayama, T., and Aono, M. (2004) Solid electrolyte nanometer switch implemented in Si LSI. International Conference on Solid State Devices and Materials, pp. 110–111.

45 Banno, N., Sakamoto, T., Hasegawa, T., Terabe, K., and Aono, M. (2006) Effect of ion diffusion on switching voltage of solid-electrolyte nanometer switch. *Jpn. J. Appl. Phys.*, **45**, 3666–3668.

46 Banno, N., Sakamoto, T., Iguchi, N., Kawaura, H., Kaeriyama, S., Mizuno, M., Terabe, K., Hasegawa, T., and Aono, M. (2006) Solid-electrolyte nanometer switch. *IEICE Trans. Electron.*, **E89-C**, 1492–1498.

47 Zhang, J.R. and Yin, J. (2008) Resistive switching devices based on Cu_2S electrolyte. IEEE International Nanoelectronics Conference, pp. 1020–1022.

48 Rahaman, S.Z., Maikap, S., Chiu, H.C., Lin, C.H., Wu, T.Y., Chen, Y.S., Tzeng, P.J., Chen, F., Kao, M.J., and Tsai, M.J. (2009) Low power operation of resistive switching memory device using novel W/$Ge_{0.4}Se_{0.6}$/Cu/Al structure. International Memory Workshop, pp. 1–4.

49 Balakrishnan, M., Thermadam, S.C.P., Mitkova, M., and Kozicki, M.N. (2006) A low power non-volatile memory element based on copper in deposited silicon oxide. IEEE Proceedings: Non-Volatile Memory Technology Symposium, pp. 104–110.

50 Schindler, C., Thermadam, S.C.P., Waser, R., and Kozicki, M.N. (2007) Bipolar and unipolar resistive switching in Cu-doped SiO_2. *IEEE Trans. Electron Dev.*, **54**, 2762–2768.

51 Schindler, C., Weides, M., Kozicki, M.N., and Waser, R. (2008) Low current resistive switching in Cu–SiO_2 cells. *Appl. Phys. Lett.*, **92**, 122910-1–122910-3.

52 Zhang, L., Huang, R., Wang, A.Z.H., Wu, D., Wang, R., and Kuang, Y. (2008) The parasitic effects induced the contact in RRAM with MIM structure. International Conference on Solid-State and Integrated Circuit Technology, pp. 932–935.

53 Schindler, C., Staikov, G., and Waser, R. (2009) Electrode kinetics of Cu–SiO_2-based resistive switching cells: overcoming the voltage–time dilemma of electrochemical metallization memories. *Appl. Phys. Lett.*, **94**, 072109-1–072109-3.

54 Sakamoto, T., Banno, N., Iguchi, N., Kawaura, H., Sunamura, H., Fujieda, S., Terabe, K., Hasegawa, T., and Aono, M. (2007) A Ta_2O_5 solid-electrolyte switch with improved reliability. Symposium on VLSI Technology, pp. 38–39.

55 Maikap, S., Rahaman, S.Z., Wu, T.Y., Chen, F., Kao, M.J., and Tsai, M.J. (2009) Low current (5 pA) resistive switching memory using high-k Ta_2O_5 solid electrolyte. Proceedings of the European Solid State Device Research Conference, pp. 217–220.

56 Rahaman, S.Z., Maikap, S., Lin, C.H., Wu, T.Y., Chen, Y.S., Tzeng, P.J., Chen, F., Lai, C.S., Kao, M.J., and Tsai, M.J. (2009) Low current and voltage resistive switching memory device using novel Cu/Ta_2O_5/W structure. International Symposium on VLSI Technology, Systems, and Applications, pp. 33–34.

57 Kozicki, M.N., Gopalan, C., Balakrishnan, M., and Mitkova, M. (2006) A low-power nonvolatile switching element based on copper–tungsten oxide solid electrolyte. *IEEE Trans. Nanotechnol.*, **5**, 535–544.

58 Lee, D., Seong, D., Jo, I., Xiang, F., Dong, R., Oh, S., and Hwang, H. (2007) Resistance switching of copper doped MoO_x films for nonvolatile memory applications. *Appl. Phys. Lett.*, **90**, 122104-1–122104-3.

59 Pyun, M., Choi, H., Park, J.B., Lee, D., Hasan, M., Dong, R., Jung, S.J., Lee, J., Seong, D., Yoon, J., and Hwang, H. (2008) Electrical and reliability characteristics of copper-doped carbon (CuC) based resistive switching devices for nonvolatile memory applications. *Appl. Phys. Lett.*, **93**, 212907-1–212907-3.

60 Choi, H., Pyun, M., Kim, T.W., Hasan, M., Dong, R., Lee, J., Park, J.B., Yoon, J., Seong, D., Lee, T., and Hwang, H. (2009) Nanoscale resistive switching of a copper–carbon-mixed layer for nonvolatile memory applications. *IEEE Trans. Electron Dev. Lett.*, **30**, 302–304.

61 Dietrich, S., Angerbauer, M., Ivanov, M., Gogl, D., Hoenigschmid, H., Kund, M., Liaw, C., Markert, M., Symanczyk, R., Altimime, L., Bournat, S., and Mueller, G. (2007) A nonvolatile 2-Mbit CBRAM memory core featuring advanced read and program control. *IEEE J. Solid-State Circuits*, **42**, 839–845.

62 Liaw, C., Kund, M., Schmitt-Landsiedel, D., and Ruge, I. (2007) The conductive bridging random access memory (CBRAM): a non-volatile multi-level memory technology. 37th European Solid State Device Research Conference, pp. 226–229.

63 Aratani, K., Ohba, K., Mizuguchi, T., Yasuda, S., Shiimoto, T., Tsushima, T., Sone, T., Endo, K., Kouchiyama, A., Sasaki, S., Maesaka, A., Yamada, N., and Narisawa, H. (2007) A novel resistance memory with high scalability and nanosecond switching. *IEDM Technical Digest*, pp. 783–786.

64 Ziegler, M.M. and Stan, M.R. (2002) Design and analysis of crossbar circuits for molecular nanoelectronics. IEEE NANO, p. 323.

65 Ziegler, M.M. and Stan, M.R. (2003) CMOS/nano co-design for crossbar-based molecular electronic systems. *IEEE Trans. Nanotechnol.*, **2**, 217.

66 Lee, M., Park, Y., Suh, D., Lee, E., Seo, S., Kim, D., Jung, R., Kang, B., Ahn, S., Lee, C., Seo, D., Cha, Y., Yoo, I., Kim, J., and Park, B. (2007). Two series oxide resistors applicable to high speed and high density nonvolatile memory. *Adv. Mater.*, **19**, 3919.

67 Likharev, K.K. (2008) Hybrid CMOS/nanoelectronic circuits: opportunities and challenges. *J. Nanoelectron. Optoelectron.*, **3**, 203.

2
Composite Solid Electrolytes
Nikolai F. Uvarov

This chapter presents a brief overview of transport and physicochemical properties of composite solid electrolytes, an attractive class of solids where fast ion diffusion is essentially governed by interfacial phenomena. The surface potential formation and the point defect equilibria at free surfaces and interfaces are analyzed in the framework of the unified Stern model. Special attention is given to true size effects caused by changing bulk characteristics of ionic salts in the nanocomposites and to the thermodynamics and stability of composite systems. The main thermodynamic property relevant for surface spreading and nanocomposite formation is the adhesion energy. Analysis of the numerous experimental data shows that nonautonomous interface phases, crystalline or amorphous, exist in some composites. The formulas proposed to describe ionic conductivity variations in such systems and the main approaches for improving or creating new composite systems of different types are reviewed.

2.1
Introduction

Composite solid electrolytes of the ionic salt–oxide type (MX–A) can be considered as a separate and wide class of ionic conductors with high diffusivity. The combination of fast ionic transport and enhanced mechanical strength, and the possibility to optimize electrolyte properties by varying the type and concentration of the heterogeneous dopant, makes these materials promising for numerous electrochemical systems, such as batteries and sensors. Since the pioneering work of Liang [1], a large number of composite ionic conductors have been studied. In fact, doping with dispersed oxides was shown to enhance the conductivity of virtually all composites based on classical (nonsuperionic) ionic salts; it is a general effect requiring scientific explanation and theoretical substantiation. Several reviews have been published devoted to the description and analysis of the ion transport in polycrystalline and composite solid electrolytes [2]. The increase in the ionic conductivity upon heterogeneous doping can be explained within the framework of the space charge model proposed by Jow and Wagner [3] and Maier [2c]. This formalism enables to interpret many phenomena observed in composite materials, but is the best suited for the explanation of experimental data on the composites containing oxides with relatively coarse grains. However, the space charge model in its classical version is correct only for ideal crystals in contact with vacuum or a structure-free medium and obviously ignores real features of the interfacial contacts, namely, changes in the structures of ionic crystals (e.g., for epitaxial contacts), the effect of elastic strains, the formation of dislocations, and so on. Moreover, if the surface concentration of defects is sufficiently high, it is impossible to ignore the interaction between the defects, which results in their ordering and the formation of superstructures and even metastable surface phases. It is known that the conductivity of composites increases as the size of dopant particles decreases. Hence, composites with nanosized grains (about 10 nm) are of particular interest for practice. Obviously, uniform mixing of such an oxide with an ionic component should produce a nanocomposite, the properties of which strongly depend on the energy of surface interaction and the peculiarities of the interface between the phases. For composites with coarse-grained additives, the presence of surfaces or interfacial contacts has virtually no effect on the bulk properties of the ionic salt; hence, the increase in the conductivity is purely of surface nature. However, in many cases, it still remains unclear whether the enhanced conductivity is primarily caused by the specific interactions at the interface or by the trivial increase in the surface conductivity as such. To answer this question, information on the conductivity of polycrystals is necessary. In nanocomposites, an ionic salt is virtually totally located at the interface. Therefore, its structure and thermodynamic characteristics can substantially change. Particularly, for ionic compounds containing high-temperature disordered phases, the latter may prove to be stable at low temperatures in nanocomposites. Therefore, it is important to understand the thermodynamic reasons for the stabilization of a disordered phase.

In this chapter, the quasichemical mechanisms of the interfacial interactions and problems of thermodynamic stability of the nanocomposites are discussed. Special emphasis is also given to the genesis of the composite morphology during its sintering. The properties of ionic salts are analyzed for a wide series of systems with an emphasis on the size effects in nanocomposites. Methods for the qualitative prediction of the conductivity and other physicochemical characteristics of the composite materials are considered and approaches for the development of novel composite systems are reviewed and discussed.

2.2
Interface Interactions and Defect Equilibria in Composite Electrolytes

The main reason responsible for the change in the physical properties of ionic salts in MX–A composites is the interface interaction between the components of the composite: ionic salt MX and oxide A. In terms of quasichemical approach [2c,g], strong ion–ion interactions may be represented as a process of the chemical

2.2 Interface Interactions and Defect Equilibria in Composite Electrolytes

adsorption of ions to the oxide surface. From the general conditions of mass and charge conservation (see Chapter 3 of the first volume), it follows that concentrations of ions are interrelated with fractions of corresponding point defects. As the conductivity in ordinary ionic salt MX occurs via migration of point defects, it is convenient to express the parameters of the chemical adsorption of ions in terms of adsorption isotherms of the corresponding defects (including impurity ions) of ith type with the adsorption energies of Δg_i. Chemical adsorption of charged species seems to be a general phenomenon typical for any polar media (including ionic crystals) and may take place for both free surfaces of MX and grain boundary MX–MX and MX–A interfaces. Therefore, the same phenomenological model can be applied to surface- or interface-related effects in polycrystalline samples and composites. Such approach is known in electrochemistry as the Stern model [4]. This model may be used for estimation of the surface potential [5], a key factor determining the concentration profile of the defects near the surface.

2.2.1
Defect Thermodynamics: Free Surface of an Ionic Crystal

In ionic salts, the lattice distortion associated with the asymmetric field at the surface results in an alteration of all characteristics of the point defects and impurity ions located on the surface with respect to the bulk defects. Their difference is the reason for the specific adsorption of defects at the crystal surface. If the adsorption energies Δg_i of oppositely charged defects are different, then an excess number of defects that have the most negative value of Δg_i appear on the surface. Defects of the opposite charge form a diffuse layer, or space charge layer (SCL), under the surface. According to the Stern model, the surface charge Q_S is determined by the sum of contributions made by all defects adsorbed on the surface. In the simplest case of the Langmuir adsorption isotherm, the surface charge is given by the expression [4]

$$Q_S = \sum_i q_i N_{S,i} \left[1 + \frac{N_i}{n_{\infty,i}} \exp\left(\frac{\Delta g_i + q_i \varphi_S}{kT}\right) \right]^{-1}, \quad (2.1)$$

where $N_{S,i}$ is the concentration of available surface sites (adsorption centers), q_i is the effective charge of the defect, $n_{\infty,i}$ is the concentration of the defects in the bulk of the crystal, N_i is the concentration of the ith type regular sites of the crystal lattice, and the summation is done over all possible charged defects of ith type. The surface charge is thus determined by both the bulk properties ($n_{\infty,i}$) and the surface-related parameters, namely, the adsorption energies Δg_i and the surface potential φ_S that is similar to the potential at inner Helmholtz layer in classical electrochemistry. The surface charge is balanced by the charge of the SCL, Q_d, formed under the crystal surface. If the probability of finding an ion at a particular point depends on the local potential through a Boltzmann distribution, $n_i(x) = n_{i,\infty} \exp(-q\varphi(x)/kT)$, then the charge density distribution and the potential gradient must satisfy the Poisson–Boltzmann equation. The solution of this equation is the Gouy–Chapman formula for the charge of SCL:

$$Q_d = A_d \left\{ \sum_i n_{\infty,i} \left[\exp\left(-\frac{q_i \varphi_{s-1}}{kT} \right) - 1 \right] \right\}^{1/2}, \qquad (2.2)$$

where $A_d = (2\varepsilon\varepsilon_0 kT)^{1/2}$ and φ_{s-1} is the potential of the outer Helmholtz layer. Assuming that in first approximation $\varphi_s \approx \varphi_{s-1}$, one may determine its value from the overall electroneutrality condition $Q_s + Q_d = 0$ and Equations 2.1 and 2.2. From these equations, it is seen that the absolute value of the surface potential in pure MX crystal depends on four independent parameters, including two values of Δg_i, total concentration of point defects in the bulk $n_{\infty,i}$ (defined only by the defect formation energy g_0), and the concentration of available surface sites N_S (determined by the assumption that $N_{S,i}$ values for both types of defects are equal to N_S). Calculated values of the surface potential for a model crystal of NaCl type with Schottky defects and superionic oxides $M^{IV}_{1-c}Me^{III}_c O_{2-c/2}$ have been reported in Ref. [5a]. Figure 2.1 shows typical temperature dependences of the surface potential for a crystal of the NaCl type with the Schottky defects doped with the impurities of bivalent metals.

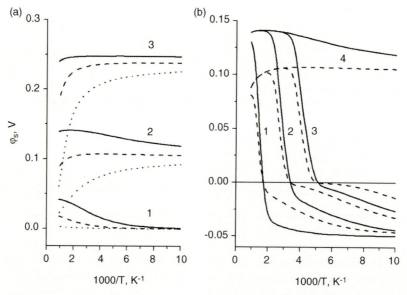

Figure 2.1 Dependences of the surface potential on the reciprocal temperature for the crystal of the NaCl type with the Schottky defects. (a) For pure crystal at $\Delta g^+ = -0.1\,\text{eV}$ and $\Delta g^- = -0.2, -0.4,$ and $-0.6\,\text{eV}$, curves 1, 2, and 3, respectively; solid, dashed, and dotted lines correspond to the concentration of the surface adsorption sites of 10^{13}, 10^{14}, and $10^{15}\,\text{cm}^{-2}$, respectively. (b) For crystal doped with the impurities of bivalent metals at $\Delta g^+ = -0.1\,\text{eV}$, $\Delta g^- = -0.4$, and zero segregation energy; the molar fractions of the dopant are 10^{-3}, 10^{-6}, and 10^{-9} for curves 1, 2, and 3, respectively; curve 4 corresponds to the ideally pure crystal; solid and dashed lines correspond to the concentration of the surface sites of 10^{15} and $10^{14}\,\text{cm}^{-2}$, respectively; defect association effects in the bulk of the crystal are neglected. The defect formation energy g_0 is equal to 1 eV; all energies, g_0, Δg^+, and Δg^- are taken to be independent of temperature.

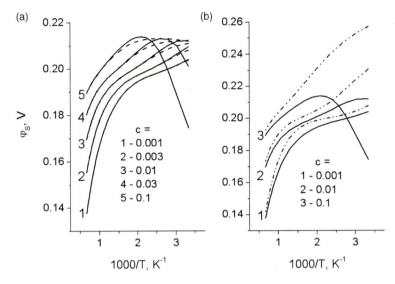

Figure 2.2 Dependences of the surface potential on the reciprocal temperature for the superionic oxide $Zr_{1-c}Me^{III}_c O_{2-c/2}$ calculated using the Stern model with Gouy–Chapman and Mott–Schottky potential distribution in the space charge at $\Delta g^- = 0$, $\Delta g^+ = -0.6\,eV$, and zero segregation energy. (a) Curves calculated with Gouy–Chapman model at different concentrations of the dopant, c; solid lines correspond to the case of no association between defects, whereas dashed lines are calculated assuming that uncharged complexes $[Me-2V_O]^{\times}$ form with the association energy of 0.4 eV. (b) The curves obtained using Gouy–Chapman and Mott–Schottky models (solid and dash-dotted lines, respectively) at different dopant concentrations; defect association effects are neglected.

Recently, the surface potential in superionic oxides was estimated using different equations for different charge distribution in SCL (Gouy–Chapman or Mott–Schottky) [5b]. The temperature dependences $\varphi_S(T)$ for these systems are presented in Figure 2.2. The obtained results may be summarized as follows:

a) The surface potential is nonzero only if $\Delta g_i < 0$ for at least one type of defects, and is governed by the difference in the defect adsorption energies Δg^- and Δg^+ of positively and negatively charged dominant defects. If $\Delta g^- = \Delta g^+$, then $\varphi_S = 0$.
b) The surface potential monotonically increases with increasing N_S.
c) In pure ionic crystal, the temperature behavior of φ_S depends on the value of Δg^α (where Δg^α energy is the most negative of Δg^+ and Δg^- ones) and differs for two cases: (i) $|\Delta g^\alpha| < g_0/4$ – the surface potential is small, and φ_S monotonically increases with the temperature and rises as a function of the defect concentration in the bulk and N_S; (ii) $|\Delta g^\alpha| > g_0/4$ – the case of high surface potentials. The values of φ_S may increase or decrease with temperature as a function of N_S values. At sufficiently high values of the adsorption energy $|\Delta g^\alpha|$ and N_S, the surface potential decreases with temperature and does not depend on the defect concentration.
d) At high N_S and $|\Delta g^\alpha|$, the surface potential tends to the value of $\varphi_S \approx (\Delta g^- - \Delta g^+)/2q$, which can be regarded as the upper limit. It is to be noted that this

expression is formally similar to the equation $\varphi_S \approx -(g^- - g^+)/2q$ obtained in frames of the approach proposed by Frenkel [6a] and later evolved by Kliewer et al. [6b,c], Poeppel and Blakely [6d], and Jamnik et al. [6e], where g^- and g^+ are the defect formation energies in the crystal bulk.

e) In MX crystal doped with MeX$_2$, a more complicated situation is observed. The φ_S values depend on the type, concentration of the impurity ions (given by the energy of dissociation of complexes (impurity ion–cation vacancy)), and their adsorption energy (or segregation energy) Δg_{Me}^+. At $|\Delta g^+| > |\Delta g^-| > |\Delta g_{Me}^+|$ or $|\Delta g_{Me}^+| > |\Delta g^+| > |\Delta g^-|$ ($\Delta g^+, \Delta g^-, \Delta g_{Me}^+ < 0$), the $\varphi_S(T)$ dependence displays an isoelectric point, that is, temperature where φ_S change its sign.

f) In superionic oxides $M_{1-c}^{IV}Me_c^{III}O_{2-c/2}$ with fluorite-type structure, the values of φ_S are generally bounded by limiting parameters given by Δg^+ and Δg_{Me}^- (the adsorption energy of oxygen vacancies and the segregation energy of Me cations, respectively). The values of φ_S obtained using Gouy–Chapman and Mott–Schottky models of SCL strongly differ in the low-temperature limit and are very close at high temperatures. Therefore, at high temperatures, the surface potential becomes practically independent of the particular form of the $\varphi(x)$ function in SCL and is determined mainly by the adsorption energies of the defects.

g) As reported by Chebotin and Perfiliev [7a], Heine [7b], and Guo et al. [7c,d], the surface potential in oxides of the $M_{1-c}^{IV}Me_c^{III}O_{2-c/2}$ family is positive; therefore, $\Delta g^+ < 0$ ($\Delta g^+ > |\Delta g_{Me}^-|$) and the surface should be enriched in anionic vacancies and depleted in cations. Within the Stern model, the adsorption of oxygen vacancies and the segregation of extrinsic cations are interrelated and even at $\Delta g_{Me}^- = 0$ a strong segregation of cations on the surface takes place. Nevertheless, the surface as a whole remains positively charged and the diffusion layer is depleted of anionic vacancies.

Representing the chemical potential of ith defects, μ_i, in the standard form $\mu_i = g_i^\pm + kT \ln[\ ]_i$, where $[\ ]_i$ is the fraction of the defects and superscript corresponds to the defect sign, and taking into account the electrical neutrality condition, one can obtain the values of $g_i^\pm$ for all the defects. From the physical point of view, each $g_i^\pm$ corresponds to the energy necessary for generation of a single isolated defect. This value is constant only within a certain temperature range corresponding to intrinsic or extrinsic conductivity regions (see Chapter 3 of the first volume). Note also that the latter regime comprises two regions, where extrinsic defects are free or associated with the complexes (defect clusters) with impurity ions. Table 2.1 lists $g_i^\pm$ values for the bulk of MX crystal doped with MeX$_2$ impurity in the three temperature regions. Using these data one can plot energy diagrams (Figure 2.3) illustrating the difference between the defect energies in the bulk of the crystal and at the surface for MeX$_2$-doped MX crystal in the different temperature ranges. This diagram differs from the diagrams reported earlier [2l,r, 6e] as in the bulk of the crystal in the intrinsic region the defect formation energies for both the defects are taken to be equal to $g_0/2$. As seen from the diagram, at negative values of adsorption energies, the surface is enriched with defects even in the case of zero surface charge (i.e., the latter hypothetical situation becomes possible at $\Delta g^+ = \Delta g^-$). In general, the surface can be considered as an independent subsystem

2.2 Interface Interactions and Defect Equilibria in Composite Electrolytes

Table 2.1 The standard values of chemical potential $g_i^{\pm}$ for cation vacancy, g^-, anion vacancy, g^+, and impurity cations, g_{Me}^+, at different temperature regions for MX crystal with Schottky defects.

Defect	Energy	Extrinsic region		Intrinsic region
		Low temperatures	High temperatures	
Cation vacancy, V'_M	g^-	$g_d/2 + (1/2)kT \ln c$	$-kT \ln c$	$g_0/2$
Anion vacancy, $V_X^{\bullet}$	g^+	$g_0 - g_d/2 - (1/2)kT \ln c$	$g_0 - kT \ln c$	$g_0/2$
Impurity cations, $Me_M^{\bullet}$	g_{Me}^+	$g_d/2 + (1/2)kT \ln c$	$-kT \ln c$	$-kT \ln c$

g_0 is the Schottky defect formation, g_d is the energy of dissociation of complexes $[V'_M - Me_M^{\bullet}]^x$, and c is the concentration of impurity MeX_2.

characterized by intrinsic surface disordering with an effective defect formation energy equal to $g_S = g_0 + (\Delta g^+ + \Delta g^-)/2$. The surface is more or less disordered than the bulk (i.e., enriched or depleted in defects on average), depending on the sign of the second term of this equation.

2.2.2
Defect Thermodynamics in Composites: Interfaces

In order to analyze defect equilibria and interactions at the composite interfaces, let us first consider the interface between ionic salt MX and oxide phase A. In this case,

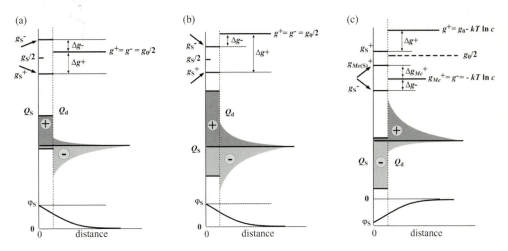

Figure 2.3 Energetic diagrams of surface and bulk defects, the potential profiles, and the charge distribution near the surface of the pure (a and b) and doped (c) ionic crystal MX with Schottky defects. The diagrams obtained in terms of the Stern model. Cases (a) and (b) correspond to the pure crystal with different Δg^+ and Δg^- values. Cases (b) and (c) relate to the MX crystal doped with bivalent impurity MeX_2 (the dopant concentration is equal to c) in intrinsic (b) and extrinsic (c) conductivity regions. Potential determining levels are indicated by arrows.

the concentration of defects in the oxide bulk is negligible and only the first oxide layer contributes to the interfacial interaction that includes pair interactions between the constituting ions of MX and A. This results in changing the defect adsorption energies, the surface potential, and, consequently, the concentration of point defects in the diffuse layer, leading to the enhancement of the conductivity of composites. Such mechanism has been proposed and discussed in detail by Maier [2c,g,h]. It has been suggested that the interfacial interactions consist of a selective chemical adsorption of M^+ cations, that is, their shift from the MX bulk to the MX–A interface. Physically, this phenomenon is equivalent to the change in the adsorption energy of positively charged defects (anionic vacancies $V_X^{\bullet}$ or interstitial cations $M_i^{\bullet}$) and the formation of high positive charge at the interface. As a result, a diffuse layer enriched with cationic vacancies forms near the interface. The energy diagram for an ideal MX–A interface is presented in Figure 2.4a. Selective adsorption of cations can be regarded as interaction of the Lewis-type acidic M^+ cations with the basic O^{2-} or OH^- centers on the oxide surface. Therefore, the adsorption energy should depend on the type of cation (acidity of the cation increases with the decrease in the ionic radius) and on the presence and strength of the basic groups on the oxide surface. The concentration of charge carriers on the surface or in the interfacial region of MX can be indeed varied due to the modification of the surface by electronic donor molecules [8a,b]. Recently, we have carried out a comparative study of electrical properties and ^{7}Li NMR data of the composites LiClO$_4$–A (A = α-Al$_2$O$_3$, γ-Al$_2$O$_3$, α-LiAlO$_2$, γ-LiAlO$_2$) [9a,b]. It was shown that the conductivity depends not only on the specific surface but also on the chemical nature and the structure of the additive. In fact, at the same value of the specific surface area, the composites with γ-Al$_2$O$_3$ and γ-LiAlO$_2$ have lower activation energy for the ionic conductivity than those containing α-Al$_2$O$_3$ and α-LiAlO$_2$ additives. Possible mechanism of such behavior may originate from the difference in the crystalline structure of oxides that influences the structure and

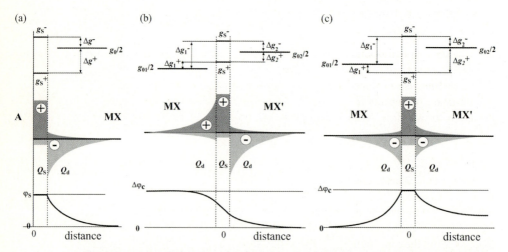

Figure 2.4 Energetic diagrams of surface and bulk defects, the potential profiles, and the charge distribution near the MX–A (a) and MX–MX′ (b and c) interfaces. The diagrams are obtained in terms of the Stern model. Cases (b) and (c) differ by the values of Δg^+ and Δg^-.

transport properties of the LiClO$_4$–A interface. The presence of a large number of mobile lithium ions in the composites was confirmed by ^{7}Li NMR.

In heterogeneous systems consisting of two ionic salts (of the MX–MX′ type), there are diffuse layers in both phases. Therefore, for the calculation of the surface potential using the Stern model, one has to take into account the defect adsorption in each phase to their common interface. Such a situation is demonstrated in the energy diagram depicted in Figure 2.4b and c for the contact of two intrinsic conductors MX and MX′ with Schottky defects. In this case, the surface potential is determined by six independent parameters: (i) the defect formation energies in the bulk of the phases MX and MX′, g_{01} and g_{02}, respectively; (ii) the energies Δg_1^+ and Δg_1^- of the defect adsorption for one of the phases (the corresponding adsorption energies for another phase are given by relations $\Delta g_2^+ = (g_{01} + g_{02})/2 - \Delta g_1^+$ and $\Delta g_2^- = (g_{01} + g_{02})/2 - \Delta g_1^-$); (iii) the number of active interface sites, N_S; and (iv) the potential difference between MX and MX′, or the Galvani potential $\Delta \varphi_c$, which cannot be measured directly. Preliminary estimations show that the following qualitatively different situations, displayed in Figure 2.4b and c, may occur:

- If $\Delta g_i^\pm > 0$, then the interface charge is close to zero, and two oppositely charged diffuse layers may be formed due to the existence of the potential drop between the phases. Such potential profile (Figure 2.4b) is typical for a classical semiconductor–liquid electrolyte boundary [10a] and has been earlier applied to the explanation of interface-related properties of MX–MX′ systems by Maier [10b], and Sato [10c].
- If $\Delta g_i^\pm < 0$, the surface potential and the interface charge are high, the potential has an extremum at the interface, and two double layers form near the interface in both phases with the space charge opposite in sign to the interface charge. Such a profile is presented in Figure 2.4c. It is qualitatively similar to that at intergrain boundaries of ionic crystals but the charge distribution is not symmetrical. A higher space charge is accumulated in the phase with lower value of the defect formation energy.

These variants correspond to some limiting cases. The real situation is intermediate and the interface potential is defined by all independent parameters mentioned above. In the case of doped ionic crystals, one can also take into account the concentration of dopants in the contacting phases and the segregation energy of dopants to the interface. As no data on the interface potential of the MX–MX′ systems are available in the literature, verification of the model is still difficult.

2.2.3
Thermodynamic Stability Criteria and Surface Spreading

The interfacial interaction in composites includes two principal terms, namely, the contribution of interatomic interactions between ions or atoms of adjacent phases MX and A and the contribution of the elastic energy of mechanical strains emerging inside the crystal lattice of the ionic salt near the interface. The latter originates from the misfit in lattice parameters of the contacting phases. The interaction is characterized by the interface energy γ_{MX-A} that is usually expressed as [11a,b]

$$\gamma_{MX-A} = \gamma_{MX} + \gamma_A - \gamma_a, \tag{2.3}$$

where γ_{MX} and γ_A are standard surface energies of MX and A, respectively, and γ_a is the adhesion energy. Free Gibbs energy of the mixture of the ionic salt MX and dispersoid A is given by the sum

$$G = \left(G^0_{MX} + \gamma_{MX}S_{MX} + G^{str}_{MX}\right) + \left(G^0_A + \gamma_A S_A + G^{str}_A\right) + \gamma_{MX-A} S_{MX-A}, \tag{2.4}$$

which includes standard values of the Gibbs energy of ith component G^0_i, the surface contribution $\gamma_i S_i$ (γ_i is the average value of specific surface energy and S_i is the total area of free surface of ith component), the excess energy due to lattice strains G^{str}_i, and the contribution of the interface energy $\gamma_{MX-A} S_{MX-A}$, where S_{MX-A} is the interface area.

There may be two variants of the changes in MX morphology on sintering [2x, 12a,b]:

a) If $dG/dS_{MX-A} > 0$, then the decrease in the MX–A interface area is favorable and trivial processes of MX recrystallization and grain coarsening take place by the Oswald "ripening" mechanism.
b) Conversely, if $dG/dS_{MX-A} < 0$, it is energetically favorable for the system to expand the interface area by spreading of MX along free A surfaces. With Equations 2.3 and 2.4, neglecting the lattice strain energies G^{str}_i, the condition $dG/dS_{MX-A} < 0$ may be represented as

$$dG/dS_{MX-A} = \gamma_{MX}(dS_{MX}/dS_{MX-A}) + \gamma_A(dS_A/dS_{MX-A}) + \gamma_{MX} + \gamma_A - \gamma_a < 0. \tag{2.5}$$

For ordinary wetting process, when liquid spreads along the plane surface with formation of a film, $dS_{MX}/dS_{MX-A} \approx 1$ and $dS_A/dS_{MX-A} \approx -1$. Consequently, Equation 2.3 reduces to the well-known Gibbs–Smith wetting criterion

$$\gamma_a > 2\gamma_{MX}, \tag{2.6a}$$

which holds for epitaxial growth of thin films on substrates. However, the surface in real composite materials has more complicated geometry and morphology. If highly dispersed oxide is taken as a heterogeneous dopant, it acts like a porous matrix easily impregnated with the ionic salt. In this case, formation of new MX–A interfaces does not lead to generation of new free MX surfaces, so $dS_{MX}/dS_{MX-A} \approx 0$, $dS_A/dS_{MX-A} \approx -1$, and the wetting may proceed at

$$\gamma_a > \gamma_{MX}, \tag{2.6b}$$

that is, much easier than in two-layer systems. In any case, Equations 2.6a and 2.6b display that an increase in the adhesion energy as well as decrease in the MX surface energy favors the wetting effect. The change in morphology of the heterogeneous system as a result of sintering for both $\gamma_a > 2\gamma_{MX}$ and $\gamma_a > \gamma_{MX}$ is schematically shown in Figure 2.5. For real systems with undefined morphology when the spreading is

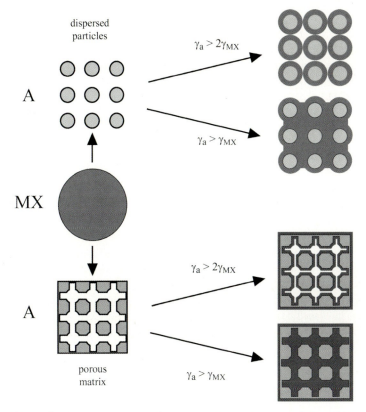

Figure 2.5 The change in morphology on sintering of the mixture of MX with dispersed particles (upper row) or porous matrix (bottom row) of A oxide phase in the cases $\gamma_a > 2\gamma_{MX}$ and $\gamma_a > \gamma_{MX}$.

accompanied by generation of strong strains, conditions (2.6a) and (2.6b) may be represented in a general form

$$\gamma_a > (2-\theta_r)\gamma_{MX} + \beta_{MX} + \beta_A, \tag{2.7}$$

where θ_r is the roughness factor ($0 \leq \theta_r \leq 1$), $\beta_{MX} = dG^{str}_{MX}/dS_{MX-A}$, $\beta_A = dG^{str}_A/dS_{MX-A}$, and β_{MX}, $\beta_A > 0$ due to a misfit between lattice parameters of MX and A. Hence, the strains hinder spreading of MX onto the surfaces of A. The last two terms seem to increase with decreasing grain size of the oxide; in particular, this follows from the analysis of X-ray diffraction (XRD) peak broadening that makes it possible to estimate both the average crystallite size and lattice deformation. In physical terms, such situation may be explained by a strong curvature of the oxide surface when its radius becomes comparable to the molecular size or lattice parameter of MX. In this case, the term β_{MX} becomes too large, condition (2.7) is not satisfied anymore, and spontaneous spreading does not occur. This effect indeed was observed by Ponomareva et al. [13a,b] in a series of composites $MHSO_4$–SiO_2 (M = Rb, Cs), where silica had different pore structures. In these experiments, the pore size of silica was varied in a wide range of 1.4–300 nm. Note that the conductivity versus pore size dependences had a maximum, and the ionic salt easily spreads onto large pores and did not penetrate the pores smaller than ∼3.5 nm.

After the salt covers all free surfaces of highly dispersed oxide ($S_{MX} \ll S_{MX-A}$), the Gibbs energy of MX changes to

$$G'_{MX} \approx G^0_{MX} + \gamma_{MX-A} S_A/2 + G^{str}_{MX} = G^0_{MX} + (\gamma_{MX-A} + 2\beta_{MX})S_A/2. \quad (2.8)$$

For physically reasonable values $\gamma_A > \gamma_a$ and $\beta_{MX} > 0$, the interface energy γ_{MX-A} is positive and G_{MX} is expected to increase. The wetting effect should proceed until all free surface of A is covered by the more mobile ionic component MX. If the A particle size is sufficiently small, $L_A \sim 10$ nm, then after prolonged sintering one can obtain the MX–A nanocomposite where the effective MX grain size of MX (L_{MX}) is close to L_A. Thus, the main reason for such spontaneous self-dispersion and changing physical properties of the ionic salts in nanocomposites MX–A is the interfacial interaction between the phases. It should be emphasized that in contrast to many nanosystems, the nanocomposite formed on the interfacial spreading is thermodynamically stable, that is, exists in a local thermodynamically equilibrium state given by the value of the specific surface area (or grain size) of the oxide.

2.3
Nanocomposite Solid Electrolytes: Grain Size Effects

When the particle size becomes smaller than 10–100 nm, physical properties of solids substantially change. During the past two decades, a significant progress was achieved in the research of various nanosystems, such as nanostructured pure and composite materials, metal nanoparticles, carbon nanotubes, mesoporous systems, and so on; a brief but thorough review of the nanoscale effects in solid-state electrochemical cells was presented in Chapter 4 of the first volume. One may note that there are two types of size effects in the heterogeneous systems: a trivial one appearing due to the decrease in the grain size without strong impact of the interfacial interactions, and a true size effect when the interfaces play a crucial role [2k,r,m,w, 14, 15]. In this section, only the true size effects in nanocomposite electrolytes are considered.

As follows from Equation 2.8, in a MX–A composite, the ionic salt has an excess Gibbs energy caused by the influence of MX–A interfaces. The temperature dependences of Gibbs free energies G^0_{MX} and G'_{MX} are schematically presented in Figure 2.6. One can see that the melting temperature of MX in the composites should be lower than that in pure MX salt. The characteristic grain size corresponding to the situation when noticeable deviations in the thermodynamic properties are observed is ∼10 nm [2m,w, 12b, 15]. The following approximate relationship for the temperature of a phase transition between high- and low-temperature MX phases, α and β, can be obtained [2x, 12b, 15c] by equating the values of Gibbs energies:

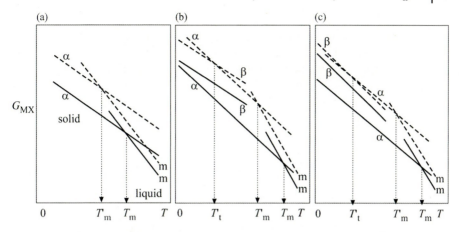

Figure 2.6 Temperature dependences of Gibbs free energies of ionic salt in standard conditions (solid lines) and in nanoparticles (dashed lines). Solid phases and melt are denoted by symbols α, β, and m, respectively. In the cases (b) and (c), the phase β does not exist in normal conditions, but appears in the nanosized particles.

$$\frac{T_t}{T_t^0} = 1 - \left[(\gamma_{MX}^\beta S_{MX}^\beta - \gamma_{MX}^\alpha S_{MX}^\alpha) + \frac{1}{2}(\gamma_{MX-A}^\beta S_{MX-A}^\beta - \gamma_{MX-A}^\alpha S_{MX-A}^\alpha) + \Delta G^{str} \right] \frac{1}{H_t}, \quad (2.9)$$

where T_t is the standard temperature of the transition, superscripts α and β correspond to high- and low-temperature phases, respectively, ΔG^{str} is the difference in the strain energies of phases α and β, and H_t is the enthalpy of the phase transition. For pure MX, S_{MX-A} is zero, whereas for composites with highly dispersed oxides $S_{MX-A} \gg S_{MX}$ and the second and third terms in brackets become prevailing. The interface energies γ_{MX} and γ_{MX-A} decrease with temperature; this explains why the transformation temperature should decrease if the phase transition is accompanied only by small changes in volume and ΔG^{str} ($S_{MX-A}^\alpha \approx S_{MX-A}^\beta \sim L_A^{-1}$; $\Delta G^{str} \approx 0$). The latter situation is indeed usually observed for thin films where the layer thickness plays the role of particle size in the above formalism. According to many experimental observations, the phase transition temperatures in films can decrease by hundreds of degrees as L decreases to 10 nm [15a–c,f]. Qualitatively similar effects have been observed in nanocomposites CsCl–Al$_2$O$_3$ [16a] and CsHSO$_4$–SiO$_2$ [16b], where the temperatures of phase transitions (melting and/or solid-state transformation) decrease with the concentration of oxide additive. The temperature of α–β transition in Li$_2$SO$_4$ decreases in Li$_2$SO$_4$–Al$_2$O$_3$ and the transition becomes diffuse [12a]. Careful calorimetric studies of $(1 - x)$AgI–xAl$_2$O$_3$ [17] nanocomposites show that as x increases, two phase transitions proceed; one occurs at the temperature characteristic of the bulk state, whereas another is observed at a different temperature and is characterized by a strong hysteresis. The smooth character of the phase transitions and strong hysteresis seem to be a general phenomenon in the nanocomposite systems.

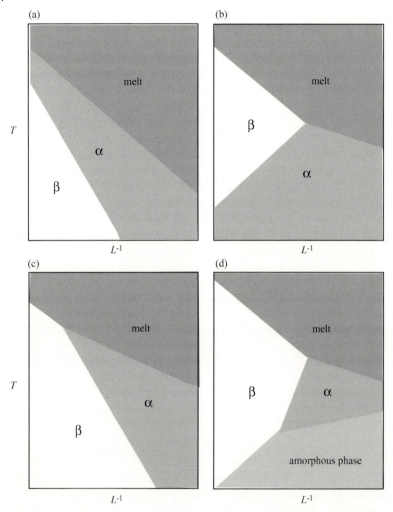

Figure 2.7 Phase diagrams represented in coordinates T_t (L^{-1}) for films of different metals deposited on supports [15a–c,f].

In general, the phase transition temperatures in a composite can either decrease or increase depending on the particular form of thermodynamic functions $G_{MX} = f(T)$ and $\Delta G^{str} = f(T)$ in different phases. The available literature contains several types of phase diagrams (Figure 2.7a–c) presented as $T_t(L^{-1})$ for different metal films [15a–c,f]. The cases c and d are of special interest as the examples of a strong size effect, when new phases unusual for the pure component are stabilized in thin films. Such effects have been observed in Al_2O_3 [18a], ZrO_2 [18b,c], and TiO_2 [18d]: crystals of these oxides smaller than 10–20 nm form metastable (γ-, δ-, θ-Al_2O_3), anatase TiO_2, or high-temperature (t-, c-ZrO_2) polymorphs. Most experimental observations in the heterogeneous systems such

2.3 Nanocomposite Solid Electrolytes: Grain Size Effects

as AgI–Al$_2$O$_3$ [17, 19a], MNO$_3$–Al$_2$O$_3$ (M = Li, Na, K) [19b], MNO$_3$–Al$_2$O$_3$ (M = Rb, Cs) [19c,d], RbNO$_3$–SiO$_2$ [19e], CsHSO$_4$–SiO$_2$ [13, 19f], and LiClO$_4$–A (A = Al$_2$O$_3$, LiAlO$_2$, SiO$_2$) [9, 19g] may be explained by the appearance of amorphous phases of the ionic salt in the composites. This is evidenced by the following effects:

i) A strong decrease in the integral intensity of XRD peaks of all crystalline phases and appearance of a wide halo on the electron diffraction patterns.
ii) Decreasing molar enthalpies of all phase transitions in MX, including the melting enthalpy. Instead, a diffuse peak appears on differential scanning calorimetry (DSC) curves at temperatures much lower than the melting point.
iii) A disappearance of abrupt conductivity changes due to MX phase transitions in the MX–A composites. The Arrhenius conductivity plots are nonlinear for some nanocomposites, where the estimated charge carrier concentration is comparable to the overall number of cations [2k], as typical for the superionic conductors or ion-conducting glasses. This is also confirmed by the absence of the conductivity changes on melting.
iv) Effects (i)–(iii) systematically become more pronounced when the total number of MX–A interfaces increases. This suggests that the amorphous phase is interface induced (nonautonomous) and is located only at the MX–A interfaces.

The volume and molar fractions of the interfacial phase (including amorphous one), f_S and x_S, respectively, its thickness (λ), and the fraction of residual bulk phase of MX, f_{bulk} (or x_{bulk}), may be estimated using the brick wall model [2k,p,x, 17, 19d]:

$$f_S = 2\beta_{cm}\left(\frac{\lambda}{L_A}\right)f(1-f), \qquad f_{bulk} = 1 - f - f_S \quad (f_{bulk} \geq 0), \tag{2.10}$$

$$x_S = 2\beta_{cm}\left(\frac{\lambda}{L_A}\right)\frac{\delta_c x(1-x)}{[1+x(\delta_c-1)]}, \qquad x_{bulk} = 1 - x - x_S \quad (x_{bulk} \geq 0), \tag{2.11}$$

where the parameter β_{cm} depends on the morphology of the composite (for cubic blocks, $\beta_{cm} \approx 3$), f (or x) is the total volume (or molar) fraction of oxide, and $\delta_c = \mu_A \rho_{MX}/\mu_{MX}\rho_A$, where μ_j and ρ_j are molecular weights and densities of the components, respectively. The values of x_{bulk} may be determined from the integral intensities of DSC peaks corresponding to phase transitions, including melting. Figure 2.8 compares the relative molar fractions of ionic salt not transformed into the amorphous state, $x_{bulk}/(1-x)$, in the composites RbNO$_3$–Al$_2$O$_3$ [19d] and AgI–Al$_2$O$_3$ [17] with a highly dispersed alumina having the specific surface area of 200–270 m^2 g^{-1}, and the data on AgI–Al$_2$O$_3$ prepared with γ-alumina (grain size of 0.06 μm) [20]. The theoretical dependences obtained from Equation 2.11 provide good fit to the data. The values of λ obtained by fitting are 3 and 4 nm for AgI–Al$_2$O$_3$ [17] and RbNO$_3$–Al$_2$O$_3$ [19d], respectively, and 7 nm for AgI–Al$_2$O$_3$ [20] composites obtained with another alumina modification. At sufficiently high concentrations of the highly dispersed alumina ($x > 0.6$–0.7 or $f > 0.5$–0.6), the effective size of MX particles is comparable to λ and virtually the whole volume of the ionic salt is in the amorphous state.

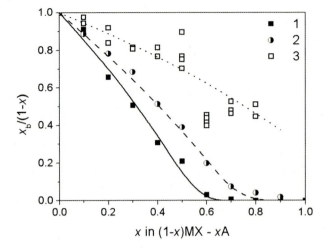

Figure 2.8 The relative molar fraction of ionic salt not transformed into the amorphous state, $x_{bulk}/(1-x)$, in composites RbNO$_3$–Al$_2$O$_3$ [19d] and AgI–Al$_2$O$_3$ [17] with highly dispersed alumina (the specific surface area of 200–270 m^2 g^{-1}), points (1) and (2), respectively, and the data reported by Lee et al. [20] for AgI–Al$_2$O$_3$ composites prepared with γ-alumina with the grain size of 0.06 μm. Lines are obtained using Equation 2.11 at λ = 4, 3, and 7.2 nm for curves 1, 2, and 3, respectively.

The self-dispersion proceeds at significant rates even at temperatures substantially below the MX melting point, that is, when the ionic salt is in the crystalline state. Due to the solid-phase spreading, the crystalline phase spontaneously transforms into the amorphous state. One possible reason for the spontaneous amorphization due to liquid- or solid-phase spreading could be the relaxation of elastic strains that arise in the MX bulk because of the salt spreading over the surface and in the pores of oxide matrix. If the lattices of the contacting phases do not match one another, the contribution of the elastic energy becomes substantial. Moreover, in contrast to thin films, the MX particles in nanopores are adhesively bound to the randomly oriented oxide surfaces surrounding a pore, which should lead to the rather active formation of microdomains and provide the excess surface energy. Apparently, the structural relaxation can occur by spontaneous amorphization of an ionic salt. Crystal phase may relax into more stable amorphous state with larger particles, and the increase in the effective particle size from L'' to L' may be due to the change in the particles shape from polyhedral to smooth one. This process may be illustrated by a schematic phase diagram shown in Figure 2.9 [2x, 21]. The concept of the spontaneous transition from the crystalline to amorphous state was discussed earlier in the context of reasons responsible for mechanical alloying [22]. Data on amorphization of alkali metal halides [23] and other inorganic crystalline hydrates [24] in nanopores are available. The amorphous hydrates incorporated into nanoporous matrices exhibit unusual thermodynamic properties relevant to changes in hydration and dehydration [24c].

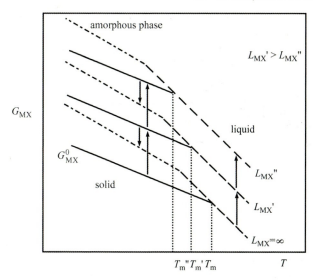

Figure 2.9 Phase diagram demonstrating the possibility of spontaneous formation of the amorphous phase in nanocomposites. As a result of the spreading, MX transforms first into the metastable crystalline state corresponding to the grain size of L''_{MX}. Then this state under the action of interfaces relaxes into more stable amorphous state with larger particles; the increase in the effective particle size from L''_{MX} to L'_{MX} may be due to the change in the shape of particles.

The composites MX–A exhibit an enhanced conductivity at temperatures below melting point or superionic phase transition in MX [2]. The conductivity goes through a maximum at 20–50 vol% of the oxide and is qualitatively described by the percolation model with two thresholds [25]. In the vicinity of these thresholds (at $f \sim 0.9$ and 0.1), the permittivity maxima are observed [17, 25d,e]. At temperatures where MX exists in molten or superionic states, the conductivity of the composites is lower than that of the individual salt and may be explained by a standard percolation model with one percolation threshold.

Most composite electrolytes reported in the literature are submicrometric systems with the effective grain size of 60–1000 nm. Transport properties of nanocomposites with the grain size on the order of 10 nm are less studied. Nevertheless, true size effects are observed only in these systems. For nanocomposites, there is a tendency to leveling conductivity parameters of the low- and high-temperature phases with the increase in the concentration of the oxide component. As a result, in $(1-x)$ AgI–xAl$_2$O$_3$ system, the conductivities of the high-temperature (superionic) and low-temperature phases become equal; that is, phase transition of the salt to the superionic state gradually disappears (Figure 2.10a) [19a, 26]. The conductivity maximum ($\sim 10^{-3}$ S cm^{-1} at 25 °C) shifts to higher concentrations, 0.5–0.6 mol % (Figure 2.10b). The activation energy in both superionic and low-temperature phases increased monotonically with x. On the temperature dependence of conductivity of the nanocomposite Li$_2$SO$_4$–Al$_2$O$_3$, there is no sharp conductivity jump at

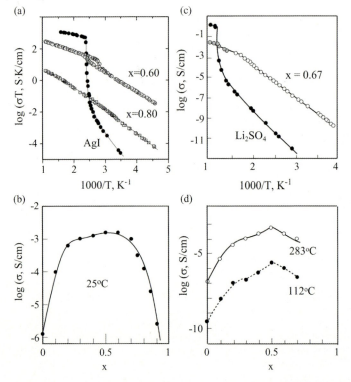

Figure 2.10 Dependences of conductivity of nanocomposites $(1-x)$AgI–xAl$_2$O$_3$ (a and b) [19a, 26] and $(1-x)$Li$_2$SO$_4$–xAl$_2$O$_3$ (c and d) [12a] on temperature (a and c) and concentration of alumina (b and d). The change of the character of the phase transition in the nanocomposites is evident. Two curves for 0.33Li$_2$SO$_4$–0.67Al$_2$O$_3$ correspond to the samples prepared in different conditions.

575 °C associated with the transition of lithium sulfate to the superionic state (Figure 2.10c) [12a]. Instead, the σ(T) dependence has a shape typical for solid electrolytes with the fluorite structure in the vicinity of the "diffuse" phase transition. Doping of alkali metal nitrates with highly dispersed alumina ($S_A = 270\,\text{m}^2\,\text{g}^{-1}$) is accompanied by a sharp increase in the conductivity (Figure 2.11) [19b,c]. The conductivity of $(1-x)$MNO$_3$–xAl$_2$O$_3$ (M = Li, Na, K, Rb, Cs) composites exhibits maxima at $x = 0.5$–0.6; the relative increase (σ/σ$_0$) depends on the cation type and varies in the range from 10^2 (CsNO$_3$–Al$_2$O$_3$) to 10^8 at 343 K (in LiNO$_3$–Al$_2$O$_3$). For $x > 0.5$–0.6, the Arrhenius plots of all composites are not linear and no conductivity jumps are observed. For example, for 0.4RbNO$_3$–0.6Al$_2$O$_3$, the Arrhenius curve is a smooth line without four sharp jumps due to phase transitions (including melting) observed in pure rubidium nitrate [19d]. It has been reported that nitrate-based composites are proton conductors in humid atmosphere [27]. We have studied in detail the transport properties of composites based on rubidium nitrate [19c,e]. The

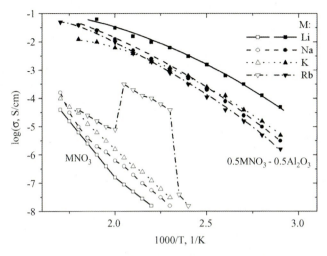

Figure 2.11 Temperature dependences of pure nitrates and the composites $0.5MNO_3$–$0.5Al_2O_3$ with high-dispersed alumina (the specific surface area $S_A = 270\,m^2\,g^{-1}$); M = Li, Na, K, Rb, Cs [19b–d].

decomposition voltage, $U_d = 4$ V, obtained by voltammetric measurements in vacuum was close to the corresponding values in fused alkali metal nitrates and halides. This makes it possible to assume that rubidium cations are the main mobile charge carriers in $RbNO_3$–Al_2O_3. Oxide additions in the composites $CsCl$–Al_2O_3 result in a gradual suppression of the abrupt conductivity change associated with the transition to the high-temperature polymorph, suggesting its stabilization at the interface [16a]. In nanocomposites $LiClO_4$–Al_2O_3, no conductivity change is observed at the melting point of lithium perchlorate [9a].

The conductivity data fairly correlate with results of calorimetric studies of the same systems. A change in temperature, a decrease in intensity, or a complete disappearance of phase transitions is observed by both electrical measurements and DSC method. The strongest effects are observed only if the specific surface area of oxide additive is sufficiently high ($S_A > \sim 100\,m^2\,g^{-1}$), oxide particles are not packed into dense aggregates, and the components have high adhesion. Moreover, even in this case one should provide conditions for effective spreading of the ionic salt MX on the oxide surfaces (thorough preliminary mixing followed by sintering). As a particular case, new metastable crystalline phases of AgI have been found in the AgI–Al_2O_3 composites. One of such phases, a 7H polytype of β-AgI [20, 28], cannot be prepared by any other method. Amorphous AgI may be obtained by fast cooling AgI–Al_2O_3 nanocomposites from 750 °C in liquid nitrogen [29a]. Amorphous AgI also forms in the composites AgI–A (A = ZrO_2, CeO_2, Sm_2O_3, MoO_3, WO_3) obtained by quenching [29b]. Recently, it was shown that nanocomposites AgI–A (A = ZrO_2, Al_2O_3 xerogels and aerogels), where oxide additives had a high specific surface area of 150–500 $m^2\,g^{-1}$, exhibited unusual transport and thermodynamic properties [29c,d].

2.4
Ionic Transport

In the composite solid electrolytes containing ionic salt and oxide phases, the ionic transport occurs mainly via interfaces. This complicates theoretical description and prediction of the transport properties in such systems. Theoretical calculations have been made on the basis of model estimations for specific morphologies [2c,g,l, 3], simplified Maxwell–Garnett equation [30], the effective medium models [31], percolation theory, and computer simulations of several model systems [32].

A most accurate calculation of the conductivity in MX–A may be carried out for limiting cases ($f \to 0$) and ($f \to 1$), when the particles of A or MX are placed in the matrix of the second phase and are isolated from each other [3, 30]. Assuming that the dispersoid particles with a size L_A are isolated by the phase MX and are covered by conducting layers with the effective thickness λ, Jow and Wagner derived an expression for the conductivity (σ) [3]:

$$\sigma = \sigma_{MX} + 3\sum_i ev_i \langle n_i \rangle K\left(\frac{\lambda}{L_A}\right)\frac{f}{(1-f)}, \tag{2.12}$$

where σ_{MX} is the bulk conductivity of MX, v_i is the mobility of ith defects, $\langle n_i \rangle$ is the average arithmetic value between the defect concentrations in the bulk and on the interface, and the summation is made over all mobile defects. This expression satisfactorily describes experimental data for CuCl–Al$_2$O$_3$, especially the conductivity dependence on the oxide grain size. However, it is only applicable for low concentrations of the oxide; at $f \to 1$, the conductivity should infinitely increase. Maier has proposed a more reasonable equation [2c,g,l]:

$$\sigma = (1-f)\sigma_{MX} + 3e\beta_L \frac{2\lambda_D}{L} fv\sqrt{n_\infty n_0}, \tag{2.13}$$

where β_L is the fraction of double layers participating in the conduction ($1/3 < \beta_L < 2/3$), λ_D is the Debye length, and n_∞ and n_0 are the defect concentrations in the MX bulk and at the interface, respectively. Using this equation and known defect mobility values, Maier quantitatively interpreted the data on conductivity in MX–Al$_2$O$_3$ (MX = AgCl, AgBr, TlCl) in the concentration range of $0 < f < 0.20$; the parameter β_L was taken as 0.5; the grain size of alumina was varied within the range of 0.06–0.3 μm. Equation 2.13 is in a better agreement with experiments but cannot explain the conductivity maximum observed in all composites. In Ref. [30a], the following expression was obtained on the basis of a simplified version of the Maxwell–Garnett equation:

$$\sigma = \left(\sigma_{MX} + \frac{3\lambda}{L_A} f\sigma_S\right)\frac{2(1-f)}{2+f}, \tag{2.14}$$

where σ_S is the conductivity of MX in the interfacial layer of thickness λ. This expression describes the conductivity maximum, despite that the initial Maxwell–Garnett model is valid only for isolated particles, that is, at low f values. Fujitsu et al.

[30b,c], however, reported that this equation did not fit experimental data for $SrCl_2–Al_2O_3$, $CaF_2–Al_2O_3$, and $CaF_2–Al_2O_3$.

The conductivity of model composites consisting of ordered isolated cubic particles A, covered by a conducting layer, has been calculated by Stoneham et al. [31a] and Wang et al. [31b]; the authors varied the conductivity profile within the conducting layers and distance between the particles. The solutions have a complex analytical form and the dependence of conductivity versus concentration has a maximum depending on the conductivity and relative thickness of the conducting layer.

For the calculations in a whole concentration range, it was proposed to use the effective medium model [31]. Accordingly, the MX–A composite was modeled as a statistical mixture of spherical particles of MX and A with the same size. For conductivity of the composite (σ), the following expression was obtained [31e]:

$$\left(\frac{f}{d_p}\right)^2 \frac{\sigma_A - \sigma}{\sigma_A + ((z_p/2)-1)\sigma} + \left(1 - \frac{f}{d_p}\right)^2 \frac{\sigma_{MX} - \sigma}{\sigma_{MX} + ((z_p/2)-1)\sigma}$$
$$+ 2\left(\frac{f}{d_p}\right)\left(1 - \frac{f}{d_p}\right) \frac{\sigma_S - \sigma}{\sigma_S + ((z_p/2)-1)\sigma} = 0, \qquad (2.15)$$

where d_p is the packing density, σ_S is the conductivity along the MX–A interface, and z_p is the packing coordination number. The concentration dependences obtained from this equation have two percolation thresholds and a smooth maximum. Qualitatively similar dependences have been obtained elsewhere [31c,d]. In Ref. [32a], the conductivity was estimated from probability of the current flow through a resistance network imitating the ensemble of the MX and A particles of the same size L, arranged in a simple cubic lattice:

$$\sigma = \sigma_{MX}\left[\frac{8\lambda}{3L}\frac{n_1}{n_0}(1-f)^3 f^2 + (1-f)^2\right], \qquad (2.16)$$

where λ is the thickness of the conducting layer, and n_1 and n_0 are the charge carrier concentrations in the conducting layer and in the bulk of MX, respectively. Equation 2.16 fairly fits the data on $LiI–Al_2O_3$ composites [1] with the conductivity maximum at 40 vol% alumina. If the particles distribution is not uniform, the maximum shifts to higher f values.

Results of computer simulations using the random resistor network model have been reported in Refs [25a–c, 32d–f]; it was shown that the concentration dependence of conductivity is described by the percolation equation with two thresholds: p_1 (insulator MX–composite ionic conductor) and p_2 (composite ionic conductor–insulator A); the values of p_1 and p_2 and the percolation exponents depend on the coordination number and dimensionality of the network [25a, 32e]. The dependences of conductivity on the grain size agree qualitatively with the experimental observations. The drawback of these simulations is a lack of analytical expression appropriate for fast estimations. Nan and Smith [25d] proposed a method for the quantitative calculation of the composite conductivity by dividing the whole concentration range into three regions, separated by some characteristic concentrations:

$$f_1 = p_1/(1+\lambda/L_A)^3, \qquad (2.17)$$

$$f_2 = 1/(1+\lambda/L_A)^3, \qquad (2.18)$$

$$f_3 = 1 - p_1 + p_2 f_2. \qquad (2.19)$$

The value of f_1 corresponds to the real percolation threshold that may differ from that for the theoretical value of $p_1 = 0.15$. At $f > f_1$, first infinite conducting cluster appears; at $f = f_2$, the infinite cluster envelopes the total volume of the composite and the conductivity reaches its maximum. The concentration f_3 corresponds to the percolation threshold from conducting to insulating state, and also differs from the ideal value of $p_2 = 0.85$. At $f < f_1$ and $f > f_3$, the conductivity is estimated using the Maxwell–Garnett equations. In the intermediate region $f_1 > f > f_3$, the effective medium model is employed with the parameters different for the cases $f < f_2$ and $f > f_2$. The obtained theoretical curves were reported to provide good quantitative description of the experimental data on Li_2SO_4–Al_2O_3 and $LiCl$–Al_2O_3 [25d]. A similar model was applied by Siekierski and Przyluski [32h] for interpreting the conductivity data in polymer composite electrolytes containing oxide additives. In general, such model is rather complicated as it includes four different equations. Moreover, there is an uncertainty in the determination of f_1, f_2, and f_3 values.

For description of the conductivity of composites of the "conductor–insulator" type, a standard mixing equation was proposed [33]:

$$\sigma^\alpha = (1-f)\sigma_{MX}^\alpha + f\sigma_A^\alpha. \qquad (2.20)$$

In traditional mixing rules, the parameter α is taken constant: $\alpha = 1$ and -1 for oriented composites consisting of parallel layers of the components when the conductivity is measured in parallel and perpendicular directions, respectively. Note that the cases when $\alpha = 0$ and $1/3$ correspond to the Lichtenecker [34a] and Landau–Lifshitz [34b] equations, respectively. Analysis shows that the parameter α is determined by the composite morphology and may vary with the concentration. In Ref. [33], it was assumed that α may be approximated by a linear dependence:

$$\alpha = (1-f)\alpha_1 + f\alpha_2, \qquad (2.21)$$

where α_1 and α_2 are determined by morphology of the composites in the dilute limits $f \to 0$ and $f \to 1$, respectively. The generalized mixing rule, Equation 2.20, with the parameter α given by Equation 2.21 provides satisfactory description of the percolation-type behavior [33].

For composite electrolytes, the mixing rule may also be rewritten in the following form:

$$\sigma^\alpha = (1-f-f_S)\sigma_{MX}^\alpha + f_S\sigma_S^\alpha + f\sigma_A^\alpha, \qquad (2.22)$$

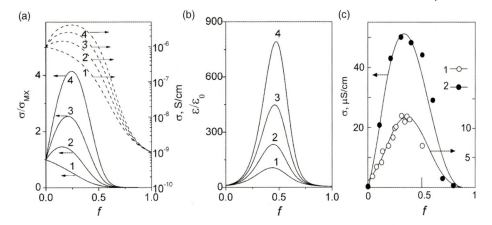

Figure 2.12 Theoretical dependences dc conductivity (a) and low-frequency dielectric permittivity (b) for a composite with $\sigma_{MX} = 1 \times 10^{-6}$ S cm^{-1}, $\sigma_A = 1 \times 10^{-9}$ S cm^{-1}, $\sigma_S = 1 \times 10^{-5}$ S cm^{-1}, and $\beta = 3$ for the ratio $\lambda/L = 0, 0.0167, 0.0333,$ and 0.05, curves 1, 2, 3, and 4, respectively. Parameters (α_1, α_2) are equal to (0.667, −0.333). The conductivity data (a) are represented in linear (solid lines) and logarithmic (dashed lines) scales. (c) Comparison of experimental conductivity data for composites LiI–Al$_2$O$_3$ [1] (1) and AgCl–Al$_2$O$_3$ [25e] (2) with theoretical curves (Equation 2.23).

where f_S and σ_S are total concentration and the conductivity of interfacial regions, the parameter $\sigma(f)$ depends on the concentration in accordance with Equation 2.21, and $f_S < f$. Evidently, this equation reduces to Equation 2.20 at $f_S \to 0$ or $\sigma_S \to 0$ when the contribution of interfaces to the total conductivity is negligible. Substituting f_S estimated from the brick wall model, Equation 2.10, yields [33b]

$$\sigma^{\alpha(f)} = (1-f)\left(1-f\frac{2\beta_{cm}\lambda}{L_A}\right)\sigma_{MX}^{\alpha(f)} + f(1-f)\frac{2\beta_{cm}\lambda}{L_A}\sigma_S^{\alpha(f)} + f\sigma_A^{\alpha(f)}. \quad (2.23)$$

The theoretical dependences σ versus f for a composite with $\sigma_{MX} = 1 \times 10^{-8}$ S cm^{-1}, $\sigma_A = 1 \times 10^{-10}$ S cm^{-1}, $\sigma_S = 1 \times 10^{-3}$ S cm^{-1}, $\beta_{cm} = 3$, and $\lambda/L = 0.1$ obtained for different α_1 and α_2 are presented in Figure 2.12. One may select the following important cases:

i) $\alpha_1 > 0, \alpha_2 > 0$. The percolation threshold values p_1 and p_2 are close to 0 and 1, respectively. At low A concentration, the oxide particles are distributed around MX grains forming the conductive cluster even at $f \to 0$, whereas at high concentrations each oxide particle is covered by thin MX layers. As a result, the percolation cluster exists practically in the whole concentration range and all composites have an enhanced conductivity.

ii) The case when $\alpha_1 > 0, \alpha_2 < 0$ is most typical. The p_1 threshold is close to 0, whereas the position of p_2 depends on the α_1/α_2 ratio and corresponds to maximum of low-frequency dielectric permittivity. At $f = p_2^* = \alpha_1/(\alpha_1 - \alpha_2)$, there is an inflection point where the maximum conductivity change is observed in logarithmic scale. This point is higher than the percolation threshold p_2 and may be called a "logarithmic" percolation threshold: at $f < p_2^*$, the composite behaves like a conductor, whereas at $f < p_2^*$, it behaves as a typical dielectric. At $\alpha_1 > 0$, the curve $\sigma(f)$ has a maximum; its position is determined by α_1 and α_2. For random mixtures, it was shown that $\alpha_1 < 2/3$ and $\alpha_2 > -1/3$ [36]. The real composites cannot be considered as statistical mixtures since the A particles at low concentrations are not distributed uniformly but located along MX grains. This leads to an increase in α_1 values up to 1.

iii) $\alpha_1 < 0$, $\alpha_2 > 0$; $|\alpha_1| < |\alpha_2|$. In this case, at low f, the conductivity decreases. At higher concentrations, the conduction increases with f and goes through a maximum at $0.8 < f < 1$. In practice, no such composites are known.

iv) $\alpha_1 < 0$, $\alpha_2 < 0$. In this case, no conductivity enhancement may be observed.

The theoretical curves, Equation 2.23, fit well experimental data for the systems LiI–Al$_2$O$_3$ [1] and AgCl–Al$_2$O$_3$ [25e] (Figure 2.12); they are in a qualitative agreement with effective medium model and percolation theory [31, 32]. The generalized mixing rule, Equation 2.23, with four fitting parameters (α_1, α_2, λ/L, and σ_S) may be used for description of the experimental data.

In the case of $(2\beta_{cm}\lambda/L)\sigma_S \gg \sigma_{MX}, \sigma_A$, Equation 2.23 reduces to the form

$$\sigma \approx \sigma_S [f(1-f)2\beta_{cm}(\lambda/L_A)]^{1/(\alpha_1(1-f)+\alpha_2 f)}, \qquad (2.24)$$

which is suitable for approximate analysis of the dependences of conductivity on concentration and grain size. According to the theoretical models reported earlier [2c,g,l, 3, 30, 31], the conductivity should increase inversely to the grain size, $\sigma \sim L_A^{-1}$, or proportionally to the specific surface area, $\sigma \sim S_A$ (which may be estimated as $S_A \approx 3(\varrho_A L_A)^{-1}$, where ϱ_A is the oxide density). This conclusion follows from Equations. The experimental dependences σ versus S_A and σ versus L_A^{-1}, obtained for LiI–Al$_2$O$_3$ [35a], AgI–Al$_2$O$_3$, and CuCl–Al$_2$O$_3$ [35b] with different grain sizes, are shown in Figure 2.13. From Equation 2.24, it follows that the conductivity dependences on the grain size or the specific surface should be described by power functions $\sigma \sim L_A^{-1/\alpha}$ and $\sigma \sim S_A^{1/\alpha}$, where the exponent $\alpha(f)$ is determined by the parameters α_1, α_2, and $\alpha = \alpha_1(1-f) + \alpha_2 f$, rather than a linear function. The power dependences $\sigma = \sigma_0 + AS_A^{1/\alpha}$ and $\sigma = \sigma_0 + AL_A^{-1/\alpha}$ with an exponent of $\alpha = 0.67 \pm 0.10$ fit the experimental data with smaller deviations than linear models ($\alpha = 1$). Substituting typical values of $\alpha_1 \approx 1$, $\alpha_2 \approx 0$, and $f \approx 0.3$–0.4 into Equation 2.21, one can obtain $\alpha \approx 0.6$–0.7 that is in agreement with the experiment. Equation 2.24 shows that linear dependences $\sigma \sim S_A$ or $\sigma \sim L_A^{-1}$ could be observed only in the limit $f \to 0$ at $\alpha_1 \approx 1$, or for a special morphology of the composites. The latter is observed, for example, when MX and A layers are oriented parallel to the direction of the electric field direction and $\alpha_1 = \alpha_2 = 1$. In general, the conductivity should obey the power function with the exponent higher than unity.

It should also be mentioned that all the equations mentioned above are valid until the surface conductivity σ_S does not depend on the effective size of MX crystals. However, generally one should take into account the σ versus L dependence, that is, the size effect for the conductivity. Analytical forms of such dependence have been obtained by Maier [2l, 37] and Khaneft [38] for pure nanocrystal under assumption

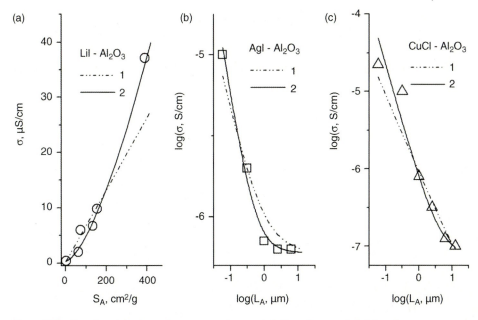

Figure 2.13 Experimental values of conductivity of composites LiI–Al$_2$O$_3$ [35a], AgI–Al$_2$O$_3$, and CuCl–Al$_2$O$_3$ [35b] with different grain sizes of alumina fitted to linear ($\sigma = \sigma_0 + A_1 S_A$ and $\sigma = \sigma_0 + A_2 L_A$) and power ($\sigma = \sigma_0 + A_3 S_A^{1/\alpha}$ and $\sigma = \sigma_0 + A_4 L_A^{-1/\alpha}$) dependences, curves 1 and 2, respectively.

that the concentration, mobility of the defects on the surface, and the defect formation energy in the bulk are constant. According to the theoretical estimations, the defect formation energy decreases on scaling down of the grain size [39]. The Debye length, being a function of the defect concentration, should also diminish. The ion mobility may also change with L. All these effects make it difficult to calculate the conductivity variations versus grain size. Moreover, at high concentration of defects, phase transitions into the disordered states may occur on the surface [40] or in the bulk [41]. In nanocomposites, the situation may be still more complicated. Nevertheless, most experimental data may be satisfactorily interpreted under the assumption that the conductivity occurs in the interface disordered phase comprising a layer of the constant thickness λ, which is characterized by a high defect concentration.

2.5
Other Properties

The addition of oxide particles to an ionic salt can greatly improve mechanical properties such as ultimate tensile strength and modulus. The mechanical properties of composite materials usually depend on structure. Thus, these properties typically depend on the shape of additive particles, their volume fraction, and the interfaces between the components. The strength of composites depends on such factors as the brittleness or ductility of the inclusions and matrix [42]. Much attention is paid to improvement of mechanical properties of polymer-based composite electrolytes for lithium batteries [43] or polymer electrolyte membranes in fuel cells (see Chapter 10 of the first volume) [44]. The most pronounced effects in both mechanical enforcement and conductivity enhancement were observed for nanocomposite systems.

There is practically no information in the literature concerning magnetic and optical properties of composite solid electrolytes. Chen et al. [45] has reported that the Morin temperature of α-Fe_2O_3 in β-AgI–α-Fe_2O_3 composites decreased compared to that of pure iron oxide. The effect was proposed to be caused by bulk interaction between the components. Results of luminescence spectroscopy studies of AgI–Al_2O_3 nanocomposites have been reported elsewhere [46]. It has been shown that after prolonged sintering of nanocomposite 0.02AgI–0.98Al_2O_3, unusual very broad nonstructured band appears instead of narrow lines in the luminescence spectra of the initial mixture. This broad band cannot be attributed to the emission of any known crystalline phases of AgI. X-ray diffraction patterns of such composite do not show any peak due to AgI phase. Hence, it is most probable that AgI occurs in the nanocomposite in the amorphous state.

2.6
Computer Simulations

Computational methods provide effective means to understand the atomic mechanisms and to model local structure near interfaces on a microscopic scale. However, despite the great progress in the application of molecular dynamics (MD) simulations in solid-state science (see Ref. [47] and Chapters 5 and 7 of the first volume), a few works have been dealing with modeling of the interface between an ionic salt and an oxide, except for the recent papers on LiI–Al_2O_3 [48], CsCl–Al_2O_3 [49], and CaF_2–BaF_2 [50] nanocomposites. In Ref. [48], an oriented nanocrystal of LiI with a thickness of 100 monolayers was placed between the surfaces of two crystals of α- or γ-alumina. The hybrid model potential function including a sum of a two-body and a three-body potential was used for MD simulations. The two-body potential consisted of Coulomb interaction, exponential repulsion, modified Born–Mayer, dispersive, quadrupole–dipole, and Born repulsive terms. The three-body potential included only the Al–O nearest-neighbor interactions. The simulations showed that the conductivity of LiI increased when it was mixed with both α- and γ-alumina. Lithium ions adsorb on vacant tetrahedral sites of the alumina surface, leaving cation vacancies under the interface. This effect is more pronounced for α-alumina and less expressed for the γ-alumina surface. The diffusion coefficient D_{Li} was higher for the LiI–α-Al_2O_3 composites, though the activation energies were similar in both cases. These results agree with the model of chemical adsorption [2c].

In Ref. [49], MD simulations of CsCl–α-Al_2O_3 using the two-layer model were reported. CsCl was placed in contact with alumina, heated up to 1100 K, and

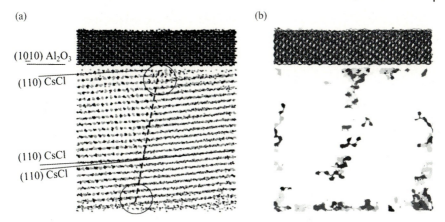

Figure 2.14 (a) The equilibrium structure of CsCl–α-Al$_2$O$_3$ nanocomposite after recrystallization simulated by MD. The crystallographic (110) planes of CsCl are denoted demonstrating the low-angle interdomain boundary (dashed line) situated between two grain boundary dislocations (marked by circles). (b) The trajectories of the most mobile ions recorded during 10^{-10} s.

transformed into the molten state. A nanocomposite was obtained after steep cooling down to 400 K, followed by crystallization of the salt. The interface was stable only for special relative orientations of crystal lattices of the contacting phases. For example, if the (10$\bar{1}$0) plane of alumina and (100) plane of CsCl contacted, a strong repulsion resulted in plastic deformation and long-term reconstruction of the interface. In order to make the system able to form energetically favorable orientation of crystal structure spontaneously, we simulated crystallization from the melt. The system after crystallization is shown in Figure 2.14a. One can see that the crystal planes (110) of CsCl are slightly inclined against the alumina plane due to the lattices misfit. The ionic salt represents a single crystal that consists of two domains disoriented by 3–3.5° in two perpendicular (110) planes and separated by a small-angle boundary. This boundary is generated by interface misfit dislocations and seems to join those located on two opposite CsCl–Al$_2$O$_3$ interfaces. The structure of alumina also undergoes changes; however, these changes are not so strong as in CsCl. The dislocation core is clouded by many defects, forming a region with an increased ion mobility. Also, incoherent grain boundaries are a source of extra free volume in the lattice and may hence act as migration paths. Figure 2.14b shows the trajectories of the most mobile ions recorded during 10^{-10} s. One can see that most mobile ions are concentrated near the interface in a core of the misfit dislocation and also in the vicinity of interdomain boundary, located in the bulk of cesium chloride and formed during crystallization. The diffusion coefficients of mobile cesium and chloride ions are close to each other, nearly $\sim 10^{-6}$ cm^2 s^{-1} at 700 K; that is, they are an order of magnitude lower than those in molten alkali halides [51].

The layered composites comprising parallel stacked arrays of two crystalline fluoride conductors, CaF_2 and BaF_2, have been recently modeled by the MD technique [50]. The CaF_2 regions in the layered system were compressed along the c-axis perpendicular to the interface. On the other hand, the BaF_2 regions were stretched along the c-axis. The diffusion coefficient and the ionic conductivity of F^- ions were shown to increase with decreasing periods in the layered fluoride conductors. Although these results agree with experiment [52], the authors mentioned that they had to fix the lattice constant at the interface to provide the stacking coherency and stability; otherwise, the heterostructure would be unstable due to great misfit between the crystal lattices.

The available results of MD simulations unambiguously show that the ion diffusion in the composites is faster than that in pure salts. However, there may be several possible reasons responsible for this effect:

i) Ion adsorption at the oxide surface with the formation of space charge layer enriched with defects.
ii) Lattice deformation near the interface resulting in an enlarging free volume.
iii) The appearance of domains and low-angle interdomain boundaries generated by the misfit dislocations. The ions located near the boundaries have an enhanced mobility.

Mechanism (i) seems to be typical for strong interfacial interactions and small misfit between the lattices MX and A when the adsorption of ions takes place with the formation of a space charge. Mechanism (ii) relates to the case of weak interfacial interaction (without ion adsorption) and relatively large lattice misfit leading to the lattice deformation. Mechanism (iii) may be similar to (i) and (ii); however, in this case the misfit is so high that the coherent interfaces become unstable and the misfit dislocations form on the surface. Thus, besides "pure" chemical adsorption, there are additional factors typical for solid-state systems, such as a misfit of the crystal lattices (given by the orientation of crystallographic planes) and mechanical properties of the components (determining the deformation, type, and concentration of dislocations and domain boundaries).

2.7
Design of the Composite Solid Electrolytes: General Approaches and Perspectives

2.7.1
Varying Chemical Nature of the Ionic Salt–Oxide Pair

This approach is traditional and may be useful, in particular, for the comparative analysis of the interfacial effects in a series of composites. To obtain composites with improved properties, one could vary ionic salt or oxides, provided that no chemical reaction takes place between the components and oxide particles are kinetically stable in the operating temperature range of the composite. The same approach is also used for optimization of properties of composite materials intended for applications under particular operation conditions. It was shown [2x] that lithium salts in the systems with alumina exhibit a stronger increase in conductivity compared to rubidium and cesium salts, while iodides more readily form nanocomposites compared to chlorides and fluorides. This suggests that the polarizing effect of the cation and the polarizability of the anion in the ionic component both play important roles in the surface interaction mechanism. The physical reason for the surface interaction in a composite of the ionic salt–oxide type originates from the tendency of both substances to decrease their surface energy due to the interaction of surface ions with the ions of neighboring phase. Owing to the difference in interionic energies and peculiarities of the crystal structures in the interfacial layers, the ideal structure inherent in individual phases should always be distorted in such a way as to provide a gain in the surface energy due to the mutual adjustment or removal of surface atoms. The relative displacement of ions from their ideal positions is determined by the balance of the interaction energies. Insofar, as in alumina, for the majority of discussed MX salts, the size of anions is larger than that of cations; it can be expected that for close packing, the interface cations have the larger free volumes and should be displaced for longer distances than the anions. As a result, an intermediate positively charged layer enriched with cations is formed in the space between the surface layers. Its charge is compensated by the cationic vacancies that constitute the diffuse layer. This process may be regarded as a chemical adsorption [2c,l] and can be presented as the following quasichemical reaction:

$$0 \leftrightarrows V'_M + (M-A)^{\bullet}_S \tag{2.25}$$

which describes the stage of the surface disordering of MX at the MX/A interface. Here the Kröger–Vink notation is used (see Chapter 3 of the first volume); the subscript S corresponds to the surface species and the last term relates to the M^+ cation adsorbed on the oxide surface. If an anion is adsorbed on the surface, another reaction occurs:

$$0 \leftrightarrows V^{\bullet}_X + (X-A)'_S \tag{2.26}$$

The isoelectric point of an oxide, pE, was proposed [53] as a measure of its surface activity. Indeed, equations similar to Equations 2.25 and 2.26 may be written for the surface interaction of an oxide with water:

$$H_2O \leftrightarrows OH' + (H-A)^{\bullet}_S \tag{2.27}$$

$$2H_2O \leftrightarrows H_3O^{\bullet} + (OH-A)'_S \tag{2.28}$$

The first reaction predominates for oxides with $pE > 7$, for instance, MgO, Al_2O_3, and CeO_2; the second prevails for the oxides with $pE < 7$ (ZrO_2, SiO_2). By analogy with aqueous solutions, one can expect that the surface reaction (2.25) should occur in the composites containing basic oxides ($pE > 7$), whereas for acidic oxides ($pE < 7$), the interface interaction should follow the mechanism (2.26). The isoelectric point of any oxide changes due to its doping with different species or by direct modification of its surface by acidic or basic agents. It enables to increase the conductivity of the composite only by the variation of its surface properties. A more general approach is to change Lewis acidity/basicity of the oxide surface. Both approaches were successfully used for the improvement of transport properties of different composites [8b]. Unfortunately, insufficient amount of available experimental data to date makes it difficult to check correctness of the direct transfer of the model of acid–base equilibria in aqueous solutions. Reference [54] reported on the conductivity enhancement in composites containing ferroelectric oxides $BaTiO_3$, $LiNbO_3$, and $KTaO_3$. The effect was proposed to be strengthened due to high dielectric permittivity of the oxide.

2.7.2
Changing the Physical State of Ionic Salt

One serious problem of crystalline composite solid electrolytes is a relatively poor contact between the crystallites of MX and A, leading to the formation of porous and brittle ceramics. To avoid this problem, the liquid, glassy, or polymer electrolytes may be used instead of crystalline ones. Among the composites of this type, the systems based on polymer or gel electrolytes are most extensively studied. They comprise solutions of ionic salts in liquids or polymers mixed with dispersed oxide fillers, such as oxides, nitrides, zeolites, ferroelectric oxides, lithium compounds, and so on [43, 44, 55]. Due to the filler presence, the composite not only maintains its mechanical stability but also has an increased conductivity. It was demonstrated that the conductivity enhancement depends on the chemical properties of dispersed ceramic component and its morphology [55b,e]. The main reason of this effect seems related to the stabilization of amorphous phase, that is, to a suppressed formation of crystalline phases with much lower ionic conductivity. This occurs owing to the surface groups of ceramic particles promoting local transformation from crystalline to amorphous state where the mobility of ions is higher [55f,g]. Another reason of the conductivity enhancement is an easier dissociation of the ionic salt in the vicinity of interface [55d–n]. The latter effect is observed even at high temperatures where polymer is in a liquid state and is accompanied by an increase in the cation transference number (t^+) [55b,c] important for the applications in lithium batteries [55b–e]. In $(PEO–LiClO_4)–Al_2O_3$ composites, the conductivity and t^+ both increase when introducing, sequentially, a series of basic, neutral, and acidic Al_2O_3 additions [55c]. Hence, it is not surprising that no conductivity enhancement was observed in some systems [55g]. The MX–A composites with inorganic glassy solid electrolytes MX were reported [56]. As mentioned in Section 2.4, the amorphous nonautonomous phases form in nanocomposites MX–A (MX = AgI, $LiClO_4$, $CsHSO_4$, and $MeNO_3$, where Me is an alkali cation and A is highly dispersed or porous Al_2O_3 or SiO_2) [2m,p,x, 9, 12a, 13a,b, 16, 17, 19, 21, 26, 29a,b]. At sufficiently high content of oxide components, these systems may be formally considered as the composites with glassy electrolytes. Solid solutions or mixtures of two salts were also suggested as solid electrolytes in MX–A systems. The composites of the (MX–MX′)–A type exhibit, in general, higher conductivities in comparison to pure MX–MX′ systems [57].

2.7.3
Changing the Physical State of the "Inert" Component

Instead of the solid oxide filler A in MX–A systems, one can use polymer network or glassy phase, either impregnated with an ion-conducting liquid phase or containing crystalline component MX. Typical examples of such systems are polymer proton-conducting membranes containing acidic end groups and easily absorbing water, for example, Nafion-type materials (see Chapter 10 of the first volume). At sufficiently high concentrations of the absorbed water, a network of liquid-phase channels provides high proton transport [58]. A qualitatively similar conductivity mechanism occurs in the gel electrolytes [59], ion exchange membranes, clays, and zeolites [60]. The polymer membranes are characterized by poor mechanical stability that may be improved by reinforcement by rigid support or solid fillers. Solid electrolytes MX–A with amorphous additive possessing a high mechanical strength may be obtained in the systems of "ionic salt–glass" type. The most known systems of the latter type are composite electrolytes with amorphous silica MX–SiO_2 (MX = LiI [61], LiBr [62], LiCl [2e], AgI [2d,i], AgCl [2c], $RbNO_3$ [19e]), which have high ionic conductivity. As silica is the chemical stable toward acids, it may be used in composite solid electrolytes with solid acids, for example, $MHSO_4$–SiO_2 (M = Cs, Rb) [13, 19f], CsH_2PO_4–SiO_2, CsH_2PO_4–SiP_2O_7, and $Cs_3H_5(SO_4)_2$–SiO_2 [63]. These and other composites of the latter type have a high proton conductivity and are promising solid electrolytes for fuel cells [63b, 64]. AgI–glass composites were obtained by controlled crystallization of AgI from supersaturated glass compositions [65]; the composites contained nanoparticles of the frozen-in high-temperature α-AgI polymorph and exhibited a fast ionic conduction [65]. Recently, the composites containing Li_2S in sulfide or oxysulfide glasses were shown to display a high lithium ion conductivity [66], which may be used in lithium solid-state batteries.

2.7.4
Alteration of the Inert Component Morphology: The Geometric Aspects

The term "morphology" (or its equivalents: micro- and mesostructure) involves all geometric parameters of the system: shape, dimension, and relative arrangement of all elements of the system, including the grains, pores, and interfaces. The conductivity of composites may be controlled in a wide range by variation of the morphology even without changing the effective grain (or pore) size. For this purpose, one can use oxide additives with particles of different dimensionality, taking into account that for one-dimensional (wires, fibers, tubes, and rods) or two-dimensional (layers, sheets, plates, and foams) systems, the orientation of individual particles has a strong effect on macro properties of the composite. Unfortunately, literature data on the influence of morphology on the conductivity of composite electrolytes are still very scarce. Dudney has studied AgCl–Al_2O_3 composites with alumina fibers as filler [67]. There is an increasing interest in the artificial micro- and nanoheterostructures prepared *in situ* using a controlled procedure that allows making of the preset architecture [2w, 68]; an example of such systems is the multilayered heterostructure comprising an array of alternating layers of two ionic salts CaF_2–BaF_2 prepared by the molecular beam epitaxy [52]. Such structures possess a high ionic conductivity depending on the thickness of the unit layer and on the heterostructure orientation. Other examples are the oxide and hydroxide layers obtained by the successive ionic layer deposition technique [69].

2.7.5
Metacomposite Materials and Nanostructured Systems

In all the cases listed above, physical and chemical properties of inert oxide additive do not change. Another situation is observed in metacomposites, that is, heterogeneous systems with the interfacial interaction changing physicochemical properties of both components [70]. Moreover, such a heterosystem exhibits properties that are not typical for both individual components. These materials constitute a new interesting class of metamaterials. The structured or nanostructured composite materials with oriented grains and coherent interfaces may also be obtained on the spinodal decomposition of solid solutions with NaCl-type structure [71] or perovskite-related systems [72]. They exhibit, again, an enhanced ionic conductivity at low temperatures caused by the interfacial phenomena.

2.7.6
Achieving the Nanoscale Level

Scaling of the systems down to nanoscale level is one of the most effective approaches to develop new materials with advanced properties. In order to prepare such systems, one has to use inert nanodispersed or nanoporous fillers or matrixes exhibiting no chemical interaction with the ionic components. The stability of the nanocomposites is determined by the adhesion energy between the components, as discussed in Section 2.3. The mechanism of the adhesion includes the same quasichemical reactions (Equations 2.25 and 2.26) that take place in ordinary composites, but the impact of the interface interaction is much stronger. The nanocomposites discussed in Section 2.4 are usually prepared using highly dispersed oxides with the characteristic size of grains or pores of ~10 nm. In these systems, strong size effects were observed and new interface-stabilized nonautonomous highly conducting phases were found.

Mesoporous oxides with ordered arrangement of pores are among the promising matrixes for the development of new composite electrolytes. Particles of such oxides contain pores of identical diameter that are long-range ordered and form one-, two-, or three-dimensional superstructures, with the lattice parameter varying from 3 to 16 nm [73]. Recently, an enhanced ionic conductivity was reported for nanocomposites of the MX–A type, LiI–A (A = Al_2O_3, SiO_2) [74], polymer composite electrolytes,

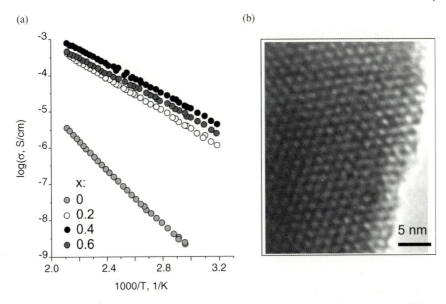

Figure 2.15 Arrhenius dependences of the conductivity of $(1-x)\text{LiClO}_4$–$x\text{SiO}_2$ nanocomposites with mesoporous silica MCM-41 at different molar concentration of the oxide (a) and high-resolution electron microscopy pattern of the composite 0.6LiClO_4–0.4SiO_2 (b).

(polymer–LiX)–A, where (polymer–LiX) is the solution of lithium salt in polymer ($X = \text{ClO}_4^-$, SO_3F^-) [75], or mesoporous proton-conducting membranes [76] with mesoporous oxide additives. Figure 2.15 shows the dependences of conductivity of nanocomposites $(1-x)\text{LiClO}_4$–$x\text{SiO}_2$ on temperature and on the concentration of additives, including the mesoporous silica and MCM-41 [77]. The sample was obtained by the impregnation of lithium salt into the mesoporous matrix. The composite with $x > 0.5$ exhibits a high conductivity and a complete absence of the conductivity jump at the melting point. This suggests that the ionic salt is in the amorphous state. In general, taking into consideration a great progress achieved during the past few years in the development of nanosystems with specific morphologies [78], the nanocomposites of new types are expected to be created in the near future. It should be mentioned that there is an optimal size of grains or pores of the oxide when the conductivity enhancement effect is maximal. On further decrease in size, the nanocomposites cannot form. This fact was first observed by Ponomareva et al. [13] and confirmed in subsequent works [74c, 79]. The effect may be explained by the contribution of strains energy, and its thermodynamic origins are discussed in Section 2.3.

All the approaches mentioned may be constructively combined in order to obtain the composites and nanocomposite electrolytes with a high performance and adopt them for particular applications in batteries, fuel cells, supercapacitors, sensors, and other electrochemical devices.

2.8
Composite Materials Operating at Elevated Temperatures

At high temperatures, both the chemical interaction between the components of any composite and the sintering processes become more intensive. In ordinary biphasic systems, these processes usually result in the material degradation due to formation of the reaction products. Nevertheless, if chemical interaction is thermodynamically unfavorable or kinetically retarded, then composites or even nanocomposites can be obtained. Such composites have superior mechanical characteristics and can be used for preparation of high-strength ceramics stable to thermal shocks. For example, addition of small amount of alumina to zirconia solid electrolytes has a minor effect on the conductivity, but enables to strongly decrease the grain boundary resistance due to the SiO_2 scavenging from the boundaries [80]. As a rule, the conductivity of solid oxide electrolytes becomes lower in composites due to the blocking effect of grain boundaries and interfaces, as briefly discussed in Section 2.2.1. The effect of conductivity enhancement was reported for nanoceramics and thin layers [81]. These phenomena are due to the particle size effect and need further detailed studies; relevant examples are found in Chapter 4 of the first volume. It is, however, critically important that such nanostructured ceramics and thin films are morphologically unstable and, in general, cannot be used for most practical applications. More promising materials are the composite ceramics. In comparison to single-phase materials, the composites are more stable toward grain coarsening process as the surface energy of each component is partially compensated by the interfacial interaction, thus decreasing the thermodynamic driving force for sintering. In particular, composite ceramics are known to be used as electrode materials in solid oxide fuel cells (SOFCs) where mixed electronic–ionic conducting systems or cermets (ceramic–metal composites) are widely used [82]. Another possible application of the composites relates to the oxygen-permeable membranes [83]. The cermets composed of solid oxide electrolytes (ZrO_2–Y_2O_3, Bi_2O_3–Y_2O_3, Bi_2O_3–Er_2O_3) and metals (Pd, Ag, Au) may be used for oxygen separation, although these are expensive. More promising are the perovskite-related oxides if these are sufficiently stable, especially in reducing atmospheres. As an example, it was shown that additions of $SrAl_2O_4$ phase to mixed conducting $SrCo_{0.8}Fe_{0.2}O_{3-\delta}$ decrease thermal expansion coefficients and improve thermal shock stability; such composites enable oxygen permeation fluxes similar to those through single-phase $SrCo_{0.8}Fe_{0.2}O_{3-\delta}$ membranes [84]. In some composites based on cuprates or vanadates, phase transition associated with wetting of the grain boundaries with eutectic melt is observed; the grain boundary wetting leads to substantially higher ionic conductivity in the composite materials [85]. Very interesting phenomena were reported in the composites formed in the systems $MeWO_4$–WO_3 (Me = Ca, Sr, Ba) [70]. These materials may be classified as metacomposites, while the formed surface phase acts as a connectivity net and possesses ambipolar surface activity and a higher mobility with respect to the parent compounds, $MeWO_4$ and WO_3. Due to the presence of this surface phase, the composites exhibit attractive electrosurface properties under the action of external electric field [70, 86].

2.9
Conclusions

In this chapter, the electrochemical, thermodynamic, and physical properties of composite solid electrolytes were briefly reviewed with an emphasis on the interfacial interactions between the components and other interfacial phenomena. The surface potential formation and the point defect equilibrium at free surfaces and interfaces were considered in the framework of the Stern model, which provides a common basis for quantitative description of the surface and interface effects. Special attention was focused on the size effects, that is, changes in the bulk characteristics of ionic salts in nanocomposites due to the influence of interface energy and the formation of interface phases. The main thermodynamic reason for the formation of nanocomposites and for the stabilization of nonautonomous interface phases is the adhesion energy γ_a. At sufficiently high γ_a values, the ionic salt in MX–A composites tends to spread along the oxide surface. If the grain size of the oxide is sufficiently small, the nanocomposite forms upon sintering. The adhesion results from the interface interaction and comprises the stage of specific adsorption of the interface ions. This leads to forming double layer in the interfacial region of ionic salt. In the case of strong adhesion, structural reconstruction or formation of the interface phases may occur. Analysis of the experimental data revealed that interface phases exist in many composites. Their structure may be either epitaxial crystalline or amorphous. The thickness of the interface phase estimated on the basis of the brick wall model is about 3–4 nm. The reason for stabilization of nonequilibrium states is complex and translates the influence of several factors, namely, the interfacial interaction between the components, particle size effect, and elastic strains in the lattice of ionic salt. The results of MD simulations showed that the main mechanisms of conductivity enhancement are the adsorption of ions to the oxide surface with formation of space charge layer, the lattice deformation near the interface, and the appearance of domains and low-angle interdomain boundaries generated by the misfit dislocations. All the above mechanisms are induced by the interfaces. The equations proposed earlier for the description of conductivity of composites were analyzed. Among them, the general mixing rule has a rather simple analytical form and provides appropriate description of the experimental conductivity data for the composite solid electrolytes in a whole concentration range. Main approaches for improving or creating new composite systems have been reviewed. Finally, composites and nanocomposites with special properties and given morphology are intensively developing. Substantial advances are seen in composite improvement in the past 5–10 years using new techniques for synthesis and new heterogeneous dopants.

Acknowledgments

Financial support from the Russian Fund for Basic Research (Grant No. 09-03-00897) and SB RAS (Integration Projects 22 and 86) is gratefully acknowledged.

References

1 Liang, C.C. (1973) *J. Electrochem. Soc.*, **120**, 1289.
2 For reviews on composite solid electrolytes, see (a) Shahi, K. and Wagner, J.B., Jr. (1981) *Solid State Ionics*, **3–4**, 295; (b) Wagner, J.B., Jr. (1985) *Solid State Batteries, NATO ASI Series E101* (eds C.A.C. Sequeira and A. Hooper), Martinus Nijhoff Publishers, p. 77; (c) Maier, J. (1985) *J. Phys. Chem. Solids*, **46**, 309; (d) Khandkar, A.C. and Wagner, J.B., Jr. (1986) *Solid State Ionics*, **18–19**, 1100; (e) Chen, L. (1986) *Materials for Solid State Batteries* (eds B.V.R. Chowdhari and S. Radhakrishna), World Scientific Publishing, Singapore, p. 69; (f) Shukla, A.K., Vaidehi, N., and Jacob, K.T. (1986) *Proc. Indian Acad. Sci. (Chem. Sci.)*, **96**, 533; (g) Maier, J. (1987) *J. Electrochem. Soc.*, **134**, 1524; (h) Maier, J. (1989) *Superionic Solids and Solid Electrolytes: Recent Trends* (eds S. Chandra and A. Laskar), Academic Press, New York, p. 137; (i) Wagner, J.B., Jr. (1989) *High Conductivity Conductors: Solid Ionic Conductors* (ed. T. Takahashi), World Scientific Publishing, Singapore, p. 102; (j) Dudney, N.J. (1989) *Annu. Rev. Mater. Sci.*, **19**, 103; (k) Uvarov, N.F., Isupov, V.P., Sharma, V., and Shukla, A.K. (1992) *Solid State Ionics*, **51**, 41; (l) Maier, J. (1995) *Prog. Solid State Chem.*, **23**, 171; (m) Uvarov, N.F. (1996) *Solid State Ionics: New Developments* (eds B.V.R. Chowdary et al..), World Scientific Publishing, Singapore, p. 311; (n) Agrawal, R.C. and Gupta, R.K. (1999) *J. Mater. Sci.*, **34**, 1131; (o) Yarostalvtsev, A.B. (2000) *Russ. J. Inorg. Chem.*, **45**, S249; (p) Uvarov, N.F. and Vanek, P. (2000) *J. Mater. Synth. Process.*, **8**, 319; (q) Maier, J. (2002) *Solid State Ionics*, **154–155**, 291; (r) Maier, J. (2003) *Solid State Ionics*, **157**, 327; (s) Jamnik, J. and Maier, J. (2003) *J. Phys. Chem. Chem. Phys.*, **5**, 5215; (t) Heitjans, P. and Indris, S. (2003) *J. Phys. Condens. Matter*, **15**, R1257; (u) Maier, J. (2004) *Solid State Ionics*, **175**, 7; (v) Schoonman, J. (2005) *Solid State Ionics*, **135–137**, 5; (w) Maier, J. (2005) *Nat. Mater.*, **4**, 805; (x) Uvarov, N.F. (2007) *Russ. Chem. Rev.*, **76**, 415; (y) Uvarov, N.F. (2008) *Composite Solid Electrolytes*, Nauka Publishers, Novosibirsk.
3 Jow, T. and Wagner, J.B., Jr. (1979) *J. Electrochem. Soc.*, **126**, 1963.
4 Stern, O. (1924) *Z. Electrochem*, **30** 508.
5 (a) Uvarov, N.F. (2007) *Russ. J. Electrochem.*, **43**, 368; (b) Uvarov, N.F. (2008) *Solid State Ionics*, **179**, 783.
6 (a) Frenkel, J. (1946) *Kinetic Theory of Liquids*, Oxford University Press, Oxford; (b) Kliewer, K.L. and Koehler, K.S. (1965) *Phys. Rev. A*, **140**, 1226; (c) Kliewer, K.L. (1966) *J. Phys. Chem. Solids*, **27**, 705; (d) Poeppel, R.B. and Blakely, J.M. (1969) *Surf. Sci.*, **15**, 507–523; (e) Jamnik, J., Maier, J., and Pejovnik, S. (1995) *Solid Sate Ionics*, **75**, 51.
7 (a) Chebotin, V.N. and Perfiliev, M.V. (1976) *Electrochemistry of Solid Electrolytes*, Russian edn, Metallurgia, Moscow; (b) Heyne, L. (1983) *Mass Transport in Solids* (eds F. Beniere and C.R.A. Catlow), Plenum Press, New York, p. 425; (c) Guo, X. and Maier, J. (2001) *J. Electrochem. Soc.*, **148**, E121; (d) Guo, X. and Waser, R. (2006) *Prog. Mater. Sci.*, **51**, 151.
8 (a) Lauer, U. and Maier, J. (1990) *Sensors Actuators*, **B2** 125; (b) Saito, Y. and Maier, J. (1995) *J. Electrochem. Soc.*, **142**, 3078.
9 (a) Ulihin, A.S., Uvarov, N.F., Mateyshina, Yu.G., Brezhneva, L.I., and Matvienko, A.A. (2006) *Solid State Ionics*, **177**, 2787; (b) Ulihin, A.S., Slobodyuk, A.B., Uvarov, N.F., Kharlamova, O.A., Isupov, V.P., Kavun, V.Ya., Uvarov, A.B., Kharlamova, N.F., Isupov, O.A., and Kavun, V.P. (2008) *Solid State Ionics*, **179**, 1740.
10 (a) Gurevich, Y.Ya. and Pleskov, Yu.V. (1983) *Photoelectrochemistry of Semiconductors*, Russian edn, Nauka, Moscow; (b) Maier, J. (1985) *Ber. Bunsen. Phys. Chem.*, **89**, 355; (c) Sato, N. (1998) *Electrochemistry at Metal and Semiconductor Electrodes*, Elsevier, Amsterdam.
11 (a) Jaycock, M.J. and Parfitt, G.D. (1981) *Chemistry of Interfaces*, John Wiley & Sons, Inc., New York; (b) Adamson, A.W. and Gast, A.P. (1997) *Physical Chemistry of Surfaces*, Wiley–Interscience.

12 (a) Uvarov, N.F., Bokhonov, B.B., Isupov, V.P., and Hairetdinov, E.F. (1994) *Solid State Ionics*, **74**, 15; (b) Uvarov, N.F. and Boldyrev, V.V. (2001) *Russ. Chem. Rev.*, **70**, 265.

13 (a) Ponomareva, V.G., Lavrova, G.V. and Simonova, L.G. (1998) *Solid State Ionics*, **107**, 120; (b) Ponomareva, V.G., Lavrova, G.V., and Simonova, L.G. (2000) *Solid State Ionics*, **136–137**, 1279.

14 Maier, J. (2003) *Z. Phys. Chem.*, **217**, 415.

15 (a) For reviews on size effects in thin films and small particles, see Palatnik, L.S. and Papirov, I.I. (1971) *Epitaxial Films*, Nauka, Moscow; (b) Morokhov, I.D., Trusov, L.I., and Chizhik, S.P. (1977) *Ultradisperse Metallic Media*, Atomizdat, Moscow; (c) Komnik, Yu.F. (1979) *Physics of Metallic Films: Size and Structural Effects*, Atomizdat, Moscow; (d) Nepiiko, S.A. (1985) *Physical Properties of Small Metallic Particles*, Naukova Dumka, Kiev; (e) Petrov, Yu.I. (1986) *Clusters and Small Particles*, Nauka, Moscow; (f) Gryaznov, V.G. and Trusov, L.I. (1993) *Prog. Mater. Sci.*, **37**, 289; (g) Khairutdinov, R.F. (1998) *Russ. Chem. Rev.*, **67**, 109; (h) Gusev, A.I. (1998) *Nanocrystalline Materials: Methods of Production and Properties*, Russian Academy of Sciences, Ekaterinburg; (i) Morris, D.G. (1998) *Materials Science Foundations*, Trans Tech Publications, Zuerich, p. 2; (j) Andrievskii, R.A. and Glezer, A.M. (1999) *Fiz. Met. Metalloved.*, **88**, 50; (k) Gusev, A.I. and Rempel, A.A. (2000) *Nanocrystalline Materials*, Fizmatlit, Moscow; (l) Gleiter, H. (2000) *Acta Mater.*, **48**, 1; (m) Klabunde, K.J. (ed.) (2001) *Nanoscale Materials in Chemistry*, John Wiley & Sons, Inc., New York; (n) Knauth, P. and Schoonman, J. (eds) (2004) *Nanostructured Materials: Selected Synthesis Methods, Properties and Applications*, Kluwer Academic Publishers, New York; (o) Sun, C.Q. (2007) *Prog. Solid State Chem.*, **35**, 1; (p) Mei, Q.S. and Lu, K. (2008) *Prog. Mater. Sci.*, **52**, 1175.

16 (a) Uvarov, N.F., Brezhneva, L.I., and Hairetdinov, E.F. (2000) *Solid State Ionics*, **136–137**, 1273; (b) Ponomareva, V.G. and Lavrova, G.V. (2001) *Solid State Ionics*, **145**, 197.

17 Uvarov, N.F., Vanek, P., Savinov, M., Zelezny, V., Studnicka, V., and Petzelt, J. (2000) *Solid State Ionics*, **127**, 253.

18 (a) Dubrovina, A.N., Akhtyamov, Yu.R., Knyazev, E.V., Ganelin, V.Ya., Trusov, L.I., Lapovok, V.N., and Terekhov, V.N. (1981) *Kristallografia*, **26**, 410; (b) Garvie, R.C. (1965) *J. Phys. Chem.*, **69**, 1238; (c) Zhang, Y.L., Jin, X.J., Ronga, Y.H., Hsu, T.Y., Jiang, D.Y., and Shi, J.L. (2006) *Mater. Sci. Eng. A*, **438–440**, 399; (d) Huber, B., Brodyanski, A., Scheib, M., Orendorz, A., Ziegler, C., and Gnaser, H. (2005) *Thin Solid Films*, **472**, 114.

19 (a) Uvarov, N.F., Khairetdinov, E.F., and Bratel, N.B. (1993) *Russ. J. Electrochem.*, **29**, 1231; (b) Uvarov, N.F., Hairetdinov, E.F., and Skobelev, I.V. (1996) *Solid State Ionics*, **86–88**, 577; (c) Uvarov, N.F., Skobelev, I.V., Bokhonov, B.B., and Hairetdinov, E.F. (1996) *J. Mater. Synth. Process.*, **4**, 391; (d) Uvarov, N.F., Vanek, P., Yuzyuk, Yu.I., Zelezny, V., Studnicka, V., Bokhonov, B.B., Dulepov, V.E., and Petzelt, J. (1996) *Solid State Ionics*, **90**, 201; (e) Lavrova, G.V. and Ponomareva, V.G. (2000) *Solid State Ionics*, **136–137**, 1285; (f) Ponomareva, V.G., Lavrova, G.V., Uvarov, N.F., and Hairetdinov, E.F. (1996) *Solid State Ionics*, **90**, 161; (g) Vinod, M.P. and Bahnemann, D. (2002) *J. Solid State Electrochem.*, **6**, 498.

20 Lee, J.S., Adams, S., and Maier, J. (2000) *J. Electrochem. Soc.*, **147**, 2407.

21 (a) Uvarov, N.F., Politov, A.A., and Bokhonov, B.B. (2000) *Solid State Ionics: Materials and Devices* (eds B.V.R. Chowdary and W. Wang), World Scientific Publishing, Singapore, p. 113; (b) Uvarov, N.F. (2000) *Z. Prikl. Khim*, **73**, 413.

22 (a) Pavlov, V.A. (1985) *Fiz. Met. Metalloved.*, **59**, 629; (b) Johnson, W.L. (1986) *Prog. Mater. Sci.*, **30**, 81; (c) Avvakumov, E.G. (1986) *Mechanical Methods of Activation of Chemical Processes*, Nauka Publishers, Novosibirsk; (d) Koch, C.C. (1989) *Annu. Rev. Mater. Sci.*, **19**, 123; (e) Serebryakov, A.V. (1991) *Metallofizika*, **13**, 115; (f) Butyagin, P.Yu. (1994) *Russ. Chem. Rev.*, **63**, 965.

23 (a) Gusev, V.A., Gagarina, V.A., Moroz, E.M., and Levitskii, E.A. (1974) *Kristallografiya*, **19**, 1289; (b) Gusev, V.A.,

Gagarina, V.A., Moroz, E.M., and Levitskii, E.A. (1976) *Kinet. Katal.*, **17**, 500.
24 (a) Aristov, Yu.I., Restuccia, G., Cacciola, G., and Parmon, V.N. (2002) *Appl. Therm. Eng.*, **22**, 191–204; (b) Gordeeva, L.G., Restuccia, G., Freni, A., and Aristov, Yu.I. (2002) *Fuel Process. Technol.*, **79**, 225–231; (c) Aristov, Yu.I., Gordeeva, L.G., and Tokarev., M.M. (2008) *Salt-in-Matrix Composite Adsorbents: Properties and Applications*, Nauka Publishers, Novosibirsk.
25 (a) Roman, H.E., Bunde, A., and Dieterich, W. (1986) *Phys. Rev. B*, **34**, 3439; (b) Roman, H.E. and Yusouff, M. (1987) *Phys. Rev. B*, **36**, 7285; (c) Roman, H.E. (1990) *J. Phys. Condens. Matter*, **2**, 3909; (d) Nan, C.W. and Smith, D.M. (1991) *Mater. Sci. Eng. B*, **10**, 99; (e) Uvarov, N.F. and Ponomareva, V.G. (1996) *Dokl. Russ. Acad. Sci.*, **351**, 358.
26 Uvarov, N.F., Hairetdinov, E.F., and Bratel, N.B. (1996) *Solid State Ionics*, **86–88**, 573.
27 Zhu, B. and Mellander, B.-E. (1994) *Solid State Ionics*, **70–71**, 285.
28 Lee, J.S. and Maier, J. (1997 Proceedings of the International Conference EUROSOLID-97 (eds A. Negro and L. Montanaro), Politecnico di Torino, p. 115–122.
29 (a) Uvarov, N.F., Shastry, M.C.R. and Rao, K.J. (1990) *Rev. Solid State Sci.*, **4**, 61; (b) Shastry, M.C.R. and Rao, K.J. (1992) *Solid State Ionics*, **51**, 311; (c) Tadanaga, K., Imai, K., Tatsumisago, M., and Minami, T. (2002) *J. Electrochem. Soc.*, **149**, A773; (d) Tadanaga, K., Imai, K., Tatsumisago, M., and Minami, T. (2000) *J. Electrochem. Soc.*, **147**, 4061.
30 (a) Phipps, J.B., Johnson, D.L., and Whitmore, D.H. (1981) *Solid State Ionics*, **5**, 393; (b) Fujitsu, S., Kobayashi, H., Koumoto, K., and Yanagida, H. (1986) *J. Electrochem. Soc.*, **133**, 1497; (c) Fujitsu, S., Miyayama, M., Koumoto, K., Yanagida, H., and Kanazawa, T. (1985) *J. Mater. Sci.*, **20**, 2103.
31 (a) Stoneham, A.M., Wade, E. and Kilner, J.A. (1979) *Mater. Res. Bull.*, **14**, 661; (b) Wang, J.C. and Dudney, N. (1986) *Solid State Ionics*, **18–19**, 112; (c) Dudney, N.J. (1985) *J. Am. Ceram. Soc.*, **68**, 538; (d) Brailsford, A.D. (1986) *Solid State Ionics*, **21**, 159; (e) Mikrajuddin, A., Shi, F.G., and Okuyama, K. (2000) *J. Electrochem. Soc.*, **147**, 3157.
32 (a) Jiang, S. and Wagner, J.B. (1995) *J. Phys. Chem. Solids*, **56**, 1113; (b) Rojo, A.G. and Roman, H.E. (1988) *Phys. Rev. B*, **37**, 3696; (c) Bunde, A. and Dieterich, W. (2000) *J. Electroceram.*, **5**, 81; (d) Bunde, A., Dieterich, W., and Roman, H.E. (1986) *Solid State Ionics*, **18–19**, 147; (e) Dieterich, W. (1989) *High Conductivity Conductors: Solid Ionic Conductors* (ed. T. Takahashi), World Scientific Publishing, Singapore, p. 17; (f) Blender, R. and Dieterich, W. (1987) *J. Phys. C*, **20**, 6113; (g) Knauth, P., Debierre, J.M., and Albinet, G. (1999) *Solid State Ionics*, **121**, 101; (h) Siekierski, M. and Przyluski, J. (1994) *Solid State Ionic Materials* (ed. B.V.R. Chowdary), World Scientific Publishing, Singapore, p. 121.
33 (a) Uvarov, N.F. (1997) *Dokl. Russ. Acad. Sci.*, **353**, 213; (b) Uvarov, N.F. (2000) *Solid State Ionics*, **136–138**, 1267.
34 (a) Lichtenecker, K. (1926) *Phys. Z.*, **27**, 115; (b) Landau, L.D. and Lifshitz, E.M. (1982) *Electrodynamics of Continuum Media*, Nauka, Moscow.
35 (a) Poulsen, F.M. (1987) *J. Power Sources*, **20**, 317; (b) Chang, M.R.W., Shahi, K., and Wagner, J.B. (1984) *J. Electrochem. Soc.*, **131**, 1213.
36 Hashin, Z. and Shtrikman, S. (1962) *J. Appl. Phys.*, **33**, 3125.
37 Maier, J. (1987) *Solid State Ionics*, **23**, 59.
38 Khaneft, A.V. (2000) *Z. Nauch. Prikl. Fotogr.*, **45**, 67.
39 (a) Qi, W.H. and Wang, M.P. (2003) *Physica B*, **334**, 432; (b) Suzdalev, I.P. and Suzdalev, P.I. (2001) *Russ. Chem. Rev.*, **70** 177.
40 (a) Bondarev, V.N. and Kuklov, A.B. (1985) *Fiz. Tverdogo Tela*, **27**, 3332.
41 (a) Gurevich, Y.Y. and Kharkats, Y.I. (1978) *J. Phys. Chem. Solids*, **38**, 751; (b) Boyce, J.B. and Huberman, A.B. (1979) *Phys. Rep.*, **51**, 189; (c) Hainovsky, N.G. and Maier, J. (1995) *Phys. Rev. B*, **51**, 15789.
42 Mishnaevsky, L.L. (2007) *Computational Mesomechanics of Composites: Numerical Analysis of the Effect of Microstructures of Composites of Strength and Damage Resistance*, Wiley–Interscience.

43 Croce, F., Curini, R. Martinelli, A., Persi, L. Ronci, F., Scrosati, B., and Caminiti, R. (1999) *J. Phys. Chem. B*, **103**, 10632–10638.

44 Herring, A.M. (2006) *Polym. Rev.*, **46**, 245–296.

45 Chen, L., Zhao, Z., and Dai, S. (1986) *Solid State Ionics*, **18–19**, 1198–1201.

46 Uvarov, N.F., Bokhonov, B.B. Politov, A.A., Vanek, P., and Petzelt, J. (2000) *J. Mater. Synth. Process.*, **8**, 327–332.

47 Catlow, C.R.A. (1997) *Computer Modeling in Inorganic Crystallography*, Academic Press, New York.

48 Huang Foen Chung, R.W.J.M. and de Leeuw, S.W. (2004) *Solid State Ionics*, **175**, 851.

49 Gainutdinov, I.I. and Uvarov, N.F. (2006) *Solid State Ionics*, **177**, 1631.

50 Nomura, K. and Kobayashi, M. (2008) *Ionics*, **14**, 131.

51 Sangster, M.J.L. and Dixon, M. (1976) *Adv. Phys.*, **25**, 247.

52 (a) Sata, N., Ebermann, K., Eberl, K., and Maier, J. (2000) *Nature*, **408**, 946; (b) Jin-Phillipp, N.J., Sata, N., Maier, J., Scheu, C., Hahn, K., Kelsch, M., and Ruhle, M. (2004) *J. Chem. Phys.*, **120**, 2375.

53 Shukla, A.K., Manoharan, R., and Goodenough, J.B. (1988) *Solid State Ionics*, **26**, 5.

54 (a) Nakamura, O. and Saito, Y. (1992) *Solid State Ionics: Materials and Applications* (eds B.V.R. Chowdary et al..), World Scientific Publishing, Singapore, p. 101; (b) Saito, Y., Asai, T., Ado, K., and Nakamura, O. (1988) *Mater. Res. Bull.*, **23**, 1661; (c) Singh, K., Lanje, U.K., and Bhoga, S.S. (1995) Tenth International Conference on Solid State Ionics, Singapore, Extended Abstracts, p. 112.

55 (a) Weston, J.E. and Steele, B.C.H. (1982) *Solid State Ionics*, **7**, 75; (b) Croce, F., Appetechi, G.B., Persi, L., and Scrosati, B. (1998) *Nature*, **394**, 456; (c) Croce, F. and Scrosati, B. (2003) *Ann. N.Y. Acad. Sci.*, **984**, 194; (d) Kumar, B. and Scanlon, L.G. (1994) *J. Power Sources*, **52**, 261; (e) Stephan, A.M. and Nahm, K.S. (2006) *Polymer*, **47**, 5952; (f) Wieczorek, W., Florjancyk, Z., and Stevens, J.R. (1995) *Electrochim. Acta*, **40**, 2251; (g) Przyluski, J., Siekierski, M., and Wieczorek, W. (1995) *Electrochim. Acta*, **40**, 2101; (h) Panero, S., Scrosati, B., and Greenbaum, S.G. (1992) *Electrochim. Acta*, **37**, 1533; (i) Capiglia, C., Mustarelli, P., Quartarone, E., Tomasi, C., and Magistris, A. (1999) *Solid State Ionics*, **118**, 73; (j) Liu, Y., Lee, J.Y., and Hong, L. (2004) *J. Power Sources*, **129**, 303; (k) Kim, J.W., Ji, K.S., Lee, J.P., and Park, J.W. (2003) *J. Power Sources*, **119–121**, 415; (l) Kim, Y.W., Lee, W., and Choi, B.K. (2000) *Electrochim. Acta*, **45**, 1473; (m) Sun, H.Y., Takeda, Y., Imanishi, N., Yamamoto, O., and Sohn, H.S. (2000) *J. Electrochem. Soc.*, **147**, 2462; (n) Kumar, B., Scanlon, L., Marsh, R., Mason, R., Higgins, R., and Baldwin, R. (2001) *Electrochim. Acta*, **46**, 1515.

56 (a) Shaju, K.M. and Chandra, S. (1995) *J. Mater. Sci.*, **30**, 3457; (b) Arof, A.K. (1994) *J. Phys. III*, **4**, 849.

57 (a) Agrawal, R.C., Kathal, K., and Gupta, R.K. (1994) *Solid State Ionics*, **74**, 137. (b) Agrawal, R.C. and Gupta, R.K. (1997) *J. Mater. Sci.*, **32**, 3327; (c) Agrawal, R.C., Verma, M.L., and Gupta, R.K. (1998) *J. Phys. D*, **31**, 2854.

58 (a) Zabolotsky, V.I. and Nikonenko, V. (1996) *Ionic Transport in Membranes*, Nauka, Moscow; (b) Yaroslavtsev, A.B., Nikonenko, V.V., and Zabolotsky, V.I. (2003) *Russ. Chem. Rev.*, **72**, 393; (c) Kreuer, K.D. (2003) *Handbook of Fuel Cells: Fundamentals, Technology and Applications, Vol. 3, Fuel Cell Technology and Applications* (eds W. Vielstich, A., Lamm, and H. Gasteiger), John Wiley & Sons, Ltd, Chichester, p. 420.

59 Mustarelli, P., Quartarone, E., Tomasi, C., and Magistris, A. (2003) *Solid State Ionics*, **135**, 81.

60 Colomban, Ph. (1992) *Proton Conductors: Solids, Membranes and Gels – Materials and Devices*, Cambridge University Press, Cambridge.

61 Phipps, J.B. and Whitmore, D.H. (1983) *Solid State Ionics*, **9–10**, 123.

62 Slade, R.C.T. and Thomson, I.M. (1988), *Solid State Ionics*, **26**, 287.

63 (a) Ponomareva, V.G. and Shutova, E.S. (2007) *Solid State Ionics*, **178**, 729; (b) Matsui, T., Kukino, T., Kikuchi, R., and Eguchi, K. (2006) *J. Electrochem. Soc.*, **153**, A339; (c) Ponomareva, V.G., Lavrova, G.V.,

and Burgina, E.B. (2005) *Solid State Ionics*, **176**, 767.

64 (a) Haile, S.M., Boysen, D.A., Chisholm, C.R.I., and Merle, R.B. (2001) *Nature*, **410**, 910; (b) Lavrova, G.V., Russkih, M.V., Ponomareva, V.G., and Uvarov, N.F. (2006) *Solid State Ionics*, **177**, 2129.

65 (a) Tatsumisago, M., Shinkuma, Y., and Minami, T. (1991) *Nature*, **354**, 217; (b) Tatsumisago, M., Saito, T., and Minami, T. (1994) *Solid State Ionics*, **70–71**, 394; (c) Minami, T., Saito, T., and Tatsumisago, M. (1996) *Solid State Ionics*, **86–88**, 415; (d) Adams, S., Hariharan, K., and Maier, J. (1995) *Solid State Ionics*, **75**, 193.

66 (a) Mizuno, F., Hayashi, A., Tadanaga, K., Minami, T., and Tatsumisago, M. (2004) *Solid State Ionics*, **175**, 699; (b) Hayashi, A., Komiya, R., Tatsumisago, M., and Minami, T. (2002) *Solid State Ionics*, **152–153**, 285; (c) Minami, T., Hayashi, A., and Tatsumisago, M. (2006) *Solid State Ionics*, **177**, 2715.

67 Dudney, N.J (1988) *Solid State Ionics*, **28–30**, 1065.

68 Despotuli, A.L. and Nikolaichik, V.I. (1993) *Solid State Ionics*, **60**, 275.

69 Tolstoy, V.P. (2006) *Russ. Chem. Rev.*, **75**, 161.

70 (a) Konisheva, E., Neiman, A., and Gorbunova, E. (2003) *Solid State Ionics*, **157**, 45; (b) Neiman, A.Ya., Pestereva, N.N., Sharafutdinov, A.R., and Kostikov, Yu. P. (2005) *Russ. J. Electrochem.*, **41**, 598; (c) Neiman, A.Ya., Uvarov, N.F., and Pestereva, N.N. (2006) *Solid State Ionics*, **177**, 3361–3369.

71 (a) Hartmann, E., Peller, V.V., and Rogalski, G.I. (1988) *Solid State Ionics*, **28–30**, 1098; (b) Hartmann, E., Peller, V.V., and Rogalski, G.I. (1990) *Solid State Ionics*, **37**, 123.

72 (a) Nemudry, A., Rudolf, P., and Schöllhorn, R. (1996) *Chem. Mater.*, **8**, 2232; (b) Nemudry, A., Goldberg, E.L., Aguirre, M., and Alario-Franco, M.A. (2002) *Solid State Sci.*, **4**, 677; (c) Nemudry, A. and Uvarov, N. (2006) *Solid State Ionics*, **177**, 2491.

73 (a) Kresge, C.T., Leonowicz, M.E., Roth, W.J., Vartuli, J.C., and Beck, J.S. (1992) *Nature*, **359**, 710; (b) Fenelonov, V.B., Derevyankin, A.Yu., Kirik, S.D., Solovyov, L.A., Shmakov, A.N., Bonardet, J.L., Gedeon, A., and Romannikov, V.N. (2001) *Microporous Mesoporous Mater.*, **44–45**, 33.

74 (a) Yamada, H., Moriguchi, I., and Kudo, T. (2005) *Solid State Ionics*, **176**, 945; (b) Maekawa, H., Tanaka, R., Sato, T., Fujimaki, Y., and Yamamura, T. (2004) *Solid State Ionics*, **175**, 281; (c) Maekawa, H., Fujimaki, Y., Shen, H., Kawamura, J., and Yamamura, T. (2006) *Solid State Ionics*, **177**, 2711.

75 (a) Subba Reddy, C.V., Wu, G.P., Zhao, C.X., Zhu, Q.Y., Chen, W., and Kalluru, R.R. (2007) *J. Non-Cryst. Solids*, **353**, 440; (b) Wang, X.L., Mei, A., Li, M., Lin, Y., and Nan, C.W. (2006) *Solid State Ionics*, **177**, 1287; (c) Kim, S., Hwang, E.J., and Park, S.J. (2008) *Curr. Appl. Phys.*, **8**, 729; (d) Xi, J., Qiu, X., Ma, X., Cui, M., Yang, J., Tang, X., Zhu, W., and Chen, L. (2005) *Solid State Ionics*, **176**, 1249.

76 Lin, Y.F., Yen, C.Y., Ma, C.C.M., Liao, S.H., Lee, C.H., Hsiao, Y.H., and Lin, H.P. (2007) *J. Power Sources*, **171**, 388.

77 Uvarov, N.F., Ulihin, A.S., Slobodyuk, A.B., Kavun, V.Ya., and Kirik, S.D. (2008) *ESC Trans.*, **11** (31), 9–17.

78 Yu, C., Tian, B., and Zhao, D. (2003) *Curr. Opin. Solid State Mater. Sci.*, **7**, 191.

79 Shigeoka, H., Otomo, J., Wen, C., Ogura, M., and Takahashi, H. (2004) *J. Electrochem. Soc.*, **151**, J76.

80 Guo, X. and Waser, R. (2006) *Prog. Mater. Sci.*, **51**, 151–210.

81 (a) Kosacki, I. and Anderson, H.U. (2001) *Encyclopedia of Materials, Science and Technology*, vol. 4, Elsevier Science Ltd., New York, p. 3609; (b) Suzuki, T., Kosacki, I., Anderson, H.U., and Colomban, Ph. (2001) *J. Am. Ceram. Soc.*, **84**, 2007; (c) Kosacki, I., Anderson, H.U., Mizutani, Y., and Ukai, K. (2002) *Solid State Ionics*, **152–153**, 431; (d) Kosacki, I., Suzuki, T., Petrovsky, V., and Anderson, H.U. (2000) *Solid State Ionics*, **136–137**, 1225; (e) Kosacki, I., Rouleau, C.M., Becher, P.F., Bentley, J. and Lowndes, D.H. (2005) *Solid State Ionics*, **176**, 1319–1326.

82 (a) Mogensen, M., Primdahl, S., Jorgensen, M.J., and Bagger, C. (2000) *J. Electroceram.*, **5**, 141–152; (b) Kim, J. and Lin, Y.S. (2000) *J. Membr. Sci.*, **167**, 123.

83 (a) Bouwmeester, H.J.M. and Burggraaf, A.J. (1997) Dense ceramic membranes for oxygen separation, in *The CRC Handbook of Solid State Electrochemistry* (eds P. J. Gellings and H.J.M. Bouwmeester), CRC Press, Boca Raton, FL, pp. 481–553; (b) Shaula, A.L., Kharton, V.V., Marques, F.M.B., Kovalevsky, A.V., Viskup, A.P., and Naumovich, E.N. (2006) *J. Solid State Electrochem.*, **10**, 28–40; (c) Sunarso, J., Baumann, S., Serra, J.M., Meulenberg, W.A., Liu, S., Lin, Y.S., and Diniz da Costa, J.C. (2008) *J. Membr. Sci.*, **320**, 13–41.

84 Yaremehenko, A.A., Kharton, V.V., Avdeev, M., Shaula, A.L., and Marques, F.B.M. (2007) *Solid State Ionics*, **178**, 1205–1217.

85 (a) Lyskov, N.V., Metlin, Yu.G., Belousov, V.V., and Tretyakov, Yu.D. (2004) *Solid State Ionics*, **166**, 207–212; (b) Belousov, V.V. (2007) *J. Eur. Ceram. Soc.*, **27**, 3459–3467.

86 Neiman, A.Ya., Pestereva, N.N., and Tsipis, E.V. (2007) *Russ. J. Electrochem.*, **43**, 672–681.

3
Advances in the Theoretical Description of Solid–Electrolyte Solution Interfaces

Orest Pizio and Stefan Sokołowski

This chapter describes theoretical approaches of the classical statistical mechanics and modern computer simulation methods to study the microscopic structure and thermodynamic and electrical properties of liquid electrolyte solutions at solid surfaces. Theoretical approaches include integral equations and density functional methods. We analyze their development to reach an accurate description of the interfacial structure and thermodynamics of nonuniform electrolyte solutions. The computer simulation techniques are reviewed. Methodological aspects concerned with the key issue of simulations, namely, accounting for the long-range electrostatic interactions, are given. Finally, we present some selected results for simple models and next for some real and experimentally important systems. Some prospects for future studies are discussed in the summary.

3.1
Introduction

At the early stage of the development of electrochemistry, the principal focus of research was on the properties of liquid electrolyte solutions. At that time, the processes occurring at electrodes were discussed rather briefly. Such a state of things reflected to a great extent the knowledge of interfacial phenomena. In the past few decades, there has been much progress in the understanding of various electrode–electrolyte interfaces. It has been mirrored in textbooks, monographs, and reviews; see, for example, Refs [1–6]. In spite of the rapid progress in this area of research, many problems have not been solved to date. In the case of systems involving liquid electrolyte solutions, this can be explained in part by an incomplete knowledge of water and aqueous solutions containing ions at microscopic level. The intrinsic complexity of these fluids in contact with various solids of interest in electrochemistry leads to additional difficulties. Moreover, one still lacks theoretical methods that involve microscopic characteristics of both participants, that is, the electrode and the electrolyte solution, at a necessary and equal level of description. This fact prohibits progress to reach a nonempirical understanding of electrochemical interfacial phenomena. Traditional theoretical methodology had received very important hints from the flourishing and seemingly unlimited in power computer simulation methods. Both their application at the classical level and their combination with quantum mechanical theoretical approaches permitted a breakthrough in interfacial electrochemistry. Consequently, our chapter is built around Section 3.2 describing dialectic development of traditional theoretical approaches initiated for point ions in a structureless dielectric medium, passing then through integral equations to modern density functional approaches. In this section, we focus both on the electric properties of interfaces and on their thermodynamic characteristics. Section 3.3 is entirely dedicated to computer simulations, their methodological aspects, and results. Whenever possible, comparisons of results coming from different tools are made and discussed. Some prospects for future studies are given in the last section of the chapter.

The reader must bear in mind that the methods involved in the type of studies we deal with are the methods of equilibrium classical statistical mechanics applied to inhomogeneous systems. Thus, the demonstrative examples are concerned with phenomena in inhomogeneous solid–fluid systems with electrostatic interactions that can be treated by this methodology. However, the methods we deal with are useful for understanding other electrochemical systems including, for example, the ones with solid electrolytes or with redox active particles in a liquid electrolyte. In particular, the density functional approaches can be extended to describe solid electrolytes and phase transformations in them. The computer simulation methods are appropriate to investigate the solid phases and liquids containing charged species.

3.2
Theoretical Approaches

3.2.1
Background

The interfacial region between a fluid (or a fluid mixture) and a solid body in electrochemistry actually represents a part of a generic problem of fluid–solid interface. However, specific features and properties of the *electrochemical* interface make this problem quite distinct from the general situation and very complex for a theoretical or simulation study.

To reach a more profound understanding of phenomena in the interface between a solid and a fluid requires accurate microscopic description of both subsystems, the solid and the fluid. We would like to comment first on the class of interfacial systems without electrostatic interactions between constituent species. It is usually assumed that the solid perturbs and changes the fluid structure close to its surface such that the properties of the interfacial region become quite specific and different from what one observes far from the solid surface. Also, it is supposed that the fluid does not exhibit any influence on the solid, in particular on its microscopic structure, even in the most exposed layers. This is true to a certain extent unless the chemical reactions involving species of two different phases occur. Moreover, this is true if the solid does not dissolve in the fluid during their contact [5]. Under common circumstances, one is interested in studying adsorption of the fluid species on a specific solid.

The adsorption given, for example, in terms of adsorption isotherms describes how the fluid behaves close to the solid. The adsorption isotherms are different depending on the energy of interaction between the fluid species and the fluid–solid energy. These isotherms have been classified into certain groups that correspond to wetting, prewetting, layering, and drying regimes according to certain phenomenological criteria; see, for example, Ref. [7]. Moreover, the criteria for classification of isotherms according to their shape have been used [8]. Characteristics of the solid phase indirectly influence the behavior of the adsorption isotherms because the fluid–solid interaction energy is formulated through the combination rules involving fluid–fluid and solid–solid interaction energy [9]. One can get a finer insight into the behavior of adsorption isotherms by taking into account the symmetry of the crystal solid surface. Again, the principal phenomena resume due to either wetting or partial wetting of a solid when a fluid comes in its contact.

In the discussion above, we assumed that the fluid and solid are in equilibrium. Thus, the temperature of the two (solid and fluid) phases is equal. The adsorption of a fluid on a single solid surface or in pores occurs from the bulk or reservoir. One can define the bulk as a portion of the fluid reasonably far from the interfacial region. The reservoir fixes the chemical potential of fluid species. The chemical potential of a solid does not enter such theoretical formulation. However, more complex equilibria must be formulated if, for example, adsorption occurs simultaneously with the dissolution of the solid into the contacting fluid phase [5].

It is common to describe the influence of a solid on the adjacent fluid in terms of an external field acting on fluid particles. In the model studies where the solid is taken just as an impermeable plane surface, the external field is one dimensional; that is, it changes only along the normal to the surface and is constant in the surface plane. If the crystalline solid surface is considered, then the concept of external field is used again. The external field results from the summation of pairwise interaction between all atoms belonging to the solid of certain symmetry with a single fluid particle and depends on three coordinates. The total solid–fluid energy then is just the sum of such external field contributions over all fluid particles [9]. To summarize, the grand thermodynamic potential of a system consisting of a solid and fluid phase in contact must be at minimum at a given fixed set of external thermodynamic parameters if we look for the description of such a system at equilibrium. Searching for such a minimum is equivalent to finding the equilibrium distribution of the fluid species at a solid. An analysis of the resulting microscopic insights permits one to investigate adsorption, phase behavior, and fine properties of the interfacial region.

Now, we would like to discuss these concepts for the systems of interest in interfacial electrochemistry. We will speak first about an electrode as a solid characterized by a certain value of the electrostatic potential that is associated with an electric field acting on an adjacent electrolyte solution. This is a very crude model but has been adopted here just as a useful starting point. Later in this chapter, finer description of the solid will be attempted.

The electrolyte solution modeling is another important issue. Several decades had passed from the pioneering developments of Debye and Hückel who showed that a two-component system of point particles with positive and negative charges is of incredible importance to construct the theory of electrolyte solutions. This model had faithfully served for years in the theory of solutions and in electrochemistry, specifically for electric double-layer problems, until quite recent times.

The microscopic, statistical mechanical theory of homogeneous and inhomogeneous fluids and electrolyte solutions, in particular, is constructed by using the concept of distribution functions. For homogeneous fluids, the most important is the pair distribution function, $g(r)$, that yields thermodynamic properties via the virial equation of state, the compressibility equation, or the energy equation. The pair distribution function describes the probability density of finding a pair of particles at a certain distance between them. It can be Fourier transformed to give a structure factor that can be measured via diffraction experiments. The pair distribution function can also be conveniently obtained by using computer simulation methods [10].

Classical homogeneous systems with electrostatic interactions have a set of specific features that are necessary to mention. An electrolyte solution is a system consisting of at least two species characterized by densities, $\varrho_{i,b}$, $i = +, -$ (the subscript b refers to bulk). Let us denote by eZ_i charges of species i where Z_i is valency. It is assumed that the fluid particles obey the Boltzmann statistics. Such systems at equilibrium must be electroneutral in total, that is, $e\sum_i \varrho_{i,b} Z_i = 0$. According to the requirement of stability of a system of many charges and existence of the statistical partition function, see, for example, Ref. [11], the bare Coulomb interaction between any two charges in a medium with the dielectric constant ε, at a distance r, becomes screened:

$$u_{\text{el},ij}(r) = Z_i Z_j e^2 / 4\pi\varepsilon_0 \varepsilon r, \tag{3.1}$$

where ε_0 is the permittivity of free space. In other words, any given charged particle is surrounded by a "cloud" composed of particles predominantly of the opposite charge. The condition of local electroneutrality is [12]

$$eZ_i = -e \sum_j Z_j \varrho_{j,b} \int [g_{ij}(r) - 1] d\mathbf{r}, \tag{3.2}$$

where $g_{ij}(r)$ is the pair distribution function for species i, j. A similar condition exists for inhomogeneous electrolytes.

The screening of Coulomb interactions between point ions yields shorter range interaction. Namely, it has the form $e^2 Z_i Z_j \exp(-\kappa r)/4\pi\varepsilon_0 \varepsilon r$, whereas the bare potential is given by Equation 3.1. Here,

$$\kappa = 1/r_D = \left(e^2 \sum_i \varrho_{i,b} Z_i^2 / \varepsilon_0 \varepsilon\, kT \right)^{1/2} \tag{3.3}$$

is the inverse of the screening length described by Debye radius r_D, k denotes the Boltzmann constant, and T is the temperature of the system. Not only the Coulomb interactions but also the interactions between multipole moments are screened. In particular, if the polar species are taken into account explicitly, the screened dipole–dipole interaction depends on the distance between two polar particles as $1/(\varepsilon r^3)$ in contrast to $1/r^3$ for bare pair interaction between two dipoles; that is, this screening results in the formation of the dielectric constant.

3.2.2
Inhomogeneous Electrolyte Solutions: Generalities

Distribution of particles in a fluid in contact with plane solid surface is not homogeneous along the axis normal to the surface, z. Statistically, such a distribution of species i is characterized by the one-particle distribution function, $g_{iw}(z)$ (the subscript w denotes the wall),

$$g_{iw}(z) = \frac{\varrho_i(z)}{\varrho_{i,b}} = 1 + h_{iw}(z), \quad (3.4)$$

where $h_{iw}(z)$ is the one-particle correlation function and $\varrho_i(z)$ is the density profile. The individual density profiles of charged particles can be combined to yield the charge density profile,

$$q(z) = e \sum_i q_i(z) = e \sum_i Z_i \varrho_i(z). \quad (3.5)$$

Consider an interface between a solid with dielectric permittivity ε_s and ionic solution (ions immersed in a medium with ε). The interaction potential of ions of species i with the surface can be written as

$$v_i(z) = v_{sh,i}(z) + v_{el,i}(z) + v_{im,i}(z), \quad (3.6)$$

where $v_{sh,i}(z)$ is the short-range nonelectrostatic potential, $v_{el,i}(z) = -eZ_iQ_sz/2\varepsilon_0\varepsilon$ (Q_s is the surface charge), and $v_{im,i}(z) = e^2Z_i^2(\varepsilon-\varepsilon_s)/[16\pi\varepsilon_0\varepsilon z(\varepsilon+\varepsilon_s)]$ is the interaction of an ion with its electrostatic image. In many models, the dielectric constants ε and ε_s are assumed equal to avoid considerations of electrostatic images. The simplest short-range nonelectrostatic potential has the hard-wall form

$$v_{sh,i}(z) = \begin{cases} 0, & \text{for } z > d_i/2, \\ \infty, & \text{otherwise,} \end{cases} \quad (3.7)$$

where d_i is the diameter of ions of type i.

The pair potential between ions, $u_{ij}(r)$, is the sum of a short-range part, usually taken as a hard-sphere interaction,

$$u_{sh,ij}(r) = \begin{cases} 0, & \text{for } r > (d_i+d_j)/2, \\ \infty, & \text{otherwise,} \end{cases} \quad (3.8)$$

and an electrostatic term, given by Equation 3.1. In the presence of a solid surface, there is an additional contribution to $u_{ij}(r)$ due to interaction of ions with their electrostatic images. In this section, we do not consider the terms coming from images; the effects due to the presence of electrostatic image charges will be discussed further in Section 3.3 dedicated to computer simulation methods.

A charged solid surface is set at $z = 0$. All the charges in the system create the electric field that is characterized by the distribution of electrostatic potential $\Psi(z)$. It satisfies the Poisson equation that is the second-order differential equation, relating the distribution of ions and $\Psi(z)$. The Poisson equation reads

$$\nabla^2 \Psi(z) = -\frac{e}{\varepsilon_0 \varepsilon} \sum_i Z_i \varrho_i(z), \quad \text{for } z \geq 0. \tag{3.9}$$

The boundary conditions to integrate this differential equation are formulated according to the geometry of the system; see, for example, Refs [13, 14].

The entire system (solid and electrolyte) must be at equilibrium; therefore, the electroneutrality holds:

$$Q_s + \sum_i e Z_i \int_0^\infty dz \, \varrho_i(z) = 0. \tag{3.10}$$

Commonly, studies on electrochemical systems are performed at a constant surface charge; thus, the electrostatic potential, $\Psi(z)$, varies upon changing thermodynamic conditions. On the other hand, the theoretical setup is conveniently determined by constancy of the electrostatic potential [15, 16]. One needs to fix electrostatic potential at a solid surface, $z = 0$, and far from the surface, $z = \infty$, the latter to provide the reference potential. Usually, one sets $\Psi(z = \infty) = 0$. In what follows, we assume that $\Psi(z = 0) = V_0$ (V_0 is referred to as the electrode potential) and choose the second integration constant according to the geometry of the system. The surface charge, Q_s, then follows from the solution of the Poisson equation for the density profiles and the electroneutrality condition, Equation 3.10.

One of the principal characteristics of the electric interfacial region is the capacitance, C, or the differential capacitance, C_d, defined by the dependence of the surface charge on the electrode potential V_0,

$$C = \frac{Q_s}{V_0}, \quad C_d = \frac{\partial Q_s}{\partial V_0}. \tag{3.11}$$

The distribution of ions corresponding to the potential V_0 does not necessarily yield nonzero surface charge. The value of V_0 corresponding to $Q_s = 0$ is called the potential of zero charge (PZC) [17]. Moreover, in studies of colloidal systems, the term ζ-potential is commonly used. It corresponds to the potential drop, measured with respect to the plane of the closest approach of ions to the surface of a colloidal particle [18].

The model of point ions ($d_i = 0$) immersed in a dielectric medium in contact with a solid surface provides a set of important results for the structure of the interfacial region [19]. We just emphasize a few of them. If, according to Gouy–Chapman (GC) approach, we apply Boltzmann distributions, $\varrho_i(z) = \varrho_{i,b}\exp[-Z_i e\Psi(z)/kT]$, in the Poisson equation (3.9), the obtained density profiles from this equation (called the Poisson–Boltzmann (PB) equation) are monotonously decaying with the distance from the wall due to effective screening of the electrostatic ion–wall interaction. The GC theory is a counterpart of the Debye–Hückel theory for uniform electrolyte solutions and, in fact, preceded it.

For 1 : 1 electrolyte ($Z_1 = -Z_2 = 1$), the density profile equation reads

$$\varrho_i(z) = \varrho_{i,b}[(1+Z_i\alpha\, e^{-\kappa z})/(1-Z_i\alpha\, e^{-\kappa z})]^2, \tag{3.12}$$

where $\alpha = \tanh(eV_0/4kT)$. Analytical solutions of the PB equation also exist for higher valency ions and for nonsymmetric electrolytes [20]. The contact values of the density profiles resulting from the GC theory, Equation 3.12, satisfy Equation 3.12 [19] and are rather successful for not very high ionic concentrations and not very high values of Q_s:

$$kT\sum_i \varrho_i(z=0) = (Q_s^2/2\varepsilon_0\varepsilon) + kT\sum_i \varrho_{i,b}. \tag{3.13}$$

It is also worth mentioning that at this level of the theory one can obtain the result showing the screening of the interactions involving electrostatic images [21].

The double-layer capacitance following from the linearized GC theory (i.e., linearized PB Equation 3.12) is (see, for example, Ref. [1])

$$C_d = \kappa\varepsilon_0\varepsilon. \tag{3.14}$$

Because of the κ dependence on temperature, C_d increases with decreasing temperature. A more general result for Z : Z electrolyte at nonzero V_0 following from Equation 3.12, a complete form of the GC equation, is

$$C_d = [\kappa\varepsilon_0\varepsilon/\cosh[ZeV_0/2kT]. \tag{3.15}$$

The capacitance has a minimum at the PZC and grows with electrolyte concentration.

The GC theory is not sufficient for high electrolyte concentration. From the analysis of experimental data, it follows that the electric double-layer capacitance is given by the equation

$$\frac{1}{C} = \frac{1}{C_{GC}} + \frac{1}{C_H}, \tag{3.16}$$

where C_{GC} is given by Equation 3.15 and the term, called the Helmholtz capacity, C_H, is independent of electrolyte concentration [1]. It has been established that C_H has contributions both from the electrode and from the electrolyte solution.

Insufficiency of the GC approach is primarily connected to the neglect of the finite sizes of ions. A step forward was considering more realistic model of electrolyte solution, namely, taking into account finite size of ions, but still describing the solvent solely via ε. The simplest of these models is the so-called restricted primitive model (RPM), which is a collection of charged hard spheres of equal diameter and equal valency; see, for example, Ref. [19] and the references therein. The RPM can be solved at the level of the Debye–Hückel theory or more sophisticated approaches, among which one can distinguish integral equations and density functional theories.

3.2.3
Integral Equations for Solid–Electrolyte Solution Problems

The most successful approaches in the treatment of one- and m-component fluids are theories that originate from the Ornstein–Zernike (OZ) equation. For a fluid mixture, this equation relates the total, $h_{ij}(r)$, and direct, $c_{ij}(r)$, correlation functions of the species i and j,

$$h_{ij}(r_{12}) = c_{ij}(r_{12}) + \sum_l \varrho_{l,b} \int dr_3\, h_{il}(r_{13}) c_{lj}(r_{23}), \quad i,j = 1, \ldots, m. \tag{3.17}$$

It was demonstrated that an m-component fluid in contact with a solid surface can be considered a special case of $(m+1)$-component mixture, in which the molecular size of one distinguished component (this may also be an ionic component) grows to infinity, while its density tends to zero. Under these conditions, the distinguished component w forms a planar surface. The set of Equation 3.17 splits into (i) equations involving the correlation functions between the component w and the remaining components (fluid components) and (ii) the equations that relate only the correlation functions between m fluid components. The first group of equations is

$$h_{iw}(z) = c_{wi}(z) + \sum_{l=1}^{m} \varrho_{l,b} \int dr'\, h_{lw}(z') c_{il}(|\mathbf{r}'|) \tag{3.18}$$

and

$$h_{iw}(z) = c_{iw}(z) + \sum_{l=1}^{m} \varrho_{l,b} \int dr'\, c_{lw}(z') h_{il}(|\mathbf{r}'|). \tag{3.19}$$

The second group (i.e., the group (ii)) is, of course, given by (3.17). There is still an equation that does not belong either to the group (i) or to the group (ii) and that can be recovered from the set (3.17) for a $(m+1)$-component mixture. It reads

$$h_{ww}(z) = c_{ww}(z) + \sum_{l=1}^{m} \varrho_{l,b} \int dr'\, c_{lw}(z') h_{lw}(|\mathbf{r}'|) \tag{3.20}$$

and relates the wall–wall total, $h_{ww}(z)$, and direct, $c_{ww}(z)$, correlation functions. It can be used to evaluate the potential of mean force between two giant particles w, that is, the solvation force between two planes. It is important for the study of the stability of colloidal suspensions.

Equations 3.18 and 3.19 are formally completely equivalent and the choice of one of them is a question of convenience. In the case of numerical calculations, Equation 3.18 is usually used because the range of the integrations is shorter than in the case of Equation 3.19. Then, Equations 3.18 and 3.4 provide a starting point for evaluation of the density profiles of ions at the surface.

All the equations (3.17)–(3.19) are exact. However, their solution requires introduction of additional relations, called closures, between the total and the direct correlation functions. The closures are approximate. Moreover, the solution of the surface integral equation (18) or (19) requires knowledge of the bulk correlation functions that are usually evaluated solving Equation 3.17 for an m-component mixture with appropriate closures. Here, we concentrate on the discussion of solutions of surface OZ equations and assume that the bulk correlation functions are known.

In order to illustrate the approach, we consider the simple case of an ionic fluid in contact with a charged hard wall. The wall is a semi-infinite half space; it is also possible to consider a wall of finite thickness, a slit-like pore, semipermeable membrane, but one should be aware that in some cases the correlations may occur between ions across the walls [6, 22].

As in the cases discussed previously, the solvent is accounted for by the dielectric constant ε. The dielectric constant of a solid is also ε, such that electrostatic image charges are not present in the system (we note that the latter assumption can be removed [23]). A general form of the closure relation is

$$h_{iw}(z) = c_{iw}(z) + \ln g_{iw}(z) + v_i(z)/kT - B_i(z), \tag{3.21}$$

where $B_i(z)$ is the so-called bridge function. The bridge function is not known in general and the design of a closure is usually equivalent to the construction of an approximation for $B_i(z)$. In particular, by setting $B_i(z) = 0$ in Equation 3.21 we obtain the hypernetted chain (HNC) approximation, which can also be written as

$$\exp[h_{iw}(z) - c_{iw}(z)] = g_{iw}(z) \exp[v_i(z)/kT]. \tag{3.22}$$

Linearization of the exponent on the left-hand side of Equation 3.22 leads to the Percus–Yevick (PY) approximation

$$h_{iw}(z) - c_{iw}(z) = 1 + g_{iw}(z) \exp[v_i(z)/kT]. \tag{3.23}$$

The PY approximation is reasonably good for hard-sphere/hard-wall system, but is not very successful for the systems with other interaction potentials. We should also mention here the mean spherical approximation (MSA) for wall–ion correlation functions. It reads

$$h_{iw}(z) = -1 \quad \text{for} \quad z < d_i/2, \qquad c_{iw}(z) = -v_i(z)/kT \quad \text{for} \quad z > d_i/2. \tag{3.24}$$

The first part of Equation 3.24 is an exact statement of the impenetrability of the solid wall; the second part of this equation is an approximation [24]. The PY and MSA permit one to obtain analytical solutions for some models of the interaction potentials.

As we have mentioned above, there are two equivalent routes to describe the double-layer structure, namely, the constant surface charge, Q_s, route, and the constant electrostatic potential, V_0, route. The constant surface charge route is easier to write down the density profile equations. For example, the HNC approximation for the ion–wall correlation functions, obtained from Equations 3.18 and 3.22, reads

$$\ln[g_{iw}(z)] = -\frac{v_i(z)}{kT} + \sum_j \varrho_{j,b} \int dz' [g_{jw}(z') - 1] C_{ij}(|z' - z|), \tag{3.25}$$

where $C_{ij}(x) = 2\pi \int dr\, r c_{ij}(\sqrt{x^2 + r^2})$ and, recall, $c_{ij}(r)$ is the bulk direct correlation function for ions i and j. In the last equation, the ion–wall potential is given by

Equation 3.6 that involves the surface charge Q_s. Its application requires integrations of long-range contribution and therefore may be tedious [22, 25].

To develop the HNC density profile equation at constant V_0, we realize that the wall–ion (and similarly ion–ion) direct correlation function can be written as a sum of the short-range part, $c_{\text{sh,iw}}(z)$, that involves the contribution from the short–range fluid–wall potential, $v_{\text{sh},i}$ (cf. Equation 3.7), and coupling between short-range and Coulombic fluid–wall interactions [19, 26],

$$c_{\text{iw}}(z) = c_{\text{sh,iw}}(z) - v_{\text{el},i}(z)/kT. \tag{3.26}$$

Similarly, the bulk direct correlation function is written as

$$c_{ij}(r) = c_{\text{sh},ij}(r) - u_{\text{el},ij}(r)/kT, \tag{3.27}$$

with $u_{\text{el},ij}(r)$ given by Equation 3.1. The HNC equation (25) becomes [19]

$$\ln[g_{\text{iw}}(z)] = -[v_{\text{sh},i}(z) + Z_i e \Psi(z)]/kT \\ + \sum_j \varrho_{j,b} \int dr' [g_{wj}(z') - 1] C_{\text{sh},ij}(|z' - z|). \tag{3.28}$$

In the above equation,

$$\Psi(z_1) = Q_s + \sum_{j=1}^{m} \int dr_2 [Z_j e \varrho_{jw}(z_2)/4\pi\varepsilon_0 \varepsilon r_{12}] \\ = -(1/2\varepsilon_0\varepsilon) \int_0^\infty |z_1 - z_2| q(z_2) dz_2 \tag{3.29}$$

and $C_{\text{sh},ij}(x) = 2\pi \int dr \, r c_{\text{sh},ij}(\sqrt{x^2 + r^2})$ and $q(z)$ is given by Equation 3.5. The last line in Equation 3.29 is obtained with the help of Equation 3.10 and is an integrated form of the Poisson equation (3.9) for a fluid in contact with a charged wall. It is worth mentioning that the GC theory results if the term containing integrals on the right-hand side of Equation 3.28 is neglected. Equation 3.28 is completely defined in terms of short-range quantities, which is not the case of Equation 3.25. The contact values of the density profiles resulting from the HNC approximation satisfy the equation [19]

$$kT \sum_i \varrho_i(z = d_i/2) = (Q_s^2/2\varepsilon_0\varepsilon) + \frac{1}{2} kT \varrho_b [1 + \partial(p/kT)/\partial\varrho_b], \tag{3.30}$$

where $\varrho_b = \sum_i \varrho_{b,i}$ and p is the bulk pressure. The last equation can be compared with the result of the GC approximation, Equation 3.13, and with the exact expression for symmetric electrolytes:

$$kT \sum_i \varrho_i(z = d_i/2) = (Q_s^2/2\varepsilon_0\varepsilon) + p. \tag{3.31}$$

The HNC approximation overestimates the sum of the contact values of the density profiles due to an inaccurate description of the excluded volume effects in the ion–wall correlations. However, the bulk HNC approximation leads to reasonably accurate predictions of the bulk pair correlation functions $h_{ij}(r)$.

The bulk direct correlation function ($c_{ij}(r)$ or $c_{sh,ij}(r)$) that must be used as an input to the theory can also be obtained from the HNC approximation. However, the resulting HNC/HNC (the first term in the last acronym points to the surface integral equation approximation, while the second to the bulk theory) yields poor results compared to computer simulations [27]. Generally, better results are obtained when the bulk correlation functions are taken from the bulk MSA. The HNC/MSA approach was used in several works; see, for example, Refs [19, 22, 25, 28]. Another way of improving the HNC approximation was proposed by D'Aguanno et al. [29]. In their method, the surface HNC approximation was used taking the bulk direct correlation functions as depending on local density, by defining "effective weighted local densities." Such an approach can be considered a precursor of the density functional methods, described below.

Despite the limited accuracy of the HNC/MSA approximation, it is still used. Recently, Ángeles and Lozada-Cassou [25] studied the so-called charge reversal at a planar charged surface. Obviously, the charged particles should be preferentially adsorbed on an oppositely charged wall. However, thermal movement of ions prevents them from "condensing" on the wall, building up a well-packed layer. Intuitively, one can expect that the amount of adsorbed counterions in the double layer is sufficient to compensate the surface charge. Under certain conditions, the adsorbed counterions can overcompensate the surface charge, yielding the surface charge reversal. Hence, an inversion of the local electric field provides that coions form a second layer. This is known as charge inversion. These phenomena are particularly important in the analysis of charge stabilization or destabilization of colloidal systems. The charge reversal can be characterized by the cumulative charge distribution

$$\bar{q}(z) = -1 - (1/Q_s) \int_{d/2}^{z} q(z') dz', \qquad (3.32)$$

where $q(z)$ is given by Equation 3.5. Figure 3.1 shows examples of $\bar{q}(z)$. For low bulk densities, $\bar{q}(z)$ monotonically increases, whereas for high densities $\bar{q}(z)$ exhibits oscillations that characterize the charge reversal.

Integral equations were also applied for nonprimitive models, that is, the models in which the solvent species are considered explicitly. These developments comprise a qualitatively new step in the investigation of the electrode–electrolyte interface. Early studies on mixtures of charged hard spheres (solutes) and hard spheres with dipole moments (solvent) were performed by Carnie and Chan [30] and by Blum and Henderson [31] by using the mean spherical approximation for particle–particle and particle–wall correlations. All angular dependent functions in the OZ equations have been expanded in series using generalized spherical harmonics. Then, the MSA problem reduces to the solution of a system of algebraic equations for certain coefficients of such expansions. The fluid model is described in terms of bulk densities of ions and dipoles, $\varrho_{i,b} = \varrho_{+,b} + \varrho_{-,b}$ and $\varrho_{d,b}$, respectively. The diameters of all species were taken equal for simplicity: $d_+ = d_- = d$; $Z_+ = |Z_-| = 1$. Also, one needs to choose the dipole moment of solvent species, μ_d, and temperature, T. It

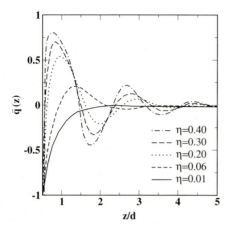

Figure 3.1 Dependence of $\bar{q}(z)$ on the distance z normalized by the species diameter (d) for the 2:2 electrolyte. The bulk packing fractions, $\eta = \pi \varrho_b d^3/6$, are given in the figure. The calculations are for $1/T^* = e^2 Z^2/(d \varepsilon k T) = 6.72$ and $\gamma^* = 2\pi e Z Q_s d/(\varepsilon k T) = 2.4$. Reprinted with permission from Ref. [25]. Copyright 2008, American Institute of Physics.

is convenient to use dimensionless parameters $d_0 = (4\pi \varrho_i e^2 / kT)^{1/2}$ and $d_2 = 4\pi \varrho_d \mu_d^2 / 3kT$ to characterize the ionic and dipolar subsystems and reduced densities, $\varrho_i^* = \varrho_i d^3$ and $\varrho_d^* = \varrho_d d^3$. Also, it is important to mention that d_0 does not contain the dielectric constant ($\kappa = d_0/(4\pi \varepsilon_0 \varepsilon)^{1/2}$, the latter is formed due to the screening of dipole–dipole interactions and depends on the parameter d_2 [32].

In the MSA, the principal output from the solution of the system of OZ integral equations is given in terms of the following parameters related to the integrals of certain harmonics of the ion–ion ($h_{ii}^{000}(r)$), ion–dipole ($h_{id}^{101}(r)$), and dipole–dipole ($h_{dd}^{112}(r)$) correlation functions:

$$b_0 = 2\pi \varrho_i \int_0^\infty dr\, r h_{ii}^{000}(r), \qquad b_1 = 2\pi (\varrho_i \varrho_d/3)^{1/2} \int_0^\infty dr\, h_{id}^{101}(r), \qquad (3.33)$$

$$b_2 = 3\pi (2/15)^{1/2} \varrho_d \int_1^\infty dr\, h_{dd}^{112}(r).$$

We introduce auxiliary combinations of the parameters b_0, b_1, and b_2 given above that conveniently provide electrostatic interfacial properties:

$$\beta_{3\cdot 2^n} = 1 + (-1)^n (1/3)(b_2/2^n), \qquad D_F = (1/2)[\beta_6(1+b_0) - b_1^2/12] \qquad (3.34)$$

and

$$a_1 = \frac{1}{2D_F^2}(-b_0 \beta_6^2 + b_1^2 \beta_{24}/3), \qquad a_2 = -\frac{b_1}{2D_F^2}\left[\beta_{12} + (b_0 \beta_3 + b_1^2/6)/2\right]. \qquad (3.35)$$

In particular, the potential drop at the interface and the differential capacitance have the following form:

$$V_0 = (E/2d_0^2 D_F)(a_1\beta_6 - b_1 a_2/2), \tag{3.36}$$

where $E = Q_s/\varepsilon_0\varepsilon$ is the electric field, and

$$C_d = \varepsilon_0\varepsilon \frac{2d_0^2 D_F}{a_1\beta_6 - b_1 a_2/2}. \tag{3.37}$$

In the limit of low and high concentrations of ions, one can perform a more detailed analysis. In particular, at low concentration, we can validate present results by comparing them with the predictions obtained previously for RPM. Namely, the expression above for the electrode potential reduces to [31]

$$V_0 = \frac{E}{\kappa}\left[1 + \frac{1}{2}\kappa d(1 + \beta_6(\varepsilon-1)/\beta_3)\right], \tag{3.38}$$

whereas the differential capacitance is

$$C_d = \varepsilon_0\varepsilon\kappa\left[1 + \frac{1}{2}\kappa d(1 + \beta_6(\varepsilon-1)/\beta_3)\right]^{-1}. \tag{3.39}$$

The counterpart of this expression in the primitive model is

$$C_d = \varepsilon_0\varepsilon 2\Gamma \approx \varepsilon_0\varepsilon\kappa\left(1 + \frac{1}{2}\kappa d\right)^{-1}. \tag{3.40}$$

The capacitance for the nonprimitive model is substantially lower compared to values obtained in the primitive model at the same values of parameters; actually, these trends lead to a more reasonable relation with experimental data. One must bear in mind, however, that it is necessary to use the same dielectric constant in the two expressions rather than to apply the dielectric constant coming from the MSA calculations for the bulk model. This issue has been discussed in detail by Blum [33].

The nonprimitive model is richer than its primitive counterpart. In particular, one obtains the values for the polarization produced by the electrode. At low ionic concentration, it reads

$$P = \frac{(\varepsilon-1)}{\varepsilon}\frac{E}{\kappa}\left[1 + \frac{1}{2}\kappa d(1 - \beta_6/\beta_3)\right]. \tag{3.41}$$

The first contribution is the well-known expression for the polarization of the dielectric continuum. However, the second term is positive by value and describes cooperative effects of the ordering of the dipole molecules in the interface region. This ordering or alignment is not restricted to the single layer of polar molecules in contact with the electrode, as it can be seen from the behavior of the relevant density profile. Moreover, this term depends on the electric field of the electrode, on the ionic concentration, and on the temperature through the value of κ. A similar analysis can be performed for high ion concentration (i.e., taking $d_2 = 0$ in the general expressions). Then,

$$V_0 = \frac{E}{2\Gamma}, \quad C_d = \varepsilon_0\varepsilon 2\Gamma. \tag{3.42}$$

These results have been obtained by Blum [34] previously.

Actually, this set of expressions for the restricted (due to equal diameters of species) nonprimitive model was obtained approximately 30 years ago. Later, there have been several developments, of which we would like to mention only two. One of them consists in the solution of the version of the bulk ion–dipole hard-sphere mixture in which the possibility of chemical association between different species is included. This analytic solution in the framework of the associative mean spherical approximation was obtained by Holovko and Kapko [35]. In our opinion, it opens the possibility to extend the temperature range for which the theory remains adequate and take into account the possible formation of ionic complexes. To the best of our knowledge, this development has so far not been extended to interfacial studies and problems of electrochemistry, in spite of much interest.

Another development has been mostly due to Schmickler and Henderson [36, 37] and focused on the finer description of the electrode subsystem. It is of much importance to electrochemistry and again seems to have room for improvement. In essence, it was attempted to describe the metal electrode subsystem explicitly rather than resort to the approximation that the electrode is an ideal conductor and just provides confinement for the electrolyte solution in the form of a charged wall.

A model for the electrode–electrolyte solution interface used in Refs [36, 37] is combined: it consists of a jellium model for metal electrode and an ensemble of hard sphere ions and dipoles for the solution. The jellium model includes the positive metal ions that are smeared out into a constant positive background of density n_+. It drops abruptly to zero at the metal surface (at $z = 0$), such that this density can be conveniently described by the step function. The electrons form a plasma with density $n_-(z)$ that spills out over the metal surface into the electrolyte solution. The surface energy E_s of the electronic plasma is considered as a functional of the electronic density profile $n_-(z)$ and can be written in a simplified manner as a sum of two terms,

$$E_s[n_-(z)] = E_{pl} + E_{cr}, \qquad (3.43)$$

where the first term, E_{pl}, includes the kinetic, the exchange, the correlation energy of the plasma, its self-interaction, its interaction with the positive background, and the inhomogeneity term due to the gradient of $n_-(z)$. The second, cross term, E_{cr}, must describe three effects, $E_{cr} = E_{ion} + E_{dip} + E_{rep}$. Namely, it gives both the electrostatic interaction energy of electrons with ions (E_{ion}) and with dipoles (E_{dip}) of the solution and the repulsion of electrons from the solution due to short-range forces, E_{rep}. One must search for the density profile $n_-(z)$ that minimizes the surface energy for a given metal (characterized by n_+) and for the electrolyte solution under given thermodynamic conditions (the solution is characterized by the same parameters as discussed above for an ion–dipole mixture).

To solve the problem for the electronic density profile, different trial functions dependent on n_+ and other parameters of the combined model have been used. In particular, the simple choice for the trial function applied in Ref. [36] was

$$n_-(z) = \begin{cases} n_+ - \dfrac{1}{2}(n_+ + \alpha Q_s/e)\exp(\alpha z), & z < 0, \\ \dfrac{1}{2}(n_+ - \alpha Q_s/e)\exp(\alpha z), & z > 0, \end{cases} \quad (3.44)$$

where Q_s is the charge of the solid surface and α is the free parameter. It is determined from the minimization of the surface energy. The algorithm to obtain the potential drop across the interface is the following. Bearing in mind that the MSA/MSA theory is adequate for low values of the surface charge, the estimate for Q_s is made by using certain density profiles for ions and dipoles from the consideration of the ion–dipole model at a charged wall. This value is obtained at given ion and dipole densities, charge of ions, and dipole moment of polar molecules, temperature, and V_0. The V_0 value follows from the solution of the Poisson–Boltzmann equation. One needs to work close to the PZC conditions.

The value of Q_s and other necessary parameters for the solution are used in the trial function for the sake of minimization of the surface energy according to Equation 3.43. Such a minimization provides α necessary to give explicitly the contribution of ions and of dipoles into the potential drop. It reads

$$V_0 = \frac{4\pi e n_+}{\alpha^2} - \frac{4\pi \mu \varrho_d}{3} + Q_s \left[\frac{d\beta_6}{2\beta_3} + \left(1 + \frac{\kappa d}{2} - \frac{\kappa d \beta_6}{2\beta_3}\right) / \kappa \varepsilon \right]. \quad (3.45)$$

The potential drop is calculated for several small values of Q_s and at $Q_s = 0$ and next the capacitance is obtained by numerical differentiation. This theory does not clarify, however, if there is any influence of the electron density of the electrode on the behavior of the density profiles of ions and dipoles close to the surface. Nevertheless, the model permits to study the Helmholtz contribution to the capacitance, cf. Equation 3.16, on the nature of metal electrode and on ion concentration. One such example is given in Figure 3.2.

This type of theory has also been developed by Badiali et al. [38]. The MSA solution is appealing because the analytic expressions for the various properties can be obtained. In an attempt to extend the linearized-type theories such as the MSA, Torrie and Patey [39] used the reference HNC approximation for a model of the solvent that includes dipoles and quadrupoles and the corresponding ion–solvent mixture. Their objective was to study solvent effects on the electric double layer. They solved the reference HNC equation for a spherical electrode that is 50 times the size of ions and solvent molecules. They found considerable structure of species near the electrode. Moreover, the differential capacitance of the double layer is in reasonable agreement with experimental data and much superior to the results of the modified GC theory [40].

The equations discussed above come from the OZ relation. This is the most popular methodology in the integral equation theory of homogeneous and inhomogeneous liquids and liquid mixtures. Another class of the integral equations for the structure of electric double layer may be derived from the Born–Green–Yvon hierarchy [27]. The application of the latter approach to double electrical layer is given in, for example, Ref. [41].

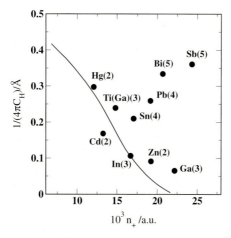

Figure 3.2 Inverse double-layer capacitance of the water–metal interphase for various sp metals. The solid line is the theoretical result explained in the text. Experimental data are given by points. Reprinted with permission from Ref. [36]. Copyright 1984, American Institute of Physics.

In general, developments of integral equations provided a big step forward toward a better description of the double-layer structure. However, the accuracy of the equations quoted above in predicting the density profiles compared to simulation results is not entirely satisfactory. Moreover, these equations are intrinsically unable to describe surface phase transitions, such as wetting transitions [42], unless a sophisticated ansatz for the bridge functions is applied. Such an ansatz is difficult to develop. Alternatively, one may attempt to consider higher order integral equations and relevant closures at a higher level, compared to the singlet theories, discussed above [43, 44]. Formally, this route would provide a complete thermodynamic description of the system in question and would also allow to study the effects due to electrostatic image charges [45]. Numerical difficulties prohibit extensive applications of such procedures. From the numerical point of view, much more convenient are the density functional (DF) approaches. They give thermodynamics directly, are easy to extend for systems with more complex molecules, and provide a possibility for describing surface phase transitions.

3.2.4
Density Functional Theories

The essence of the DF approaches relies on the development of approximations for the free energy $F[\varrho_i]$ as a functional of the density profiles $\varrho_i(\mathbf{r})$ [46]. Commonly, these theories are formulated and applied in the grand canonical ensemble. We remind the reader that the grand canonical thermodynamic potential is

$$\Omega[\varrho_i] = F[\varrho_i] + \sum_i \int d\mathbf{r}[\mu_i - v_i(\mathbf{r})]\varrho_i(\mathbf{r}), \tag{3.46}$$

where μ_i is the chemical potential of species (ions) i and the density profiles follow from the variational principle

$$\delta\Omega/\delta\varrho_i(\mathbf{r}) = 0. \tag{3.47}$$

The accuracy of the DF approaches is determined by the accuracy of approximation for the free energy. Usually, free energy approximations are constructed in the spirit of perturbational treatments for interparticle interactions.

At present, theoretical descriptions of the electric double layer concentrate on simple models for the electrolyte solution and solid surface. Among these models is the already discussed RPM. In many applications, the solid surface is an infinite planar hard wall with a uniform surface charge density Q_s. Other shapes for solid surfaces (e.g., spherical or cylindrical) were studied as well, but they are not discussed here for the sake of brevity. Usually, the dielectric constant of the space occupied by the solid is taken equal to that for an ionic solution, ε.

For the RPM, the Helmholtz free energy functional $F[\varrho_i]$ comprises ideal, $F_{id}[\varrho_i]$, and excess, $F_{ex}[\varrho_i]$, terms. The ideal term is known exactly,

$$F_{id}[\varrho_i] = \sum_i \int d\mathbf{r}[\ln\varrho_i(\mathbf{r}) - 1]\varrho_i(\mathbf{r}). \tag{3.48}$$

The functional $F[\varrho_i]$ is a generating functional for the family of the direct correlation functions (dcfs)[46]. Specifically, the one- and two-particle dcfs follow from the following functional derivatives:

$$c_i(r_1) = \frac{\beta \delta F[\varrho_i]}{\delta \varrho_i(r_1)} \quad \text{and} \quad c_{ij}(\mathbf{r}_1, \mathbf{r}_2) = \frac{\beta \delta^2 F[\varrho_i]}{\delta \varrho_i(r_1) \delta \varrho_j(r_2)}. \tag{3.49}$$

Note that here $c_{ij}(r_1, r_2)$ is the two-particle direct correlation function of a *nonuniform* fluid. We have already seen that dcfs are conveniently written as a sum of a short-range and a Coulomb term, cf. Equations 3.26 and 3.27.

Now, it is necessary to show how the electrostatic potential enters the expression for the grand thermodynamic potential. For this purpose, let us assume that the excess free energy is known for the state characterized by the density profiles $\varrho_{i,0}(\mathbf{r})$. To get the excess free energy at any arbitrary state with densities $\varrho_i(\mathbf{r})$, we functionally integrate the free energy along the linear density path [46]

$$\varrho_i(\mathbf{r}) = \varrho_{i,0}(\mathbf{r}) + \alpha[\varrho_i(\mathbf{r}) - \varrho_{i,0}(\mathbf{r})], \tag{3.50}$$

where $\alpha \in [0, 1]$ is the so-called "charging" parameter. Then,

$$\begin{aligned}
-\beta F_{ex}[\varrho_i] =\ & -\beta F_{ex}[\varrho_{i,0}] + \sum_i \int d\mathbf{r}[\varrho_i(\mathbf{r}) - \varrho_{i,0}(\mathbf{r})]c_i(\mathbf{r}) \\
& + \frac{1}{2}\sum_{i,j} \int d\mathbf{r}\, d\mathbf{r}'[\varrho_i(\mathbf{r}) - \varrho_{i,0}(\mathbf{r})][\varrho_i(\mathbf{r}') - \varrho_{i,0}(\mathbf{r}')] \\
& \times \int_0^1 d\alpha \int_0^\alpha d\alpha'\, c_{ij}(\mathbf{r}, \mathbf{r}'; \alpha'),
\end{aligned} \tag{3.51}$$

where $c_{ij}(r,r';\alpha')$ must be calculated for all intermediate states, characterized by α'. Assuming that the initial state is just the bulk electrolyte at densities $\varrho_{i,0}(\mathbf{r}) = \varrho_{i,b}$, from Equations 3.9, 3.26, 3.27 and 3.51 we obtain (cf. Ref. [47])

$$\Omega[\varrho_i] - \Omega_b = \sum_i \left\{ \int d\mathbf{r}\, \varrho_i(\mathbf{r}) v_{sh,i}(z) + kT \int d\mathbf{r}\, \varrho_i(\mathbf{r}) \ln[\varrho_i(\mathbf{r})/\varrho_{i,b}] \right.$$
$$\left. - eZ_i \int d\mathbf{r}\, \varrho_i(\mathbf{r}) \Psi_i(\mathbf{r}) \right\} + F_{hs}[\varrho_i]$$
$$- \frac{kT}{2} \sum_{i,j} \int d\mathbf{r}\, d\mathbf{r}'\, [\varrho_i(\mathbf{r}) - \varrho_{i,b}][\varrho_j(\mathbf{r}') - \varrho_{j,b}] \int_0^1 d\alpha \int_0^\alpha d\alpha'\, c_{sr,ij}(\mathbf{r},\mathbf{r}';\alpha'), \quad (3.52)$$

where Ω_b is the grand potential for the bulk system and F_{hs} is the hard-sphere contribution to the free energy functional; in the last equation, $c_{sr,ij}(r,r';\alpha')$ is the short-range part of the *nonuniform* two-particle dcf.

Condition (3.47) yields the density profile equation

$$kT \ln[\varrho_i(\mathbf{r})/\varrho_{i,b}] = -v_{sh,i}(z) - eZ_i \Psi_i(\mathbf{r}) + kT\tilde{c}_i(\mathbf{r}) + \frac{\delta F_{hs}}{\delta \varrho_i(\mathbf{r})}, \quad (3.53)$$

where

$$\tilde{c}_i(r) = \sum_j \int d\mathbf{r}'\, [\varrho_j(\mathbf{r}') - \varrho_{j,b}] \int_0^1 d\alpha\, c_{sr,ij}(\mathbf{r},\mathbf{r}';\alpha). \quad (3.54)$$

A commonly made approximation is to assume $c_{sr,ij}(\mathbf{r},\mathbf{r}';\alpha)$ to be equal to its bulk counterpart and thus independent of the "charging" parameter α. Moreover, it can be taken from the analytical solution of the MSA for the RPM [48].

In the original derivation of the DF theory [47] for RPMs, the hard-sphere contribution to the free energy functional was calculated from the Carnahan–Starling equation of state, according to the nonlocal density functional prescription of Tarazona (cf. Ref. [46]). This approach relies on the introduction of a weighted density, defined as an integral of the local density with an appropriate weight function. Tarazona's approach is quite accurate for nonuniform uncharged hard spheres; see Ref. [46] and the references therein.

In Figure 3.3, we present examples of the ionic density profiles, calculated for 2:2 (Figure 3.3a) and 1:1 (Figure 3.3b) electrolytes using this theory. All the calculations were carried out assuming $\varepsilon = 78.5$ and $d = 0.425$ nm at $T = 298$ K and compared with the results of computer simulations [49]. In the case of divalent electrolyte (Figure 3.3a), we observe the already mentioned charge reversal that consists in the formation of a second layer of coions next to the first layer of counterions. Nearly all the charge from counterions is concentrated in a thin layer at the wall, such that a "dipole"-like (i.e., the wall and the counterions) structure is formed. The response of the system to these dipoles is in the formation of a layer of coions between $d < z < 2d$. The theory describes the double-layer structure reasonably well, although it underestimates the magnitude of the maximum for the coions profile. At a higher bulk ionic concentration and for a high surface charge density

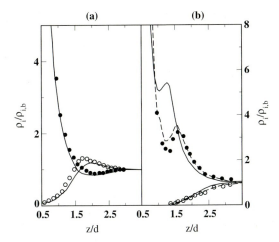

Figure 3.3 Comparison of the DF theory density profiles with computer simulation results [49] (symbols). Part (a) is for 2:2 electrolyte at the bulk density corresponding to 0.5 M ion concentration and for $Q_s^* = Q_s d^2/e = 0.1704$; part (b) is for 1:1 at 1 M and for $Q_s^* = 0.7$. Solid lines give the results of DF theory and dashed lines in (b) give the results of Henderson and Plischke [43]. Reprinted with permission from Ref. [47]. Copyright 1990, American Institute of Physics.

(Figure 3.3b), the agreement of the theory and simulations is worse. The theory overestimates the counterions' local density within the second layer and incorrectly predicts the position of its maximum. The second-order integral equation approach, mentioned in the preceding section [43], appears to be reasonably accurate in this case. However, it is computationally much more expensive compared to the DF approach.

As we have noted above, the theory developed by Mier-y-Teran et al. [47] was based on the Tarazona's nonlocal approach. An alternative DF theory was proposed by Patra and Ghosh [50–53]. Their method is based on the Curtin–Ashcroft theory for uncharged confined fluids [46]. The approach of Patra and Ghosh was also applied to double layers at nonplanar surfaces [18, 54].

However, the most successful in the description of nonuniform hard-sphere systems is undoubtedly the fundamental measure theory (FMT), which was proposed originally by Rosenfeld [55]. According to FMT, the term F_{hs} is written as the integral of the free energy density, $F_{hs} = \int d\mathbf{r}\, \Phi_{hs}(\mathbf{r})$, and the free energy density results from very accurate Mansoori–Carnahan–Starling–Leland–Boublik equation of state for hard-sphere fluid mixtures:

$$\Phi_{hs} = -n_0 \ln(1-n_3) + \frac{n_1 n_2 - \mathbf{n}_1 \cdot \mathbf{n}_2}{1-n_3}$$
$$+ n_2^3(1-\xi^2)^3 \frac{n_3 + (1-n_3)^2 \ln(1-n_3)}{36\pi n_3^2 (1-n_3)^2}. \tag{3.55}$$

In the above, $\xi(\mathbf{r}) = |\mathbf{n}_2(\mathbf{r})|/n_2(\mathbf{r})$. The scalar, n_β, for $\beta = 0, 1, 2, 3$ and vector, $\mathbf{n}_\beta$, for $\beta = 1, 2$ weighted densities are given by

$$n_\beta(\mathbf{r}) = \sum_i \int d\mathbf{r}'\, \varrho_i(\mathbf{r}') w_\beta^{(i)}(\mathbf{r}, \mathbf{r}'), \tag{3.56}$$

where $w_\beta^{(i)}$ are the weight functions. Similar relations hold for the vector weighted densities, $\mathbf{w}_\beta^{(i)}$; see the original work [55].

The above method is appealing due to its reasonable level of simplicity of the description of electrostatic contribution to the free energy. This term is developed as a perturbation with respect to the homogeneous reference system, that is, the bulk electrolyte. However, under certain circumstances, the method is not satisfactory. Namely, it fails to describe vapor–liquid phase transition due to electrostatic interactions. On the other hand, description of ionic distribution close to the solid surface is not always very accurate, as it follows from a comparison of the density profiles obtained in this theory with available computer simulation data.

Recently, a new approach to account for electrostatic free energy, F_{el}, in inhomogeneous RPM and related models has been proposed by Gillespie et al. [56, 57]. The principal idea of the method, called reference fluid density (RFD) approach according to the original work, is to use an inhomogeneous electrolyte as a reference model to study an inhomogeneous system of ions in question. The excess free energy is $F_{ex} = F_{hs} + F_{el}$. The hard-sphere contribution is described within the FMT, whereas the RFD concerns the electrostatic term written in the form

$$\begin{aligned} F_{el}[\varrho_i] = {} & F_{el}[\varrho_{i,\text{ref}}] - kT \sum_j \int \mathbf{r} c_{i,el}[\varrho_{j,\text{ref}}] \Delta \varrho_j(\mathbf{r}) \\ & - \frac{kT}{2} \sum_{i,j} \int d\mathbf{r} d\mathbf{r}'\, c_{ij,el}(\mathbf{r}, \mathbf{r}') \Delta \varrho_i(\mathbf{r}) \Delta \varrho_j(\mathbf{r}'), \end{aligned} \tag{3.57}$$

where $\Delta \varrho_i(\mathbf{r}) = \varrho_i(\mathbf{r}) - \varrho_{i,\text{ref}}(\mathbf{r})$ and the electrostatic contributions to the dcfs are evaluated for reference density profiles, $\varrho_{i,\text{ref}}$. The electrostatic contribution to the free energy F_{el} is the generating functional for the corresponding dcfs, for example,

$$c_{i,el}(\mathbf{r}) = -\frac{\delta \beta F_{el}}{\delta \varrho_i(\mathbf{r})}, \quad c_{ij,el}(\mathbf{r}, \mathbf{r}') = -\frac{\beta \delta^2 F_{el}}{\delta \varrho_i(\mathbf{r}), \delta \varrho_j(\mathbf{r}')}. \tag{3.58}$$

The function $c_{i,el}(\mathbf{r})$ can be expanded into a series to obtain

$$\begin{aligned} c_{1,el}(\mathbf{r}, \{\varrho_i(\mathbf{r})\}) \approx {} & c_{1,el}(\mathbf{r}, \{\varrho_{i,\text{ref}}(\mathbf{r})\}) \\ & + \sum_j \int d\mathbf{r}'\, c_{ij,el}(\mathbf{r}, \mathbf{r}'; \{\varrho_{i,\text{ref}}(\mathbf{r}')\}) \Delta \varrho_i(\mathbf{r}'). \end{aligned} \tag{3.59}$$

The RFD approach considers the reference fluid densities as functionals of the particle densities. The densities $\varrho_{i,\text{ref}}(r)$ must be chosen such that both the first- and the second-order electrostatic dcfs can be evaluated. According to the original work [56], the reference fluid densities have the form

$$\varrho_{i,\text{ref}}(\mathbf{r}; \{\varrho_i(\mathbf{r})\}) = \int d\mathbf{r}'\, \alpha_i(\mathbf{r}') \varrho_i(\mathbf{r}') w(\mathbf{r}, \mathbf{r}'), \tag{3.60}$$

where the scaling factor $\alpha_i(\mathbf{r}) = P(\mathbf{r})$ for $Z_i \geq 0$ and $\alpha_i(\mathbf{r}) = P(\mathbf{r})Q(\mathbf{r})$ for $Z_i < 0$. The functions introduced just above are

$$P(\mathbf{r}) = \sum_i Z_i^2 \varrho_i(\mathbf{r}) / \left[\sum_{Z_i \geq 0} \varrho_i(\mathbf{r}) + Q(\mathbf{r}) \sum_{Z_i < 0} Z_i^2 \varrho_i(\mathbf{r}) \right] \qquad (3.61)$$

and

$$Q(\mathbf{r}) = \sum_{Z_i \geq 0} Z_i \varrho_i(\mathbf{r}) / \sum_i |Z_i| \varrho_i(\mathbf{r}). \qquad (3.62)$$

The choice of scaling factors ensures that the fluid with densities $\alpha_i(\mathbf{r})\varrho_i(\mathbf{r})$ is electroneutral at any point $\mathbf{r}$. Moreover, it has the same ionic strength as the fluid with densities $\varrho_i(r)$. Finally, the weighting function, $w(\mathbf{r}, \mathbf{r}')$ in Equation 3.60, is given by

$$w(\mathbf{r}, \mathbf{r}') = \frac{1}{4\pi R_{el}^3/3} \theta(|\mathbf{r}' - \mathbf{r}| - R_{el}(\mathbf{r})), \qquad (3.63)$$

where $\theta(x)$ denotes the step function and $R_{el}(\mathbf{r})$ denotes the radius of the sphere over which averaging is performed. It is not a well-defined quantity. However, a reasonable approximation is to take it as a sum of average ion radius and the local screening length, $s(\mathbf{r})$, as follows:

$$R_{el}(\mathbf{r}) = \sum_i (d/2) \alpha_i(\mathbf{r}) \varrho_i(\mathbf{r}) / \sum_i \alpha_i(\mathbf{r}) \varrho_i(\mathbf{r}) + s(\mathbf{r}). \qquad (3.64)$$

We recall that for RPM all d_i values are equal and therefore the subscript i in the above equation is dropped.

The local screening length can be chosen extending the MSA result for the screening length of the bulk electrolyte in the form, $s(\mathbf{r}) = [2\Gamma(\mathbf{r})]^{-1}$, where

$$2\Gamma(\mathbf{r})d = \sqrt{1 + 2\kappa(\mathbf{r})d} - 1. \qquad (3.65)$$

Here, $\kappa(\mathbf{r})$ is the inverse Debye radius (cf. Equation 3.3), calculated at each point $\mathbf{r}$ for the set of reference fluid densities $\varrho_{i,\text{ref}}(\mathbf{r})$. Technical issues of the application of the procedure were discussed in detail by Gillespie et al. [56, 57].

It can be shown [56, 57] that the equation for the density profiles from the RFD approach has the form of Equation 3.53, but where $\tilde{c}_i(\mathbf{r}) \equiv \tilde{c}_{i,\text{el}}(\mathbf{r})$,

$$\tilde{c}_{i,\text{el}}(\mathbf{r}) = \sum_j \int d\mathbf{r}' \, \Delta\varrho_i(\mathbf{r}') c_{ij,\text{sr}}(\mathbf{r}, \mathbf{r}'; \{\varrho_{j,\text{ref}}(\mathbf{r})\}). \qquad (3.66)$$

One of the principal achievements of the method by Gillespie et al. [56, 57] is the proposed flexible weighting procedure for the evaluation of electrostatic free energy contribution. This procedure permits to develop an alternative density functional method. Before presenting it, we emphasize that the two approaches outlined above follow the so-called compressibility route to thermodynamics. The compressibility route for the RPM does not permit to capture vapor–liquid phase transition. This is in contrast to available computer simulation data predicting vapor–liquid transition in the bulk RPM with confidence. Moreover, surface phase transitions expected for the class of interfacial systems of our interest would not be captured by those theories. Recently, we have attempted to develop the approach avoiding the compressibility route [58].

The electrostatic contribution to the free energy can be written similarly as the hard-sphere free energy in the FMT. Namely, the starting point is

$$F_{el}[\{\varrho_i(\mathbf{r})\}] = \int d\mathbf{r}\, \Phi_{el}[\{\varrho_{i,\mathrm{ref}}(\mathbf{r})\}], \qquad (3.67)$$

where Φ_{el} is the electrostatic free energy density evaluated for the reference fluid densities. It can be well approximated by using the solution of the MSA for the RPM in the framework of the energy route [58]:

$$\Phi_{el}\left[\{\varrho_{i,\mathrm{ref}}(\mathbf{r})\}\right] = -\frac{e^2}{T^*}\sum_i Z_i^2 \varrho_{i,\mathrm{ref}}(\mathbf{r}) \frac{\Gamma(\mathbf{r})}{1+\Gamma(\mathbf{r})d} + \frac{\Gamma(\mathbf{r})^3}{3\pi}. \qquad (3.68)$$

The expression for Γ is given by Equation 3.65. The equation for the density profiles is given by Equation 3.53 with $\tilde{c}_i(\mathbf{r}) \equiv \tilde{c}_{i,el}(\mathbf{r})$, where now

$$\tilde{c}_{i,el}(\mathbf{r}) = \frac{\delta F_{el}[\{\varrho_i(\mathbf{r})\}]}{\delta \varrho_i(\mathbf{r})} \qquad (3.69)$$

and where $F_{el}[\{\varrho_i(\mathbf{r})\}]$ is given by Equations 3.67 and 3.68.

The final expression for the grand thermodynamic potential reads

$$\Omega = kT \sum_i \int d\mathbf{r}\, \varrho_i(\mathbf{r})[\ln \varrho_i(\mathbf{r}) - 1] + \int d\mathbf{r}\, \Phi_{hs}(n_\alpha(\mathbf{r})) + \int d\mathbf{r}\, \Phi_{el}(\varrho_i(\mathbf{r})) \\ + e \sum_i Z_i \int d\mathbf{r}\, \varrho_i(\mathbf{r}) \Psi(\mathbf{r}) + \sum_i \int d\mathbf{r}\, \varrho_i(\mathbf{r})[v_{sh,i}(\mathbf{r}) - \mu_i]. \qquad (3.70)$$

The first term is the ideal free energy. The second term, resulting from FMT, describes the hard-sphere contribution, while the third term takes into account electrostatic contribution to the free energy functional. The inhomogeneous solution is in equilibrium with its bulk counterpart. Thus, to be consistent one chooses the chemical potential for the bulk fluid as follows: $\mu_i = \mu_{i,hs} + \mu_{i,el}$, where the Mansoori–Carnahan–Starling–Leland–Boublik equation of state and the MSA energy route solution are used in calculations.

Up to now our principal focus was on the development of adequate tools to describe hard-sphere and electrostatic contributions. There is, however, one more concept of much importance in the classical statistical mechanics of interfacial systems. This is the concept of chemical association. It is important in several aspects. It is known that with decreasing temperature the strength of electrostatic ion–ion interaction increases and the formation of ion pairs and larger aggregates is expected. On the other hand, in many systems of interest ions can chemically bond to the solid surface [59]. These phenomena can alter all the properties of the interfacial region. Therefore, improvements and extensions of the theory of solid–electrolyte interfaces can benefit from the associative concept applied either to describe the electrostatic attraction more adequately or to deal with additional forces yielding chemical association.

The fundamentals of the description of chemical association in the theory of uniform uncharged fluids were developed by Wertheim. This method has been applied to inhomogeneous fluids in several works (see Ref. [60] for a more detailed description and references therein). In the framework of the perturbational density functional theory, the effects of association are accounted for by an additional term in the free energy, $F_{as,el}$ [61]. We restrict our attention to the effects of association that result in the formation of dimers of oppositely charged ions. The free energy due to association can be written as [61]

$$F_{as,el} = \int d\mathbf{r}\, \Phi_{as,el}[\{\varrho_{i,ref}(\mathbf{r})\}]$$
$$\beta\Phi_{as,el} = \sum_i \varrho_{i,ref}(\mathbf{r})\left[\ln[\chi(\{\varrho_{i,ref}(\mathbf{r})\})] + \frac{1}{2} - \frac{1}{2}\chi(\{\varrho_{i,ref}(\mathbf{r})\})\right] \quad (3.71)$$

where χ is the fraction of nonbonded ions, that is, the degree of dissociation following from the local mass action law for inhomogeneous system. For the specific case under study, this law reads

$$K(T^*, \{\varrho_{i,ref}(\mathbf{r})\}) = 2[1-\chi(\{\varrho_{i,ref}(\mathbf{r})\})]/\chi(\{\varrho_{i,ref}(\mathbf{r})\}) \sum_i \varrho_{i,ref}(\mathbf{r}), \quad (3.72)$$

where $K(T^*, \{\varrho_{i,ref}(\mathbf{r})\})$ is the association constant, which can be written as a product of temperature-dependent term and the function of temperature and density,

$$K(T^*, \{\varrho_{i,ref}(\mathbf{r})\}) = K^0(T^*)K^\gamma(T^*, \{\varrho_{i,ref}(\mathbf{r})\}). \quad (3.73)$$

The first multiplier, $K^0(T^*)$, is the thermodynamic association constant,

$$K^0(T^*) = 8\pi d^3 \sum_{m\geq 2} \frac{(T^*)^{-2m}}{(2m)!(2m-3)}, \quad (3.74)$$

which follows from Ebeling and Grigo's work [62]. The second multiplier, $K^\gamma(T^*, \{\varrho_{i,ref}(\mathbf{r})\})$, was successfully approximated by Zhou and Stell [63],

$$K^\gamma(T^*, \{\varrho_{i,ref}(\mathbf{r})\}) = \frac{(1-\eta_{ref}(\mathbf{r})/2)}{(1-\eta_{ref}(\mathbf{r}))^3} \exp\left[-\frac{\Gamma d(2+\Gamma d)}{T^*(1+\Gamma d)^2}\right], \quad (3.75)$$

where $\eta_{ref} = \pi \sum_i \varrho_{i,ref}(\mathbf{r})d^3/6$. Due to the additional term in the excess free energy, the density profile equation undergoes modification, namely, the term $\delta F_{as,el}/\delta \varrho_i(\mathbf{r})$ arises on the right-hand side of Equation 3.53.

To be specific, we assume below that the inhomogeneity of the system is along normal to the solid surface, that is, along z-axis. The density profiles determine the value of charge on the solid surface, Q_s, if the electrostatic potential, V_0, is the external variable. Consequently, the differential capacitance of the electric double layer can be obtained.

Let us proceed now to some illustrative examples concerning recent developments. The principal outcome of the DF theory is given in terms of the density profiles. Their accuracy determines quality of the theory. We saw in Figure 3.3 that a version of the density functional theory by Mier-y-Teran et al. [47] has not always been quite successful in describing the double-layer structure. The theory developed by Patra and Ghosh [50–52] is, in general, much more accurate. However, all the DF theories employing the compressibility route have failed to predict a possible depletion of counterions at a charged wall at low temperatures; that is, they do not take into account the interion correlations under these conditions properly. Such depletion has been observed in computer simulations [64], however. The theory that uses the energy route is able to describe such a phenomenon [58].

In Figure 3.4, we show examples of the density profiles resulting from the DF approach proposed in Ref. [58]. Figure 3.4a shows the profiles for $Z_1 = |Z_2| = Z = 1$ electrolyte at a quite low temperature at a charged wall, $Q_s^* = 0.00765$. Despite that the wall is charged and attracts the counterions, the counterion profile is depleted. Figure 3.4b shows an example of the counterion and coion profiles obtained under the same thermodynamic conditions as the profiles in Figure 3.3b. Now, the agreement of the theory [58] with simulations [49] is much better.

One of the important problems in the double-layer theory is to explain the temperature dependence of the double-layer capacitance for the RPM at a charged surface. Computer simulations revealed that the capacitance has a negative slope at high temperatures and a positive slope at low temperatures [65]. The first successful approach explaining such phenomenon was proposed by Holovko et al. [66], who incorporated chemical association into the MSA/MSA integral equation theory. However, all the DF approaches were unable to reproduce the computer simulation

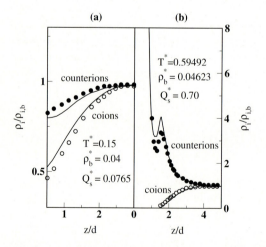

Figure 3.4 Comparison of DF results [58] (lines) with computer simulations (symbols, taken from Ref. [64] in (a) and from Ref. [49] in (b)) for the density profiles of an RPM at a charged hard wall. The bulk reduced densities, $\varrho_b^* = \varrho_b d^3$, are given in the figure. Reprinted with permission from Ref. [58]. Copyright 2004, American Institute of Physics.

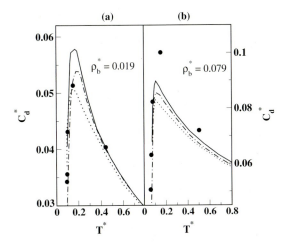

Figure 3.5 Dependence of $C_d^* = C_d d/\varepsilon$ on temperature. Symbols denote the results of simulations [65], dash-dotted lines result from the theory of Holovko et al. [66], and solid and dotted lines were obtained from the DF approach with [67] and without [68] association, respectively. Reprinted with permission from Ref. [67]. Copyright 2005, American Institute of Physics.

results. In contrast, the DF theory that follows from the energy route [58, 61] has successfully reproduced the simulation results.

Figure 3.5 shows a comparison of the temperature dependence of the double-layer capacitance obtained from simulations [65], from the theory of Holovko et al. [66] and from the DFT [58, 61]. The calculations are for 1:1 RPM at a charged wall ($Q_s^* = 0.005$) at two bulk densities, $\varrho_b^* = 0.019$ (Figure 3.5a) and 0.079 (Figure 3.5b). Two versions of the DF approach were employed: one takes into account the association between ions to form ion pairs [61, 67] and the other neglects such association [58, 68]. The DF theory without association reproduces simulation data quite well. Incorporation of association into the free energy does not improve the agreement with simulations for the temperature dependence of capacitance in the present DF approach. In contrast, in the theory of Holovko et al. based on the compressibility route (common MSA/MSA) improvement of $C_d(T^*)$ dependence comes solely from the associative concept. In our approach, the energy route is sufficient to describe $C_d(T^*)$ very well.

The bulk counterpart of the DF following from the MSA energy route predicts the existence of liquid–vapor phase transition. It is of interest to study how confinement of an RPM influences the liquid–vapor phase diagram. This problem was tackled in Refs [58, 61]. In Figure 3.6a, we show examples of the phase diagrams of an RPM in a slit-like pore with uncharged walls in the T^*–$\langle \varrho \rangle$ plane. Here, $\langle \varrho \rangle$ is the average density of the confined electrolyte, $\langle \varrho \rangle = \sum_i \int_0^H \varrho_i(z) dz / H$, where H is the pore width. Inclusion of the chemical association into the theory provides additional mechanism for the formation of ionic pairs besides correlations described by the free energy via the energy route. The fraction of pairs depends both on density and on the

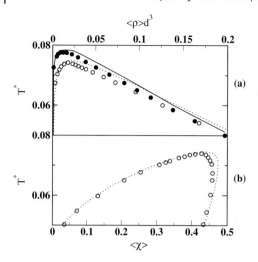

Figure 3.6 (a) Comparison of the phase diagrams for bulk (solid and dotted line) and confined (solid and empty circles) fluid. Solid line and filled symbols correspond to the system without association, whereas dotted line and open circles to the system with association. Part (b) displays the average dissociation degrees along the liquid–vapor coexistence lines. The abbreviations for bulk and confined RPM are the same as in (a). The pore width is $H = 21d$. Reprinted with permission from Ref. [61]. Copyright 2005, American Institute of Physics.

electrolyte reduced temperature; see Figure 3.6b where the average dissociation degree $\langle \chi \rangle = \int_0^H \chi(z) \mathrm{d}z / H$ has been plotted.

The formation of pairs of oppositely charged ions modifies effective interactions between all the particles and, consequently, lowers the critical temperature. Also, the confinement of the fluid results in decrease in the critical temperature. The phenomenon observed here is *capillary evaporation* rather than capillary condensation; that is, the *bulk* fluid in an equilibrium with the confined one has higher density than the confined fluid. Figure 3.6b shows the average dissociation degree along the liquid–vapor coexistence lines shown in Figure 3.6a. The gas-like phase for confined system is characterized by a lower degree of dissociation compared to the bulk RPM; however, the values of $\langle \chi \rangle$ for confined and bulk liquid-like phases are similar. How good are the envelopes for the ionic vapor–ionic liquid-phase transition in confined systems is an open question. At present, there are no computer simulation data validating theoretical predictions at a quantitative level. Nevertheless, the results presented in Figure 3.6 demonstrated for the first time a possibility of description of the phase transitions for confined fluids within the DF approach.

There have been many theoretical developments for complex electric double layers, for example, for mixtures involving molecules of complex architecture and charged groups, polymer/polyelectrolyte brushes on solid surfaces, and so on [69–73]. In general, methodological elements for the description of such systems in the framework of integral equations and density functional approaches are similar to what we have discussed above. Again, due to space constraints we do not discuss them in detail. However, one example seems to be appropriate because it shows the property that is measured experimentally and can be obtained both by using computer simulations and by using theoretical methods. Namely, we would like to refer to adsorption isotherms. The model in question in this example is a mixture of polyions (chains composed of tangentially bonded charged spherical segments) and both the positive and negative spherical monovalent ions ($Z_+ = 1$, $Z_- = -1$) in contact with a charged solid surface described by Q_s^*. The diameter of the segments and of ions is equal ($d_s = d_+ = d_-$), and the segments are monovalent with $Z_s = -1$. The model has been described by the density functional approach [74] and results in the density profiles of ions and in the profiles of individual segments, $\varrho_{i,s}(z)$. The latter can be combined to provide the polyion density profile, $\varrho_p(z)$,

$$\varrho_p(z) = \sum_{i=1}^{M} \varrho_{i,s}(z), \tag{3.76}$$

where M is the number of segments in a chain. Theoretical density profiles of species provide the excess adsorption isotherm straightforwardly,

$$\Gamma_{ex,i} = \int_0^\infty dz [\varrho_i(z) - \varrho_{i,b}]. \tag{3.77}$$

The excess adsorption isotherm shown in Figure 3.7 is for polyions on bulk density, $\varrho_{p,b}$, at different values of the surface charge density, Q_s^*. It is worth mentioning that the theory is able to reproduce computer simulation data. The excess adsorption increases with increasing surface charge density. However, unlike Langmuir adsorption isotherm, the excess adsorption in the present case exhibits nonmonotonous variation with increasing polyion density. The reduction in the surface excess is attributed to the restriction of chain configurations and by already mentioned charge

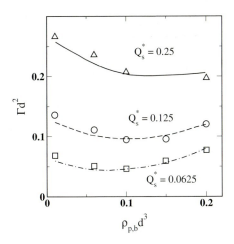

Figure 3.7 Adsorption isotherms for polyions at different surface charge densities, Q_s^*. Theoretical results [74] are given by lines, whereas the simulation data are denoted by the points [72]. The model is at 298 K, the Bjerrum length is 0.714 nm, and $M = 10$. Reprinted with permission from Ref. [74]. Copyright 2006, American Chemical Society.

inversion effects in the region of very low polyion densities. A weak increase in adsorption at high polyion densities was attributed to the excluded volume effects.

To summarize briefly this section, it is worth mentioning, however, that the present stage of development of DF methods for systems of charged particles is still not satisfactory in many aspects [75] and there exists much room for several improvements. In particular, a wide area concerning the role of polar species in the interfacial phenomena is in the initial stage of development. Further theoretical progress for model systems is in many aspects inspired by the experimental data concerning real electrode–electrolyte systems and to much extent by the results of computer simulation methods employing interaction potentials with different degrees of sophistication.

3.3
Computer Simulations

Molecular dynamics (MD) and Monte Carlo (MC) computer simulations are valuable tools that help understand the structure and thermodynamics of electric double layers. They provide data for testing theoretical approaches and yield important insight into molecular origin of several phenomena, particularly in the cases when theoretical description would require introduction of substantial simplifications. Computer simulations are based on statistical mechanics and form a link between the microscopic motion of atoms (in some situations of electrons [76] as well) and macroscopic properties that can be measured experimentally directly or indirectly.

Computer simulations were initially applied to the primitive model of electrolytes and simple models of solid surfaces, in which ions are charged hard spheres, solvent is a dielectric continuum, and the surface is a charged hard wall. Later on, the simulations of the electric double layers were carried out employing more sophisticated molecular models, for example, in which solvent molecules are described explicitly, the interactions between the species take into account dispersion forces, and the microscopic structure of a surface may be considered. More recently, the simulations have been used to study complex systems involving quite big molecules such as surfactants, biomolecules, membranes, and so on [77].

Present computational facilities have advanced to a stage where the so-called *ab initio* simulations, which calculate the electronic structure "on the fly," are able to treat nonuniform systems involving electrolytes. These methods, due to their quantum mechanical nature, are able to describe electronic rearrangement in the interfacial region; thus, they lead to a better understanding of the mechanisms of electrochemical reactions. Undoubtedly, such approaches will continue to develop. Despite a significant progress in computer modeling, to date essential insight into "global" double-layer properties, for example, concentration dependence, differential capacitance, temperature dependence, conductivity, and, in particular, phase behavior and criticality, still comes from the simple models.

Before computer simulations are carried out, one needs to select a statistical ensemble appropriate for a given problem and specify all the details of the microscopic model and interaction potentials, in particular. The simulations of equilibrium systems comprising ionic solution at an electrode are usually carried out in "classical" statistical ensembles, for example, in the microcanonical ensembles, canonical ensembles, grand canonical ensembles, and so on. In all the above cases, a fixed charge on the electrode is assumed. In experiments, when an electrical field is applied to the electrodes, the electrical potential difference or the voltage is the control parameter rather than the charge density on the electrode. The voltage cannot be considered an interaction potential because it is related to the charge density distribution via the Poisson equation. Therefore, in order to mimic such external conditions in simulations, a constant voltage ensemble for the Monte Carlo simulation was developed [15, 16].

The constant voltage ensemble is formulated as an extension of the canonical ensemble by introducing the voltage between the anode and the cathode. In this ensemble, the charge density induced on the electrode fluctuates as the charges are transferred between the anode and the cathode. The charge density on the electrode at equilibrium is calculated as the ensemble average. In performing MC simulations of ions in the constant voltage grand canonical ensemble, there are three kinds of moves: (1) translation of ions, (2) charge transfer between the anode and the cathode, and (3) insertion and deletion of ions. In a trial translational move of ions, the conventional method is used. An ion is randomly chosen from all the anions and cations and the position of the ion is randomly changed within a fixed maximum distance. In a trial charge transfer move for the impermeable electrodes, the charge density on the anode is changed by $+\xi Q_{s,m}$, where $Q_{s,m}$ denotes the maximum charge density to be transferred in a move and ξ is the random number from the range $[-1, 1]$. At the same time, the charge density on the cathode is changed by $-\xi Q_{s,m}$. In a trial insertion, a minimal set of ions is inserted in randomly chosen positions in the simulation cell. In a trial deletion, a minimal set of ions that is randomly chosen from the system is deleted from the simulation cell. After each trial move, the ratio of the Boltzmann factors for the new and old configurations is calculated and the new trial configuration is either accepted or rejected according to the standard Metropolis method [10]. A detailed description of the algorithm and theoretical backgrounds is given by Kiyohara and Asaka [15, 16].

The existing literature on the application of computer simulation methods in the study of ionic solutions at surfaces is too voluminous to be even briefly mentioned in this chapter. Therefore, below we outline only some methodological aspects, that is, concerned with accuracy of calculations of long-range electrostatic interactions. This step is crucial in the application of simulation techniques to electrochemical interfaces. Then, we will present some characteristic and important results.

3.3.1
Summation of Long-Range Forces

A long-range force is usually defined as one in which the spatial interaction falls not faster than r^{-D}, where D is the dimensionality of the system. In this category are the Coulomb interaction between ions ($\sim r^{-1}$) and the dipole–dipole interaction between molecules ($\sim r^{-3}$). The range of these forces is usually much greater than the simulation box length. The brute force solution to the problem would be to increase the size of the simulation box, but even with modern computers the time of simulations would then prohibitively increase. This approach is still used [78].

The problem of evaluating the long-range Coulomb forces has a long history [79]. There exist two main groups of methods that can be used to tackle the problem of long-range forces. The first group comprises lattice summation methods, known as the Ewald summation, including the interaction of an ion or a molecule with all its periodic replicas or copies (usually called images); see, for example, Ref. [10] for the principles of the computer simulation methods. The second group of methods is known as the reaction field methods. Both methods assume that the interaction of molecules beyond a cutoff distance can be handled in an average way, using macroscopic electrostatics [80], and require an a priori estimate of the relative permittivity. In particular, a charge distribution within a spherical cavity polarizes the surrounding medium. This polarization, depending on the relative permittivity of the medium, affects the charge distribution in the cavity. The Ewald method provides an efficient procedure of summation of the interactions between an ion and all its periodic replicas. Formally, the potential energy of a system comprising the basic simulation cell of the linear dimension L with N_L charges and all its periodic replicas is

$$u = \frac{1}{8\pi\varepsilon_1\varepsilon_0} \sideset{}{'}\sum_{\mathbf{n}} \sum_{i,j=1}^{N_L} Z_i Z_j e^2 |\mathbf{r}_i - \mathbf{r}_j + \mathbf{n}|^{-1}, \qquad (3.78)$$

where ε_1 is the dielectric constant of the surrounding medium. The first sum is over all the cells of replicas (images) $\mathbf{n} = (n_x L, n_y L, n_z L)$ with n_i ($i = x, y, z$) being integers. The prime indicates that $i \neq j$ for $\mathbf{n} = \mathbf{0}$. The sum (3.78) is conditionally convergent; that is, the result depends on the order in which one adds up the terms [10]. The usual choice is to take boxes in order of their proximity to the basic simulation box. In the case of nonuniform systems, the order of the summation must be consistent with the imposed periodic boundary conditions. For example, in the slab geometry the periodic boundary conditions are used only in the directions parallel to the wall. Thus, the first term is $\mathbf{n} = (0,0)$, the second term comprises the boxes at $\mathbf{n} = (\pm L, 0), (0, \pm L)$, and so on.

Adopting this approach, we must specify the nature of the medium surrounding the constructed sphere (circle in the case of slab geometry). The results for a sphere surrounded by an ideal conductor ($\varepsilon_1 = \infty$) and for a sphere surrounded by vacuum ($\varepsilon_1 = 1$) are different [81]. The original method was derived for $\varepsilon_1 = \infty$ and then from that result the Ewald sum for a sphere surrounded by a vacuum, $\varepsilon_1 = 1$, is computed [81, 82]. In the Ewald method, each charge i is surrounded by a charge distribution of equal magnitude and opposite sign. This distribution is assumed to be Gaussian:

$$p_i(\mathbf{r}) = Z_i K_w^3 \exp(-K_w^2 r^2)/\pi^{3/2}, \qquad (3.79)$$

where K_w is the width of the distribution. The value of K_w depends on the size of the simulation box, and usually $K_w \approx 5/L$ is adopted. The distribution (3.79) acts like an ionic atmosphere, to screen the interaction between neighboring charges. The screened interactions are short-ranged, and the total energy is calculated by summing over all the molecules in the central box and all their replicas. This summation is carried out by employing the Fourier transform technique. All the details, together with the code of the computer program, can be found in Refs [10, 83].

The original procedure of Ewald can be readily extended to dipolar systems [81]. A simple method for modeling dipoles (and also higher multipole moments) is to represent them as partial charges within a molecule. Then, the Ewald method may be applied directly to each partial charge at a particular site. The only complication in this case is in the self-term, that is, the term that describes the interaction between charges within the same molecule. In a simulation, one should subtract this term. The method for efficient subtraction was given by Heyes [84].

The standard Ewald method is facing two principal difficulties. First, the calculation time essentially increases due to the summation over all the replicas. In order to make these calculations faster, several modifications to the original procedure have been proposed [85–87]. The second difficulty is that the combination of long-range Coulomb forces with periodic boundaries invokes a nonisotropic electric field having the symmetry of the crystalline lattice composed of a main cell as an elementary unit; that is, the Ewald procedure tends to overemphasize the periodic nature of the model fluid. In order to remove this nonphysical effect, Yakub and Ronchi [87] proposed a modification of the Ewald method suitable for simulation of ionic fluids. They introduced an effective electrostatic interaction potential of two ions that is preaveraged over all orientations of the main cell. Explicit expressions for this preaveraged potential were developed.

The second approach to account for the long-range forces is reaction field method or, in the case of nonuniform systems, the charged sheet method. Its application depends on the system geometry. We describe it briefly for the case of a slab geometry [49]. It is assumed that the fluid is confined between two charged walls at a distance H. The basic simulation box has the dimensions $L \times L \times H$. In the central cell, the calculation of energy is carried out explicitly using the intermolecular potentials. The influence of the lateral charges (and also of dipoles if the solvent is present explicitly) surrounding the central cell is taken into account by introducing sheets parallel to the charged walls. Each ion at z_i (and dipole) has its own sheet at z_i. A charged sheet of dimension $L_s \times L_s$ corresponds to each ion in the box and similarly the polarized sheet corresponds to each dipole. A charged sheet carries a uniform charge density $Z_i e / L^2$, where $Z_i e$ is the charge of the ion i in the box, and a polarized sheet has a uniform surface polarization density of μ/L^2, where μ is the dipole moment of the dipole in the box. Each particle interacts with each sheet. Determining the electrostatic potential energy, one must subtract the interactions in the central cell. The potential energy of an ion $Z_i e$ at z_i above the center of a charged sheet (csh) corresponding to an ion $Z_j e$ at z_j is

$$u_{\text{csh}}(|z_i - z_j|; L_s) = (Z_i Z_j e^2 / L_s^2) \int_{-L_s/2}^{L_s/2} \int_{-L_s/2}^{L_s/2} dx\, dy / r. \tag{3.80}$$

Similar equations can be written for ion–dipole and dipole–dipole [88]. Thus, the total two-particle electrostatic ion–ion energy is [88]

$$U_{\text{el,ion-ion}} = \sum_{i<j}^{N} (Z_i Z_j e^2 / r_{ij}) \\ + \sum_{i,j=1}^{N} [u_{\text{csh}}(|z_i - z_j|; L_s = \infty) - u_{\text{csh}}(|z_i - z_j|; L_s = L)], \quad (3.81)$$

where N is the total number of ions confined between two walls in the central box.

This method provides an essential improvement over the procedure proposed by Torrie and Valleau [49]. The charged sheet method of Boda *et al.* [88] was next applied in several works, for example, Refs [89–91]. A modification to the case of an ionic fluid in cylindrical pores was described in Ref. [92]. In comparison to the Ewald procedure, the charged sheet method is computationally very fast.

There are two other important techniques to handle the long-range forces. The first is due to Ladd [93, 94] for studying dipolar systems. The interaction of a dipole with all its minimum image neighbors is considered explicitly. The dipole of interest from the main box is considered as the center of a "new" periodic array of replicated boxes. Ladd expands the electric field due to the particles in each of the neighboring boxes in a multipole expansion with respect to the center of the simulation box. The energy of a molecular dipole at **r** is the sum of the interactions with multipoles from neighboring boxes plus the explicit minimum image neighbor interaction.

The second method is the particle–particle and particle–mesh (abbreviated as PPPM, or as P^{3}M) algorithm [95]. Similar to the Ewald procedure, this method separates the total force acting on an ion i into short-range and long-range parts. The former is handled directly by using the Coulomb potential. The latter is calculated using the following steps: (i) the charge density in the fluid is approximated by assigning charges to a finely spaced mesh in the simulation box; (ii) the Fourier transform is used to solve the Poisson equation for the electrostatic potential due to the charge distribution on the mesh – this gives the potential at each mesh point; and (iii) the field at each mesh point is calculated by numerically differentiating the potential and then the force on a particular particle is calculated from the mesh field by interpolation.

The P^{3}M method was employed in the simulation of melting of ionic crystals [96]. However, recently Crozier [97] found that the original P^{3}M method is inadequate for correctly modeling the strong long-range Coulombic interactions for systems with discrete solvent molecules. There were several modifications of this approach [98–102]. A modified P^{3}M procedure was used to investigate the interfacial electrochemical properties of an aqueous electrolyte solution with explicit water molecules between charged atomistic electrodes [102].

3.3.2
Simulations of Simple Models for Electric Double Layer

One of the first electric double-layer simulations, and one of the most often used in numerous tests of theoretical approaches (cf. Section 3.2.4), was carried out by Torrie and Valleau [49]. What might also be important for the reader is that the above-cited work provides a nice description of the algorithm and the methodology of RPM double-layer computer experiments. In their simulations, Torrie and Valleau assumed that the dielectric constants of the solid and an 1:1 electrolyte solution, modeled as RPM fluid, are equal. The fluid is confined between two charged walls, separated by a rather large distance ($H = 24.5d$ or $H = 36.51d$). Using Grand Canonical MC simulations, they evaluated the ionic profiles, the profile of electrostatic potential, and the potential drop V_0. The results of this work have already been presented in Figures 3.3 and 3.4.

The double-layer primitive model simulations have been reported in numerous works; see, for example, Refs [103–106]. For example, Bhuiyan et al. [103] applied MC simulation and the modified PB equation to investigate the planar electric double layer for the primitive model at low temperatures. Capacitance as a function of temperature at low surface charge was determined for 1:1, 2:2, 2:1, and 3:1 electrolytes. They observed that for 1:1 electrolyte the double-layer potential as a function of surface charge exhibits a maximum at low densities. At high densities, the double-layer potential is negative with a negative slope. Hou et al. [105] studied the formation of the double layer of various electrolytes with ions of the same or different diameters in slit-type pores via MC simulation. In large pores, the distribution of ionic species is similar to that observed in an isolated planar double layer. Simulations demonstrated that double-layer overlap is not only a function of ionic strength and surface charge density but also a function of the diameters and charges of ionic species involved. The overlap results from strong ion–ion correlation effects and the asymmetries in size and charge of ions, and is most pronounced in the case of trivalent counterions with large diameter. Wang et al. [106] applied a density functional approach and canonical MC simulations to describe the ionic structure around the DNA molecule (modeled as infinitely long cylinder) immersed in mixed size counterion solutions.

Computer simulations were helpful in discovering several new phenomena. A nice example obtained almost 30 years ago is the work of Torrie [107]. Introduction of a surface carrying a charge density Q_s into an electrolyte solution results in the potential difference V_0 between the surface and the bulk solution. Suppose the surface charge is increased. Is it possible for V_0 to decrease, that is, could the differential capacitance, $(\partial V_0/\partial Q_s)^{-1}$, be negative? The classical theory of Landau and Lifshitz rules out such a possibility [108]. However, subsequent considerations [109, 110] demonstrated that there exists no rigorous argument that the double-layer capacitance at a single electrode must always be positive. However, the total capacitance of the system composed of a cell between a cathode and an anode must be positive [110].

Torrie [107] performed MC simulations of the double layer for a primitive model of a 0.05 M 1:2 electrolyte and found that for this system V_0 decreases as Q_s is increased. This was the first evidence of the existence of a negative capacitance for a double layer at a single wall. Similar MC simulations, but carried out for 1:1 and 2:2 RPMs in contact with a spherical charged colloidal particle, were carried out by González-Tovar et al. [111]. While for a 1:1 electrolyte an increase in Q_s led to an

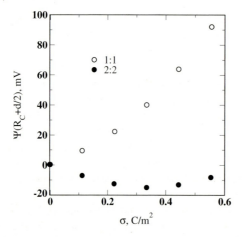

Figure 3.8 The electrostatic potential $\Psi(R_C + d/2)$ versus the charge of a colloidal particle of the radius $R_C = 3d$ at $T = 298.15$ K in the 1 : 1 and 2 : 2 electrolytes at the concentration of 0.5 M ($d = 0.714$ nm). Reprinted with permission from Ref. [111]. Copyright 2004, American Institute of Physics.

increase in V_0, the situation for a 2: 2 electrolyte was different, see Figure 3.8. In addition to computer simulations, González-Tovar et al. [111] solved the singlet HNC/MSA equation for the system in question. The results obtained by them clearly demonstrated that the capacitance could be negative. Moreover, they obtained the next unexpected result from simulation and theory, namely, that the second derivative of the potential drop with respect to the charge can be positive; that is, the curve V_0 versus Q_s can be concave. They explained the observed artifacts on the grounds of ionic adsorption and layering of the ionic species and of the compactness of the diffuse double layer. An interplay between electrostatic and ionic size correlation effects is responsible for the observed behavior.

A nonmonotonic behavior of the double-layer capacitance with temperature, already discussed in Section 3.2.4 (cf. Figure 3.5), was also studied in Refs [65, 103, 112, 113]. Lamperski et al. [113] investigated the dependence of the differential capacitance, C_d, on Q_s in the neighborhood of PZC. They showed that for aqueous electrolytes at room temperature and at zero surface charge, the differential capacitance not only has a minimum but can also have a maximum for molten salts and ionic liquids. Similar results were obtained recently by Fawcett et al. [114] on the basis of MC simulations.

Usually, it is assumed that a wall at which a double layer is formed is uniformly charged, that is, the surface charge is "smeared out." However, in several experiments, the surface charge is created during dissociation of some surface groups [2, 5]. In such cases, the surface charge is not uniformly distributed. Recently, Madurga et al. [115] carried out MC simulations aimed at a comparison of the structure of the double layer for a "smeared out" charge and for models with a discrete surface charge distribution. Different discretization models were considered. The effect of discreteness was analyzed in terms of charge density profiles. For point surface groups, a complete equivalence in the case of uniformly distributed charge was found if the profiles were analyzed as functions of the distance to the charged surface. However, the differences between the two cases are observed in planes parallel to the surface. In contrast, significant discrepancies with approaches that do not account for discreteness were found if charge sites of finite size were placed on the surface.

The ionic density profiles have some characteristic properties. One of them is the contact value of the profiles at a charged hard wall. The expression predicting the local density contact value of symmetric electrolytes was developed a long time ago [19] (cf. Section 3.2.3, Equation 3.31). However, quite recently Bhuiyan et al. [116] proposed a generalization of the contact theorem to include not only symmetric but also asymmetric (e.g., 2:1/1:2) electrolytes. The MC simulations they performed for different concentrations and temperatures present convincing evidence that the generalization is valid at least for low electrode charges.

Torrie et al. [23] considered the effects of surface polarization, that is, of the electrostatic image force due to the dielectric discontinuity, on the structure of the electric double layer in the case of nonuniform 1:1 electrolyte. The dielectric constant of the solid, ε_s, was taken to be different from the dielectric constant of the solution, ε. We already know that the presence of the discontinuity at $z = 0$ yields a set of fictitious electrostatic image charges in the region $z < 0$. With each ion of charge $q_i = Z_i e$ at x_i, y_i, z_i, there is an associated image charge of the strength $q'_i = q_i(\varepsilon - \varepsilon_s)/(\varepsilon + \varepsilon_s) \equiv q_i \alpha_c$ located at $x_i, y_i, -z_i$ [117]. During the MC simulations, the difference between the number of cations and anions, $\Delta N = N_+ - N_-$, is kept constant. The electroneutrality condition for the entire system is assured by imposing the wall charge, ΔNe, located at $z = 0$. The *imposed total mean* surface charge density per unit surface area, Q_s, then consists of the contribution of mean density (due to image charges), $-\alpha_c Q_s$, *superimposed* on the contribution $(1 + \alpha_c)Q_s$. Each ion interacts with all other ions in the central simulation box, with the images of all ions (including its own image) and with the component $(1 + \alpha)Q_s$ of the surface charge. When the charged sheet method is employed to handle the long-range forces, it must also take into account the existence of image charges [23]. Two kinds of walls were considered: (i) the conducting material, $\varepsilon_s = \infty$, then $\alpha = -1$, and (ii) a vacuum, $\varepsilon_s = 1$, then $\alpha = 0.97484$ ($\varepsilon = 78.5$) [23, 118]. They referred to the cases (i) and (ii) as "attractive" and "repulsive" images, respectively.

The profiles for both attractive and repulsive electrostatic images are shown in Figure 3.9. The first series of the grand canonical MC simulation was carried out for a very low surface charge density, $Q_s^* = 0.01$, at a concentration 0.01 M and at $T = 298$ K; see Figure 3.9a. The effect of electrostatic images on the structure of the double layer is rather strong. When $\varepsilon_s = 1$, the repulsion is strong enough to create the maximum in counterion density near $z = 1.5d$. On the other hand, strong attraction when $\varepsilon_s = \infty$ causes a minimum in the coion density at about the same distance from the wall. However, as the surface charge increases, the effect of the forces due to electrostatic images on the ion density profiles becomes less pronounced, as shown for $Q_s^* = 0.08$ at 0.1 M and at the same temperature in Figure 3.9b. Torrie et al. [23] concluded that at low surface charges, the effects of surface polarization on the ionic local densities in the diffuse layer are quite large and

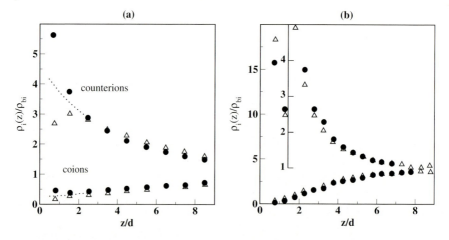

Figure 3.9 Ionic local densities for repulsive (triangles) and attractive (circles) walls from grand canonical MC simulations. Dotted lines show the results for a model system for which $\varepsilon_s = \varepsilon$ (no electrostatic images). Part (a) is for $Q_s^* = 0.01$ and 0.01 M electrolyte, whereas part (b) is for $Q_s^* = 0.098$ and 0.1 M solution. Reprinted with permission from Ref. [23]. Copyright 1982, American Institute of Physics.

lead to extrema in the ionic densities near the wall. This can have important implications for colloidal stability in which a component of the force between colloids is proportional to the ionic densities at the surface. The simulations discovered that the effect of electrostatic images on the total potential drop of the layer was fairly small; however, it increases with the surface charge density. They also successfully tried to describe the simulation results using the singlet HNC equation and showed that this theory is able to accurately estimate the effects of electrostatic images.

The effect of electrostatic images on the double-layer structure was also simulated in Refs [119–123]. In particular, Messina [124] investigated that effect on the adsorption of ions on spherical colloidal particle. The simulations were carried out at a room temperature of $T = 298$ K, assuming $\varepsilon = 80$, $\varepsilon_s = 2$ (or 80), the radius of colloidal particle $7.5d$, and different valencies and concentrations of electrolytes. One example of the obtained profiles is shown in Figure 3.10. The bulk 2:2 electrolyte concentration is rather low, $\varrho_b d^3 = 0.014$. Figure 3.10a shows the results for moderate charge on the colloidal particle, $Q_s = 60e$, whereas Figure 3.10b for a high charge, $Q_s = 180e$. The role of the images is similar to the already discussed case of a planar wall.

The results presented above clearly demonstrate the importance of accounting for electrostatic image charges in simulations of electric double layer formed at the interface between two phases of different dielectric permittivities. The efficient calculation of induced charges in an inhomogeneous dielectric is particularly important in molecular biology, chemical physics, and electrochemistry. Therefore, development of an efficient method for their evaluation is very important. One such method was proposed by Allen et al. [125]. More recently, Boda et al. [126, 127] generalized and extended the work of Allen et al. to models of the double layer that

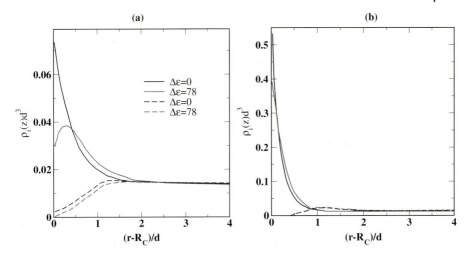

Figure 3.10 Divalent salt–ion distribution with colloidal particle charge of $Q = 60e$ (a) and of $Q = 180e$ (b). The solid and dashed lines correspond to counterion and coion densities, respectively. $\Delta\varepsilon = \varepsilon - \varepsilon_s$. Reprinted with permission from Ref. [124]. Copyright 2002, American Institute of Physics.

include the different dielectric coefficients of the electrode, of the commonly postulated "inner" layer, and of the electrolyte. They reported MC simulation results for the density profiles of 1:1, 2:2, and 2:1 electrolytes that were based on the developed method and compared the results with an extended PB theory.

3.3.3
Solvent Effects on Electric Double Layer: Simple Models

The principal deficiency of the primitive models discussed above is that they ignore the presence of a molecular solvent. To take into consideration the effect of the solvent molecules and to examine the solvent structure at the interface, more realistic models are necessary. For this purpose, the models of ion–dipole mixtures, where the solvent molecules are modeled as hard spheres (or van der Waals molecules) with embedded point dipoles, were proposed [128]. A more sophisticated model, where the solvent particles have a quadrupole tensor describing the water molecule, in addition to the point dipole, was investigated by theory [39] (cf. Section 3.2.3). To our knowledge, for the confined systems such a model was not used in simulations.

Recently, a more realistic double-layer modeling was used by Spohr [129] in MD simulations. The author investigated aqueous NaCl and CsF solutions near a model electrode. The ions were considered as charged Lennard–Jones particles, whereas the water molecules were described by the rigid SPC/E model; see, for example, Ref. [130]. The electrode potential, in addition to the electrostatic and image–charge interactions, contains a nonelectrostatic contribution describing the effect of surface corrugation. This study showed different behavior of various ions at the interface.

Specifically, smaller ions tend to form rigid solvation shells preventing them from contact adsorption. The larger ones exhibit stronger contact adsorption. Contact adsorption did not occur on the uncharged electrodes. Glosli and Philpott [131] studied a similar model and observed the same behavior.

The computer simulation of ion–dipole mixtures is difficult even in the bulk phase, cf. [132]. Such mixtures have many low-energy configurations separated by strong barriers. Examples of such minima are ion pairing, a fully solvated ion, and, in the case of the double-layer systems, layering at the charged surface. The standard Metropolis technique [10] is slow to take the system out from these minima and common sampling becomes very inefficient.

To avoid at least some of these problems, several authors [88, 133] used a simple model, called the solvent primitive model (SPM), in which the solvent molecules are neutral hard spheres and their polar nature is implicitly taken into account by assuming the dielectric constant of the medium equal to ε (specifically $\varepsilon = 78.5$). This investigation showed that the presence of the solvent molecules induces strong structures, as evidenced by the ion and solvent distributions. Several layers of ions and solvent particles were found close to both the charged and uncharged walls. These features are absent in the primitive model simulations. Of course, the SPM allows only a very crude description of the solvent, including only the excluded volume effects.

In Figure 3.11, we show the co- and counterion profiles and the profile of a primitive solvent of hard spheres [88]. The profiles of ionic species shown in Figure 3.11 differ greatly from the profiles presented in Figures 3.9 and 3.10. First of all, we can distinguish well-developed layers that result from the presence of a dense solvent and from entropic packing effects of solvent particles at the walls. Because the

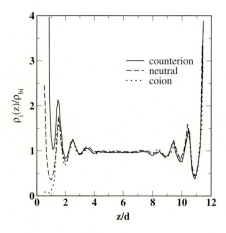

Figure 3.11 Normalized density profiles for the solvent primitive model system in a slit $H = 12d$. The reduced charge on the left wall is $Q_s^* = 0.42$; the right wall is uncharged. The calculations are for 1 M 1:1 electrolyte at $T = 300$ K, $\varepsilon = 78.5$, and $d = 0.425$ nm. The reduced bulk solvent density is 0.627. Reprinted with permission from Ref. [88]. Copyright 1998, American Institute of Physics.

diameters of all the species were taken as equal, the positions of consecutive maxima and minima for all species coincide.

Lamperski and Zydor [134] considered a similar system to those investigated by Boda et al. [88]. They carried out MC simulations of symmetric 1 : 1 and asymmetric 1 : 2 electrolytes using SPM and observed that the presence of the solvent molecules in the system induces layering of ions and solvent molecules in the vicinity of the electrode surface. The presence of the solvent molecules reduces the thickness of the electric double layer, lowers the value of the mean electrostatic potential, and raises the capacitance compared to the relevant primitive model. The SPM was also applied to other systems [54, 135].

We would like to comment on results for a 1 : 1 electrolyte in a dipolar solvent. The solvent molecules are hard spheres with point dipoles (the dipole moment was that of water, $|\mu| = 1.8$ Debye). The system with nominal charges and this dipole moment exhibits "practically nonergodic" behavior [88]. To restore "practical ergodicity," Boda et al. [88] introduced parameters that describe switching on the interactions, λ_i and λ_μ. The "intermediate" models with moderated interaction strengths were studied in detail. In Figure 3.12a and b, we show the density profile of ions and dipolar solvent. The calculations were carried out for the bulk reduced density of each ionic species of 0.056 (Figure 3.12 a) and 0.057 (Figure 3.12b) and for the bulk density of the dipolar solvent of 0.607. The parameter λ_μ was 0.5 and the parameter $\lambda_i = 0.2$. All remaining parameters (pore width, diameters of species, and temperature) are the same as in Figure 3.11. The structure of the solvent and of ions at the surface is similar to that shown in Figure 3.11. However, when the value of λ_i increases drastic changes in the structure occur, including separation of ionic species at high values of λ_i [88]. This study of ion–dipole mixture implies that more sophisticated models, especially for

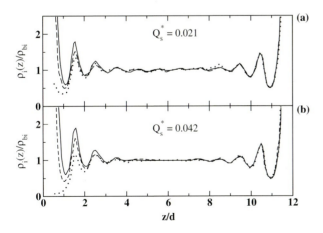

Figure 3.12 Normalized density profiles for ion–dipole mixtures confined between walls. The solid, the dotted, and the dashed lines represent coion, counterion, and averaged over orientation dipole density profiles, respectively. All the parameters are given in the text. Reprinted with permission from Ref. [88]. Copyright 1998, American Institute of Physics.

the solvent (water), are necessary for a more adequate description of electrochemical interfacial problems.

3.3.4
Solvent Effects on Electric Double Layer: Models of Water

Of course, all the above-discussed models are very simple in comparison with experimental systems. Deeper insights can be obtained applying more sophisticated modeling for the solvent. In particular, Yang et al.[89] modeled water molecules by the TIP4P-FQ potential. According to the model, a water molecule has four sites (one describing oxygen, two representing hydrogens, and one representing fictitious charged site) [89]. The nonelectrostatic ion–ion and ion–water interactions are described by the Lennard–Jones potentials parameterized for aqueous sodium chloride solution. The solution is in a slit-like pore with charged hard walls. To evaluate the ion–wall and water–wall energies, ion and water molecules are considered as spheres of radii equal to 0.1 and 0.125 nm, respectively. Because the TIP4P-FQ allows for the fluctuation of the charges within water molecule, this intramolecular energy must also be taken into account. In Figure 3.13, the density

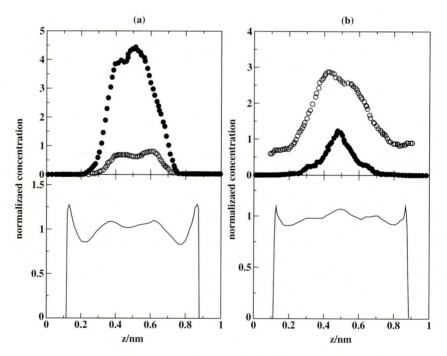

Figure 3.13 Ion and water normalized (i.e., divided by the corresponding values for bulk fluid) concentration profiles in a 1 nm slit-like nanopore with the surface charge density of $-0.05\,C\,m^{-2}$ (a) and of $0.05\,C\,m^{-2}$ (b) at $T=300$ K. The upper panels show the profiles of ions: solid and open circles are the profiles of counterions and coions, respectively. The lower panels show the profiles for water species based on the location of oxygen atoms. Reprinted with permission from Ref. [89]. Copyright 2002, American Institute of Physics.

profiles of species for this solution in a narrow slit-like pore obtained from MC canonical ensemble simulations are shown. The upper panels display the profiles of ionic species, whereas the lower panels show the profiles of the center of water molecules. For a negatively charged surface (Figure 3.13a), the maximum of counterion (Na^+) concentration is in the center of the pore and decays toward the surfaces. This behavior is explained because an adsorbed water layer at the charged surfaces is formed. Water molecules compete with ions at the charged surfaces, and in the considered case the molecules are more favorably adsorbed at the pore walls comparing to the ionic species.

When the pore walls are positively charged, the counterions (Cl^- in this case) are present at the surface (Figure 3.13b). The chloride ion has a larger diameter and has a lower hydration energy compared to the sodium ion. The chloride ions adsorbed at the surfaces yield lower water concentration there. The height of the first water local density peaks is lower in the case of positive charged walls (Figure 3.13b) than for negatively charged walls (Figure 3.13a).

Quite recently, Willard et al. [136] reported results of MD simulations of water in an aqueous ionic solution at an interface with a model crystalline platinum electrode. In the simulations, the electrode was an ideally polarizable hydrophilic metal, supporting image–charge interactions for charged species. The electrode was maintained at a constant electrical potential with respect to the solution. The water–water interactions were modeled using the SPC/E potential; however, to describe interactions of water and ions (Li^+, Ru^{2+}, Ru^{3+}, Cl^-) with the electrode, parameterized potentials derived from *ab initio* calculations are used. Strong attraction and preferential planar orientation of water molecules at the electrode surface were observed. This ordering is different from the structure that might be deduced for continuous models of electrode interfaces, that is, the models where the electrode lacks explicit structure. Furthermore, this ordering significantly affects the probability of ions to reach the surface. Willard et al. [136] described the concomitant motion and configurations of water molecules and ions as functions of the electrode potential. Moreover, the length scales over which ionic atmospheres fluctuate were analyzed. The statistics of these fluctuations depends on the surface structure and ionic strength of the solution. The fluctuations are sufficiently large such that the mean ionic atmosphere appears to be a poor descriptor of the aqueous environment near the metal surface. These findings are important both for the description of electrochemical reactions and for the description of charge transfer between the electrode and the redox species in the solution. A similar study, but for molten salt mixtures confined between model metallic electrodes that were maintained at a constant electrical potential, was carried out by Reed et al. [137].

In the above, we described the application of computer simulations to systems with increasing complexity. One of the most complicated systems is the one that involves water–ionic solutions in contact with surfaces of minerals. Although several works dedicated to the problem have been published, we discuss just one example [138]. The simulations are usually carried out using MD methods. In order to give insight into the problems and demonstrate applicability of the simulations, we briefly outline the results of Kerisit et al. [138] who applied MD to study the structure of NaCl aqueous solution at the (100) surface of a goethite.

Before carrying out MD simulations, the (ideal) structure of the surface plane must be determined. This was done using the Born model of solids [139] with parameterized quantum mechanical expressions describing nonelectrostatic potentials between the species (iron, oxygen, hydroxyl oxygen, and hydrogen).

A slab of simulated goethite crystal was cut exposing the (100) surface, which is comprised of alternating layers that contain either hydroxide groups or iron and oxygen atoms. There are, therefore, two possible terminations and according to the calculations, the most stable surface is the one terminated by a hydroxide layer. It is shown in the left panel of Figure 3.14. The surface hydroxyl ions are bonded to two cations from the Fe−O layer below. In this layer, the iron atoms are in a fivefold coordination, whereas the oxygen ions remain in their bulk coordination and are bonded to three cations. If the surface is in contact with vacuum, the hydroxyl ions rotate to point away from the surface and thus the surface becomes corrugated, with grooves in between hydroxyl rows.

After preparing the surface of goethite, canonical ensemble MD simulations were carried out. The simulational cell comprised 1.6 nm thick two identical goethite slabs that were set at the bottom and top of the simulation cell. At each goethite surface, a thick (up to 18 nm in some runs) slab of water containing dissolved NaCl was deposited. These slabs were separated by 10 nm wide vacuum (during simulations, some water molecules entered this space, so, in fact, two interfaces were simulated). The first is the goethite–solution interface; the second one is the solution–vapor interface. The runs were carried out at 300 K; flexible shell model potential for water–water interaction [140] and sophisticated potentials for the nonelectrostatic interactions between all remaining species and solute species with the atoms of the solid were applied; for details see Ref. [138].

Figure 3.15a shows the profiles of the salt concentration (in mol dm^{-3}) and the density profile of water species (the bulk water density, that is, the density on the plateau of the profile is equal to 55 mol dm^{-3}). The local density of water exhibits oscillations up to 1.5 nm from the surface (the origin of the coordination system is

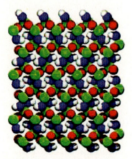 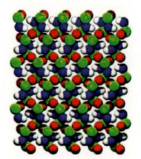

Figure 3.14 Snapshots of two possible terminating planes of the (100) surface of goethite. Iron, oxygen, hydroxyl oxygen, and hydrogen are colored green, red, blue and white, respectively. Left panel gives the most stable surface. Reproduced from Ref. [138] with permission of The Royal Society of Chemistry.

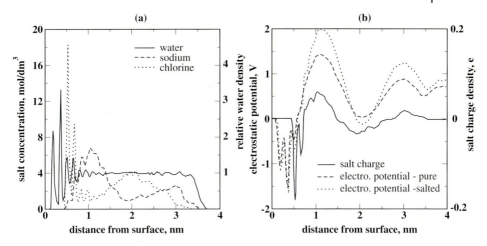

Figure 3.15 (a) Variation in salt concentration and water density as a function of distance from the goethite surface. (b) Variation in electrostatic potential as a function of distance from the goethite surface in comparison to the salt charge density. The results of the electrostatic potential for salt-free water are also included. Reproduced from Ref. [138] with permission of The Royal Society of Chemistry.

defined as the plane passing through the center of the uppermost layer of iron atoms). The first peak corresponds to H_2O adsorbed to the surface's grooves at a height of 0.18 nm above the surface. The second peak corresponds to those water molecules that are bonded to the surface hydroxyls. Each layer contains on average one water molecule per surface cation and thus the period of oscillation is similar to the diameter of H_2O molecule. The calculation of water residence times shows that molecules in the first layer remain adsorbed to the surface's groove for a rather long time (half a nanosecond). Also, one can see small oscillations of the water density profiles at the liquid–vapor interface. What is important to note is that although the presence of ions causes local disruptions of the water hydrogen bond structure, the shape of the water density profile is not affected by the presence of electrolyte ions.

Figure 3.15a also shows the sodium and chloride distributions. Both the Na^+ and the Cl^- profiles exhibit layering close to the surface. The oscillations persist to about 1.2 nm and the period for Cl^- oscillations coincides with that for H_2O (in the case of Na^+, a shift in the profiles is observed). More important is that there is a clear buildup of negative charge near the surface due to the adsorption of Cl^- in the third and fourth layers and depletion of Na^+ profiles in this region. Beyond about 0.8 nm, the Cl^- concentration decreases sharply and there is a large excess of Na^+ in the next layer of 1 nm width. Due to oscillations of ionic density, the overall salt charge distribution exhibits oscillations, as shown in Figure 3.15b. The calculated distributions differ significantly in shape from the classical view of the electrical double layer in the vicinity of a mineral surface, whereby, beyond the Stern layer, the net electrical charge is thought to decay exponentially. Recent molecular dynamics simulations of molten and gaseous NaCl by Keblinski et al. [141] and by Spohr [129] confirm the microscopic structure of the interfacial region discussed above.

Of course, the electrostatic potential (in the direction normal to the surface) was calculated from the Poisson equation. The calculations were carried out for the simulations concerning pure water and salt solution. The electrostatic potential was found to be very similar in both cases, cf. Figure 3.15b. The narrow peaks close to the surface (up to 0.6 nm) are due to the specific orientation of water molecules near the surface. Beyond this region, the peaks become very broad and their amplitude is slightly affected by the presence of ions. In this region, there is a striking correlation with the salt charge distribution, which would imply that the modification of the water structure caused by the mineral surface controls the distribution of electrolyte ions in the vicinity of this charge-neutral surface.

Kerisit et al. [138] also studied the effect of the surface charge on the structure of the solution at the goethite surface. In order to model a positive surface charge, a hydroxyl group was removed from each surface of the slab and the system was neutralized by adding chloride ions to each water slab. However, to obtain negatively charged surface a hydroxyl group from each surface was replaced by an oxygen atom and the system was charge compensated by adding a sodium ion. The charges at the positive and negative surfaces were 4.5 and 24.5 $\mu C\ cm^{-2}$, respectively. The negatively charged surface still showed not only the same chloride peak near the surface but also a slight increase in sodium density near the interface. On the positively charged surface, the distribution is identical to that of the neutral surface except for a higher first chloride peak. However, in both cases these changes are small and the overall shape of the salt distribution is not affected by the charge on the surfaces. Obviously, an increase in the surface charges would magnify these effects. The last results provide a nice example of possibilities of computer simulation methods. In fact, simulations are very helpful in understanding the molecular mechanism of the formation of double layers at solid surfaces.

A more accurate description of the forces in molecular dynamics simulations is still possible. Indeed, to describe ionic solutions in contact with solid surfaces first-principles MD simulation can be employed. These methods are usually based on Car–Parrinello method, in which the forces are "on-fly" calculated from quantum mechanical density functional theory [142]. Examples of the application of such approaches in simulations of ionic solutions near solid surface have been presented in Refs [143–147]. However, these methods in spite of being promising for interface studies at present are at the initial stage of development.

3.3.5
Summary

This chapter reviewed important methodological elements for the description of phenomena at the interfaces between solid electrodes and liquid electrolyte solutions. We discussed entirely theoretical aspects and computer simulation methods. The principal focus was on the microstructural characteristics of interfaces and thermodynamic and electrical properties resulting from the microscopic structure. The theoretical part begins with some generalities concerning simple models of solid and solution and the Poisson–Boltzmann approach. Next, more sophisticated integral equation approaches and density functional methods are discussed. These methods are discussed quite comprehensively. The next part of the chapter deals with the application of the Monte Carlo and molecular dynamics methodology to interfacial problems in electrochemistry. There we discussed a crucial issue, namely, the account of long-range electrostatic interactions in the interfacial systems. The applications of methods to different models of augmenting complexity then followed. Some illustrative examples involving theoretical and computer simulation results were given throughout the manuscript. General approaches and models considered throughout the chapter are useful for the description of a wide class of electrochemical systems. Basic methodology is essentially the same but requires certain extensions depending on the system and properties of interest.

Due to space limitations, we were unable to cover several specific topics, phenomena, and classes of systems of different level of complexity. In particular, the solutions of complex molecules (polyelectrolytes and other biomolecules) were not discussed either in theory or in simulations. This type of molecules has been used for surface modification of the solid surfaces in chemistry and electrochemistry. Consequently, the electric double layers with desired properties can be designed. Our principal focus was on planar interfaces. The problems related to double layers of other shape, as well as the electric interfaces formed on the surface of mesoscopic and nanoscopic objects, remained out of our attention. Also, the progress achieved in understanding the electric double layers in microporous materials and complex ionic channels by means of theoretical and simulation approaches has not been discussed. Our focus mostly was on simple equilibria. Complex problems of solubility of solids put in contact with electrolyte solutions combined with other types of equilibria still lack well-established theoretical and simulation methods. The same concerns a variety of chemical reactions. Transport and electrokinetic phenomena that accompany adsorption and thermodynamic aspects of the behavior of electrolyte solutions at electrodes were not included. We were restricted to the problems where classical statistical mechanics still reigns. Quantum mechanical approaches mentioned in Section 3.3.4 are of much interest, but would require a separate chapter to provide a reasonable understanding of the state of art for the interested reader. Still we think that the chapter may be useful both for the experts in this and related areas and for the less experienced readers.

Acknowledgment

We are grateful to Professor W.R. Fawcett for carefully reading the manuscript and providing valuable suggestions.

References

1. Schmickler, W. (1996) *Interfacial Electrochemistry*, Oxford University Press, New York.
2. Fawcett, W.R. (2004) *Liquids, Solutions, and Interfaces*, Oxford University Press, New York.
3. Hamann, C.H., Hamnett, A., and Vielstich, W. (2007) *Electrochemistry*, Wiley-VCH Verlag GmbH, Weinheim.
4. Wieczkowski, A. (ed.) (1999) *Interfacial Electrochemistry: Theory, Experiment and Application*, Marcel Dekker, New York.
5. Tombacz., E. (2002) Adsorption from electrolyte solutions, in *Adsorption: Theory, Modeling, and Analysis Surfactant Science Series*, vol. 107 (ed. J. Toth), Marcel Dekker, New York, pp. 711–742.
6. Attard, P. (1996) Electrolytes and the electric double layer. *Adv. Chem. Phys.*, **92**, 1–159.
7. Pandit, R., Schick, M., and Wortis, M. (1982) Systematics of multilayer adsorption phenomena on attractive substrates. *Phys. Rev. B*, **26**, 5112–5140.
8. Gregg, S.J. and Sing, K.S.W. (1982) *Adsorption, Surface Area, and Porosity*, Academic Press, New York.
9. Steele, W.A. (1973) *The Interaction of Gases with Solid Surfaces*, Pergamon Press, Oxford.
10. Allen, M.P. and Tildesley, D. (1991) *Computer Simulation of Liquids*, Clarendon Press, Oxford.
11. Falkenhagen, H., Ebeling, W., and Hertz, H.G. (1971) *Theorie der Electrolytes*, Hizzel Verlag, Leipzig.
12. Stillinger, F.H. and Lovett, R. (1968) Ion-pair theory of concentrated electrolytes. I. Basic concepts. *J. Chem. Phys.*, **48**, 3858–3868.
13. Hribar, B., Vlachy, V., Bhuiyan, L.B., and Outhwaite, C.W. (2000) Ion distributions in a cylindrical capillary as seen by the modified Poisson–Boltzmann theory and Monte Carlo simulations. *J. Phys. Chem. B*, **104**, 11522–11527.
14. Duval, J.F.L., Leermakers, F.A.M., and van Leeuwen, H.P. (2004) Electrostatic interactions between double layers: influence of surface roughness, regulation, and chemical heterogeneities. *Langmuir*, **20**, 5052–5065.
15. Kiyohara, K. and Asaka, K. (2007) Monte Carlo simulation of electrolytes in the constant voltage ensemble. *J. Chem. Phys.*, **126**, 214704-1–214704-9.
16. Kiyohara, K. and Asaka, K. (2007) Monte Carlo simulation of porous electrodes in the constant voltage ensemble. *J. Phys. Chem. C*, **111**, 15903–15909.
17. Frumkin, A.N. (1979) *Potentials of Zero Charge*, Nauka, Moscow.
18. Goel, T. and Patra, C.N. (2007) Structure of spherical electric double layer: a density functional approach. *J. Chem. Phys.*, **127**, 034501-1–034501-7.
19. Blum, L. and Henderson, D. (1992) Statistical mechanics of electrolytes at interfaces, in *Fundamentals of Inhomogeneous Fluids* (ed. D. Henderson), Marcel Dekker, New York, Chapter 6.
20. Levine, S. (1951) The free energy of the double layer of a colloidal particle. *Proc. Phys. Soc. A*, **64**, 781–790.
21. Yukhnovsky, I.R. and Kuryliak, I.I. (1976) Generalization of the collective variables method for confined system of charged particles. *Ukr. Phys. J.*, **21**, 1772–1781.
22. Lozada-Cassou, M. (1992) Fluids between walls and in pores, in *Fundamentals of Inhomogeneous Fluids* (ed. D. Henderson), Marcel Dekker, New York, Chapter 8.
23. Torrie, G.M., Valleau, J.P., and Patey, G.N. (1982) Electrical double layer. II. Monte Carlo and HNC studies of image effects. *J. Chem. Phys.*, **76**, 4615–4622.
24. Henderson, D. and Plischke, M. (1988) Pair and singlet correlation functions of inhomogeneous fluids calculated using the Ornstein–Zernike equation. *J. Phys. Chem.*, **92**, 7177–7185.
25. Ángeles, F.J. and Lozada-Cassou, M. (2008) On the regimes of charge reversal. *J. Chem. Phys.*, **128**, 174701-1–174701-7.
26. Waisman, E. and Lebowitz, J.L. (1972) Mean spherical model integral equation for charged hard spheres. I. Method of solution. *J. Chem. Phys.*, **56**, 3086–3093.

27 Hansen, J.-P. and McDonald, I.R. (2006) *Theory of Simple Liquids*, 3rd edn, Academic Press, New York.

28 Lozada-Cassou, M. and Henderson, D. (1983) Application of the hypernetted chain approximation to the electrical double layer: comparison with Monte Carlo results for 2:1 and 1:2 salts. *J. Phys. Chem.*, **87**, 2821–2828.

29 D'Aguanno, B., Nielaba, P., Alts, T., and Forstmann, F. (1986) A local HNC/HNC approximation for the 2:2 restricted primitive model electrolyte near a charged wall. *J. Chem. Phys.*, **85**, 3476–3481.

30 Carnie, S.L. and Chan, D.Y.C. (1980) The structure of electrolytes at charged surfaces: ion–dipole mixtures. *J. Chem. Phys.*, **73**, 2949–2957.

31 Blum, L. and Henderson, D. (1981) Mixtures of hard ions and dipoles against a charged hard wall: Ornstein–Zernike equation, some exact results and the mean spherical approximation. *J. Chem. Phys.*, **74**, 1902–1910.

32 Wertheim, M.S. (1971) Exact solutions of the mean spherical model for fluids of hard spheres with permanent electric dipole moments. *J. Chem. Phys.*, **54**, 4291–4298.

33 Blum, L. (1974) Solution of a model for the solvent–electrolyte interactions in the mean spherical approximation. *J. Chem. Phys.*, **61**, 2129–2133.

34 Blum, L. (1977) Theory of electrified interfaces. *J. Phys. Chem.*, **81**, 136–147.

35 Holovko, M.F. and Kapko, V.I. (2007) Ion–dipole model for electrolyte solutions: application of the associative mean spherical approximation. *Condens. Matter Phys.*, **10**, 397–406.

36 Schmickler, W. and Henderson, D. (1984) The interphase between jellium and a hard sphere electrolyte: a model for the electric double layer. *J. Chem. Phys.*, **80**, 3381–3386.

37 Schmickler, W. and Henderson, D. (1986) The interphase between jellium and a hard sphere electrolyte: capacity–charge characteristics and dipole potentials. *J. Chem. Phys.*, **85**, 1650–1657.

38 Badiali, J.P., Rosinberg, M.L., Vericat, F., and Blum, L. (1983) A microscopic model for the liquid metal–ionic solution interface. *J. Electroanal. Chem.*, **158**, 253–267.

39 Torrie, G.M. and Patey, G.N. (1991) Molecular solvent models of electrical double layers. *Electrochim. Acta*, **36**, 1677–1684.

40 Carnie, S.L. and Torrie, G.M. (1984) The statistical mechanics of the electrical double layer. *Adv. Chem. Phys.*, **546**, 141–253.

41 Caccamo, C., Pizzimenti, G., and Blum, L. (1986) An improved closure for the Born–Green–Yvon equation for the electrical double layer. *J. Chem. Phys.*, **84**, 3327–3335.

42 Denesyuk, N.A. and Hansen, J.-P. (2004) Wetting transitions of ionic solutions. *J. Chem. Phys.*, **121**, 3613–3624.

43 Henderson, D. and Plischke, M. (1988) Pair and singlet correlation functions of inhomogeneous fluids calculated using the Ornstein–Zernike equation. *J. Phys. Chem.*, **92**, 7177–7185.

44 Eaton, A.C. and Haymet, A.D.J. (2001) Electrolytes at charged interfaces: pair integral approximations for model 2-2 electrolytes. *J. Chem. Phys.*, **114**, 10938–10947.

45 Kjellander, R. (1988) Inhomogeneous Coulomb fluids with image interactions between planar surface. II. On the anisotropic hypernetted chain approximation. *J. Chem. Phys.*, **88**, 7129–7137.

46 Evans, R. (1992) Density functionals in the theory of nonuniform fluids, in *Fundamentals of Inhomogeneous Fluids* (ed. D. Henderson), Marcel Dekker, New York, Chapter 3.

47 Mier-y-Teran, L., Suh, S.H., White, H.S., and Davis, H.T. (1990) A nonlocal free-energy density-functional approximation for the electrical double layer. *J. Chem. Phys.*, **92**, 5087–5098.

48 Waisman, E. and Lebowitz, J.L. (1972) Mean spherical model integral equation for charged hard spheres. II. Results. *J. Chem. Phys.*, **56**, 3093–3099.

49 Torrie, G.M. and Valleau, J.P. (1980) Electrical double layers. I. Monte Carlo study of a uniformly charged surface. *J. Chem. Phys.*, **73**, 5807–5816.

50 Patra, C.N. and Ghosh, S.K. (1993) Weighted-density-functional theory of nonuniform ionic fluids: application to electric double layers. *Phys. Rev. E*, **47**, 4088–4097.

51 Patra, C.N. (1999) Structure of electric double layers: a simple weighted density functional approach. *J. Chem. Phys.*, **111**, 9832–9838.

52 Patra, C.N. and Ghosh, S.K. (2002) Structure of electric double layers: a self-consistent weighted-density-functional approach. *J. Chem. Phys.*, **117**, 8938–8943.

53 Patra, C.N. and Ghosh, S.K. (2003) Weighted-density-functional approach to the structure of nonuniform fluids. *J. Chem. Phys.*, **118**, 8326–8330.

54 Goel, T., Patra, C.N., Ghosh, S.K., and Mukherjee, T. (2008) Molecular solvent model of cylindrical electric double layers: a systematic study by Monte Carlo simulations and density functional theory. *J. Chem. Phys.*, **129**, 154707-1–154707-10.

55 Rosenfeld, Y. (1993) Free-energy model for inhomogeneous fluid mixtures: Yukawa-charged hard spheres, general interactions, and plasmas. *J. Chem. Phys.*, **98**, 8126–8148.

56 Gillespie, D., Nonner, W., and Eisenberg, R.S. (2003) Density functional theory of charged, hard-sphere fluids. *Phys. Rev. E*, **68**, 031503–031513.

57 Gillespie, D., Valisko, N., and Boda, D. (2005) Density functional theory of the electrical double layer: the RFD functional. *J. Phys. Condens. Matter*, **17**, 6609–6626.

58 Pizio, O., Patrykiejew, A., and Sokołowski, S. (2004) Phase behavior of ionic fluids in slitlike pores: a density functional approach for the restricted primitive model. *J. Chem. Phys.*, **121**, 11957–11964.

59 Borówko, M., Pizio, O., and Sokołowski, S. (2000) Nonuniform associating fluids, in *Computational Methods in Surface and Colloid Science* (ed. M. Borówko), Marcel Dekker, New York, Chapter 4.

60 Pizio, O. and Sokołowski, S. (2006) On the effects of ion–wall chemical association on the electric double layer: a density functional approach for the restricted primitive model at a charged wall. *J. Chem. Phys.*, **125**, 024512-1–024512–9.

61 Pizio, O. and Sokołowski, S. (2005) Phase behavior of the restricted primitive model of ionic fluids with association in slitlike pores. Density-functional approach. *J. Chem. Phys.*, **122**, 144707-1–144707-7.

62 Ebeling, W. and Grigo, M. (1980) An analytical calculation of the equation of state and the critical point in a dense classical fluid of charged hard spheres. *Ann. Phys.*, **37**, 21–30.

63 Stell, G. and Zhou, Y. (1989) Chemical association in simple models of molecular and ionic fluids. *J. Chem. Phys.*, **91**, 3618–3623.

64 Boda, D., Henderson, D., Mier-y-Teran, L., and Sokołowski, S. (2002) The application of density functional theory and the generalized mean spherical approximation to double layers containing strongly coupled ions. *J. Phys. Condens. Matter*, **14**, 11945–11954.

65 Boda, D., Henderson, D., and Chan, K.Y. (1999) Monte Carlo study of the capacitance of the double layer in a model molten salt. *J. Chem. Phys.*, **110**, 5346–5350.

66 Holovko, M., Kapko, V., Henderson, D., and Boda, D. (2001) On the influence of ionic association on the capacitance of an electrical double layer. *Chem. Phys. Lett.*, **341**, 363–368.

67 Reszko-Zygmunt, J., Sokołowski, S., and Pizio, O. (2005) Temperature dependence of the double layer capacitance for the restricted primitive model: the effect of chemical association between ions. *J. Chem. Phys.*, **123**, 016101–1–016101–2.

68 Reszko-Zygmunt, J., Sokołowski, S., Henderson, D., and Boda, D. (2005) Temperature dependence of the double layer capacitance for the restricted primitive model of an electrolyte solution from a density functional approach. *J. Chem. Phys.*, **122**, 084504-1–084504–6.

69 Smagala, T.G., Patrykiejew, A., Sokołowski, S., Pizio, O., and Fawcett, W.R. (2008) Restricted primitive model for electrolyte solutions in contact with

solid surface modified by grafted chains: a density functional approach. *J. Chem. Phys.*, **128**, 024907-1–024907-10.

70 Pizio, O., Bucior, K., Patrykiejew, A., and Sokołowski, S. (2005) Density-functional theory for fluid mixtures of charged chain particles and spherical counterions in contact with charged hard wall: adsorption, double layer capacitance, and the point of zero charge. *J. Chem. Phys.*, **123**, 214902-1–214902-9.

71 Patra, C.N. and Yethiraj, A. (2000) Density functional theory for the nonspecific binding of salt to polyelectrolytes: thermodynamic. *Biophys. J.*, **78**, 699–706.

72 Patra, C.N., Chang, R., and Yethiraj, A. (2004) Structure of polyelectrolyte solutions at a charged surface. *J. Phys. Chem. B*, **108**, 9126–9132.

73 Reddy, G., Chang, R., and Yethiraj, A. (2006) Adsorption and dynamics of a single polyelectrolyte chain near a planar charged surface: molecular dynamics simulations with explicit solvent. *J. Chem. Theory Comput.*, **2**, 630–636.

74 Li, Z. and Wu, J. (2006) Density functional theory for planar electric double layers: closing the gap between simple and polyelectrolytes. *J. Phys. Chem. B*, **110**, 7473–7484.

75 Durand-Vidal, S., Simonin, J.-P., and Turq, P. (2000) *Electrolytes at Interfaces*, Kluwer, Dordrecht.

76 Spohr, E. (2003) Some recent trends in computer simulations of aqueous double layers. *Electrochim. Acta*, **49**, 23–27.

77 Messina, R. (2009) Electrostatics in soft matter. *J. Phys. Condens. Matter*, **21**, 113102-1–113102-18.

78 Rose, D.A. and Benjamin, I. (1993) Adsorption of Na^+ and Cl^- at the charged water–platinum interface. *J. Chem. Phys.*, **99**, 2283–2290.

79 Ewald, P.P. (1921) Die Berechnung optischer und elektrostatischer Gitterpotentiale. *Ann. Phys.*, **369**, 253–287.

80 Fröhlich, H. (1949) *Theory of Dielectrics: Dielectric Constant and Dielectric Loss*, Oxford University Press, Oxford.

81 De Leeuw, S.W., Perram, J.W., and Smith, E.R. (1980) Simulation of electrostatic systems in periodic boundary conditions. I. Lattice sums and dielectric constant. *Proc. R. Soc. Lond.*, **A373**, 27–56.

82 Heyes, D.M. (1981) Electrostatic potentials and fields in infinite point charge lattices. *J. Chem. Phys.*, **74**, 1924–1929.

83 Rappaport, D.C. (1995) *The Art of Molecular Dynamics Simulation*, Cambridge University Press, Cambridge, UK.

84 Heyes, D.M. (1983) MD incorporating Ewald summations on partial charge polyatomic systems. *CCP5 Quarterly*, **8**, 29–36.

85 Martyna, G.J. and Tuckerman, M.E. (1999) A reciprocal space based method for treating long range interactions in *ab initio* and force-field-based calculations in clusters. *J. Chem. Phys.*, **110**, 2810–2821.

86 Duan, Z.-H. and Krasny, R. (2000) An Ewald summation based multipole method. *J. Chem. Phys.*, **113**, 3492–3495.

87 Yakub, E. and Ronchi, C. (2003) An efficient method for computation of long-ranged Coulomb forces in computer simulation of ionic fluids. *J. Chem. Phys.*, **113**, 11556–11560.

88 Boda, D., Chan, K.-Y., and Henderson, D. (1998) Monte Carlo simulation of an ion–dipole mixture as a model of an electrical double layer. *J. Chem. Phys.*, **109**, 7362–7371.

89 Yang, K.-L., Yiacoumi, S., and Tsouris, C. (2002) Monte Carlo simulations of electrical double-layer formation in nanopores. *J. Chem. Phys.*, **117**, 8499–8507.

90 Lee, M., Chan, K.-Y., and Tang, Y.W. (2002) Forces between charged surfaces in a solvent primitive model electrolyte. *Mol. Phys.*, **100**, 2201–2211.

91 Quesada-Pérez, M., Martín-Molina, A., and Hidalgo-Álvarez, R. (2004) Simulation of electric double layers with multivalent counterions: ion size effect. *J. Chem. Phys.*, **121**, 8618–8626.

92 Boda, D., Busath, D.D., Henderson, D., and Sokołowski, S. (2000) Monte Carlo simulations of the mechanism for channel selectivity: the competition

between volume exclusion and charge neutrality. *J. Phys. Chem. B*, **104**, 8903–8910.
93 Ladd, J.A.C. (1977) Monte Carlo simulation of water. *Mol. Phys.*, **33**, 1039–1050.
94 Ladd, A.J.C. (1978) Long-range dipolar interactions in computer simulations of polar liquids. *Mol. Phys.*, **36**, 463–474.
95 Eastwood, J.W., Hockney, R.W., and Lawrence, D. (1980) P3M3DP: the three dimensional periodic particle–particle/particle–mesh program. *Comput. Phys. Commun.*, **19**, 215–261.
96 Hockney, R.W. and Eastwood, J.W. (1981) *Computer Simulation Using Particles*, McGraw-Hill, New York.
97 Crozier, P.S., Rowley, R.L., Spohr, E., and Henderson, D. (2000) Comparison of charged sheets and corrected 3D Ewald calculations of long-range forces in slab geometry electrolyte systems with solvent molecules. *J. Chem. Phys.*, **112**, 9253–9257.
98 Deserno, M. and Holm, C. (1998) How to mesh up Ewald sums. I. A theoretical and numerical comparison of various particle mesh routines. *J. Chem. Phys.*, **109**, 7678–7693.
99 Deserno, M. and Holm, C. (1998) How to mesh up Ewald sums. II. An accurate error estimate for the particle–particle-particle–mesh algorithm. *J. Chem. Phys.*, **109**, 7694–7701.
100 Pollock, E.L. and Glosli, J. (1996) Comments on P^3M, FMM, and the Ewald method for large periodic Coulombic systems. *Comput. Phys. Commun.*, **95**, 93–99.
101 Crowley, M., Darden, T., Cheatham, T., and Deerfield, D. (1997) Adventures in improving the scaling and accuracy of a parallel molecular dynamics program. *J. Supercomput.*, **11**, 255–278.
102 Crozier, P.S., Rowley, R.L., and Henderson, D. (2000) Molecular dynamics calculations of the electrochemical properties of electrolyte systems between charged electrodes. *J. Chem. Phys.*, **113**, 9202–9207.
103 Bhuiyan, L.B., Outhwaite, C.W., and Henderson, D. (2006) Planar electric double layer for a restricted primitive model electrolyte at low temperatures. *Langmuir*, **22**, 10630–10634.
104 Hou, C.H. (2008) Electrical double layer formation in nanoporous carbon materials. PhD dissertation, Georgia Institute of Technology (http://hdl.handle.net/1853/22698)
105 Hou, C.-H., Taboada-Serrano, P., Yiacoumi, S., and Tsouris, C. (2008) Monte Carlo simulation of electrical double-layer formation from mixtures of electrolytes inside nanopores. *J. Chem. Phys.*, **128**, 044705-1–044705-8.
106 Wang, K., Yu, Y.X., Gao, G.H., and Luo, G.S. (2005) Density-functional theory and Monte Carlo simulation study on the electric double layer around DNA in mixed-size counterion systems. *J. Chem. Phys.*, **123**, 234904-1–234904-8.
107 Torrie, G.M. (1992) Negative differential capacities in electrical double layers. *J. Chem. Phys.*, **96**, 3772–3774.
108 Landau, L.D. and Lifshitz, E.M. (1960) Chapter III, in *Electrodynamics of Continuous Media*, Pergamon Press, Oxford.
109 Attard, P., Wei, D., and Patey, N.G. (1992) On the existence of exact conditions in the theory of electrical double layers. *J. Chem. Phys.*, **96**, 3767–3771.
110 Partenskii, M.B. and Jordan, P.C. (1993) The admissible sign of the differential capacity, instabilities, and phase transitions at electrified interfaces. *J. Chem. Phys.*, **99**, 2992–3002.
111 González-Tovar, E., Jiménez-Angeles, F., Messina, R., and Lozada-Cassou, M. (2004) A new correlation effect in the Helmholtz and surface potentials of the electrical double layer. *J. Chem. Phys.*, **120**, 9782–9792.
112 Ballone, P., Pastore, G., Tosi, M.P., Painter, K.R., Grout, P.J., and March, N.H. (1984) Dependence of capacitance of metal–molten salt interface on local density profiles near electrode. *Phys. Chem. Liq.*, **13**, 269–277.
113 Lamperski, S., Outhwaite, C.W., and Bhuiyan, L.B. (2009) The electric double-layer differential capacitance at and near zero surface charge for a restricted primitive model electrolyte. *J. Phys. Chem. B*, **113**, 8925–8929.

114 Fawcett, W.R., Ryan, P.J., and Smagala, T.G. (2009) The properties of the diffuse double layer at high electrolyte concentrations. *J. Phys. Chem. B*, **113** (43), 14310–143104.

115 Madurga, S., Martín-Molina, A., Vilaseca, E., Mas, F., and Quesada-Pérez, M. (2007) Effect of the surface charge discretization on electric double layers: a Monte Carlo simulation study. *J. Chem. Phys.*, **126**, 234703-1–234703–11.

116 Bhuiyan, L.B., Outhwaite, C.W., and Henderson, D. (2009) Evidence from Monte Carlo simulations for a second contact value theorem for a double layer formed by 2:1/1:2 salts at low electrode charges. *Mol. Phys*, **107**, 343–347.

117 Corson, D. and Lorrain, P. (1962) Chapter 4, in *Introduction to Electromagnetic Fields and Waves*, Freeman, San Francisco, CA.

118 Torrie, G.M., Valleau, J.P., and Outhwaite, C.W. (1984) Electrical double layers. IV. Image effects for divalent ions. *J. Chem. Phys.*, **81**, 6296–6300.

119 Croxton, T., McQuarrie, D.A., Patey, G.N., Torrie, G.M., and Valleau, J.P. (1981) Ionic solution near an uncharged surface with image forces. *Can. J. Chem.*, **59**, 1998–2003.

120 Moreira, A.G. and Netz, R.R. (2002) Counterions at charge-modulated substrates. *Europhys. Lett.*, **57**, 911–917.

121 Saad, A. and Tobazeon, R. (1982) Study of the double layer at an insulator/liquid interface by step voltage transients. *J. Phys. D: Appl. Phys.*, **15**, 2505–2512.

122 Greberg, H., Kjellander, R., and Åkesson, T. (1996) Ion–ion correlations in electric double layers from Monte Carlo simulations and integral equation calculations. *Mol. Phys.*, **87**, 407–422.

123 Alawneh, M. and Henderson, D. (2007) Monte Carlo simulation of the double layer at an electrode including the effect of a dielectric boundary. *Mol. Simul.*, **33**, 541–547.

124 Messina, R. (2002) Image charges in spherical geometry: application to colloidal systems. *J. Chem. Phys.*, **117**, 11062–11074.

125 Allen, R., Hansen, J.-P., and Melchionna, S. (2001) Electrostatic potential inside ionic solutions confined by dielectrics: a variational approach. *Phys. Chem. Chem. Phys.*, **3**, 4177–4186.

126 Boda, D., Gillespie, D., Nonner, W., Henderson, D., and Eisenberg, B. (2004) Computing induced charges in inhomogeneous dielectric media: application in a Monte Carlo simulation of complex ionic systems. *Phys. Rev. E*, **69**, 046702-1–046702–10.

127 Henderson, D., Gillespie, D., Nagy, T., and Boda, D. (2005) Monte Carlo simulation of the electric double layer: dielectric boundaries and the effects of induced charge. *Mol. Phys.*, **103**, 2851–2861.

128 Benjamin, I. (1997) Molecular dynamic simulations in interfacial electrochemistry, in *Modern Aspects of Electrochemistry* (eds J.O.M. Bockris, R.E. White, and B.E. Conway), Plenum Press, New York, pp. 115–179, Chapter 3.

129 Spohr, E. (1998) Computer simulation of the structure of the electrochemical double layer. *J. Electroanal. Chem.*, **450**, 327–334.

130 Pusztai, L., Pizio, O., and Sokołowski, S. (2008) Comparison of interaction potentials of liquid water with respect to their consistency with neutron diffraction data of pure heavy water. *J. Chem. Phys.*, **129**, 184103-1–184103–6.

131 Philpott, M.R. and Glosli, J.N. (1997) *Molecular Dynamics Simulation of Interfacial Electrochemical Processes: Electric Double Layer Screening in Solid–Liquid Electrochemical Interfaces*, ACS Symposium Series, vol. 656 (eds G. Jerkiewicz, M.P. Soriaga, K. Uosaki and A. Wieckowski) ACS, Washington, DC, Chapter 2, pp. 13–30.

132 Eggebrecht, J. and Peters, G.H. (1993) Multipolar electrolyte solution models. II. Monte Carlo convergence and size dependence. *J. Chem. Phys.*, **98**, 1539–1545.

133 Zhang, L., Davis, H.T., and White, H.S. (1993) Simulations of solvent effects on confined electrolytes. *J. Chem. Phys.*, **98**, 5793–5799.

134 Lamperski, S. and Zydor, A. (2007) Monte Carlo study of the electrode|solvent primitive model electrolyte interface. *Electrochim. Acta*, **52**, 2429–2436.

135 Boda, D., Henderson, D., Patrykiejew, A., and Sokołowski, S. (2000) Simulation and density functional study of a simple membrane. II. Solvent effects using the solvent primitive model. *J. Chem. Phys.*, **113**, 802–806.

136 Willard, A.P., Reed, S.K., Madden, P.A., and Chandler, D. (2009) Water at an electrochemical interface: a simulation study. *Faraday Discuss.*, **141**, 423–441.

137 Reed, S.K., Madden, P.A., and Papadopoulos, A. (2008) Electrochemical charge transfer at a metallic electrode: a simulation study. *J. Chem. Phys.*, **128**, 124701-1–124701-10.

138 Kerisit, S., Cooke, D.J., Marmier, A., and Parker, S.C. (2005) Atomistic simulation of charged iron oxyhydroxide surfaces in contact with aqueous solution. *Chem. Commun.*, **24**, 3027–3029.

139 Born, M. and Huang, K. (1954) *Dynamical Theory of Crystal Lattices*, Oxford University Press, Oxford, UK.

140 de Leeuw, N.H. and Parker, S.C. (1998) Molecular-dynamics simulation of MgO surfaces in liquid water using a shell-model potential for water. *Phys. Rev. B*, **58**, 13901–13908.

141 Keblinski, P., Eggebrecht, J., Wolf, D., and Phillpot, R. (2000) Molecular dynamics study of screening in ionic fluids. *J. Chem. Phys.*, **113**, 282–291.

142 Car, R. and Parinello, M. (1985) Unified approach for molecular dynamics and density-functional theory. *Phys. Rev. Lett.*, **55**, 2471–2474.

143 Spohr, E. (1999) Molecular simulation of the electrochemical double layer. *Electrochim. Acta*, **44**, 1697–1705.

144 Izvekov, S., Mazzolo, A., Van Opdorp, K., and Voth, G.A. (2001) *Ab initio* molecular dynamics simulation of the Cu (110)–water interface. *J. Chem. Phys.*, **114**, 3248–3257.

145 Guymon, C.G., Rowley, R.L., Harb, J.N., and Wheeler, D.R. (2005) Simulating an electrochemical interface using charge dynamics. *Condens. Matter Phys.*, **8**, 335–356.

146 Netz, R.R. (2004) Water and ions at interfaces. *Curr. Opin. Colloid Interface Sci.*, **9**, 192–197.

147 Suter, J.L., Boek, E.S., and Sprik, M. (2008) Adsorption of a sodium ion on a smectite clay from constrained *ab initio* molecular dynamics simulations. *J. Phys. Chem. C*, **112**, 18832–18839.

4
Dynamical Instabilities in Electrochemical Processes
István Z. Kiss, Timea Nagy, and Vilmos Gáspár

Electrochemical cells often exhibit instabilities that can result in temporal current/potential oscillations or large spatial variations of reaction rate over the electrode surface. The complexity arises from the interplay of nonlinear chemical reactions and various physical processes. In this chapter, the characterization, description, and control of these temporal and spatiotemporal phenomena are reviewed. The use of recent experimental and data processing tools is illustrated with different types of self-organized behavior on single- and multielectrode systems. Development in the engineering and control of dynamical instabilities of electrochemical cells is summarized. Examples are provided for uniform and metastable pitting corrosion, electrocatalytic reactions, solid-state and polymer electrolyte membrane fuel cells, sensors, and some other solid-state electrochemical applications.

4.1
Introduction

Electrochemical cells are excellent examples of far-from-equilibrium systems in which irreversible processes take place under the action of coupled nonlinear processes due to chemical reactions, electrical effects (double-layer charging, potential drops, migration), and mass transport. The description of complex dynamic responses of the cells is limited with classical tools of equilibrium, reversible thermodynamics, and simple linear evolution laws. Advances in irreversible thermodynamics [1] and in the qualitative theory of differential equations during the second half of the last century led to a general understanding of how nonlinear evolution laws may result in different forms of dynamical instabilities (bistability, oscillations, and chaos) in far-from-equilibrium systems. Wojtowicz [2] and 20 years later Hudson and Tsotsis [3, 4] pointed out in their comprehensive reviews on electrochemical oscillations that there are perhaps more examples of such behavior in electrochemical systems than in any other area of chemical kinetics. Therefore, analysis of nonlinear behavior is necessary for understanding important dynamical features of electrochemical systems that

Solid State Electrochemistry II: Electrodes, Interfaces and Ceramic Membranes.
Edited by Vladislav V. Kharton.
© 2011 Wiley-VCH Verlag GmbH & Co. KGaA. Published 2011 by Wiley-VCH Verlag GmbH & Co. KGaA.

operate in a far-from-equilibrium regime. Still, until the 1990s, there have been only a small number of systematic investigations aimed at developing the general framework within which dynamical instabilities can be interpreted.

The excellent reviews by Koper [5] and Krischer [6, 7] gave detailed accounts on the progress (until about 10 years ago) in understanding the true physicochemical origin of dynamical instabilities in electrochemical systems resulting, for example, in a standard method for a classification of electrochemical oscillators [8]. Here, we do not intend to give a complete account of all experimental and theoretical results regarding dynamical instabilities in electrochemistry. For the interested reader, we refer to some excellent reviews published during the past few decades [2–5, 7–19].

In this chapter, we summarize what has been achieved by rigorous applications of the advanced theories in quantitative characterization, engineering, and control of electrochemical oscillations and chaos. We also show how simple scaling relations can hold for certain situations allowing the design of dynamical behavior. Technological advances in instrumentation interfaced with fast personal computers allowed the investigations of spatiotemporal dynamics of single and large arrays of electrodes. With these tools at hand, new strategies have been developed to study the various forms of a collective behavior emerging through coupling of nonlinear units [20], for example, synchronization [21]. The methods and strategies considered in this chapter are general and can be applied to any nonlinear electrochemical system, either solid or liquid. The available information on solid-state systems is, however, more scarce, so most of the particular examples discussed later were selected from systems with liquid electrolyte.

We start with a theoretical introduction to the origin and classification of dynamical instabilities in electrochemical systems and follow by the characterization of nonlinear phenomena occurring at different conditions. Basic experimental methods and data processing algorithms are discussed briefly. We then give an overview of the literature reporting on the investigation of bistability, oscillations, chaos, pattern formation, and their control in different electrochemical systems. We close with a brief overview of applications in understanding corrosion, designing fuel cells, and solid-state devices.

4.2
Origin and Classification of Dynamical Instabilities in Electrochemical Systems

In this section, we elaborate on the origin of instabilities in electrochemical systems leading to the appearance of critical behavior such as bistability, oscillations, chaos, and pattern formation. First, we discuss the crucial processes that may result in nonlinear dependence of the essential system variables on some bifurcation parameters.

We then present a systematic method for dynamical classification of electrochemical systems. The definition of bistability, oscillations, and chaos is given and the origin of these phenomena is discussed in the framework of bifurcations theory and the linear stability analysis of the differential equations describing the evolution laws.

4.2.1
Classification Based on Essential Species

The classification of dynamical instabilities was originally developed for oscillatory electrochemical reactions [8]. The classification scheme is general in nature and thus can be applied to other nonlinear phenomena (e.g., bistability and excitability). In the following, we will present the classification assuming oscillatory behavior.

In an oscillatory chemical system, the concentrations of many species exhibit periodic variations in time. A common feature is the existence of a number of species, which are "essential" to the oscillatory mechanisms [22–24]. When the concentration of an essential species is (externally) kept constant, oscillations will not occur. In contrast, nonessential species simply indicate the presence of oscillations. For example, pH often varies in many chemical oscillatory systems; however, there is a well-defined set of reactions, "pH oscillators" [25, 26], in which the variations would not occur in buffered solutions. From a modeling perspective, the concentration/surface coverage of the essential species will play a role of essential variables needed for mathematical description of the dynamics.

An important step in characterization of nonlinear behavior is the identification of essential variables. Usually the number of variables is relatively low; basic electrochemical mechanisms often consider only two to four essential species [5, 27, 28]. In electrochemistry, in addition to traditional "chemical" variables (e.g., concentration/surface coverage of molecules), the electrode potential (potential drop across the double layer driving the reaction) could also be considered as a variable. Therefore, the electrode potential plays the role of pseudospecies.

There are two important characteristics of essential variables:

- **Timescales:** Essential variables vary on typical timescales; it is customary to think of variables as "fast" and "slow" [23]. For example, the variations of the electrode potential are often determined by the double-layer charging process having a characteristic timescale of RC_dA, where R is the uncompensated (series) resistance of the cell (Ω), C_d is double-layer capacitance (F cm^{-2}), and A is the surface area of the electrode (cm^2). Because the double-layer capacitance is typically a small number, the timescale of electrode potential is small; therefore, in a large number of systems, electrode potential is a fast variable.
- **Feedback loop:** Nonlinear behavior in chemical systems is often related to some special kinetic features of the reactants and products. Positive feedback is considered when the rate of reaction is larger than that expected without the special kinetic effect; examples include autocatalysis and self-inhibition. In negative feedback loops, the reaction slows down as it progresses. It is important to point out that feedback loops are considered from a general perspective where not only chemical reactions but also physicochemical processes (e.g., potential drops in electrolyte, mass transfer) are taken into account. The types of feedback loops in which essential species participate can be determined by stoichiometric network analysis [22, 23] and by analysis of the signs of the Jacobian matrix elements of the underlying ordinary differential equations describing the system [8].

4 Dynamical Instabilities in Electrochemical Processes

The electrochemical oscillators can be classified into two major categories: strictly potentiostatic and negative differential resistance (NDR) systems.

4.2.1.1 "Truly" or "Strictly" Potentiostatic Systems

In strictly potentiostatic systems, cell instabilities are caused by the feedback loops in the reaction network. In these systems, the electrode potential is simply a parameter that affects the rate constant of reactions that involve charge transfer. Setting of proper electrode potential is thus needed to satisfy parametric conditions for instabilities.

4.2.1.2 Negative Differential Resistance Systems

In NDR systems, the electrode potential is an essential dynamical variable and crucially involved in establishing the necessary positive and negative feedback loops. In an NDR oscillator when the electrode potential is kept constant, the oscillations disappear. As we shall see below, this will imply the presence of a potential region where for a reaction the Faradaic current decreases with potential and thus creates a negative differential resistance. Because the electrode potential (e) is dynamical variable, an ordinary differential equation should be derived for the temporal evolution. Figure 4.1a shows a simple equivalent circuit [29] that considers three major processes:

- Double-layer charging is represented by a capacitance C_d. For simplicity, it is assumed that C_d does not depend on the electrode potential.
- IR drop in electrolyte and external circuitry. The potential (IR) drop in the electrolyte and external circuitry (e.g., resistance externally added to the electrode) can be modeled with a series resistance R_s. The series resistance depends on cell geometry (size and placement of electrodes).
- Charge transfer process. The electrochemical reaction is represented by an impedance Z_F. Often this circuit element is simplified with an ohmic circuit element, whose resistance depends on electrode potential.

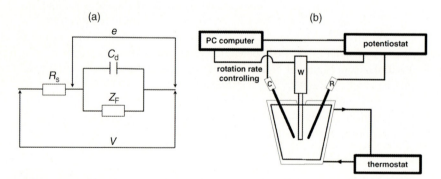

Figure 4.1 (a) Simple equivalent circuit for analysis of electrochemical instabilities. (b) Standard three-electrode electrochemical cell with a liquid electrolyte and working (W), reference (R), and counter (C) electrodes. The potentiostat is interfaced with a computer that may also regulate the rotation rate of a rotating disk working electrode.

4.2 Origin and Classification of Dynamical Instabilities in Electrochemical Systems

In a typical potentiostatic electrochemical experiment, the circuit potential V is kept constant. By writing a charge balance for the total current ($i_{tot} = (V-e)/R_s$) obtained by summing the charging ($i_c = AC_d(de/dt)$) and Faradaic [$i_F(e)$] currents,

$$i_{tot} = \frac{(V-e)}{R_s} = AC_d \frac{de}{dt} + i_F(e). \tag{4.1}$$

(Note that it is assumed in this simple derivation that near-surface concentrations of electroactive species are the same as bulk concentrations; thus, they are constant and the Faradaic current will depend only on potential.) The equation for variation of electrode potential from this equation [28] is

$$C_d \frac{de}{dt} = \frac{(V-e)}{AR_s} - \frac{i_F(e)}{A}. \tag{4.2}$$

Negative slope in $i_F(e)$ can be obtained by two major routes illustrated in Figure 4.2. The NDR systems are therefore further classified into two subgroups.

N-NDR Systems In N-NDR systems, the negative differential resistance appears through an N-shaped polarization curve (Figure 4.2a). In this example, $i_F(e)$ decreases in Equation 4.2 as e is increased. Because of the negative sign in front of the $i_F(e)$ term in Equation 4.2, with increasing electrode potential de/dt decreases resulting in self-inhibition. This self-inhibition will result in positive feedback loop that includes the electrode potential.

Conditions for cell instabilities can be derived by stability analysis of Equation 4.2 [5]. The steady-state solution (e_{ss}) of Equation 4.2,

$$\frac{de}{dt} = 0 = \frac{(V-e_{ss})}{C_d A R_s} - \frac{1}{C_d A} i_F(e_{ss}), \tag{4.3}$$

is stable only if the Jacobian of Equation 4.2 is negative:

$$\frac{-1}{C_d A R_s} - \frac{1}{C_d A} \frac{di_F(e)}{de} < 0. \tag{4.4}$$

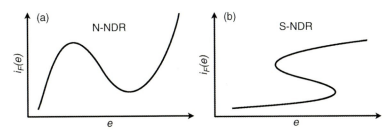

Figure 4.2 Classes of negative differential resistance systems. (a) N-shaped polarization curve (N-NDR). (b) S-shaped polarization curve (S-NDR).

Because A, C_d, and R are always positive, this inequality implies that instability cannot be induced by the electrode potential on the positive slope of the polarization curve. However, on the negative slope ($di_F(e)/de < 0$), the stationary state can be unstable if

$$\frac{1}{R_s} < \left|\frac{di_F(e)}{de}\right|. \tag{4.5}$$

Equation 4.5 shows that in N-NDR systems with large enough series resistance, the stationary steady solution can become unstable.

The Faradaic current can be expressed more generally as

$$i_F(e) = zFAck_r(e), \tag{4.6}$$

where z is the number of electrons involved in the charge transfer process, F is the Faraday constant, $k_r(e)$ is the potential-dependent rate constant, and c is the (near) surface concentration of the electroactive species. Koper identified [5] three major mechanisms that can produce negative slopes: (1) $dA/de < 0$, (2) $dk_r/de < 0$, and (3) $dc/de < 0$.

1) The available electrode surface area decreases with increasing electrode potential. Most prominent example is the anodic passivation of metals.
2) The decrease of rate constant with increasing electrode potential. The rate constant is considered as an apparent rate constant. Because of the potentially complicated reaction mechanism, it is possible that adsorption of an inhibitor or desorption of a catalyst occurs with increasing potential. Extensively studied examples include reduction of metal ions in the presence of organic agents [6].
3) Decrease of concentration of the electroactive species with increasing overpotential. Concentrations of the species at the reaction plane might not be equal to the near-surface concentration due to the complicated structure of the double layer. This Frumkin effect can cause negative differential resistance in many anion reduction reactions in low ionic strength solutions [6].

When there are many electrochemical reactions, it is possible to have parallel reactions in which N-NDR character of one reaction is hidden by the positive slope of other "traditional" reactions. Such systems are called hidden N-NDR (HN-NDR) systems [6, 30]; for example, the negative slope observed in transpassive dissolution of Ni in sulfuric acid due to the adsorption of blocking bisulfate ions can be hidden by the chemical dissolution of the oxide layer [5, 31]. HN-NDR systems exhibit unique dynamical features; for example, they can exhibit oscillations under galvanostatic conditions [6].

S-NDR Systems Negative differential resistance can also develop through an S-shaped polarization curve shown in Figure 4.2b. In this example, the positive feedback loop in chemical variables causes bistability (upper and lower branches of the S-curve) and a middle branch with NDR. In contrast to the N-NDR systems, bistability occurs at very low resistances. (In fact, at large resistance, bistability disappears.) The S-NDR system was identified, for example, with certain deposition reactions of bivalent cations (e.g., Zn^{2+} where the reaction step

$$Me^{2+} + Me_{(a)}^+ + e^- \rightleftharpoons 2Me_{(a)}^+ \qquad (4.7)$$

results in autocatalytic positive feedback with the adsorbed metal ion species [32]).

4.2.2
Classification Based on Nonlinear Dynamics

In the previous section, we have seen how negative differential resistance may result in an unstable steady-state solution for the electrode potential (e_{ss}). The electrode potential is, however, only one of the essential variables affecting the dynamics. Equally important variables are the surface or near-surface concentration of the electroactive species as discussed earlier. The time evolution of such a multivariable system at a given set of parameters can be calculated by solving the *system of ordinary differential equations* (ODEs) that follows from the physical laws describing the system. Standard ODE solvers are readily available to perform such task [33]. Other important goals are to find the location of the stable and unstable steady states (and periodic orbits) and to investigate the properties of these states as parameters are changed. *Bifurcation* means a sudden change in the number or in the character of the steady-state solutions resulting in new dynamical behavior of the system. In bifurcation diagrams, the position and the stability of steady (stationary) and oscillatory states are shown as a function of a system parameter. The stability or instability and the character of the steady states can be determined by *linear stability analysis*. This is also an essential tool in understanding the wealth of dynamical phenomena occurring in electrochemical systems. There are several excellent textbooks on the relevant mathematical machinery [34–36]. Here, we shall give only a brief introduction to dynamical systems theory and define the basic nomenclature that is necessary to the classification of nonlinear behavior in electrochemical systems.

Since in many cases the dynamics of multivariable systems can be well approximated by simple two-variable models (after reducing the number of variables by using quasi-steady-state approximation), we shall explain the most important concepts of linear stability analysis by assuming two dynamical variables only. In this case, the time evolution of the system is described by ODEs as follows:

$$\frac{dx_i(t)}{dt} = f_i(\mathbf{x}), \qquad i = 1, 2, \qquad (4.8)$$

where $\mathbf{x} = [x_1(t), x_2(t)]$ gives the coordinates of the phase point at time $t > 0$. We now define the Jacobian matrix (**J**) at a steady-state (stationary) point $\bar{\mathbf{x}} = (\bar{x}_1, \bar{x}_2)$:

$$\mathbf{J} = \begin{pmatrix} \partial f_1/\partial x_1 & \partial f_1/\partial x_2 \\ \partial f_2/\partial x_1 & \partial f_2/\partial x_2 \end{pmatrix}_{\bar{\mathbf{x}}}. \qquad (4.9)$$

When the phase point is in the small neighborhood of the stationary point, functions f_i can be approximated by Taylor expansion. Thus, the change in the distance from the stationary point can be well described by a set of *linear differential equations* written in matrix form as follows:

$$\frac{d\delta x}{dt} = J\delta x, \qquad (4.10)$$

where $\delta \mathbf{x} = [x_1(t) - \bar{x}_1, x_2(t) - \bar{x}_2]^T$. A standard result of linear algebra is that the stability and character of the stationary point are uniquely determined by the eigenvalues (λ_1, λ_2) of the Jacobian matrix (**J**) that can be calculated as the solution to the following characteristic equation:

$$\det(\mathbf{J} - \lambda \mathbf{I}) = 0, \qquad (4.11)$$

where **I** is the 2×2 identity matrix.

Generally, the eigenvalues can be real or complex but from the point of view of stability, it is enough to investigate the sign of the real part of the complex number. If the signs of the real parts of the eigenvalues are all negative, the stationary point is *stable*, and the phase point will approach the steady-state solution from every direction; an initial perturbation will always die out. On the other hand, if there is at least one eigenvalue with a positive real part, the stationary point is *unstable*. In this case, an initial perturbation will grow and the phase point will depart from the stationary point. Table 4.1 gives the classification of the type of stationary points that are possible in two-variable systems based on the nature of the eigenvalues.

In an n-variable dynamical system, linear stability analysis will result in n eigenvalues. The more the variables define the dynamics of a system, the more the possibilities are for the combination of the eigenvalues. However, the basic statements about the stability and instability of the steady state also hold for multivariable systems. If all eigenvalues have negative real parts, the stationary point will be stable, while one eigenvalue with a positive real part is enough for instability.

As mentioned above, a *bifurcation* could occur upon changing the value of a system parameter μ. Bifurcation is generally associated with the change in the sign of at least one of the eigenvalues. The bifurcation can occur at the critical value μ_{crit}, where the

Table 4.1 Classification of steady states in two-variable dynamical systems.

Steady state	Eigenvalues	Local time behavior
Stable node	$\lambda_1, \lambda_2 < 0$	Monotonic decay
Unstable node	$\lambda_1, \lambda_2 > 0$	Monotonic growth
Saddle	$\lambda_1 < 0, \lambda_2 > 0$	Monotonic decay and growth in "opposite" directions
Stable focus	Re $\lambda \pm i$ Im λ, Re $\lambda < 0$	Oscillatory decay
Unstable focus	Re $\lambda \pm i$ Im λ, Re $\lambda > 0$	Oscillatory growth

eigenvalue becomes zero. There are two types of bifurcations that are essential to most of the studied electrochemical systems:

a) **Saddle–node (SN) bifurcation.** It occurs when one of the stability directions is reversed and a (stable or unstable) node changes into a saddle. This bifurcation may also occur when at μ_{crit} a stable node coalesces with a saddle resulting in the formation of a saddle–node point. The SN bifurcation is closely related to the appearance (or disappearance) of *bistability*. When a system has more than one stable solution, the final state will depend on the initial conditions. The fact that two steady states may coexist at a given set of parameters could result in the so-called *hysteresis*: the switch between the two steady states will occur at different values of the bifurcation parameter upon increasing or decreasing its value.

b) **Hopf (H) bifurcation.** It can occur when the real part of a complex pair of eigenvalues becomes zero. Such condition often indicates the onset of sustained oscillations. As the bifurcation parameter is varied so that the system passes through a Hopf bifurcation, a new periodic orbit or *limit cycle* develops. It is a closed curve in the phase space surrounding the steady state. The bifurcation is called *supercritical* when the new limit cycle is a stable object. At the bifurcation point, the steady state loses its stability; it turns into an unstable focus. In a *subcritical* Hopf bifurcation, the new limit cycle is unstable; further bifurcation (usually a saddle–node bifurcation of periodic orbits) is required for experimental observation of the oscillations.

Bifurcation scenarios are often represented in the so-called reconstructed phase space (or state space). This technique is regularly applied to most experimental systems, since it is almost impossible to follow the time evolution of all dynamical variables. In electrochemical systems, for example, one can readily measure the current (or potential); however, information on the surface or near-surface concentration of the electroactive species is often not available. According to Ref. [37], the phase space can be reconstructed from the time series values of only one dynamical variable $x(t)$ as follows: one should plot $x(t)$, $x(t-\tau)$, $x(t-2\tau)$, ..., $x[t-(m-1)\tau]$ in an m-dimensional phase space, where τ is an arbitrarily chosen *delay time*. An empirical rule of thumb is that the optimal value for τ is between one-third and one-tenth of the oscillatory period. It has been proven that a reconstructed attractor is topologically equivalent to the original n-dimensional attractor if $m \geq (2n+1)$.

Hopf bifurcation is not the only way by which a limit cycle can be born or disappear. It can also happen when a growing limit cycle runs into a saddle point. At the critical value μ_{crit}, there is a so-called *homoclinic orbit*. It is a closed-loop trajectory that starts and ends on exactly the same saddle point. This type of bifurcation is called *saddle–loop (SL) bifurcation*. A third way for the bifurcation of periodic solutions is when stable and unstable limit cycles coexist and they collapse onto each other at μ_{crit}, resulting in mutual disappearance. This bifurcation is referred to as *saddle node of periodic (SNP) orbits*.

In dynamical systems of more than two variables, the topological restrictions of the plane are removed and, therefore, more complex solutions can evolve. Most important difference is that in three- or more dimensional phase space, the trajectories can cross over each other without violating the uniqueness theorem. This results in more complex periodic orbits leading to mixed-mode oscillations (MMOs), quasiperiodicity (irregular periodicity), or chaos.

Deterministic chaos is a very special state of a dynamical system. In the phase space, chaos can be identified by a special object (chaotic attractor) that has an infinite periodicity; although the motion on the attractor consists of cyclic variations, the cycle never repeats itself. In the long term, the motion of a system point on the chaotic attractor is unpredictable as the dynamics is very sensitive to the initial conditions. Small error in the definition of the initial state will result in an exponentially growing difference in the predicted values. An often observed, simple bifurcation scenario that may lead to chaos is the *cascade of period-doubling bifurcations* [38, 39]. At a bifurcation of a periodic orbit of period T (that turns unstable), a new stable biperiodic ($2T$) orbit can be formed. This period doubling may repeat as the bifurcation parameter is changed resulting in even higher periodic orbits. The distance between the critical parameter values $\Delta \mu_k = \mu_k - \mu_{k-1}$ corresponding to the subsequent period-doubling events becomes smaller and smaller as k is increased, leading to a limit point where the period of the attractor becomes infinite. Beyond this point, the system is in the state of chaos. The sequence of period-doubling bifurcations leading to chaotic oscillations on the copper–phosphoric acid system is shown in Figure 4.3a. Note that the transition back from chaotic to periodic behavior occurs through a cascade of period-halving bifurcations. The chaotic attractor is represented in a reconstructed phase space in Figure 4.3b.

The transition from chaotic to periodic motion may also occur through *intermittency*, in which a periodic behavior is interspersed with chaotic bursts. In general, this bifurcation (also called tangent bifurcation) occurs when a chaotic attractor hits a saddle node of periodic orbits. A third route to chaos involves a *secondary Hopf* or *torus bifurcation* of a simple periodic orbit leading to modulated oscillations. Details of these scenarios are discussed in many excellent textbooks [34, 40, 41].

Nonlinear behavior is not restricted to homogeneous (well-stirred) systems discussed so far. In spatially extended systems, where transport processes (the most common form is diffusion) are coupled to nonlinear kinetics, some type of pattern formation is almost inevitable [42]. Patterns that often resemble some geometrical form are due to spatial variation in system variables such as concentration, current, potential, and so on. The pattern can be stationary (structures) or dynamical (traveling and spiral waves, oscillatory fronts).

Stationary structures are often interpreted by the Turing mechanism in which the steady state of an activator–inhibitor system is unstable to inhomogeneous perturbations but stable to all spatially homogeneous perturbations [36]. The spatial structure possesses an intrinsic wavelength that is determined by kinetic/mass transfer constants but is independent of the size of the system as long as the system size is larger than the pattern wavelength. Critical difference in the transport properties of the activator or inhibitor species plays a decisive role in the dynamics and stability of the structures.

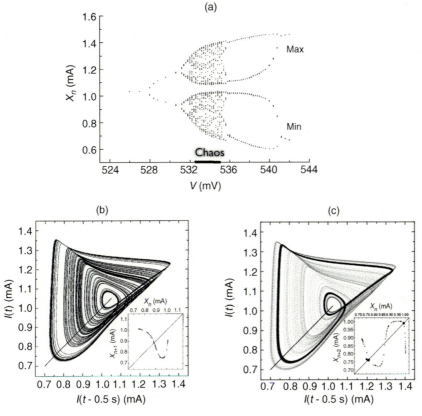

Figure 4.3 Chaotic behavior of Cu electrodissolution in phosphoric acid solution in a rotating disk experiment (diameter 5 mm, rotation rate 1850 rpm) with a total series resistance of 202 Ω and at −17.5 °C. (a) Bifurcation diagram showing both maxima and minima of current oscillations as a function of the applied anodic potential. (b) Reconstructed chaotic attractor in the phase space of time delay coordinates (anodic potential 532.0 mV, rotation rate 1800 rpm). The next-return map x_{n+1} versus x_n has been generated by using successive current values on the Poincaré section. (c) Stabilized period-2 orbit embedded in the chaotic attractor (anodic potential 527.0 mV, rotation rate 1900 rpm). The next-return map x_{n+2} versus x_n has been generated by using successive current values on the Poincaré section [81].

Dynamical patterns of oscillatory systems close to a Hopf bifurcation are often described with the complex Ginzburg–Landau equation (CGLE) and its variants [43]. A vast variety of nonlinear physical problems can be described with CGLE that includes electrochemical pattern formation [44]; the patterns are often related to uniform oscillations losing stability to some oscillating Fourier modes. A review on waves and patterns in electrochemical systems where, in addition to diffusion, migration plays an important role has been given by Krischer *et al.* [45].

4.3
Methodology

In this section, we briefly summarize the basic experimental and data processing techniques that are generally applied to investigate temporal and spatiotemporal phenomena in electrochemical systems.

4.3.1
Experimental Techniques

4.3.1.1 Techniques for Temporal Dynamics

The basic tool for studying nonlinear dynamics of electrochemical processes is the standard three-electrode electrochemical cell. Figure 4.1b shows a jacketed cell holding working (W), reference (R), and counter (C) electrodes. The working electrode can be a standing electrode (e.g., in reactions under kinetic control), a rotating disk, or a ring electrode. In many studies, the effect of potential drop in the cell is studied either by changing cell geometry (working-to-reference electrode placement) or by an external resistance attached in series with the working electrode. In potentiostatic (galvanostatic) experiments, the circuit potential (current) is kept constant and the current (potential) is measured with chronoamperometry (chronopotentiometry). Therefore, typical parameters that affect the dynamics in a potentiostatic experiment are circuit potential, temperature, rotation rate, external resistance, electrolyte composition, and cell geometry (placement and size of the electrodes).

In addition to potentiostatic measurement, linear sweep voltammetry is also often applied to detect the bifurcations that may lead to oscillations by varying the potential, and so on. The standard method of obtaining a bifurcation diagram under potentiostatic conditions is a very slow scan (0.01–10 mV s^{-1}) of the circuit potential allowing the system to relax onto the new attractor (stationary point, limit cycle, and so on).

Because of the large number of experimental parameters, it is often difficult to find the oscillatory conditions. *Impedance spectroscopy* can be applied to determine the necessary potentials/resistance conditions for the appearance of bifurcations leading to instabilities without the need to generate a nonequilibrium phase diagram by systematically varying the parameters over large ranges [46]. Impedance spectroscopy can be considered as a tool that is capable of performing an experimental linear stability analysis of the system without knowing the detailed kinetic model for the electrochemical processes. The analysis of the negative real part of the impedance spectrum gives valuable information (see Section 4.4) about the bistability and oscillatory region of circuit potential and resistance.

4.3.1.2 Techniques for Spatial Dynamics

Characterization of spatiotemporal self-organization effects requires imaging an essential species (or a corresponding physical property) of the system. This task is very challenging since many patterns are nonstationary and a compromise needs to be made to find optimal spatial and temporal resolutions.

In some reactions, optical microscopy can be applied to detect the spatiotemporal dynamics. In metal dissolution/passivation reactions, there could be a visual contrast between the active (shiny metal) and passive (dark oxide film) electrode surfaces. In an experimental setup utilizing upward facing electrode [47–49], the spatial changes can be followed with a video camera with high frame capture rate. This technique was successfully used with Fe [47, 48] and Co [49] electrodissolution reactions.

Because the electrode potential is an essential variable in majority of reactions, the potential distribution over the electrode provides valuable information about the dynamics. A set of potential probes can be placed close to the (standing) electrode surface. This technique was applied in a range of systems, including formic acid oxidation [50], Ni electrodissolution [51], and Co dissolution [52]. In a design with rotating ring electrode, only one (sensing) reference electrode is positioned under the ring; the time series data obtained from the reference electrode are converted to spatial positions and thus temporal resolution is determined by the usually relatively fast rotation rate [53–56]. This technique was very successfully applied to a range of reactions, including CO oxidation [57] and H_2 oxidation [54–56]. Two-dimensional imaging of electrode potential with high spatial and temporal resolution is possible based on potential dependence of resonance conditions for the excitation of surface plasmons [58]. In an innovative surface plasmon microscopy setup, 2D imaging of potential waves and stationary patterns was accomplished in reduction of peroxodisulfate [58] and CO electrooxidation reactions [59], respectively.

In reactions where properties of surface films vary in the reaction, ellipsometry can be applied. *In situ* observations of spreading of pitting corrosion of stainless steel were accomplished [60] with combination of two different imaging methods simultaneously, namely, ellipsomicroscopy for visualizing changes of surface film properties and contrast-enhanced microscopy for monitoring nucleation and reactivation of metastable corrosion pits. Other applications include semiconductor interfaces [61].

Full kinetic description of the system is greatly facilitated by determining concentrations of electroactive species directly. Spatially resolved *in situ* infrared spectroscopy is a powerful tool that was applied to image CO coverage in CO electrooxidation [62] and interfacial chemical composition of semiconductor space charge layer [63].

Multielectrode array configurations [64] have been used in studies that required high resolution in temporal behavior (e.g., oscillations and chaos) at the expense of spatial resolution. The electrode arrays with small wires and close spacing exhibited dynamical behavior similar to that observed with one large electrode in iron electrodissolution [64]. Typically, metal electrode wires are embedded in an insulating material (Teflon or epoxy) so that the reaction takes place only at the ends (see Figure 4.4a and b); the currents of individual electrodes are determined by multichannel current meters. The establishment of well-defined mass transfer conditions for providing fresh electrolyte solutions to the surface and the flexible design of electrodes of various sizes and spacings are major experimental difficulties of electrode array studies. When the number of electrodes is small (e.g., two) or local measurements are not required, rotating disks can be used. An impinging jet system

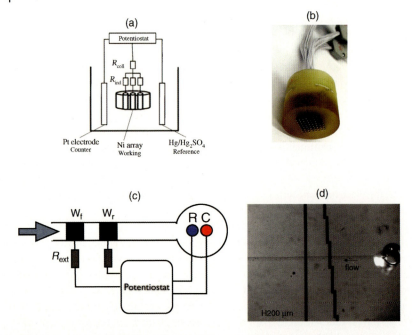

Figure 4.4 Experimental designs for studying dynamics of coupled electrodes using electrode arrays. (a) Schematic of the experimental setup. (b) An 8 × 8 Ni electrode array embedded in epoxy. (c) Schematic of experimental setup for on-chip fabricated microfluidic flow cell [69]. W_f, W_r, C, R: front and rear working, counter, and reference electrodes, respectively. (d) Microfluidic dual-electrode setup with 100 μm Pt band electrodes over which a 100 μm wide flow channel is placed [69].

was developed for the investigation of oscillations on the mass transfer limited region of dissolution of arrays of relatively large (>10) number of electrodes [65]. However, investigators have chosen electrochemical reactions under kinetic control where critical phenomena occur at stagnant or weakly stirred conditions because of the difficulty of providing laminar flow in macroelectrode cells [66–68]. Recently, on-chip fabrication technologies with microfluidic flow cells (Figure 4.4c and d) provided alternative means for flexible cell design with well-controlled mass transfer conditions [69].

4.3.2
Data Processing

Large amount of data can be collected in experiments that represent the state of the electrode with an experimentally observable quantity (e.g., current) as a function of time. Methods based on classical digital signal processing can be applied to obtain information about frequency/amplitude of the signals. In addition, nonlinear dynamics developed a set of tools with which characterization of various complexity features is possible.

4.3.2.1 Digital Signal Processing

With oscillatory reactions, a standard tool for the analysis of times series data obtained from current, electrode potential, or concentrations is the power spectrum [4]. Instead of looking at the data in time domain, the data are analyzed in the frequency domain when the amplitude (A_m) or the power (A_m^2) of each frequency component is determined by the fast Fourier transform (FFT) method. Oscillations with harmonic waveform are characterized by a single line (delta peak) in the power spectrum corresponding to the frequency (ω) of the oscillations. Because of noise in the measurements, instead of a single line a narrow peak will occur. Simple periodic oscillations exhibit frequency components at superharmonic components (at frequencies $= \omega k$, where $k = 1, 2, \ldots$ is an integer). Quasiperiodic oscillations are characterized by two (or more) incommensurate peaks in the power spectrum. Chaotic data exhibit broad power spectrum that indicates some universal features of the transition (e.g., presence of strong sub- and superharmonic components with base frequency of the cycles, ω) [70]. Distinction between the three types of motion based only on the power spectra is quite subjective because of the presence of noise and thus further tests are required [41].

The FFT method can be used for the entire time series data or for a short time window. The latter is used when the frequency of the data is required as a function of time. When applied to short time, windowing techniques have to be applied to avoid spectral leaking. Even with proper windowing, the continuous FFT spectrum has serious fundamental limitations on the temporal and frequency resolution. Wavelet transform provides improved temporal and frequency (timescale) resolution by replacing the harmonic basis set with a mother wavelet, which is a wavelike oscillation whose amplitude starts with zero, increases, and then decreases back to zero (e.g., Mexican hat shape) [71]. With wavelet analysis, instead of power spectrum, the amplitude is plotted as a function of the "timescale" (equivalent to inverse frequency) of the wavelet. Wavelet analysis was successfully applied to pitting corrosion [72] and electrodissolution systems [73, 74] to reveal time-dependent dynamical features of the electrochemical processes.

Several methods have been developed to characterize dynamical features based on time-dependent phase ($\phi(t)$) and amplitude ($A_m(t)$) of the oscillations $x(t)$. These two quantities can be obtained by plotting the data in a 2D state space where at each time the system is represented as a point; the phase and amplitude of the oscillations are the angle and the magnitude of the state vector pointing from the origin to the phase point [75]. The 2D state space is often constructed using the Hilbert transform

$$H(t) = \frac{1}{\pi} PV \int_{-\infty}^{\infty} \frac{x(\tau) - \langle x \rangle}{t - \tau} d\tau, \tag{4.12}$$

where $\langle \rangle$ denotes temporal average and the integral should be evaluated in the sense of Cauchy principal value (PV) [75]. The state space is thus constructed from the $H(t)$ versus $x(t) - \langle x \rangle$ plots and the phase and amplitude are obtained as follows:

$$\phi(t) = \arctan \frac{H(t)}{x(t) - \langle x \rangle}, \quad A_m(t) = \sqrt{H(t)^2 + (x(t) - \langle x \rangle)^2}. \tag{4.13}$$

These quantities are meaningful only when the phase space trajectories have proper rotation around the origin. Characterization of experimental signal with the concept of phase is a powerful technique that has found many applications [75]. For example, the frequency of *any* (including chaotic) oscillator can be obtained from the slope of the phase versus time plot. Other methods also exist that use derivative signals [76, 77], wavelets [78], peak finding [75], recurrence analysis [79], and locking-based measurement [80] for characterization of phase dynamics of nonlinear systems.

The precision of the oscillations can be characterized by the half width of the dominant peak for simple periodic oscillations [75]. For chaotic time series, the precision can be characterized by realizing that the $\phi(t) - 2\pi\omega t$ quantity exhibits random walk; diffusion coefficient for the random walk can be defined and used as a measure of the precision [75].

4.3.2.2 Analysis of Time Series Data with Tools of Nonlinear Dynamics

Analyzing time series data of electrochemical systems (obtained from current, electrode potential, and other measurements) allows the characterization of the dynamics by constructing bifurcation diagrams as system parameters are varied. It is an easy task in case of stationary or simple oscillatory behavior. However, it is becoming increasingly harder for more complex dynamics, such as mixed-mode and quasiperiodic oscillations, and especially for chaos. In order to distinguish and characterize genuine chaos from noise, one should start with *reconstructing the attractor* by applying the embedding technique of Takens described in Section 4.2.2. In most experimental situations, it is enough to construct two-dimensional (Figure 4.3b) [81] or three-dimensional (see Figure 4.5b) [82] portraits of the attractor. The coherent structure of the emerging objects shown in these figures proves the deterministic nature of the chaotic dynamics. In other examples, such as the attractor shown in Figure 4.6b [77], higher dimensional embeddings are required.

A method that is often applied to analyze the chaotic attractor involves the construction of a first recurrence map or *Poincaré map* from the intersection of an n-dimensional attractor with an $(n-1)$-dimensional hyperplane (the *Poincaré plane*) transversal to the flow of the system. The intersecting points constitute the so-called *Poincaré section*. This section will be just one point for a simple limit cycle, a cluster of points for an orbit of higher periodicity, and, seemingly, a continuous curve for a chaotic attractor. The Poincaré map or *next-return map* is a simple plot of the values of two consecutive intersection points on the section. As a consequence, a Poincaré map turns the original continuous dynamical system into a discrete one with one dimension less than that of the original system. As an example, Figure 4.3b [81] shows the next-return map calculated from the time series data of chaotic current oscillations of Cu dissolution in phosphoric acid electrolyte. Note that since the reconstructed attractor is two dimensional, the next-return map is one dimensional. Obviously, it is also possible to plot, for example, the values of every second, third, and so on return to the section. Figure 4.3c [81] shows the second next-return map of the chaotic Cu dissolution. This higher order mapping will be utilized in chaos control (see later).

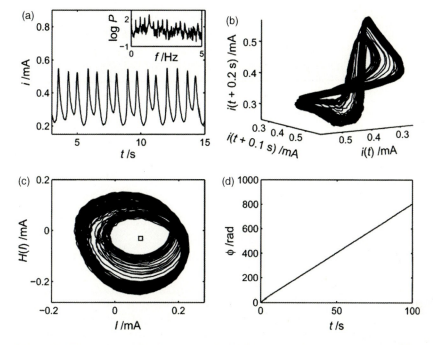

Figure 4.5 Phase-coherent chaotic behavior of Ni electrodissolution in sulfuric acid with 909 Ω external resistance [82]. (a) Time series of current and power spectrum (inset). (b) Reconstructed attractor using time delay coordinates. (c) Phase portrait obtained with Hilbert transform. (d) Phase versus time.

The chaotic attractor is described as a strange attractor when it has noninteger (fractal) dimension. For example, the correlation dimension of the chaotic attractor shown in Figure 4.3b [81] has been found to be 2.25. For details on calculating the correlation dimension of a chaotic attractor, see the seminal paper by Grassberger and Procaccia [83].

Another quantitative measure of the chaotic attractor is the Lyapunov exponent. It measures the rate of separation of infinitesimally close trajectories. It is assumed that two trajectories in the phase space with an initial separation $\varepsilon_i(0)$ will diverge as follows:

$$\varepsilon_i(t) \approx \varepsilon_i(0)\exp(\lambda_i t), \tag{4.14}$$

where λ_i is a Lyapunov exponent. Since the rate of separation can be different for different orientations, there is a series of Lyapunov exponents. At least one positive Lyapunov exponent is a strong indicative of chaos. For details on calculating the Lyapunov exponents, see Ref. [84]. An important application of Lyapunov exponents is to calculate the Kolmogorov entropy of the chaotic attractor [40]. It is perhaps the most accurate measure of the chaotic motion: its value gives the rate by which information about the state of the system is lost in time.

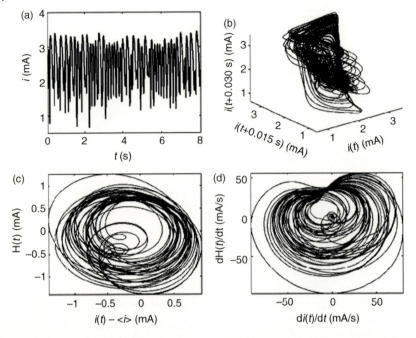

Figure 4.6 Nonphase-coherent chaotic behavior of Fe electrodissolution in sulfuric acid [77]. (a) Time series of the current. (b) Attractor using time delay coordinates. (c) Phase space using the Hilbert transform. (d) Phase space using the derivative coordinates. Black circle at the origin denotes the center of rotation.

There exist many computer software packages that implement the methods described above. Most notably, Nonlinear Time Series Analysis (TISEAN) software provides uniform interface for a comprehensive set of analysis tools [85].

The analysis of pattern formation is a very challenging task. An effective method to analyze stable patterns is the two-dimensional fast Fourier transform [33], which is the series expansion of an image function in terms of "cosine" image (orthonormal) basis functions. The Karhunen–Loève (KL) decomposition (a variant of principal component analysis) [86] has been found to be a powerful tool for analyzing spatiotemporal dynamics in electrochemical systems. The numerical method [87] considers several hundreds of discrete snapshots of a spatially extended system characterized by a scalar variable (e.g., current), depending on time and position, and calculates the eigenfunctions and temporal amplitudes of the principal modes (patterns) constituting the overall dynamics.

4.4
Dynamics

In this section, we shall encounter the most important *experimental* reports on the application of the tools presented earlier for the quantitative characterization of the dynamical instabilities in electrochemical systems. The discussion will start with the simple bistability, continue with the most often observed oscillations, and end with the complex dynamical behavior of coupled electrodes and spatiotemporal pattern formation.

4.4.1
Bistability

One of the simplest nonlinear phenomenon is *bistability* that can be observed under both potentiostatic and galvanostatic control. It is the result of the coexistence of two stable steady states (in most cases, nodes) that are separated in the phase space by a separatrix running through a third steady state, which is a saddle point. Bistability manifests itself experimentally in a hysteresis loop: the stable steady state that is attained by the system depends on the history of how the control parameters are changed. Reviews by Krischer [6, 7] enlist the necessary conditions for the appearance of bistability for all types of NDR systems and define how these conditions could be determined by impedance spectroscopy.

In an NDR system, in which the electrode potential plays the role of a fast activator, bistable behavior may occur under both galvanostatic and potentiostatic control with large ohmic resistance R_Ω. As an example, Figure 4.7 shows experimental bifurcation diagrams of Ni–sulfuric acid electrodissolution at two different external resistance values. The SN symbol denotes the saddle–node bifurcation points (predicted by impedance spectroscopy) that are responsible for the appearance of bistability at large potential values. The larger the series resistance, the larger the range for bistability. Dashed lines show the estimated position of the saddle points assuming a traditional Z-shaped curve representing the bistability. The saddle-type steady states could be stabilized and tracked in a bifurcation diagram by an adaptive control algorithm [88]. Figure 4.7c–m shows the changes in the reconstructed phase space for the Ni–sulfuric acid system at a total resistance of 300 Ω as the circuit potential is varied. To obtain these diagrams, two-dimensional state–space reconstruction was used with $i(t)$ and $i(t − 0.5$ s$)$ values. Traditional bistability (coexistence of two nodes and a saddle) can be observed in panel (i) at $V = 1.774$ V; above this potential, the saddle point slowly approaches the upper steady state (panels (j) and (k)). After the collision of the saddle and the node at $V = 1.841$ V (panel (l)), only the lower steady state can be observed in panel (m). Between the potentials corresponding to panels (f) and (h), one can also observe the coexistence of a stable limit cycle and a stable node resulting in a nontraditional bistability.

Rare appearance of *tristability* due to the existence of two NDR regions in a polarization curve has also been observed, for example, during methanol oxidation on Pt [89] and for the Ni(II)–N_3^- electroreduction at streaming Hg electrode [19]. Tristability can be explained by the coexistence of five different steady states in the phase space, for example, three stable nodes and two saddle points.

In an S-NDR system, in which the electrode potential plays the role of a slow inhibitor, the polarization curve is S-shaped at vanishing ohmic resistance R_Ω giving rise to bistability. As an example, Figure 4.8a shows an experimentally determined polarization curve of Zn deposition on a rotating Zn disk electrode.

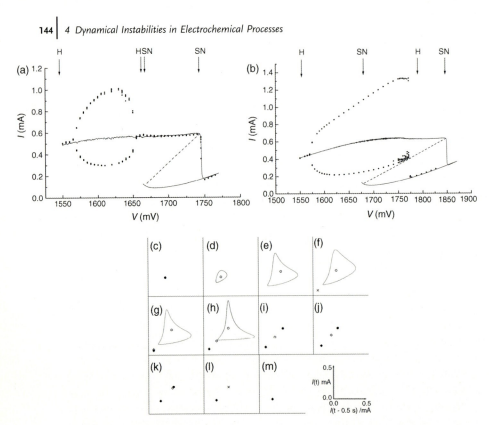

Figure 4.7 Experimental bifurcation diagrams of the electrodissolution of a Ni wire (diameter 1 mm) in sulfuric acid solution at 10 °C at total series resistance of (a) 150 Ω and (b) 300 Ω. Symbols H and SN denote Hopf and saddle–node bifurcations. (c–m) Reconstructed phase space of the Ni–sulfuric acid system at a total series resistance of 300 Ω obtained at various circuit potentials: (c) 1.565 V, (d) 1.575 V, (e) 1.660 V, (f) 1.676 V, (g) 1.680 V, (h) 1.746 V, (i) 1.774 V, (j) 1.788, (k) 1.841 V, (l) 1.848 V, and (m) 1.856 V. Solid (open) circles represent the upper stable (unstable) steady states, open and solid diamonds represent the saddle and lower stable points, respectively, and × denotes the saddle–node points. The solid curves correspond to limit cycle oscillations [74].

4.4.2
Oscillations

Oscillations in electrochemical systems are perhaps the oldest and most studied nonlinear phenomena. Almost two centuries ago, Fechner [90] reported the first observation of current oscillations during electrodissolution of an iron wire in nitric acid. Since then, the experimental evidence on oscillations in different processes (electrochemical reduction at mercury electrode, metal electrodissolution, electrocatalytic oxidation at metal surfaces, reduction at semiconductor electrodes, etc.) has been "exponentially" accumulating. A comprehensive review on the several hundreds of systems showing oscillatory dynamics has been given by Hudson and Tsotsis [4] that covers almost all relevant literature up to 1993. Reviews by Koper [5]

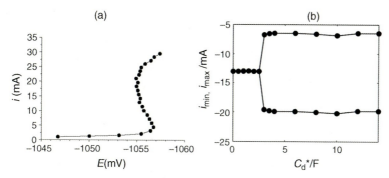

Figure 4.8 Bistability and oscillations in S-NDR Zn electrodeposition system. (a) IR-compensated ($R_s = 3.0\ \Omega$) S-shaped polarization curve of Zn electrodeposition. Experimental conditions include 7 mm diameter Zn disk electrode rotated at 1000 rpm in 0.72 mol dm^{-3} ZnCl$_2$ + 2.67 mol dm^{-3} NH$_4$Cl buffer (pH = 5.2) at 26 °C, scan rate of -4.0 mV s^{-1} (Z. Kazsu, 2002, unpublished results). (b) One-parameter bifurcation diagram of Zn electrodeposition at $V_0 = -1100$ mV and $R_\Omega = 9.7\ \Omega$ showing the minima and maxima of current oscillations as a function of the pseudocapacitance C_d^* [91].

and Krischer [6, 16] give an account of later findings and also provide a systematic way to classify electrochemical oscillators as detailed earlier.

In this review, we shall focus mainly on current oscillations, which may appear during polarization scans (Figure 4.9a), or at different fixed potentials resulting in different waveforms, for example, nearly sinusoidal or relaxational, as shown in Figure 4.9b and c, respectively. In an N-NDR system, current oscillations appear around a branch of *negative* slope under potentiostatic conditions but for intermediate values of ohmic resistance R_Ω. (These systems show only bistability under galvanostatic control.) In an HN-NDR system, current oscillations occur around a branch of *positive slope* if the ohmic series resistance R_Ω is large enough. (Under galvanostatic control, potential oscillations may also occur.) In an S-NDR system, current oscillations may develop under potentiostatic control, but because of extreme parametric conditions – very large values of specific double-layer capacitance C_d (F cm^{-2}) – the polarization curve could exhibit oscillations only with very slow reactions. With the use of *differential controller* [91], a specific pseudocapacitance C_d^* can be introduced into the system, resulting in the appearance of current oscillations through a Hopf bifurcation in Zn electrodissolution (Figure 4.8b). For the experimental implementation of the controller, the circuit potential V is varied around V_0 according to the following equation:

$$V = V_0 + \gamma \frac{dV}{dt} - \gamma R_\Omega \frac{di}{dt}, \qquad (4.15)$$

where γ is the control gain. The proposed differential controller can also be applied to experimentally identify essential dynamical variables in oscillatory systems and to explore their role in the feedback loops. For electrochemical systems, the controller allows the identification of the type of electrochemical oscillations based on whether the added pseudocapacitance induces or suppresses current oscillations.

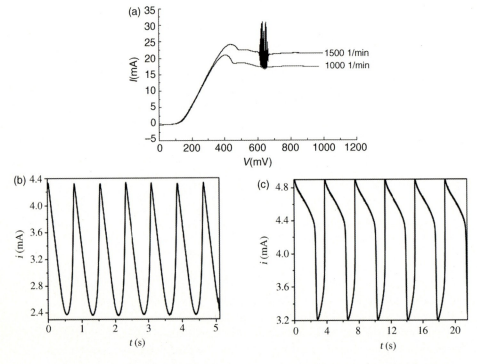

Figure 4.9 Oscillations in N-NDR Cu electrodissolution system. (a) Experimental anodic polarization curves (scan rate 10 mV s^{-1}) of a rotating copper disk electrode of 7 mm diameter in o-phosphoric acid at different rotation rates. For experimental conditions, see Ref. [93]. (b) Nearly sinusoidal-type current oscillations during electrodissolution of a rotating copper disk electrode (5 mm diameter) at $V = 67$ mV, $R_s = 85\,\Omega$, rotation rate 1500 rpm, and temperature $-5\,°C$. (c) Relaxation-type current oscillations by using the same cell as in (b), but $V = 590$ mV, $R_s = 130\,\Omega$, rotation rate 1000 rpm, and temperature $5\,°C$. In both experiments (b) and (c), a Hg/Hg$_2$SO$_4$ reference electrode has been used (T. Nagy, 2009, unpublished results).

As we have pointed out in Section 4.1, oscillation is related to the appearance of a limit cycle in the phase space of the dynamical variables. In Figure 4.7c–h, we can follow the birth and death of a limit cycle in the Ni–sulfuric acid electrodissolution system. With increasing potential, the upper steady state (c) loses its stability (Hopf bifurcation) and a stable limit cycle appears around the unstable focus (d); the size of the limit cycle increases with increasing potential (e–g) until the limit cycle collides with the saddle point (h) leading to the death of the limit cycle by further increasing the potential (i). This bifurcation is called *saddle–loop bifurcation* resulting in a so-called *homoclinic orbit*. Since the closed loop trajectory starts and ends on the same saddle point, it is a periodic orbit of infinite period.

Appearance of current oscillations depends on many features of the electrochemical systems. Chemical properties include the type of electrochemical reaction, the electrode material, the composition of the electrolyte, and so on, while physical properties include the solution resistance, the cell constant, the electrode size, the rotation rate, the external resistance, and so on. In Figure 4.9a, one can observe, for example, how a simple increase of the rotation rate of a disk electrode may lead to the appearance of current oscillations during polarization scan. Generally, most of the constraints are fixed, and the behavior is mapped in the parameter space of external resistance versus the electrode potential (current) under potentiostatic (galvanostatic) conditions. These *traditional phase diagrams* show the ranges of bistability and oscillations in connection with SN, SL, and Hopf bifurcations, respectively, at the given conditions.

To determine these bifurcations, impedance spectroscopy has been found to be a convenient tool. As an example, Figure 4.10a shows the impedance spectra of the Cu–sodium acetate–glacial acid (SAGA) oscillator measured at different electrode potentials under potentiostatic control [92]. The plots start close to the origin of the complex plane since all measurements were performed with full IR compensation

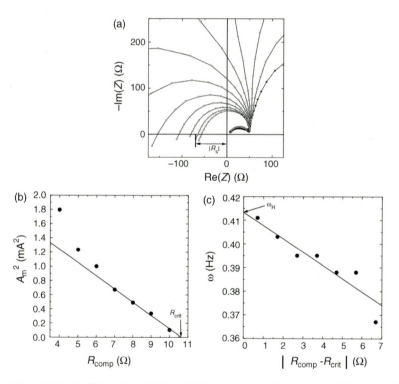

Figure 4.10 Oscillations in N-NDR Cu–SAGA system: Hopf bifurcation and impedance analysis. (a) Impedance spectra of the Cu–SAGA system at different electrode potentials. The first spectrum (solid circles) has been measured at 280.0 mV by gradually decreasing the frequency from 17.0 kHz (8 points per decade). Other spectra (open circles) have been determined at $280.0 + n \times 10.0$ mV true anodic potentials ($n = 1, 2, \ldots, 11$). (b) Square of the amplitude of current oscillations as a function of the compensated resistance $R_{comp} < R_s = 70.0\ \Omega$; $R_{crit} = 10.7 \pm 0.8\ \Omega$. (c) Frequency of oscillations as a function of $|R_{comp} - R_{crit}|$; $\omega_H = 0.413 \pm 0.003$ Hz [92].

($R_{comp} = 70.0\,\Omega$). By increasing the potential, the spectra would cross the negative part of the real axis at finite frequencies. Since turning off the IR compensation results in only a horizontal shift of the impedance plot (in this case, 70 Ω to the right), one can conclude that a Hopf bifurcation would occur in the uncompensated system at a true potential value between 370 and 380 mV resulting in oscillations. According to the theory, during Hopf bifurcation, the oscillatory amplitude scales with the square root of the bifurcation parameter and the frequency decreases to a minimal, nonzero value. These relationships are shown, for example, in Figure 4.10b and c, respectively, for a Hopf bifurcation taking place during Cu electrodissolution with a critical IR compensation and resulting in the loss of oscillations.

Impedance studies of electrochemical oscillators led to the application of cell geometry-independent phase diagrams to define the oscillatory regions [92]. These *nontraditional phase diagrams* are constructed from the reciprocal value of the uncompensated series resistance at the Hopf point R_H and the "true" electrode potential. Figure 4.11a shows an experimentally determined diagram for the Cu–SAGA system. The horizontal line corresponds to an arbitrary chosen value of the total series resistance R_s of the circuit. Its crossing with the bell-shaped curve gives the potential range for oscillations in the given system. Recently, a *scaling relationship* was derived for such nontraditional phase diagrams as follows:

$$R_H A - C_r(e_{ss}) = \frac{D_r(e_{ss})}{d_{rr}^{0.5}}, \tag{4.16}$$

where R_H is the solution resistance at a given Hopf point, A is the surface area of the electrode, and C_r and D_r are parameters that should be determined from the bell-shaped diagrams measured at different rotation rates (d_{rr}) (Figure 4.11b). It was shown that all scaled data points characterizing the onset of oscillations should fall – independent of the size of the electrode and the rotation rate – on a single plot (Figure 4.11c) [93]. Scaling relations like this could allow the "engineering" of the dynamical behavior of oscillating electrochemical systems.

Oscillations are typically characterized by waveform and frequency that are expected to have complicated nonlinear dependence on experimental parameters such as concentrations, resistance, rate constants, temperature, rotation rate, surface area, and so on. Electrochemical systems exhibit a wide range of waveforms such as smooth (nearly sinusoidal) (Figure 4.9b) or relaxation type (Figure 4.9c), periodic, quasiperiodic, and chaotic. In general, it is very difficult to predict these characteristics, for example, the frequency. However, according to the principle of critical simplification, around bifurcation points the underlying mathematical structure simplifies and often relatively simple formulas can be obtained [94]. Such approximate formula has been derived and *experimentally verified* for the frequency dependence of NDR-type oscillators [95]. Two crucial timescales affect the frequency of an electrochemical oscillator: the electrical timescale τ_e that is related to the charging of the double layer ($\tau_e = C_d R A$) and the chemical timescale τ_c that is related to the rate of the electrochemical reactions ($\tau_c = a_t / 2k_r(e^*)$). In these definitions, a_t is the thickness of the Nernst diffusion layer, $k_r(e^*)$ is the potential-dependent (pseudo)

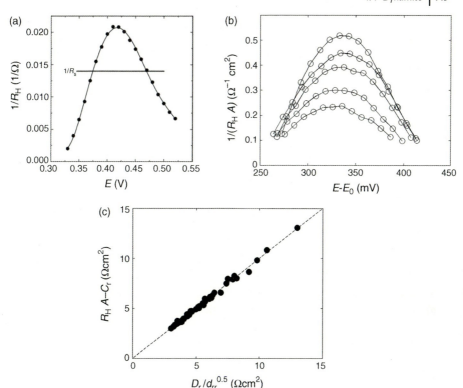

Figure 4.11 Nontraditional phase diagram for oscillatory N-NDR system and their rotation rate dependence. (a) Nontraditional phase diagram of the Cu–SAGA system. The plot shows the reciprocal value of the minimal uncompensated series resistance R_H as a function of the true electrode potential E. The horizontal line corresponds to the reciprocal series resistance R_s of the applied electrochemical cell [92]. (b) Nontraditional phase diagram of the Cu–phosphoric acid system as a function of rotation rate from top to bottom: 3000, 2500, 2000, 1500, and 1000 rpm, while the diameter of the disk electrode is kept constant (3 mm). (c) Scaling relationship $(R_H A - C_r)$ versus $D_r/d_{rr}^{0.5}$ for all bifurcation data shown in (b) [93].

first-order rate constant of the reaction, and e^* is the electrode potential at the Hopf point. The frequency of the oscillations can be expressed [95] with a combination of the two frequencies (inverse timescales):

$$\omega^* \approx \sqrt{\omega_c \omega_e} = \sqrt{\frac{2k_r(e^*)}{a_t R C_d A}}. \tag{4.17}$$

The equation also interprets the often observed Arrhenius-type temperature dependence of the frequency under conditions close to Hopf bifurcation point [95]. (There are counterexamples: formic acid oxidation exhibits temperature overcompensation; decrease of frequency with temperature [96].) The experimental validity of the frequency equation indicates that "apparent" rate constants can be extracted from frequency measurements of electrochemical oscillations, which can aid further modeling and engineering of complex responses of electrochemical cells.

4.4.3
Chaos

Chaotic oscillations have been reported in a relatively large number of electrochemical systems [3, 4]. Here, we focus on how the methodologies in Section 4.3 can be applied to experimental systems in some of the best documented accounts of electrochemical chaos.

The complex oscillatory feature of the anodic electrodissolution of copper in phosphoric acid was investigated in great detail [97]; the complex oscillatory behavior was carefully mapped as function of the rotation rate of the disk electrode, circuit potential, and external resistance [81, 97, 98]. A bifurcation diagram showing the period-doubling route to chaos [81] is shown in Figure 4.3a. The chaotic behavior spans a relatively small potential range of about 3 mV. The chaotic attractor shown in a 2D state space with delay coordinates (Figure 4.3b) has low-dimensional character ($D_2 = 2.25 \pm 0.1$) and the corresponding next-return map using the diagonal Poincaré section is nearly one dimensional.

Koper and Gaspard proposed [27] an interesting model for interpretation of electrochemical oscillations. Because a chaotic ordinary differential equation system needs at least three variables, they augmented the standard two-variable (electrode potential and near-surface concentration) model with a third variable representing the concentration of the electroactive species in a second diffusion layer. It was shown that chaotic behavior occurs because of the slow perturbation of the oscillatory system by a third variable due to the slower diffusion. It is quite probable that the large number of electrochemical chaotic oscillations and their similarities are caused by some common mechanism. A candidate for the mechanism is this mass transfer-induced chaos in the Koper–Gaspard model.

Ni electrodissolution in sulfuric acid also exhibits chaotic behavior under both galvanostatic [31, 99] and potentiostatic conditions [100, 101]. The chaotic current oscillations (see Figure 4.5a) develop through period-doubling bifurcation under potentiostatic condition; the information dimension of the reconstructed attractor in Figure 4.5b was found to be 2.3 [102]. The phase of the chaotic Ni dissolution system was analyzed in great detail [101, 103]. The Hilbert transform method was capable of providing a 2D state space shown in Figure 4.5c where the oscillations exhibited a unique center of rotation; the angle to the phase points provides phase as a function of time (Figure 4.5d). The slope of the line fitted to the phase versus time curve gives the (angular) frequency of the oscillations. Phase analysis provided an interesting duality [104] of stochastic and deterministic features of the phase-coherent chaotic behavior. It was shown that as a result of the period-doubling route to chaotic behavior, short-period oscillations are often followed by long-period oscillations for about 80% of the cycles. Because of this regularity of the phase dynamics, the frequency of the oscillations can be obtained from relatively short time series.

However, the long-term behavior of the deterministic phase dynamics of the chaotic oscillator can be considered as stochastic [75]; the $\varphi - \omega t$ quantity exhibits random walk for which diffusion coefficient can be defined. The precision of the chaotic oscillations, measured by the phase diffusion coefficient, deteriorates with increase in temperature; for example, the chaotic behavior at 35 °C is nonphase coherent [105]. In contrast to the Cu dissolution system, the Ni dissolution is under kinetic control, and thus it does not require well-defined mass transfer; because of the kinetic control, the Ni dissolution system is an excellent choice for studying pattern formation with single electrodes and electrode arrays [66, 67].

The chaotic iron electrodissolution in sulfuric acid exhibits higher dynamical complexity than that observed for Cu and Ni dissolution [3, 106, 107]. The dynamic behavior strongly depends on the surface area of the iron rotating disk electrode. With increasing electrode size, transitions were observed from periodic oscillations to low-order chaos to higher order chaos [107]. This higher order chaos exhibits strong nonphase-coherent character, as shown in Figure 4.6 [77]. The Hilbert transform does not produce unique center of rotation; instead, a method based on derivative Hilbert transform [76, 77] can be used for phase definition.

Some other well-studied chaotic systems include the reduction of In^{3+} in the presence of SCN^- ions [108] where low-order chaos similar to that observed with Cu dissolution in phosphoric acid can be obtained; H_2 oxidation in the presence of $CuCl_2$, where interior crisis (collision of a "small" chaotic attractor with a stable manifold of a saddle-type limit set) was observed [109]; and Cu dissolution in acidic chloride-containing media where Shilnikov chaos was shown to occur [110]. It is interesting to point out that the addition of halides to simple oscillatory systems often results in increase in complexity of dynamics: addition of Br^- and Cl^- ions in H_2O_2 reduction [111] and formic acid oxidation [112], respectively, induced chaotic behavior.

4.4.4
Bursting

Bursting oscillations are characterized by a complex waveform in which slow, "silent" dynamics are interrupted by fast, repetitive spiking [113]. A mathematical description and a formal classification of bursting dynamics in neurons were proposed by Rinzel [113] and later extended by Izhikevich [114]. The essential variables describing the bursting dynamics can be divided into slow and fast variables depending on the timescales over which they vary. The fast subsystem, described by a minimum of two variables, is responsible for the fast spiking. The slow subsystem is responsible for the slow modulation causing the periodic appearance of the fast spiking behavior.

Although bursting was originally observed and studied in biology (e.g., "parabolic" bursting of neuronal pacemakers [115], "square wave" bursting of insulin secreting pancreatic β-cells [116], or the "elliptic" bursting of rodent trigeminal interneurons [117]), it also occurs in electrochemical systems. Electrochemical bursting oscillations have been reported during H_2O_2 reduction on platinum [118, 119], iron dissolution in sulfuric acid with halogen additives [120, 121] and with dichromate ions coupled with graphite or zinc electrodes [122]. A possible mechanism of bursting in electrochemical systems has been proposed based on a polarization scan with two negative slopes and a (hypothetical) slow variable [5]. Chaotic [74] and periodic [123] bursting can also occur in anodic iron electrodissolution (see Figure 4.12). The relatively slow (1–10 Hz) oscillations alternate with very fast (about 100 Hz) spiking. The separation of the timescales can be visualized with wavelet transform. The time-resolved spectrum shows how the instantaneous amplitudes of

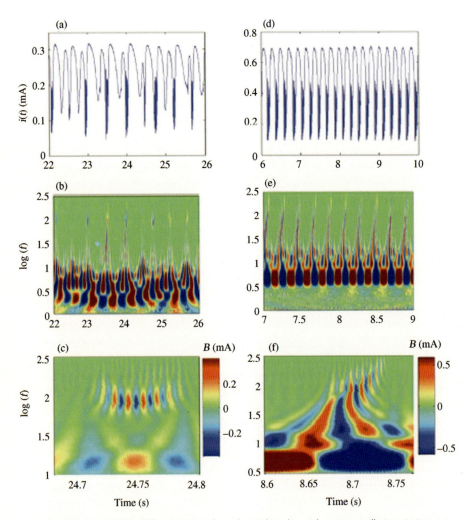

Figure 4.12 Irregular (left column) and regular (right column) bursting oscillations in iron electrodissolution in sulfuric acid [74]. (a and d) Current versus time. (b-c and e-f) Wavelet transform of current in a color-coded plot (color denotes amplitudes (B) of the wavelets).

the slow and fast frequency components alternate with time. The periodic bursting oscillations were successfully modeled with the standard iron electrodissolution model ("fast" subsystem with essential variables: electrode potential and surface H^+ and Fe^{2+} concentrations) augmented with a slow fourth variable of surface film thickness that affects the series resistance of the cell [123].

Bursting oscillations in iron electrodissolution can also be observed in the active–passive oscillatory region by addition of halides: elliptic and square wave types of bursting oscillations were observed [124].

4.4.5
Dynamics of Coupled Electrodes

The use of more than one working electrode in an electrochemical cell allows the investigation of effects of interactions on dynamical features of the reactions. Two types of interactions should be considered:

- Electrical interactions due to the potential drop in the electrolyte
- Chemical interactions due to the diffusion of chemical species.

In many reactions, electrical interactions dominate over chemical interactions [6] because potential changes of one electrode practically instantaneously affect the potential of other electrodes (i.e., the electrical coupling is fast compared to the slow diffusion) and potential drops in the electrolyte are very difficult to avoid. Cell geometry (placement and size of working, reference, and counter electrodes) greatly affects the potential drops in the cell, both the local dynamics through changes in solution resistance (or electrode composition in case of solid electrolytes) and the strength/length scales of global/long-range electric coupling [6].

Consider first a dual (working) electrode setup with a liquid electrolyte. The relative effects of chemical and electric coupling were explored by cell designs where the position of the electrodes could be changed and the working electrodes could be isolated from chemical interactions (e.g., with a graphite plate). In a typical macrocell, the working-to-reference electrode distance was found to be a dominating coupling factor. With increasing distance, the large potential drop on the series resistance of the cell induces a strong electrical coupling between the working electrodes. The variation in electric coupling strength with working-to-reference electrode placement was interpreted theoretically [6, 125] and demonstrated in synchronization of dual-electrode H_2O_2 reduction [126] and active–passive iron oscillations [127] in macrocells. This effect can also be observed in microfluidic flow cells where the working-to-reference electrode distance is often large [69]. Figure 4.13 shows that as the reference-to-working electrode distance is increased, the drifting oscillations (Figure 4.13a) become synchronized (Figure 4.13b).

The effect of reference electrode position on dynamics in dual-electrode setups is difficult to interpret because both the local dynamics (e.g., frequency and waveform of oscillations) and the coupling strength are affected through changes of series resistance. The effect of electric coupling can be studied by applying a combination of

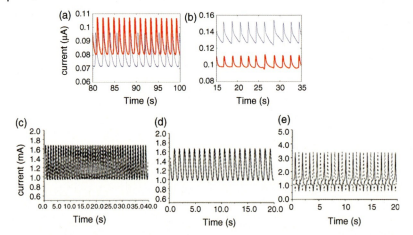

Figure 4.13 Effect of global coupling in dual-electrode setup on synchronization. *Top row*: Effect of working-to-reference electrode placement in oscillatory dual-electrode formic acid oxidation on Pt [69]. (a) Desynchronized current oscillation with near (2.4 mm) working-to-reference electrode placements. (b) Synchronized current oscillations with far (11 mm) working-to-reference electrode placements. *Bottom row*: Effect of collective (series) resistance in oscillatory dual-electrode Ni electrodissolution in sulfuric acid [66]. (c) Desynchronized current oscillations without added electric coupling with smooth oscillations ($K_{gs}=0$). (d) In-phase synchronized current oscillations with added electric coupling with smooth oscillations ($K_{gs}=0.03$). (e) Antiphase synchronized current oscillations with added electric coupling with relaxation oscillations ($K_{gs}=1.50$).

parallel (R_p) and series (R_s) resistance (Figure 4.4a). The series resistance induces global coupling [66] whose strength can be quantified by a global coupling strength K_{gs}:

$$K_{gs} = \frac{N_{el} R_s}{R_p}, \qquad (4.18)$$

where N_{el} is the number of electrodes. The experiments can be carried out by varying K_{gs} and keeping the total cell resistance ($R_{tot} = R_p + N_{el} R_s$) constant.

The effects of electrical coupling strength on the dynamical behavior have been investigated for small (2–4) sets [66, 102, 103, 128] and large populations (64) of electrodes [67, 100, 129–132] in Ni electrodissolution in sulfuric acid for both periodic and chaotic oscillators.

Without added coupling ($K_{gs} = 0$) in a dual-electrode ($N_{el} = 2$) setup, the oscillators interact only very weakly through the electrolyte; because there is a small difference between the two electrodes (e.g., surface area and conditions), the individual oscillators exhibit periodic oscillations with slightly different (e.g., 1%) frequencies [66]. Figure 4.13c shows the current oscillations just above Hopf bifurcation point where the waveform is close to harmonic (smooth oscillators).

A small amount of coupling is capable of bringing the frequencies of the two oscillators together (phase or frequency synchronization) [66, 128]; Figure 4.13d shows the in-phase synchronized behavior of the two local oscillators. Synchronization theories predict dephasing behavior where the synchronization could occur in antiphase configuration [133]; with relaxation oscillators (close to a homoclinic bifurcation), whose waveform is more similar to spikes, antiphase synchrony was achieved in the experiments (see Figure 4.13e) [66, 128].

A population of oscillators can exhibit collective dynamical features. In the 1960s and 1970s, Winfree and Kuramoto predicted that in a large population of oscillators, the transition to synchrony will take place through a second-order phase transition: there exists a critical coupling strength (K_c) below which the population will be fully desynchronized [134, 135]. Above the coupling strength, the synchrony will quickly increase by forming a group of synchronized oscillators; the number of elements in the synchronized group is expected to increase with increase in K_{gs}. The Kuramoto transition was experimentally confirmed with an array of 64 Ni electrodes (8 × 8 configuration) [67]. Figure 4.14 shows that as the coupling is increased, the second-order phase transition takes place such that the Kuramoto order parameter r_0 (related to the amplitude of the mean current oscillations) starts to increase above K_c. With strongly relaxation oscillators instead of single synchronized group, two or three synchronized groups are observed that have constant phase differences [66, 128].

The effect of electric coupling on chaotic dynamics is more complicated. With two electrodes without any added coupling ($K_{gs} = 0$) (Figure 4.15a), the phase analysis

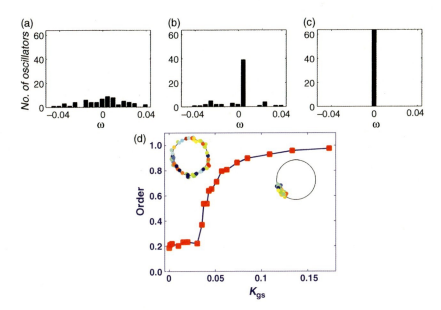

Figure 4.14 Emergence of coherence (Kuramoto transition) in oscillatory Ni electrodissolution with an 8 × 8 electrode array [67, 223]. (a) Frequency distribution without added electric coupling ($K_{gs} = 0$). (b) Frequency distribution just above the phase transition point ($K_{gs} = 0.29$). (c) Frequency distribution at strong electric coupling strength ($K_{gs} = 0.52$). (d) Kuramoto order parameter as a function of coupling strength. Insets illustrate behavior in 2D state space below and above the phase transition. The frequencies in panels (a)–(c) are renormalized with the mean value of the natural frequency distribution.

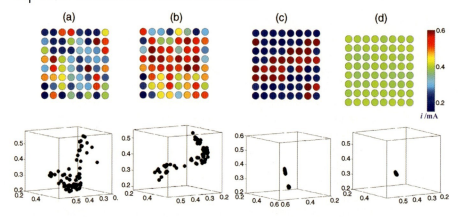

Figure 4.15 Synchronization of chaotic oscillations in chaotic Ni electrodissolution in sulfuric acid on an 8 × 8 electrode array [82, 100]. *Top*: Snapshots of currents on the electrodes. *Bottom*: Snapshot of position in state space constructed from the currents (mA) of all 64 electrodes. (a) Desynchronized chaos, $K_{gs}=0$. (b) Phase-synchronized chaos with weak added electrical coupling, $K_{gs}=0.12$. (c) Dynamical differentiation (clustering) with moderate electrical coupling, $K_{gs}=3$. (d) Identical synchronization with strong added electrical coupling, $K_{gs}=\infty$.

reveals that similar to that observed with periodic oscillators the frequencies are slightly different [103]. With very weak added coupling, phase synchronization sets in where the frequencies become identical but the amplitudes are not correlated (Figure 4.15b). As shown in Figure 4.15d, at very strong coupling, the currents of the electrodes exhibit identical variation (similar to that of a single electrode) and thus identical synchronization sets in.

In a population of chaotic oscillators, in addition to the desynchronized, phase synchronized, and identically synchronized states, there exists a coupling strength region below identical synchronization where chaotic clustering (or dynamical differentiation) takes place [100]. The array splits into groups; the elements in each group have identical dynamics different from that of the other group. Two clusters with a large number of possible cluster configurations have been observed with chaotic Ni dissolution. A representative cluster configuration is shown in Figure 4.15c. At coupling strengths slightly weaker and stronger than that required for clustering dynamics, "itinerant clustering" was observed [132]. The cluster configurations varied with time: spontaneous changes in the number of clusters and their configurations were detected.

In iron electrodissolution in sulfuric acid, because of the large current density there is a strong coupling between the iron wires even without added external resistance. In-phase, out-of-phase, and drifting synchrony were observed in a two-electrode setup [68, 127]. In 1D ring and 2D arrays, traveling waves develop in both mass transfer and active–passive oscillatory regions [64, 65].

Stationary (nonoscillatory) patterns can also be studied with electrode arrays. The CO electrooxidation S-NDR system exhibits chemical autocatalysis [136] in which electrical (inhibitor) coupling can induce pattern formation in dual-electrode setup.

One electrode without added external resistance exhibits the S-NDR bistability [136]. Dual-electrode experiments [137] with individual resistances eliminate bistability (Figure 4.16a) as expected because of the large IR drop in the cell. However, when collective resistance is applied, nonuniform currents can be recorded, as shown in Figure 4.16b, with one electrode having larger and the other electrode having lower currents than those observed without electrical coupling. Pattern formation induced by electrical coupling makes catalysis design very difficult because overall performance of the catalysis depends on the interactions and inherent catalytic property of the electrode.

In numerical simulations of a globally coupled population of electrochemical S-NDR oscillators, several different scenarios for complete synchronization, partial synchronization, and dynamical cluster formation have been observed [138].

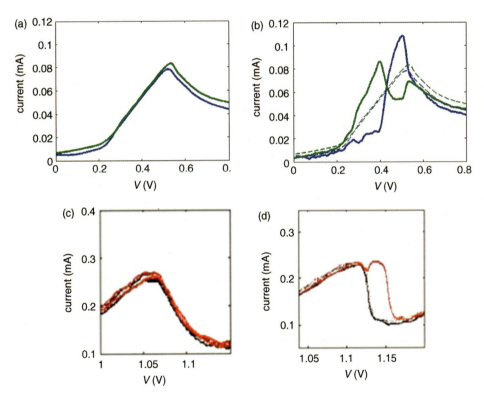

Figure 4.16 Stationary pattern formation via symmetry breaking in S-NDR (top, with positive coupling) and N-NDR (bottom, with negative coupling) systems. *Top*: CO electrooxidation in sulfuric acid/saturated Na_2SO_4 on Pt in a dual-electrode setup [137]. (a) Polarization curve with 4 kΩ individual resistors ($K_{gs}=0$). (b) Asymmetric current states with added positive coupling ($K_{gs}=\infty$). *Bottom*: Ni electrodissolution in sulfuric acid in dual-electrode setup [151]. (c) Polarization curve without added resistance exhibiting N-NDR. (d) Asymmetric current states with individual resistors (602 Ω) compensated fully by IR compensation.

4.4.6
Dynamics of Pattern Formation

Pattern formation in electrochemical system is dominated by the electric coupling of migration currents [6]. In contrast to patterns in reaction diffusion systems where the coupling is mainly local, in electrochemical systems the coupling was found to consist of positive long-range coupling superimposed by global coupling [6, 125]. Distant and near reference electrode placements impose positive or negative global coupling, respectively.

If a bistable system is in a passive state and a small disturbance is imposed on the electrode surface, an activation wave will sweep through the surface. When the reference electrode is far from the working electrode, accelerating fronts [53, 139] are obtained, as illustrated in Figure 4.17a. With negative coupling (e.g., close, pointwise reference electrode) [140], the wave could be initiated at a remote location.

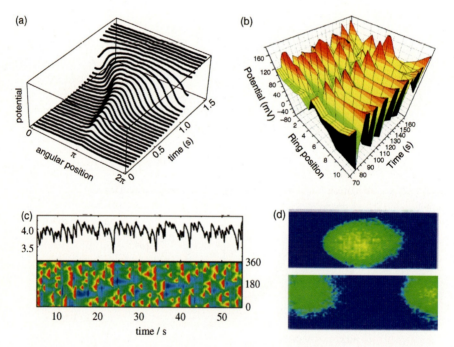

Figure 4.17 Representative patterns in electrochemical systems. (a) Accelerating potential front during the reduction of $S_2O_8^{2-}$ on a silver ring electrode [53]. (b) Standing wave potential oscillations in the electrocatalytic oxidation of formic acid on Pt ring electrode [142]. (c) Spatiotemporal turbulence in H_2 oxidation reaction on Pt in the presence of Cu^{2+} and Cl^- ions. *Top panel*: Total current. *Bottom panel*: Space–time plot of interfacial potential [56]. (d) Stationary potential (and adsorbate) patterns during the periodate reduction on Au(111) electrodes in the presence of camphor adsorbate (blue: camphor free area; orange and yellow: high camphor coverage) [59]. Panels (a), (c), and (d) are by courtesy of Katharina Krischer. Panel (b) is by courtesy of Peter Strasser.

Remote triggering and accelerated fronts occur in electrochemical systems naturally; such patterns in reaction–diffusion systems are rare.

In oscillatory systems, a wide variety of patterns have been observed [6]. In addition to clustering (similar to those observed on electrode arrays) [55], standing waves [54], asymmetric target patterns [141], and fully developed turbulence (chaos in both space and time, as shown in Figure 4.17c) [56] can be observed in H_2 oxidation reaction on Pt ring electrode with poison (Cu^{2+} and Cl^- ions). Standing waves (Figure 4.17b), rotating pulses, and in-phase active or passive oscillations were also observed in formic acid oxidation on Pt ring electrode [50, 142, 143]. Active–passive oscillations of iron electrodissolution on a ring electrode exhibit cluster formation where the ring splits into two equally sized region oscillating in antiphase [144]. Galvanostatic conditions impose very strong positive coupling [145] that is expected to destroy nonuniform patterns; nonetheless, patterns are still possible because of the complicated phase and amplitude clusters; Ni electrodissolution exhibited standing wave oscillations under galvanostatic conditions [51].

Two distinct types of stationary patterns have been observed in cell geometries favoring negative and positive coupling. With negative coupling, stationary patterns develop where propagation of the activation front is inhibited because of the presence of negative coupling; stationary domains were found in $S_2O_8^{2-}$ reduction on Ag electrode [146] and H_2 oxidation on Pt in sulfuric acid [147]. These types of patterns have wavelengths that strongly depend on the size of the electrode. With strong positive coupling of S-NDR systems, stationary Turing patterns are possible [148]. Turing patterns often occur in the activator–inhibitor system where the inhibitor is strongly coupled; in S-NDR systems, the electrode potential being the inhibitor can be strongly (positively) coupled in usual cell geometry with far reference-to-working electrode placement. Turing patterns were experimentally confirmed in periodate reduction reaction on Au(111) in the presence of camphor adsorbate (Figure 4.17d) [59] and in CO electrooxidation [57, 62]. In agreement with the theory on Turing structures, the wavelength of the patterns does not depend on the size of the electrode.

Electrodeposition reactions often exhibit spiral and target patterns that are very common in homogeneous reaction diffusion systems and heterogeneous catalysis. For example, in codeposition of silver and antimony, spiral structure was observed that consisted of concentration patterns of silver/antimony ratio [149].

4.5
Control of Dynamics

Control of critical behavior (most importantly, bistability, oscillations, and pattern formation) by perturbations of a system parameter serves two purposes:

- In classical kinetic studies, such behavior is usually avoided because extraction of qualitative and quantitative information about mechanism can be quite complicated. The goal of the control is to eliminate the critical behavior and allow the application of standard methods.

- In studies with emphasis on nonlinear science control perturbation adds to the complexity of the system. Therefore, the existing critical behavior can be modified or new types of self-organized phenomena can be observed in the "controlled" system. Control can also stabilize unstable dynamical behavior (steady states and oscillations) that can provide valuable information about the system dynamics and bifurcations.

The control can be open loop (when a predefined waveform is superimposed on a system parameter) or closed loop (when the perturbation is determined as a function of essential variables of the system). We give a few examples of the effects of open- and closed-loop control action applied to electrochemical systems.

4.5.1
IR Compensation

N-NDR systems do not exhibit oscillations/bistability with low value of the series resistance resulting in small IR drop in the cell. The goal of IR compensation is to diminish the series resistance of a cell by a feedback circuit that basically simulates a "negative" resistance attached in series with the working electrode [29]. By selecting proper (negative) resistance value for the IR compensation, bistability and oscillations have been suppressed in many examples [92].

However, recent investigations pointed out a major weakness of using IR compensation for kinetic studies: IR compensation can introduce negative coupling of the electrode potential variable in the spatially extended system (e.g., among reacting sites) [150] or in an electrode array among the electrodes [151]. This negative global coupling is the direct consequence of the negative series (collective) resistance implemented by the IR compensation. Krischer *et al.* have shown [150] that the negative global coupling-induced dynamical behavior through IR compensation is similar to those observed with close reference-to-working electrode placements (e.g., stationary domains, clustering). Similarly, in dual-electrode arrays, IR compensation (applied to both electrodes based on total current) was shown to induce symmetry breaking bifurcation in Ni electrodissolution (Figure 4.16c and d) and antiphase oscillations and multiple antisymmetric (active/passive) steady states in Fe electrodissolution [151].

These studies suggest that the application of IR compensation should be made with great care because of the induced pattern formation.

4.5.2
Periodic Forcing

In classical impedance spectroscopy [29], periodic forcing of the circuit potential is applied to an electrochemical cell exhibiting a stable stationary solution. The description of the effect of a periodic forcing signal to an oscillatory system requires application of theoretical tools of nonlinear science [75]. The periodic forcing signal (of frequency ω_f) above a critical forcing amplitude can entrain the oscillatory system: the natural frequency of the system (ω_0) becomes adjusted to the forcing frequency. The phase difference between the forcing signal and the system variable is a complicated function of the forcing amplitude but often in the range of 0–$\pi/2$ rad. In the forcing amplitude versus forcing frequency phase diagram, the entrained states form long vertical lines at resonant frequencies ($k\omega_f = m\omega_0$, where k and m are integers) called Arnold tongues. In addition, in many periodically forced systems, bifurcations to chaotic behavior were predicted from simple oscillator models [152].

The appearance of Arnold tongues has been confirmed in periodically forced iron electrodissolution [153]. In iron dissolution, harmonic forcing of periodic electrochemical oscillators resulted in entrainment, spike generation, quasiperiodicity [154], and harmonic, subharmonic, and superharmonic entrainment [153]. Regular oscillations were transformed to chaotic by periodic forcing of the potential in reduction of $Fe(CN)_6^{3-}$ on glassy carbon electrode [155]. With harmonic forcing of periodic Ni electrodissolution, complex oscillation waveforms were observed [156].

In Ni electrodissolution, periodic forcing was also applied to a single chaotic oscillator [101]. At large forcing amplitude, the chaotic behavior was suppressed and period-1 and period-2 oscillations were observed. A bifurcation diagram showing an inverse period-doubling sequence with increase in forcing amplitude is shown in Figure 4.18a. (Similar results were also obtained with Cu electrodissolution [157].) When the forcing frequency was similar to the natural frequency of the oscillator, a critical forcing amplitude was observed above which the chaotic oscillations were entrained. The entrained states in the forcing amplitude versus frequency space, as shown in Figure 4.18b, formed an Arnold tongue. Chaotic phase synchronization and suppression of chaos are considered two general effects of periodic forcing signal on a chaotic system.

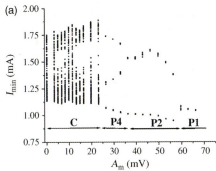
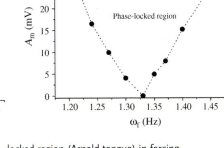

Figure 4.18 Effect of periodic forcing on chaotic dissolution of Ni in sulfuric acid [101]. (a) Bifurcation diagram of the forced system showing the minima of the oscillations (I_{min}) as a function of the amplitude of the forcing (A_m). C: chaos, Pn: period-n oscillations. (b) Phase-locked region (Arnold tongue) in forcing amplitude–forcing frequency parameter space. The sinusoidal forcing waveform of amplitude A_m and frequency ω_f is superimposed on the circuit potential.

Local forcing was also applied to iron electrodissolution for a single electrode (with laser perturbation) [158, 159] and to one element in an array [160]. Pacemaker activity was recorded in both examples when the entire system could be entrained by the local perturbations.

Global forcing applied to electrode arrays resulted in global phase synchronization [80], stabilization of various cluster configurations [161], and appearance of resonance cluster states [162].

4.5.3
Chaos Control

Control of chaotic systems has been the subject of intense research during the past decades. The goal is typically the stabilization of unstable periodic orbits (UPO) embedded in a chaotic time signal. Since chaos is prevalent in electrochemical systems, they proved to be ideal for experimentally testing different control strategies proposed by theoreticians. Electrochemical systems are ideal systems since the reproducibility of experiments is good, the level of internal noise is low, the period of chaotic oscillations may be very short, the behavior can be monitored by simple current or voltage measurements, and the control parameter (potential or current) is easily attainable. The idea has been put into practice first by Parmananda et al. [163]. They reported on controlling the chaotic current oscillations during the anodic dissolution of a rotating copper disk electrode in SAGA buffer by applying a feedback-type control method developed by Ott–Grebogi–Yorke (OGY) [164]. It is based on the sensitivity of a chaotic system to initial conditions and utilizes the short-term predictability of the deterministic dynamics: a desired periodic orbit embedded in the chaotic attractor is being stabilized by applying *small*, time-dependent (discrete) perturbations to a control parameter, for example, the circuit potential. A simplified version of the OGY method, the so-called proportional feedback (SPF) algorithm, has been applied to stabilize periodic orbits in the chaotic current oscillations of the copper–phosphoric acid system [81]. The control formula is as follows:

$$\delta V_n = K_{SPF}(x_n - x_f), \tag{4.19}$$

where δV_n is the potential perturbation applied at the nth crossing of the Poincaré section by the chaotic trajectory, K_{SPF} is feedback gain, x_n is the current value on the Poincaré map at the nth return, and x_f is the current value of the fixed point on the map corresponding to the UPO. The desired fixed point can be found as the crossing of the next-return map with the diagonal, as shown in Figure 4.3b. Results of SPF control are shown in Figure 4.19a. The potential perturbations δV_n (right axis) applied at the successive returns to the Poincaré section are shown in the inset. Figure 4.3c shows that not only period-1 but also period-2 (or higher order periodic orbits) can be stabilized by successive perturbations. In this case, the Poincaré map had been constructed from every second return values to the Poincaré section.

In addition to the simple [81] and recursive proportional feedback methods [165], adaptive learning algorithm [166], artificial neural network [167], resonant control [157], and the so-called Pyragas method [168] have been applied successfully to controlling chaotic current oscillations in electrochemical systems. The Pyragas method is based on a delayed feedback formula

$$\delta V(t) = C_P[i(t) - i(t-\tau)], \qquad (4.20)$$

where $\delta V(t)$ is the continuously applied potential perturbation, $i(t)$ is the actual current value, and $i(t-\tau)$ is the current value at delay time τ earlier.

4.5.4
Delayed Feedback and Tracking

Delay methods have also been proven to be very effective in achieving synchronization and control of chaos on electrode arrays [169]. Figure 4.19b shows chaotic oscillations of the total current of an array of $N_{el} = 4$ Ni electrodes. Synchronization of the subsystems has been achieved by simultaneously perturbing each individual resistance r_k according to

$$\delta r_k(t) = K_S \left(i_k(t) - \frac{i_{tot}(t)}{N_{el}} \right), \qquad (4.21)$$

where K_S is a coupling constant. Once the system is synchronized, a simple delayed feedback formula is applied to stabilize the desired unstable periodic orbit of period τ by using the Pyragas method (Equation 4.20) for the perturbation of the circuit potential but now based on the total current values. Figure 4.19b shows that by an appropriate choice of the control parameters, the Pyragas method can be applied to control uniform higher periodic unstable orbits, in this case period-2 current oscillations.

An extended time delay autosynchronization (ETDAS) and delay optimization with a descent gradient method were also applied for tracking unstable stationary states and unstable periodic orbits in the experimental bifurcation diagrams, respectively [170]. Results of such tracking are shown in Figure 4.7a and b for Ni electrodissolution. The method could be successfully applied, for example, for tracking period-2 oscillations through a period-doubling cascade of the chaotic current oscillation during Cu electrodissolution [170].

4.5.5
Adaptive Control

Control techniques often require an exact target dynamics, for example, location of the steady states to be stabilized. Adaptive control techniques [171] attempt to find and stabilize unstable steady states and oscillations without prior knowledge about their locations.

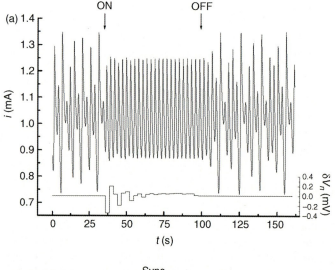

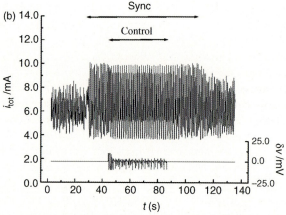

Figure 4.19 Control of electrochemical chaos. (a) Current (left axis) versus time for an interval when control for stabilizing period-1 oscillations has been switched on at 35.2 s and then switched off at 99.3 s. The potential perturbations δV_n (right axis) applied at the successive returns to the Poincaré section are shown in the inset. For experimental conditions, see Figure 4.3b [81]. (b) Time series of the total current (left axis) and perturbations of the circuit potential (right axis) during stabilization of period-2 current oscillations of four Ni electrodes without global coupling. The individual resistors are perturbed according to Equation 4.21 ($K_S = -100\,\Omega\,\text{mA}^{-1}$ and $\delta r_{max} = 20\,\Omega$) during a period of length "Sync" while the circuit potential is perturbed according to Equation 4.20 ($\tau = 1.52$ s, $C_P = 0.00956\,\text{V}\,\text{mA}^{-1}$, and $\delta v_{max} = 10$ mV) during a period of length "Control." For other details of the experiment, see Ref. [169].

The control procedure is illustrated with a potentiostatic system where the control parameter is the circuit potential (V) and the experimental observable is the current (i). The circuit potential can be set to a value of $V(t) = V_0 + \delta V(t)$, where V_0 is a base potential and $\delta V(t)$ is the perturbation:

$$\delta V(t) = k_{\text{fg}}[i(t)-y(t)], \quad dy/dt = \lambda[y(t)-i(t)], \tag{4.22}$$

where k_{fg} is the feedback gain and y is an auxiliary variable that adapts to the current with a rate of λ. Such adaptive feedback can stabilize unstable focus, node, and saddle point with appropriate values of k_{fg} and λ. Figure 4.20 shows successful stabilization of an unstable focus and a saddle with negative and positive values of λ, respectively, in oscillatory Ni electrodissolution in sulfuric acid. The adaptive controller can stabilize steady states in large parameter region, and thus can map entire phase diagrams showing both stable and unstable phase space objects. Extension of the method can also be applied to periodic orbits [172].

4.5.6
Synchronization Engineering

Recent advances in nonlinear science dealt with the design of far-from-equilibrium self-organized structure [173], for example, synchronization structure of a large population of oscillator assemblies. A major question of both theoretical and practical importance is how to bring the collective behavior of a rhythmic system to a desired condition or, equivalently, how to avoid a deleterious condition without destroying the inherent behavior of its constituent parts. The efficient design of a complex dynamic structure is a formidable task that requires simple yet accurate models incorporating integrative experimental and mathematical approaches that can handle hierarchical complexities and predict emergent, system-level properties.

The kinetics–mass transfer-type models often used in electrochemistry are generally not detailed enough for use in design of collective behavior. It was shown that

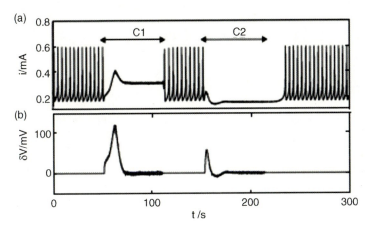

Figure 4.20 Adaptive control of unstable focus (region C1) and saddle (region C2) stationary points in oscillatory Ni electrodissolution in sulfuric acid [172]. (a) Current versus time. (b) Perturbation of circuit potential versus time. The circuit potential is perturbed according to Equation 4.22 during the control periods with negative (C1) and positive (C2) values of λ, respectively. C1: stabilization of an unstable focus steady state with stable controller. C2: stabilization of saddle steady state with unstable controller.

nonlinear feedback loops can be rigorously designed using experiment-based phase models [128] to "dial up" a desired collective behavior without requiring detailed knowledge of the underlying physiochemical properties of the target system [174]. Weak feedback signals can be designed so as to have a minimal impact on the dynamics of the individual electrodes while producing a collective behavior of the population that is both qualitatively and quantitatively different from the dynamic behavior of an uncontrolled system.

The method, termed as "synchronization engineering" [174–176], was demonstrated to create phase-locked oscillators with arbitrary phase difference, subtle dynamical structures such as itinerant cluster dynamics, desynchronization, and various cluster states. For example, Figure 4.21 shows a "slow switching" state where under the feedback the four-oscillator system itinerates among two-cluster states along heteroclinic orbits.

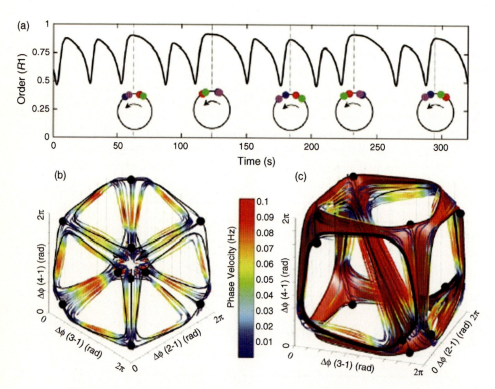

Figure 4.21 Synchronization engineering: designed sequential cluster patterns in Ni electrodissolution in sulfuric acid in a four-electrode array with a cubic feedback of electrode potential to the circuit potential [174]. (a) Time series of the Kuramoto order parameter along with selected cluster configurations. (b and c) Trajectory in state space of phase differences. The black lines represent theoretically calculated heteroclinic connections between cluster states (black fixed points). The red surface in (c) is the set of trajectories traced out by a heterogeneous phase model. The experimental trajectory is colored according to its phase velocity.

4.5.7
Effect of Noise

Intuitively, destructive effects are expected when a dynamical system is exposed to noisy environment: noise destroys ordered temporal and spatial structures. However, studies with general equations describing nonlinear systems have shown that noise can play a constructive role that promotes self-organized behavior in certain situations [75]. Some of these constructive roles were demonstrated in electrochemical systems by superimposing a noisy perturbation on the cell potential.

In coherence resonance [177], noise induces oscillations in an otherwise stable steady system. Figure 4.22a shows that without noise the Ni electrodissolution exhibits a stable steady state; however, small amount of noise can induce oscillatory behavior (see Figure 4.22b) [178]. This effect can be interpreted by the "noisy precursor" mechanism: there exists an oscillatory state in nearby parameter space and noise drives the system intermittently into this state [179].

During iron dissolution in sulfuric acid, coherence resonance can be observed close to a homoclinic bifurcation [180]. Moreover, the presence of Cl^- in the system results in effects similar to those with external noise and thus the authors proposed that oscillations in the presence of Cl^- ions are due to intrinsic noise effects stemming from localized corrosion [181, 182]. This internal noise effect was also used to improve the regularity of the spatial distribution of porous silicon structures [183].

Noise can also affect dynamic spatial structures [184, 185]. Figure 4.22c and d shows chaotic oscillations in Ni electrodissolution in a 64-electrode array setup. There is no obvious synchronization; therefore, the total current exhibits small fluctuations due to finite size effects. Small amount of global noise induces chaotic phase synchronization resulting in more ordered space–time plots and strong oscillations of the mean current (Figure 4.22e and f). Noise can also induce complete chaotic synchronization [186], aperiodic stochastic resonance [182], and replicable aperiodic spike trains [187].

4.6
Toward Applications

Electrochemistry has many applications; thus, the question naturally arises: What is the role of nonlinear phenomena in some practically important electrochemical systems? In previous examples (Section 4.5), we considered some electrochemical reactions from academic point of view; here, we show some examples of nonlinear behavior that arise in systems that are closer to important applications: corrosion, fuel cells, semiconductors, and solid-state sensors.

Commonly used stainless steels (and other passive film-forming metals), which are designed to be corrosion resistant, can nevertheless undergo localized pitting corrosion, which rapidly leads to their failure [188]. Pitting corrosion shows a sharp rise in corrosion rate that occurs with only a small change in conditions, for example,

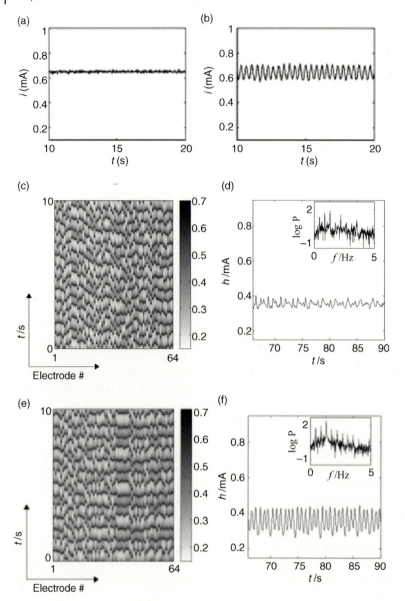

Figure 4.22 Noise-induced dynamics in Ni electrodissolution in sulfuric acid. (a and b) Coherence resonance [178]. (a) Stationary current of a single electrode slightly below Hopf bifurcation point. (b) Noise-induced oscillations with weak noise superimposed on stationary circuit potential. (c–f) Noise-induced phase synchronization of chaotic oscillations on an 8 × 8 electrode array [185]. (c) Space–time plot for weakly electrically coupled ($K_{gs} = 0.014$) electrode array; desynchronized chaos. (d) Corresponding mean current in power spectrum (inset) showing very weak oscillations. (e) Space–time plot with weak noise superimposed on circuit potential for system shown in panel (c); phase synchronized chaos. (f) Corresponding mean current and power spectrum (inset) showing noise-induced collective oscillations.

applied potential [189]. A combination of ellipsometry, optical enhanced microscopy, and parallel current measurements revealed that the onset of pitting corrosion represents a cooperative critical phenomenon [60]. Electrochemical reactions at a metastable pit change ion concentrations and weaken the protective film over defect sites. Each pit enhances the probability of appearance of further pits at defect sites within a wide zone of weakened film around it: an autocatalytic reproduction of pits can take place. Sudden transitions are thus associated with an explosive growth in the number of active pits. Similar cooperative metastable pit formation was observed in electrode array studies with Al where the individual metastable pits were visualized by using pH-sensitive agar gels [190]. Note that there is no obvious NDR character in these reactions; therefore, the pattern formation exhibits a distinct mechanism from those in Section 4.5. There are corrosion patterns that are related to the positive feedback mechanism due to NDR: two-dimensional wave patterns (propagating pulses, rotating spirals, and serpentine structures) have been observed on corroding steel plates in nitric acid with propagation velocity on the order of millimeters per second [191].

Fuel cells often accommodate reactions with NDR characteristics [16]. On the anode, oxidation of H_2 (especially, in the presence of poisons) and various C1 substrates all exhibit nonlinear phenomena [6]. On the cathode, the reduction of O_2 can take place through the reduction of H_2O_2 under certain conditions [16]; the reduction of H_2O_2 on Pt has rich dynamics with at least five distinct regions of oscillations [6, 192, 193]. The role of nonlinear effects in fuel cells is system specific: complex responses [194] can give rise to increased power generation by as much as 100% [195] or lead to cell failure due to nonuniformities and pattern formation of internal potential differences [196]. In polymer electrolyte membrane fuel cells (PEMFCs), water management introduces further complications [197]; steady-state multiplicity can develop because of the autocatalytic nature of the interplay between water and the reaction rate, which is enhanced through membrane humidification. Water, the (overall) reaction product in the PEMFC, autocatalytically accelerates the reaction rate by enhancing proton transport through the PEM. Externally forced fuel cells can be applied to limit the inhibition of the electrocatalyst by CO [198]; in the presence of periodic current pulses, the cell can attain in the presence of CO contamination up to 70% of the cell voltage obtained when the anode is fed with pure hydrogen. Nonlinear effects on steady-state behavior were also observed in solid oxide fuel cells (SOFCs) [199]. The temperature dependence of the electrolyte conductivity was shown to influence the occurrence of multiple (3–5) steady states, instabilities, and the formation of hot spots.

A classical solid-state process is the anodic dissolution of silicon in fluoride media. Damped oscillations were observed without external resistance; when external resistance is present (e.g., from contact of the back of the electrode), the oscillations can become sustained [200]. The oscillations at microscopic domains are described by the current burst model: fast oxide growth takes place during an active time, and slow oxide dissolution takes place during an idle time. Two detailed microscopic models have been elaborated by the groups of Foell [200, 201] and Lewerenz [202, 203]. In the former current-burst model, the origin of the microscopic oscillation is electrical. It is assumed that the current through the interfacial oxide film is switched on when the electric field gets larger than some critical value, leading to fast increase of oxide thickness, and it is switched off when the electric field decreases below another (lower) critical value; these two threshold values are distributed with two distinct probability laws. In the latter current-burst model, the sudden increase in current density is triggered by stress-induced mechanical breakdown of the oxide film. Macroscopic oscillations can be obtained only if the local domains become synchronized [204]. This can be achieved by a series resistance in the external circuitry. Complicated subharmonic cluster patterns can form local oscillatory domains [205]. The entrained oscillators in the different domains have the same frequency but exhibit irregular distribution of amplitudes.

In addition to temporal complexity, the silicon dissolution process exhibits rich spatial features. Electrochemically etched pores can form where, for example, macropores and octahedrally shaped pores can grow simultaneously [206]. The current-burst model can describe the pore formation under the assumption that the system self-organizes and switches the pore morphologies to that mode that optimally consumes the available electronic holes in the reactions.

Nanoprecipitation-assisted ion current oscillations [207] can occur in conical nanopores: the addition of small amounts of divalent cations to a buffered monovalent ionic solution results in an oscillating ionic current through the nanopore. The behavior is caused by the transient formation and redissolution of nanoprecipitates, which temporarily block the ionic current through the pore.

Nonlinear behavior of reduction of H_2O_2 on semiconductor surfaces (e.g., GaAs), which plays a role in the semiconductor etching process, was also thoroughly investigated [208, 209]. These systems hold the potential for new phenomena related to the difference of semiconductor/electrolyte from metal/electrolyte interface and to the presence of photocurrent oscillations and oscillatory light emissions. Periodic, MMO, and chaotic current oscillations accompanied with similar electroluminescence were observed with H_2O_2 reduction on GaAs [208, 209]. The oscillations occur due to the combined effects of correlated H_2O_2 reduction consuming an e^- and creating a hole, direct chemical etching, and parallel (competing) H^+ reduction. Result of the analysis is that for oscillations at large enough ohmic drops, there should be a potential region where the electron trapping becomes less effective with increase of potential, for example, through anomalous band bending effect. Real-time *in situ* infrared spectroscopy and ellipsometry revealed that the oscillations are accompanied by oscillations in thickness of a porous layer of solid arsenic hydride, with typical variations of a few tens of nanometers [61]. The current peaks in the oscillations coincide with the sudden dissolution of protecting arsenic hydride and thus H_2O_2 adsorbs at an elevated rate. Current oscillations and luminescence were also observed during electrochemical pore etching of n-type GaP(100) in HF and HBr electrolytes [210].

Oscillations can also occur with electrodeposition of metals or nonmetals on semiconducting surfaces, for example, open-circuit potential oscillations on p-Si during electronless Cu deposition from the HF + $CuSO_4$ solution when the Cu deposit formed a continuous porous film composed of submicrometer-sized particles [211]. The oscillations were explained by the autocatalytic shift in the flat-band potential of Si, caused by the change in the coverage of the Si oxide and the connection and disconnection of the Cu film with the Si surface.

Electromotive force (emf) of solid electrolyte concentration cells can be used to detect oscillatory surface reactions. In heterogeneous gas-phase CO oxidation, the emf of Y_2O_3-stabilized ZrO_2 electrolyte, on which the 0.1 μm thick Pt catalyst was deposited by electron beam evaporation, was shown to follow the rate of reaction and surface CO concentration [212]. Surface CO could be detected even under conditions where CO was very scarce compared to oxygen. The emf was generated by mixed electrode potential involving electrochemical reactions of O^{2-} with CO and oxygen adsorbed on Pt during the reaction. In this example, the oscillations are not generated by an electrochemical process, but are produced by the heterogeneous catalytic CO oxidation reaction to CO_2; the emf acts as a "high-impedance" probe of modulations in the relative concentrations of adsorbed CO and oxygen species reacting on the Pt surface. However, if a current is passed through the device, oxygen is transported through the electrolyte to or from the active electrode, thereby modifying the surface concentrations. Such electrical perturbation can alter the nature of the oscillation; for example, periodic modulation of the current was observed to cause complete repetitive quenching and activation of the emf oscillation and a corresponding modulation in the CO production [213]. The effect on oscillations of electrochemical pumping of O^{2-} was extensively studied with a Pt catalyst film that was used both as a catalyst and as an electrode of the solid electrolyte cell CO, O_2, Pt|ZrO_2 (8 mol% Y_2O_3)| Pt, O_2 [214]. The electrochemical O^{2-} pumping had dramatic non-Faradaic effect on the behavior: the rate could be increased by as much as 500% under severely reducing conditions. Reaction rate oscillations could be induced or stopped by adjusting the potential of the catalyst electrode. The frequency of the electrochemically induced oscillations was a linear function of the applied current.

Many other solid-state gas sensors could operate under similar nonequilibrium conditions; oscillations with resistive-type oxygen sensors consisting of TiO_2 porous ceramics were observed [213]. In this case, oscillations in surface coverage modulate the degree of reduction of the specimen and produce a corresponding oscillation in the TiO_2 electronic resistance. The authors [213] proposed that oscillations in the resistance [215] of ThO_2-doped SnO_2 structures exposed to CO very likely have the same origin.

The emf decay curves exhibited oscillations after the cathodic polarization of the oxidized sample in Pt, Co|ZrO_2 (+ CaO)|air, Pt solid electrolyte galvanic cell when the Co was completely consumed at the metal–electrolyte interface [216].

4.7
Summary and Outlook

The research area of nonlinear behavior in electrochemical systems was rejuvenated in the 1980s and 1990s by application of nonlinear science. Many of the previously described "irregular" behaviors were rigorously characterized and new systems were discovered. General stability characteristics of operation of electrochemical cells were revealed. Since the mid-1990s, tremendous progress has been made in theoretical description and experimental characterization of surface patterns. Large amount of new kinetic data have become available in which concentrations of several chemical species are obtained with temporal and spatial resolution. We attempt to outline some recent exciting developments in both electrochemistry and nonlinear science that can fuel future explorations in the field.

4.7.1
Electrochemistry of Small Systems

Nanoscale structures in all fields of science challenge experimentation and theoretical description. Fluctuation effects similar to those observed at macroscales in Section 4.5.7 are expected to play a role. In addition, novel effects are also expected to arise [217]: the discreteness and stochasticity of an electron transfer event cause fluctuations of the electrode potential that render all elementary electrochemical reactions to be faster on a nanoelectrode than predicted by the macroscopic (Butler–Volmer) electrochemical kinetics. With development of low-current devices, measurement of attoampere currents becomes possible that makes investigation of single-electron transfer processes a reality. Exciting progress is being made in single-molecule electrochemistry [218] that could open the Pandora's box of experimental molecular dynamics. Molecular structures are often described with equations similar to self-organized dissipative structures [134]; however, conservative self-organized dynamics received little attention in chemistry because of the lack of experimental examples. In parallel, nonlinear science has already been developing techniques to deal with many-body effects in mesoscopic and nanoscopic systems [219].

4.7.2
Description of Large, Complex Systems

At the other end of length scales, efficient description of large, multiscale systems is also a challenge. With enough chemical/physical knowledge, one can build large, detailed models such as those developed for lead–acid battery [220]. Application of tools of nonlinear science, however, is well developed for relatively low-dimensional systems. Two approaches can be applied [221]: model reduction (e.g., through manifolds) and subsequent analysis of the simple model or a recent coarse graining method. Both approaches need further refinement and test for applications.

4.7.3
Engineering Structures

Systematic design of temporal and spatial structures could be advantageous in certain situations, for example, for enhancing reactions or creating structure templates. The emergent structures could be achieved by inherent or externally imposed feedbacks/interactions. General design principles are needed.

4.7.4
Integration of Electrochemical and Biological Systems

Both electrochemical and biological systems exhibit a broad spectrum of nonlinear features. Interfacing these two systems, for example, through electrochemical and biological "nerve" propagation, could create further unforeseen complexities.

This strongly subjective list will probably change as soon as novel materials (e.g., memristors [222]), theoretical methods, and experimental techniques will become readily available. However, as nonlinear effects in electrochemical systems are inherent, it is certain that nonlinear dynamics will be a useful tool in the hands of electrochemists.

References

1. Nicolis, G. and Prigogine, I. (1977) *Self-Organization in Nonequilibrium Systems*, John Wiley & Sons, Inc., New York.
2. Wojtowicz, J. (1973) Oscillatory behavior in electrochemical systems, in *Modern Aspects of Electrochemistry* (eds J.O.M. Bockris and B.E. Conway), Plenum Press, New York, pp. 47–120.
3. Hudson, J.L. and Bassett, M.R. (1991) *Rev. Chem. Eng.*, **7**, 109–170.
4. Hudson, J.L. and Tsotsis, T.T. (1994) *Chem. Eng. Sci.*, **49**, 1493–1572.
5. Koper, M.T.M. (1996) *Adv. Chem. Phys.*, **92**, 161–298.
6. Krischer, K. (1999) Principles of temporal and spatial pattern formation in electrochemical systems, in *Modern Aspects of Electrochemistry* (eds B.E. Conway, O.M. Bockris, and R.E. White), Kluwer Academic Publishers, New York, p. 1.
7. Krischer, K. (2002) Nonlinear dynamics in electrochemical systems, in *Advances in Electrochemical Science and Engineering* (ed. R.C. Alkire), Wiley-VCH Verlag GmbH, Berlin, pp. 90–203.
8. Strasser, P., Eiswirth, M., and Koper, M.T.M. (1999) *J. Electroanal. Chem.*, **478**, 50–66.
9. Keizer, J. (1977) Mechanisms of electrochemical oscillations, in *Special Topics in Electrochemistry* (ed. P.A. Rock), Elsevier, Amsterdam, pp. 111–127.
10. Poncet, P., Braizaz, M., Pointu, B., and Rousseau, J. (1977) *J. Chim. Phys. Physicochim. Biol.*, **74**, 452–458.
11. Fahidy, T.Z. and Gu, Z.H. (1995) Recent advances in the study of the dynamics of electrode processes, in *Modern Aspects of Electrochemistry* (eds R.E. White, J.O.M. Bockris, and B.E. Conway), Plenum Press, New York, pp. 383–409.
12. Scheeline, A., Kirkor, E.S., Kovacs-Boerger, A.E., and Olson, D.L. (1995) *Mikrochim. Acta*, **118**, 1–42.
13. Inzelt, G. (1996) *Stud. Univ. Babes-Bolyai Chem.*, **41**, 47–69.
14. Koper, M.T.M. (1998) *J. Chem. Soc. Faraday Trans.*, **94**, 1369–1378.
15. Dini, D. (2002) *Recent Res. Dev. Electrochem.*, **5**, 47–62.
16. Krischer, K. and Varela, H. (2003) Oscillations and other dynamic instabilities, in *Handbook of Fuel Cells: Fundamentals, Technologies and Applications* (eds W. Vielstich, A. Lamm, and A.H. Gasteiger), John Wiley & Sons, Ltd., Chichester, pp. 679–701.
17. Ramaswamy, R., Scibioh, M.A., and Viswanathan, B. (2004) *Proc. Indian Natl. Sci. Acad. A*, **70**, 515–519.
18. Fahidy, T.Z. (2006) *Russ. J. Electrochem.*, **42**, 506–511.
19. Orlik, M. (2009) *J. Solid State Electrochem.*, **13**, 245–261.
20. Winfree, A.T. (2002) *Science*, **298**, 2336–2337.

21 Strogatz, S. (2004) *Sync: The Emerging Science of Spontaneous Order*, Penguin Books Ltd., UK.
22 Clarke, B.L. (1980) *Adv. Chem. Phys.*, **43**, 1.
23 Eiswirth, M., Freund, A., and Ross, J. (1991) *Adv. Chem. Phys.*, **80**, 127–199.
24 Chevalier, T., Schreiber, I., and Ross, J. (1993) *J. Phys. Chem.*, **97**, 6776–6787.
25 Rabai, G., Orban, M., and Epstein, I.R. (1990) *Acc. Chem. Res.*, **23**, 258–263.
26 Rabai, G., (1998) *ACH Models Chem.*, **135**, 381–392.
27 Koper, M.T.M. and Gaspard, P. (1992) *J. Chem. Phys.*, **96**, 7797–7813.
28 Koper, M.T.M. and Sluyters, J.H. (1991) *J. Electroanal. Chem.*, **303**, 73–94.
29 Bard, A.J. and Faulkner, L.R. (1980) *Electrochemical Methods*, John Wiley & Sons, Inc., New York.
30 Koper, M.T.M. and Sluyters, J.H. (1994) *J. Electroanal. Chem.*, **371**, 149–159.
31 Lev, O., Wolfberg, A., Pismen, L.M., and Sheintuch, M. (1989) *J. Phys. Chem.*, **93**, 1661–1666.
32 Suter, R.M. and Wong, P.Z. (1989) *Phys. Rev. B*, **39**, 4536–4540.
33 Press, W.H., Teukolsky, A.A., Vetterling, W.T., and Flannery, B.P. (2007) *Numerical Recipes: The Art of Scientific Computing*, Cambridge University Press, Cambridge.
34 Strogatz, S.H. (2000) *Nonlinear Dynamics and Chaos*, Westview Press, Cambridge.
35 Acheson, D. (1998) *From Calculus to Chaos: An Introduction to Dynamics*, Oxford University Press, New York.
36 Epstein, I.R. and Pojman, J.A. (1998) *An Introduction to Nonlinear Chemical Dynamics: Oscillations, Waves, Patterns, and Chaos*, Oxford University Press, Oxford.
37 Takens, F. (1981) Detecting strange attractors in turbulence, in *Dynamical Systems and Turbulence* (eds D.A. Rand and L.-S. Young), Springer, Heidelberg.
38 Feigenbaum, M.J. (1979) *J. Stat. Phys.*, **21**, 669–706.
39 Feigenbaum, M.J. (1980) *Los Alamos Sci.*, **1**, 4–27.
40 Ott, E. (1993) *Chaos in Dynamical Systems*, Cambridge University Press, New York.
41 Marek, M. and Schreiber, I. (1988) *Chaotic Behaviour of Deterministic Dissipative Systems*, Cambridge University Press, Cambridge.
42 Kapral, R. and Showalter, K. (1995) *Chemical Waves and Patterns (Understanding Chemical Reactivity)*, Kluwer Academic Publishers, Dordrecht.
43 Aranson, I.S. and Kramer, L. (2002) *Rev. Mod. Phys.*, **74**, 99–143.
44 Garcia-Morales, V., Hoelzel, R.W., and Krischer, K. (2008) *Phys. Rev. E*, **78**, 026215.
45 Krischer, K., Mazouz, N., and Grauel, P. (2001) *Angew. Chem., Int. Ed.*, **40**, 851–869.
46 Koper, M.T.M. (1996) *J. Electroanal. Chem.*, **409**, 175–182.
47 Hudson, J.L., Tabora, J., Krischer, K., and Kevrekidis, I.G. (1993) *Phys. Lett. A*, **179**, 355.
48 Sayer, J.C. and Hudson, J.L. (1995) *Ind. Eng. Chem. Res.*, **34**, 3246–3251.
49 Otterstedt, R.D., Plath, P.J., Jaeger, N.I., and Hudson, J.L. (1996) *J. Chem. Soc. Faraday Trans.*, **92**, 2933–2939.
50 Lee, J. et al. (2001) *J. Chem. Phys.*, **115**, 1485–1492.
51 Lev, O., Sheintuch, M., Pismen, L.M., and Yarnitzky, C. (1988) *Nature*, **336**, 458–459.
52 Otterstedt, R.D., Plath, P.J., Jaeger, N.I., and Hudson, J.L. (1996) *Phys. Rev. E*, **54**, 3744–3751.
53 Flätgen, G. and Krischer, K. (1995) *Phys. Rev. E*, **51**, 3997–4004.
54 Grauel, P., Varela, H., and Krischer, K. (2001) *Faraday Discuss.*, **120**, 165–178.
55 Varela, H., Beta, C., Bonnefont, A., and Krischer, K. (2005) *Phys. Chem. Chem. Phys.*, **7**, 2429–2439.
56 Varela, H., Beta, C., Bonnefont, A., and Krischer, K. (2005) *Phys. Rev. Lett.*, **94**, 174104.
57 Bonnefont, A., Varela, H., and Krischer, K. (2003) *ChemPhysChem*, **4**, 1260–1263.
58 Flatgen, G. et al. (1995) *Science*, **269**, 668–671.
59 Li, Y.J. et al. (2001) *Science*, **291**, 2395–2398.
60 Punckt, C. et al. (2004) *Science*, **305**, 1133–1136.
61 Erne, B. et al. (2000) *J. Phys. Chem. B*, **104**, 5974–5985.

62 Morschl, R., Bolten, J., Bonnefont, A., and Krischer, K. (2008) *J. Phys. Chem. C*, **112**, 9548–9551.
63 Chazalviel, J., Erne, B., Maroun, F., and Ozanam, F. (2001) *J. Electroanal. Chem.*, **502**, 180–190.
64 Fei, Z., Kelly, R., and Hudson, J.L. (1996) *J. Phys. Chem.*, **100**, 18986–18991.
65 Fei, Z. and Hudson, J.L. (1998) *Ind. Eng. Chem. Res.*, **37**, 2172–2179.
66 Kiss, I.Z., Wang, W., and Hudson, J.L. (1999) *J. Phys. Chem. B*, **103**, 11433–11444.
67 Kiss, I.Z., Zhai, Y.M., and Hudson, J.L. (2002) *Science*, **296**, 1676–1678.
68 Karantonis, A., Pagitsas, M., Miyakita, Y., and Nakabayashi, S. (2004) *J. Phys. Chem. B*, **108**, 5836–5846.
69 Kiss, I.Z., Munjal, N., and Martin, R.S. (2009) *Electrochim. Acta*, **55**, 395–403.
70 Farmer, J.D. (1981) *Phys. Rev. Lett.*, **47**, 179–182.
71 Grossman, A. and Morlet, J. (1984) *SIAM J. Math. Anal.*, **15**, 723–736.
72 Planinsic, P. and Petek, A. (2008) *Electrochim. Acta*, **53**, 5206–5214.
73 Darowicki, K., Krakowiak, Λ., and Zielinski, A. (2002) *Electrochem. Commun.*, **4**, 158–162.
74 Kiss, I.Z., Lv, Q., Organ, L., and Hudson, J.L. (2006) *Phys. Chem. Chem. Phys.*, **8**, 2707–2715.
75 Pikovsky, A.S., Rosenblum, M., and Kurths, J. (2001) *Synchronization: A Universal Concept in Nonlinear Science*, Cambridge University Press, Cambridge.
76 Osipov, G. *et al.* (2003) *Phys. Rev. Lett.*, **91**, 024101.
77 Kiss, I.Z., Lv, Q., and Hudson, J.L. (2005) *Phys. Rev. E*, **71**, 035201.
78 Lachaux, J., Rodriguez, E., Martinerie, J., and Varela, F. (1999) *Hum. Brain Mapp.*, **8**, 194–208.
79 Romano, M.C. *et al.* (2005) *Europhys. Lett.*, **71**, 466–472.
80 Rosenblum, M.G. *et al.* (2002) *Phys. Rev. Lett.*, **89**, 264102.
81 Kiss, I.Z., Gáspár, V., Nyikos, L., and Parmananda, P. (1997) *J. Phys. Chem. A*, **101**, 8668–8674.
82 Kiss, I.Z., Zhai, Y.M., and Hudson, J.L. (2002) *Ind. Eng. Chem. Res.*, **41**, 6363–6374.
83 Grassberger, P. and Procaccia, I. (1983) *Physica D*, **9**, 189–208.
84 Wolf, A., Swift, J.B., Swinney, H.L., and Vastano, J.A. (1985) *Physica D*, **16**, 285–317.
85 Hegger, R., Kantz, H., and Schreiber, T. (1999) *Chaos*, **9**, 413.
86 Breuer, K.S. and Sirovich, L. (1991) *J. Comput. Phys.*, **96**, 277.
87 Armbruster, D., Heiland, R., and Kostelich, E.J. (1994) *Chaos*, **4**, 421.
88 Pyragas, K., Pyragas, V., Kiss, I.Z., and Hudson, J.L. (2004) *Phys. Rev. E*, **70**, 026215.
89 Chen, S. and Schell, M. (1999) *Electrochim. Acta*, **44**, 4773–4780.
90 Fechner, A.T. (1828) *Schweigg. J. Phys. Chem.*, **53**, 61–76.
91 Kiss, I.Z., Kazsu, Z., and Gáspár, V. (2005) *J. Phys. Chem. A*, **109**, 9521–9527.
92 Kiss, I.Z., Gáspár, V., and Nyikos, L. (1998) *J. Phys. Chem. A*, **102**, 909–914.
93 Kiss, I.Z., Kazsu, Z., and Gáspár, V. (2009) *Phys. Chem. Chem. Phys.*, **11**, 7669–7677.
94 Yablonsky, G.S., Mareels, I.M.Y., and Lazman, M. (2003) *Chem. Eng. Sci.*, **58**, 4833–4842.
95 Kiss, I.Z., Pelster, L.N., Wickramasinghe, M., and Yablonsky, G.S. (2009) *Phys. Chem. Chem. Phys.*, **11**, 5720–5728.
96 Nagao, R., Epstein, I.R., Gonzalez, E.R., and Varela, H. (2008) *J. Phys. Chem. A*, **112**, 4617–4624.
97 Albahadily, F.N. and Schell, M. (1988) *J. Chem. Phys.*, **88**, 4312–4319.
98 Schell, M. and Albahadily, F.N. (1988) *J. Chem. Phys.*, **90**, 822–828.
99 Lev, O., Wolffberg, A., Sheintuch, M., and Pismen, L.M. (1988) *Chem. Eng. Sci.*, **43**, 1339–1353.
100 Wang, W., Kiss, I.Z., and Hudson, J.L. (2000) *Chaos*, **10**, 248–256.
101 Kiss, I.Z. and Hudson, J.L. (2001) *Phys. Rev. E*, **64**, 046215.
102 Kiss, I.Z., Wang, W., and Hudson, J.L. (2000) *Phys. Chem. Chem. Phys.*, **2**, 3847–3854.

103 Kiss, I.Z. and Hudson, J.L. (2002) *Phys. Chem. Chem. Phys.*, **4**, 2638–2647.
104 Davidsen, J., Kiss, I.Z., Hudson, J.L., and Kapral, R. (2003) *Phys. Rev. E*, **68**, 026217.
105 Nawrath, J. et al. (2010) *Phys. Rev. Lett.*, **104**, 038701.
106 Diem, C.B. and Hudson, J.L. (1987) *AIChE J.*, **33**, 218–224.
107 Wang, Y. and Hudson, J.L. (1991) *AIChE J.*, **37**, 1833–1843.
108 Koper, M.T.M., Gaspard, P., and Sluyters, H.J. (1992) *J. Chem. Phys.*, **97**, 8250–8260.
109 Krischer, K. et al. (1991) *Ber. Bunsen. Phys. Chem.*, **95**, 820–823.
110 Bassett, M.R. and Hudson, J.L. (1988) *J. Phys. Chem.*, **92**, 6963–6966.
111 Mukouyama, Y., Kikuchi, M., and Okamoto, H. (2005) *J. Solid State Electrochem.*, **9**, 290–295.
112 Okamoto, H., Kikuchi, M., and Mukouyama, Y. (2008) *J. Electroanal. Chem.*, **622**, 1–9.
113 Rinzel, J. (1987) A formal classification of bursting mechanisms in excitable systems, in *Mathematical Topics in Population Biology, Morphogenesis and Neurosciences* (eds E. Teramoto and M. Yamaguti), Springer, Berlin, pp. 267–281.
114 Izhikevich, E.M. (2000) *Int. J. Bifurcat. Chaos*, **10**, 1171–1266.
115 Plant, R.E. (1981) *J. Math. Biol.*, **11**, 15–32.
116 Chay, T.R. and Keizer, J. (1983) *Biophys. J.*, **42**, 181–190.
117 Pernarowski, M., Miura, R.M., and Kevorkian, J. (1992) *SIAM J. Appl. Math.*, **52**, 1627–1650.
118 Fetner, N. and Hudson, J.L. (1990) *J. Phys. Chem.*, **94**, 6506–6509.
119 van Venrooij, T.G.J. and Koper, M.T.M. (1995) *Electrochim. Acta*, **40**, 1689–1696.
120 Sazou, D., Pagitsas, M., and Georgolios, C. (1992) *Electrochim. Acta*, **37**, 2067–2076.
121 Sazou, D., Pagitsas, M., and Gerogolios, C. (1993) *Electrochim. Acta*, **38**, 2321–2332.
122 D'Alba, F. and Lucarini, C. (1995) *Bioelectrochem. Bioenerg.*, **38**, 185–189.
123 Organ, L., Kiss, I.Z., and Hudson, J.L. (2003) *J. Phys. Chem. B*, **107**, 6648–6659.
124 Koutsaftis, D., Karantonis, A., Pagitsas, M., and Kouloumbi, N. (2007) *J. Phys. Chem. C*, **111**, 13579–13585.
125 Christoph, J. and Eiswirth, M. (2002) *Chaos*, **12**, 215–230.
126 Mukouyama, Y. et al. (1996) *Chem. Lett.*, **6**, 463–464.
127 Karantonis, A., Pagitsas, M., Miyakita, Y., and Nakabayashi, S. (2003) *J. Phys. Chem. B*, **107**, 14622–14630.
128 Kiss, I.Z., Zhai, Y.M., and Hudson, J.L. (2005) *Phys. Rev. Lett.*, **94**, 248301.
129 Zhai, Y., Kiss, I.Z., Daido, H., and Hudson, J.L. (2005) *Physica D*, **205**, 57–69.
130 Kiss, I.Z., Zhai, Y., and Hudson, J.L. (2002) *Phys. Rev. Lett.*, **88**, 238301.
131 Mikhailov, A.S. et al. (2004) *Proc. Natl. Acad. Sci. USA*, **101**, 10890–10894.
132 Kiss, I.Z. and Hudson, J.L. (2003) *Chaos*, **13**, 999–1009.
133 Han, S., Kurrer, C., and Kuramoto, Y. (1995) *Phys. Rev. Lett.*, **75**, 3190–3193.
134 Kuramoto, Y. (1984) *Chemical Oscillations, Waves and Turbulence*, Springer, Berlin.
135 Winfree, A.T. (1967) *J. Theor. Biol.*, **16**, 15–42.
136 Koper, M.T.M., Schmidt, T.J., Markovic, N.M., and Ross, P.N. (2001) *J. Phys. Chem. B*, **105**, 8381–8386.
137 Kiss, I.Z., Brackett, A.W., and Hudson, J.L. (2004) *J. Phys. Chem. B*, **108**, 14599–14608.
138 Birzu, A. and Gáspár, V. (2009) *Electrochim. Acta*, **55**, 383–394.
139 Otterstedt, R.D. et al. (1996) *Chem. Eng. Sci.*, **51**, 1747–1756.
140 Christoph, J., Strasser, P., Eiswirth, M., and Ertl, G. (1999) *Science*, **284**, 291–293.
141 Plenge, F., Varela, H., and Krischer, K. (2005) *Phys. Rev. Lett.*, **94**, 198301.
142 Strasser, P. et al. (2000) *J. Phys. Chem. A*, **104**, 1854–1860.
143 Lee, J. et al. (2003) *Phys. Chem. Chem. Phys.*, **5**, 935–938.
144 Green, B. and Hudson, J. (2001) *Phys. Rev. E*, **63**, 026214.
145 Mazouz, N., Flatgen, G., Krischer, K., and Kevrekidis, I.G. (1998) *J. Electrochem. Soc.*, **145**, 2404–2411.
146 Grauel, P., Christoph, J., Flatgen, G., and Krischer, K. (1998) *J. Phys. Chem. B*, **102**, 10264–10271.

147 Grauel, P. and Krischer, K. (2001) *Phys. Chem. Chem. Phys.*, **3**, 2497–2502.
148 Mazouz, N. and Krischer, K. (2000) *J. Phys. Chem. B*, **104**, 6081–6090.
149 Krastev, I. and Koper, M.T.M. (1995) *Physica A*, **213**, 199–208.
150 Krischer, K. *et al.* (2003) *Electrochim. Acta*, **49**, 103–115.
151 Jain, S., Kiss, I.Z., Breidenich, J., and Hudson, J.L. (2009) *Electrochim. Acta*, **55**, 363–373.
152 Scott, S.K. (1991) *Chemical Chaos*, Clarendon Press, Oxford.
153 Pagitsas, M., Sazou, D., Karantonis, A., and Georgolios, C. (1992) *J. Electroanal. Chem.*, **327**, 93–108.
154 Karantonis, A., Pagitsas, M., and Sazou, D. (1993) *Chaos*, **3**, 243–255.
155 Varma, V.S. and Upadhyay, P.K. (1989) *J. Electroanal. Chem.*, **271**, 345–349.
156 Berthier, F. (2004) *J. Electroanal. Chem.*, **572**, 267–281.
157 Parmananda, P. *et al.* (2000) *J. Phys. Chem. B*, **104**, 11748–11751.
158 Karantonis, A., Shiomi, Y., and Nakabayashi, S. (2001) *Chem. Phys. Lett.*, **335**, 221–226.
159 Shiomi, Y., Karantonis, A., and Nakabayashi, S. (2001) *Phys. Chem. Chem. Phys.*, **3**, 479–488.
160 Fei, Z. and Hudson, J.L. (1997) *J. Phys. Chem. B*, **101**, 10356–10364.
161 Wang, W., Kiss, I.Z., and Hudson, J.L. (2001) *Phys. Rev. Lett.*, **86**, 4954–4957.
162 Kiss, I.Z., Zhai, Y., and Hudson, J.L. (2008) *Phys. Rev. E*, **77**, 046204.
163 Parmananda, P., Sherard, P., Rollins, R.W., and Dewald, H.D. (1993) *Phys. Rev. E*, **47**, R3003–R3006.
164 Ott, E., Grebogi, C., and Yorke, J.A. (1990) In *Chaos: Soviet–American Perspectives on Nonlinear Science*, (ed. D.K. Campbell) American Institute of Physics, New York, pp. 153–172.
165 Rollins, R.W., Parmananda, P., and Sherard, P. (1993) *Phys. Rev. E*, **47**, 780–783.
166 Rhode, M.A., Rollins, R.W., and Dewald, H.D. (1997) *Chaos*, **7**, 653.
167 Kiss, I.Z. and Gáspár, V. (2000) *J. Phys. Chem. A*, **104**, 8033–8037.
168 Parmananda, P. *et al.* (1999) *Phys. Rev. E*, **59**, 5266–5271.
169 Kiss, I.Z., Gáspár, V., and Hudson, J.L. (2000) *J. Phys. Chem. B*, **104**, 7554–7560.
170 Kiss, I.Z., Kazsu, Z., and Gáspár, V. (2006) *Chaos*, **16**, 033109.
171 Pyragas, K., Pyragas, V., Kiss, I.Z., and Hudson, J.L. (2002) *Phys. Rev. Lett.*, **89**, 0244103.
172 Pyragas, K. (2001) *Phys. Rev. Lett.*, **86**, 2265–2268.
173 Mikhailov, A.S. and Showalter, K. (2006) *Phys. Rep.*, **425**, 79–194.
174 Kiss, I.Z., Rusin, C.G., Kori, H., and Hudson, J.L. (2007) *Science*, **316**, 1886–1889.
175 Rusin, C.G., Kiss, I.Z., Kori, H., and Hudson, J.L. (2009) *Ind. Eng. Chem. Res.*, **48**, 9416–9422.
176 Kori, H., Rusin, C.G., Kiss, I.Z., and Hudson, J.L. (2008) *Chaos*, **18**, 026111.
177 Gang, H., Ditzinger, T., Ning, C.Z., and Haken, H. (1993) *Phys. Rev. Lett.*, **71**, 807.
178 Kiss, I.Z., Hudson, J.L., Santos, G.J.E., and Parmananda, P. (2003) *Phys. Rev. E*, **67**, 035201.
179 Wiesenfeld, K. (1985) *J. Stat. Phys.*, **38**, 1071.
180 Santos, G., Rivera, M., Eiswirth, M., and Parmananda, P. (2004) *Phys. Rev. E*, **70**, 021103.
181 Rivera, M., Santos, G., Uruchurtu-Chavarin, J., and Parmananda, P. (2005) *Phys. Rev. E*, **72**, 030102.
182 Santos, G.J.E., Rivera, M., Escalona, J., and Parmananda, P. (2008) *Philos. Trans. R. Soc. Lond. A*, **366**, 369–380.
183 Escorcia-Garcia, J., Agarwal, V., and Parmananda, P. (2009) *Appl. Phys. Lett.*, **94**, 133103.
184 Zhou, C.S., Kurths, J., Kiss, I.Z., and Hudson, J.L. (2002) *Phys. Rev. Lett.*, **89**, 014101.
185 Kiss, I.Z. *et al.* (2003) *Chaos*, **13**, 267–278.
186 Kiss, I.Z., Hudson, J.L., Escalona, J., and Parmananda, P. (2004) *Phys. Rev. E*, **70**, 026210.
187 Parmananda, P., Santos, G., Rivera, M., and Showalter, K. (2005) *Phys. Rev. E*, **71**, 031110.
188 Frankel, G. (1998) *J. Electrochem. Soc.*, **145**, 2970.

189 Szklarska-Smialowska, Z. (1986) *Pitting Corrosion of Metals*, National Association of Corrosion Engineers, Houston, TX.
190 Sasaki, K. and Isaacs, H. (2004) *J. Electrochem. Soc.*, **151**, B124–B133.
191 Agladze, K. and Steinbock, O. (2000) *J. Phys. Chem. A*, **104**, 9816–9819.
192 Mukouyama, Y. *et al.* (2001) *J. Phys. Chem. B*, **105**, 10905–10911.
193 Mukouyama, Y. *et al.* (2001) *J. Phys. Chem. B*, **105**, 7246–7253.
194 Benziger, J., Chia, J.E., Kimball, E., and Kevrekidis, I.G. (2007) *J. Electrochem. Soc.*, **154**, B835–B844.
195 Zhang, J.X. and Datta, R. (2004) *Electrochem. Solid State Lett.*, **7**, A37–A40.
196 Benziger, J.B. *et al.* (2006) *J. Power Sources*, **155**, 272–285.
197 Chia, E., Benziger, J., and Kevrekidis, I. (2004) *AIChE J.*, **50**, 2320–2324.
198 Carrette, L., Friedrich, K., Huber, M., and Stimming, U. (2001) *Phys. Chem. Chem. Phys.*, **3**, 320–324.
199 Mangold, M., Krasnyk, M., and Sundmacher, K. (2006) *J. Appl. Electrochem.*, **36**, 265–275.
200 Foca, E., Carstensen, J., and Foell, H. (2007) *J. Electroanal. Chem.*, **603**, 175–202.
201 Carstensen, J., Prange, R., and Foll, H. (1999) *J. Electrochem. Soc.*, **146**, 1134–1140.
202 Grzanna, J., Jungblut, H., and Lewerenz, H. (2000) *J. Electroanal. Chem.*, **486**, 181–189.
203 Grzanna, J., Jungblut, H., and Lewerenz, H. (2000) *J. Electroanal. Chem.*, **486**, 190–203.
204 Chazalviel, J.-N. and Ozanam, F. (2010) *Electrochim. Acta*, **55**, 656–665.
205 Miethe, I., Garcia-Morales, V., and Krischer, K. (2009) *Phys. Rev. Lett.*, **102**, 194101.
206 Lolkes, S. *et al.* (2003) *Mater. Sci. Eng. B*, **101**, 159–163.
207 Powell, M.R. *et al.* (2008) *Nat. Nanotechnol.*, **3**, 51–57.
208 Koper, M.T.M., Meulenkamp, E.A., and Vanmaekelbergh, D. (1993) *J. Phys. Chem.*, **97**, 7337–7341.
209 Koper, M.T.M. and Vanmaekelbergh, D. (1995) *J. Chem. Phys.*, **99**, 3687–3696.
210 Wloka, J., Lockwood, D., and Schmuki, P. (2005) *Chem. Phys. Lett.*, **414**, 47–50.
211 Nagai, T. *et al.* (2006) *Chaos*, **16**, 037106.
212 Okamoto, H., Kawamura, G., and Kudo, T. (1983) *J. Catal.*, **82**, 322–331.
213 Hetrick, R.E. and Logothetis, E.M. (1979) *Appl. Phys. Lett.*, **34**, 117–119.
214 Yentekakis, I.V. and Vayenas, C.G. (1988) *J. Catal.*, **111**, 170–188.
215 Nitta, M., Kanefusa, S., Taketa, Y., and Haradome, M. (1978) *Appl. Phys. Lett.*, **32**, 590.
216 Enoki, K., Hagiwara, S., Kaneko, H., and Saito, Y. (1977) *Nippon Kinzoku Gakkaishi*, **41**, 505–510.
217 Garcia-Morales, V. and Krischer, K. (2010) *Proc. Natl. Acad. Sci.*, **107**, 4528–4532.
218 Ulgut, B. and Abruna, H.D. (2008) *Chem. Rev.*, **108**, 2721–2736.
219 Ullmo, D. (2008) *Rep. Prog. Phys.*, **71**, 026001.
220 Boovaragavan, V., Methakar, R.N., Ramadesigan, V., and Subramanian, V.R. (2009) *J. Electrochem. Soc.*, **156**, A854–A862.
221 Gorban, A.N. *et al.* (2007) *Model Reduction and Coarse-Graining Approaches for Multiscale Phenomena*, Springer, New York.
222 Strukov, D.B., Snider, G.S., Stewart, D.R., and Williams, R.S. (2008) *Nature*, **453**, 80–83.
223 Zhai, Y.M., Kiss, I.Z., and Hudson, J.L. (2004) *Ind. Eng. Chem. Res.*, **43**, 315–326.

5
Fuel Cells: Advances and Challenges
San Ping Jiang and Xin Wang

This chapter reviews the status, development, and challenges of various fuel cell technologies, ranging from the most advanced alkaline fuel cells, proton exchange membrane fuel cells, and solid oxide fuel cells to the emerging new members of the fuel cell families such as microbial fuel cells, single-chamber solid oxide fuel cells, and direct carbon fuel cells. Common to all new and emerging technologies, most technical barriers for the commercial viability and wide use of fuel cell technologies are the cost and durability, briefly considered in the chapter. The principles, classification, applications, and fuels for the electrochemical cells are also briefly reviewed.

5.1
Introduction

The excessive and accelerated use of fossil fuels, especially coal, oil, and gas, in the past 100 years has triggered a global energy crisis, and carbon dioxide emission from power generation using fossil fuels is considered a key contributor to climate change and related environmental problems. With increasing energy demand and depleting fossil fuel reserves, the current power generation from fossil fuels will not be sustainable.

Consequently, there are urgent needs to increase electricity generation efficiency and to develop renewable energy sources. For example, coal is believed to be the bridging energy source with the diminishing of crude oil, and currently, half of electricity produced in the United States comes from coal-fired power plants. Coal's share of electricity production in many developing countries exceeds this amount reaching over 70% in China and India. Coal generates more CO_2 emissions than any other conventional energy source [1]. However, the CO_2 and other pollutant emission can be reduced by integration of coal gasification and fuel cells (integrated gasification fuel cell, IGFC). The energy efficiency of the IGFC power plant can be 56–60% depending on gasification and fuel cell operating conditions [2]. With the increased efficiency, CO_2 emission will dramatically decrease.

Solid State Electrochemistry II: Electrodes, Interfaces and Ceramic Membranes.
Edited by Vladislav V. Kharton.
© 2011 Wiley-VCH Verlag GmbH & Co. KGaA. Published 2011 by Wiley-VCH Verlag GmbH & Co. KGaA.

Fuel cell is an energy conversion device to electrochemically transform the chemical energy of fuels such as hydrogen, methanol, ethanol, natural gas, and light hydrocarbons to electricity, and hence, fuel cells inherently have a significantly higher efficiency than that of conventional energy conversion technologies such as internal combustion engine (ICE), the efficiency of which is limited by the Carnot cycle. Fuel cell is considered to be the most efficient, and less polluting power generating technology, and is a potential and viable candidate to moderate the fast increase in power requirements and to minimize the impact of the increased power consumption on the environment. Fuel cells are also versatile devices ranging from room-temperature fuel cells such as proton exchange membrane fuel cells (PEMFCs) to high-temperature (500–1000 °C) solid oxide fuel cells (SOFCs). SOFCs allow internal reforming, promote rapid kinetics with nonprecious materials, and offer high flexibilities in fuel choice. The other advantages of fuel cells are the quiet operation and flexible modulability.

The fuel cell technology is not new and was invented more than 160 years ago by Sir William Grove. In fact, this is one of the oldest electrical energy conversion technologies. Surprisingly, the technological development of fuel cells has lagged far behind the more well-known energy conversion technologies such as steam and internal combustion engine. The reasons can be reduced mainly to economic factors, materials problems, and certain inadequacies in the operation of electrochemical devices. During the past 20 years, fuel cells have received enormous attention worldwide as alternative electrical energy conversion systems because of their huge potential for power generation in portable, transport, and stationary applications. One of the major factors that have influenced the development and demonstration of fuel cells in past few decades is the worldwide concern about the finite reserves of fossil fuels and environmental consequences of the increasing consumptions of fossil fuel in electricity production and for the propulsion of vehicles.

This chapter reviews the status, development, and challenges of various fuel cell technologies. The status and development of emerging new members of fuel cells communities such as microbial fuel cells (MFCs), direct carbon fuel cells (DCFCs), and so on will also be briefly reviewed. The principles, classifications, and fuels for the electrochemical cells will be briefly discussed.

5.1.1
Principle and Classification of Fuel Cells

A fuel cell is an electrochemical cell and thus the fuel cell reaction can be generally represented by two-electrode reactions (anode and cathode) with an overall reaction described in Equation 5.1.

$$A_{ox1} + B_{red1} \rightarrow C_{red2} + D_{ox2} \tag{5.1}$$

An ideal fuel cell can produce current while sustaining a steady voltage as long as the fuel is supplied. The theoretical voltage of the cell, E_{thermo}, can be thermodynamically predicted by the Nernst equation.

$$E_{thermo} = E^0 - \frac{RT}{n_e F} \ln(\Pi), \qquad (5.2)$$

where E^0 is the standard cell potential (V), R is the gas constant (8.314 J mol^{-1} K^{-1}), T is the temperature (K), n_e is the number of electrons transferred in the reaction, F is the Faraday's constant (96 485 C mol^{-1}), and Π is the product of chemical activities of effluents divided by those of reactants.

For H_2/O_2 fuel cells with proton exchange membrane (PEM) electrolyte, hydrogen is oxidized at the anode and protons enter the electrolyte and are transported to the cathode:

$$H_2 \rightarrow 2H^+ + 2e^- \qquad (E_a^0 \sim 0 \text{ V versus SHE}) \qquad (5.3)$$

where SHE stands for standard hydrogen electrode and E_a^0 is the standard potential for the hydrogen oxidation reaction. Electrons flow in the external circuit to the cathode, where the supplied oxygen reacts according to

$$\frac{1}{2}O_2 + 2H^+ + 2e^- \rightarrow H_2O \qquad (E_c^0 \sim 1.23 \text{ V versus SHE}) \qquad (5.4)$$

where E_c^0 is the standard potential for the oxygen reduction reaction (ORR). Thus, the overall reaction for the H_2/O_2 fuel cells is the combination of oxygen ions with protons to form water:

$$\frac{1}{2}O_2 + H_2 \rightarrow H_2O \qquad (5.5)$$

The standard cell potential, E^0, for a H_2/O_2 fuel cell is 1.23 V at 25 °C. In practice, the actual voltage of a fuel cell is less than the predicted thermodynamic voltage due to irreversible losses when the current is higher than zero. The three major irreversible losses that affect fuel cell performance are activation losses or activation polarization (η_{act}), ohmic losses (η_Ω), and mass transport losses or concentration polarization (η_{conc}); see also Chapter 12 of the first volume. The extent of these losses varies from one system to another. The actual operational voltage output (E_{op}) of a fuel cell can be determined by subtracting the voltage losses associated with each processes from the E_{thermo} value as follows:

$$E_{op} = E_{thermo} - [(\eta_{act} + \eta_\Omega + \eta_{conc})_{cathode} + (\eta_{act} + \eta_\Omega + \eta_{conc})_{anode}]. \qquad (5.6)$$

Current generation in a fuel cell depends largely on the kinetics of the reaction that takes place at the anode and the cathode. The reaction kinetics is limited by an activation energy barrier that impedes the reaction. When current is drawn from a fuel cell, a portion of the potential at the anode and cathode is then lost to overcome this activation barrier, that is, η_{act}, characterized by an exponential loss of potential on the current–voltage curve at low current densities. The η_Ω represents the voltage lost in order to drive the charge transport (i.e., electrons, oxygen ions, or protons). This loss generally follows the Ohm's law. The η_{conc} is due to the reactant depletion or product accumulation and usually occurs at high currents. In addition, there are parasitic losses either due to the reactant crossover that causes the mixed potentials

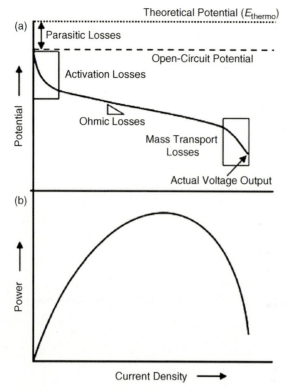

Figure 5.1 Schematic representation of the current–potential and power versus current dependence in a fuel cell, illustrating the energy dispersion.

or due to the internal shorting caused by the mixed ionic and electronic conduction of the electrolyte membrane. Figure 5.1 shows schematically the losses taking place in a fuel cell.

Fuel cells are usually classified according to their working temperature or to the electrolyte employed. There are, thus, low- and high-temperature fuel cells. Low-temperature fuel cells include alkaline fuel cells (AFCs), proton exchange membrane fuel cells, direct methanol fuel cells (DMFCs), and phosphoric acid fuel cells (PAFCs). The high-temperature fuel cells operate at 500–1000 °C, including molten carbonate fuel cells (MCFCs) and SOFCs. Due to their high operating temperature, MCFCs and SOFCs have a high tolerance to typical catalyst poisons such CO, produce high-quality heat for reforming hydrocarbons, and offer the possibility of direct utilization of hydrocarbon fuels. The electrolytes in fuel cells can be solid (polymer or ceramic) or liquid (aqueous or molten), and must have high ionic (primarily O^{2-}, OH^-, H^+, or CO_3^{2-}) conductivity with negligible electrical conductivity. A second grouping is based on fuels used in fuel cells, such as DMFC, direct formic acid fuel cells (DFAFCs), and DCFCs. Areas of application include battery replacement in portable electronic devices, transportation, prime movers, and/or auxiliary power units in vehicles, residential combined heat and power (CHP), and large-scale megawatt stationary power generation.

Table 5.1 lists the different types of fuel cells, classified according to the electrolyte employed, together with their main characteristics [3–6].

5.1.2
Fuels for Electrochemical Cells

5.1.2.1 Hydrogen, Methanol, Formic Acid, and Liquid Fuels

Hydrogen is the most attractive and cleanest fuel for fuel cells as its only by-product is water with no greenhouse gas emission. At present, steam reforming, partial oxidation, and autothermal reforming of hydrocarbons as well as biofeed stocks are the major processes for hydrogen generation, but all these methods produce a large amount of CO and/or CO_2. Low-temperature fuel cells, such as AFCs and PEMFCs, are extremely sensitive toward CO and CO_2 impurities: PEMFCs are poisoned by very low (ppm) levels of CO and the tolerance level for CO_2 is extremely low in the case of AFCs. Thus, hydrogen generated from fossil fuel or hydrocarbons needs to be cleaned to reduce the CO level to the acceptable ppm level. Preferential oxidation of CO is one of the most effective methods for trace CO cleanup from the reformate stream [7].

For transportation and portable applications of fuel cells, the challenge is the storage and transportation of hydrogen. There are several options in the development of storage technologies for hydrogen: liquid hydrogen, pressurized hydrogen, metal hydrides, borohydrides, carbon nanotubes (CNTs), zeolites, and metal oxide frameworks. Despite the enormous effort in search for a material that could store appropriate amount of hydrogen, the US Department of Energy (DoE) targets of 6% by mass appear to be very tough to meet [8]. State-of-the-art compressed hydrogen storage consists of lightweight tanks using polymer and carbon fibers containing hydrogen compressed to 700 bar. The infrastructure for hydrogen fuel station is also very costly and, for safety reasons, a hydrogen filling station will be very different from the gasoline station. Also from an energy point of view, large-scale transportation of hydrogen and the necessary compression or liquefaction of hydrogen can be highly unattractive. Thus, effective and safe on-site or onboard generation of hydrogen could speed up the introduction of fuel cell system without the need of a widespread hydrogen infrastructure.

The direct use of liquid fuels in fuel cells is of significant importance due to potentially higher energy density and higher maximum thermodynamic efficiencies. Liquid fuels such as methanol and ethanol have several advantages with respect to hydrogen. They are relatively cheap, easily handled, transported, and stored and have a high theoretical energy density. However, apart from the energy density, the toxicological–ecological hazards of the liquid fuels and the environmental effects of the by-product of the liquid fuel oxidation reactions should also be taken into account when selecting a particular fuel. For example, methanol, the most studied liquid fuel in fuel cells, predominantly produced by steam reforming of natural gas, is also highly flammable and toxic. On the other hand, ethanol would be a preferred

Table 5.1 Types of fuel cells according to the electrolyte employed and their main characteristics.

	AFC	PEMFC	PAFC	MCFC	SOFC
Temperature (°C)	60–90	80–90	160–200	600–650	500–1000
Electrolyte	KOH	Polymer	H_3PO_4	$LiCO_3/K_2CO_3$	$Y_2O_3\text{–}ZrO_2$
Anode	Pt/C	Pt/C	Pt/C	Ni alloy	Ni/YSZ
Cathode	Pt/C	Pt/C	Pt/C	NiO	LSM
Primary fuel	H_2	H_2	H_2/reformate	H_2/CO reformate	H_2/CO/CH_4 reformate
Oxidant	O_2	O_2/air	O_2/air	CO_2/O_2/air	O_2/air
Efficiency (%)	35–40	40–50	32–42[a]	47–57[a]	60–65[a]
Power range (W)	1000	1–100 000	50 000–200 000	200 000–300 000	1000–125 000
Main applications	Space transportation	Portable transportation	Stationary	Stationary	Stationary

a) The production of additional electric energy by means of thermal energy cogeneration is not considered.

liquid fuel for portable fuel cells as ethanol can be easily produced by hydration of acetylene or can be derived by fermentation of sugar-rich raw materials and is less harmful compared to methanol.

The electrochemical oxidation of majority liquid fuels in low-temperature fuel cells is hardly a complete reaction. Thus, the environmental effects of the by-products need to be considered in the design of fuel cell systems. For example, the main by-products of the electrooxidation of methanol are formaldehyde and formic acid [9]. Formaldehyde is highly irritant, corrosive, carcinogenic, and toxic. In the case of ethanol fuel, the main challenge is the breakage of the C−C bond. Final products of the ethanol oxidation are acetaldehyde and acetic acid [10], which are also highly flammable and carcinogenic. Dimethyl ether (DME) is also used as liquid fuel for fuel cells as DME is the simplest ether with no C−C bonds and low toxicity compared to methanol. However, as reported by Mizutani *et al.* [11], the main by-products of direct dimethyl ether fuel cells are methanol and methyl formate and the formation of methanol does not depend on the current but increases with increasing temperature. Wang *et al.* [12] studied the oxidation of 1-propanol and 2-propanol in a PEMFC using online mass spectrometry. Propanal is the main product of the electrooxidation of 1-propanol, while the electrooxidation of 2-propanol mainly yields acetone. Thus, treatment processes must be incorporated into fuel cell system or be created as the by-products of the oxidation reactions of the liquid fuels cannot be directly discharged without proper cleaning treatment. Demirci [13] gave a detailed thermodynamic and environmental analysis of potential liquid fuels in fuel cells. Table 5.2 lists specific energy density, hazard, environmental effects, and main by-products of the common liquid fuels used in fuel cells.

5.1.2.2 Methane, Hydrocarbons, and Biomass-Derived Renewable Fuels

Given the lack of a supply infrastructure and difficulties in storing hydrogen, the ability of fuel cells to operate at high efficiency on hydrocarbon fuels is considered a major advantage of high-temperature fuel cells such as MCFCs and SOFCs. Natural gas (primary methane) is seen as an ideal fuel due to its abundance, existing distribution infrastructure, and low cost.

The steam reforming of methane and hydrocarbons is an established process used on an industrial scale for the production of hydrogen. The first step of the steam reforming results in the formation of a mixture of CO and H_2 (syngas) and further reaction via water–gas shift (WGS) reaction converts CO and steam into more H_2. During internal reforming, these two reactions occur simultaneously, and the equilibrium composition of the gas is dictated by steam to carbon ratio, temperature, and pressure. Methane can also be directly or indirectly used as fuel in high-temperature MCFCs and SOFCs.

Renewable fuels derived from biomass (particularly from nonfood biomass) are an attractive alternative for fuel cells as the electricity generation from biomass-based renewable fuels through fuel cells not only has high efficiency but is also carbon neutral. The typical product gas of the biomass gasification consists of H_2, CO, CH_4, CO_2, H_2O, N_2, and some impurities such as sulfur, chlorine, and alkali metals [14]. The fuel from the biomass gasifier can also contain considerable amounts of tar

Table 5.2 Theoretical specific energy density, hazard, environmental effects, and main by-products of the most common liquid fuels used in fuel cells [10, 12, 13].

Name	Chemical formula	n_e	Specific energy (kWh kg^{-1})	Health effects	Fire hazard	Main by-product
Ethanol	C_2H_5OH	12	8.028	Irritant	Highly flammable	Formic acid, acetaldehyde
Ethylene glycol	C_2H_6O	10	5.268	Irritant, harmful if swallowed	Flammable	Oxalic acid
Dimethyl ether	CH_3OCH_3	12	8.377		Highly flammable	Methanol, methyl formate
Methanol	CH_3OH	6	6.073	Toxic	Highly flammable	Formic acid, formaldehyde
Formic acid	$HCOOH$	2	1.63	Irritant, harmful	Flammable	CO
1-Propanol	$CH_3CH_2CH_2OH$	18	9.07	Irritant	Highly flammable	Propanal
2-Propanol	$CH_3CH(OH)CH_3$	18	8.99	Irritant	Highly flammable	Acetone

n_e: number of electrons involved in the chemicals of the liquid fuels.

depending on the type of gasifier used. Tar is a complex of aromatics and can be represented by a mixture of toluene, naphthalene, phenol, and pyrene. Thus, certain prereforming and cleaning of biomass gasification fuels are needed for fuel cell applications.

5.2
Alkaline and Alkaline Membrane Fuel Cells

5.2.1
Alkaline Fuel Cells

AFCs use an aqueous solution of alkali metal hydroxides, primarily KOH, as the electrolyte, with typical concentration of ~30%. Hydrogen is introduced into the anodic compartment, while oxygen or purified air is provided to the cathode compartment. The process generates water, electrical energy, and heat. The by-product water and heat have to be removed. This is usually achieved by recirculating the electrolyte, while water is removed by evaporation [15, 16]. Figure 5.2 shows the operating principle of an AFC. The overall reaction in an AFC is the recombination of H_2 and O_2 to form water (Equation 5.5).

AFCs have a long history of successful operation in the space program; however, they have not been used extensively in transportation and other applications, mostly as a result of perceived problems of the electrolyte carbonation and corrosion. Yet, there has been a resurgence of interest in AFCs recently [17–19] partly because they have the potential for lower cost in mass production than other types of low-temperature fuel cells. Here, the historic background and development, and issues in the key components of AFCs are briefly described.

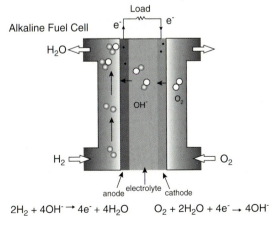

Figure 5.2 Operating principle of an alkaline fuel cell.

5.2.1.1 Historic Development and Prospects

AFCs were first described in 1902 by James H. Reid [20] and in 1955 by Francis T. Bacon [21, 22] who constructed a 5 kW AFC with 30% KOH as electrolyte operated in H_2/O_2 at 200 °C under a pressure of 5 MPa to prevent boiling of the electrolyte. In these fuel cells, a porous nickel electrode and a lithiated nickel oxide electrode were used as anode and cathode, respectively. The performance of these AFCs reached 1 A cm^{-2} at 0.8 V [23]. AFCs developed by Pratt & Whitney Co. and UTC Fuel Cells were used by NASA for Apollo moon flights and Space Shuttle missions to power onboard systems and create drinking water [22, 24, 25]. In parallel, Justi and Winsel [26] achieved similar results at lower pressures and temperatures by increasing the electrode internal area using Raney nickel and Raney silver electrodes.

Almost at the same time, Union Carbide (UC) developed AFCs for terrestrial mobile applications based on liquid caustic electrolytes and polytetrafluoroethylene (PTFE)-bonded electrodes. From the early 1970s, Elenco in Belgium also developed PTFE-bonded carbon electrodes. The company focused on hydrogen–air fuel cell power plants (5–60 kW) for small stations and city buses. In 1986, the European Space Agency (ESA) started the development of a reusable spacecraft for their manned missions. In Russia, the "Photon" AFC was built for manned space flights with high Pt catalyst loadings [25].

In the long history of the development, research, and promotion of AFC technologies, one should mention Professor Kordesch who played a leading role in driving the commercialization of the technology and is best known for developing an AFC-powered Austin A40 car in the early 1970s [27, 28].

5.2.1.2 Gas Diffusion Cathode, Electrolyte Carbonation, and Corrosion

The life span and output power of AFCs greatly depend on the polarization loss at the cathode due to the slow reaction kinetics of the ORR. The cathodes of AFC are generally composed of several PTFE-bonded carbon layers applied to a metal mesh that is used as the current collector. The catalyst is supported on a carbon substrate, usually a high surface area acetylene black or carbon black (e.g., Vulcan XC72R). PTFE is commonly used as binder for the electrodes to enhance the electrode hydrophobicity. The hydrophobicity helps prevent the flooding in the diffusion layer due to the penetration of the electrolyte. In general, carbon cloth or porous carbon paper is used as backing material [17, 29]. The catalyst layer provides a three-phase reaction zone for the gaseous fuel, the electrolyte, and the catalyst. Apart from carbon black, other catalyst supports have been tried, including carbon nanotubes, CNT–perovskite composite, and CNT–active carbon black composite. As for other fuel cells, the manner of production of electrodes for AFCs has significant impact on the performance of the cell.

AFC stacks have generally favored monopolar construction. Nickel mesh is predominantly used as an electrode substrate in AFCs because it is highly conductive and corrosion resistant to KOH. On the other hand, nickel foam with a lower weight to surface ratio and thus less expensive than nickel mesh is shown to be a good substrate for AFC electrodes [30]. However, nickel passivates in alkaline solutions, leading to the formation of a surface oxide layer and increase in the electrode resistance. This can be overcome by either coating nickel with gold, silver, or graphite or embedding it into the diffusion layer [31]. The issues of the electrocatalysts for AFCs are similar to that in alkaline membrane fuel cells (AMFCs) and will be discussed in Section 5.2.2.2.

Carbonation is considered as the major technical challenge in the development of terrestrial AFC running with air. CO_2 in the air and CO_2 formed by the oxidation of carbon support interact with the electrolyte, forming carbonate/bicarbonate (CO_3^{2-}/HCO_3^-) in the liquid alkaline electrolyte:

$$CO_2 + 2OH^- \rightarrow CO_3^{2-} + H_2O \tag{5.7}$$

$$CO_2 + 2KOH \rightarrow K_2CO_3 + H_2O \tag{5.8}$$

Carbonation causes a decrease in ionic conductivity of the electrolyte, a slower ORR kinetics due to a lower OH^- concentration, a reduced oxygen solubility, and an increased electrolyte viscosity. The precipitated carbonate can also block the electrolyte pathways and pores of the catalyst and diffusion layers [31], though there are significant disagreements regarding the exact poisoning and contamination mechanism. The solubility of the carbonate compounds depends on the temperature, the concentration, and the type of electrolyte. KOH solution is almost exclusively used as the electrolyte as it has a higher ionic conductivity than NaOH solution and K_2CO_3 has a higher solubility and is less likely to precipitate. A method to solve carbonation problem was suggested early by circulating the electrolyte instead of using a stabilized matrix. However, electrolyte leakage and parasitic power losses are among the most challenging problems and require the development of improved stack designs.

To prevent electrolyte carbonation, air needs to be pretreated to reduce CO_2 content to between 5 and 30 ppm before it enters the fuel cell [17]. Pure hydrogen can be introduced directly into the anode compartment. However, if hydrogen is produced from carbon-containing fuel sources, it also needs to be purified. Several ways are used to separate CO_2 from fuel streams, such as chemical absorption (using soda lime) and physical absorption (using molecular sieves).

Corrosion is another issue facing AFCs. The corrosion of the carbon support and the degradation of the PTFE binder decrease the performance of the electrodes in AFCs. Carbon can be oxidized by O_2 and HO_2^- radicals formed as intermediates during the ORR. HO_2^- radicals formed will attack and break the polymer chains, leading to the loss of mechanical stability (crack formation) and hydrophobicity (weeping) of the electrodes. Increase in operating temperature accelerates the loss of the electrode activity. This would cause the flooding of cathode structure by the electrolyte due to the penetration of alkaline solution through the electrode. Incorporating radical scavengers in the polymer chains or using other catalysts with lower peroxide formation was proposed to alleviate this problem [24, 31].

In addition to the requirement of very pure gases, durability of AFC systems remains a key issue, especially when using non-noble metal catalysts and air as oxidant.

5.2.2
Alkaline Membrane Fuel Cells

As discussed above, AFCs suffer from carbonation and corrosion problems that cause the performance deterioration of AFCs. To eliminate the adverse effect of CO_2, anion exchange membranes have been developed to replace the liquid alkaline electrolyte. Fuel cells made of such alkaline polyelectrolytes (AMFCs) should in principle have the advantages of AFCs and PEMFCs (see Section 5.3). In particular, the typical AFC issue of electrolyte leakage and carbonate precipitate in the cathode can be avoided because, unlike KOH solutions, there is no dissociated cation in the phase of the polymer electrolyte, thus no salt precipitates would be formed.

5.2.2.1 Alkaline Anion Exchange Membranes

The grand challenge in the development of AMFCs is the alkaline anion exchange membranes (AAEMs). The main issues in the development of AAEMs are the anion conductivity and stability of the membrane in alkaline solution particularly at elevated temperatures. Increase in the ionic conductivity is challenging because the mobility of OH^{-1} is only one third to half of that of H^+ ions. Most of the AAEMs are based on quaternized polymer.

Similar to the acid-doped polybenzimidazole (PBI), alkaline-doped PBI membrane was used by Xing and Savadogo for H_2/O_2 AMFCs [32, 33]. PBI doped with KOH, NaOH, and LiOH showed conductivities in the range of 5×10^{-5} to 10^{-1} S cm^{-1}. The conductivity is 10^{-1} S cm^{-1} when the PBI is treated with 6 M KOH at 70–90 °C. A H_2/O_2 AMFC based on the KOH-doped PBI showed a power density of 370 mW cm^{-2} at 0.6 V, similar to that of a H_2/O_2 PEMFC based on Nafion 117 (see Chapter 10 of the first volume). An alkaline-doped PBI membrane was also developed for direct methanol AMFCs by Hou et al. [34]. At room temperature, methanol permeability through this membrane was one order of magnitude lower than that of Nafion membrane. For a direct methanol AMFC at 90 °C based on this KOH-doped PBI membrane, the peak power density was ∼31 mW cm^{-2}.

Varcoe and Slade et al. [35] developed a quaternary ammonium-functionalized radiation-grafted poly(ethylene-co-tetrafluoroethylene) (ETFE) AAEM. With this membrane as the electrolyte, the cell achieved a maximum power of 130 mW cm^{-2} in H_2/O_2 and 8.5 mW cm^{-2} in methanol/O_2, though the methanol permeability of the AAEM is substantially reduced compared to that of Nafion 115 membranes [36]. The interfacial resistance between the AAEM and the electrodes and the mass transport in the electrodes are considered to be the main factors affecting the cell performance.

Fang and Shen [37] inducted chloromethyl groups and then quaternary ammonium groups into poly(phthalazinone ether sulfone ketone) as AAEMs for AMFCs. The membrane exhibited a thermal stability below 150 °C and an ionic conductivity of 0.14 S cm^{-1} in 2 M KOH solution at room temperature. Wang et al. [38] made quaternary ammonium AAEMs of high ionic conductivities from polysulfone, chloromethyl groups, and different amines. The membrane showed an ionic conductivity of 3.1×10^{-2} S cm^{-1} at room temperature and remained stable in

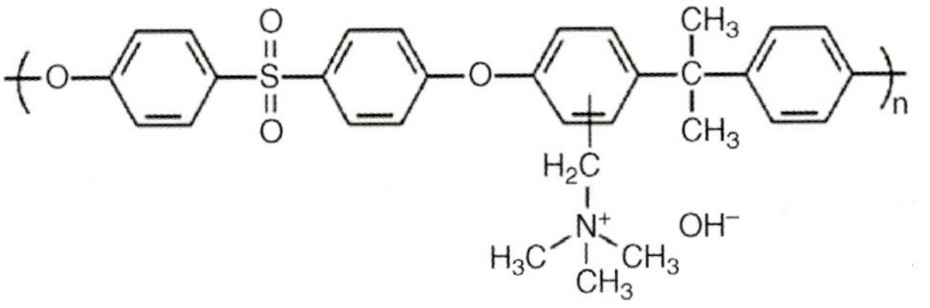

Figure 5.3 The structure of quaternary ammonium polysulfone (QAPS) with hydroxide anions. (Reproduced with permission from Ref. [45].)

alkaline solutions up to 8 M KOH at room temperature. AAEMs synthesized from chloroacetylated poly(2,6-dimethyl-1,4-phenylene oxide) (CPPO) and bromomethylated poly(2,6-dimethyl-1,4-phenylene oxide) (BPPO) [39, 40], cross-linked quaternized poly(vinyl alcohol) membranes (QAPVA) [41], quaternized cardo polyetherketone (QPEK-C) [42], cross-linked PVA/sulfosuccinic acid membrane [43], and vinylbenzyl chloride-grafted poly(vinylidene fluoride) (PVDF) and poly(tetrafluoroethene-co-hexafluoropropylene) (FEP) [44] have also been evaluated as AAEMs in hydrogen and/or methanol AMFCs.

Recently, Zhuang and coworkers [45] reported the synthesis of quaternary ammonium polysulfone (QAPS) (see Figure 5.3). The results showed that QAPS is highly stable in 2 M KOH solution at room temperature and is thermally stable in air below 120 °C. The ionic conductivity is on the order of 10^{-2} S cm^{-1} at room temperatures. A H_2/O_2 AFC based on a QAPS membrane with an ion exchange capacity (IEC) of 1.18 mmol g^{-1} and Pt anode and cathode achieved an open-circuit voltage (OCV) of 1.08 V and a maximum power density of 110 mW cm^{-2}.

To enhance the mechanical, thermal, and chemical stability of AAEMs for fuel cell application, the development of hybrid and composite materials has been attempted. Huang et al. [46] synthesized a composite AAEM from a pyridinium-type polymer and a fibrous woven structure for direct methanol AMFC. It was shown that the woven component considerably improved the tensile strength of the membrane, and the membrane showed an acceptable conductivity but a poor performance in the direct methanol AMFC. Wu et al. [47, 48] prepared polyethylene oxide (PEO)–SiO$_2$ hybrid membranes from alkoxysilane-functionalized PEO-1000, N-triethoxysilylpropyl-N,N,N-trimethylammonium iodine, and other alkoxysilanes. The membranes have good flexibility and good hydrophilicity. However, their ion exchange capacities are not high (1.10–1.65 mmol g^{-1}) and not stable under alkali conditions. Xiong et al. [49] developed hybrid organic–inorganic membranes based on quaternized poly(vinyl alcohol) (PVA) and tetraethoxysilanes as AAEM. The presence of silica in the membranes improves the thermal stability and methanol permeability. The membrane with 5% silica showed the lowest methanol permeability (8.45×10^{-7} cm^2 s^{-1} at 30 °C) and highest ionic conductivity (1.4×10^{-2} S cm^{-1} at 60 °C). Direct methanol, ethanol, and isopropanol AMFCs with PVA/TiO$_2$ membrane as the electrolyte showed maximum power densities of 9.25, 8.00, and 5.45 mW cm^{-2}, respectively [50].

The chemical and thermal stability of AAEMs is a major concern, especially at elevated temperatures [51]. In order to convert the Cl^{-1} anion in the as-synthesized

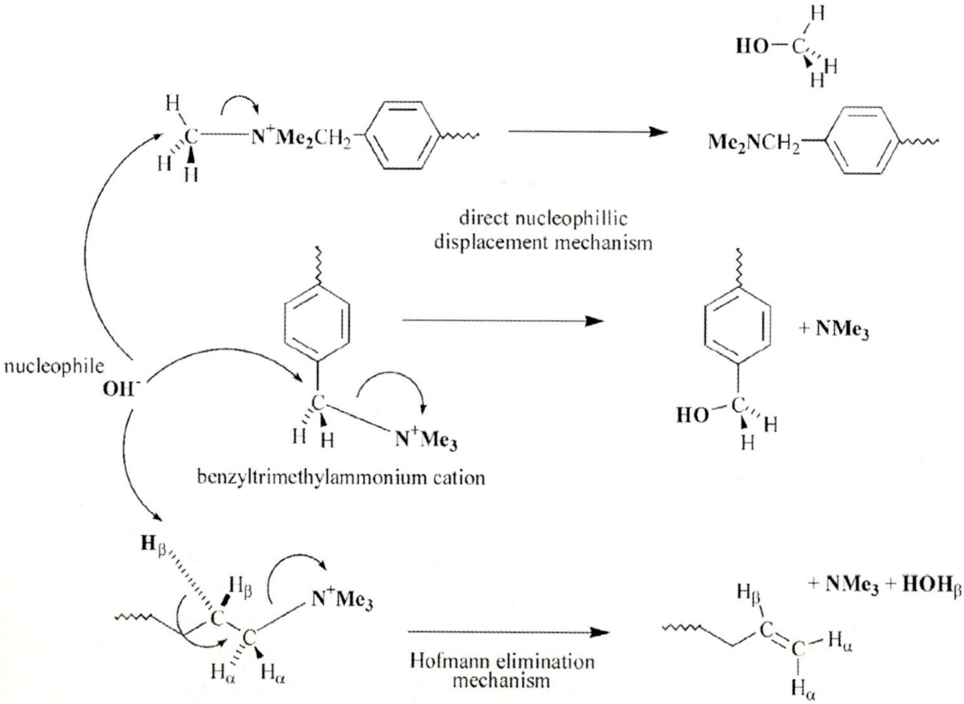

Figure 5.4 Two alternative mechanisms of displacement of the trimethylammonium groups by hydroxide anions in AAEMs at elevated temperatures. (Reproduced with permission from Ref. [33].)

polymer into an OH^{-1} anion, the membrane has to be immersed in KOH solution for anion replacement. Thus, chemical stability in KOH solution is essential. Instability is mainly due to the displacement of the ammonium group by the OH^- anions (an excellent nucleophile) via a direct nucleophilic displacement and/or a Hoffmann elimination reaction when β-hydrogens are present (see Figure 5.4).

5.2.2.2 Electrocatalysts in Alkaline Medium

Interest in electrocatalysis in alkaline media has been increasing in recent years largely due to latest developments in AFCs and AMFCs. There are several distinctive advantages of electrocatalysis in alkaline media over that in acidic media. In general, most electrocatalytic processes would be more facile in alkaline solutions than in acidic solutions due to the lower degree of anion adsorption in alkaline solutions. Hsueh et al. [52] found that the ORR proceeds at lower overpotentials in KOH solutions than in H_2SO_4 and H_3PO_4 solutions. In contrast to the acidic conditions in which very few electrode materials are stable, a much wider range of materials are stable in alkaline environment, including much less expensive materials such as Ni and Ag. Recently, Spendelow and Wieckowski gave a critical review on the electrocatalysis of ORR in alkaline solutions [53].

Pt-based materials are used as ORR electrocatalyst in alkaline solutions, and the activity of Pt can be further improved by formation of Pt alloys such PtPd [54], PtCo [55], and PtAu [56]. Ag, which is much cheaper and more abundant than Pt, has displayed promising activity for ORR in alkaline media [57]. The concentration of alkaline solutions has positive effect on the activity of Ag catalysts for ORR and in very high pH solutions the activity of Ag catalysts is comparable to Pt catalysts [58]. Carbon-supported Ag cathodes have been reported to be more stable than Pt cathodes during long-term operation [59]. The combination of reasonably high activity, good stability, and relatively low price makes Ag attractive as an alkaline ORR cathode material. The activity of Ag for ORR can also be improved by incorporating elements such as Mg and Co [60, 61] and tungsten carbide [62]. Besides Ag-based catalysts, a variety of catalytic materials have been investigated for ORR in alkaline media. These include carbon [63], manganese, and cobalt oxides [64–66], metal macrocycles such as phthalocyanine [67] and porphyrins [68], carbon-supported Pd [69], perovskite oxides (e.g., $La_{1-x}Ca_xCoO_3$) [70], and Mn- and Co-containing spinels [71–73].

There have been extensive studies on Pt as electrocatalysts for the small alcohol oxidation in alkaline media. The methanol oxidation on Pt is a complicated process that involves the formation of intermediate such as CO or HCO, and the rate of CO/HCO oxidation is determined by the coverage of adsorbed CO/HCO and OH [74]. Carbonate and, under some conditions, formate may also form in alkaline media for the methanol oxidation on Pt catalysts [75]. In the case of ethanol oxidation reaction in alkaline media, the prevalence of partial oxidation to acetaldehyde and acetate [76, 77] makes Pt an unattractive anode catalyst for direct ethanol fuel cells (DEFCs). The electrocatalytic activity of Pt can be enhanced by adding oxides such as Pt–CeO_2/C, Pt–MgO/C, Pt–ZrO_2/C, and Pt–NiO/C [78, 79].

Pd-based catalysts show high oxidation activities for methanol, ethanol, glycerol, and ethylene glycol in alkaline media [80–84]. Pd modified with various oxides, including CeO_2, NiO, Co_3O_4, and Mn_3O_4, displayed better activity and stability than the conventional PtRu/C catalysts. Among them, Pd–NiO (6: 1 by weight)/C showed the highest activity for the ethanol oxidation, while Pd–Co_3O_4 had the best activity for the oxidation of methanol, ethylene glycol, and glycerol. It was proposed that OH species formed on the surface of oxide at lower potential are beneficial for the oxidation of the intermediate species on Pd [80, 82, 83]. Incorporation of tungsten carbide into Pd by an intermittent microwave heating technique has shown to enhance the activity of Pd for ethanol oxidation [85]. Pd-based alloys such as Pd–Ag were also found to exhibit higher current density and better stability than Pd/C and Pt/C. The enhancement effect of Ag on Pd for ethanol oxidation reaction was suggested as the result of the promoted adsorption of OH to the alloy catalysts due to an upshift of Pd d-band center [54, 86]. Pd–(Ni–Zn)/C and Pd–(Ni–Zn–P)/C are shown to have better activity for ethanol oxidation than Pd/C [87, 88]. Nanotechnology plays an important role in electrocatalytic activity. For example, Pd nanowire arrays (NWAs) exhibit a much higher ethanol oxidation activity than Pd film and E-TEK PtRu/C catalyst [89].

The Ni catalysts commonly used in AFCs could not work well for AMFC where the hydrogen anode made from Ni catalysts shows an electrode potential characteristic of nickel oxides instead of the potential of a hydrogen electrode. Lu et al. [90] addressed this problem by decorating Ni surface with Cr oxide. Using the density functional theory (DFT) analysis, it was shown that decoration of the Ni surface with Cr oxide may weaken the Ni—O bond with no influence on the formation of surface Ni—H bonds. An AMFC fabricated with QAPS membrane, Cr-decorated Ni anode, and Ag cathode achieved a power density of $50\,\mathrm{mW\,cm^{-2}}$ at $60\,°C$ in H_2/O_2, an encouraging result with an unoptimized electrode/electrolyte interface.

5.3
Polymer Electrolyte Membrane Fuel Cells

5.3.1
Technology Development of PEMFCs

The core component of a PEMFC is membrane electrode assembly (MEA), which consists of a proton exchange membrane sandwiched between anode and cathode; detailed description can be found in Chapter 10 of the first volume and Chapter 7 of this book. Hydrogen is oxidized at the anode and the proton is transported to the cathode through the PEM. The electrons produced flow to a cathode through an external load, reducing ambient oxygen to water. Figure 5.5 illustrates the operating principle of a PEMFC.

To form a stack, MEAs are connected in series by bipolar plates. For the past decades, extensive research has been conducted on all the components in PEMFCs and significant progresses have been made in the power density specifications close to meet the requirement for automotive and stationary applications. The focus of the R&D activities is shifting to the cost reduction and the improvement in the durability and efficiency of the PEMFC system.

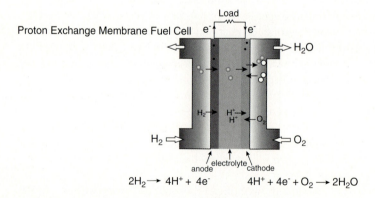

Figure 5.5 Operating principle of a proton exchange membrane fuel cell.

5.3.1.1 Nafion and Other Perfluorinated Membranes

The earliest development of PEM for fuel cell applications involves the phenol sulfonic and polystyrene sulfonic membranes [91, 92]. However, the as-prepared membranes have relatively low mechanical strength, short life span, and insufficient proton conductivity (see also Chapter 10 of the first volume). A major breakthrough in the field of PEMFCs came with the employment of perfluorosulfonic acid (PFSA) polymer membranes commercialized under trademark, Nafion, by DuPont. The Nafion membranes possess a high acidity and high conductivity ($\sim 0.1\,\text{S cm}^{-1}$ at 30 °C, 100% RH for Nafion 117 [93]) and are far more stable than the polystyrene sulfonate membranes. PFSA consists of a PTFE-based structure that is chemically inert in reducing and oxidizing environments, and can be operated at temperatures between 0 and 80–90 °C (see Figure 5.6).

However, Nafion membranes must be fully hydrated to exhibit good proton conductivity. Dehydration of this type of membrane must be prevented as its ionic conductivity dramatically decreases when its water content drops below full saturation [94]. The requirement of the auxiliary humidification unit complicates the system design and lowers power density of the system. Therefore, research activities have been carried out to find alternatives or modify the Nafion membrane for the operation under elevated temperatures or low relative humidity (RH). To realize this goal, various approaches have been developed, one of which is to synthesize Nafion-based hybrid composite membranes by the incorporation of hygroscopic metal oxide particles including SiO_2 [95, 96], ZrO_2 [97], TiO_2 [98], and so on into the hydrophilic domains of the Nafion membrane to enhance their water retention properties and thermal stability.

Another approach is to develop self-humidifying membranes. Watanabe et al. [99] fabricated Nafion 112-based self-humidifying membranes by uniformly incorporating Pt nanoparticles. The origin of this self-humidifying action is the recombination of permeated H_2 and O_2 on the Pt catalytic sites. Yang et al. [100] prepared self-humidifying membrane by recasting the Nafion and Pt/C catalyst to form a Pt/C–Nafion layer followed by hot pressing together with the pure Nafion membrane. The disadvantage of these methods is that the presence of Pt nanoparticles or Pt/C catalyst may cause a short circuit of the membrane. Reinforced membranes commercialized by Gore consist of a porous PTFE matrix filled with a Nafion-type electrolyte [101]. The benefit of this composite membrane is that the membrane can be very thin, typically 30 μm or even less, due to the strength of the PTFE matrix. Thin membranes have a low resistance and it is much easy to keep them hydrated than a thick membrane.

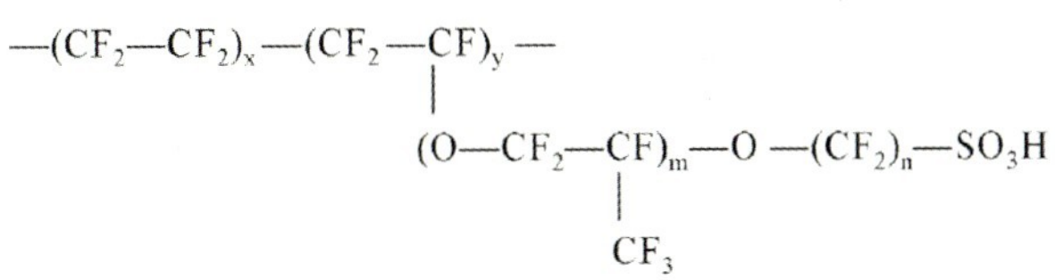

$$-(CF_2-CF_2)_x-(CF_2-CF)_y-$$
$$|$$
$$(O-CF_2-CF)_m-O-(CF_2)_n-SO_3H$$
$$|$$
$$CF_3$$

Figure 5.6 Structure of a PFSA polymer.

The high cost of PFSA-based membranes has been the driving force for the development of alternative membranes for the PEMFCs. Many of them are based on aromatic hydrocarbon, such as sulfonated poly(ether ether ketone) [102], polybenzimidazoles [103], sulfonated polystyrene [104], and sulfonated polyimides [105]. It has been proven to be difficult to develop cheaper alternatives that can meet the requirements of conductivity and durability [106].

5.3.1.2 Pt, PtRu, and Nonprecious Metal Composite Electrocatalysts

Only platinum-based catalysts have sufficient activity in the 40–80 °C temperature range to meet performance targets of PEMFCs. The major challenges associated with Pt catalyst are (i) its low tolerance to CO if reformate or methanol is used as the fuel, (ii) the low electrocatalytic efficiency for ORR at the cathode side, and (iii) the high cost. The activity improvement of Pt electrocatalyst toward fuel cell applications is usually realized by the appropriate design of bimetallic structure.

Bimetallic Alloy Catalysts The most classical bimetallic electrocatalysts for CO tolerance are PtRu catalysts. The significantly enhanced CO tolerance of PtRu catalysts is explained by the bifunctional mechanism, that is, the oxygen-containing species such as OH can be adsorbed on Ru surface at a lower potential than Pt, which may react with CO adsorbed on Pt surface, releasing the active sites for further reaction [107]:

$$Pt\text{-}CO + Ru\text{-}OH \rightarrow Pt + Ru + CO_2 + H^+ + e^- \tag{5.9}$$

PtSn bimetallic catalysts also show enhanced electrocatalytic activity for the methanol oxidation in DMFCs. However, the proposed origin of enhanced activity on PtSn is different from that on PtRu due to the significant difference in the electronegativity of Sn and Ru. The charge transfer from Sn to Pt would increase the electron density around the Pt atoms and downshift their d-band center, which thus leads to the weakened chemisorption energy with adsorbate, such as CO intermediate generated from the methanol oxidation in DMFCs [107]. Therefore, the activity enhancement of PtSn electrocatalysts is mainly due to the modification of electronic environment of the Pt sites (electronic effect).

The cathode reaction involves the electrochemical reduction of oxygen, and thus the chemisorption energy of oxygen species on Pt surface would have a significant effect on its electrocatalytic activity. Until now, the most extensively investigated cathode bimetallic electrocatalysts are PtNi and PtCo. Stamenkovic et al. [108] have demonstrated that the $Pt_3Ni(111)$ surface is 10-fold more active for ORR than the corresponding monometallic Pt(111) surface. They found that the $Pt_3Ni(111)$ surface has an unusual electronic structure (d-band center shift). The outermost atomic layer consists of pure Pt and the second layer is Ni-rich under operating conditions relevant to fuel cells. The d-band center of Pt on the surface was downshifted due to the electronic modification by the underlying Ni, which results in the weakened chemisorption of OH^- on Pt surface, thus leading to increased number of active sites for O_2 adsorption and reaction.

Bimetallic alloy electrocatalysts can also be synthesized with core–shell structure with the Pt shell as the electrocatalytic surface site for the reaction while the solid core part of Pt nanoparticles is replaced by other inexpensive metal elements [109, 110]. Adzic and coworkers [111] studied the electrocatalytic activity for ORR of Pt monolayer on Ru(0001), Ir(111), Rh(111), Au(111), and Pd(111) in a 0.1 M $HClO_4$ solution on a disk electrode and found out that Pt monolayer on Pd(111) electrode is the most active catalyst for ORR. This core–shell approach provides a potential way to significantly reduce the usage of precious Pt. Besides, the appropriate selection of core metal could modify the electronic structure of Pt shell, which would allow further improvement in the electrocatalytic activity of Pt.

Kibler et al. [112] presented an electrochemistry and theoretical study with pseudomorphic palladium monolayer on different single-crystal electrode surfaces, showing the significant effect of lateral strain on the electrochemical properties of the surface Pd monolayer. As illustrated in Figure 5.7, the direct effect of the lateral strain on the d-band center is the shift in energy with nearest-neighbor separations, which would subsequently affect the binding energies of adsorbates. The origin of the shift in the d-band center is a change in the interatomic distances within the surface monolayer. If the d-band center of a metal is more than half-filled (tensile strain), an expanded pseudomorphic monolayer would lead to an upshift of the d-band center due to band narrowing and energy conservation, reflecting increased chemisorption energy with adsorbates. In contrast, if the d-band center of a metal is of compressive strain, its d-band center would downshift, resulting in weakened adsorption. Therefore, during the design of core–shell bimetallic electrocatalysts, the choice of different metal cores would lead to significantly different modification on the electronic structure of the shell.

The electrocatalytic activity and performance of Pt-based catalysts critically depend on the deposition, distribution, and size of Pt nanoparticles supported on high

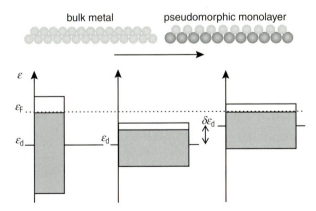

Figure 5.7 The illustration of the shift in the d-band center caused by the interatomic distance change within and overlayer. If the d-band center of a metal is more than half-filled, an expanded pseudomorphic monolayer will lead to an upshift of d-band center due to band narrowing and energy conservation. (Reproduced with permission from Ref. [112].)

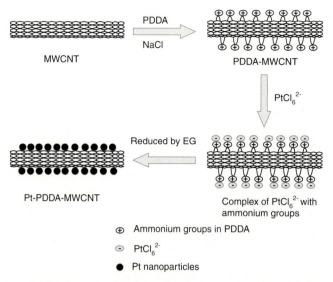

Figure 5.8 PDDA functionalization of MWCNTs and the *in situ* synthesis of Pt nanoparticles on PDDA-functionalized MWCNTs. (Reproduced with permission from Ref. [185].)

surface carbon or CNTs [113]. And the activity of Pt nanoparticles is also significantly affected by the nature of their interaction with carbon supports. Wang et al. [114] presented a novel and highly effective method for the deposition of Pt nanoparticles on poly(diallyldimethylammonium chloride) (PDDA) polyelectrolyte-functionalized multiwalled carbon nanotubes (MWCNTs) (see Figure 5.8). The strong adsorption of the positively charged PDDA on MWCNTs is believed due to the σ–π interaction between PDDA and the basal plane of graphite of MWCNTs [115]. Pt nanoparticles are then synthesized *in situ* on the PDDA-functionalized MWCNTs via the self-assembly between negatively charged Pt precursors and positively charged functional groups of PDDA. The subsequent reduction by ethylene glycol (EG) gives Pt nanoparticles with uniform distribution and high density. High-density and uniformly distributed Pt nanoparticle catalysts with Pt loading as high as 93 wt% have been successfully synthesized on PDDA-functionalized MWCNTs [116].

Non-Pt Electrocatalysts The high cost of Pt serves as a constant drive for researchers to develop non-Pt catalysts based on heat-treated macrocyclic compounds of transition metals, ruthenium and palladium alloy catalysts [117–120]. The most promising one appears to be Pd alloys and high loaded RuSe/C. Shao et al. [121] synthesized Pd–Fe alloy nanoparticles as cathode electrocatalyst for ORR and the results indicate that the Pd_3Fe/C electrocatalyst shows a higher catalytic activity than the monometallic Pd and the state-of-the-art commercial Pt/C electrocatalysts. The significantly improved electrocatalytic activity for ORR on Pd–Fe electrocatalyst was attributed to the modification of the electronic structure of Pd by Fe via orbital overlapping, which, in turn, alters the d-band center. Recent study on self-assembled Pd/heteropolyacid, tungstophosphoric acid ($H_3PW_{12}O_{40}$, HPW) on polyelectrolyte

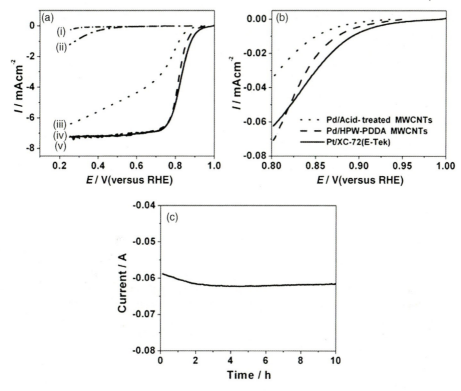

Figure 5.9 (a) Polarization curves for the ORR on raw MWCNT (i), HPW–PDDA–MWCNTs (ii), Pd/acid-treated MWCNTs (iii), Pd/HPW–PDDA–MWCNTs (iv), and E-TEK Pt/C (v) electrodes in a 0.5 M O_2-saturated H_2SO_4 solution with a scan rate of 5 mV s^{-1} and a rotation speed of 2500 rpm. (b) The kinetic current density (I_k) obtained from RDE experiment normalized by the electrochemical surface area of the catalysts. (c) Stability test of Pd/HPW–PDDA–MWCNT catalyst at 0.6 V (versus RHE) using gas diffusion electrode at room temperature; electrode area: 2 cm^2; Pd loading: 1 mg cm^{-2}. (Reproduced with permission from Ref. [122].)

(PDDA)-functionalized MWCNTs shows that the Pd/HPW–PDDA–MWCNT catalyst is stable and its activity is comparable to the conventional Pt/C electrocatalysts for the ORR in acidic media (see Figure 5.9) [122], indicating the promising potential of the Pd/HPW–PDDA–MWCNT catalysts for ORR in fuel cell applications.

Tungsten- and molybdenum-based materials were investigated as hydrogen oxidation catalysts for PEMFCs. However, the performance of non-Pt catalysts for hydrogen oxidation reaction is generally much lower than that of the Pt.

5.3.1.3 Self-Humidification and Electrode Structure Optimization

The electrodes used in PEMFC are gas diffusion electrodes, which consist of a gas diffusion layer (GDL) to function as a support layer, and a catalyst layer where the electrochemical reactions occur. GDL is typically a dual-layer carbon-based porous material, including a macroporous carbon fiber paper or carbon cloth substrate covered by a thin microporous layer (MPL) consisting of carbon black powder and a hydrophobic agent. A GDL must also have the sufficient ability of transporting the reactant gas from flow channels to catalyst layer, draining out produced liquid water, and collecting the current produced in the catalyst layer. Inefficient removal of product water has an extremely strong effect on the fuel cell performance as it can completely block the transport of oxygen to the reaction sites [123].

The most commonly used materials for GDL are a mixture of carbon black and PTFE supported on hydrophobic treated carbon cloth or carbon paper. The microstructure and composition (i.e., the carbon and PTFE contents) of the GDL not only affect its electrical resistance but also affect the Pt utilization of the catalyst layer [124]. For the catalyst layer, Pt-based catalysts supported on carbon are commonly used, and the catalyst layer is usually impregnated with a Nafion solution to introduce the proton conduction path to the membrane. The key to an efficient electrode in a PEMFC is that the Pt nanoparticles in the catalyst layer have to have simultaneous access to the gas (hydrogen for anode or oxygen for cathode), the electron to/from external load circuit, and the proton to/from the PEM. The thickness of the catalyst layer is kept small ($\sim$10 µm) to minimize resistance for reactants and protons [125]. Another variation of this structure is to coat this hydrophilic catalyst layer directly on the membrane to ensure more intimate interface between the catalyst layer and the membrane [126].

For the PEMFCs operated at elevated temperatures and low humidity, the tendency to lose water in the catalyst layer, in particular the anode side of the cell, remains. Similar to self-humidifying membranes, addition of nanosized hygroscopic materials such as silica [127, 128] in the catalyst layer has been adopted to retain water in the electrode. However, because the additives are not proton and electron conductive, the addition of excess hydrophilic particles would have detrimental effect on the cell performance. This could be overcome by using mesoporous carbon and mesoporous carbon–silica to replace high surface area carbon as support in the catalyst layer [129]. The water condensation inside the nanochannels of highly ordered mesoporous supports enhances the water retention of the electrodes.

5.3.1.4 Bipolar Plates and Stack Design

Bipolar plate is a multifunctional component of the fuel cell stack, acting as a separator between the fuel, oxidant, and coolant, distributing reactant and product streams, and collecting the current by electrochemical reaction. To fulfill the required functions, bipolar plates must have a high electrical conductivity, low gas permeability, good corrosion resistance, sufficient strength, substantially high thermal conductivity, and low cost. Bipolar plates are conventionally made from high-density graphite. Materials such as moldable graphite, carbon–polymer composites, or thin metal sheets are also evaluated and developed as bipolar plates. Figure 5.10 presents the classification of the bipolar plates for PEMFCs.

Composites normally consist of a polymer resin used as a binder and a conductive filler such as graphite, carbon nanotube, metal-coated graphite, or low melting metal alloys [130]. Although the composites have a somewhat lower conductivity and thermal conductivity compared to graphite and metal, these offer the advantage of

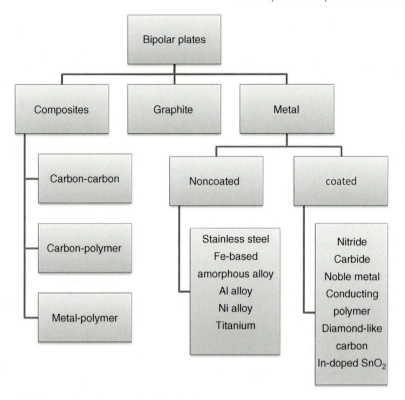

Figure 5.10 Classification of bipolar plates.

reducing manufacturing time and cost, since they can be fabricated to any shape by extrusion or molding and no postmachining process to form the flow channel is needed. The polymer binder to conductive filler ratio is a compromise between the electrical conductivity and the mechanical strength of the composites.

An alternative to graphite and graphite–polymer composite plates is metallic plates such as Ai, Ti, Ni, and stainless steel. Metals, as thin sheets, offer potential benefits as they have excellent mechanical properties, have high electrical and thermal conductivity, and can be easily stamped to a desired shape. The major concern with the metallic plates is the contact resistance between the bipolar plates and the GDL, attributed to resistant oxide films formed on the surface. The corrosion of metallic materials such as stainless steel leads to the formation of multivalent cations (e.g., Fe^{3+}, Ni^{2+}, and Cr^{3+}), which can seriously affect durability of the membrane and catalyst. Better interfacial contact and corrosion resistance can be obtained either by using special alloys [131] or by coating stainless steel plates with nitride, carbide, and conducting metal oxides. The major research focus in bipolar plates is mainly on the use of graphite/polymer composites, metallic materials coated with noble metals or various nitride- or carbide-based alloys to reduce contact resistance and to prevent corrosion [130, 132–134].

While single cell design is usually done based on the desired experimental purposes, the stack design would be much more complicated since the desired overall voltage and power and durability of cell performance should be considered. Furthermore, additional parameters such as specific geometrical requests and cooling system must be taken into account also for the final application of stacks. Not too much information about stack design is available in the open literature mainly because most of the R&D work is conducted by industrial companies. Nevertheless, several reviews are available for the outline of the basic principles [135, 136].

One critical issue in the operation and performance of a PEMFC is the design of flow fields, which provide channels for transporting reactant gases, liquid water, and water vapor. The configuration, dimension parameters, and aspect ratio significantly impact the performance of PEMFCs. Li and Sabir [137] reviewed the impact of the design of several types of flow fields, including pin-type, parallel straight pattern, interdigitated flow, serpentine pattern, and multiple serpentine flow. As noted by Li and Sabir [137], although the multiple serpentine flow field ensures adequate water removal, mitigates stagnant area formation, and reduces the reactant pressure drop relative to single serpentine design, the reactant pressure drop through each of the serpentines remains relatively high due to the relatively long path of each channel. The pin-type flow, on the other hand, can result in a low pressure drop, but the reactants flowing through such flow fields tend to follow the path of least resistance, which may lead to the formation of stagnant areas and uneven reactant distribution.

5.3.1.5 Heat, Water Management, and Modeling

Heat and water are generated as by-products during the fuel cell operation. Their management in a fuel cell system is normally coupled and remains a big challenge [138]. The main heat sources in a PEMFC include the entropic heat of reactions, irreversibilities of the electrochemical processes, and the ohmic resistances. Since PEMFC is normally operated below 100 °C, the temperature difference between the cell and the environment is too small to effectively evolve the heat to the environment or provide heat utilization. Thus, air cooling, liquid cooling, and heat exchangers are commonly used. Liquid cooling is the most common cooling method adopted by practically all developers. This involves the design of bipolar plates with internal cooling channels between the anode and the cathode. On the other hand, heat exchangers are generally employed for high-power PEMFC systems, such as those used in transportation and stationary power stations.

Water management is one of the major issues in PEMFC technology. Although water, in gaseous or liquid form, is a product of the reactions at the cathode, it is difficult to keep the water in the membrane. The water balance depends on the operating parameters, such as pressure, temperature, humidity, and gas flow rates. Sufficient amounts of water should be present in the PEM to maintain high proton conductivity. Membrane and catalyst layer dehydration tends to occur when PEMFC is operated at high temperature, low relative humidity, low current density, or high airflow rate due to the insufficient water supply. Electroosmotic drag also causes membrane dehydration at the anode side. On the other hand, excess water would lead to flooding in the cathode, blocking the oxygen transport to the catalytic sites. One way to improve water management is to humidify the gases entering the fuel cell. An integrated stack can use cooling water flowing through the stack to humidify the gases [139].

Better water management can also be achieved by optimization of the flow field design. Designing the flow field pattern to have a higher internal pressure drop will remove the excess liquid water from the fuel cell more efficiently. Heat and water generated from fuel cell stack could be partially reused. For example, if water is used as a coolant in fuel cell system, the heat and water management could be integrated into one single handling system. In this way, water removes the generated heat from the stack; the same water and heat are used to humidify the reactant gases.

Extensive modeling studies have been conducted in an attempt to understand the phenomena inside MEA and stack. The major focus has been on the transport and distribution of various species, such as reactants, water, and heat (temperature), within the single fuel cell domain. Two good reviews on the modeling have been published by Weber and Newman [140] and Djilali [141]. Simulation studies both on the stack level and on the system level have also been reported, providing an analysis on full system performance including auxiliary components such as compressors and pumps [142, 143].

5.3.1.6 Durability and Degradation of Electrocatalysts, Membrane and Cell Performance

Although significant development of PEMFCs has been made, the large-scale application of this technology is still impeded by its high cost and limited life span. Both the electrode and the membrane are susceptible to aging effects that lead to performance loss of PEMFCs. Extensive studies indicate that performance loss of PEMFCs is mainly caused by the loss of the electrochemically active surface area (ECSA) of the electrocatalysts and the degradation of polymer electrolyte membranes [144]. The loss of ECSA of Pt electrocatalysts in the fuel cell electrode is mainly related to the coarsening, dissolution of Pt nanoparticles, and the corrosion of the catalyst carbon support under the fuel cell operation conditions [145]. Carbon corrosion is believed to be induced mainly in the transient regimes (e.g., startup and shutdown cycles) and by fuel starvation either from uneven flow or from gas flow blockage.

Membrane degradation can be caused by mechanical, thermal, and chemical/electrochemical effects/processes. The proton-conducting polymer can lose its conductivity due to dehydration [146], due to contamination with trace metal ions originated from corrosion of metallic bipolar plates, or due to physical degradation such as membrane creep, microcrack formation, and dry/wet cycling [147]. Figure 5.11 shows the SEM micrographs of a Nafion membrane treated for 48 h in H_2O_2/metal ion-containing solution [147]. The small bubbles gradually became pinholes most likely due to decomposition of the repeating units in the membrane. The dimensional change of membrane due to relative humidity cycling or change during fuel cell operation [148] is also detrimental to mechanical durability.

Various strategies have been employed to prepare more durable electrocatalysts for the application in PEMFCs. Zhang *et al.* [149] demonstrated significantly improved Pt stability against dissolution under potential cycling regimes by modifying

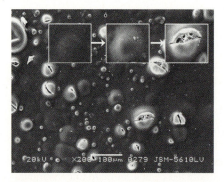

Figure 5.11 SEM micrograph of the surface of the Nafion membrane treated in H_2O_2/metal ion-containing solution for 48 h. The metal ion solution was obtained by treating SS316L stainless steel in 0.5 mol l^{-1} nitric acid. (Reproduced with permission from Ref. [147].)

Pt nanoparticles with small gold clusters. It is believed that Au atoms are preferentially deposited on high-energy surface sites of Pt and reduce the dissolution of these unstable sites. Instead of conventional carbon black, more corrosion-resistant materials, such as carbon nanotubes, have been proposed as electrocatalyst support and enhanced durability has been observed [150, 151]. Corrosion of carbon support can also be alleviated by enhancing water retention on the anode. To completely eliminate the corrosion problem associated with carbon, Yan *et al.* [152] studied the supportless Pt and PtPd nanotubes as electrocatalysts for ORR in fuel cells with high durability, high utilization, and high activity.

To mitigate membrane degradation, efforts have been made by modifying the membrane structure to improve its chemical stability by incorporating regenerative radical scavengers [153] in the membrane and by using PTFE-reinforced membrane [154, 155]. To prevent mechanical failure of the membrane, the MEA and flow field structure must be carefully designed to avoid local drying and to ensure the uniform mechanical stresses across the MEA. The durability of PEMFCs also depends on the operational parameters such as startup/shutdown cycle, fuel starvation, and freeze–thaw cycle. Detailed discussion on the durability and degradation issues of PEMFC components can be found in Refs [156, 157].

5.3.2
PEMFCs with Liquid Fuels

Other than the cost and durability issues, the widespread application of hydrogen/air-based PEMFCs is also limited by the difficulty in the storage and transportation of hydrogen fuel, as well as the lack of refueling infrastructure. To overcome these problems, PEMFCs with various liquid fuels have been proposed. Among them, single carbon-containing compounds have been extensively studied, such as methanol and formic acid, as they have relatively clean oxidative pathways. More complex fuels such as ethanol are also attractive due to their higher energy density, lower toxicity, and greater availability. The PEMFCs with liquid fuel have three main members, direct methanol fuel cells, direct formic acid fuel cells, and direct ethanol fuel cells, and will be discussed here. In addition, other liquid fuels such 1-propanol and 2-propanol [158], dimethyl ether [11], ethylene glycol [159], and so on were also tested as fuels in PEMFCs.

5.3.2.1 Direct Methanol Fuel Cells

Interest in DMFCs stems mainly from the high volumetric energy density (4.8 kWh l^{-1}) compared to liquid hydrogen (2.6 kWh l^{-1}), and the relative ease of handling and distribution. The system design is also relatively simple. A DMFC is different from the conventional hydrogen PEMFC only by the anode reaction:

$$CH_3OH + H_2O \rightarrow CO_2 + 6H^+ + 6e^- \tag{5.10}$$

However, the performance obtained in a DMFC is inferior compared to a PEMFC supplied with pure hydrogen. There are three main reasons for this: (i) the electrochemical activities of Pt- and Pt-based alloy catalysts for the methanol oxidation reaction (MOR) are low due to the sluggish electrokinetics of the reaction involving six electrons to fully oxidize methanol to CO_2 and H_2O; (ii) methanol migration from the anode to cathode, leading to a mixed potential at the cathode with a consequential lowering of the OCV and poisoning the cathode catalysts; and (iii) electroosmosis of water from the anode to the cathode (up to 19 water molecules per proton) in the case of diluted methanol at the anode, causing severe flooding at the cathode. The high methanol crossover through Nafion membranes allows only low concentrations (generally 1–2 M) to be fed to a DMFC.

Similar to the electrocatalysts in PEMFCs, the most significant challenge for DMFCs is the development of active, robust, and low-cost catalysts [9]. The most common anode catalysts in DMFCs are PtRu binary alloys with the molar ratio of 1:1 because carbon monoxide is the intermediate in the MOR that will poison Pt-only catalysts. The high, usually unsupported, catalyst loading (up to 4 mg cm^{-2}) at the anode is also used to maximize methanol oxidation, although the cost of such high Pt load would be prohibitively high for widespread commercial applications. Another approach is to raise the DMFC operating temperature to >150 °C where the electrokinetics of MOR are more facile. This approach is feasible with acid-doped PBI membrane and inorganic proton exchange membranes (see Section 5.6.5).

Extensive research has been carried out on the modification of Nafion-based membranes to reduce the methanol crossover by blocking methanol transport pathways while maintaining proton conductivity. For example, composite membranes such as Nafion–silica [160, 161] and Nafion–zirconium phosphate [162] have been developed and show lowered methanol permeability. Sandwiching a Pd thin film between Nafion membranes, depositing a Pd- or Pd-based alloy thin film on the surface of a Nafion membrane, or depositing Pd nanoparticles through ion exchange followed by chemical reduction has been shown to reduce the methanol crossover [163–165]. Unfortunately, the overall cell resistance increases with the introduction of the Pd thin film.

Pore-filling membranes proposed by Yamaguchi and coworkers [166–168] also show reduced methanol crossover. As methanol crossover takes place mainly through the polymer electrolyte filler, the fact that composite membrane mechanically prevents the electrolyte from swelling can lead to decreased methanol crossover due to the confined passage. Various substrates were attempted, including polytetrafluoroethylene [169], cross-linked polyethylene [166], and commercial porous polyimide [167]. However, the reported performance of DMFC based on this type of membrane is either lacking or not satisfactory mainly because of the difficulty to completely fill the porous substrate to form a workable membrane with sufficient proton conductivity and mechanical strength.

Various multilayer membrane structures have also been investigated. For example, Wang et al. [170] directly polymerized protonated polyaniline (PANI) on Nafion membranes, forming a composite membrane, in which a thin PANI layer (methanol barrier) is sandwiched between the Nafion membrane and the anodic catalyst layer, with PANI facing the anode. The methanol permeability of the as-fabricated PANI/Nafion composite membrane is reduced to 59% of that of the as-received Nafion membrane. Using layer-by-layer (LbL) self-assembly technique to modify Nafion membrane is also shown to be effective in reducing the methanol permeability of Nafion [171, 172]. Figure 5.12 shows the LbL self-assembly of PDDA polyelectrolytes

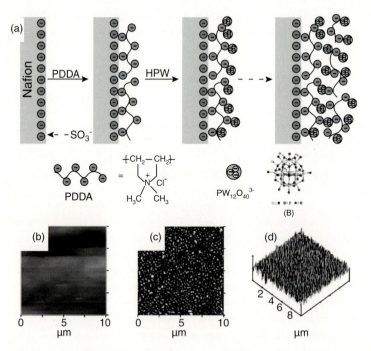

Figure 5.12 (a) Scheme of the LbL self-assembly of PDDA/HPW bilayers on Nafion membranes, (b) AFM images of N212 membrane, (c) AFM image of N212 membrane modified with four self-assembled PDDA–PWA bilayers, and (d) three-dimensional AFM image of (c). (Reproduced with permission from Ref. [172].)

and Keggin-type phosphotungstic acid (HPW) molecules on a Nafion membrane. Owing to the existence of a negatively charged sulfonic acid group, SO_3^-, on the surface of the Nafion membrane, self-assembly of positively charged polycation PDDA occurs as a result of the electrostatic interaction. Dipping the self-assembled PDDA–Nafion in a solution containing negatively charged HPW molecules would lead to the self-assembly of a HPW monolayer, forming a PDDA–HPW bilayer structure that is stabilized by the strong electrostatic force. The best performance was obtained for the cell that used an N212 membrane with four self-assembled PDDA–HPW bilayers, achieving a power density of 30.9 mW cm^{-2}. This is an increase of 103% in power density compared to the cell using unmodified N212 membrane (15.2 mW cm^{-2}). Power density as high as 200 mW cm^{-2} was reported for a DMFC when operated at 120 °C in a methanol concentration of 0.5 mol l^{-1}, catalyst loadings of >2 mg cm^{-2}, Nafion 117 membrane (178 μm), and a back pressure of 2 atm at the cathode [173].

However, to reduce the cost of DMFCs, a more radical approach is required. One approach is to replace expensive Nafion-based fluorinated membranes with hydrocarbon membranes [174]. The hydrocarbon membranes such as sulfonated polythion-ether sulfone (sPTES), sulfonated poly(ether ether ketone) (sPEEK), and sulfonated polyphosphazene (sPPZ) are relatively cheap and are potentially more effective for DMFCs than Nafion.

DMFCs that have no auxiliary liquid pumps and gas blowers/compressors are known as "passive" and are developed specifically for powering portable electronic devices. Figure 5.13 shows a typical design of a passive DMFC [175]. The MEA for the passive DMFC is similar to that for "active" DMFCs and is composed of diffusion and catalyst layers on both anode and cathode sides of a PEM. Since the anode is directly in contact with the methanol solution in the fuel reservoir,

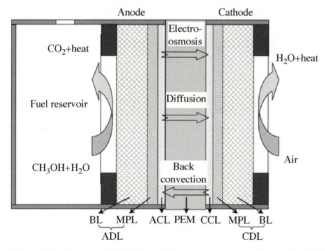

Figure 5.13 Schematic of MEA and heat/mass transport in passive DMFC. (Reproduced with permission from Ref. [175].)

methanol and water are transported from the reservoir to the anode catalytic layer mainly by diffusion due to the concentration difference between the fuel reservoir and the catalyst layer [176]. Cathode is directly exposed to ambient air, also known as air breathing cathode.

Methanol concentration is one of the few adjustable parameters that influence the performance of passive DMFCs. In contrast to active DMFCs, high concentration of methanol improves the cell performance due to the improved mass transfer of methanol from the fuel reservoir to the anode catalyst layer. In general, passive DMFC stacks adopt the monopolar arrangement to completely make use of the air breathing cathode. Kim et al. [177] designed and fabricated a passive DMFC stack, reaching a power of $\sim$1.0 W and a power density of 37 mW cm^{-2}. The most important issue in the design of passive DMFCs is how to effectively enhance the efficiency of the supply/removal of reactants/products without invoking active means so that the fuel cell performance and stability can be increased.

5.3.2.2 Direct Formic Acid Fuel Cells

DFAFCs have attracted considerable interest recently as portable power sources because of lower crossover through the Nafion-based membrane and a higher kinetic activity than methanol [178]. The anode reaction of a DFAFC is described as

$$HCOOH \rightarrow CO_2 + 2H^+ + 2e^- \quad (E_a^0 \sim -0.25 \text{ V versus SHE}) \quad (5.11)$$

Thus, the overall reaction with the oxygen as oxidant is

$$HCOOH + \frac{1}{2}O_2 \rightarrow CO_2 + H_2O \quad (5.12)$$

The OCV of a DFAFC is $\sim$1.48 V, higher than either hydrogen or direct methanol fuel cells. Although formic acid has a lower theoretical energy density (2.1 kWh l^{-1}) compared to 4.8 kWh l^{-1} for methanol fuel, the smaller permeability through Nafion membrane allows the use of highly concentrated fuel solutions and thinner membranes in DFAFCs.

The most commonly accepted mechanism for formic acid oxidation is the so-called "dual pathway mechanism": direct oxidation through a dehydrogenation reaction without the formation of CO and formation of adsorbed CO intermediate via dehydration. Thus, to enhance the DFAFC efficiency and to avoid poisoning of the catalyst, anode catalyst selection is pivotal in directing the oxidation pathway to form CO_2 and H_2O without the formation of CO intermediates.

Although platinum is a commonly used electrocatalyst in fuel cell applications, the electrocatalytic activity of formic acid oxidation on Pt is relatively low. Pt-based bimetallic catalysts such as PtPd, PtRu, PtPb, and PtAu were developed for DFAFCs [179–181]. Addition of a second metal not only reduces the use of Pt but also enhances the activity of the catalysts for the oxidation reaction of formic acid. Recently, Wang and coworkers [178] synthesized PtAu bimetallic catalysts with a decoration structure of Pt on Au cores and significant enhancement in the electrocatalytic activity toward formic acid oxidation has been observed. This enhancement

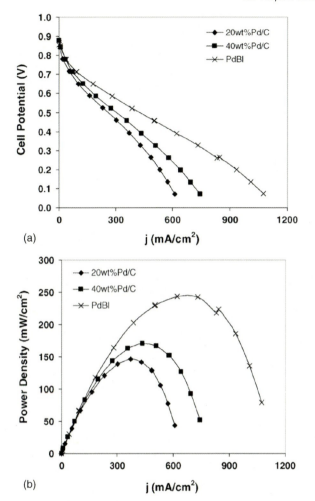

Figure 5.14 (a) Fuel cell polarization plots and (b) power density curves with 3 M formic acid at 30 °C. The flow rate of formic acid was 1 ml min^{-1}. Dry air was supplied to the cathode at a flow rate of 390 sccm without back pressure. (Reproduced with permission from Ref. [183].)

was attributed to the suppression of CO formation on the Pt surface in Pt-decorated Au catalysts. Ordered intermetallic electrocatalysts, such as PtBi, PtIn, and PtPb, also showed significant enhancement for the formic acid oxidation [182].

Ha et al. [183, 184] demonstrated that Pd-based catalysts possess superior performance for formic acid oxidation compared to Pt. Figure 5.14 shows performance curves of DFAFCs with 20 wt% Pd/C, 40 wt% Pd/C, and unsupported Pd black (PdB1) [183]. Unsupported Pd black was used as cathode. The best performance was obtained on the DFAFCs with unsupported Pd black catalysts with a maximum power density of 243 mW cm^{-2}. However, DFAFC with 20 wt% Pd/C outperforms the DFAFC with Pd black based on the Pd catalyst loading. The electrocatalytic activity of Pd catalysts also depends strongly on the morphology of the Pd nanoparticles [185]. The performance of the DFAFCs with Pd-based catalysts deteriorates with operation time, but the lost activity can be largely recovered by applying a high anodic potential [183].

Similar to DMFCs, "passive" operation of DFAFCs is also possible. For passive DFAFCs, the air breathing cathode is used and the anode flow field is replaced by a fuel reservoir. Formic acid is delivered from fuel reservoir to the anode catalyst layer simply by diffusion and capillary forces [186].

5.3.2.3 Direct Ethanol Fuel Cells

Although methanol is widely used as the fuel for PEMFCs, its toxicity as a harmful chemical limits its widespread application to some extent. In contrast, ethanol has no toxicity and can be produced in large quantity from fermentation process. In addition, ethanol possesses a higher theoretical energy density than methanol and is easy to store and handle. Therefore, direct ethanol fuel cell (DEFC) is particularly attractive for portable power sources. Considering the similar molecular structure of methanol and ethanol, both binary and ternary Pt-based catalysts were also tested for the ethanol oxidation reaction (EOR). Among the binary catalysts, PtM/C (M = Sn, Ru, Pd, W), PtSn is the best binary catalysts for EOR [187, 188]. The electrocatalytic activity is related to the Pt/Sn atomic ratios and the best activity and stability of PtSn catalysts were observed for the PtSn catalysts with a Pt/Sn atomic ratio of 2:1 (see Figure 5.15) [189]. Ternary catalysts were also extensively studied for EOR. Zhou et al. [190] studied the activity of PtRuW and PtRuMo ternary catalysts for EOR. PtRuW (1:1:1) and PtRuMo (1:1:1) showed an excellent activity. The power output of the DEFCs with PtRuW and PtRuMo anode catalysts was 31–38 mW cm^{-2} at 90 °C, higher than 29 mW cm^{-2} for the cell with PtRu anode, but still inferior to 52 mW cm^{-2} obtained on the cell with PtSn (1:1) anode catalysts.

Recently, Adzic and coworkers [191] developed a ternary electrocatalyst, Pt/Rh/SnO$_2$, for the electrooxidation of ethanol. The electrocatalyst was prepared with a cation adsorption–reduction–galvanic displacement method with the deposition of Pt and Rh atoms on SnO$_2$. It was reported that the as-prepared electrocatalysts could effectively split the C–C bond in ethanol at room temperature in acid solutions, leading to the direct formation of CO$_2$, which has never been realized with existing catalysts, such as Pt/C. The authors demonstrated that the enhanced electrocatalytic activity contributes to the specific properties of each of the constituents, induced by their interactions. Substantial progress has been made in the areas ranging from fundamentals to the development of electrode materials and fuel cell stack assembly [10, 192, 193]. However, in order to realize the direct use of ethanol as the fuel for fuel cells at low temperatures, the breakthrough in the electrolyte and electrocatalyst development is still needed. The formation of highly flammable and carcinogenic byproducts due to the incomplete oxidation of ethanol is a serious concern in the development of DEFCs.

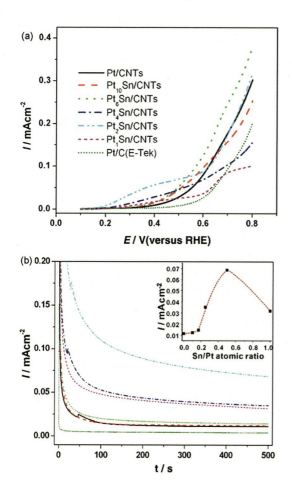

Figure 5.15 (a) Linear sweep voltammetry and (b) chronoamperometry curves for the ethanol oxidation on Pt/CNTs and PtSn/CNTs with different Pt/Sn atomic ratios in a N_2-saturated 0.5 M H_2SO_4 + 1.0 M ethanol solution at room temperature. LSV was measured at a scan rate of 1 mV s^{-1} and CA was measured at 0.5 V (versus RHE). Inset in (b) shows the plot of current measured at 0.5 V (RHE) after 500 s test versus the Sn/Pt atomic ratios of PtSn/CNT catalysts. (Reproduced with permission from Ref. [189].)

5.4 Phosphoric Acid Fuel Cells and Molten Carbonate Fuel Cells

5.4.1 Phosphoric Acid Fuel Cells

The PAFC is the first fuel cell technology that has been widely used in civilian applications. With the same half cell reactions, it is very similar to PEMFC, except it operates at higher temperature (~200 °C) and uses concentrated phosphoric acid as the electrolyte. Because of the elevated temperature, PAFC shows certain tolerance toward fuel impurity and the activation overpotential loss at cathode is also significantly reduced.

In a PAFC, the liquid electrolyte between the anode and the cathode is embedded in a porous matrix made of chemically stable and electronic insulating materials. The matrix combined with electrolyte is used to provide the ionic conductivity, to minimize crossover of the reactant gases, and to avoid short circuit of the electrodes. The electrolyte should have good chemical, thermal, and electrochemical stability. At present, 100% H_3PO_4 is widely used as electrolyte in PAFCs [194]. The operating principle of a PAFC is the same as that of PEMFCs in the case of H_2/O_2 (see Figure 5.5).

5.4.1.1 Electrocatalysts, Electrolyte, and Gas Diffusion Electrode

Electrocatalysts Similar to PEMFCs, the cathodic electrocatalysts used in PAFCs are carbon-supported Pt-based nanoparticle electrocatalysts. Typically, there are two main ways for Pt nanoparticle deposition: (1) colloidal sol method and (2) impregnation method. Pt-based alloy catalysts are also shown to possess significant advantages compared to the pure Pt catalysts for PAFCs. For example, Seo et al. [195] developed a chemical vapor deposition (CVD) method to modify Pt/C catalyst with Cr as the cathodic catalyst for a PAFC. The Cr-modified Pt/C catalyst showed a higher electrocatalytic activity for ORR than the original one.

Among the various noble metal catalysts that are stable under acidic conditions, Pt is still the most active electrocatalyst for the hydrogen oxidation reaction. Due to the fact that the hydrogen oxidation on Pt is about 10^5 times faster than ORR in acidic media, fewer efforts have been devoted to study anode electrocatalysts compared to cathode electrocatalysts. However, the presence of impurities and contaminants such as S and CO in the hydrogen source could easily poison Pt catalyst, resulting in low reaction efficiency and fast catalyst deactivation [194].

Electrolyte and Matrix In a PAFC system, the SiC-based matrix (0.1–0.2 mm thick) has been commonly utilized. The thickness of SiC matrix should be small in order to reduce the ohmic resistance. However, such thin SiC matrix could lead to the poor mechanical strength, which limits the maximum pressure difference between the anode and the cathode. Low ionic resistance and low gas crossover through the matrix are also necessary. The properties of the matrix should be appropriate to ensure that the liquid electrolyte is immobilized inside the cell permanently. Various techniques have been used to optimize the matrix. For example, mixtures of SiC with other metal carbides or silicides were used to fabricate a thin layer of matrix [196]. For the fabrication of the matrix/electrolyte composite, SiC can be mixed with a small amount of PTFE and applied to the catalyst layer of the electrode by a screen printing method, then followed by subsequent filling with the electrolyte.

Gas Diffusion Electrodes A typical gas diffusion electrode is made of mixtures of catalyst particles (Pt on carbon black, Pt/C) and PTFE, forming the electrolyte network and gas network in the electrodes [197]. The relative percentage of electrolyte and gas network in the gas diffusion electrode could be tuned by the PTFE content. The particle size of PTFE is generally much bigger than that of carbon black particles, which may lead to insufficient cover of the carbon black surface by PTFE in the gas diffusion electrode. Watanabe and coworkers [197] proposed a design of a gas diffusion electrode by wet-proofing of carbon black with fluorinated polyethylene (FPE/CB), where Pt/CB and FPE/CB function as electrolyte network and gas network, respectively.

The degradation explanation of Pt-based catalysts in PEMFC as discussed in Section 5.3.1.6 is also partly applicable to the PAFC system. However, the degradation of Pt-based catalyst is more severe in PAFCs because of the highly corrosive (acidic) electrolyte and high operating temperatures [144]. Similarly, degradation of electrocatalysts could also be evaluated by the ECSA loss. The ECSA loss in a PAFC system has been attributed to several factors: (1) the dissolution of Pt particles, especially at the cathode when operating at above 0.8 V; (2) formation of larger particles by the redeposition of Pt atoms via the electrochemical Ostwald ripening; and (3) migration of Pt from cathode to anode through the electrolyte.

5.4.1.2 Sustainability and Challenges

PAFC is one of the most stable and successful types in the entire fuel cell family. However, it also suffers from some severe problems, especially the high overall cost and the limited life span of the PAFC plants. The high overall cost is contributed by many components, including the separator, bipolar plate, and Pt-based electrocatalyst. The technology developed to fabricate bipolar plates must be suitable for the mass production of PAFC for the extensive worldwide utilization of this technology. The cost of Pt-based electrocatalysts is another challenge. Recent research has been focused on finding alternative Pt-free electrocatalysts, for example, macrocycle compounds. However, the active sites of these Pt-free electrocatalysts could be easily destroyed by peroxide produced during the oxygen reduction at the cathode. Meanwhile, alloying Pt with other suitable and nonprecious metal could enhance the electrocatalytic activity and reduce the cost of the catalyst. Finally, the life span of a PAFC could be enhanced using novel stable supports, such as carbon nanotubes and TiO_2.

5.4.2
Molten Carbonate Fuel Cells

An attractive feature of MCFCs is their expected clean and efficient generation of electricity from hydrocarbon fuels due to their high operating temperature of ~650 °C. A typical MCFC consists of Ni/Cr anodes, lithium-doped NiO cathodes, and lithium aluminate, $LiAlO_2$, matrix filled with lithium and potassium carbonate mixture (e.g., 62 mol% Li_2CO_3/38 mol% K_2CO_3) as the electrolyte. The operating principle of an MCFC is based on the transfer of carbonate ions (CO_3^{2-}) from the cathode to the anode through the molten carbonate immobilized in the ceramic matrix. The fuel gas is a humidified mixture of H_2 and CO, and the oxidant is a mixture of O_2 and CO_2 that may contain water vapor. Note that the latter may enable

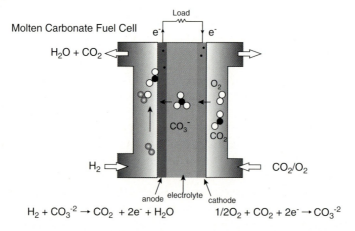

Figure 5.16 Schematic representation of the operating principle of a molten carbonate fuel cell.

transport of protons, increasing the overall performance. Figure 5.16 explains the basic operating principle of an MCFC.

In the cell construction, the anode, electrolyte, and cathode are thin, flat plates (typical thickness 0.4–1.8 mm). A bilayer nickel/stainless steel such as austenitic 310S and 316L separator plate serves as a current collector and provides the anode–cathode interconnection. A major determinant for satisfying requirement to life span of at least 40 000 h (4.5 years) is the stability and behavior of various cell components. In this aspect, the stability of NiO cathode plays a vital role [198].

5.4.2.1 NiO Cathode

The cathode material widely used in MCFCs is porous NiO. NiO is a p-type semiconductor and when used as the cathode in the fuel cell, its conductivity is enhanced by doping with lithium provided by Li_2CO_3 in the molten electrolyte. However, the slow dissolution of the NiO cathode into the electrolyte is one of the major technical obstacles to the commercialization of the MCFCs. NiO has a small degree of solubility in the carbonate electrolyte. The transport of dissolved Ni species from the cathode to the anode surface and the reduction of the dissolved Ni species and precipitation of metallic Ni into the electrolyte matrix are considered to be the main cathode dissolution mechanism [199]. The NiO cathode dissolution not only leads to the loss of the cathode's surface area for ORR but also possibly results in short circuiting.

To overcome these problems, various approaches have been developed mainly based on two different strategies: to use alternative cathode materials or to adjust the composition or the acidity of the carbonates to reduce the dissolution of NiO. For the former case, as a typical example, $LiCoO_2$ has been shown to be a promising component for cathode because of its higher stability and lower solubility than nickel oxide. However, the problem associated with $LiCoO_2$ is its high manufacturing cost and low electrical conductivity. The use of additives in the electrolyte such as alkaline earth metal oxides or carbonates based on Ba or Sr has been reported to decrease the acidity of the electrolyte, thus improving the stability of the NiO cathode [200, 201]. Unfortunately, the use of additives would lead to the decrease in cell performance as a trade-off.

Alternatively, modification of the NiO cathode composition by coating has been attempted. In this case, only a small amount of stabilizer is deposited on the internal surface of NiO without any detrimental effect on electrical conductivity, resulting in the suppression of the NiO dissolution. A variety of stabilizers have been investigated as coatings, such as $LiCoO_2$, $La_{0.8}Sr_{0.2}CoO_3$, $LiFeO_2$, CeO_2, $La_{0.8}Sr_{0.2}MnO_3$, Co_3O_4, TiO_2, and so on [202–207]. For example, TiO_2-coated NiO cathode shows a remarkable decrease in solubility, and changes in the network structure, pore size, or particle size of TiO_2-coated cathode before and after the solubility testing are negligible [204].

5.4.2.2 Anode, Stability and Retention of Electrolyte, Corrosion of Separate Plate, and Component Stability

Unalloyed porous Ni anodes used by early developers were found to shrink during operation under stack compressive load. Alloying with chromium and/or aluminum strengthens oxide dispersion resulting in adequate creep strength. Excellent mechanical and chemical stability of the Ni–Al anode was demonstrated in an 18 000 h field operation [208].

In the MCFC system, the electrolyte matrix is used to support the molten carbonate electrolyte and to provide the electronic insulation and ionic transport between anode and cathode. The MCFC performance strongly depends on the amount of molten carbonate in the electrolyte matrix, and this would require that the electrolyte matrix should have high porosity and narrow pore size distribution. At present, the most commonly used electrolyte matrix is made of $LiAlO_2$ in the form of submicron powder. It has been shown that $LiAlO_2$ may exist in three different crystallographic forms, α, β, and γ. Among them, γ-$LiAlO_2$ has been considered as the most suitable form for MCFCs [209]. The main issue related to the electrolyte matrix is its degradation caused by corrosion of the molten carbonate and possible transformation of the γ-$LiAlO_2$ in the matrix to α-$LiAlO_2$ under the MCFC operating conditions [209, 210]. Thermal cycling of the stack is considered the major cause for matrix cracking. Addition of crack arrestors such as large and rod-shaped γ-$LiAlO_2$ particles was found to improve the matrix strength and the stability of the cell performance [211, 212]. Recently, Lee et al. [213] studied the aluminum-reinforced γ-$LiAlO_2$ matrices for MCFCs. After the reinforcement, the porosity was increased from 54% for the pure γ-$LiAlO_2$ matrix to 62%, while the pore size decreased to 0.1 μm. The cell performance on aluminum-reinforced γ-$LiAlO_2$ matrix was found superior to that on nonreinforced matrix.

In addition to the NiO cathode dissolution, the corrosion of stainless steel-based separator plates in MCFCs due to direct contact with the corrosive molten carbonate is also the most critical problem for the MCFC performance and life span [214–217]. The corrosion of the separator plate mainly takes place in three regions of the MCFC: the cathode, the anode, and the wet seal region [217]. In the cathode area, the separator corrosion is mainly related to the growth of the product layer, and this would consume the electrolyte material and lead to increase in electrical resistance between the separator and the electrode. The corrosion rate is proportional to the square root of time, indicating that the corrosion is diffusion limited [217]. The corrosion at the anode side involves rapid formation of thick multilayered oxide scales. Usually, Ni-clad stainless steels are used to prepare corrosion-resistant separator plates due to their thermodynamic stability in the anode-side environment. The carburization layer formed below the Ni coating resulted in the mechanical failure and cell performance loss. The corrosion problem in the wet seal area of the MCFC separator plate is critically serious since the wet seal area is exposed to a highly corrosive environment.

Coating the separator plate with a suitable material possessing high corrosion resistance, good electrical conductivity, good high-temperature resistance, and acceptable cost has been reported effective for protection [214, 217, 218]. Several coating techniques have been investigated, such as chemical vapor deposition of TiN, TiC, and TiCN, ion vapor deposition of Al followed by the heat treatment, thermal spraying of Al, electrodeposition of Ni and Al, sol-gel coating $LiCoO_2$ or perovskite, and so on [215, 216, 218, 219]. For example, Kawabata et al. [216] electroplated Ni and Al on the stainless steel plates followed by heat treatment and the high corrosion resistance was obtained due to the formation of a $LiAlO_2$ layer on the Al–Ni intermetallic compound after heat treatment.

Rare earth (RE) metals are used as additives in MCFCs to improve the creep resistance, corrosion resistance, and high-temperature resistance of materials. However, as pointed out by Wee and Lee [201], efforts to enhance the properties of MCFC materials using RE have not yielded marked effects as their use in SOFCs.

5.4.2.3 Direct Internal Reforming MCFCs

The MCFC system operates at relatively high temperatures enabling an acceptable reforming reaction kinetics for natural gas and other light hydrocarbons in the fuel cell. Thus, the thermal and chemical features of the fuel cell and reforming reactions are uniquely complementary to an efficient integration of both these reactions inside the anode compartment. Figure 5.17 shows the principle of the direct internal reforming MCFC (DIR-MCFC) developed by FuelCell Energy Inc. [4].

A hydrocarbon fuel such as natural gas is introduced into the anode compartment along with steam. Unused fuel from the cell is oxidized with fresh air and is introduced to the cathode side. In the presence of a catalyst in the anode compartment, the fuel and water react to form hydrogen. The heat required for the reaction is provided by the fuel cell. Hydrogen reacts electrochemically with CO_3^{2-} at the anode to release water, CO_2, and heat (which is partly consumed for the reforming reaction). This "one-step" process leads to a simpler, more efficient, and cost-effective energy conversion system compared to external reforming fuel cells. The main issue in the case of DIR-MCFC appears to be deactivation of the catalysts due to sintering of the support or of Ni crystallites, the most commonly used reforming catalyst. The migration of electrolyte may lead to filling the pores of the carrier and thus deactivate the catalyst.

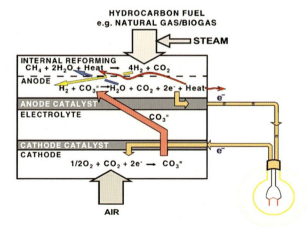

Figure 5.17 Schematic representation of the operating principle of a DIR-MCFC. (Reproduced with permission from Ref. [4].)

5.5
Solid Oxide Fuel Cells

An SOFC consists of a porous anode, a fully dense solid electrolyte, and a porous cathode. Driven by the differences in oxygen chemical potentials, oxygen ions migrate through the electrolyte to the anode where they are consumed by oxidation of fuels such as hydrogen, methane, and hydrocarbons (C_nH_{2n+2}). Figure 5.18 illustrates the operating principle of an SOFC; see also Chapter 12 of the first volume.

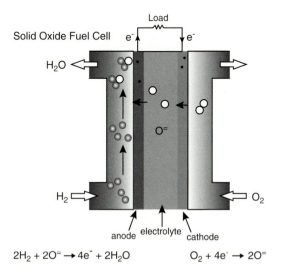

Figure 5.18 Schematic representation of operating principle of a solid oxide fuel cell.

In Kröger–Vink notation (see Chapter 3 of the first volume), the ORR can also be written as

$$O_2 + 2V_O^{\bullet\bullet} + 4e^- = 2O_O^X \tag{5.13}$$

where $V_O^{\bullet\bullet}$ is vacant oxygen site and O_O^X is an oxygen ion on a regular oxygen site in the solid electrolyte lattice.

Yttria-stabilized zirconia (YSZ) is the most commonly used solid electrolyte, while lanthanum–strontium manganite (LSM) and Ni/YSZ often serve as the cathode and anode, respectively (see Chapter 12 of the first volume). Traditional SOFCs operate at high temperatures, 900–1000 °C, because of the relatively low oxygen ion conductivity and high activation energy of oxide electrolytes such as YSZ. However, lowering the operating temperature of SOFCs to intermediate range of 500–800 °C brings substantial technical and economic benefits. The cost of an SOFC system can be substantially reduced by using less costly metal alloys as interconnect and compliant temperature gaskets [220]. Furthermore, as the operating temperature is reduced, thermodynamic efficiency, system reliability, and durability of cell performance increase. This increases the possibility of using SOFCs for a wide variety of applications, including residential and automotive applications. On the other hand, reduction in operating temperature results in a significant increase in the electrolyte and electrode resistivity and the polarization losses. To compensate for the performance losses, the thickness of electrolyte layer has to be reduced in order to lower the ohmic resistance of the cell. However, when using a thin electrolyte layer, the electrolyte can no longer mechanically support the cell. Thus, anode- or cathode-supported structures need to be employed. There are significant activities in the research and development in materials, fabrication technologies, and stack designs for intermediate temperature SOFCs or IT-SOFCs. The reader may refer to numerous reviews covering essentially all topics, namely, electrodes [221–224], electrolytes [225, 226], interconnects [220, 227], and seals [228].

5.5.1
Development of Key Engineering Materials

5.5.1.1 Anode
The basic requirements for anode materials are that they must be stable in the reducing environment and should have sufficient electronic conductivity and activity for electrochemical oxidation of fuel gases under operating conditions. Since the SOFCs operate at high temperatures (up to ~1000 °C), the anode must also be chemically and thermally compatible with other cell components during cell fabrication and operation.

The state-of-the-art anode materials are based on the Ni/YSZ cermets. The function of the oxide component is primarily to reduce the sintering of the Ni metal phase, to decrease the thermal expansion coefficient (TEC), and to improve the electrochemical performance of the anode. Thus, the composition and phase distribution are important for the microstructure and performance of the anodes.

For example, with addition of 30 vol% YSZ, the TEC of the Ni/YSZ cermet is $\sim 12.5 \times 10^{-6}$ K^{-1}, substantially less than that of pure Ni. In a simple two-phase system, the theory predicts the percolation threshold at $\sim$30 vol% of the phase with higher conductivity for the transition from dominant ionic conductivity to dominant electronic conductivity. This is generally true for the electrical conductivity of the Ni/YSZ cermet systems, as shown by Dees *et al.* [229]. The YSZ phase in the cermet plays a significant role in the creation of additional reaction sites by extension of the three-phase boundary (TPB).

The microstructure and distribution of Ni and YSZ phases in the Ni/YSZ cermets depend strongly on the characteristics of starting NiO and zirconia powders (i.e., the particle size and size distribution), the heat treatment of the raw powders, the Ni/YSZ ratio, sintering temperature, the fabrication process, and so on. Jiang [230] studied the effect of sintering temperature on the activity and performance of Ni/YSZ anodes. Good YSZ to YSZ connectivity was evident for anodes sintered at 1400 °C, the same temperature at which the lowest electrode ohmic resistance and polarization resistance were observed. The sintering of the YSZ phase in the cermet has significant effect not only on the formation of Ni-to-Ni electrical contact network in the cermet but also on the formation of a good bonding between the YSZ phase in the cermet and the YSZ electrolyte. A comprehensive review of the effect of the powder characteristics and fabrication processes on the performance and microstructure of Ni/YSZ cermet anodes has been published [221].

The microstructure and performance of Ni/YSZ cermet anodes can also be improved by impregnation of nanosized ionic conductive and active particles such as doped ceria, $Sm_{0.20}Ce_{0.80}O_{1.9}$, by wet impregnation method. Jiang *et al.* [231] studied the effect of the SDC and YSZ impregnation on the performance of Ni/YSZ cermet anodes. As shown in Figure 5.19, fine and nanosized YSZ and SDC particles in the range of 100–300 nm were uniformly distributed in Ni/YSZ cermet structure. The formation of fine oxide particles inhibits sintering, grain growth, and agglomeration of zirconia and Ni phases during processing at high temperatures [231]. For H_2 oxidation on Ni/YSZ cermet anode without impregnation, the anode overpotential (η) was 234 mV at 250 mA cm^{-2} and 800 °C. After impregnation with $Sm_{0.2}Ce_{0.8}O_{1.9}$, η was reduced down to 56 mV. Using a 0.5 µm thick yttria-doped ceria (YDC) interlayer between anode and electrolyte, the electrode polarization resistance for Ni/YSZ anode for methane oxidation was $\sim$1.2 Ω cm^2 at 600 °C, which is much smaller than 6.6 Ω cm^2 for the Ni/YSZ cermet anode without the interlayer under the same conditions [232]. This is due to the enlarged sites for the reaction at the electrode/electrolyte interface via the electrochemically active and mixed conducting interlayer, such as YDC. The segregation of impurities such as Si at the Ni/YSZ and YSZ electrolyte interface regions can have significant effect on the electrode behavior of Ni anodes [233].

An important issue regarding the Ni/YSZ cermet anodes is the degradation under SOFC operating conditions. The predominant microstructure change is agglomeration and coarsening of the Ni phase, primarily due to the poor wettability between the metallic Ni and YSZ oxide phase [234]. At SOFC operation temperatures, coarsening kinetics is fast and agglomeration of Ni occurs relatively

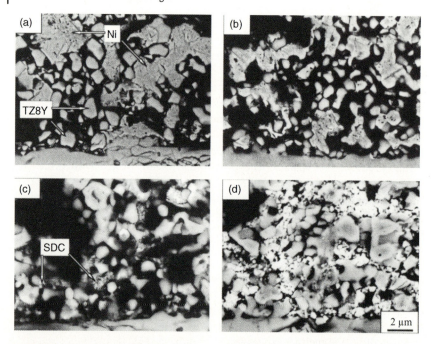

Figure 5.19 SEM micrographs of Ni/YSZ cermet anodes with and without impregnation after the fuel cell testing: (a) none, (b) impregnation with 3 M $Y_{0.03}Zr_{0.97}(NO_3)_x$, (c) impregnation with 3 M $Sm_{0.2}Ce_{0.8}(NO_3)_x$, and (d) impregnation three times with 3 M $Sm_{0.2}Ce_{0.8}(NO_3)_x$ solution. The scale bar applies to all SEM micrographs. (Reproduced with permission from Ref. [231].)

rapidly [235]. Thus, a deep microstructural optimization is required to achieve good stability.

Coarsening processes can be simulated in three dimensions using phase-field models, analyzing a 3D Ni/YSZ anode measured by the focus ion beam scanning electron microscopy (FIB-SEM) [236] as the starting structure (see Figure 5.20). The simulations assumed that the YSZ phase did not change whereas the Ni phase was allowed to coarsen via Ni surface diffusion. Considerable coarsening, as well as smoothening of surfaces, is evident comparing the results in Figure 5.20. From the simulation, the TPB density in the anode versus time can be obtained. The results showed, in particular, that the TPB density initially decreases rapidly with time, before decreasing more gradually at longer times. Overall, the TPB density decreased by ~15%, and the Ni surface area decreases in a similar manner. Note that comparison with experimental data suggested that the total coarsening time corresponded to ~400 h at 1000 °C, and a Ni surface diffusion coefficient of $4 \times 10^{-14}\,m^2\,s^{-1}$.

For hydrocarbon fuels ranging from natural gas to diesel, the most abundant impurity is sulfur, existing as gaseous hydrogen sulfide, H_2S, after reforming. Sulfur is a major impurity in coal, so sulfur tolerance is a major issue for SOFC power plants

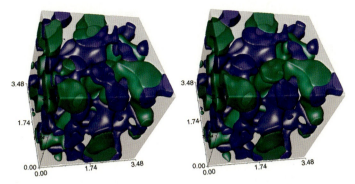

Figure 5.20 3D image renderings of a Ni/YSZ anode functional layer in the as-prepared state (a) and after coarsening using a phase field model (b). Green represents Ni, blue represents pore, and the YSZ phase is transparent. (Courtesy of Professor Katsuyo Thornton, University of Michigan, MI)

designed to utilize gasified coal. Nickel- and alloy-based metals can corrode in high carbon activity gases by a process known as metal dusting. Metal dusting involves carbon deposition and incorporation into the metal that results in disintegration of bulk metals and alloys into metal particles at high temperatures (300–850 °C). Degradation of the metallic structure can be avoided by providing a high enough steam to carbon (S/C) ratio in fuel gas, for example, by mixing the fuel with H_2O [237]. However, high S/C ratio is not attractive for fuel cells as it lowers their electrical efficiency by diluting the fuel. The endothermic nature of the steam reforming reaction can also cause local cooling and steep thermal gradients potentially capable of mechanically damaging the cell stack. Thus, development of anodes with high tolerance toward carbon deposition and sulfur poisoning is critical for SOFCs utilizing hydrocarbon fuels.

Replacing Ni with carbon-inert Cu to form Cu/YSZ cermet anodes [238] and using Ni–Cu/YSZ cermet anodes prepared by impregnation of porous YSZ layer with copper and nickel nitrate solution [239] were found to improve the resistance toward carbon deposition. Gadolinia-doped ceria (GDC) impregnation can substantially enhance the electrocatalytic activity of Ni anodes in methane and suppress carbon deposition even at open circuit in 97% CH_4/3% H_2O at 800 °C [240]. Ni/YSZ-based cermet anodes have a very limited tolerance to H_2S. As shown in Figure 5.21 for 50 ppm H_2S, the sulfur poisoning is characterized by a two-stage behavior: a rapid drop in the cell performance upon exposure to H_2S, followed by a gradual but persistent deterioration in performance [241]. At open circuit or low current density, the poisoning effect of H_2S on the electrocatalytic activity of Ni/YSZ cermet anodes is most severe. Upon switching from an H_2S-containing fuel back to a clean fuel, the recovery is slow enough that the degradation is not completely reversible over a reasonable time frame. One speculation is that sulfur adsorption changes the surface energies of different crystallographic planes, such that the Ni particle surface planes gradually change to less electrochemically active ones [241]. Another study [242] showed that the slow degradation at high H_2S levels (hundreds of ppm) was

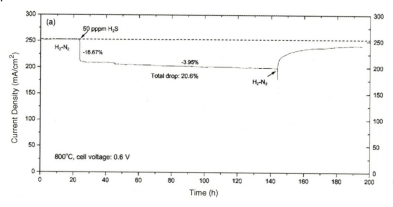

Figure 5.21 Sulfur poisoning and regeneration (or desulfurization) processes of Ni/YSZ anodes in a fuel mixture with 50 ppm H_2S. (Reproduced with permission from Ref. [241].)

accompanied by transport of Ni from the Ni/YSZ anode, resulting in a reduced anode conductivity that could explain the slow degradation. A number of groups have shown that perovskite oxides, for example, (La,Sr)(Cr,Mn)O_3 (LSCM) and (La,Sr)VO_3 (LSV), exhibit better sulfur tolerance [243–246]; LSV makes it possible to work with more than 10% H_2S in the fuel [245]. Unfortunately, LSV appears to be insufficiently stable for use in SOFCs.

Oxide-based materials based on perovskite-related structures have been extensively investigated as alternative anode for SOFCs. Examples include titanate-based perovskites [247], yttria-stabilized zirconia–terbia (YSZT) such as Y_2O_3–ZrO_2–TiO_2 and Sc_2O_3–Y_2O_3–ZrO_2–TiO_2 [248], perovskite $SrTiO_3$ doped with La or Nb and niobium-based tetragonal tungsten bronzes [249, 250], $La_{0.75}Sr_{0.25}Cr_{0.5}Mn_{0.5}O_3$ (LSCM) [251], and double perovskites $Sr_2Mg_{1-x}Mn_xMoO_{6-\delta}$ [252]. Among the perovskite-based materials, LSCM appears to be most promising. As shown in Figure 5.22, the maximum power density of a cell with $(La_{0.75}Sr_{0.25})_{0.9}Cr_{0.5}Mn_{0.5}O_3$ anode and LSM cathode measured was 0.47 W cm^{-2} at 900 °C, which is considered to be compatible with the cell based on Ni/YSZ cermet anodes, considering that the YSZ electrolyte used in the cell had a thickness of 0.3 mm [251]. However, the electrical conductivity of LSCM is very low, ~0.22 S cm^{-1} in 10% H_2/N_2 [253]. This factor is a serious concern for practical SOFC electrodes as low electrical conductivity will result in the high ohmic losses and the high contact resistance between the electrode and the current connector. Review on the selected new electrode materials for SOFCs is found in Chapter 6.

5.5.1.2 Cathode

LSM is still among the materials of choice for cathodes of SOFCs with YSZ electrolytes [223]. LSM shows a high thermal stability and compatibility with YSZ; more important, if compared to other cathode materials, LSM has an excellent microstructural stability and its long-term performance stability has been well established. For undoped $LaMnO_3$, the TEC is in the range of 11.33×10^{-6} to

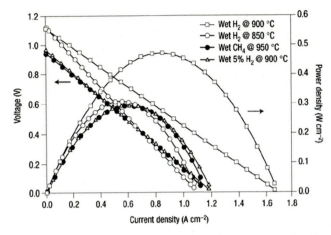

Figure 5.22 Performance of a model SOFC with $(La_{0.75}Sr_{0.25})_{0.9}Cr_{0.5}Mn_{0.5}O_3$ anode and LSM cathode measured at different temperatures in humidified fuel gases. YSZ electrolyte thickness was 0.3 mm. (Reproduced with permission from Ref. [251].)

12.4×10^{-6} K^{-1} [254], close to that of most commonly used YSZ electrolyte, which is approximately 10.3×10^{-6} K^{-1} in the temperature range from 50 to 1000 °C in air or in H$_2$ atmosphere.

LaMnO$_3$ is a perovskite material with intrinsic p-type conductivity (see also Chapters 3, 9 and 12 of the first volume). Electronic conductivity is enhanced by substitution of the La^{3+} site with divalent ions such as Sr^{2+} and Ca^{2+}. Of the alkaline earth dopants, Sr substitution is preferred for SOFC applications because the resultant perovskite forms stable compounds with high conductivity in the oxidizing atmosphere [255]. Figure 5.23 shows the Arrhenius plots of the total conductivity, log σT against $1/T$, for La$_{1-x}$Sr$_x$MnO$_3$ ($0 \leq x \leq 0.7$) in pure oxygen [256]. The electronic conductivity increases with increase in x; the maximum is observed for $x = 0.5$. However, extensive tests showed that La$_{1-x}$Sr$_x$MnO$_3$, where $x = 0.1$–0.2, still provides high conductivity while maintaining mechanical and chemical stability with YSZ [255, 257].

As the incorporation and bulk diffusion of oxygen inside LSM cannot be expected to occur to a significant degree [258], the TPB areas become the reaction sites for the O$_2$ reduction. The very low oxygen ionic conductivity of LSM is considered to be the main factor in the high polarization losses of LSM cathode for the O$_2$ reduction reaction in reduced SOFC operating temperatures [259]. Various strategies have been developed to improve the electrocatalytic activity of the LSM-based cathodes. Murray and Barnett [260, 261] showed that addition of YSZ and GDC to LSM significantly reduced the electrode polarization resistance. The electrode polarization resistance decreased as the GDC concentration increased to 50 wt%. The electrochemical performance of an LSM cathode can also be enhanced substantially by introducing catalytically active nanoparticles, such as doped CeO$_2$, into the LSM porous structure [262]. The use of composite cathodes can also significantly reduce the effect of the Cr poisoning [263].

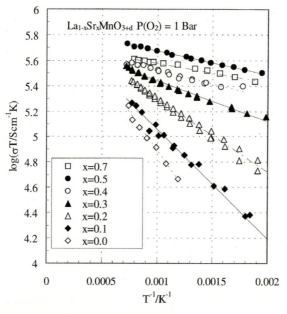

Figure 5.23 Temperature dependence of total conductivity (σ) of $La_{1-x}Sr_xMnO_3$ ($x = 0–0.7$) in pure oxygen. (Reproduced with permission from Ref. [256].)

Another family of materials that has been extensively studied for SOFC cathode is $LaCoO_3$-based perovskites. $LaCoO_3$, like $LaMnO_3$, shows p-type electronic conductivity. Acceptor-doped cobaltites are characterized by enhanced lattice oxygen vacancy formation and hence high ionic conductivity. However, the TECs of Co-rich perovskites are generally too high for both YSZ and doped CeO_2 electrolytes. The average TEC of undoped $LaCoO_3$ is about 20×10^{-6} K^{-1}. An alternative to the Co-rich perovskites is the Sr-doped $LaFeO_3$ that has a lower TEC and a good chemical compatibility with doped CeO_2 electrolyte. Compositions in the system (La,Sr)(Co, Fe)O_3, such as $La_{0.6}Sr_{0.4}Co_{0.2}Fe_{0.8}O_3$ or LSCF6428 [264], may have desirable properties for IT-SOFC cathode applications. The oxygen self-diffusion coefficient of cobaltite-based materials is several orders of magnitude higher than that of manganites [258]. Addition of ionically conductive phase, such as GDC, and noble metal, such as Pt and particularly Pd, also promotes the electrocatalytic activity of LSCF cathodes [265, 266].

Replacing lanthanum with barium at the A-site of LSCF substantially enhances its electrochemical activity for the O_2 reduction reaction at intermediate temperatures. Shao and Haile [267] applied $Ba_{0.5}Sr_{0.5}Co_{0.8}Fe_{0.2}O_3$ (BSCF) as cathode to a doped ceria electrolyte cell and achieved the power densities of 1.01 W cm^{-2} and 402 mW cm^{-2} at 600 and 500 °C, respectively, when operated with hydrogen as fuel and air as cathode gas. The electrode polarization resistance is very low, 0.51–0.6 Ω cm^2 at 500 °C. However, the TEC of BSCF is $\sim 20 \times 10^{-6}$ K^{-1} and the conductivity is relatively low, ~ 25 S cm^{-1} at 800 °C [268].

A number of other oxide materials have also been investigated as cathodes of SOFCs. Examples are lanthanum nickelate, $La_2NiO_{4+\delta}$ (LN) [269–271], La(Ni, M)O_3 (M = Al, Cr, Mn, Fe, Co, and Ga) perovskites [272], and lanthanum–strontium manganese chromite, $La_{0.75}Sr_{0.25}Cr_{0.5}Mn_{0.5}O_3$ [273]. However, their long-term stability under SOFC operating conditions is generally unproven.

One important issue with the SOFC cathodes is their chemical compatibility and reactivity with other fuel cell components especially with yttria–zirconia electrolyte and chromium-containing interconnect materials at both processing and operating temperatures. The reactivity between LSM and the electrolyte (usually YSZ) has been extensively studied [274–276]. Mitterdorfer and Gauckler [277] investigated the mechanism of the formation of $La_2Zr_2O_7$ as a function of stoichiometry composition of LSM. Excess lanthanum oxide within the perovskite reacts immediately with YSZ to form dense $La_2Zr_2O_7$ blocking layer at the interface. The conductivity of $La_2Zr_2O_7$ is more than 100 times lower than that of zirconia. Nevertheless, the reactivity of LSM with YSZ can be reduced by using A-site or lanthanum-deficient manganite, $La_{1-x}MnO_3$ or $(La,Sr)_{1-x}MnO_3$ ($x = 0.1–0.2$) [278, 279].

$La_{1-x}Sr_xMO_3$ (M = Co, Ni, and Fe) perovskites are relatively unstable compared to their manganite counterparts. The former readily reacts with zirconia electrolytes, leading to the formation of secondary phases at temperature as low as 800 °C in air. The interaction between the nanostructured thin-film cathodes of $La_{0.5}Sr_{0.5}CoO_3$ and YSZ could even occur at 650–700 °C [280]. Using interlayer such as doped ceria can prevent the interaction, but the long-term stability of such double layer structure is largely unknown.

The stringent requirements for the IT-SOFC cathodes could also be fulfilled by the combination of highly active MIEC materials such as BSCF and the structurally stable and highly conductive LSM [281]. In such nanostructured BSCF-impregnated LSM composite cathode, the uniformly distributed BSCF nanoparticles could significantly enhance the electrochemical activity, while the LSM skeleton would provide an effective electron transfer and matching of the effective TEC. The interfacial reaction between the impregnated BSCF and YSZ can be avoided or minimized due to the low processing temperature. Figure 5.24 shows the performance of Ni/YSZ anode-supported YSZ film cells with pure LSM and 1.1 mg cm^{-2} BSCF-impregnated LSM cathodes at different temperatures, under H_2/air gradient.

Understanding the interaction between the metallic interconnect and the cathode is very important for developing IT-SOFCs. Metallic interconnects, such as chromia-forming alloys (e.g., stainless steel), generate volatile Cr-containing species in oxidizing atmospheres. These species may lead to a rapid degradation of the SOFC performance [263, 282–284]. The latest studies showed, however, that it is possible to develop novel cathodes with high tolerance toward Cr poisoning [285, 286].

5.5.1.3 Electrolytes

Figure 5.25 compares the oxygen ion conductivity of selected solid oxide electrolytes considered for SOFCs [226]; more detailed information is available in Chapters 2, 3 and 9 of the first volume.

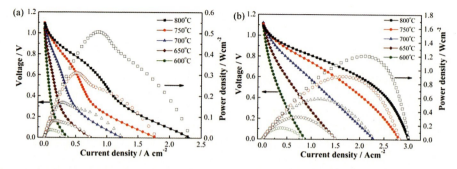

Figure 5.24 Performance of the Ni/YSZ anode-supported YSZ electrolyte film cells with (a) pure LSM cathode and (b) 1.1 mg cm^{-2} BSCF-impregnated LSM composite cathode at different temperatures in H_2/air. H_2 flow rate: 200 ml min^{-1}; oxidant: stationary air. (Reproduced with permission from Ref. [281].)

Any SOFC electrolyte must meet the requirements of fast ionic transport, negligible electronic conduction, and thermodynamic stability over a wide range of temperature and oxygen partial pressure. In addition, they must have thermal

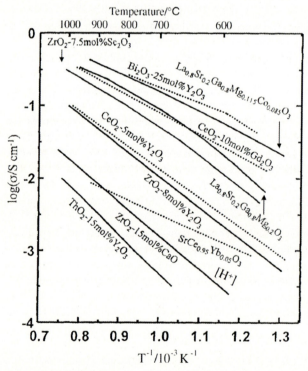

Figure 5.25 Oxygen ion conductivity of selected electrolyte materials. (Reproduced with permission from Ref. [226].)

expansion compatible with the electrodes and other cell components, good mechanical properties, and negligible interaction with electrode materials under operation and processing conditions. These requirements limit the choice of electrolyte materials to zirconia and doped ceria and $LaGaO_3$ [226, 287, 288].

In the Y_2O_3–ZrO_2 system, a maximum of the conductivity of ZrO_2 electrolyte with respect to the dopant content occurs around 8–9 mol% Y_2O_3 [289]. This maximum corresponds to the minimum dopant level required to fully stabilize the high-temperature cubic phase.

Scandia-doped zirconia (ScSZ) provides a higher conductivity, attributed to the smaller mismatch in size between Zr^{4+} and Sc^{3+} ions [290]. ScSZ with 9.0 mol% Sc_2O_3 (9ScSZ) has the conductivity of $0.34\,S\,cm^{-1}$ at $1000\,°C$ [291]. Another issue of zirconia-based electrolytes is the decrease in ionic conductivity due to aging during operation at high temperatures. For example, the initial conductivity of the 8.0 mol% Sc_2O_3-doped ZrO_2 (8ScSZ) was $0.13\,S\,cm^{-1}$ at $800\,°C$ and dropped to $0.012\,S\,cm^{-1}$ after annealing at $1000\,°C$ for $1000\,h$ [292]. Such a degradation might be partly reduced by increasing the Sc_2O_3 content [293]. Also, grain boundaries in stabilized zirconia exhibit a blocking effect with regard to the ionic transport [294, 295]. The segregation of impurities such as silica [296] is detrimental, but can be improved by minor oxide additives [297].

Pure ceria is not a good oxygen ion conductor, but its conductivity can be increased significantly by substituting Ce^{4+} with divalent alkaline earth or trivalent rare earth ions [298]. Highest conductivities have been reported for either Gd- or Sm-doped ceria (GDC and SDC) with maximum conductivity obtained around 10–20 mol% Gd_2O_3 or Sm_2O_3 [299–301]. Ceria-based electrolytes suffer from the partial reduction of Ce^{4+} to Ce^{3+}. However, the leakage current decreases substantially with decreasing temperature [302].

Another major class of SOFC electrolyte materials is doped $LaGaO_3$ perovskites, in particular the lanthanum–strontium magnesium gallate (LSGM) systems [226, 303]; see Chapter 9 of the first volume. Ishihara and coworkers [304] fabricated 5 μm thick $La_{0.8}Sr_{0.2}Ga_{0.8}Mg_{0.2}O_3$ thin electrolyte cells with 400 nm thick $Gd_{0.2}Ce_{0.8}O_2$ interlayer by pulse laser deposition, achieving a power density as high as 1.95 and $0.61\,W\,cm^{-2}$ at 600 and $500\,°C$, respectively.

Table 5.3 lists the typical values of the oxygen ion conductivity and its activation energies in the low- and high-temperature ranges for the main electrolyte materials employed in SOFCs. Note that higher conductivity values are sometimes observed in thin films and other nanostructured systems (see Chapter 4 of the first volume). Kosacki et al. [305] investigated the ionic conductivity of highly textured YSZ films deposited on MgO substrate and observed exceptionally high ionic conductivity for a film thickness of more than 60 nm. Recently, Garcia-Barriocanal et al. [306] reported a huge ionic conductivity enhancement at interfaces of epitaxial $YSZ/SrTiO_3$ heterostructures. This indicates that nanoscale effects at the interface could be manipulated to enhance the ionic conductivity of zirconia-based electrolytes for low- and intermediate-temperature SOFC applications.

Apatite-type oxides with a general formula $(La/M)_{10-x}(Si/GeO_4)_6O_{2\pm\delta}$ (M = Mg, Ca, Sr, Ba) are attracting considerable attention [309–311]. The drawback of the

Table 5.3 Conductivity and activation energy for some electrolyte materials employed in SOFCs.

Composition	E_a (eV) 400–500 °C	E_a (eV) 850–1000 °C	σ (S cm^{-1}) 1000 °C	σ (S cm^{-1}) 600 °C	Reference
3 mol% Y_2O_3–97 mol% ZrO_2	0.95	0.80	0.056	3.62×10^{-3}	[287]
8 mol% Y_2O_3–92 mol% ZrO_2	1.10	0.91	0.16	8.73×10^{-3}	[287]
10 mol% Y_2O_3–90 mol% ZrO_2	1.09	0.83	0.13		[287]
12 mol% Y_2O_3–88 mol% ZrO_2	1.20	1.04	0.068		[287]
9.0 mol% Sc_2O_3–91 mol% ZrO_2	1.30	0.72	0.34		[291]
$Gd_{0.2}Ce_{0.8}O_{1.9}$		0.86		16.9×10^{-3}	[307]
$Sm_{0.2}Ce_{0.8}O_{1.9}$		1.0		9.7×10^{-3}	[301]
$Y_{0.2}Ce_{0.8}O_{1.9}$		0.68		14×10^{-3}	[308]
$La_{0.8}Sr_{0.2}Ga_{0.8}Mg_{0.2}O_3$		0.63		17×10^{-3}	[226]

apatite-based oxides is the very high sintering and densification temperatures (e.g., as high as 1650–1700 °C) [312, 313]. Bismuth oxide-based materials, such as yttria- and erbia-stabilized bismuth oxides, showed highest conductivities [314]. However, the high oxygen mobility is a result of weak metal–oxygen bonds and thus Bi_2O_3-based materials have lower stability under reduced partial pressure of oxygen at the anode side, resulting in decomposition to metallic Bi.

Detailed information on the proton-conducting electrolytes used in SOFCs is available in Chapters 7 and 12 of the first volume. As an important example, one can note that Iwahara [315] observed appreciable proton conductivity of $BaCeO_3$-based oxides at high temperatures, 1×10^{-2} S cm^{-1} at 1000 °C. However, one major concern with these materials is the chemical stability when exposed to CO_2 under long-term fuel cell operation conditions.

5.5.1.4 Interconnect, Sealing, and Balance of Plant

Only a few oxide materials could be used as interconnect materials in SOFCs. Alkaline earth (AE)-doped $MCrO_3$ (M = La, Y, and Pr) are the most studied ceramic interconnect materials for the high-temperature SOFCs [316, 317]. The increase in AE content results in a higher TEC, which causes thermal stresses and thus decreases long-term stability [318]. The praseodymium in $PrCrO_3$-based oxides exists in the two valence states, Pr^{3+} and Pr^{4+} [319], which decreases chemical and dimensional stability. $La_{0.7}Ca_{0.3}CrO_3$-doped CeO_2 composite [320] and $Nd_{0.75}Ca_{0.25}Cr_{0.98}O_{3-\delta}$ [321] were also considered for interconnects. However, the decrease in their electrical and thermal conductivities with decreasing temperature is a major challenge in the developments of ceramic interconnect for IT-SOFCs.

Metallic materials based on transition metal-based oxidation-resistant alloys have been considered to be the primary candidates as the interconnect materials of IT-SOFCs due to the economic and easy processing benefits in addition to the high electrical and thermal conductivities. These include Ni(–Fe)–Cr-based heat-resistant alloys, Cr alloys, and chromia-forming ferric stainless steels [220, 227]. The alloys with the formation of a protective and semiconductive chromia scale to minimize further environmental attack during the high-temperature operation and with TECs of 11.0 to 12.5 × 10^{-6} K^{-1} are the preferred candidates. The conductivity of chromia oxides is $\sim 10^{-2}$ S cm^{-1} at 800 °C in air [322]. A good example in this category is Plansee Ducralloy with a composition of 94% Cr, 5% Fe, and 1% Y_2O_3 (as $Cr_5FeY_2O_3$) [323]. To further increase the electrical conductivity of the scale and to reduce the chromium vaporization, a new alloy that contains 0.5% Mn (Crofer 22 APU) was developed [324]. The oxide scale consisting of a $(Mn,Cr)_3O_4$ spinel top layer shows a higher electrical conductivity [325].

However, without effective protective coatings, the vaporization of chromium species from chromia scale poisons the cathodes and seriously degrades the cell performance [326]. To reduce the growth rate of the oxide scale and the vaporization of chromium species, a thin and dense oxide coating with high electrical conductivity such as LSM and LSCo is often deposited on the metallic interconnect. The chromium volatility can also be suppressed by modifying the metallic interconnect materials. Recently, Hua et al. [327] reported a novel Ni–Mo–Cr alloy with a TEC value of 13.92 × 10^{-6} K^{-1} between 35 and 800 °C. After oxidation treatment at 750 °C for 1000 h, the area-specific resistance (ASR) of this alloy was 4.48 mΩ cm^2. The poisoning study using LSM cathode indicates that the Cr deposition and poisoning of the Ni–Mo–Cr alloy is remarkably reduced compared to the conventional Fe–Cr alloy [328].

The sealing material has been regarded as one of the most significant technical challenges in the development of planar SOFCs, while sealing is much less of a problem for tubular SOFCs. The sealants can be broadly classified into rigid bonded seals, compressive seals, and compliant bonded seals. Each offers advantages and limitations. In rigid bonded sealing, the sealant forms a joint that is nondeformable at room temperature. Because the final joint is brittle, it is critical for the sealant to match the TEC of the adjacent substrates. High-temperature glass and ceramic glass such as alkali silica glasses and $BaO–CaO–SiO_2$ [329] are among the most important rigid bonded sealants employed in joining SOFC stacks. These materials have acceptable stability in reducing and oxidizing atmospheres, are generally inexpensive, and can be readily applied to the sealing surfaces as a powder paste or a tape cast sheet. They are electrically insulating and their TEC can be adjusted to those of electrolyte and metallic interconnect. However, the brittle nature of glasses and ceramic glasses makes these seals vulnerable to cracking, and they tend to transform in phases and react with the cell components and interconnect materials under SOFC operation conditions in a long run due to their intrinsic thermodynamic instability [330, 331].

Compressive seals have been developed to avoid the disadvantages of the rigid bonded seals, with the merit of flexibility and compressibility, allowing the cells and interconnects to expand and contract freely during thermal cycles and operation. This type of sealing relies on the compressive load of the stack. So far, two kinds of compressive seals have been considered, that is, the deformable metallic seals and the mica-based seals. The deformable metallic seals include ductile silver [332] and corrugated or C-shaped superalloy gaskets [333], but their application is limited due to their high electronic conductivity. The most common compressive sealing material is based on mica. Mica belongs to a class of layered minerals known as phyllosilicates and is composed of cleavable silicate sheets. By incorporating a compliant interlayer such as a deformable metal or glass at the interface to form the hybrid mica-based seals, the sealing properties of mica are significantly improved [334, 335]. Compliant bonded seals are based on metallic braze. Metallic materials have lower stiffness compared to ceramics and can undergo plastic deformation, which allow for accommodation of thermal and mechanical stresses. Silver and gold are stable in air and are commonly used as metal braze materials [336]. A major challenge in obtaining a good metal–ceramic joint is adequate wetting of the ceramic by the braze metal. Comprehensive review articles have recently been published on sealant materials used for SOFCs [228, 337].

Balance of plant (BoP) consists of the components that support the operation of the fuel cell stack. This includes the manifolds, piping, heat exchangers, burners, blowers, and so on. Since the exhaust gas is at the same temperature as the stack itself, the BoP components that come into contact with the gas steam need to be able to operate at the same temperature and chemically and thermally stable. Refractory ceramics, high-temperature alloy (e.g., Inconel), and conventional austenistic stainless steels can be used manifolds and piping for SOFCs.

5.5.2
SOFC Structures and Configurations

5.5.2.1 Cell Structures

There are a number of cell support structures and each is classified according to the layer that mechanically supports the cell. These include electrolyte-supported, anode- or cathode-supported, and porous substrate- or metal-supported structures (see Figure 5.26) [338].

Due to the thick electrolyte (typically in the range of ~100 μm) needed to mechanically support other cell components, the electrolyte-supported SOFCs are primarily developed for operation at high temperatures (~900 °C or above) [339]. However, reduction in the operating temperature to intermediate range (500–800°) would greatly reduce the degradation of SOFC components, widen the materials selection, lessen the sealing problem, and enable the use of low-cost metallic interconnects and BoP components. On the other hand, the overall performance

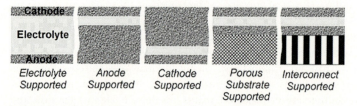

Figure 5.26 Illustration of different types of cell support architectures for SOFCs. (Reproduced with permission from Ref. [338].)

of an SOFC would decrease because of the reduced ionic conductivity of the solid electrolytes and the increased polarization resistance of the electrodes.

Both the rapid progress in the thin-film technology [340] and the identification of alternative electrolyte materials with higher ionic conductivity such as GDC have significantly reduced the ohmic losses associated with solid electrolyte. The use of thin electrolyte layers requires the electrolyte to be supported on an appropriate substrate. The substrate can be a functional anode or cathode, or a porous support providing gas diffusion and transportation for fuel cell reactions. As the substrate is the principal structural component in these cells, it is necessary to optimize the conflicting requirements of mechanical strength and high gas permeability.

Up to now, the most popular supported cell structure is the anode-supported thin-film SOFCs. The state-of-the-art anode-supported SOFC is based on porous Ni/YSZ cermets as a support. The anode support usually consists of a relatively thick porous supporting substrate (200–500 μm) and a thin and fine-structure electrode layer, the anodic function layer (AFL). To reduce the electrolyte ohmic resistance and to enhance the cell efficiency, the electrolyte layer deposited should be as thin as possible. As a general role, the film thickness is inversely proportional to the pore size and/or propagated roughness of the surface, which means that the larger the pore size, the more difficult it is to get a thinner electrolyte. Thus, the AFL is also required to have proper pore structure with low surface roughness. Tape casting and tape calendering processes are the common techniques in the fabrication of anode-supported structure for thin-film SOFCs [341, 342]. Power density as high as 1.8 W cm^{-2} at 800 °C was reported for such anode-supported cells [343]. However, the porous composite anode support is relatively weak mechanically and can have difficulty withstanding the thermal and mechanical stresses generated by rapid temperature fluctuation. Moreover, Ni/NiO redox cycling induced by air diffusion into the anode compartment during the loss of fuel supply and other operational excursion can disrupt the anode microstructure, leading to irreversible degradation [344].

Metal-supported structures are often used to overcome the problems associated with anode-supported cells [345, 346], in which the Ni/YSZ anode support is replaced by a metal (usually stainless steel). Metal support improves thermal shock resistance, reduces temperature gradients due to the high thermal conductivity of metals, and thus enhances the robustness of SOFCs. The metal-supported cells are particularly suitable for operating at 600 °C or lower as the corrosion constraints on the stainless steel materials would be far less severe. Other supported thin-film cell structures include thin-film SOFCs supported by cathode [347], and gas diffusion porous substrate typified by the Rolls-Royce's segmented in series design [348].

5.5.2.2 Stack Design and Configurations

There are two main SOFC configurations, that is, tubular and planar (see Chapter 12 of the first volume). Performance of planar SOFCs is theoretically higher than that of tubular because of the reduced in-plane ohmic resistance. In addition, tape casting and other mass production techniques, for example, screen printing and plasma spray, can be easily applied for planar SOFC component production, thus making it possible to substantially reduce production cost. On the other hand, the tubular configuration, because of its geometry, is capable of solving the problems related to cracking, thermocycling, startup time, and sealing.

5.6
Emerging Fuel Cells

5.6.1
Protic Ionic Liquid Electrolyte Fuel Cells

Protic ionic liquids (PILs) belong to ionic liquid groups and are low-temperature molten salts. The salts are characterized by weak interactions, owing to the combination of a large cation and a charge-delocalized anion. This results in a low tendency to crystallize due to flexibility (anion) and dissymmetry (cation). The archetype of ionic liquids is formed by the combination of a 1-ethyl-3-methylimidazolium cation and an N,N-bis(trifluoromethane)sulfonamide anion. This combination gives a fluid with an ionic conductivity comparable to many organic electrolyte solutions and an absence of decomposition or significant vapor pressure up to $\sim$300–400 °C.

PILs, consisting of a combination of Brønsted acids and bases that could form hydrogen bonds and act as proton carriers, have been considered effective proton transfer carriers for HT-PEMFCs. PILs have the unique properties of ionic liquids of high proton conductivity, nonvolatility, and high thermal and chemical stability. PIL-based membranes could be approached by immobilizing ionic liquids by polymerization of the components or gelification by a neutral macromolecule with an architecture allowing a water-independent proton conductivity associated with a high thermal stability (see Figure 5.27) [349].

PIL-based PEMs have been extensively studied [350–352]. Similar to phosphoric acid (H_3PO_4)-doped PBI composite membranes, the release of PIL component is the major challenge for stability and durability [350]. Retention of PILs in PEMs can be enhanced by incorporation of silica and mesoporous silica nanoparticles [353]. Silica hybrid membranes based on a PIL, N-ethylimidazolium trifluoromethane-sulfonate ([EIm][TfO]), and copolymer, poly(styrene-co-acrylonitrile) (SAN), provide proton conductivity of up to 1×10^{-2} S cm^{-1} at 160 °C under anhydrous conditions [353].

5.6.2
Microbial Fuel Cells

Microbial fuel cells use bacteria and microorganisms as catalysts for converting the chemical energy of feedstock such as glucose and organic matters in wastewaters into electricity. The MFC concept was first demonstrated by Potter in 1912 using living cultures of *Escherichia coli* and *Saccharomyces* and Pt electrodes [354]. Similar to any fuel cells, MFCs consist of anode and cathode compartments, which are often separated by a PEM such as Nafion. The anode compartment contains microorganisms that oxidize the available substrate (i.e., the electron donor). Substrates used in

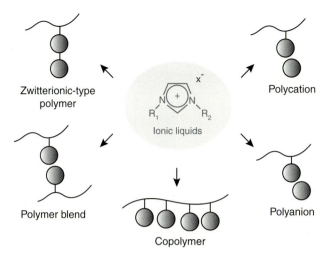

Figure 5.27 Strategies to immobilize ionic liquids in polymeric architectures to develop ionic liquid-based proton-conducting electrolyte membranes. (Reproduced with permission from Ref. [349].)

MFC vary from sugars and organic acids such as glucose or acetate to complex polymers such as starch and cellulose. Domestic, industrial, and animal waste streams have been used as feedstock for generating electricity in MFCs. Marine sediment, soil, wastewater, fresh water sediment, and activated sludge are all rich sources for microorganisms in MFCs [355, 356].

However, it is extremely difficult to utilize the electrons generated by the reaction occurring inside the cell. One solution is via the electron mediators. A mediator in an oxidized state can be easily reduced by capturing the electrons from within the membrane. The mediator then moves across the membrane and releases the electrons to the anode and becomes oxidized again in the anodic chamber. This cyclic process or mediated electron transfer (MET) facilitates the transfer of electrons and thus increases the power output. Figure 5.28 shows the operating principle of an MFC [357]. Chemical mediators such as neutral red (NR) or anthraquinone-2,6-disulfonate (AQDS) can be added to the system to allow electricity production by bacteria [358]. Unfortunately, the toxicity and instability of synthetic mediators limit their applications in MFCs. However, some microbes can either use naturally occurring compounds such as microbial metabolites such as humic acid, anthraquinone, and the oxyanions of sulfur as mediators [359] or directly transfer electrons to the anode [360].

The anode materials in MFCs must be conductive, biocompatible, and chemically stable in the reactor solution. Metal anodes consisting of noncorrosive stainless steel mesh can be utilized, but copper is not useful due to the toxicity of even trace copper ions to bacteria. The most versatile electrode material is carbon. Kim et al. [361] used *Proteus vulgaris* in suspension within a thionine mediator solution to generate power from a variety of carbon forms. Using a mixed bacterial culture originated from an anaerobic sludge and a hexacyanoferrate mediator, an MFC produced a power output

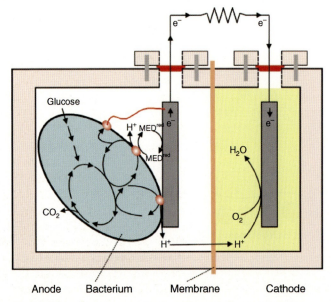

Figure 5.28 Operating principle of a microbial fuel cell. (Reproduced with permission from Ref. [357].)

of up to $360\,\mu W\,cm^{-2}$ using a glucose fuel [362]. Instead of immobilizing the microbe, immobilizing the mediator is effective in facilitating the electron transfer in MFCs [363].

The reaction at the cathode compartment can be classified into aerobic or anaerobic reactions depending on the source of the final electron acceptor available. In aerobic cathode, oxygen is the most suitable electron acceptor. To increase the rate of oxygen reduction, Pt catalysts are usually used. Non-Pt materials such as pyrolyzed iron(II) phthalocyanine or CoTMPP have also been used as MFC cathodes [364]. Mediators that undergo reversible redox reactions (e.g., ferricyanide) can also reduce the cathodic overpotential in MFCs [365]. However, the use of ferricyanide is not sustainable in practice since ferricyanide must be chemically regenerated and is potentially toxic. The cathode compartment can also be maintained under anaerobic conditions. In this case, microorganisms transfer the electrons from the cathode to the final electron acceptor, for example, nitrate [366]. Nevertheless, the performance of biocatalysts for the cathodic reaction is constrained by high activation overpotentials [366]. As noted by Rismani-Yazdi et al. [367], the performance of an MFC is also significantly affected by the operation parameters and electrode design.

Power generation of MFCs is still very low (e.g., in the range of several $\mu W\,cm^{-2}$ level). There are significant energy barriers to the anodic oxidation of substrates and cathodic reduction of oxygen in MFCs, in addition to the significant internal ohmic resistance and concentration losses [368]. Thus, MFCs are particularly suitable for powering small telemetry systems and wireless sensors that require only low power to transmit signals (e.g., temperature) to receivers in remote areas [369]. In combination with wastewater treatment process, the power produced by MFCs could potentially halve the electricity needed in a conventional treatment process. However, in addition to the challenges in the improving efficiency and long-term stability of the MFC stack and systems, the success of specific MFC application in wastewater treatment will depend on the concentration and biodegradability of the organic matter in the influent, the wastewater temperature, and the absence of toxic chemicals. A breakthrough in the development of inexpensive electrodes that resist fouling is also needed.

5.6.3
Biofuel Cells

Biofuel cells follow the same operating principle as the MFCs. In biofuel cells, enzymes are used to catalyze biomass or biofuel substrate to produce electricity, or biological fluids are used as fuel sources for the electrical activation of implantable electronic medical devices or prosthetic aids. The development of enzyme-based cells (or enzymatic biofuel cells) began in the 1960s and received extensive attention due to the high turnover rates associated with enzymes that lead to a high biocatalysis rate. Enzymes such as bilirubin oxidase can have higher catalytic activity and lower overpotential than the Pt catalysts [370]. Enzymes allow a wide variety of fuels to be utilized without the need for expensive and time-consuming purification because enzymes typically do not react with or get passivated by impurities in the fuel. The selectivity of enzymes also allows for the elimination of the PEMs used to separate cathodic and anodic solutions by simultaneous use of a biocathode and a bioanode.

Similar to MFCs, the spatial separation of the biocatalytic redox sites from the electrode prevents the electrical contacting of the enzyme with the electrode (the dimensions of redox proteins are in the range of 7–20 nm). Thus, the electrical communication between the redox centers of enzymes and the electrodes needs to be established. Mediators have also been employed in biofuel cell systems through polymerization on the electrode surface prior to enzyme immobilization, coimmobilization of enzyme and mediator simultaneously, or simply allowing the mediator to be free in solution. Many low molecular weight redox-active compounds and polymers have been incorporated to mediate the electron transfer, including organic dyes such as methylene green, phenazines, and azure dyes along with other redox-active components such as ferrocene, ferrocene derivatives, and conducting salts [371]. The problems associated with the mediated system are that mediators are often not biocompatible or have short life spans. However, some enzymes are capable of direct electron transfer (DET) via their active site. DET occurs through the enzyme's ability to act as a "molecular transducer" that converts the chemical signal to an electrical one through the transfer of charge to a stable redox species that is in turn capable of transferring this charge to another molecule or electrode surface. Many of these enzymes contain redox-active metal centers that perform the catalytic electron transfer. Enzymes such as laccase can catalyze the four-electron reduction of O_2 to water through electron transfer from the electrode surface directly to the active

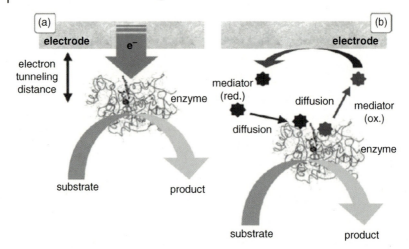

Figure 5.29 Electron transfer mechanisms utilized in biofuel cells: (a) direct electron transfer and (b) mediated electron transfer. (Reproduced with permission from Ref. [375].)

sites and through to the substrate. Figure 5.29 explains electron transfer mechanism through the MET and DET.

However, DET is correlated directly to enzyme proximity and orientation to the electrode surface in order for electron tunneling to occur, allowing only the biocatalytic reaction to be the limiting process. These issues can be addressed by implementing immobilization techniques such as cross-linking, wiring, sandwich, and encapsulation. The term "wiring" refers to chemically binding or attracting the enzyme of interest to the anode or cathode surface through covalent or ionic interactions in such a way that there is an electron pathway to the electrode. The structure of carbon electrodes has significant effect on the cell performance. When laccase was immobilized on superdispersed colloid graphite and acetylene black electrodes, the rate of ORR per enzyme molecule at the graphite electrode was found to be five times higher than that at the acetylene black electrode [372]. Kim and Grate [373] examined the utilization of CNTs for direct noncovalent immobilization of cytochrome, and glucose oxidase showed a significant improvement in the life span of the cells.

Pizzariello et al. [374] constructed a biofuel cell using a solid binding matrix based on graphite particles combined with enzymes and ferrocene-based mediators, which is spray-printed on inert polymer film as the electrodes. Glucose oxidase and horseradish peroxidase were included in the anode and cathode, respectively, to give a cell with a power output of $0.15\,\mu W\,cm^{-2}$ at an operating potential of 0.021 V (1 mM H_2O_2 and 1 mM glucose). Other limitations to the system efficiency include incomplete oxidation of fuels, short life spans, and reduced performance due to slow DET and/or problems associated with the stability or thermodynamics of redox mediator. Maintaining the integrity and performance of the three-dimensional protein structure of both the enzyme active sites and the macromolecules as a whole requires accurate control of temperature, pH, and chemical components of the solvent environment. Moehlenbrock and Minteer [375] summarized recently the advances made in the immobilization and stabilization of enzymes at biocathodes and bioanodes of biofuel cells, and Willner et al. [376] gave an excellent review on the methods of constructing the integrated anode and cathode units in biofuel cells.

The development of biofuel cells for practical applications is a field that is still in its infancy, and there is much potential for further improvement [377]. With the improvements in immobilization techniques, biofuel cells can also be stacked [378]. Mano et al. [379, 380] constructed a miniaturized biofuel cell (0.26 mm^2 footprint of electrodes and reaction volume of 0.0026 mm^3) and generated a power of 4.4 μW under continuous operation for 6 days in a physiological, glucose-enriched buffer solution (pH 7.2, 0.14 M NaCl, 20 mM phosphate, 30 mM glucose, 37 °C), indicating the feasibility of implantable biofuel cells. The major challenge seems to be the increase in the power output of biofuel cell stacks.

5.6.4
Microfluidic Fuel Cells

Different from the conventional classifications for fuel cells, a microfluidic fuel cell is a fuel cell where the fluid delivery and removal, reaction sites, and electrode structures are all confined to a microfluidic channel [381]. Microfluidic fuel cells typically operate in a colaminar flow configuration without a physical barrier, such as a membrane, to separate the anode and the cathode. Thus, microfluidic fuel cells are also called membraneless fuel cells. Two aqueous streams, one containing the fuel (anolyte) and the other containing the oxidant (catholyte), flow side by side via a single microfluidic channel. The laminar flow region is characterized by low Reynold's numbers.

Microfluidic fuel cells can be manufactured by inexpensive, well-established micromachining methods for electronics and microfluidic chips. Typically, these devices consist of a microchannel, two electrodes, and a liquid-tight support structure. Figure 5.30 shows a silicon-based microfluidic fuel cell developed by Cohen et al. [382]. Microchannel stencils were created by silicon etching, using a standard photolithographic procedure. Pt deposited on a polyamide film by electron beam evaporation technique was used as electrodes. Using 0.5 M formic acid as anolyte and O$_2$-saturated 0.1 M H$_2$SO$_4$ solution as catholyte, an OCV of 0.53 V and maximum power density of ~86 μW cm^{-2} were obtained for a single 1 mm wide, 380 μm thick, and 5 cm long microchannel.

In addition to formic acid and oxygen, a variety of fuels and oxidants have been tested on microfluidic fuel cells, including hydrogen [383, 384], methanol [385, 386], hydrogen peroxide [387], vanadium redox species [388, 389], and potassium permanganate [390]. Microfluidic fuel cells were also incorporated into biofuel cells, for example, based on glucose fuel [391, 392]. The microfluidic fuel cell design avoids many of the issues encountered in conventional PEM-based fuel cells and is significantly simpler as it requires no auxiliary humidification and water management. However, there are many challenging issues facing the development of microfluidic fuel cells. The power output of a single planar microfluidic fuel cell

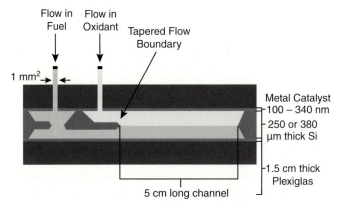

Figure 5.30 Side view of a silicon-based microfluidic fuel cell. (Reproduced with permission from Ref. [382].)

is still very low compared to conventional PEMFCs, and the feasibility of enlarging a planar cell is limited by the structural constraints. Due to the low level of power output, the task to develop a microfluidic fuel cell system with auxiliary equipment and fluid storage and circulation would not be an easy one [381].

5.6.5
High-Temperature Proton Exchange Membrane Fuel Cells

Two important factors have led to an increase in research and development effort toward high-temperature (120–200 °C) proton exchange membranes (HT-PEMs): low CO tolerance and poor heat transfer on a system level associated with the PEMFC operated at temperatures of 70–80 °C. In the state-of-the-art PEMFC, a CO level of 10 ppm could lead to a loss in fuel cell power of 20–50% [3]. The key benefit of fuel cell operation at temperatures above 100 °C is the ability to use direct reformate since operation up to 200 °C can afford the presence of CO up to 3 wt% [393]. Other advantages include faster electrode reaction kinetics, simplified water and heat management, higher energy efficiency, and reduced usage of precious Pt and Pt alloy catalysts.

The most significant challenge in HT-PEMFCs is the development of proton exchange membranes that can operate at elevated high temperature under anhydrous or low RH conditions; see also Chapter 10 of the first volume. Another challenge associated with HT-PEMFCs is the deficiency in HT-specific fuel cell architecture test station design, testing, and design protocols [394].

5.6.5.1 HT-PEMFCs Based on Phosphoric Acid-Doped PBI PEM
Much work has been devoted to the development of phosphoric acid (H_3PO_4)-doped PBI composite membranes, which can be operated at a temperature between 125 and 200 °C [395]. PBI with a glass transition temperature of 425–435 °C is a basic polymer ($pK = 5.5$) containing imidazole groups that are protonated after treatment with

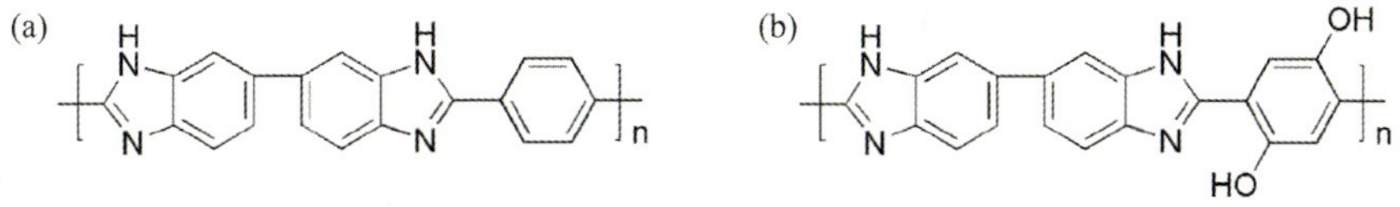

Figure 5.31 Chemical structure of (a) *para*-PBI and (b) dihydroxy-PBI.

phosphoric acid, resulting in highly conducting membranes [396]. Figure 5.31 shows the chemical structure of PBI with *para*-PBI and dihydroxy-PBI. Composite membranes with high H_3PO_4 content of 85 wt% are obtained through a sol-gel process [397]. The heterocycle of the PBI is involved in the proton transport process as it provides free electron pairs for proton binding. The conductivity of hybrid PBI membrane is reported to be 0.25–0.35 S cm^{-1} at 160 °C for H_3PO_4 dopant content of 40 wt% [398]. The mechanical properties of the doped membranes are greatly affected by the amount of phosphoric acid due to the plasticization effect.

The degradation mechanism of H_3PO_4-doped PBI membrane fuel cells is similar to that discussed in PEMFCs including the acid loss from the membrane, fast catalyst dissolution in hot acid medium, Pt catalyst sintering, thermal stress, and thermal degradation of the carbon support [399, 400]. The transition through the liquid water regime below 100 °C during startup and shutdown could lead to a leaching out of the acid if water condensation is allowed to happen. However, Yu *et al.* [398] performed detailed durability tests of PBI-based PEMFC operated at 160 °C under steady-state, thermal cycle, and shutdown–startup conditions. The fuel cell voltage degradation rate was 4.9 μV h^{-1} for steady-state operating condition, indicating the possibility of a long-term operation (>10 000 h) without any significant performance degradation. Dynamic tests (thermal cycle and shutdown–startup) also show that the H_3PO_4 loss from the MEAs was relatively small.

5.6.5.2 HT-PEMFCs Based on Inorganic and Ceramic PEM

Great efforts have been dedicated to develop HT-PEMs based on mesoporous or nanoporous inorganic materials. Mesoporous inorganic materials have a pore size range of 2–50 nm and are characterized by high specific surface area, nanosized channels or frameworks with an ordered or disordered interconnected internal structure, and high structural stability that allow their potential applications as proton exchange membranes operating at elevated temperatures. Lu and coworkers [401] reported sol-gel-derived mesostructured zirconium phosphates with proton conductivities of about 10^{-8}–10^{-6} S cm^{-1}. Vichi *et al.* [402] synthesized nanoporous anatase thin film with conductivity values from 10^{-5} to 10^{-3} S cm^{-1} in the range of 33–81% RH at room temperature. However, the conductivity of these pure mesoporous materials is too low to be practical for fuel cell applications. Recently, higher proton conductivities were reported for mesoporous thin films. Yamada *et al.* [403] reported TiO_2–P_2O_5 mesoporous nanocomposite with a proton conductivity of 2×10^{-2} S cm^{-1} at 160 °C. Halla *et al.* [404] synthesized meso-SiO_2–$C_{12}EO_{10}OH$–CF_3SO_3H as a new proton-conducting electrolyte and observed a conductivity of 1×10^{-3} S cm^{-1} at room temperature and 90% RH. But no fuel cell performance is reported in both cases.

Uma and Nogami [405] synthesized inorganic glass composite membrane consisting of a mixture of phosphotungstic acid (HPW) and phosphomolybdic acid (HPM), and reported very high conductivity values, 1.014 S cm^{-1} at 30 °C and 85% RH for a mesostructured HPW/HPM–P$_2$O$_5$–SiO$_2$ glass and 1.01×10^{-1} S cm^{-1} at 85 °C under 85% RH for a mesostructured HPW–P$_2$O$_5$–SiO$_2$ glass [406]. The cell performance based on these inorganic PEMs is 35–42 mW cm^{-2} in H$_2$/O$_2$ at ~30 °C under 30% RH.

Recently, Lu et al. [407] developed a novel inorganic proton exchange membrane based on highly ordered mesoporous MCM-41 silica with assembled HPW nanoparticles by vacuum-assisted impregnation method. The proton conductivity of HPW/MCM-41 mesoporous silica inorganic PEM is 0.018 and 0.045 S cm^{-1} at 25 and 150 °C, respectively. Most significantly, the PEMFCs based on the HPW/MCM-41 mesosilica membrane showed an impressive performance, achieving a maximum power density of 95 mW cm^{-2} in H$_2$/O$_2$ at 100 °C and 100% RH and 90 mW cm^{-2} in methanol/O$_2$ at 150 °C and 0.67% RH of cathode (see Figure 5.32). The TEM image shows that HPW is uniformly distributed inside the silica mesoporous channels (shown as black dots in Figure 5.32b).

Another group of oxides that show promising potential as HT-PEMs for fuel cells are solid acids based on oxyanion groups (SO$_4^{2-}$, PO$_4^{3-}$, SeO$_4^{2-}$, AsO$_4^{3-}$), linked together by hydrogen bonds and charge balanced by large cation species (Cs$^+$, Rb$^+$, NH$^+$, K$^+$); see also Chapter 7 of the first volume. Some of these acids undergo a polymorphic, structural transition from an ordered state to a highly disordered state, in which the oxyanion groups undergo almost free reorientation. Accompanying this order–disorder transition is a drastic increase in proton conductivity, by as much as four orders of magnitude in the case of CsHSO$_4$. Above the transition temperatures, the conductivity is typically ~10^{-2} S cm^{-1} [408], adequate for fuel cell applications.

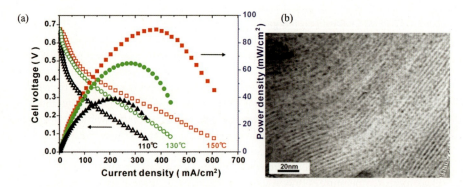

Figure 5.32 (a) Polarization curves and power density of a cell with a 30 wt% HPW/MCM-41 mesosilica electrolyte membrane in methanol/O$_2$ at 110 °C/2.2% RH, 130 °C/1.1% RH, and 150 °C/0.67% RH. (b) TEM image of the 30 wt% HPW/MCM-41 mesosilica electrolyte membrane. O$_2$ flow rate: 300 ml min^{-1}; back pressure: 1.5 atm; methanol solution (2.0 M) flow rate: 1 ml min^{-1}; cathode Pt loading: 1.0 mg cm^{-2} 50 wt% Pt/C; anode Pt loading: 4.0 mg cm^{-2} 50 wt% PtRu/C. The membrane thickness was ~0.5 mm. (Reproduced with permission from Ref. [407].)

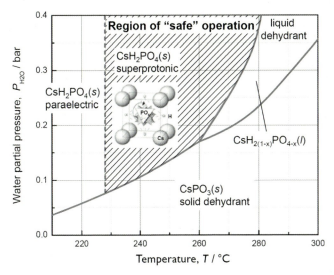

Figure 5.33 Phase stability diagram of CsH_2PO_4. (Reproduced with permission from Ref. [410].)

Among the solid acids studied, CsH_2PO_4 has been identified to be most appropriate for fuel cell electrolyte applications [409]. As shown in Figure 5.33 for the phase diagram of CsH_2PO_4 [410], the superprotonic transition temperature of this compound is 228 °C. Thus, the cell must be heated above this phase transition temperature before the cell is operable. To prevent the dehydration at the high operating temperatures, the gas streams at both anode and cathode must be humidified (>0.2 atm). Despite the structural constrains on the application of CsH_2PO_4 proton exchange membranes, high performance and power densities were reported; for example, 250 mW cm^{-2} in methanol fuel at ~250 °C [411].

5.6.6
Single-Chamber Solid Oxide Fuel Cells

In single-chamber solid oxide fuel cells (SC-SOFCs), both anode and cathode are exposed to the same mixture of fuel and oxidant gas. As a result, the gas-sealing problem can be inherently avoided since no separation between fuel and air is required. In addition, carbon deposition becomes less problematic due to the presence of a large amount of oxygen in the mixture. Figure 5.34 shows three types of possible geometries for SC-SOFCs [412].

In all three arrangements, the two electrodes should be highly selective: (i) one electrode (anode) has to be electrochemically active for the oxidation of the fuel but inert to ORR and (ii) the other electrode (cathode) has to show the opposite properties. The concept of SC-SOFC was first demonstrated by Hibino and Iwahara in 1993 [413]. Using YSZ as electrolyte, Ni/YSZ as anode, and Au as cathode, the cell achieved an OCV of 350 mV and power density of 2.3 mW cm^{-2} at 950 °C with a methane–air mixture (methane/oxygen = 2 : 1). Replacing the Au cathode with conventional LSM

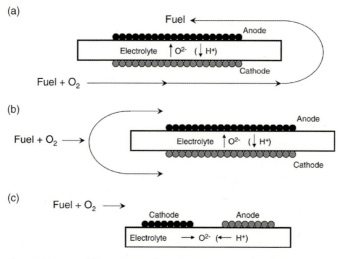

Figure 5.34 Possible geometries for SC-SOFCs. (Reproduced with permission from Ref. [412].)

resulted in the significant increase in the OCV to 795 mV and maximum power density to 121 mW cm^{-2} at 950 °C. Doped ceria can also be used as an electrolyte in SC-SOFCs. Using SDC as electrolyte, Ni/SDC as anode, and Sm$_{0.5}$Sr$_{0.5}$CoO$_3$ as cathode, the cell achieved an OCV of 800 mV and a peak power density of 644 mW cm^{-2} at 550 °C with a methane–air mixture [414]. SC-SOFCs also exhibit better performance with higher hydrocarbon fuels. Due to the fact that the gas at the electrodes in SC-SOFCs is a fuel–air mixture, the electrolyte does not need to be in dense form [415]. The use of a porous electrolyte allows to prepare thin electrolyte films by low-cost fabrication methods such as screen printing. LSCF and BSCF can also be considered as cathodes in SC-SOFCs [267, 416].

Though the heat generated by the partial oxidation of hydrocarbons at the anode is able to sustain the cell temperature without external heating [417], the efficiency of the fuel cell was estimated to be ∼1%. Thus, the energy conversion efficiency of SC-SOFCs needs to be improved substantially. Challenges still remain in the development of highly selective anode and cathode materials for fuel oxidation and ORR, respectively, for SC-SOFCs. However, due to the simplified cell structure (i.e., the compact design) and thus due to the enhanced mechanical and thermal tolerance of these cells, SC-SOFCs can be used for the energy recovery from waste fuels such as the engine exhaust [418], where the use of conventional SOFCs would be problematic.

5.6.7
Microsolid Oxide Fuel Cells

To date, miniaturized fuel cells utilizing proton exchange membranes and liquid methanol fuels (i.e., DMFCs) have been the primary focus of interests. However, high loading of precious metal catalysts such as Pt and PtRu is required for DMFCs to obtain the beneficial energy output due to the CO poisoning, and under operating conditions methanol crossover is still a serious problem [419, 420]. Due to the persistent challenges with polymer-based microfuel cells, there is a growing interest in the development of micro-SOFCs (μ-SOFCs) for portable power sources. With these systems, hydrocarbon fuels, in addition to hydrogen, can be used directly at the anode and this reduces the need for fuel preforming [417]. μ-SOFC systems are predicted to have higher specific energy and energy density, compared to μ-DMFCs, μ-PEMFCs, and existing battery technologies [421]. In contrast to large SOFC systems, the operating temperature of μ-SOFCs can be reduced to below 600 °C.

There are two possible designs for μ-SOFCs: tubular and planar. Tubular μ-SOFCs consist of small, needle-like tubes bundled together into a stack. In the tubular configuration, microtubular cells are obtained by extrusion and dip coating processes of very small ceramic tubes (diameter <0.4 mm). The tubular configuration is well suited to rapid startup/shutdown due to the high thermal shock resistance and low mass [422]. The electrolyte surface area to volume ratio increases with reduced diameters of microtubes, leading to an increased power density for these tubes [423]. The decrease in tube diameter also enables to reduce wall thickness without any degradation of cell's mechanical properties [424]. Roy et al. [425] showed that reduction of the cell and a decrease in the wall thickness lead to an increase in hoop strength of microtubular SOFCs.

The most common stack configuration of tubular μ-SOFCs is, probably, planar multicell array (PMA) [426]. In this stack arrangement, the anode represents the internal layer of the tube, while the external surface is the cathode. The current collector have two cylinders, one is in contact with the inner part of the cell (i.e., the anode) and the other with the outer (i.e., the cathode), and thus every contiguous cell of the PMA is connected in series. One problem with PMA design is the long electron path along both the anode and the cathode, which could lead to high electrical resistance and low power density. Funahashi et al. [427] proposed "cubic bundle" configuration for microtubular cells. In this structure, all the cells are connected electrically in parallel through current collectors of LSCF and Ag powder, and porous MgO matrices are used for the arrangement and support of tubular cells. Figure 5.35 shows the procedures of the arrangement of cubic bundle-based tubular cells [427]. A tubular μ-SOFC stack with three series-connected bundles produced a maximum power of 0.6 W at 500 °C, corresponding to a power density of 0.22 W cm^{-2}. The tubular cell consisted of extruded Ni/GDC anode-supported tube with dip-coated GDC electrolyte and LSCF cathode.

MEMS microfabrication techniques such as lithography and etching in combination with thin-film deposition process are commonly used in the design and fabrication of planar μ-SOFCs on Ni, silicon wafers, or glass–ceramic substrates. Magnetron sputtering, pulsed laser deposition (PLD), atomic layer deposition (ALD), and spray pyrolysis (SP) are some of the common thin-film techniques employed in planar μ-SOFCs. Chen et al. [428] fabricated a μ-SOFC based on a thin-film electrolyte deposited on a nickel foil substrate by a PLD method. The Ni foil substrate was then processed into a porous anode by photolithographic patterning and wet etching to

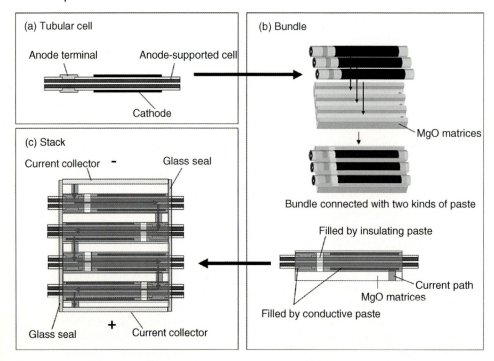

Figure 5.35 Concepts of (a) tubular cell structure, (b) bundle structure arranged tubular cell in the groove of MgO matrices, and (c) series-connected stack structure. (Reproduced with permission from Ref. [427].)

develop pores for gas transport into the fuel cell. An LSC thin-film cathode was then deposited on the electrolyte, and a porous NiO/YSZ cermet layer was added to the anode to improve the electrode performance. The μ-SOFC yielded a maximum output power density of 110 mW cm^{-2} at 570 °C. Huang et al. [429] fabricated a thin-film SOFC containing 50–150 nm thick free-standing YSZ or GDC electrolyte and 80 nm porous Pt cathode and anode, using sputtering, lithography, and etching techniques. The unit cell area is 0.06 mm^2 and the peak power is 200 and 400 mW cm^{-2} at 350 and 400 °C, respectively. The high power densities achieved were attributed to the ultrathin electrolyte and the high charge transfer reaction rates at the interfaces between the nanoporous electrodes (cathode and/or anode) and the nanocrystalline thin electrolyte.

Su et al. [430] fabricated μ-SOFC with free-standing ultrathin corrugated YSZ electrolyte cells using a sequence of MEMS processing steps. Figure 5.36 shows the process flow for fabrication of the corrugated planar μ-SOFC. The thin-film electrolyte is generated by patterning the silicon wafer with standard photolithography and deep reactive ion etching (DRIE) to create cup-shaped trenches. The electrolyte thin film (∼70 nm) is then deposited on the silicon template via ALD technique. Etching with KOH and sputtering of both Pt anode and cathode on both sides of the free-standing YSZ membrane lead to total cell thickness of ∼300 nm. A high power

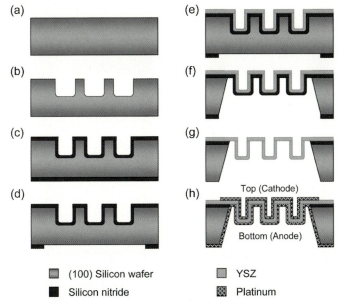

Figure 5.36 Process flow diagram for fabrication of the corrugated thin-film SOFC. The (100) silicon substrate is etched by DRIE to generate the template for pattern transfer (a and b). A 100 nm thick silicon nitride layer is deposited with LPCVD on both sides of wafer (c). The backside of silicon nitride is patterned with openings (d), followed by ALD deposition of YSZ on template (e). Silicon template is etched in KOH (f) and silicon nitride etch stop is removed by plasma etching (g). The free-standing corrugated electrolyte is deposited with porous platinum on top (cathode) and bottom (anode) sides acting as both electrode and catalyst. (Reproduced with permission from Ref. [430].)

density, 677 mW cm^{-2} at 400 °C, was obtained [430]. However, photolithography in combination with the etching process is complicated and often difficult since the etching process could damage the fuel cell materials. In addition, the resolution of wet etching is generally limited to ~10 μm. Dry etching would allow much finer patterns but is time consuming for a multilayered Ni-based substrate.

Joo and Choi [431] fabricated μ-SOFCs based on a porous and thin Ni substrate using screen printing technique. The cell achieved a maximum power density of 26 mW cm^{-2} at 450 °C in H_2/air. The advantage of screen printing is its simplicity and low cost in comparison to the lithography and etching processes. Muecke et al. [432] reported the fabrication of planar μ-SOFCs on a photostructurable glass–ceramic substrate by thin-film and micromachining techniques. The anode is a sputtered Pt film and the cathode is made of a SP-deposited LSCF, a sputtered Pt film, and Pt paste. YSZ deposited by PLD and SP is used as electrolytes. The total thickness of all layers is less than 1 μm. The contribution of the electrolyte resistance to the total cell resistance was found negligible for all cells. The best performance was observed on cells with a bilayer electrolyte of PLD-YSZ and SP-YSZ with a maximum power density of 152 mW cm^{-2} at 550 °C.

To develop μ-SOFCs with high performance and stability, the design of the support structure, the deposition techniques for the thin-film electrode and electrolyte, and the good and stable electrode/electrolyte interfaces are the areas of challenge.

5.6.8
Direct Carbon Fuel Cells

DCFC is the only fuel cell type using solid fuel. In DCFC, solid carbon is directly fed into the anode compartment and electrochemically oxidized to CO_2, generating electricity. Figure 5.37 shows the operating principle of a DCFC based on solid electrolyte. The overall cell reaction is given by Equation 5.14.

$$C + O_2 = CO_2 \tag{5.14}$$

The theoretical electrochemical conversion efficiency based on Equation 5.14 slightly exceeds 100%. This is because the entropy change for the cell reaction is positive ($\Delta S = 1.6\,J\,K^{-1}\,mol^{-1}$ at 600 °C), which results in a slightly larger standard Gibbs free energy change ($\Delta G = -395.4\,kJ\,mol^{-1}$ at 600 °C) than the standard enthalpy change ($\Delta H = -394.0\,kJ\,mol^{-1}$ at 600 °C) [433]. The reacting carbon and the product (CO_2) exist as pure substances in separate phases; therefore, their chemical potentials are fixed and independent of the extent of the fuel conversion or position within the cell.

Electrochemical oxidation of carbon requires high temperature because of the sluggish kinetics. Thus, both MCFC and SOFC configurations can be used for DCFCs. Despite the high electrochemical efficiency, the performance reported for DCFCs is far below that of MCFCs and SOFCs. Cooper and Berner [434] used mixed molten carbonate (Li_2CO_3–K_2CO_3) as the electrolyte and carbon paste as the anode with open-foam Ni as the current collector. The maximum power density varied in the range of 80–90 mW cm^{-2} at 800 °C. Using carbon black as anode with Pt as the

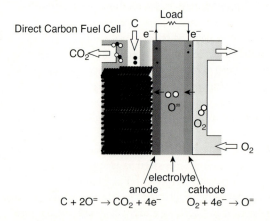

Figure 5.37 Operating principle of a direct carbon fuel cell based on solid electrolyte.

current collector, a power density of $10\,mW\,cm^{-2}$ at 0.25 V was obtained on a SOFC configuration with a 0.12 mm thick YSZ electrolyte [435]. The low performance is partially related to the poor contact between the solid carbon anode and the electrolyte.

Apart from the difficulties in continuous feeding of the solid carbon fuels, the stacking and scaling up of DCFCs will be the most challenging. The compact bipolar design with large electrode areas developed for gaseous fuels seems impossible for a DCFC. Raw coal might not be directly used as fuel in DCFCs without cleaning because impurities such as sulfur and chloride will not only degrade the cell performance but also cause a significant atmospheric pollution if the exhaust is directly released.

5.7
Applications of Fuel Cells

As reviewed in this chapter, there exist a wide range of different fuel cell technologies, each with its own characteristics and suited to different applications. The range of fuel cell applications and the size of potential markets for fuel cell-based energy devices are enormous.

PEMFCs have been the technology of choice for transportation because of their low-temperature operation and rapid startup capability. The past decades have seen a significant progress in the power density and durability close to the target requirements for automotive and stationary applications. However, the cost of PEMFCs and the system complexity are still too high. Tsuchiya and Kobayashi [436] estimated in 2004 that the cost for the low-volume production of PEMFCs is $1833 kW^{-1} and the lion's share of the cost is in the MEA and bipolar plate manufacturing. In addition to the high cost of hydrogen fuel infrastructure, the successful entry of hybrid and battery-powered vehicles suggests that they might achieve prominence before we might expect fuel cell vehicles to become widespread. Thus, it makes sense to consider the opportunities for PEMFCs in other markets. For example, fuel cells can compete favorably with lead–acid batteries for forklift applications.

To date, tubular SOFC-based power generation systems of up to 250 kW size have been produced and operated. A 100 kW size power generation system using tubular SOFCs was fabricated by Siemens-Westinghouse and operated for 2 years in the Netherlands on desulfurized natural gas without any significant performance degradation [437]. The system provided up to 108 kW of AC electricity at an efficiency of 47% and ~85 kW of hot water for the local heating system. With the combination of pressurized SOFC–gas turbine power system, the efficiency is expected to reach ~70%. The challenges for a widespread application of tubular SOFCs systems are the low power densities (0.25–$0.30\,W\,cm^{-2}$) and high manufacturing cost.

Planar SOFCs offer, in principle, a much higher power density. One of the significantly growing application areas for planar SOFC systems (1–5 kW) is the residential combined heat and power (CHP) systems operating on natural gas, developed by several companies such as Ceramic Fuel Cells Ltd (CFCL) in Australia and Ceres Power in the United Kingdom. For example, the BlueGen units with power output of up to 2 kW and 60% electrical efficiency produced by CFCL can produce up to 17 000 kWh of electricity per annum, more than sufficient to power an average home. Lowering the operation temperature, shortening the startup and shutdown time, and providing more rugged construction due to compliant seals and metallic interconnects make IT-SOFCs a viable technology for mobile applications. IT-SOFCs have been demonstrated for use as auxiliary power units (APUs) and traction power in vehicles [438].

The power density requirement for portable power sources is ever increasing, and power consumption is forecast to pose long-term technical challenges for the portable electronics industries, which are working to find ways to extend the running time of mobile devices such as portable computers, MP3 players, and mobile phones. The power range is 0.1–1 W for MP3 player, 2–5 W for mobile phones, and 15–30 W for laptop computers. In fact, the performance, mass and volumes, and life span of power supplies limit most applications of microelectromechanical systems (MEMS) technology [439]. Miniaturized fuel cells or microfuel cells are particularly very attractive for powering portable electronic devices. The important potential benefit over batteries is a longer usage time between recharges, given the considerably greater energy density of liquid fuels than that of state-of-the-art batteries. DMFCs have been the primary technology of interest for high-energy portable power sources. In this application, Pt black is the preferred catalyst for the cathode and unsupported PtRu catalyst at the loading level of 2–4 mg cm^{-2} is needed to achieve practical power densities in DMFCs. There are numerous developers of small DMFC systems intending to reduce the size and to increase the power density, including Smart Fuel Cell Inc, Toshiba Corporation, MIT, Samsung, and so on.

Significant and growing military interests in the fuel cell technology for various applications include submarine power, unmanned aerial and underwater vehicles, base camp power, and soldiers' personal power suppliers. The ability of high-temperature SOFCs and μ-SOFCs to operate directly on logistic fuels (e.g., JP-8) would be particularly attractive for such applications. Fuel cell technologies offer the potential for extended mission length.

The integration of biofuels such as bioethanol, or biomass gasification fuel cells, is a promising and forthcoming technology for electricity and heat cogeneration along with profound environmental and socioeconomic benefits. SOFC creates many synergies in the integrated system. The high-quality heat eluted from SOFCs can be used either in external thermal cycles, increasing the total yield, or to cover the thermal demands of the integrated process (reforming of the fuel, thermochemical treatment of biomass). Figure 5.38 shows a simplified flow diagram of the integrated biomass gasification SOFC system. Omosun et al. [440] showed that for a biomass-fueled SOFC system, the overall theoretical efficiency for the cogeneration of electricity and heat can be as high as 60%. Such integrated systems are well suited for the distributed power generation for remote areas due to the wide availability of biomass and flexible modulability of SOFCs. One of the main focus areas of the Solid State Energy Conversion Alliance (SECA) program is to develop large (>100 MW) integrated coal gasification SOFC power systems [2].

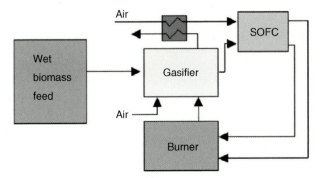

Figure 5.38 An integrated biomass gasifier SOFC system. (Reproduced with permission from Ref. [440].)

In addition to the electricity and heat generation, fuel cells can also be used for chemical and energy cogeneration [441]. In the cogeneration mode, the fuel cells are characterized by their current efficiency, selectivity with respect to the product obtained, and/or current and power densities.

5.8
Final Remarks

Despite the fact that the concept of fuel cell was almost as old as the steam engine and IC engines, majority of fuel cell technologies are still either in the field trial stage or are in the early stage of development in the case of emerging fuel cells, though some technologies are moving close to entering into specific niche market such as cogeneration, forklift traction power, backup power, and industrial battery replacement. The issues related to the development and commercialization of the fuel cell technologies are truly complex, while the technologies developed in 1970–1990s were less feasible with respect to other technical solutions.

Common to all new and emerging technologies, two of the most technical barriers for the commercial viability and acceptance of the fuel cell technologies are the cost and durability. The high cost makes the fuel cell-based power sources not commercially competitive to existing types of power applications. Though material costs would remain essentially constant, it is anticipated that the production costs will come down with volume. Fabrication process takes a lion's share of the cost of fuel cells. Besides cost reduction, durability is the most important issue to be addressed before any fuel cell technology can be commercially successful. For different applications, the requirements for fuel cell life span vary significantly; for example, for PEMFCs, it ranges from 5000 h for cars to 20 000 h for buses and 40 000 h of continuous operation for stationary applications. The performance of a fuel cell or stack is affected by many internal and external factors, such as fuel cell design and assembly, material degradation, operational conditions, and impurities or contaminants. Thus, performance degradation is unavoidable, but can be minimized through a comprehensive understanding of degradation and failure mechanisms and development of appropriate solutions. Unfortunately, the present understanding of these mechanisms on fuel cell components and particularly on the stack level is far from sufficient. One of the reasons is the long period of time needed and the high cost of the durability and life tests. For example, it is estimated that for testing a fuel cell bus system (275 kW) for 20 000 h, the fuel expense alone would be approximately $2 million (3.8 billion liters of hydrogen at $5.3 m^{-3}) [157].

In addition, the infrastructure for the production, storage, and delivery and transportation of fuels needs to be provided to support fuel cell-based energy devices and systems. In order not to be bound by the uncertain future of "hydrogen economy," fuel cells need to be able to operate effectively on alternative fuels such as natural gas, alcohols, biofuels, LPG, military logistic fuels, and so on. High-temperature fuel cells, such as MCFCs and SOFCs, offer an attractive option that is much more fuel flexible than low-temperature fuel cells such as PEMFCs and is suitable for a wide range of applications.

Fuel cells are considered environment friendly. However, they are not emission free and the fuels (in particular the liquid fuels) are not completely safe. The carbon-containing fuels produce CO_2, which is a greenhouse gas. Liquid fuels such as methanol and the other alcohols produce hazardous by-products and some of them are toxic and highly flammable. For example, hydrazine (H_2NNH_2, theoretical mass energy density is 2.6 kWh kg^{-1}) is alkaline in aqueous solution and has been shown to be compatible with AMFCs with good power output (power density of 617 mW cm^{-2} at a 4 M hydrazine hydrate concentration in 1 M KOH solution [442]). However, hydrazine is carcinogenic, very hazardous to health and environment, unstable, and should be generally avoided as a fuel. Thus, the environment and safety issues of the fuel and the products or by-products emitted from the cells should be considered.

The cost of the fuel cell systems can be significantly reduced by the breakthroughs in the new materials and catalysts. However, as the materials and catalysts for fuel cells become increasingly complex, development of new characterization techniques, especially those allowing *in situ* and quantitative measurements on the atomic level, are also critical for a quantum leap in understanding many fundamental issues. For example, the synchrotron-based *in situ* X-ray absorption spectroscopy is a powerful technique to yield structural and electronic information, and the technique was used to correlate the adsorbed coverage of species, such as CO, on PtRu surface during the potential cycling [443]. In the case of high-temperature SOFCs, the 3D image techniques based on the FIB-SEM [236, 444] demonstrate the feasibility of visualization of the electrochemically active, inactive, and unknown TPBs at the electrode/electrolyte interface and their correlation with the electrochemical activity and degradation of the electrodes. Further developments in this very important area will substantially enhance our fundamental understanding of the electrochemical reaction mechanisms.

Meeting the world's gigantic energy demands in a sustainable manner with low environmental impact is undoubtedly one of the most important challenges of this century. Fuel cells as one of the most efficient energy conversion technologies can and will play an increasingly important role in meeting the challenges. However, the questions still remain: can fuel cells fulfill the promise of clean and efficient energy systems for transportation? Can fuel cells sustain, in a trouble-free manner, the limitless hunger for energy by portable electronics? We may not have the answers today, but increasing and widespread adoption of fuel cell technologies is anticipated to contribute to a reduced dependence on fossil fuels, lessen CO_2 and noxious pollutant emissions, improve security of energy supply, and enable sustainable development of the society, though the process may take longer than anticipated.

Abbreviations

AFCs	alkaline fuel cells
AFL	anode function layer
AMFCs	alkaline membrane fuel cells
ASR	area-specific resistance
BoP	balance of plant
CHP	combined heat and power
CNTs	carbon nanotubes
DCFCs	direct carbon fuel cells
DEFCs	direct ethanol fuel cells
DET	direct electron transfer
DFAFCs	direct formic acid fuel cells
DIR-MCFCs	direct internal reforming molten carbonate fuel cells
DMFCs	direct methanol fuel cells
EOR	ethanol oxidation reaction
GDC	gadolinia-doped ceria or $(Gd,Ce)O_{2-\delta}$
GDL	gas diffusion layer
HT-PEMs	high-temperature proton exchange membranes
HT-PEMFCs	high-temperature proton exchange membrane fuel cells
ICE	internal combustion engine
IT-SOFCs	intermediate-temperature solid oxide fuel cells
LSCF	lanthanum–strontium cobalt ferrite or $(La,Sr)(Co,Fe)O_3$
LSCo	lanthanum–strontium cobaltite or $(La,Sr)CoO_3$
LSGM	lanthanum–strontium manganese gallate or $(La,Sr)(Mg,Ga)O_3$
LSM	lanthanum–strontium manganite or $(La,Sr)MnO_3$
MET	mediated electron transfer
MFCs	microbial fuel cells
MOR	methanol oxidation reaction
MPL	microporous layer
MWCNTs	multiwalled carbon nanotubes
OCV	open-circuit voltage
ORR	oxygen reduction reaction
PAFCs	phosphoric acid fuel cells
PBI	polybenzimidazole

PDDA	poly(diallyldimethylammonium chloride)
PEMFCs	proton exchange membrane fuel cells
PFSA	perfluorosulfonic acid
PILs	protic ionic liquids
PTFE	polytetrafluoroethylene
RH	relative humidity
SC-SOFCs	single-chamber solid oxide fuel cells
ScSZ	scandia-doped zirconia or Sc_2O_3–ZrO_2
SDC	samaria-doped ceria or $(Sm,Ce)O_{2-\delta}$
SOFCs	solid oxide fuel cells
TEC	thermal expansion coefficient
TPB	three-phase boundary
YDC	yttria-doped ceria or $(Y,Ce)O_{2-\delta}$
YSZ	yttria-doped zirconia or Y_2O_3-ZrO_2
μ-SOFCs	microsolid oxide fuel cells

Acknowledgment

We thank Dr. Shuangyin Wang for the technical drawings.

References

1. http://news.mongabay.com/2008/0614-co2_emissions.html (accessed on 30 June 2010).
2. Surdoval, W.A. (2009) *ECS Trans.*, **25** (2), 21–28.
3. de Bruijn, F. (2005) *Green Chem.*, **7**, 132–150.
4. Farooque, M. and Maru, H.C. (2006) *J. Power Sources*, **160**, 827–834.
5. Singhal, S.C. (2002) *Solid State Ionics*, **152**, 405–410.
6. Haile, S.M. (2003) *Mater. Today*, **6**, 24.
7. Choudhary, T.V. and Goodman, D.W. (2002) *Catal. Today*, **77**, 65–78.
8. Ross, D.K. (2006) *Vacuum*, **80**, 1084–1089.
9. Arico, A.S., Srinivasan, S., and Antonucci, V. (2001) *Fuel Cells*, **1**, 133–161.
10. Lamy, C., Lima, A., LeRhun, V., Delime, F., Coutanceau, C., and Leger, J.M. (2002) *J. Power Sources*, **105**, 283–296.
11. Mizutani, I., Liu, Y., Mitsushima, S., Ota, K.I., and Kamiya, N. (2006) *J. Power Sources*, **156**, 183–189.
12. Wang, J.T., Wasmus, S., and Savinell, R.F. (1995) *J. Electrochem. Soc.*, **142**, 4218–4224.
13. Demirci, U.B. (2007) *J. Power Sources*, **169**, 239–246.
14. Yu, J., Tian, F.J., McKenzie, L.J., and Li, C.Z. (2006) *Process Saf. Environ. Protect.*, **84**, 125–130.
15. De Geeter, E., Mangan, M., Spaepen, S., Stinissen, W., and Vennekens, G. (1999) *J. Power Sources*, **80**, 207–212.
16. McLean, G.F., Niet, T., Prince-Richard, S., and Djilali, N. (2002) *Int. J. Hydrogen Energy*, **27**, 507–526.
17. Bidault, F., Brett, D.J.L., Middleton, P.H., and Brandon, N.P. (2009) *J. Power Sources*, **187**, 39–48.
18. Duerr, M., Gair, S., Cruden, A., and McDonald, J. (2007) *J. Power Sources*, **171**, 1023–1032.
19. Gulzow, E., Schulze, M., and Gerke, U. (2006) *J. Power Sources*, **156**, 1–7.
20. Reid, J.H. (1902) Process of generating electricity. US Patent 736016.

21 Andujar, J.M. and Segura, F. (2009) *Renew. Sust. Energy Rev.*, **13**, 2309–2322.

22 Barak, M. (1966) *Adv. Energy Conversion*, **6**, 29–32, IN21–IN24, 33–55.

23 Bockris, J.O.M. and Appleby, A.J. (1986) *Energy*, **11**, 95–135.

24 Tomantschger, K., Findlay, R., Hanson, M., Kordesch, K., and Srinivasan, S. (1992) *J. Power Sources*, **39**, 21–41.

25 Kordesch, K. and Cifrain, M. (2003) *204th Meeting of the Electrochemical Society*, The Electrochemical Society, Inc., Orlando, FL.

26 Appleby, A.J. (1990) *J. Power Sources*, **29**, 3–11.

27 Kordesch, K., Gsellmann, J., Cifrain, M., Voss, S., Hacker, V., Aronson, R.R., Fabjan, C., Hejze, T., and Daniel-Ivad, J. (1999) *J. Power Sources*, **80**, 190–197.

28 Kordesch, K. and Oliveira, J.C.T. (1988) *Int. J. Hydrogen Energy*, **13**, 411–427.

29 Tomantschger, K., McClusky, F., Oporto, L., Reid, A., and Kordesch, K. (1986) *J. Power Sources*, **18**, 317–335.

30 Bidault, F., Brett, D.J.L., Middleton, P.H., Abson, N., and Brandon, N.P. (2009) *Int. J. Hydrogen Energy*, **34**, 6799–6808.

31 Cifrain, M. and Kordesch, K.V. (2004) *J. Power Sources*, **127**, 234–242.

32 Xing, B. and Savadogo, O. (2001) *Hydrogen/Oxygen Polymer Electrolyte Membrane Fuel Cells (PEMFCS) Based on Alkaline-Doped Polybenzimidazole (PBI)*, Ecole Polytechnique de Montreal, Montreal, Canada, pp. 321–323.

33 Varcoe, J.R. and Slade, R.C.T. (2005) *Fuel Cells*, **5**, 187–200.

34 Hou, H., Sun, G., He, R., Sun, B., Jin, W., Liu, H., and Xin, Q. (2008) *Int. J. Hydrogen Energy*, **33**, 7172–7176.

35 Varcoe, J.R. and Slade, R.C.T. (2006) *Electrochem. Commun.*, **8**, 839–843.

36 Varcoe, J.R., Slade, R.C.T., Lam How Yee, E., Poynton, S.D., Driscoll, D.J., and Apperley, D.C. (2007) *Chem. Mater.*, **19**, 2686–2693.

37 Fang, J. and Shen, P.K. (2006) *J. Membr. Sci.*, **285**, 317–322.

38 Wang, G., Weng, Y., Chu, D., Chen, R., and Xie, D. (2009) *J. Membr. Sci.*, **332**, 63–68.

39 Wu, L. and Xu, T. (2008) *J. Membr. Sci.*, **322**, 286–292.

40 Wu, L., Xu, T., Wu, D., and Zheng, X. (2008) *J. Membr. Sci.*, **310**, 577–585.

41 Xiong, Y., Fang, J., Zeng, Q.H., and Liu, Q.L. (2008) *J. Membr. Sci.*, **311**, 319–325.

42 Xiong, Y., Liu, Q.L., and Zeng, Q.H. (2009) *J. Power Sources*, **193**, 541–546.

43 Yang, C.-C., Chiu, S.-J., and Chien, W.-C. (2006) *J. Power Sources*, **162**, 21–29.

44 Danks, T.N., Slade, R.C.T., and Varcoe, J.R. (2003) *J. Mater. Chem.*, **13**, 712–721.

45 Pan, J., Lu, S.F., Li, Y., Huang, A.B., Zhuang, L., and Lu, J.T. (2008) *Adv. Funct. Mater.*, **20**, 312–319.

46 Huang, A., Xia, C., Xiao, C., and Zhuang, L. (2006) *J. Appl. Polym. Sci.*, **100**, 2248–2251.

47 Wu, Y., Wu, C., Yu, F., Xu, T., and Fu, Y. (2008) *J. Membr. Sci.*, **307**, 28–36.

48 Wu, Y., Wu, C., Xu, T., Yu, F., and Fu, Y. (2008) *J. Membr. Sci.*, **321**, 299–308.

49 Xiong, Y., Liu, Q.L., Zhu, A.M., Huang, S.M., and Zeng, Q.H. (2009) *J. Power Sources*, **186**, 328–333.

50 Yang, C.-C., Chiu, S.-J., Lee, K.-T., Chien, W.-C., Lin, C.-T., and Huang, C.-A. (2008) *J. Power Sources*, **184**, 44–51.

51 Neagu, V., Bunia, I., and Plesca, I. (2000) *Polym. Degrad. Stabil.*, **70**, 463–468.

52 Hsueh, K.L., Gonzalez, E.R., and Srinivasan, S. (1983) *Electrochim. Acta*, **28**, 691–697.

53 Spendelow, J.S. and Wieckowski, A. (2007) *Phys. Chem. Chem. Phys.*, **9**, 2654–2675.

54 Lima, F.H.B., Zhang, J., Shao, M.H., Sasaki, K., Vukmirovic, M.B., Ticianelli, E.A., and Adzic, R.R. (2007) *J. Phys. Chem. C*, **111**, 404–410.

55 Xiong, L. and Manthiram, A. (2004) *J. Mater. Chem.*, **14**, 1454–1460.

56 Luo, J., Njoki, P.N., Lin, Y., Wang, L., and Zhong, C.J. (2006) *Electrochem. Commun.*, **8**, 581–587.

57 Demarconnay, L., Coutanceau, C., and Leger, J.M. (2004) *Electrochim. Acta*, **49**, 4513–4521.

58 Chatenet, M., Genies-Bultel, L., Aurousseau, M., Durand, R., and Andolfatto, F. (2002) *J. Appl. Electrochem.*, **32**, 1131–1140.

59 Okajima, K., Nabekura, K., Kondoh, T., and Sudoh, M. (2005) *J. Electrochem. Soc.*, **152**, D117–D120.

60 Lee, H.K., Shim, J.P., Shim, M.J., Kim, S.W., and Lee, J.S. (1996) *Mater. Chem. Phys.*, **45**, 238–242.

61 Lima, F.H.B., de Castro, J.F.R., and Ticianelli, E.A. (2006) *J. Power Sources*, **161**, 806–812.

62 Meng, H. and Shen, P.K. (2006) *Electrochem. Commun.*, **8**, 588–594.

63 Xu, J., Huang, W.H., and McCreery, R.L. (1996) *J. Electroanal. Chem.*, **410**, 235–242.

64 Calegaro, M.L., Lima, F.H.B., and Ticianelli, E.A. (2006) *J. Power Sources*, **158**, 735–739.

65 Yang, J. and Xu, J.J. (2003) *Electrochem. Commun.*, **5**, 306–311.

66 Jiang, S.P., Lin, Z.G., and Tseung, A.C.C. (1990) *J. Electrochem. Soc.*, **137**, 764–769.

67 Zhang, D., Chi, D., Okajima, T., and Ohsaka, T. (2007) *Electrochim. Acta*, **52**, 5400–5406.

68 Xie, X.-Y., Ma, Z.-F., Ma, X.-X., Ren, Q., Schmidt, V.M., and Huang, L. (2007) *J. Electrochem. Soc.*, **154**, B733–B738.

69 Yang, Y.-F., Zhou, Y.-H., and Cha, C.-S. (1995) *Electrochim. Acta*, **40**, 2579–2579.

70 Hammouche, A., Kahoul, A., Sauer, D.U., and De Doncker, R.W. (2006) *J. Power Sources*, **153**, 239–244.

71 Sugawara, M., Ohno, M., and Matsuki, K. (1997) *J. Mater. Chem.*, **7**, 833–836.

72 De Koninck, M. and Marsan, B. (2008) *Electrochim. Acta*, **53**, 7012–7021.

73 Heller-Ling, N., Prestat, M., Gautier, J.L., Koenig, J.F., Poillerat, G., and Chartier, P. (1997) *Electrochim. Acta*, **42**, 197–202.

74 Spendelow, J.S., Goodpaster, J.D., Kenis, P.J.A., and Wieckowski, A. (2006) *J. Phys. Chem. B*, **110**, 9545–9555.

75 Morallon, E., Rodes, A., Vazquez, J.L., and Perez, J.M. (1995) *J. Electroanal. Chem.*, **391**, 149–157.

76 Wang, Q., Sun, G.Q., Jiang, L.H., Xin, Q., Sun, S.G., Jiang, Y.X., Chen, S.P., Jusys, Z., and Behm, R.J. (2007) *Phys. Chem. Chem. Phys.*, **9**, 2686–2696.

77 Lopezatalaya, M., Morallon, E., Cases, F., Vazquez, J.L.K., and Perez, J.M. (1994) *J. Power Sources*, **52**, 109–117.

78 Xu, C., Shen, P.K., Ji, X., Zeng, R., and Liu, Y. (2005) *Electrochem. Commun.*, **7**, 1305–1308.

79 Bai, Y., Wu, J., Xi, J., Wang, J., Zhu, W., Chen, L., and Qiu, X. (2005) *Electrochem. Commun.*, **7**, 1087–1090.

80 Xu, C., Shen, P.K., and Liu, Y. (2007) *J. Power Sources*, **164**, 527–531.

81 Antolini, E. (2007) *J. Power Sources*, **170**, 1–12.

82 Shen, P.K. and Xu, C. (2006) *Electrochem. Commun.*, **8**, 184–188.

83 Xu, C., Tian, Z., Shen, P., and Jiang, S.P. (2008) *Electrochim. Acta*, **53**, 2610–2618.

84 Hu, F., Chen, C., Wang, Z., Wei, G., and Shen, P.K. (2006) *Electrochim. Acta*, **52**, 1087–1091.

85 Hu, F.P. and Shen, P.K. (2007) *J. Power Sources*, **173**, 877–881.

86 Greeley, J., Nørskov, J.K., and Mavrikakis, M. (2002) *Annu. Rev. Phys. Chem.*, **53**, 319–348.

87 Bambagioni, V., Bianchini, C., Filippi, J., Oberhauser, W., Marchionni, A., Vizza, F., Psaro, R., Sordelli, L., Foresti, M.L., and Innocenti, M. (2009) *ChemSusChem*, **2**, 99–112.

88 Bianchini, C., Bambagioni, V., Filippi, J., Marchionni, A., Vizza, F., Bert, P., and Tampucci, A. (2009) *Electrochem. Commun.*, **11**, 1077–1080.

89 Xu, C., Wang, H., Shen, P.K., and Jiang, S.P. (2007) *Adv. Mater.*, **19**, 4256–4259.

90 Lu, S.F., Pan, J., Huang, A.B., Zhuang, L., and Lu, J.T. (2008) *Proc. Natl. Acad. Sci. USA*, **105**, 20611–20614.

91 Kundu, P.P., Sharma, V., and Shul, Y.G. (2007) *Crit. Rev. Solid State*, **32**, 51–66.

92 Smitha, B., Sridhar, S., and Khan, A.A. (2005) *J. Membr. Sci.*, **259**, 10–26.

93 Ma, C.S., Zhang, L., Mukerjee, S., Ofer, D., and Nair, B.D. (2003) *J. Membr. Sci.*, **219**, 123–136.

94. Anantaraman, A.V. and Gardner, C.L. (1996) *J. Electroanal. Chem.*, **414**, 115–120.
95. Tang, H.L. and Pan, M. (2008) *J. Phys. Chem. C*, **112**, 11556–11568.
96. Tang, H., Wan, Z., Pan, M., and Jiang, S.P. (2007) *Electrochem. Commun.*, **9**, 2003–2008.
97. Ren, S., Sun, G., Li, C., Song, S., Xin, Q., and Yang, X. (2006) *J. Power Sources*, **157**, 724.
98. Traversa, E. and Licoccia, S. (2006) *J. Power Sources*, **159**, 12.
99. Watanabe, M., Uchida, H., and Emori, M. (1998) *J. Phys. Chem. B*, **102**, 3129.
100. Yang, B., Fu, Y.Z., and Manthiram, A. (2005) *J. Power Sources*, **139**, 170.
101. Liu, W., Ruth, K., and Rusch, G. (2001) *J. New Mater. Electrochem. Syst.*, **4**, 227–232.
102. Jiang, R., Kunz, H.R., and Fenton, J.M. (2005) *J. Power Sources*, **150**, 120–128.
103. Li, Q., Jensen, J.O., Savinell, R.F., and Bjerrum, N.J. (2009) *Prog. Polym. Sci.*, **34**, 449–477.
104. Chen, S.-L., Krishnan, L., Srinivasan, S., Benziger, J., and Bocarsly, A.B. (2004) *J. Membr. Sci.*, **243**, 327–333.
105. Bai, H. and Ho, W.S.W. (2008) *J. Membr. Sci.*, **313**, 75–85.
106. Buchi, F.N., Gupta, B., Haas, O., and Scherer, G.G. (1995) *Electrochim. Acta*, **40**, 345–353.
107. Arico, A.S., Srinivasan, S., and Antonucci, V. (2001) *Fuel Cells*, **1**, 1–29.
108. Stamenkovic, V.R., Fowler, B., Mun, B.S., Wang, G.J., Ross, P.N., Lucas, C.A., and Markovic, N.M. (2007) *Science*, **315**, 493.
109. Wang, S.Y., Kristian, N., Jiang, S.P., and Wang, X. (2009) *Nanotechnology*, **20**.
110. Luo, J., Wang, L., Mott, D., Njoki, P.N., Lin, Y., He, T., Xu, Z., Wanjana, B.N., Lim, I.I.S., and Zhong, C.J. (2008) *Adv. Mater.*, **20**, 4342–4347.
111. Zhang, J.D., Vukmirovic, M.R., Xu, Y., Mavrikakis, M., and Adzic, R.R. (2005) *Angew. Chem., Int. Ed.*, **44**, 2132.
112. Kibler, L.A., El-Aziz, A.M., Hoyer, R., and Kolb, D.M. (2005) *Angew. Chem., Int. Ed.*, **44**, 2080–2084.
113. Tian, Z.Q., Jiang, S.P., Liang, Y.M., and Shen, P.K. (2006) *J. Phys. Chem. B*, **110**, 5343–5350.
114. Wang, S., Jiang, S.P., and Wang, X. (2008) *Nanotechnology*, **19**.
115. Yang, D.Q., Rochette, J.F., and Sacher, E. (2005) *J. Phys. Chem. B*, **109**, 4481–4484.
116. Wang, S.Y., Jiang, S.P., White, T.J., Guo, J., and Wang, X. (2009) *J. Phys. Chem. C*, **113**, 18935–18945.
117. Lefevre, M., Proietti, E., Jaouen, F., and Dodelet, J.P. (2009) *Science*, **324**, 71–74.
118. Medard, C., Lefevre, M., Dodelet, J.P., Jaouen, F., and Lindbergh, G. (2006) *Electrochim. Acta*, **51**, 3202–3213.
119. Shukla, A.K. and Raman, R.K. (2003) *Annu. Rev. Mater. Res.*, **33**, 155–168.
120. Mustain, W.E. and Prakash, J. (2007) *J. Power Sources*, **170**, 28–37.
121. Shao, M.-H., Sasaki, K., and Adzic, R.R. (2006) *J. Am. Chem. Soc.*, **128**, 3526–3527.
122. Wang, D.L., Lu, S.F., and Jiang, S.P. (2010) *Chem. Commun.*, **46**, 2058–2060.
123. Wilson, M.S., Valerio, J.A., and Gottesfeld, S. (1995) *Electrochim. Acta*, **40**, 355–363.
124. Han, M., Chan, S.H., and Jiang, S.P. (2006) *J. Power Sources*, **159**, 1005–1014.
125. Wilson, M.S. and Gottesfeld, S. (1992) *J. Appl. Electrochem.*, **22**, 1–7.
126. Wilson, M.S., Valerio, J.A., and Gottesfeld, S. (1995) *Electrochim. Acta*, **40**, 355–363.
127. Jung, U.H., Park, K.T., Park, E.H., and Kim, S.H. (2006) *J. Power Sources*, **159**, 529–532.
128. Han, M., Chan, S.H., and Jiang, S.P. (2007) *Int. J. Hydrogen Energy*, **32**, 385–391.
129. Tang, H. and Jiang, S.P. (2008) *J. Phys. Chem. C*, **112**, 19748–19755.
130. Hermann, A., Chaudhuri, T., and Spagnol, P. (2005) *Int. J. Hydrogen Energy*, **30**, 1297–1302.
131. Makkus, R.C., Janssen, A.H.H., de Bruijn, F.A., and Mallant, R. (2000) *J. Power Sources*, **86**, 274–282.
132. Tawfik, H., Hung, Y., and Mahajan, D. (2007) *J. Power Sources*, **163**, 755–767.

133 Heras, N.D.L., Roberts, E.P.L., Langton, R., and Hodgson, D.R. (2009) *Energy Environ. Sci.*, **2**, 206–214.

134 Tawfik, H., Hung, Y., and Mahajan, D. (2007) *J. Power Sources*, **163**, 755–767.

135 Squadrito, G., Barbera, O., Giacoppo, G., Urbani, F., and Passalacqua, E. (2008) *Int. J. Hydrogen Energy*, **33**, 1941–1946.

136 Rajalakshmi, N., Pandiyan, S., and Dhathathreyan, K.S. (2008) *Int. J. Hydrogen Energy*, **33**, 449–454.

137 Li, X.G. and Sabir, M. (2005) *Int. J. Hydrogen Energy*, **30**, 359–371.

138 Kandlikar, S.G. and Lu, Z. (2009) *Appl. Therm. Eng.*, **29**, 1276–1280.

139 Choi, K.H., Park, D.J., Rho, Y.W., Kho, Y.T., and Lee, T.H. (1998) *J. Power Sources*, **74**, 146–150.

140 Weber, A.Z. and Newman, J. (2004) *Chem. Rev.*, **104**, 4679–4726.

141 Djilali, N. (2005) Computational modelling of polymer electrolyte membrane (PEM) fuel cells: challenges and opportunities. 18th International Conference on Efficiency, Costs, Optimization, Simulation, and Environmental Impact of Energy Systems, Trodheim, Norway, June 22–24, pp 269–280.

142 Cownden, R., Nahon, M., and Rosen, M.A. (2001) *Int. J. Hydrogen Energy*, **26**, 615–623.

143 De Francesco, M. and Arato, E. (2002) *J. Power Sources*, **108**, 41–52.

144 Borup, R., Meyers, J., Pivovar, B., Kim, Y.S., Mukundan, R., Garland, N., Myers, D., Wilson, M., Garzon, F., Wood, D., Zelenay, P., More, K., Stroh, K., Zawodzinski, T., Boncella, J., Mcgrath, J.E., Inaba, M., Miyatake, K., Hori, M., Ota, K., Ogumi, Z., Miyata, S., Nishikata, A., Zyun, S., Uchimoto, Y., Yasuda, K., Kimijima, K.-I., and Iwashita, N. (2007) *Chem. Rev.*, **107** (10), 3904–3951.

145 Shao-Horn, Y., Sheng, W.C., Chen, S., Ferreira, P.J., Holby, E.F., and Morgan, D. (2007) *Top. Catal.*, **46**, 285–305.

146 Handley, C., Brandon, N.P., and van der Vorst, R. (2002) *J. Power Sources*, **106**, 344–352.

147 Tang, H.L., Shen, P.K., Jiang, S.P., Fang, W., and Mu, P. (2007) *J. Power Sources*, **170**, 85–92.

148 Huang, X.Y., Solasi, R., Zou, Y., Feshler, M., Reifsnider, K., Condit, D., Burlatsky, S., and Madden, T. (2006) *J. Polym. Sci. B*, **44**, 2346–2357.

149 Zhang, J., Sasaki, K., Sutter, E., and Adzic, R.R. (2007) *Science*, **315**, 220–222.

150 Wang, X., Li, W.Z., Chen, Z.W., Waje, M., and Yan, Y.S. (2006) *J. Power Sources*, **158**, 154–159.

151 Wang, S.Y., Wang, X., and Jiang, S.P. (2008) *Langmuir*, **24**, 10505–10512.

152 Chen, Z.W., Waje, M., Li, W.Z., and Yan, Y.S. (2007) *Angew. Chem., Int. Ed.*, **46**, 4060–4063.

153 Trogadas, P., Parrondo, J., and Ramani, V. (2008) *Electrochem. Solid State Lett.*, **11**, B113–B116.

154 Li, M.Q., Shao, Z.G., Zhang, H.M., Zhang, Y.M., Zhu, X.B., and Yi, B.L. (2006) *Electrochem. Solid State Lett.*, **9**, A92–A95.

155 Liu, Y.H., Nguyen, T.H., Kristian, N., Yu, Y.L., and Wang, X. (2009) *J. Membr. Sci.*, **330**, 357–362.

156 de Bruijn, F.A., Dam, V.A.T., and Janssen, G.J.M. (2008) *Fuel Cells*, **8**, 3–22.

157 Wu, J.F., Yuan, X.Z., Martin, J.J., Wang, H.J., Zhang, J.J., Shen, J., Wu, S.H., and Merida, W. (2008) *J. Power Sources*, **184**, 104–119.

158 Qi, Z.G. and Kaufman, A. (2002) *J. Power Sources*, **112**, 121–129.

159 Peled, E., Duvdevani, T., Aharon, A., and Melman, A. (2001) *Electrochem. Solid State Lett.*, **4**, A38–A41.

160 Miyake, N., Wainright, J.S., and Savinell, R.F. (2001) *J. Electrochem. Soc.*, **148**, A905.

161 Staiti, P., Arico, A.S., Baglio, V., Lufrano, F., Passalacqua, E., and Antonucci, V. (2001) *Solid State Ionics*, **145**, 101.

162 Yang, C., Srinivasan, S., Arico, A.S., Baglio, V., and Antonucci, V. (2001) *Electrochem. Solid State Lett.*, **4**, A31.

163 Pu, C., Huang, W., Ley, K.L., and Smotkin, E.S. (1995) *J. Electrochem. Soc.*, **142**, L119.

164 Yoon, S.R., Hwang, G.H., Cho, W.I., Oh, I.H., Hong, S.A., and Ha, H.Y. (2002) *J. Power Sources*, **106**, 215.

165 Kim, Y.J., Choi, W.C., Woo, S.I., and Hong, W.H. (2004) *Electrochim. Acta*, **49**, 3227.

166 Yamaguchi, T., Miyata, F., and Nakao, S. (2003) *Adv. Mater.*, **15**, 1198.

167 Yamaguchi, T., Zhou, H., Nakazawa, S., and Hara, N. (2007) *Adv. Mater.*, **19**, 592.

168 Yamauchi, A., Ito, T., and Yamaguchi, T. (2007) *J. Power Sources*, **174**, 170–175.

169 Yamaguchi, T., Kuroki, H., and Miyata, F. (2005) *Electrochem. Commun.*, **7**, 730–734.

170 Wang, C.H., Chen, C.C., Hsu, H.C., Du, H.Y., Chen, C.P., Hwang, J.Y., Chen, L.C., Shih, H.C., Stejskal, J., and Chen, K.H. (2009) *J. Power Sources*, **190**, 279–284.

171 Jiang, S.P., Liu, Z., and Tian, Z.Q. (2006) *Adv. Mater.*, **18**, 1068–1072.

172 Yang, M., Lu, S.F., Lu, J.L., Jiang, S.P., and Xiang, Y. (2010) *Chem. Commun.*, **46**, 1434–1436.

173 Thomas, S.C., Ren, X.M., Gottesfeld, S., and Zelenay, P. (2002) *Electrochim. Acta*, **47**, 3741–3748.

174 Neburchilov, V., Martin, J., Wang, H.J., and Zhang, J.J. (2007) *J. Power Sources*, **169**, 221–238.

175 Zhao, T.S., Chen, R., Yang, W.W., and Xu, C. (2009) *J. Power Sources*, **191**, 185–202.

176 Bae, B., Kho, B.K., Lim, T.H., Oh, I.H., Hong, S.A., and Ha, H.Y. (2006) *J. Power Sources*, **158**, 1256–1261.

177 Kim, D.J., Cho, E.A., Hong, S.A., Oh, I.H., and Ha, H.Y. (2004) *J. Power Sources*, **130**, 172–177.

178 Kristian, N., Yan, Y., and Wang, X. (2007) *Chem. Commun.*, 353–355.

179 Waszczuk, P., Barnard, T.M., Rice, C., Masel, R.I., and Wieckowski, A. (2002) *Electrochem. Commun.*, **4**, 599–603.

180 Rice, C., Ha, S., Masel, R.I., and Wieckowski, A. (2003) *J. Power Sources*, **115**, 229–235.

181 Choi, J.H., Jeong, K.J., Dong, Y., Han, J., Lim, T.H., Lee, J.S., and Sung, Y.E. (2006) *J. Power Sources*, **163**, 71–75.

182 Casado-Rivera, E., Volpe, D.J., Alden, L., Lind, C., Downie, C., Vazquez-Alvarez, T., Angelo, A.C.C., Disalvo, F.J., and Abruna, H.D. (2004) *J. Am. Chem. Soc.*, **126**, 4043–4049.

183 Ha, S., Larsen, R., and Masel, R.I. (2005) *J. Power Sources*, **144**, 28–34.

184 Larsen, R., Zakzeski, J., and Masel, R.I. (2005) *Electrochem. Solid State Lett.*, **8**, A291–A293.

185 Wang, S.Y., Wang, X., and Jiang, S.P. (2008) *Nanotechnology*, **19**, 265–601.

186 Ha, S., Dunbar, Z., and Masel, R.I. (2006) *J. Power Sources*, **158**, 129–136.

187 Song, S.Q., Zhou, W.J., Zhou, Z.H., Jiang, L.H., Sun, G.Q., Xin, Q., Leontidis, V., Kontou, S., and Tsiakaras, P. (2005) *Int. J. Hydrogen Energy*, **30**, 995–1001.

188 Tsiakaras, P.E. (2007) *J. Power Sources*, **171**, 107–112.

189 Wang, D.L., Lu, S.F., and Jiang, S.P. (2010) *Electrochim. Acta*, **55**, 2964–2971.

190 Zhou, W.J., Zhou, Z.H., Song, S.Q., Li, W.Z., Sun, G.Q., Tsiakaras, P., and Xin, Q. (2003) *Appl. Catal. B*, **46**, 273–285.

191 Kowal, A., Li, M., Shao, M., Sasaki, K., Vukmirovic, M., Zhang, J., Marinkovic, N.S., Liu, P., Frenkel, A.I., and Adzic, R.R. (2009) *Nat. Mater.*, **8**, 325.

192 Antolini, E. (2007) *Appl. Catal. B*, **74**, 337–350.

193 Song, S.Q. and Tsiakaras, P. (2006) *Appl. Catal. B*, **63**, 187–193.

194 Sammes, N., Bove, R., and Stahl, K. (2004) *Curr. Opin. Solid State Mater. Sci.*, **8**, 372–378.

195 Seo, S.J., Joh, H.I., Kim, H.T., and Moon, S.H. (2005) Properties of Pt/C catalyst modified by chemical vapor deposition of Cr as a cathode of phosphoric acid fuel cell. 56th Annual ISE Meeting, Busan, South Korea, September 25–30, 2005, pp. 1676–1682.

196 Caires, M.I., Buzzo, M.L., Ticianelli, E.A., and Gonzalez, E.R. (1997) *J. Appl. Electrochem.*, **27**, 19–24.

197 Hara, N., Tsurumi, K., and Watanabe, M. (1996) *J. Electroanal. Chem.*, **413**, 81–88.

198 Minh, N.Q. (1988) *J. Power Sources*, **24**, 1–19.

199 Freni, S., Barone, F., and Puglisi, M. (1998) *Int. J. Energy Res.*, **22**, 17–31.

200 Ota, K.I., Matsuda, Y., Matsuzawa, K., and Mitsushima, S., and Kamiya, N. (2005) Effect of rare earth oxides for improvement of MCFC. International Workshop on Molten Carbonate Fuel Cells and Related Science and Technology, Toulouse, France, August 29–31, 2005, pp. 811–815.

201 Wee, J.H. and Lee, K.Y. (2006) *J. Mater. Sci.*, **41**, 3585–3592.

202 Ganesan, P., Colon, H., Haran, B., and Popov, B.N. (2003) *J. Power Sources*, **115**, 12–18.

203 Han, J., Kim, S.G., Yoon, S.P., Nam, S.W., Lim, T.H., Oh, I.H., Hong, S.A., and Lim, H.C. (2001) Performance of LiCoO2-coated NiO cathode under pressurized conditions. 7th Grove Fuel Cell Symposium, London, UK, September 11–13, 2001, pp. 153–159.

204 Hong, M.Z., Lee, H.S., Kim, M.H., Park, E.J., Ha, H.W., and Kim, K. (2006) *J. Power Sources*, **156**, 158–165.

205 Huang, B., Wang, S.R., Yu, Q.C., Liu, Y., and Hu, K.A. (2005) *J. Appl. Electrochem.*, **35**, 1145–1156.

206 Lee, H., Hong, M.Z., Bae, S.C., Park, E., and Kim, K. (2003) *J. Mater. Chem.*, **13**, 2626–2632.

207 Park, E., Hong, M.Z., Lee, H., Kim, M., and Kim, K. (2005) *J. Power Sources*, **143**, 84–92.

208 Yuh, C., Colpetzer, J., Dickson, K., Farooque, M., and Xu, G. (2006) *J. Mater. Eng. Perform.*, **15**, 457–462.

209 Suski, L. and Tarniowy, M. (2001) *J. Mater. Sci.*, **36**, 5119–5124.

210 Terada, S., Nagashima, I., Higaki, K., and Ito, Y. (1998) *J. Power Sources*, **75**, 223–229.

211 Hyun, S.H., Cho, S.C., Cho, J.Y., Ko, D.H., and Hong, S.A. (2001) *J. Mater. Sci.*, **36**, 441–450.

212 Kim, S.D., Hyun, S.H., Lim, T.H., and Hong, S.A. (2004) *J. Power Sources*, **137**, 24–29.

213 Lee, J.J., Choi, H.J., Hyun, S.H., and Im, H.C. (2008) *J. Power Sources*, **179**, 504–510.

214 Durante, G., Vegni, S., Capobianco, P., and Golgovici, F. (2005) High temperature corrosion of metallic materials in molten carbonate fuel cells environment. Fuel Cell Seminar 2004, San Antonio, TX, 2004, pp. 204–209.

215 Zhou, L., Lin, H.X., Yi, B.L., and Zhang, H.M. (2007) *Chem. Eng. J.*, **125**, 187–192.

216 Kawabata, Y., Fujimoto, N., Yamamoto, M., Nagoya, T., and Nishida, M. (2000) *J. Power Sources*, **86**, 324–328.

217 Frangini, S. (2007) Corrosion of metallic stack components in molten carbonates: critical issues and recent findings. International Workshop on Degradation Issues in Fuel Cells, Crete, Greece, September 19–21, 2007, pp. 462–468.

218 Perez, F.J., Duday, D., Hierro, M.P., Gomez, C., Aguero, A., Garcia, M.C., Muela, R., Pascual, A.S., and Martinez, L. (2002) *Surf. Coat. Technol.*, **161**, 293–301.

219 Keijzer, M., Hemmes, K., De Wit, J.H.W., and Schoonman, J. (2000) *J. Appl. Electrochem.*, **30**, 1421–1431.

220 Fergus, J.W. (2005) *Mater. Sci. Eng. A*, **397**, 271–283.

221 Jiang, S.P. and Chan, S.H. (2004) *J. Mater. Sci.*, **39**, 4405–4439.

222 Zhu, W.Z. and Deevi, S.C. (2003) *Mater. Sci. Eng. A*, **362**, 228–239.

223 Jiang, S.P. (2008) *J. Mater. Sci.*, **43**, 6799–6833.

224 Tsipis, E.V. and Kharton, V.V. (2008) *J. Solid State Electrochem.*, **12**, 1039–1060.

225 Fergus, J.W. (2006) *J. Power Sources*, **162**, 30–40.

226 Ishihara, T. (2006) *Bull. Chem. Soc. Jpn.*, **79**, 1155–1166.

227 Zhu, W.Z. and Deevi, S.C. (2003) *Mater. Sci. Eng. A*, **348**, 227–243.

228 Fergus, J.W. (2005) *J. Power Sources*, **147**, 46–57.

229 Dees, D.W., Claar, T.D., Easler, T.E., Fee, D.C., and Mrazek, F.C. (1987) *J. Electrochem. Soc.*, **134**, 2141–2146.

230 Jiang, S.P. (2003) *J. Electrochem. Soc.*, **150**, E548–E559.

231 Jiang, S.P., Duan, Y.Y., and Love, J.G. (2002) *J. Electrochem. Soc.*, **149**, A1175–A1183.

232 Murray, E.P., Tsai, T., and Barnett, S.A. (1999) *Nature*, **400**, 649–651.

233 Jensen, K.V., Primdahl, S., Chorkendorff, I., and Mogensen, M. (2001) *Solid State Ionics*, **144**, 197–209.

234 Tsoga, A., Naoumidis, A., and Nikolopoulos, P. (1996) *Acta Mater.*, **44**, 3679–3692.

235 Jiang, S.P. (2003) *J. Mater. Sci.*, **38**, 3775–3782.

236 Wilson, J.R., Kobsiriphat, W., Mendoza, R., Chen, H.Y., Hiller, J.M., Miller, D.J., Thornton, K., Voorhees, P.W., Adler, S.B., and Barnett, S.A. (2006) *Nat. Mater.*, **5**, 541–544.

237 Eguchi, K., Kojo, H., Takeguchi, T., Kikuchi, R., and Sasaki, K. (2002) *Solid State Ionics*, **152–153**, 411–416.

238 Kim, H., da Rosa, C., Boaro, M., Vohs, J.M., and Gorte, R.J. (2002) *J. Am. Ceram. Soc.*, **85**, 1473–1476.

239 Kim, H., Lu, C., Worrell, W.L., Vohs, J.M., and Gorte, R.J. (2002) *J. Electrochem. Soc.*, **149**, A247–A250.

240 Wang, W., Jiang, S.P., Tok, A.I.Y., and Luo, L. (2006) *J. Power Sources*, **159**, 68–72.

241 Zha, S.W., Cheng, Z., and Liu, M.L. (2007) *J. Electrochem. Soc.*, **154**, B201–B206.

242 Lussier, A., Sofie, S., Dvorak, J., and Idzerda, Y.U. (2008) *Int. J. Hydrogen Energy*, **33**, 3945–3951.

243 Zha, S., Tsang, P., Cheng, Z., and Liu, M. (2005) *J. Solid State Chem.*, **178**, 1844–1850.

244 Mukundan, R., Brosha, E.L., and Garzon, F.H. (2004) *Electrochem. Solid State Lett.*, **7**, A5–A7.

245 Aguilar, L., Zha, S., Li, S., Winnick, J., and Liu, M. (2004) *Electrochem. Solid State Lett.*, **7**, A324–A326.

246 Marina, O.A. and Pederson, L.R. (2002) *The Fifth European Solid Oxide Fuel Cell Forum (ESOFC-V)* (ed. J. Huijismans), European Fuel Cell Forum, Lucerne, Switzerland, pp. 481–489.

247 Marina O.A., Canfield, N.L., and Stevenson, J.W. (2002) *Solid State Ionics*, **149**, 21–28.

248 Tao, S. and Irvine, J.T.S. (2002) *J. Solid State Chem.*, **165**, 12–18.

249 Slater, P.R., Fagg, D.P., and Irvine, J.T.S. (1997) *J. Mater. Chem.*, **7**, 2495–2498.

250 Slater, P.R. and Irvine, J.T.S. (1999) *Solid State Ionics*, **120**, 125–134.

251 Tao, S.W. and Irvine, J.T.S. (2003) *Nat. Mater.*, **2**, 320–323.

252 Huang, Y.H., Dass, R.I., Xing, Z.L., and Goodenough, J.B. (2006) *Science*, **312**, 254–257.

253 Jiang, S.P., Liu, L., Ong, K.P., Wu, P., Li, J., and Pu, J. (2008) *J. Power Sources*, **176**, 82–89.

254 Aruna, S.T., Muthuraman, M., and Patil, K.C. (1997) *J. Mater. Chem.*, **7**, 2499–2503.

255 Kuo, J.H., Anderson, H.U., and Sparlin, D.M. (1990) *J. Solid State Chem.*, **87**, 55–63.

256 Mizusaki, J., Yonemura, Y., Kamata, H., Ohyama, K., Mori, N., Takai, H., Tagawa, H., Dokiya, M., Naraya, K., Sasamoto, T., Inaba, H., and Hashimoto, T. (2000) *Solid State Ionics*, **132**, 167–180.

257 Yokokawa, H., Sakai, N., Kawada, T., and Dokiya, M. (1990) *Solid State Ionics*, **40–41**, 398–401.

258 Carter, S., Selcuk, A., Chater, R.J., Kajda, J., Kilner, J.A., and Steele, B.C.H. (1992) *Solid State Ionics*, **53-6**, 597–605.

259 Jiang, S.P. (2002) *Solid State Ionics*, **146**, 1–22.

260 Murray, E.P., Tsai, T., and Barnett, S.A. (1998) *Solid State Ionics*, **110**, 235–243.

261 Murray, E.P. and Barnett, S.A. (2001) *Solid State Ionics*, **143**, 265–273.

262 Jiang, S.P. and Wang, W. (2005) *J. Electrochem. Soc.*, **152**, A1398–A1408.

263 Jiang, S.P., Zhen, Y.D., and Zhang, S. (2006) *J. Electrochem. Soc.*, **153**, A1511–A1517.

264 Esquirol, A., Brandon, N.P., Kilner, J.A., and Mogensen, M. (2004) *J. Electrochem. Soc.*, **151**, A1847–A1855.

265 Hwang, H.J., Ji-Woong, M.B., Seunghun, L.A., and Lee, E.A. (2005) *J. Power Sources*, **145**, 243–248.

266 Sahibzada, M., Benson, S.J., Rudkin, R.A., and Kilner, J.A. (1998) *Solid State Ionics*, **115**, 285–290.

267 Shao, Z.P. and Haile, S.M. (2004) *Nature*, **431**, 170–173.

268 Wei, B., Lu, Z., Li, S.Y., Liu, Y.Q., Liu, K.Y., and Su, W.H. (2005) *Electrochem. Solid State Lett.*, **8**, A428–A431.

269 Jennings, A.J. and Skinner, S.J. (2002) *Solid State Ionics*, **152**, 663–667.
270 Skinner, S.J. (2003) *Solid State Sci.*, **5**, 419–426.
271 Laberty, C., Zhao, F., Swider-Lyons, K.E., and Virkar, A.V. (2007) *Electrochem. Solid State Lett.*, **10**, B170–B174.
272 Chiba, R., Yoshimura, F., and Sakurai, Y. (1999) *Solid State Ionics*, **124**, 281–288.
273 Bastidas, D.M., Tao, S.W., and Irvine, J.T.S. (2006) *J. Mater. Chem.*, **16**, 1603–1605.
274 Yokokawa, H., Sakai, N., Kawada, T., and Dokiya, M. (1991) *J. Electrochem. Soc.*, **138**, 2719–2727.
275 Yokokawa, H. (2003) *Annu. Rev. Mater. Res.*, **33**, 581–610.
276 Jiang, S.P., Zhang, J.P., and Foger, K. (2003) *J. Eur. Ceram. Soc.*, **23**, 1865–1873.
277 Mitterdorfer, A. and Gauckler, L.J. (1998) *Solid State Ionics*, **111**, 185–218.
278 Kenjo, T. and Nishiya, M. (1992) *Solid State Ionics*, **57**, 295–302.
279 Jiang, S.P., Love, J.G., Zhang, J.P., Hoang, M., Ramprakash, Y., Hughes, A.E., and Badwal, S.P.S. (1999) *Solid State Ionics*, **121**, 1–10.
280 Peters, C., Weber, A., and Ivers-Tiffee, E. (2008) *J. Electrochem. Soc.*, **155**, B730–B737.
281 Ai, N., Jiang, S.P., Lu, Z., Chen, K.F., and Su, W.H. (2010) *J. Electrochem. Soc.*, **157**, B1033–B1039.
282 Jiang, S.P., Zhang, S., and Zhen, Y.D. (2006) *J. Electrochem. Soc.*, **153**, A127–A134.
283 Badwal, S.P.S., Deller, R., Foger, K., Ramprakash, Y., and Zhang, J.P. (1997) *Solid State Ionics*, **99**, 297–310.
284 Taniguchi, S., Kadowaki, M., Kawamura, H., Yasuo, T., Akiyama, Y., Miyake, Y., and Saitoh, T. (1995) *J. Power Sources*, **55**, 73–79.
285 Zhen, Y.D., Tok, A.I.Y., Boey, F.Y.C., and Jiang, S.P. (2008) *Electrochem. Solid State Lett.*, **11**, B42–B46.
286 Zhen, Y.D., Tok, A.I.Y., Jiang, S.P., and Boey, F.Y.C. (2007) *J. Power Sources*, **170**, 61–66.
287 Badwal, S.P.S. (1992) *Solid State Ionics*, **52**, 23–32.
288 Kharton, V.V., Marques, F.M.B., and Atkinson, A. (2004) *Solid State Ionics*, **174**, 135–149.
289 Hull, S. (2004) *Rep. Prog. Phys.*, **67**, 1233–1314.
290 Nomura, K., Mizutani, Y., Kawai, M., Nakamura, Y., and Yamamoto, O. (2000) *Solid State Ionics*, **132**, 235–239.
291 Badwal, S.P.S., Ciacchi, F.T., Rajendran, S., and Drennan, J. (1998) *Solid State Ionics*, **109**, 167–186.
292 Yamamoto, O., Arati, Y., Takeda, Y., Imanishi, N., Mizutani, Y., Kawai, M., and Nakamura, Y. (1995) *Solid State Ionics*, **79**, 137–142.
293 Badwal, S.P.S., Ciacchi, F.T., and Milosevic, D. (2000) *Solid State Ionics*, **136**, 91–99.
294 Guo, X. and Maier, J. (2001) *J. Electrochem. Soc.*, **148**, E121–E126.
295 Guo, X. and Waser, R. (2006) *Prog. Mater. Sci.*, **51**, 151–210.
296 Badwal, S.P.S. and Rajendran, S. (1994) *Solid State Ionics*, **70**, 83–95.
297 Liu, Y. and Lao, L.E. (2006) *Solid State Ionics*, **177**, 159–163.
298 Mogensen, M., Sammes, N.M., and Tompsett, G.A. (2000) *Solid State Ionics*, **129**, 63–94.
299 Steele, B.C.H. (2000) *Solid State Ionics*, **129**, 95–110.
300 Zhan, Z.L., Wen, T.L., Tu, H.Y., and Lu, Z.Y. (2001) *J. Electrochem. Soc.*, **148**, A427–A432.
301 Jung, G.B., Huang, T.J., and Chang, C.L. (2002) *J. Solid State Electrochem.*, **6**, 225–230.
302 Zhang, X., Robertson, M., Deces-Petit, C., Qu, W., Kesler, O., Maric, R., and Ghosh, D. (2007) *J. Power Sources*, **164**, 668–677.
303 Ishihara, T., Tabuchi, J., Ishikawa, S., Yan, J., Enoki, M., and Matsumoto, H. (2006) *Solid State Ionics*, **177**, 1949–1953.
304 Yan, J.W., Matsumoto, H., Enoki, M., and Ishihara, T. (2005) *Electrochem. Solid State Lett.*, **8**, A389–A391.
305 Kosacki, I., Rouleau, C.M., Becher, P.F., Bentley, J., and Lowndes, D.H. (2005) *Solid State Ionics*, **176**, 1319–1326.

306 Garcia-Barriocanal, J., Rivera-Calzada, A., Varela, M., Sefrioui, Z., Iborra, E., Leon, C., Pennycook, S.J., and Santamaria, J. (2008) *Science*, **321**, 676–680.

307 Zha, S.W., Xia, C.R., and Meng, G.Y. (2003) *J. Power Sources*, **115**, 44–48.

308 Vanherle, J., Horita, T., Kawada, T., Sakai, N., Yokokawa, H., and Dokiya, M. (1996) *J. Eur. Ceram. Soc.*, **16**, 961–973.

309 Vincent, A., Savignat, S.B., and Gervais, F. (2007) *J. Eur. Ceram. Soc.*, **27**, 1187–1192.

310 Leon-Reina, L., Losilla, E.R., Martinez-Lara, M., Bruque, S., and Aranda, M.A.G. (2004) *J. Mater. Chem.*, **14**, 1142–1149.

311 Slater, P.R., Sansom, J.E.H., and Tolchard, J.R. (2004) *Chem. Rec.*, **4**, 373–384.

312 Tao, S.W. and Irvine, J.T.S. (2001) *Mater. Res. Bull.*, **36**, 1245–1258.

313 Jiang, S.P., Zhang, L., He, H.Q., Yap, R.K., and Xiang, Y. (2009) *J. Power Sources*, **189**, 972–981.

314 Verkerk, M.J. and Burggraaf, A.J. (1981) *Solid State Ionics*, **3–4**, 463–467.

315 Iwahara, H. (1988) *Solid State Ionics*, **28**, 573–578.

316 Mori, M., Yamamoto, T., Itoh, H., and Watanabe, T. (1997) *J. Mater. Sci.*, **32**, 2423–2431.

317 Fergus, J.W. (2004) *Solid State Ionics*, **171**, 1–15.

318 Sakai, N., Yokokawa, H., Horita, T., and Yamaji, K. (2004) *Int. J. Appl. Ceram. Technol.*, **1**, 23–30.

319 Liu, Z.G., Zheng, Z.R., Huang, X.Q., Lu, Z., He, T.M., Dong, D.W., Sui, Y., Miao, J.P., and Su, W.H. (2004) *J. Alloys Compd.*, **363**, 60–62.

320 Zhou, X.L., Deng, F.J., Zhu, M.X., Meng, G.Y., and Liu, X.Q. (2007) *J. Power Sources*, **164**, 293–299.

321 Shen, Y., Liu, M.N., He, T.M., and Jiang, S.P. (2010) *J. Power Sources*, **195**, 977–983.

322 Holt, A. and Kofstad, P. (1994) *Solid State Ionics*, **69**, 137–143.

323 Quadakkers, W.J., Hansel, M., and Rieck, T. (1998) *Mater. Corros.*, **49**, 252–257.

324 Blum, L., Buchkremer, H.P., Gross, S., Gubner, A., de Haart, L.G.J., Nabielek, H., Quadakkers, W.J., Reisgen, U., Smith, M.J., Steinberger-Wilckens, R., Steinbrech, R.W., Tietz, F., and Vinke, I.C. (2007) *Fuel Cells*, **7**, 204–210.

325 Yang, Z.G., Hardy, J.S., Walker, M.S., Xia, G.G., Simner, S.P., and Stevenson, J.W. (2004) *J. Electrochem. Soc.*, **151**, A1825–A1831.

326 Jiang, S.P., Zhang, S., and Zhen, Y.D. (2005) *J. Mater. Res.*, **20**, 747–758.

327 Hua, B., Pu, J., Zhang, J.F., Lu, F.S., Chi, B., and Jian, L. (2009) *J. Electrochem. Soc.*, **156**, B93–B98.

328 Chen, X.B., Hua, B., Pu, J., Li, J., Zhang, L., and Jiang, S.P. (2009) *Int. J. Hydrogen Energy*, **34**, 5737–5748.

329 Eichler, K., Solow, G., Otschik, P., and Schaffrath, W. (1999) *J. Eur. Ceram. Soc.*, **19**, 1101–1104.

330 Yang, Z.G., Stevenson, J.W., and Meinhardt, K.D. (2003) *Solid State Ionics*, **160**, 213–225.

331 Jiang, S.P., Christiansen, L., Hughan, B., and Foger, K. (2001) *J. Mater. Sci. Lett.*, **20**, 695–697.

332 Duquette, J. and Petric, A. (2004) *J. Power Sources*, **137**, 71–75.

333 Bram, M., Reckers, S., Drinovac, P., Monch, J., Steinbrech, R.W., Buchkremer, H.P., and Stover, D. (2004) *J. Power Sources*, **138**, 111–119.

334 Chou, Y.S., Stevenson, J.W., and Chick, L.A. (2002) *J. Power Sources*, **112**, 130–136.

335 Chou, Y.S. and Stevenson, J.W. (2002) *J. Power Sources*, **112**, 376–383.

336 Khan, T.I. and Al-Badri, A. (2003) *J. Mater. Sci.*, **38**, 2483–2488.

337 Weil, K.S. (2006) *JOM*, **58**, 37–44.

338 Brett, D.J.L., Atkinson, A., Brandon, N.P., and Skinner, S.J. (2008) *Chem. Soc. Rev.*, **37**, 1568–1578.

339 Foger, K. and Love, J.G. (2004) *Solid State Ionics*, **174**, 119–126.

340 Jiang, S.P. (2010) *Handbook of Nanostructured Film Devices and Coatings*, vol. **3** (ed. S. Zhang), CRC Press, pp. 155–187.

341 Minh, N.Q. (2004) *Solid State Ionics*, **174**, 271–277.

342 Srivastava, P.K., Quach, T., Duan, Y.Y., Donelson, R., Jiang, S.P., Ciacchi, F.T., and Badwal, S.P.S. (1997) *Solid State Ionics*, **99**, 311–319.

343 Kim, J.W., Virkar, A.V., Fung, K.Z., Mehta, K., and Singhal, S.C. (1999) *J. Electrochem. Soc.*, **146**, 69–78.

344 Sarantaridis, D. and Atkinson, A. (2007) *Fuel Cells*, **7**, 246–258.

345 Lang, M., Henne, R., Schaper, S., and Schiller, G. (2001) *J. Therm. Spray Technol.*, **10**, 618–625.

346 Brandon, N.P., Blake, A., Corcoran, D., Cumming, D., Duckett, A., El-Koury, K., Haigh, D., Kidd, C., Leah, R., Lewis, G., Matthews, C., Maynard, N., Oishi, N., McColm, T., Trezona, R., Selcuk, A., Schmidt, M., and Verdugo, L. (2004) *J. Fuel Cell Sci. Technol.*, **1**, 61–65.

347 Liu, Y., Hashimoto, S., Nishino, H., Takei, K., and Mori, M. (2007) *J. Power Sources*, **164**, 56–64.

348 Cassidy, M., Boulfrad, S., Irvine, J., Chung, C., Jorger, M., Munnings, C., and Pyke, S. (2009) *Fuel Cells*, **9**, 891–898.

349 Armand, M., Endres, F., MacFarlane, D.R., Ohno, H., and Scrosati, B. (2009) *Nat. Mater.*, **8**, 621–629.

350 Fernicola, A., Panero, S., and Scrosati, B. (2008) *J. Power Sources*, **178**, 591–595.

351 Yan, F., Yu, S.M., Zhang, X.W., Qiu, L.H., Chu, F.Q., You, J.B., and Lu, J.M. (2009) *Chem. Mater.*, **21**, 1480–1484.

352 Cho, E., Park, J.S., Sekhon, S.S., Park, G.G., Yang, T.H., Lee, W.Y., Kim, C.S., and Park, S.B. (2009) *J. Electrochem. Soc.*, **156**, B197–B202.

353 Lin, B.C., Cheng, S., Qiu, L.H., Yan, F., Shang, S.M., and Lu, J.M. (2010) *Chem. Mater.*, **22**, 1807–1813.

354 Potter, M.C. (1912) *Proc. R. Soc. Lond. B*, **84**, 260–276.

355 Niessen, J., Harnisch, F., Rosenbaum, M., Schroder, U., and Scholz, F. (2006) *Electrochem. Commun.*, **8**, 869–873.

356 Zhang, E.R., Xu, W., Diao, G.W., and Shuang, C.D. (2006) *J. Power Sources*, **161**, 820–825.

357 Logan, B.E., Hamelers, B., Rozendal, R.A., Schrorder, U., Keller, J., Freguia, S., Aelterman, P., Verstraete, W., and Rabaey, K. (2006) *Environ. Sci. Technol.*, **40**, 5181–5192.

358 Bond, D.R., Holmes, D.E., Tender, L.M., and Lovley, D.R. (2002) *Science*, **295**, 483–485.

359 Lovley, D.R. (1993) *Annu. Rev. Microbiol.*, **47**, 263–290.

360 Chaudhuri, S.K. and Lovley, D.R. (2003) *Nat. Biotechnol.*, **21**, 1229–1232.

361 Kim, N., Choi, Y., Jung, S., and Kim, S. (2000) *Biotechnol. Bioeng.*, **70**, 109–114.

362 Rabaey, K., Lissens, G., Siciliano, S.D., and Verstraete, W. (2003) *Biotechnol. Lett.*, **25**, 1531–1535.

363 Park, D.H. and Zeikus, J.G. (2003) *Biotechnol. Bioeng.*, **81**, 348–355.

364 Zhao, F., Harnisch, F., Schroder, U., Scholz, F., Bogdanoff, P., and Herrmann, I. (2005) *Electrochem. Commun.*, **7**, 1405–1410.

365 Oh, S.E. and Logan, B.E. (2006) *Appl. Microbiol. Biotechnol.*, **70**, 162–169.

366 Clauwaert, P., Rabaey, K., Aelterman, P., De Schamphelaire, L., Ham, T.H., Boeckx, P., Boon, N., and Verstraete, W. (2007) *Environ. Sci. Technol.*, **41**, 3354–3360.

367 Rismani-Yazdi, H., Carver, S.M., Christy, A.D., and Tuovinen, I.H. (2008) *J. Power Sources*, **180**, 683–694.

368 Du, Z.W., Li, H.R., and Gu, T.Y. (2007) *Biotechnol. Adv.*, **25**, 464–482.

369 Shantaram, A., Beyenal, H., Raajan, R., Veluchamy, A., and Lewandowski, Z. (2005) *Environ. Sci. Technol.*, **39**, 5037–5042.

370 Mano, N., Fernandez, J.L., Kim, Y., Shin, W., Bard, A.J., and Heller, A. (2003) *J. Am. Chem. Soc.*, **125**, 15290–15291.

371 Chaubey, A. and Malhotra, B.D. (2002) *Biosens. Bioelectron.*, **17**, 441–456.

372 Tarasevich, M.R., Bogdanovskaya, V.A., and Kapustin, A.V. (2003) *Electrochem. Commun.*, **5**, 491–496.

373 Kim, J. and Grate, J.W. (2003) *Nano Lett.*, **3**, 1219–1222.

374 Pizzariello, A., Stred'ansky, M., and Miertus, S. (2002) *Bioelectrochemistry*, **56**, 99–105.

375 Moehlenbrock, M.J. and Minteer, S.D. (2008) *Chem. Soc. Rev.*, **37**, 1188–1196.

376 Willner, I., Yan, Y.M., Willner, B., and Tel-Vered, R. (2009) *Fuel Cells*, **9**, 7–24.

377 Davis, F. and Higson, S.P.J. (2007) *Biosens. Bioelectron.*, **22**, 1224–1235.

378 Teodorescu, S.G., Gellett, W.L., Kesmez, M., and Schumacher, J. (2006) *209th ECS Meeting Abstract*, The Electrochemical Society, Pennington, Denver.

379 Mano, N. and Heller, A. (2003) *J. Electrochem. Soc.*, **150**, A1136–A1138.

380 Mano, N., Mao, F., and Heller, A. (2003) *J. Am. Chem. Soc.*, **125**, 6588–6594.

381 Kjeang, E., Djilali, N., and Sinton, D. (2009) *J. Power Sources*, **186**, 353–369.

382 Cohen, J.L., Westly, D.A., Pechenik, A., and Abruna, H.D. (2005) *J. Power Sources*, **139**, 96–105.

383 Cohen, J.L., Volpe, D.J., Westly, D.A., Pechenik, A., and Abruna, H.D. (2005) *Langmuir*, **21**, 3544–3550.

384 Mitrovski, S.M. and Nuzzo, R.G. (2006) *Lab Chip*, **6**, 353–361.

385 Sung, W. and Choi, J.W. (2007) *J. Power Sources*, **172**, 198–208.

386 Jayashree, R.S., Egas, D., Spendelow, J.S., Natarajan, D., Markoski, L.J., and Kenis, P.J.A. (2006) *Electrochem. Solid State Lett.*, **9**, A252–A256.

387 Hasegawa, S., Shimotani, K., Kishi, K., and Watanabe, H. (2005) *Electrochem. Solid State Lett.*, **8**, A119–A121.

388 Kjeang, E., McKechnie, J., Sinton, D., and Djilali, N. (2007) *J. Power Sources*, **168**, 379–390.

389 Kjeang, E., Michel, R., Harrington, D.A., Djilali, N., and Sinton, D. (2008) *J. Am. Chem. Soc.*, **130**, 4000–4006.

390 Choban, E.R., Markoski, L.J., Wieckowski, A., and Kenis, P.J.A. (2004) *J. Power Sources*, **128**, 54–60.

391 Togo, M., Takamura, A., Asai, T., Kaji, H., and Nishizawa, M. (2007) *Electrochim. Acta*, **52**, 4669–4674.

392 Togo, M., Takamura, A., Asai, T., Kaji, H., and Nishizawa, M. (2008) *J. Power Sources*, **178**, 53–58.

393 Li, Q.F., He, R.H., Gao, J.A., Jensen, J.O., and Bjerrum, N.J. (2003) *J. Electrochem. Soc.*, **150**, A1599–A1605.

394 Zhang, J.L., Xie, Z., Zhang, J.J., Tanga, Y.H., Song, C.J., Navessin, T., Shi, Z.Q., Song, D.T., Wang, H.J., Wilkinson, D.P., Liu, Z.S., and Holdcroft, S. (2006) *J. Power Sources*, **160**, 872–891.

395 Savinell, R., Yeager, E., Tryk, D., Landau, U., Wainright, J., Weng, D., Lux, K., Litt, M., and Rogers, C. (1994) *J. Electrochem. Soc.*, **141**, L46–L48.

396 Li, Q.F., He, R.H., Jensen, J.O., and Bjerrum, N.J. (2003) *Chem. Mater.*, **15**, 4896–4915.

397 Xiao, L.X., Zhang, H.F., Scanlon, E., Ramanathan, L.S., Choe, E.W., Rogers, D., Apple, T., and Benicewicz, B.C. (2005) *Chem. Mater.*, **17**, 5328–5333.

398 Yu, S., Xiao, L., and Benicewicz, B.C. (2008) *Fuel Cells*, **8**, 165–174.

399 Zhai, Y.F., Zhang, H.M., Liu, G., Hu, J.W., and Yi, B.L. (2007) *J. Electrochem. Soc.*, **154**, B72–B76.

400 Stevens, D.A. and Dahn, J.R. (2005) *Carbon*, **43**, 179–188.

401 Hogarth, W.H.J., da Costa, J.C.D., Drennan, J., and Lu, G.Q. (2005) *J. Mater. Chem.*, **15**, 754–758.

402 Vichi, F.M., Tejedor-Tejedor, M.I., and Anderson, M.A. (2000) *Chem. Mater.*, **12**, 1762–1770.

403 Yamada, M., Li, D.L., Honma, I., and Zhou, H.S. (2005) *J. Am. Chem. Soc.*, **127**, 13092–13093.

404 Halla, J.D., Mamak, M., Williams, D.E., and Ozin, G.A. (2003) *Adv. Funct. Mater.*, **13**, 133–138.

405 Uma, T. and Nogami, M. (2008) *Anal. Chem.*, **80**, 506–508.

406 Uma, T. and Nogami, M. (2007) *Chem. Mater.*, **19**, 3604–3610.

407 Lu, S.F., Wang, D.L., Jiang, S.P., Xiang, Y., Lu, J.L., and Zeng, J. (2010) *Adv. Mater.*, **22**, 971.

408 Haile, S.M., Boysen, D.A., Chisholm, C.R.I., and Merle, R.B. (2001) *Nature*, **410**, 910–913.

409 Boysen, D.A., Uda, T., Chisholm, C.R.I., and Haile, S.M. (2004) *Science*, **303**, 68–70.

410 Chisholm, C.R.I., Boysen, D.A., Papandrew, A.B., Zecevic, S., Cha, S.,

Sasaki, K.A., Varga, A., Giapis, K.P., and Haile, S.M. (2009) *Electrochem. Soc. Interface*, **Fall 2009**, 53–59.

411 Haile, S.M., Chisholm, C.R.I., Sasaki, K., Boysen, D.A., and Uda, T. (2007) *Faraday Discuss.*, **134**, 17–39.

412 Yano, M., Tomita, A., Sano, M., and Hibino, T. (2007) *Solid State Ionics*, **177**, 3351–3359.

413 Hibino, T. and Iwahara, H. (1993) *Chem. Lett.*, **7**, 1131–1134.

414 Hibino, T., Hashimoto, A., Yano, M., Suzuki, M., Yoshida, S., and Sano, M. (2002) *J. Electrochem. Soc.*, **149**, A133–A136.

415 Riess, I. (2006) *Solid State Ionics*, **177**, 1591–1596.

416 Suzuki, T., Jasinski, P., Anderson, H.U., and Dogan, F. (2004) *J. Electrochem. Soc.*, **151**, A1678–A1682.

417 Shao, Z.P., Haile, S.M., Ahn, J., Ronney, P.D., Zhan, Z.L., and Barnett, S.A. (2005) *Nature*, **435**, 795–798.

418 Nagao, M., Yano, M., Okamoto, K., Tomita, A., Uchiyama, Y., Uchiyama, N., and Hibino, T. (2008) *Fuel Cells*, **8**, 322–329.

419 Jiang, S.P., Liu, Z.C., and Tian, Z.Q. (2006) *Adv. Mater.*, **18**, 1068.

420 Wasmus, S. and Kuver, A. (1999) *J. Electroanal. Chem.*, **461**, 14–31.

421 Evans, A., Bieberle-Hutter, A., Rupp, J.L.M., and Gauckler, L.J. (2009) *J. Power Sources*, **194**, 119–129.

422 Bujalski, W., Dikwal, C.A., and Kendall, K. (2007) *J. Power Sources*, **171**, 96–100.

423 Pusz, J., Mohammadi, A., and Sammes, N.M. (2006) *J. Fuel Cell Sci. Technol.*, **3**, 482–486.

424 Sarkar, P., Yamarte, L., Rho, H.S., and Johanson, L. (2007) *Int. J. Appl. Ceram. Technol.*, **4**, 103–108.

425 Roy, B.R., Sammes, N.M., Suzuki, T., Funahashi, Y., and Awano, M. (2009) *J. Power Sources*, **188**, 220–224.

426 Sammes, N.M., Du, Y., and Bove, R. (2005) *J. Power Sources*, **145**, 428–434.

427 Funahashi, Y., Shimamori, T., Suzuki, T., Fujishiro, Y., and Awano, M. (2009) *Fuel Cells*, **9**, 711–716.

428 Chen, X., Wu, N.J., Smith, L., and Ignatiev, A. (2004) *Appl. Phys. Lett.*, **84**, 2700–2702.

429 Huang, H., Nakamura, M., Su, P.C., Fasching, R., Saito, Y., and Prinz, F.B. (2007) *J. Electrochem. Soc.*, **154**, B20–B24.

430 Su, P.C., Chao, C.C., Shim, J.H., Fasching, R., and Prinz, F.B. (2008) *Nano Lett.*, **8**, 2289–2292.

431 Joo, J.H. and Choi, G.M. (2008) *J. Power Sources*, **182**, 589–593.

432 Muecke, U.P., Beckel, D., Bernard, A., Bieberle-Hutter, A., Graf, S., Infortuna, A., Muller, P., Rupp, J.L.M., Schneider, J., and Gauckler, L.J. (2008) *Adv. Funct. Mater.*, **18**, 3158–3168.

433 Cao, D.X., Sun, Y., and Wang, G.L. (2007) *J. Power Sources*, **167**, 250–257.

434 Cooper, J.F. and Berner, K. (2005) Fuel Cell Seminar, Palm Springs, CA.

435 Tao, T. (2005) Fuel Cell Seminar, Palm Springs, CA.

436 Tsuchiya, H. and Kobayashi, O. (2004) *Int. J. Hydrogen Energy*, **29**, 985–990.

437 Singhal, S.C. (2000) *MRS Bull.*, **25**, 16–21.

438 Mukerjee, S., Haltiner, K., Klotzbach, D., Vordonis, J., and Iyer, A. (2009) *ECS Trans.*, **25** (2), 59–63.

439 Cook-Chennault, K.A., Thambi, N., and Sastry, A.M. (2008) *Smart Mater. Struct.*, **17**, 043001.

440 Omosun, A.O., Bauen, A., Brandon, N.P., Adjiman, C.S., and Hart, D. (2004) *J. Power Sources*, **131**, 96–106.

441 Alcaide, F., Cabot, P.L., and Brillas, E. (2006) *J. Power Sources*, **153**, 47–60.

442 Asazawa, K., Sakamoto, T., Yamaguchi, S., Yamada, K., Fujikawa, H., Tanaka, H., and Oguro, K. (2009) *J. Electrochem. Soc.*, **156**, B509–B512.

443 Scott, F.J., Mukerjee, S., and Ramaker, D.E. (2007) *J. Electrochem. Soc.*, **154**, A396–A406.

444 Chen, H.-Y., Yu, H.-C., Cronin, J.S., Wilson, J.R., Barnett, S.A., and Thornton, K. (2011) *Journal of Power Sources*, **196**, 1333–1337.

6
Electrodes for High-Temperature Electrochemical Cells: Novel Materials and Recent Trends

Ekaterina V. Tsipis and Vladislav V. Kharton

Following Chapter 5 where state-of-the-art fuel cell technologies are reviewed, this chapter presents a short survey on the novel electrode materials for the solid oxide fuel cells (SOFCs) with oxygen ion- and proton-conducting electrolytes. Attention is primarily drawn to most recent research studies published during the past few years. Numerous literature data on the electrode performance and key properties of the electrode materials are systematically compared, including the total electrical conductivity, thermal expansion, and reactivity with solid electrolytes. A comparison of the cell performance with various fuels and cell components is also presented. In addition to the fuel cells, related applications such as high-temperature electrolyzers of steam and carbon dioxide are briefly considered.

6.1
Introduction

As illustrated by many examples discussed in the first volume [1] and in the preceding chapters, the progress in the field of solid-state electrochemical devices is significantly limited by the materials science-related factors. Particularly, the development of high- and intermediate-temperature (IT) solid-state electrochemical devices, such as solid oxide fuel cells (SOFCs), electrolyzers (SOEC) of carbon dioxide and water vapor, oxygen and hydrogen pumps, and various ceramic reactors and sensors are associated with a search for novel cathode and anode materials with superior electrocatalytic activity, optimization of their fabrication and processing technologies, and efforts to deeper understand the electrochemical reaction mechanisms (see Refs [1–8] and references therein, Chapters 12 and 13 of the first volume and Chapters 5 and 9–11 of this book). For example, commercialization of SOFCs requires reducing costs and enhancing their reliability and long-term stability. These problems can be partially solved by decreasing the SOFC operation temperatures down to the so-called intermediate-temperature range, 770–1070 K, which makes it possible to use less expensive construction materials, to suppress degradation caused by high operating temperatures and by thermal cycling, to facilitate miniaturization, and to improve efficiency of the kW-scale generators [5–10]. On the other hand,

Solid State Electrochemistry II: Electrodes, Interfaces and Ceramic Membranes.
Edited by Vladislav V. Kharton.
© 2011 Wiley-VCH Verlag GmbH & Co. KGaA. Published 2011 by Wiley-VCH Verlag GmbH & Co. KGaA.

lowering operation temperature also leads to a greater role of electrode polarization that may become critical for the overall performance. In fact, approaches elaborated to improve the electrode properties are similar for SOFC systems and other electrochemical devices based on oxygen ion- or proton-conducting solid electrolytes.

This chapter is centered on the comparative analysis of electrochemical behavior and properties relevant to the electrode applications, such as thermal expansion, stability, and mixed conductivity of the oxide and composite materials, recently reported as promising for the electrochemical devices, primarily fuel cells and electrolyzers. In this chapter, no attempt has been made to provide a complete overview of all novel electrode materials, relevant phase equilibria, and microstructural design, or a thorough analysis of the microscopic electrode reaction mechanisms. Furthermore, since the major emphasis is on novel rather than on well-established materials and approaches, the scarce information and preliminary reports available in the literature may lead to a need for additional studies and validation employing complementary experimental methods. The references in this chapter were selected in order to show typical relationships between the properties of various electrode materials reported during the past 2–4 years and, in general, to reflect state of the art in this attractive field.

6.2
General Comments

The anode and cathode are supposed to possess a high electronic conduction and catalytic activity toward oxidation and reduction processes, respectively, and to have an appropriate microstructure minimizing mass transport limitations (see Chapter 12 of the first volume [1] and Chapter 5 of this book). Moreover, all the components of high-temperature electrochemical cells should be chemically and thermomechanically compatible with each other, and be stable under the operation and fabrication conditions. These stringent and often conflicting requirements have resulted in a continuous search for optimum electrode materials and their preparation methods. In addition to the key role of electrode microstructure and the effects of solid oxide electrolyte on the electrode performance, the performance is primarily governed by the properties of electrode materials determined, in turn, by their chemical and phase compositions and crystal structure.

The most widely studied materials for potential use as oxygen electrodes are based on the perovskite-related oxide compounds (Chapters 9 and 12 of the first volume [1] and Chapter 5 of this book). These are often more tolerant to extensive cation substitution and possess better transport properties with respect to other known families. The lower overpotentials (η) and polarization resistances (R_η) are usually observed for the electrode materials made of mixed electronic–ionic conductors (MIECs) owing to effective spatial expansion of the electrochemical reaction zone beyond the triple-phase boundary (TPB) (see Refs [5, 7, 8] and references cited therein). In this respect, the cobaltite- and nickelate-based phases are of primary interest for the IT range, although many Fe- and Cu-containing materials attract serious attention as well. Whilst the electronic and oxygen ionic conductivities of manganites, still considered as state-of-the-art cathode materials for SOFCs operating at 1070–1270 K, are lower compared to their Fe-, Co-, and Ni-containing analogues, the latter families exhibit other serious disadvantages, including excessively high thermal and/or chemical expansion and limited thermodynamic stability.

For operation in reducing environments, ceramic–metal composites (cermets) containing stabilized zirconia and/or doped ceria fluorites and Ni metal are widely employed [8, 10]. The presence of nickel, however, provokes carbon deposition at the anode surface of hydrocarbon-fueled cells and may induce various types of microstructural degradation, all leading finally to the electrode destruction; therefore, a number of alternative metal and oxide materials, which may be used without metallic phase, are being considered. The major groups of ceramic compositions for the fuel electrodes include, in particular, perovskite-related titanates, chromites, and molybdates, such as $(Sr,R)TiO_{3\pm\delta}$ (R = lanthanide cations or Y), $LaCrO_3$-based solid solutions, or $Sr_2MoMO_{6-\delta}$ (M = Mg, Mn, Cr). Note that under reducing conditions, many transition metal-containing oxide materials exhibit a limited thermodynamic stability, a poor electronic transport, or insufficient catalytic activity, limiting their use.

6.3
Novel Cathode Materials for Solid Oxide Fuel Cells: Selected Trends and Compositions

The perovskite-related cobaltites and their derivatives show a relatively high electrochemical activity compared to other groups of the cathode materials (Tables 6.1 and 6.2 and Figures 6.1–6.4). Consequently, these materials are being widely appraised as promising IT SOFC cathodes, despite the excessively high volume changes that may be induced by temperature changes (thermal expansion) and/or oxygen chemical potential and overpotential variations (chemical expansion) [11–19]. For the relevant information, readers may refer to Chapter 3 of the first volume [1] and Chapter 9 of this book; Table 6.1 and Figure 6.2 present several examples. The thermal and chemical expansivity limit compatibility with common solid oxide electrolytes and long-term cell operation. For instance, the polarization resistance of porous $Sm_{0.5}Sr_{0.5}CoO_{3-\delta}$ cathode in contact with $Ce_{0.8}Sm_{0.2}O_{2-\delta}$ (CSO20) electrolyte was found to undergo fast degradation both on thermal cycling and during an isothermal operation [11]. Partial substitution of cerium for cobalt in $Sm_{1-x}Sr_xCoO_{3-\delta}$ was shown to slightly suppress the lattice expansion, decrease the conductivity, and improve the cathode performance; however, Ce solubility limit is quite low, ≤ 5 mol% [12]. The highest electrochemical performance of single-chamber solid oxide fuel cells (SC-SOFCs), fabricated with $R_{0.6}Sr_{0.4}Co_{1-x}Fe_xO_{3-\delta}$ (R = La, Nd; x = 0, 0.5) perovskite cathodes, $Ce_{0.9}Gd_{0.1}O_{2-\delta}$ (CGO10) interlayer, thin 8 mol% yttria-stabilized zirconia (8YSZ) electrolytes, and Ni–8YSZ anodes by tape casting, cofiring, and screen printing, was achieved for $La_{0.6}Sr_{0.4}CoO_{3-\delta}$ cathode providing a maximum power density (P_{max}) of ~ 550 mW cm^{-2} at 1073 K [13]. The

Table 6.1 Properties of selected cobaltite- and ferrite-based cathode materials in air.

Composition	$\sigma^{1073\,K}$ (S cm^{-1})	Average TECs		Chemical interaction		$R_\eta^{1073\,K}$ (Ω cm^2)	References
		T (K)	$\bar{\alpha}$ ($\times 10^6$ K^{-1})	Electrolyte	T (K)		
SrCo$_{0.95}$Sb$_{0.05}$O$_{3-\delta}$	178	673–1273	19.35	CNO20	>1273	0.02	[14]
Sr$_2$Co$_{0.9}$Fe$_{0.1}$NbO$_{6-\delta}$	5.7			CGO10	>1473	0.34$^{a)}$	[16]
La$_{0.5}$Sr$_{0.5}$Co$_{0.8}$Fe$_{0.2}$O$_{3-\delta}$				LSGM(10 20)	>1173	0.23	[20]
PrBaCo$_2$O$_{5+\delta}$				CSO20	≥1373		[26]
PrBa$_{0.5}$Sr$_{0.5}$Co$_2$O$_{5+\delta}$		1173	21.9	CGO10		0.05	[28]
SmBa$_{0.5}$Sr$_{0.5}$Co$_2$O$_{5+\delta}$		973	22.9	CGO10	≥1473	0.02	[28]
		1173		8YSZ	>1173	1.35	[28]
				8YSZ/CGO10 interlayer		0.01	
Sm$_{0.5}$Sr$_{0.5}$CoO$_{3-\delta}$				LSGM(10 20)	>1173	0.18	[20]
Sm$_{0.5}$Sr$_{0.5}$Co$_{0.8}$Fe$_{0.2}$O$_{3-\delta}$				LSGM(10 20)	>1173	0.26	[20]
GdBa$_{0.5}$Sr$_{0.5}$Co$_2$O$_{5+\delta}$		1173	18.7	CGO10		0.05	[28]
Ba$_2$CoMo$_{0.5}$Nb$_{0.5}$O$_{6-\delta}$	1.2	873–1173		CSO20	>1323	0.05$^{a)}$	[25]
Ba$_{0.5}$Sr$_{0.5}$Co$_{0.8}$Fe$_{0.2}$O$_{3-\delta}$				CSO20	≥1273	0.04	[25]
Ba$_{1.2}$Sr$_{0.8}$CoO$_{4+\delta}$				LSGM(10 20)	>1373	0.3	[20]
YBaCo$_4$O$_{7+\delta}$	74	750–1273	7.61	CGO10	>1273	0.08	[33, 34]
YBaCo$_{3.2}$Fe$_{0.8}$O$_{7+\delta}$		750–1273	10	LSGM(9.8 20)		0.11	[33]
YBaCo$_3$ZnO$_{7+\delta}$	19	293–973	7.69	YSZ	≥1273	0.05	[37]
		293–1273	7.43	CGO20	>1373	0.26	
				CGO20			[36]
				LSGM(20 20)	≥1273		
TbBaCo$_3$ZnO$_{7+\delta}$		298–1173	9.45	CGO20	≥1073	0.07	[36]
Ca$_3$Co$_4$O$_{9-\delta}$		1093	10	YSZ	>1073	0.68$^{a)}$	[38]
				CGO10			

Composition		T range	$\bar{\alpha}$	Electrolyte	T range	$R_\eta^{1073\,K}$	Ref.
$Sr_{0.9}K_{0.1}FeO_{3-\delta}$	26			LSGM(20 17)	>1273	$0.2^{b)}$	[45]
$LaSrFeO_{4+\delta}$				CSO10	≥1373		[47]
$(La_{0.8}Sr_{0.2})_{0.95}FeO_{3-\delta}$	44			8YSZ			[52]
$(La_{0.8}Sr_{0.2})_{0.95}Fe_{0.8}Cu_{0.2}O_{3-\delta}$	120	293–1273	10.65				[52]
$(La_{0.8}Sr_{0.2})_{0.95}Fe_{0.8}Ni_{0.2}O_{3-\delta}$	175	293–1273	10.47	8YSZ	≥1273		[52, 53]
				CGO	>1373		[53]
$(La_{0.8}Sr_{0.2})_{0.95}Fe_{0.8}Ni_{0.2}O_{3-\delta}$				CGO10		0.5	[50]

Notes: 8YSZ: 8 mol% Y_2O_3-stabilized ZrO_2 $(Zr_{0.85}Y_{0.15}O_{1.93})$; CGO10: $Ce_{0.9}Gd_{0.1}O_{1.95}$; CGO20: $Ce_{0.8}Gd_{0.2}O_{1.9}$; CSO20: $Ce_{0.8}Sm_{0.2}O_{1.9}$; CNO20: $Ce_{0.8}Nd_{0.2}O_{1.9}$; LSGM(10 20): $La_{0.9}Sr_{0.1}Ga_{0.8}Mg_{0.2}O_{3-\delta}$; LSGM(20 20): $La_{0.8}Sr_{0.2}Ga_{0.8}Mg_{0.2}O_{3-\delta}$; LSGM(20 17): $La_{0.8}Sr_{0.2}Ga_{0.83}Mg_{0.17}O_{3-\delta}$; LSGM(9.8 20): $(La_{0.9}Sr_{0.1})_{0.98}Ga_{0.8}Mg_{0.2}O_{3-\delta}$; LSAO: $La_{10}Si_5AlO_{26.5}$. $\sigma^{1073\,K}$ is the total electrical conductivity at 1073 K; $\bar{\alpha}$ is the apparent thermal expansion coefficient averaged in the given temperature range; $R_\eta^{1073\,K}$ is the electrode polarization resistance at 1073 K.

a) Extrapolated.
b) Estimated from polarization data.

Table 6.2 Properties[a] of selected Ni- and Mn-containing cathode materials in air.

Composition	$\sigma^{1073\,K}$ (S cm^{-1})	j^{1173} (×10^8 mol s^{-1} cm^{-2})	Average TECs		Polarization resistance		References
			T (K)	$\bar{\alpha}$ (×10^6 K^{-1})	Electrolyte	$R_\eta^{1073\,K}$ (Ω cm^2)	
La$_2$NiO$_{4+\delta}$	59		548–1173	13.8	LSGM(10 20)	2.0	[55, 56]
	76.4	10.0	400–1290	13.6			[57, 66]
La$_3$Ni$_2$O$_{7-\delta}$	52[b]		548–1173	13.7	LSGM(10 20)	1.5	[55, 56]
La$_4$Ni$_3$O$_{10-\delta}$	86[b]		548–1173	13.5	LSGM(10 20)	0.9	[55, 56]
La$_2$Ni$_{2.9}$Cu$_{0.1}$O$_{10-\delta}$					LSAO	2.2	[66]
La$_{3.95}$Sr$_{0.05}$Ni$_2$CoO$_{10-\delta}$					LSAO	1.6	[66]
La$_2$Ni$_{0.9}$Co$_{0.1}$O$_{4\pm\delta}$					LSAO	2.7	[66]
La$_2$Ni$_{0.8}$Cu$_{0.2}$O$_{4+\delta}$	64.6	4.7	400–1240	13.3	LSGM(10 20)	0.6	[55]
			323–1273	13.9	LSGM(9.8 20)	0.62	[61, 65]
					LSGM(10 20)	2.8	[60]
					8YSZ	6.1	[60]
					LSAO	1.54	[65]
La$_2$Ni$_{0.5}$Cu$_{0.5}$O$_{4+\delta}$	31.9		300–900	13.4	LSGM(9.8 20)	1.6	[66]
			900–1370	14.5	LSAO	2.2	
Pr$_2$Ni$_{0.8}$Cu$_{0.2}$O$_{4+\delta}$	167.9		300–1030	13.2	LSGM(9.8 20)	0.29	[57]
			1030–1150	17.5			
			1210–1280	12.9			
La$_{1.9}$Sm$_{0.1}$NiO$_{4\pm\delta}$	82				LSGM(20 20)	3.1	[55]
La$_{0.95}$Sr$_{0.05}$NiO$_{4\pm\delta}$					CGO10	0.65	[62]
LaSr$_2$Mn$_{1.6}$Ni$_{0.4}$O$_{7-\delta}$	22.2	0.01	300–900	12.5	LSAO	24	[66]

$Gd_{0.6}Ca_{0.4}Mn_{0.9}Ni_{0.1}O_{3-\delta}$	137.7		900–1370	14.4		
			550–900	10.8	LSAO	20	[66]
			900–1370	12.1		
$Sr_{0.7}Ce_{0.3}Mn_{0.9}Cr_{0.1}O_{3-\delta}$	296.8	0.02	550–950	12.0	LSAO	10.2	[66]
			950–1370	15.1		
$SrMn_{0.6}Nb_{0.4}O_{3-\delta}$	18.9		300–800	11.3	LSAO	31	[66]
			800–1200	12.3		

a) j is the steady-state oxygen permeation flux through 1.0 mm thick membrane under oxygen partial pressure gradient of 0.21/0.021 atm. Other symbols are explained in the notes to Table 6.1.
b) Porous samples.

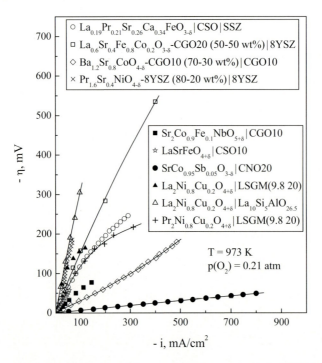

Figure 6.1 Comparison of the overpotentials of various cathode layers at 973 K in air [14, 16, 29, 43, 47, 57, 61, 65, 241, 242]. 8YSZ is 8 mol% Y_2O_3-stabilized ZrO_2; SSZ is Sc_2O_3-stabilized ZrO_2; CGO10 and CGO20 correspond to $Ce_{0.9}Gd_{0.1}O_{1.95}$ and $Ce_{0.8}Gd_{0.2}O_{1.9}$, respectively; CSO10 is $Ce_{0.9}Sm_{0.1}O_{1.95}$, CNO20 is $Ce_{0.8}Nd_{0.2}O_{1.9}$, and LSGM(9.8 20) is $(La_{0.9}Sr_{0.1})_{0.98}Ga_{0.8}Mg_{0.2}O_{3-\delta}$. A 4–6 μm thick samaria-doped ceria interlayer between the electrode and the electrolyte was used in Ref. [43].

substitution of Nd for La and Fe for Co, which might provide better compatibility of the materials, reduces cell performance, and the effect of iron doping is much more pronounced. The thermal and chemical expansion of these materials remains, however, very high [8].

Among the $SrCo_{1-x}Sb_xO_{3-\delta}$ ($x = 0$–0.2) cathodes applied onto $Ce_{0.8}Nd_{0.2}O_{2-\delta}$ (CNO20) electrolyte, the composition with $x = 0.05$ displays the highest conductivity and lowest polarization resistance (Figure 6.2), ranging from 0.009 to 0.23 Ω cm² at 873–1173 K [14]. Stabilization of tetragonal ($x = 0.05$–0.15) or cubic ($x = 0.2$) structure was reported to suppress abrupt changes in the lattice expansion at elevated temperatures, observed for the parent cobaltite due to phase transitions associated with thermally induced disordering. The apparent average linear thermal expansion coefficient (TEC) tends, however, to increase with x, reaching 29.3×10^{-6} K^{-1} at 673–1273 K and $x = 0.15$ [14]. Similar effects were found for the $SrCo_{1-x}Nb_xO_{3-\delta}$ system considered for cathode materials in lanthanum gallate-based cells [15]. For the tetragonal double perovskite $Sr_2Co_{1-x}Fe_xNbO_{5+\delta}$, the total conductivity and cathode performance decrease on increasing x [16]. Doping strontium cobaltite with iron, stabilizing the perovskite structure, does not reduce the lattice expansion on heating

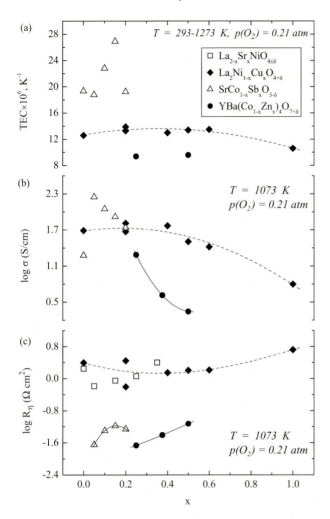

Figure 6.2 Composition dependences of the average linear thermal expansion coefficients (a), total electrical conductivity (b), and cathode polarization resistance (c) of SrCo$_{1-x}$Sb$_x$O$_{3-\delta}$ [14], YBa(Co$_{1-x}$Zn$_x$)$_4$O$_{7+\delta}$ [37], La$_{2-x}$Sr$_x$NiO$_{4\pm\delta}$ [62], and La$_2$Ni$_{1-x}$Cu$_x$O$_{4+\delta}$ [60, 61, 66]. R_η values were obtained in contact with ceria-based solid electrolytes, except for La$_2$Ni$_{1-x}$Cu$_x$O$_{4+\delta}$ cathodes applied onto LSGM. In the case of YBa(Co$_{1-x}$Zn$_x$)$_4$O$_{7+\delta}$ and La$_{2-x}$Sr$_x$NiO$_{4\pm\delta}$, the polarization resistance values correspond to porous (YBa(Co$_{1-x}$Zn$_x$)$_4$O$_{7+\delta}$–Ce$_{0.8}$Gd$_{0.2}$O$_{1.9}$ (50–50 wt%) composite and cone-shaped electrodes, respectively [37, 62].

down to the level appropriate for electrode applications [8]. In addition to thermo-mechanical strains, Sr(Co,Fe)O$_{3-\delta}$ exhibit typical ordering processes in the oxygen sublattice on cooling, which often result in the formation of brownmillerite-type phases and cause significant decrease in the oxygen ionic conductivity. As the partial substitution of Ba for Sr suppresses vacancy ordering and lowers the oxygen content variations, a significant attention was drawn to the (Ba,Sr)(Co,Fe)O$_{3-\delta}$ system,

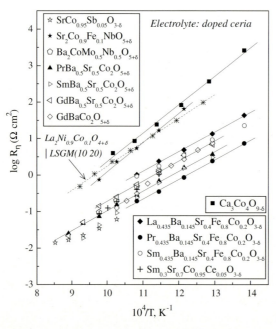

Figure 6.3 Temperature dependences of the polarization resistance of Co-containing cathode layers in contact with ceria-based solid electrolytes in air [12, 14, 16, 25, 27, 28, 38, 41]. The data on $La_2Ni_{0.9}Co_{0.1}O_{4-\delta}$ applied onto $La_{0.9}Sr_{0.1}Ga_{0.8}Mg_{0.2}O_{3-\delta}$ electrolyte [55] are shown for comparison.

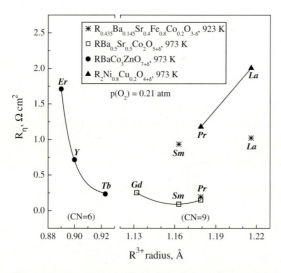

Figure 6.4 Comparison of the polarization resistances of ferrite-, cobaltite- and nickelate-based cathodes with various A-site cations, in contact with CGO (gadolinia-doped ceria) [28, 36, 41] and $(La_{0.9}Sr_{0.1})_{0.98}Ga_{0.8}Mg_{0.2}O_{3-\delta}$ [57, 61] solid electrolytes.

particularly $Ba_{0.5}Sr_{0.5}Co_{0.8}Fe_{0.2}O_{3-\delta}$ (BSCF) [17, 18, 20–24]. At the same time, problems of interaction with CO_2 and water vapor important in the IT range increase on barium incorporation [21]. Moreover, the cubic BSCF perovskite phase was demonstrated to undergo gradual decomposition into a perovskite phase mixture on cooling in air below 1173 K [22–24], making the use of BSCF in the IT SOFCs questionable. The apparent TECs of $Ba_{0.5}Sr_{0.5}Co_{1-x}Fe_xO_{3-\delta}$ ($x=0.2$–0.8) contributed by chemical expansivity may achieve $29.8 \times 10^{-6} K^{-1}$ at 298–1273 K and $p(O_2) = 10^{-5}$ atm, depending on sample prehistory [24]. This may be responsible for the minimum polarization resistance observed at rather low (1173 K) sintering temperature of such electrodes applied onto CGO10 electrolyte [18].

Substantially lower TECs, relatively high mixed conductivity, and fast oxygen exchange kinetics drew a significant interest to the layered cobaltites where the state of Co cations is often more stable with respect to disordered perovskite analogues. Important compositional series are $A_2CoBO_{5+\delta}$ (A = Sr, Ba; B = Nb, Mo), $RBaCo_2O_{5+\delta}$ (R = Pr, Gd–Ho or Y), and $RBaCo_4O_{7+\delta}$ (R = Dy–Yb or Y) [8, 16, 25–37]; the electrochemical performance of selected cathodes is illustrated in Table 6.1 and Figures 6.1–6.7. Compared to BSCF, the B-site cation-ordered double perovskite $Ba_2CoMo_{0.5}Nb_{0.5}O_{5+\delta}$ shows a considerably lower total conductivity, but a similar electrochemical activity owing to the effect of Mo doping, higher structural stability, and reduced reactivity with CSO20 [25]. The electrode layers of the A-site cation-ordered $PrBaCo_2O_{5+\delta}$ double perovskite applied onto CSO20 exhibit an area-specific resistance of 0.4Ω cm^2 at 873 K in air; the corresponding SOFC with thin-film electrolyte was reported to have $P_{max} = 620$ mW cm^{-2} [26].

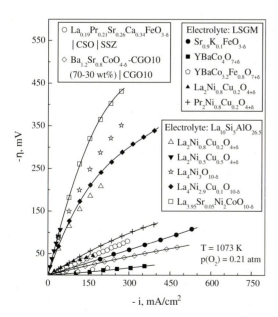

Figure 6.5 Comparison of the cathodic overpotentials of various half-cells with porous mixed conducting electrodes at 1073 K in air [29, 33, 34, 43, 45, 57, 61, 65, 66].

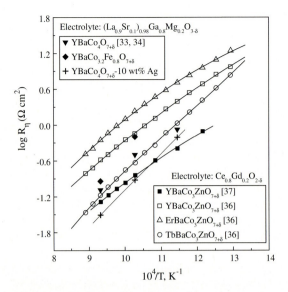

Figure 6.6 Temperature dependences of the polarization resistances of porous RBa(Co,M)$_4$O$_7$-based cathode layers in contact with (La$_{0.9}$Sr$_{0.1}$)$_{0.98}$Ga$_{0.8}$Mg$_{0.2}$O$_{3-\delta}$ and Ce$_{0.8}$Gd$_{0.2}$O$_{1.9}$ solid electrolytes, in air [33, 34, 36, 37].

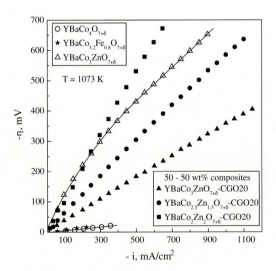

Figure 6.7 Cathodic polarization curves for YBa(Co,Fe)$_4$O$_{7+\delta}$|LSGM(9.8 20) [33, 34], YBaCo$_3$ZnO$_{7+\delta}$|CGO20|LSGM(20 20) [37], and YBaCo$_{1-x}$Zn$_x$O$_{7+\delta}$|YBaCo$_{1-x}$Zn$_x$O$_{7+\delta}$–CGO20 (50–50 wt%)|CGO20|LSGM(20 20) [37] cells in air at 1073 K. CGO20 is Ce$_{0.8}$Gd$_{0.2}$O$_{1.9}$, LSGM(20 20) is La$_{0.8}$Sr$_{0.2}$Ga$_{0.8}$Mg$_{0.2}$O$_{3-\delta}$, and LSGM(9.8 20) corresponds to (La$_{0.9}$Sr$_{0.1}$)$_{0.98}$Ga$_{0.8}$Mg$_{0.2}$O$_{3-\delta}$.

High oxygen surface exchange rate and a reasonable level of the oxygen ion diffusion in the IT range were also reported for GdBaCo$_2$O$_{5+\delta}$, leading to an excellent cathode performance [27] (Figure 6.3). The electrochemical properties of RBa$_{0.5}$Sr$_{0.5}$Co$_2$O$_{5+\delta}$ (R = Pr, Sm, Gd) as cathode materials for IT-SOFC were analyzed in Ref. [28]. A very low R_η value, 0.092 Ω cm^2 at 973 K, was observed for SmBa$_{0.5}$Sr$_{0.5}$Co$_2$O$_{5+\delta}$ electrode sintered on CGO10 at 1273 K to form symmetrical cells (Figures 6.3 and 6.4). Ba$_{1.2}$Sr$_{0.8}$CoO$_{4+\delta}$ with K$_2$NiF$_4$-type structure [29] shows a worse performance compared to the perovskite-type cobaltites in contact with ceria-based electrolytes. Note that in many literature reports on the behavior of novel cobaltite-based electrode materials, their phase stability and/or expansivity critical for the practical applications were not assessed. At the same time, the apparent TECs of double-perovskite cobaltites are still high; the cobaltites with other layered structures usually possess a lower stability compared to the perovskite-type analogues and often undergo various phase transitions [8, 31–35]. In most cases, solid electrolyte additions to the cathode layers lower the strains and enable to enlarge the electrochemical reaction zone, resulting in better performance (Figure 6.8).

Oxide materials derived from recently discovered RBaCo$_4$O$_7$ (R = lanthanide or Y), whose crystal structure consists of closely packed alternating triangular and Kagomé layers of corner-sharing Co tetrahedra [30, 31], also attract a growing interest due to their unusual oxygen sorption and transport properties [32–37]. Porous YBaCo$_4$O$_7$-based cathodes show very high electrochemical activity in contact with LaGaO$_3$- and ceria-based solid electrolytes at 873–1073 K (Table 6.1 and Figures 6.2 and 6.4–6.8). However, at atmospheric oxygen pressure, YBaCo$_4$O$_7$-based compounds appear thermodynamically stable only above approximately 1123–1173 K and metastable at lower temperatures. A slow oxygen uptake at 1000–1200 K causes complete decomposition of YBaCo$_4$O$_{7+\delta}$ into a mixture of perovskite-like phases and binary

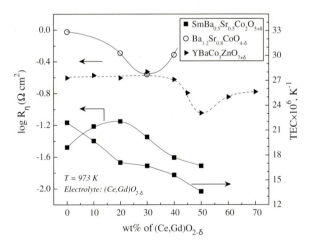

Figure 6.8 Polarization resistances of composite cobaltite–CGO cathodes in contact with Ce$_{0.9}$Gd$_{0.1}$O$_{1.95}$ (CGO10) [28, 29] and Ce$_{0.8}$Gd$_{0.2}$O$_{1.9}$ (CGO20) [37] solid electrolytes, and TECs of SmBa$_{0.5}$Sr$_{0.5}$Co$_2$O$_{5+\delta}$–CGO10 composite ceramics [28], at 973 K in air.

oxides when the average cobalt oxidation state approaches $+3$; decomposition is accompanied by a conductivity jump and dramatic volume contraction presenting a serious drawback to electrochemical applications. In the RBa(Co,M)$_4$O$_{7+\delta}$ (R = Y, Er, Tb, Ca, In; M = Zn, Fe, Al) series, several compositions such as YBaCo$_{4-x}$Zn$_x$O$_{7+\delta}$ ($1 \leq x \leq 2$) were found stable at the SOFC operating temperatures and thermomechanically compatible with the electrolyte materials [36, 37]. Although the substitution of Zn for Co decreases the electronic conductivity and cathode performance (Figures 6.2 and 6.6), under open-circuit conditions the YBaCo$_3$ZnO$_{7+\delta}$ and composite YBaCo$_3$ZnO$_{7+\delta}$–CGO20 (50–50 wt%) cathodes exhibit low R_η values [37], comparable to those of materials derived from the perovskite-type cobaltites, while the cathodic overpotentials in air seem rather high (Figure 6.7). The electrochemical activity of porous RBaCo$_3$ZnO$_{7+\delta}$ layers increases in a sequence Er < Y < Tb [36] (Figures 6.4 and 6.6). The misfit compound Ca$_3$Co$_4$O$_{9-\delta}$, where the crystal structure built of alternating CdI$_2$-type and rock salt slabs, also shows moderate expansion and is proposed as a promising cathode material [38]. The performance of porous Ca$_3$Co$_4$O$_{9-\delta}$ electrodes is similar to that of layered nickelates (Figures 6.1 and 6.3) and may be further improved by CGO10 additions.

As for the cobaltite-based materials, perovskite-type ferrites exhibit very high thermal and chemical expansion limiting their electrode applications. In general, however, incorporation of Fe cations into the cobaltite perovskites decreases both the thermal and chemical induced strains and the electrochemical activity. In the most widely studied perovskite-type ferrite system, La$_{1-x}$Sr$_x$FeO$_{3-\delta}$, the highest level of electronic and oxygen ionic transport is observed at $x \approx 0.5$ [39], while further acceptor-type doping promotes vacancy ordering and hole localization processes, with a negative impact on the transport properties. Increasing difference of the R^{3+} and A^{2+} cation radii in R$_{0.5}$A$_{0.5}$FeO$_{3-\delta}$ (R = La, Pr, Nd, Sm; A = Sr, Ba) results in higher oxygen deficiency and lower oxygen ionic and p-type electronic conductivities [40]. Contrary to the ionic conductivity variations, the role of surface exchange kinetics in the oxygen permeation processes tends to decrease on Ba^{2+} doping and on decreasing R^{3+} size in R$_{0.5}$Sr$_{0.5}$FeO$_{3-\delta}$ series. In correlation with this behavior, the incorporation of Ba^{2+} reduces polarization resistance of R$_{0.58-x}$Ba$_x$Sr$_{0.4}$Fe$_{0.8}$Co$_{0.2}$O$_{3-\delta}$ (R = La, Pr, Sm) electrodes in contact with ceria-based electrolyte [41]; the best performance achieved for $x = 0.145$ and R = Pr is close to the level of undoped cobaltites (Figures 6.3 and 6.4). The total electrical conductivity and cathode performance of rhombohedral R$_{0.5}$A$_{0.5}$FeO$_{3-\delta}$ (R = La, Pr; A = Sr, Ca, Ba) was reported to increase with average A-site cation radius [42]. Very interesting results [43] were obtained on the mixed solid solutions R$_{1-x}$A$_x$FeO$_{3-\delta}$ (R = La, Pr, Nd; A = Ca, Sr), synthesized fixing the average A-site cation radius and cation size mismatch in order to isolate the effect of divalent dopant concentration from the steric effects. The maximum total conductivity, predominantly electronic in air, was found for La$_{0.19}$Pr$_{0.31}$Sr$_{0.26}$Ca$_{0.24}$FeO$_{3-\delta}$ ceramics, while the La$_{0.19}$Pr$_{0.21}$Sr$_{0.26}$Ca$_{0.34}$FeO$_{3-\delta}$ cathode layer showed minimum overpotentials in contact with scandia-stabilized zirconia (SSZ) and CSO [43], thus indicating correlations with the oxygen ionic conduction and/or oxygen vacancy ordering processes rather than with electronic transport. For La$_{0.8}$A$_{0.2}$FeO$_{3-\delta}$

(A = Ca, Sr, Ba), the highest ionic conductivity and lowest polarization resistance of the composite cathodes calcined at 1373 K were found for $La_{0.8}Sr_{0.2}FeO_{3-\delta}$ [44].

For porous layers formed of Co-free $Sr_{0.9}K_{0.1}FeO_{3-\delta}$ perovskite [45], quite low overpotentials (Figure 6.5) and high power densities (Table 6.3) were observed in the cells based on $La_{0.8}Sr_{0.2}Ga_{0.83}Mg_{0.17}O_{3-\delta}$ (LSGM(20 17)) solid electrolyte. The effects of potassium doping on the phase stability, interaction with solid electrolytes, and thermal expansion, and also potential volatilization require, however, further evaluation. The polarization resistances of dense thin-film $SrTi_{1-x}Fe_xO_{3-\delta}$ ($x = 0.05$–0.80) layers in contact with single-crystal YSZ solid electrolyte were reported comparable to those of LSFC and BSCF thin films [46]. It should also be mentioned that as for cobaltite-based materials, attractive combinations of physicochemical and thermomechanical properties may also be expected for layered ferrites. As a particular example, at 973 K the $LaSrFeO_{4+\delta}$ porous electrode layers sintered on CSO10 at 1273 K show the open-circuit polarization resistance of $3.95\ \Omega\ cm^2$ and cathodic overpotential of -57 mV at current density of -55 mA cm^{-2} [47]. These values are close to the other materials with K_2NiF_4-type structure (Table 6.1 and Figure 6.1).

The incorporation of Ni or Cu cations in the perovskite materials is considered as a possible strategy to optimize their properties relevant for electrode applications. In general, such doping often makes it possible to improve the thermal expansion and electronic conductivity, but worsens phase stability; its effects on ionic transport are more complex [8]. As an example, the total polarization resistance of porous $(La_{0.6}Sr_{0.4})_{0.99}Co_{0.9}Ni_{0.1}O_{3-\delta}$ and $(La_{0.8}Sr_{0.2})_{0.99}Co_{0.8}Ni_{0.2}O_{3-\delta}$ electrodes in contact with CGO10 electrolyte is approximately $0.1\ \Omega\ cm^2$ at 973 K in air [48]. The incorporation of Sr^{2+} in perovskite-type $La_{1-y}Sr_yFe_{1-x}Ni_xO_{3-\delta}$ increases the oxygen deficiency and ionic transport at elevated temperatures, but leads to a lower thermodynamic stability as reflected by narrowing the solid solution domain at 1373 K down to $x \approx 0.25$ at $y = 0.10$ and $x \approx 0.12$ at $y = 0.20$ in air [49]. The average thermal expansion coefficients of $La_{1-y}Sr_yFe_{1-x}Ni_xO_{3-\delta}$ ($x = 0.1$–0.4, $y = 0.1$–0.2) ceramics vary in the ranges $(12.4$–$13.4) \times 10^{-6}\ K^{-1}$ at 700–1150 K and $(14.2$–$18.0) \times 10^{-6}\ K^{-1}$ at 1150–1370 K, rising with strontium and nickel content. These additives also lower the temperature of orthorhombic $\rightarrow$ rhombohedral phase transition and increase the total conductivity. Doping with nickel was found to have a weak negative effect on the ionic transport, probably due to defect cluster formation involving oxygen vacancies and Ni^{2+} [49]. $(La_{1-x}Sr_x)_zFe_{1-y}M_yO_{3-\delta}$ ($x = 0.2$–0.4; $y = 0.2$–0.8; $z = 0.8$–1.1; M = Ni, Cu) perovskites were proposed as alternative SOFC cathode materials [50–52]. However, the initial performance and long-term stability of $(La_{0.8}Sr_{0.2})_{0.95}Fe_{0.8}Ni_{0.2}O_{3-\delta}$ were found to be inferior to LSFC cathodes [50]. For $(La_{0.8}S_{0.2})_{0.95}Fe_{1-y}M_yO_{3-\delta}$ system, the substitution level is limited by dopant solubility, close to 20 mol% Ni or Cu. Doping with copper and decreasing z in $(La_{1-x}Sr_x)_zFe_{1-y}M_yO_{3-\delta}$ lower the melting point and consequently the sintering temperature [52]. These factors may not only enable to decrease interaction with solid electrolyte ceramics but also promote microstructural instability of the porous electrodes at the SOFC operating temperature, particularly their sintering. Similar to many other perovskite electrodes [8], the A-site cation deficiency in

Table 6.3 Examples of the maximum power density in various single SOFCs with alternative anodes.

Anode	Electrolyte/thickness	Cathode	T (K)	Fuel	P_{max} (W cm^{-2})	Reference
Ni$_{9.6}$Fe (10 wt% Fe$_2$O$_3$-coated NiO)	LSGM(10 20), 6 μm |CSO20, 0.5 μm	Sm$_{0.5}$Sr$_{0.5}$CoO$_{3-\delta}$	973	Wet H$_2$	1.79	[83]
Ni–CGO10 (50 wt% NiO), Ni$_6$Fe$_7$ support	CGO10/25 μm	La$_{0.6}$Sr$_{0.4}$CoO$_{3-\delta}$–CGO10 (50–50 wt%)	923	H$_2$	0.68	[84]
Ni$_{3.8}$Fe–CSO20 (50–50 wt%)	LSGM(20 17)/0.3 mm, CSO20 interlayer	SrCo$_{0.8}$Fe$_{0.2}$O$_{3-\delta}$	1073 973	Wet H$_2$	1.43 0.62	[86]
CoFe–CSO20 (50–50 wt%)	LSGM(20 17)/0.3 mm, CSO20 interlayer	SrCo$_{0.8}$Fe$_{0.2}$O$_{3-\delta}$	1073	Wet H$_2$	1.07	[87]
Ce$_{0.8}$Fe$_{0.2}$O$_{2-\delta}$	LSGM/0.88 mm	Sm$_{0.5}$Sr$_{0.5}$Fe$_{0.8}$Co$_{0.2}$O$_{3-\delta}$	1073	Wet CH$_4$	0.05	[88]
La$_{0.2}$Sr$_{0.7}$TiO$_{3-\delta}$ impregnated with CGO20 and Cu	8YSZ/75 μm	La$_{0.6}$Sr$_{0.4}$CoO$_{3-\delta}$	1023	Wet H$_2$	0.50	[90]
Sr$_{0.88}$Y$_{0.08}$TiO$_{3-\delta}$–8YSZ (50–50 wt%), infiltrated with CeO$_{2-\delta}$ and Ru	8YSZ/10 μm	La$_{0.65}$Sr$_{0.3}$MnO$_{3-\delta}$–8YSZ (50–50 wt%)	1073	H$_2$ 10 ppm H$_2$S–H$_2$	0.51 0.47	[91]
La$_{0.4}$Sr$_{0.6}$Ti$_{0.4}$Mn$_{0.6}$O$_{3-\delta}$–8YSZ (65–35 vol%)	8YSZ/0.2 mm	La$_{0.65}$Sr$_{0.3}$MnO$_{3-\delta}$–8YSZ (50–50 wt%)|LSM	1129	Wet H$_2$ Wet CH$_4$	0.43[a] 0.06	[94]
La$_{0.8}$Sr$_{0.2}$Cr$_{0.82}$Ru$_{0.18}$O$_{3-\delta}$	LSGM(10 20)/0.4 mm	La$_{0.6}$Sr$_{0.4}$Fe$_{0.8}$Co$_{0.2}$O$_{3-\delta}$–CGO10|LSFC	1073	Wet H$_2$	0.53	[116]
(La$_{0.75}$Sr$_{0.25}$)$_{0.95}$Cr$_{0.5}$Mn$_{0.5}$O$_{3-\delta}$	LSGMCo/0.6 mm	Gd$_{0.4}$Sr$_{0.6}$CoO$_{3-\delta}$	1123	Wet H$_2$	0.30[a]	[99]
La$_{0.75}$Sr$_{0.25}$Cr$_{0.5}$Mn$_{0.5}$O$_{3-\delta}$	LSGM(20 17)/0.3 mm, CLO interlayer	SrCo$_{0.8}$Fe$_{0.2}$O$_{3-\delta}$	1123	Dry H$_2$ Dry CH$_4$	0.21 0.10	[109]
La$_{0.75}$Sr$_{0.25}$Cr$_{0.5}$Mn$_{0.5}$O$_{3-\delta}$	LSGM(20 17)/0.25 mm	SrCo$_{0.8}$Fe$_{0.2}$O$_{3-\delta}$	1073 1123	Dry H$_2$ Dry CH$_4$	0.18 0.25	[102]
La$_{0.75}$Sr$_{0.25}$Cr$_{0.5}$Mn$_{0.5}$O$_{3-\delta}$	LSGM(10 20)/1.5 mm	La$_{0.75}$Sr$_{0.25}$Cr$_{0.5}$Mn$_{0.5}$O$_{3-\delta}$	1073	Wet 5% H$_2$	0.05	[20]
		Ba$_{0.5}$Sr$_{0.5}$Co$_{0.8}$Fe$_{0.2}$O$_{3-\delta}$	1073	Wet 5% H$_2$	0.12	[20]

6.3 Novel Cathode Materials for Solid Oxide Fuel Cells: Selected Trends and Compositions

Composition	Electrolyte	Cathode	T (K)	Fuel	Value	Ref.
$La_{0.75}Sr_{0.25}Cr_{0.5}Mn_{0.5}O_{3-\delta}$	LSGM(10 20)/0.12 mm	$La_{0.8}Sr_{0.2}MnO_{3\pm\delta}$	1073	Wet H_2	0.57	[100]
$La_{0.75}Sr_{0.25}Cr_{0.5}Mn_{0.5}O_{3-\delta}$	8YSZ/0.25 mm	$(La,Sr)MnO_3$	1173	Wet H_2	0.28	[98]
$La_{0.75}Sr_{0.25}Cr_{0.5}Mn_{0.5}O_{3-\delta}$	YSZ/83 μm	$La_{0.8}Sr_{0.2}MnO_{3\pm\delta}$ impregnated with CSO20	1123	Dry H_2	0.20	[96]
				Dry CH_4	0.03	
$La_{0.75}Sr_{0.25}Cr_{0.5}Mn_{0.5}O_{3-\delta}$ impregnated with CSO20			1123	Dry H_2	0.40	
				Dry CH_4	0.08	
$La_{0.75}Sr_{0.25}Cr_{0.5}Mn_{0.5}O_{3-\delta}$ impregnated with Ni			1123	Dry H_2	1.00	
				Dry CH_4	0.36	
$La_{0.75}Sr_{0.25}Cr_{0.5}Mn_{0.5}O_{3-\delta}$ impregnated with CSO20 + Ni			1123	Dry H_2	1.14	
				Dry CH_4	0.42	
$La_{0.75}Sr_{0.25}Cr_{0.5}Mn_{0.5}O_{3-\delta}$–YSZ graded	YSZ/0.3 mm, CGO20 interlayer	$La_{0.8}Sr_{0.2}MnO_{3\pm\delta}$ graded	1173	Wet H_2	0.47	[95]
				Wet CH_4	0.2	
$La_{0.75}Sr_{0.25}Cr_{0.5}Mn_{0.5}O_{3-\delta}$–YSZ (50–50 wt%)	YSZ/1 mm	Pt	1073	CH_4	0.02	[105]
				C_2H_5OH	0.01	
$La_{0.75}Sr_{0.25}Cr_{0.5}Mn_{0.5}O_{3-\delta}$–YSZ (50–50 wt%), impregnated with Pd	YSZ/1 mm	Pt	1073	CH_4	0.05	
				C_2H_5OH	0.11	
$La_{0.8}Sr_{0.2}Cr_{0.5}Mn_{0.5}O_{3-\delta}$–YSZ (45–65 wt%)	YSZ/60 μm	$La_{0.8}Sr_{0.2}FeO_{3-\delta}$–YSZ (40–60 wt%)	973	Wet H_2	0.11	[107]
$La_{0.8}Sr_{0.2}Cr_{0.5}Mn_{0.5}O_{3-\delta}$–YSZ (45–65 wt%), impregnated with Pd	YSZ/60 μm	$La_{0.8}Sr_{0.2}FeO_{3-\delta}$–YSZ (40–60 wt%)	973	Wet H_2	0.50	
8YSZ impregnated with	8YSZ/50 μm	$La_{0.8}Sr_{0.2}MnO_{3\pm\delta}$–YSZ (50–50 wt%)	1073	Dry H_2	0.21	[112]
Cu–$La_{0.75}Sr_{0.25}Cr_{0.5}Mn_{0.5}O_{3-\delta}$ (35–65 wt%)				CH_4	0.06	
				C_4H_{10}	0.13	
Cu–$La_{0.75}Sr_{0.25}Cr_{0.5}Mn_{0.5}O_{3-\delta}$ (20–80 wt%)	LSGM(20 17)/0.3 mm, CLO interlayer	$SrCo_{0.8}Fe_{0.2}O_{3-\delta}$	1123	Dry H_2	0.86	[109]
				Dry CH_4	0.48	
Cu–$La_{0.75}Sr_{0.25}Cr_{0.5}Mn_{0.5}O_{3-\delta}$ (20–80 wt%), sputtered Pt	LSGM(20 17)/0.25 mm	$SrCo_{0.8}Fe_{0.2}O_{3-\delta}$	1123	Dry H_2	0.85	[102]
				Dry CH_4	0.52	

(Continued)

Table 6.3 (Continued)

Anode	Electrolyte/thickness	Cathode	T (K)	Fuel	P_{max} (W cm^{-2})	Reference
$Pr_{0.7}Ca_{0.3}Cr_{0.6}Mn_{0.4}O_{3-\delta}$	8YSZ/0.37 mm	$Pr_{0.7}Ca_{0.3}Cr_{0.6}Mn_{0.4}O_{3-\delta}$	1223	Wet H_2	0.25	[111]
			1223	Wet CH_4	0.16	
$La_{0.8}Sr_{0.2}Sc_{0.2}Mn_{0.8}O_{3-\delta}$	10SSZ/0.3 mm	$La_{0.8}Sr_{0.2}Sc_{0.2}Mn_{0.8}O_{3\pm\delta}$	1173	Wet H_2	0.31	[119]
			1173	Wet CH_4	0.13	
$LaSr_2Fe_2CrO_{9-\delta}$–CGO10 (50–50 wt%)	CGO10/0.3 mm	$La_{0.6}Sr_{0.4}Fe_{0.8}Co_{0.2}O_{3-\delta}$–CGO10 (50–50 wt%)	1073	Wet H_2	0.30	[118]
$Sr_2MgMoO_{6-\delta}$	LSGM(10 20)/0.4 mm	$Sr_{0.9}K_{0.1}FeO_{3-\delta}$	1073	Wet H_2	0.37	
	LSGM(20 17)/0.3 mm		1073	H_2	0.68	[45]
$Sr_2MgMoO_{6-\delta}$	LSGM(20 17)/0.3 mm, CLO4 interlayer	$SrCo_{0.8}Fe_{0.2}O_{3-\delta}$	1073	Dry H_2	0.84	[120]
				Wet H_2	0.81	
				Dry CH_4	0.44	
				Wet CH_4	0.35	
$Sr_2MgMoO_{6-\delta}$	LSGM(20 20)/0.6 mm With CGO20 interlayer	$La_{0.6}Sr_{0.4}Fe_{0.2}Co_{0.8}O_{3-\delta}$	1073	Wet H_2	0.27	[122]
				Wet H_2	0.33	
$Sr_2MnMoO_{6-\delta}$	LSGM(20 17)/0.3 mm, CLO4 interlayer	$SrCo_{0.8}Fe_{0.2}O_{3-\delta}$	1073	Dry H_2	0.65	[120]
				5 ppm $H_2S–H_2$	0.57	
$Sr_2Mg_{0.9}Cr_{0.1}MoO_{6-\delta}$		$SrCo_{0.8}Fe_{0.2}O_{3-\delta}$	1073	Dry H_2	0.79	
				5 ppm $H_2S–H_2$	0.61	
$Sr_{1.2}La_{0.8}MgMoO_{6-\delta}$	LSGM(20 17)/0.3 mm CLO5 interlayer	$SrCo_{0.8}Fe_{0.2}O_{3-\delta}$	1073	Dry CH_4	0.47	[121]
				Wet CH_4	0.49	
				Wet C_2H_6	0.19	
				Wet C_3H_8	0.17	

Notes: In most cases, "wet H_2" and "wet CH_4" refer to the gas humidified under ambient conditions (~3% H_2O); 10SSZ is 10 mol% Sc_2O_3-stabilized ZrO_2; CLO5 and CLO4 are $Ce_{0.4}La_{0.5}O_{2-\delta}$ and $Ce_{0.6}La_{0.4}O_{2-\delta}$, respectively; LSGMCo is $La_{0.8}Sr_{0.2}Ga_{0.8}Mg_{0.15}Co_{0.05}O_{3-\delta}$.
a) Extrapolated.

$(La_{1-x}Sr_x)_zFe_{1-y}M_yO_{3-\delta}$ was found favorable to decrease thermal expansion and reactivity with 8YSZ. The Ni-substituted materials exhibit the highest conductivity in these compositional series and also have a greater reactivity with YSZ if compared to the parent ferrites, a result of reduced thermodynamic stability of the doped perovskite phases; no indication of chemical reaction or cation interdiffusion between $(La_{0.8}Sr_{0.2})_{0.95}Fe_{0.8}Ni_{0.2}O_{3-\delta}$ and gadolinia-doped ceria (CGO) was revealed by X-ray diffraction (XRD) analysis [52, 53]. At the same time, despite the lower electrical conductivity, $La_{0.8}Sr_{0.2}FeO_{3-\delta}$ porous layers were reported to possess higher performance than $La_{0.7}Sr_{0.3}Fe_{0.8}Ni_{0.2}O_{3-\delta}$ and $LaNi_{0.6}Fe_{0.4}O_{3-\delta}$ applied onto YSZ electrolyte with CSO20 interlayer [51]. The effects of A-site nonstoichiometry on the electrochemical activity of $La_{1-x}Fe_{0.4}Ni_{0.6}O_{3-\delta}$ were investigated using model cone-shaped electrodes and CGO10 electrolyte [54]; increasing x results in a separation of NiO, inhibiting oxygen reduction reaction.

During the past decade, materials with Ruddlesden–Popper (RP) structure built of n perovskite-like layers alternating with one rock salt sheet, in particular the nickelate series $(R,A)_{n+1}Ni_nO_{3n+1}$, have attracted significant attention among most promising groups of the cathode materials (Figures 6.1, 6.2, 6.4–6.5 and 6.9 and Table 6.2). Their advantages include a relatively high level of oxygen ionic transport, reasonable electronic conductivity, and moderate thermal expansion [8, 9, 55–62]. Increasing the number of perovskite layers in the RP nickelates (n) usually leads to higher ionic and electronic conduction [55–57]. These effects are primarily associated with increasing concentration of Ni—O—Ni bonds responsible for the electronic transport, progressive delocalization of the p-type electronic charge carriers, and increasing vacancy migration contribution to the oxygen ion diffusivity. Note that except for oxygen hyperstoichiometric $La_2NiO_{4+\delta}$ and its analogues, where anion diffusion is essentially dominated by the interstitial migration, most RP nickelates are oxygen deficient at elevated temperatures. As a particular consequence, the electrochemical activity of $La_{n+1}Ni_nO_{3n+1-\delta}$ cathodes becomes substantially higher when n increases from 1 to 3 [55, 56]. Analogous results were also obtained for praseodymium nickelate electrodes where $Pr_4Ni_3O_{10-\delta}$ phase is formed due to oxidative decomposition of undoped [58] or Cu-doped [57] $Pr_2NiO_{4+\delta}$ (Figures 6.4–6.5 and 6.9 and Table 6.2). Modest A-site deficiency was found to decrease the polarization resistance of $Nd_{1-x}NiO_{4+\delta}$ porous layers in contact with 8 mol% 8YSZ solid electrolyte [58]. An opposite trend was, however, reported in Ref. [59] where the $La_3Ni_2O_{7-\delta}$ cathode displayed higher overpotentials than $La_2NiO_{4+\delta}$ in contact with CSO20–2 mol% Co-doped ceramics.

Another necessary comment is related to the higher concentration of A-site cations in the RP phases in comparison with their perovskite-type analogues. As for the other factors including thermodynamic stability of the electrode material phase or the presence of high-diffusivity additives, the concentration of A-site cations may substantially affect reactivity with the solid oxide electrolytes. For example, chemical reaction between $La_2Ni_{1-x}Cu_xO_{4+\delta}$ ($x = 0–1$) and 8YSZ leads to the formation of the insulating $La_2Zr_2O_7$ pyrochlore phase and is observed at relatively low temperatures, such as 1173 K [60]. This effect becomes more pronounced with increasing x, leading to a worse cathode performance. Therefore, the layered nickelate–cuprate electrodes

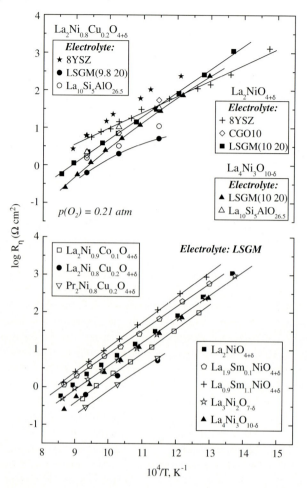

Figure 6.9 Temperature dependences of the polarization resistance of porous [55–58, 60, 61, 65, 66] and cone-shaped [62] nickelate cathodes in contact with various solid electrolytes in air.

should be used in contact with doped ceria or lanthanum gallate electrolytes rather than with YSZ (Figure 6.9). At 973–1073 K, porous $Pr_2Ni_{0.8}Cu_{0.2}O_{4+\delta}$ electrodes deposited on $(La_{0.9}Sr_{0.1})_{0.98}Ga_{0.8}Mg_{0.2}O_{3-\delta}$ (abbreviated as LSGM(9.8 20)) electrolyte substrates exhibit lower polarization resistances compared to $La_2Ni_{0.8}Cu_{0.2}O_{4+\delta}$ (Figures 6.4 and 6.9 and Table 6.2), while cathodic reduction decreases their performance [57, 61]. Under oxidizing conditions, the extensive oxygen uptake at temperatures below 1073–1223 K leads to reversible decomposition of Pr_2NiO_4-based solid solutions into $Pr_4Ni_3O_{10}$ and praseodymium oxide phases [57]. The substitution of copper for nickel decreases the oxygen content and phase transition temperature, while the incorporation of iron cations has an opposite effect. Both types of doping tend to decrease stability in reducing and even inert environment. The steady-state oxygen permeability of $Pr_2NiO_{4+\delta}$ ceramics at 1173–1223 K, limited by both surface exchange kinetics and bulk ionic conduction, is similar to that of $La_2NiO_{4+\delta}$. The phase transformation on cooling results in considerably higher electronic conductivity and oxygen permeation and is also associated with significant volume changes revealed by dilatometry [57]. The area-specific resistance of cone-shaped $La_{2-x}Sr_xNiO_{4+\delta}$ electrodes in contact with CGO10 electrolyte [62] is close to that of perovskite electrodes, being yet much higher than for the best ferrite–cobaltite materials. The lowest value, 23.8 $\Omega\,cm^2$ in air at 873 K, was found for $La_{1.75}Sr_{0.25}NiO_{4+\delta}$ [62].

Apatite-type $La_{10-x}(SiO_4)_6O_{2\pm\delta}$ silicates and their derivatives possess a substantially high oxygen ionic conductivity, moderate thermal expansion, low electronic transport in a wide range of oxygen chemical potentials, and relatively low costs, and may thus be considered for potential use as SOFCs electrolytes [7, 63, 64]. The performance of mixed conducting cathodes in contact with silicate ceramics is, however, usually lower compared to other solid electrolytes; the high polarization resistance originates primarily from the surface diffusion of silica, partially blocking the electrochemical reaction zone, without formation of secondary phases detectable by XRD [65, 66]. This explains very poor performance of $La_{0.75}Sr_{0.25}Mn_{0.8}Co_{0.2}O_{3-\delta}$-based cathode layers cosintered with $La_9SrSi_6O_{26.5}$ electrolyte at 1673 K, while no diffusion of the electrode components into the apatite ceramics was detected [67]. The electrochemical activity of porous $La_{0.8}Sr_{0.2}Fe_{0.8}Co_{0.2}O_{3-\delta}$–$Ce_{0.8}Gd_{0.2}O_{2-\delta}$ (CGO20), $La_{0.7}Sr_{0.3}MnO_{3-\delta}$–CGO20, $SrMn_{0.6}Nb_{0.4}O_{3-\delta}$, $Sr_{0.7}Ce_{0.3}Mn_{0.9}Cr_{0.1}O_{3-\delta}$, $Gd_{0.6}Ca_{0.4}Mn_{0.9}Ni_{0.1}O_{3-\delta}$, $La_2Ni_{1-x}Cu_xO_{4+\delta}$ ($x=0.2$, 0.5), $La_2Ni_{0.8}Cu_{0.2}O_{4+\delta}$–Ag, $LaSr_2Mn_{1.6}Ni_{0.4}O_{7-\delta}$, $La_4Ni_{3-x}Cu_xO_{10-\delta}$ ($x=0$–0.1), and $La_{3.95}Sr_{0.05}Ni_2CoO_{10-\delta}$ electrodes was studied at 873–1073 K in contact with apatite-type $La_{10}Si_5AlO_{26.5}$ electrolyte [65, 66]; note that the level of oxygen ionic conductivity in the latter electrolyte is higher than that of YSZ in the intermediate temperature range [64]. In all cases, however, the polarization resistances and overpotentials of nickelate-based cathodes applied onto silicate electrolyte are substantially higher compared to similar layers applied onto LSGM(9.8 20); see Figures 6.1, 6.5, 6.9 and 6.10 and Table 6.2. The electrochemical activity of porous nickelate-based layers was found to correlate with the concentration of mobile ionic charge carriers and bulk oxygen transport, thus decreasing in the series $La_4Ni_{2.9}Cu_{0.1}O_{10-\delta} > La_4Ni_3O_{10-\delta} > La_{3.95}Sr_{0.05}Ni_2CoO_{10-\delta}$ and on copper doping in K_2NiF_4-type $La_2Ni_{1-x}Cu_xO_{4-\delta}$ (Figure 6.5). Compared to the intergrowth nickelate materials, the manganite-based electrodes exhibit substantially worse electrochemical properties (Figure 6.10), in correlation with the level of oxygen ionic and electronic conduction in Mn-containing phases [66]. Due to the effects of cation interdiffusion between the cell components, the performance of mixed conducting cathodes applied onto $La_{10}Si_5AlO_{26.5}$ can be improved by reducing electrode fabrication temperature [65]. Qualitatively similar results were obtained in [68] for $La_{0.8}Sr_{0.2}MnO_{3\pm\delta}$, $La_{0.7}Sr_{0.3}FeO_{3-\delta}$, $La_{0.6}Sr_{0.4}Co_{0.2}Fe_{0.8}O_{3-\delta}$, and $La_{0.6}Sr_{0.4}Co_{0.8}Fe_{0.2}O_{3-\delta}$ cathodes and for $La_{0.75}Sr_{0.25}Cr_{0.5}Mn_{0.5}O_{3-\delta}$, NiO–CGO, and $Sr_2MgMoO_{6-\delta}$ anodes applied onto $La_{10}Si_{5.5}Al_{0.5}O_{26.75}$ solid electrolyte. The information on electrochemical behavior of various electrode materials in contact with apatite-type solid electrolytes is, however, still scarce and nonsystematic.

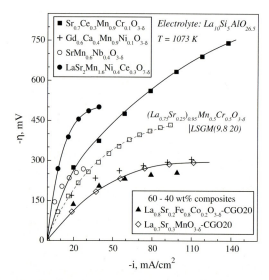

Figure 6.10 Cathodic polarization curves of Fe-, Cr-, and Mn-containing electrode layers in contact with apatite-type $La_{10}Si_5AlO_{26.5}$ solid electrolyte [65, 66]. The data on $(La_{0.75}Sr_{0.25})_{0.95}Cr_{0.5}Mn_{0.5}O_{3-\delta}$ applied onto $(La_{0.9}Sr_{0.1})_{0.98}Ga_{0.8}Mg_{0.2}O_{3-\delta}$ [104] are shown for comparison. Pt mesh current collectors were used in all cases.

The electrochemical performance of Ca-, Ag-, and Sr-doped pyrochlore-type $Bi_2Ru_2O_7$ and their composites with 20 mol% erbia-stabilized δ-Bi_2O_3, which exhibits a very high level of the oxygen ionic transport (Chapters 2 and 9 of the first volume), was evaluated in contact with ceria-based electrolyte [69]. At 973 K, the area-specific resistances of undoped, 5 mol% Ca-doped, and 5 mol% Sr-doped bismuth ruthenate electrode were 1.45, 1.24, and 1.41 $\Omega\,cm^2$, respectively. Addition of erbia-stabilized bismuth oxide up to 31–44 wt% made it possible to further reduce the electrode resistance down to 0.08–0.11 $\Omega\,cm^2$ [69]. Due to high reactivity and thermal expansion of Bi_2O_3-containing materials, this family of electrode compositions may be considered for potential use in the electrochemical cells based on doped $Bi_4V_2O_{11}$ (BIMEVOX) solid electrolytes; their stability in ceria-based cells is rather problematic.

6.4
Oxide and Cermet SOFC Anodes: Relevant Trends

Systematic information on the anode performance available in the literature is essentially limited to the electrochemical cells with zirconia-, ceria-, and lanthanum gallate-based electrolytes (see Chapters 9 and 12 of the first volume and Chapter 5 of this book). The latter group of solid oxide electrolytes refers primarily to $La_{1-x}Sr_xGa_{1-y}Mg_yO_{3-\delta}$ (LSGM); the long-term stability of transition metal cation-doped gallates under anodic conditions still requires further investigations. The two major approaches, which can be identified in the developments of SOFC anode materials, are an optimization of cermet compositions and a search for novel oxide phases with high electrochemical activity and sufficient thermodynamic stability both under oxidizing and reducing conditions [5, 8, 10, 70–72]. Continuous attention is drawn to the Ni-ceria-based cermets. The advantages of ceria as the anode component originate, first of all, from a very high catalytic activity for the combustion reactions involving oxygen, particularly to carbon oxidation beneficial for the fuel cells operating on hydrocarbons and biogas. In addition, reduced $CeO_{2-\delta}$ and its derivatives possess a substantial mixed oxygen ionic and n-type electronic conductivity; the transport properties and reducibility can be enhanced by acceptor-type doping (Chapter 9 of the first volume). Nevertheless, the use of ceria-based compositions without any metallic phase results in quite poor anode performance, even with current collectors formed of Au ink and Pt paste [20, 88]. The electrocatalytic activity of $Ni–Ce_{0.8}Ti_{0.2}O_{2-\delta}$ (60 wt% NiO) was found higher compared to $Ni–CeO_{2-\delta}$ cermet anode under 10% CH_4 fuel, a result of the higher conductivity of doped ceria [73]. A $Ni–Cu–CeO_{2-\delta}$ (14–16–70 vol%) anode fired at reduced temperature, 1173 K, onto CGO10 electrolyte, had $R_\eta < 0.1 \, \Omega \, cm^2$ at 873 K under wet 50% H_2 [74]. However, as the cermets with ceria-based phases suffer from dimensional instabilities caused by the local $p(O_2)$ variations and minor TEC mismatch, the presence of one redox-stable phase with moderate thermal expansion, such as YSZ, is still desirable [8, 75]. Taking into account the durability issues related to carbon deposition and sulfur poisoning, efforts are being undertaken to use other transition metal components or lower Ni concentrations (Figures 6.11– 6.13 and Table 6.3). The anode assemblies consisting of one Ni–CGO10 functional layer and a $Ni–La_{0.9}Mn_{0.8}Ni_{0.2}O_{3-\delta}$ contact layer were proposed to reduce nickel

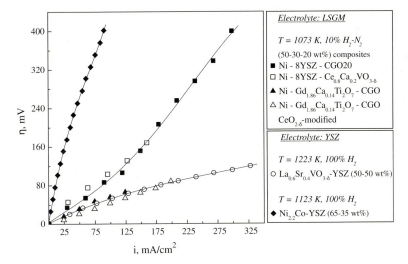

Figure 6.11 Anodic overpotentials of $La_{0.6}Sr_{0.4}VO_{3-\delta}$–YSZ composite [125] and various Ni-containing cermets with current collectors made of Pt mesh [75] and Pt–Rh mesh embedded in the electrode layer [82] in H_2-containing environments. Experimental conditions are given in the legend.

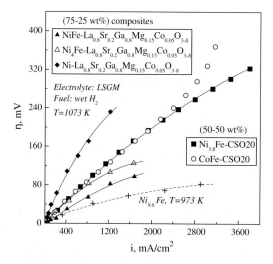

Figure 6.12 Anodic overpotentials of various cermets at 1073 K and one Ni–Fe alloy at 973 K, with $Ce_{0.8}Sm_{0.2}O_{2-\delta}$ (CSO20) interlayer [83, 86, 87] and current collector made of Pt mesh [86, 89] and Au mesh [87]. H_2 humidified under ambient conditions (~3% H_2O) was used in all cases.

content in the anode [76]. The redox tolerance was investigated at 1123 K, displaying that the driving force behind the performance degradation is nickel agglomeration that occurs even when the nickel content is below percolation threshold. Although nickel should be reduced under the anode operation conditions and segregate from

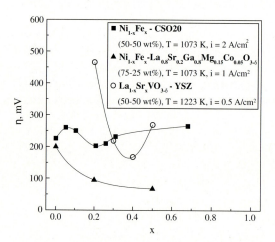

Figure 6.13 Composition dependences of the anodic overpotentials of the cermet and oxide composite electrodes under a fixed current density, in contact with $La_{0.8}Sr_{0.2}Ga_{0.83}Mg_{0.17}O_{3-\delta}$, LSGM(20 17)/$Ce_{0.8}Sm_{0.2}O_{1.9}$ (CSO20) interlayer [86], $La_{0.8}Sr_{0.2}Ga_{0.8}Mg_{0.15}Co_{0.05}O_{3-\delta}$ [89], and yttria-stabilized zirconia [125] solid electrolytes. The measurements were performed in pure H_2 [125] or H_2 humidified under ambient conditions (~3% H_2O) with Pt mesh current collectors [86, 89].

$La_{0.9}Mn_{0.8}Ni_{0.2}O_{3-\delta}$, the perovskite structure was retained after 100 redox cycles using 50% H_2O-H_2 [76]. Quantum mechanical calculations employing density functional theory (DFT) to calculate the stability of surface-adsorbed hydrogen atoms, oxygen atoms, and hydroxyl radicals for a variety of metals (Mn, Fe, Co, Ni, Cu, Ru, Rh, Pd, Ag, Pt, Au) showed that the anode electrochemical activity should be highest for Ni that has an intermediate oxygen adsorption energy, while it decreases for both increasing (Ru, Co, Fe, Mn) and decreasing (Rh, Pd, Pt, Au) oxygen binding energies; the activity was found uncorrelated with the stability of surface-adsorbed hydrogen [77]. The copper-based anodes, which may be of interest for hydrocarbon-fueled SOFCs in combination with ceria catalysts due to low catalytic activity of Cu toward the C—C bond formation, however, exhibit poor electrochemical properties and undergo fast degradation [8, 72, 78, 79].

One interesting solution is based on the use of alloys as the anode components. Changes in the electronic structures of supported catalysts, induced by the formation of surface alloys, were measured and related to the chemical activities and materials' catalytic performance in accordance with the d-band model [80]. A particular conclusion was that the antibonding adsorbate–Ni states for Sn/Ni catalysts are populated to a higher degree than the antibonding states for monometallic Ni, decreasing the strength of interaction between the adsorbates (O, C, OH, CO, CH_x, etc.) and the Ni sites on the Sn/Ni substrate. The catalytic behavior of Cu and bimetallic CuNi composites with CeO_2-based oxides (20–80 and 40–60 wt%) for direct oxidation of dry methane up to 1173 K was analyzed by temperature-programmed reduction and oxidation (TPR/TPO) and X-ray photoelectron spectroscopy (XPS) [81]; the catalytic activity appears to depend not only on the presence of nickel but also on the nature of dopants in ceria (e.g., Gd or Tb). Despite the drawbacks primarily associated with the metal oxidation under high current densities, attempts were made to introduce metallic Co and Fe in the cermets. For example, performance of bimetallic $Ni_{1-x}Co_x$–YSZ and $Ni_{1-x}Cu_x$–YSZ anodes was tested in H_2, CH_4, and H_2S-CH_4 gas mixtures [82]. Copper additions were shown to result in large metal particle sizes and lower SOFC performance, while the use of cobalt has an opposite effect. In both cases, the performance tends to quickly degrade in dry CH_4 due to carbon deposition and anode delamination from YSZ electrolyte. However, $Ni_{2.22}$Co–YSZ (70 wt% oxide) cermets demonstrated a higher activity in 10% H_2S-CH_4 than in H_2 fuel due to formation of a sulfidated alloy (Ni–Co–S) under operating conditions [82].

Ni–Fe alloys were proposed for the anode layers [83], anode supports [84], or cermet components [85, 86]; selected parameters are summarized in Figures 6.12 and 6.13 and Table 6.3. The maximum power density of an anode-supported single cell consisting of a Ni–Fe alloy produced by reduction of 10 wt% Fe_2O_3-coated NiO was reported to decrease after one thermal cycle from 1790 to 1620 mW cm^{-2} at 973 K in humidified H_2 fuel [83]. A composite $Ni_{8.56}$Fe (Ni–Fe 90–10 wt%)–MgO–LSGM(10 20) (87.2–2.5–10 wt%) anode was reported to exhibit a sufficient stability under CH_4 atmosphere at 973 K, a high tolerance against carbon deposition, and lower overpotentials compared to MgO-free $Ni_{8.56}$Fe–LSGM (90–10 wt%) cermet, particularly due to an increased basicity and surface activity [85]. Among various compositions of the Ni–Fe alloys, $Ni_{3.81}Fe$–CSO20 (50–50 wt%) cermet displayed the best performance, being essentially stable during about 160 h operation; under open-circuit conditions, the polarization resistance was 0.105 $\Omega\,cm^2$ at 1073 K [86]. Despite ordering processes on cooling below 1000 K, CoFe was tested as a component of Ni-free cermet with CSO20 [87]. Ni–Fe–$La_{0.8}Sr_{0.2}Ga_{0.8}Mg_{0.15}Co_{0.05}O_{3-\delta}$ composite materials were also reported to be promising for the SOFC applications [89], though cobalt is likely to segregate from the latter phase under the anodic conditions and decomposition of the lanthanum gallate phase may be expected to occur.

The progress achieved in the field of oxide anodes is still substantially driven by the problems associated with carbon deposition and sulfur poisoning. Due to the stability requirements, the choice of parent oxide systems for the ceramic anodes is essentially limited to titanates, chromites, molybdates, and vanadates; as for the oxide cathodes, the most extensively studied are perovskite-related compounds. Representative examples and typical properties are illustrated by Figures 6.11 and 6.13–6.19 and Tables 6.3–6.5. It should be separately mentioned that prereduction of the oxide materials, primarily titanates [90], has often a very positive impact on the final anode performance [90]. The anodic activity therefore depends on the porous layer prehistory in addition to other factors (see, in particular, Chapter 12 of the first volume and Ref. [8]); this often leads to contradicting results reported in the literature. Irrespective of these factors, a relatively good sulfur tolerance in

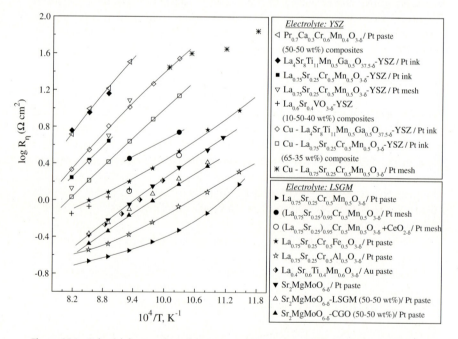

Figure 6.14 Selected data on the polarization resistances of various cermet and oxide anode layers [20, 79, 92, 94, 101, 104, 111, 122, 125]. The measurements were performed under 100% H_2 environment [125], diluted H_2 [20, 79, 92, 94, 104, 111, 122], or humidified (3% H_2O) CH_4 [101]. The current collectors are specified in the legend after anode composition.

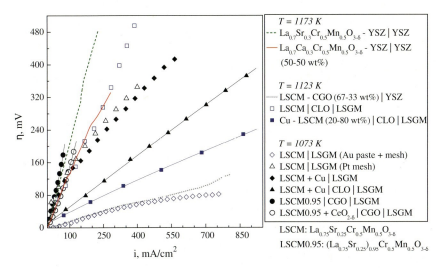

Figure 6.15 Polarization curves of $(La,Sr)_{1-x}Cr_{0.5}Mn_{0.5}O_{3-\delta}$-based anodes in contact with various solid electrolytes with or without ceria-based interlayers, in dry H_2 [102, 109], wet H_2 [100, 106, 113], and wet 10% H_2 [104]. The current collector types are indicated in parentheses for selected cells.

10–40 ppm H_2S-humidified H_2 was reported for the anode formed using $Sr_{0.88}Y_{0.08}$-$TiO_{3-\delta}$–8YSZ porous electrode backbone infiltrated with ceria and Ru (particle sizes of 30–200 nm) [91]. Copper additions were shown to improve the performance of $La_4Sr_8Ti_{11}Mn_{0.5}Ga_{0.5}O_{37.5-\delta}$–YSZ anodes [92]. The total conductivity and catalytic

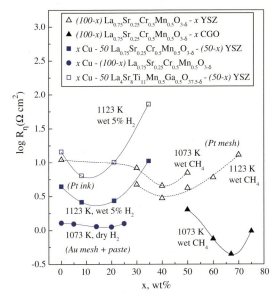

Figure 6.16 Comparison of the polarization resistances of various composite anodes in contact with YSZ electrolyte [92, 101, 106] and LSGM with $(Ce,La)O_{2-\delta}$ interlayer [109].

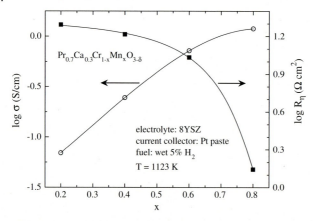

Figure 6.17 Total electrical conductivity of dense ceramics and anode polarization resistance of porous electrode layers of $Pr_{0.7}Ca_{0.3}Cr_{1-x}Mn_xO_{3-\delta}$, at 1123 K in humidified 5% H_2–Ar [111].

activity of the pure titanates is, however, quite low, limiting their applicability. For instance, the electrochemical activity of electrodes based on a redox stable Sr vacancy compensated Nb-doped $SrTiO_3$ with the nominal composition $Sr_{0.94}Ti_{0.9}Nb_{0.1}O_3$, which is essentially governed by the concentrations of Ti^{3+} and oxygen vacancies in the vicinity of the triple-phase boundaries, was found insufficient [93]. At 1123 K in 97% H_2–3% H_2O, the polarization resistance of porous $Sr_{0.94}Ti_{0.9}Nb_{0.1}O_3$–8YSZ (50–50 wt%) layer in contact with 8YSZ electrolyte and Pt paste current collector was 16.3 $\Omega\,cm^2$, increasing on further redox cycling [93]; the role of Pt/electrode interface was assumed negligible based on the comparison with literature data on

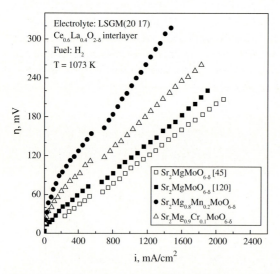

Figure 6.18 Polarization curves of the $Sr_2(Mg,M)MoO_{6-\delta}$ anode layers in contact with $La_{0.8}Sr_{0.2}Ga_{0.83}Mg_{0.17}O_{3-\delta}$ (LSGM(20 17)) solid electrolyte at 1073 K in H_2 [45, 120]. Pt mesh with a small amount of Pt paste was used as the current collector.

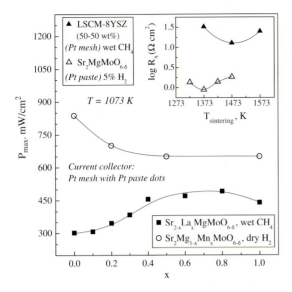

Figure 6.19 Maximum power density of the model SOFCs with Sr_2MgMoO_6-based anodes [120–122]. The inset shows polarization resistances of $La_{0.75}Sr_{0.25}Cr_{0.5}Mn_{0.5}O_{3-\delta}$–YSZ and $Sr_2MgMoO_{6-\delta}$ layers as a function of anode sintering temperature [101, 122]. YSZ [101] and LSGM [120–122] electrolytes were used with $(Ce,La)O_{2-\delta}$ interlayer [120, 121]; the current collector type is indicated in italic.

Pt electrodes. Notice that taking into account the surface spreading phenomena and microstructural alterations during long-term operation (see Chapter 9 and references cited therein), the latter assumption may be questionable in many situations. For the $La_{0.4}Sr_{0.6}Ti_{1-x}Mn_xO_{3-\delta}$ ($x=0$–0.6) series, the composition with $x=0.6$ displays highest conductivity [94] (Table 6.4). This material also exhibits a moderate TEC, chemical expansion of 0.5% on redox cycling, no essential interaction with 8YSZ up to 1673 K, and a significant electrochemical activity with Au paste current collector under humidified H_2, which becomes much worse when switching to methane (Table 6.3).

Another attractive group of the potential anode materials is based on perovskite-like $La_{1-x}Sr_xCr_{1-y}Mn_yO_{3-\delta}$ ($x=0$–0.3; $y \leq 0.6$). Important advantages of this family include a substantial electrochemical activity in reducing and oxidizing atmospheres (Figures 6.14–6.15 and 6.17 and Tables 6.5–6.7), compatibility with various solid electrolytes, a good dimensional stability (Table 6.4), and sulfur tolerance [95–114]. Porous $La_{0.75}Sr_{0.25}Cr_{0.5}Mn_{0.5}O_{3-\delta}$ (LSCM)-based layers exhibit reasonably low overpotentials in contact with YSZ, CGO, and LSGM electrolytes. While a poor adherence was observed for LSCM deposited on zirconia, the electrochemical performance and adhesion both can be improved by YSZ additions [92], which also enhance ionic conduction in the anode materials. Cu–LSCM (~70 wt% CuO) cermets were found to exhibit lower polarization resistances than Cu–CeO_2 under 5% H_2 and to have a better adhesion to the YSZ surface [79]. Since additions of Pd to the LSCM dense anode film led only to a small decrease in polarization resistance in H_2-containing

Table 6.4 Properties of selected anode materials.

Composition	Total conductivity			Average TECs			Reference
	T (K)	$p(O_2)$ (atm)	σ (S cm^{-1})	T (K)	$p(O_2)$ (atm)	$\bar{\alpha}$ ($\times 10^6$ K^{-1})	
$La_{0.2}Sr_{0.7}TiO_{3-\delta}$	1173	5% H_2	25$^{a)}$	300–1273	0.21	8.6	[90]
				300–1273	5% H_2	9.5	
$La_{0.4}Sr_{0.6}Ti_{0.4}Mn_{0.6}O_{3-\delta}$	1083	10^{-18}	1.5	303–1373	0.21	11.9	[94]
		0.21	22.6				
$LaCr_{0.5}Mn_{0.5}O_{3-\delta}$	1073	100% H_2	3.7				[114]
$La_{0.8}Sr_{0.2}Cr_{0.5}Mn_{0.5}O_{3-\delta}$	1073	100% H_2	1.3	337–708	0.21	8.9	[95]
				793–1229	0.21	10.1	
$(La_{0.75}Sr_{0.25})_{0.95}Cr_{0.5}Mn_{0.5}O_{3-\delta}$	1173	0.21	30.6	373–923	0.21	10.8	[104]
		10^{-18}	2.2	923–1223	0.21	12.7	
				1223–1523	0.21	14.1	
				923–1223	6×10^{-4}	12.5	
				923–1223	5×10^{-21}– 3×10^{-14}	11.7	
$La_{0.75}Sr_{0.25}Cr_{0.5}Mn_{0.5}O_{3-\delta}$	1173	10^{-21}	1.5	434–1173	0.21	13.7	[95, 101]
$La_{0.75}Sr_{0.25}Cr_{0.5}Mn_{0.5}O_{3-\delta}$–8YSZ (50–50 wt%)				564–1173	0.21	12.2	[101]
$Sr_2MgMoO_{6-\delta}$	1073	0.21	3×10^{-3}	382–633	0.21	11.7	[122]
		5% H_2	0.8				
		5% H_2	4.3				[120]

Material							
$Sr_{1.4}La_{0.6}MgMoO_{6-\delta}$	1073	10^{-18}	8.5	691–1074	0.21	12.7	[121]
$Sr_2MnMoO_{6-\delta}$	1073	5% H_2	5.0				
		10^{-18}	5.7				
		10^{-20}	7.9				
$CaV_{0.5}Mo_{0.5}O_{4-\delta}$	973–1173	0.21	10^{-5}–10^{-4}	473–1173	0.21	10	[124]
$CaV_{0.5}Mo_{0.5}O_{3-\delta}$	1123	10^{-18}	320	473–1173	10% H_2	13	
	1123	10^{-20}	445				
$La_{0.8}Sr_{0.2}VO_{3-\delta}$	873–1273	10% H_2	79–158				[125]

a) Prereduced sample.

Table 6.5 Total polarization resistance at 1073 K and activation energy for the area-specific electrode conductivity of various anode layers under open-circuit conditions in H_2-containing environments.

Composition	Surface modification	Current collector	Electrolyte	$R_\eta^{1073\,K}$ ($\Omega\,cm^2$)	E_a (kJ mol^{-1}) (T (K))	Reference
$Sr_{0.88}Y_{0.08}TiO_{3-\delta}$–8YSZ (50–50 wt%)	As-prepared	Pt mesh + paste	8YSZ	2.6		[91]
	$CeO_{2-\delta}$			1.3		
	Ru			0.8		
	$CeO_{2-\delta}$ + Ru			0.5		
$La_{0.4}Sr_{0.6}Ti_{0.4}Mn_{0.6}O_{3-\delta}$–8YSZ (65–35 vol%)	As-prepared	Au paste	8YSZ	0.5$^{a)}$	103 (1003–1129)	[94]
$Sr_{0.94}Ti_{0.9}Nb_{0.1}O_{3-\delta}$–8YSZ (50–50 wt%)	As-prepared	Pt paste	8YSZ	16.3		[93]
$LaCr_{0.5}Mn_{0.5}O_{3-\delta}$	As-prepared	Pt mesh	YSZ	3.3		[114]
$La_{0.8}Sr_{0.2}Cr_{0.5}Mn_{0.5}O_{3-\delta}$	As-prepared	Pt mesh	YSZ	8.6		
$(La_{0.75}Sr_{0.25})_{0.95}Cr_{0.5}Mn_{0.5}O_{3-\delta}$	As-prepared	Pt wire	LSGM(9.8 20)	35.6		[104]
	Ni			5.1		
	Ni + $CeO_{2-\delta}$			4.8		
$(La_{0.75}Sr_{0.25})_{0.95}Cr_{0.5}Mn_{0.5}O_{3-\delta}$	As-prepared	Pt mesh	LSGM(9.8 20)	2.8	66 (973–1073)	[104]
	$CeO_{2-\delta}$			1.3	87 (973–1073)	
$La_{0.75}Sr_{0.25}Cr_{0.5}Mn_{0.5}O_{3-\delta}$	As-prepared	Au mesh + paste	LSGM(10 20)	0.16		[100]
$La_{0.75}Sr_{0.25}Cr_{0.5}Mn_{0.5}O_{3-\delta}$	As-prepared	Au mesh + paste	8YSZ	0.6		[95]
$La_{0.75}Sr_{0.25}Cr_{0.5}Mn_{0.5}O_{3-\delta}$	As-prepared	Pt paste	LSGM(10 20)	0.3	45 (973–1073)	[20]
$La_{0.75}Sr_{0.25}Cr_{0.5}Fe_{0.5}O_{3-\delta}$	As-prepared	Pt paste	LSGM(10 20)	1.6	66 (973–1073)	[20]
$La_{0.75}Sr_{0.25}Cr_{0.5}Al_{0.5}O_{3-\delta}$	As-prepared	Pt paste	LSGM(10 20)	0.4	62 (973–1073)	[20]
$Pr_{0.7}Ca_{0.3}Cr_{0.5}Mn_{0.5}O_{3-\delta}$	As-prepared	Pt paste	8YSZ	32	130 (1073–1223)	[111]
$La_{0.75}Sr_{0.25}Cr_{0.5}Mn_{0.5}O_{3-\delta}$–YSZ (50–50 wt%)	As-prepared	—	YSZ	1.3		[105]
	Pd			0.9		

Material	Condition	Current collector	Electrolyte			Ref.
$La_{0.75}Sr_{0.25}Cr_{0.5}Mn_{0.5}O_{3-\delta}$–YSZ (50–50 wt%)	As-prepared	Pt ink	8YSZ	7.5[a]	105 (1123–1223)	[92]
Cu–$La_{0.75}Sr_{0.25}Cr_{0.5}Mn_{0.5}O_{3-\delta}$–YSZ (10–50–40 wt%)	As-prepared	Pt ink	8YSZ	4.4	110 (973–1223)	[92]
$La_{0.75}Sr_{0.25}Cr_{0.5}Mn_{0.5}O_{3-\delta}$–CSO15 (75–25 wt%)	As-prepared	Pt paste	LSGM(10 20)	0.4	62 (973–1073)	[20]
$LaCr_{0.5}Mn_{0.5}O_{3-\delta}$–CGO (50–50 wt%)	As-prepared	Pt mesh	YSZ	0.21		[114]
$La_{0.8}Sr_{0.2}Cr_{0.5}Mn_{0.5}O_{3-\delta}$–CGO (50–50 wt%)	As-prepared	Pt mesh	YSZ	0.74		
$LaSr_2Fe_2CrO_{9-\delta}$–CGO10 (50–50 wt%)	As-prepared	Sputtered Au grid	LSGM(10 20)	0.3		[118]
$Sr_2MgMoO_{6-\delta}$	As-prepared	Pt paste	LSGM(20 20)	0.9	77 (848–1173)	[122]
$Sr_2MgMoO_{6-\delta}$–LSGM(20 20) (50–50 wt%)	As-prepared	Pt paste	LSGM(20 20)	0.8		[122]
$Sr_2MgMoO_{6-\delta}$–CGO20 (50–50 wt%)	As-prepared	Pt paste	LSGM(20 20)	0.7		[122]

a) Extrapolated.

Table 6.6 Total polarization resistance at 1073 K and activation energy for the area-specific electrode conductivity of $(La_{0.75}Sr_{0.25})_{1-x}Cr_{0.5}M_{0.5}O_{3-\delta}$-based layers in contact with LSGM, under open-circuit conditions in air.

Composition	Surface modification	Current collector	Electrolyte	$R_\eta^{1073\,K}$ ($\Omega\,cm^2$)	T (K)	E_a (kJ mol^{-1})	Reference
$(La_{0.75}Sr_{0.25})_{0.95}Cr_{0.5}Mn_{0.5}O_{3-\delta}$	As-prepared	Pt wire	LSGM(9.8 20)	24.1	873–1073	27	[104]
$(La_{0.75}Sr_{0.25})_{0.95}Cr_{0.5}Mn_{0.5}O_{3-\delta}$	PrO$_x$	Pt wire	LSGM(9.8 20)	40.1	973–1223	33	
Ag-$(La_{0.75}Sr_{0.25})_{0.95}Cr_{0.5}Mn_{0.5}O_{3-\delta}$ (10–90 wt%)	As-prepared	Pt wire	LSGM(9.8 20)	22.3			
	PrO$_x$	Pt wire	LSGM(9.8 20)	8.2			
$(La_{0.75}Sr_{0.25})_{0.95}Cr_{0.5}Mn_{0.5}O_{3-\delta}$	As-prepared	Pt mesh	LSGM(9.8 20)	13.5			
$La_{0.75}Sr_{0.25}Cr_{0.5}Mn_{0.5}O_{3-\delta}$	As-prepared	Pt paste	LSGM(10 20)	17.6	873–923 923–1073	47 137	[20]
$La_{0.75}Sr_{0.25}Cr_{0.5}Fe_{0.5}O_{3-\delta}$	As-prepared	Pt paste	LSGM(10 20)	34.8	873–973 973–1073	14 98	[20]
$La_{0.75}Sr_{0.25}Cr_{0.5}Al_{0.5}O_{3-\delta}$	As-prepared	Pt paste	LSGM(10 20)	1.4	873–923 923–1073	118 185	[20]

Table 6.7 Comparison of cathodic overpotentials at $i = -30\,\text{mA}\,\text{cm}^{-2}$ for porous $(La_{0.75}Sr_{0.25})_{0.95}Cr_{0.5}Mn_{0.5}O_{3-\delta}$ and $Pr_2Ni_{0.8}Cu_{0.2}O_{4+\delta}$ electrodes in contact with LSGM (9.8 20) and total conductivity of dense ceramics at 1073 K in air [57, 104].

Ceramics	σ (S cm^{-1})	Cathode	Current collector	η (mV)
$(La_{0.75}Sr_{0.25})_{0.95}Cr_{0.5}Mn_{0.5}O_{3-\delta}$	25.0	Ag-modified	Pt wire	−669
			Pt gauze	−254
$Pr_2Ni_{0.8}Cu_{0.2}O_{4+\delta}$	167.9	As-prepared	Pt wire	−11
			Pt gauze	−10

atmospheres, the oxygen ionic conduction was suggested as an electrochemical reaction rate-determining factor [108]. The ionic transport in LSCM is indeed very low, in spite of the oxygen vacancy formation under reducing conditions [104]. Consequently, CeO_2- and ZrO_2-based additives and interlayers decrease the electrode polarization [92, 95, 96, 102–107, 113, 114], though no essential effects were observed in Ref. [20]. An optimum composition of the LSCM–CGO anodes was observed for the component ratio of 33:67 wt% [106]. The most crucial factor relevant for the majority of chromite- and titanate-based anodes is, however, a relatively low electronic conduction. In particular, the total conductivity of LSCM lies in the range of 20–40 S cm^{-1} and is essentially independent of $p(O_2)$ in oxidizing and moderately reducing environments at 1000–1250 K, but decreases down to 1–5 S cm^{-1} on reduction [95, 98, 104]. For model cells, the limiting effects of electronic transport can be avoided using specific current collectors, such as Au paste and mesh [95, 99, 100, 109], Ag paint [96, 107], or Pt ink and paste [92, 107, 111]. Analogously, sputtering of even small amounts of Pt over the Cu–LSCM layer improved the anode performance [102]. These approaches cannot, however, be used in real SOFCs where the current collectors made of stainless steels or $LaCrO_3$-based ceramics possess considerably higher contact and/or bulk resistance compared to the noble metals. The electronic transport limitations make it necessary to use LSCM in combination with metal additives, such as Ni or Cu [79, 92, 96, 102, 104, 109, 112], and to consider these phases among other promising components of the anode cermets. On the other hand, it seems necessary to include catalytically active metal or oxide phases in the LSCM-based anodes, especially when operating in hydrocarbon fuels, since the catalytic activity of LSCM is also relatively low [110]. Addition of 0.5–9 wt% Pd, Rh, Ni, Fe, or ceria made it possible to substantially reduce the polarization of $(La_{0.75}Sr_{0.25})_{1-z}Cr_{0.5}Mn_{0.5}O_{3-\delta}$ anodes in contact with YSZ- and LSGM-based electrolytes in H_2- and CH_4-containing environments [96, 104, 107]. Surface deposition of Pd nanoparticles was found to significantly enhance the electrocatalytic activity of porous LSCM-based layers in methane and ethanol, and had a minor effect in H_2 [105, 108, 109]. The incorporation of nickel leads to a stronger effect on the anode performance compared to ceria-based phases [96, 104].

In comparison with $La_{0.75}Sr_{0.25}Cr_{0.5}Mn_{0.5}O_{3-\delta}$, Sr-free $LaCr_{0.5}Mn_{0.5}O_{3-\delta}$ was reported to have better properties, namely, higher thermochemical stability, electronic conductivity, and anode performance under humidified H_2 [114]. For $La_{0.7}A_{0.3}Cr_{0.5}Mn_{0.5}O_{3-\delta}$, the highest electrochemical activity was claimed for A = Ca, whereas the barium solubility limit was below 30 mol% [113]. Moderate (10 mol%) doping of LSCM with cerium improves the anode performance without any essential effect on the total conductivity. At 1173 K, $La_{0.65}Ce_{0.1}Sr_{0.25}Cr_{0.5}Mn_{0.5}O_{3-\delta}$|8YSZ anode displayed polarization resistances of 0.2 and 1.6 Ω cm^2 in humidified H_2 and CH_4, respectively. [115].

The anode polarization resistance of $La_{0.8}Sr_{0.2}Cr_{1-x}Ru_xO_{3-\delta}$ ($x = 0$–0.25)–CGO10 (50–50 wt%) composites applied onto LSGM(10 20) was shown to decrease during the initial period of SOFC operation before reaching a minimum [116]. The rate of these changes and the electrode performance both increase with increasing x, a result of higher density of Ru surface nanoparticles observed by transmission electron microscopy (TEM). The measured initial surface coverage by Ru nanoparticles was consistent with a model where ruthenium supply to the surface is limited by the Ru bulk out-diffusion, but the coverage saturation requires longer times [116]. Another alternative anode composition, $La_{0.75}Sr_{0.25}Cr_{0.5}Fe_{0.5}O_{3-\delta}$, was reported to be active for methane reforming and total oxidation, depending on the oxygen nonstoichiometry level [117]. At 1123 K, its polarization resistances in contact with 8YSZ electrolyte and Au mesh (fixed with a small amount of Au paste), about 1.79 and 1.15 Ω cm^2 in wet 5% H_2 and wet 100% H_2, respectively are, however, still too high for a viable SOFC electrode system [117]. The R_η values of $La_{0.75}Sr_{0.25}Cr_{0.5}Fe_{0.5}O_{3-\delta}$ anodes, applied onto LSGM and coated with Pt paste, are found higher compared to those of LSCM [20]. The $LaSr_2Fe_2CrO_{9-\delta}$ material that belongs to the same family appears stable under the SOFC anode conditions up to approximately 1073 K [118]. The performance of $LaSr_2Fe_2CrO_{9-\delta}$–CGO10 (50–50 wt%) composite anode was slightly better in the cells with LSGM(10 20) electrolyte in comparison to CGO10, under humidified H_2 at 1073 K [118]. $La_{0.8}Sr_{0.2}Sc_{0.2}Mn_{0.8}O_{3-\delta}$, showing higher electrical conductivity and electrocatalytic activity than LSCM under both anodic and cathodic conditions, was also reported to have a sufficient chemical and structural stability due to the backbone effect of Sc^{3+} in the perovskite lattice [119]. In all these cases, however, detailed experimental appraisal of the long-term stability and possible surface segregation phenomena under highly reducing environments is necessary.

The double $Sr_2Mg_{1-x}M_xMoO_{6-\delta}$ (M = Mn, Cr) perovskites were recently proposed [120] as the candidates that meet the requirements for long-term stability with a tolerance to sulfur and display a superior electrochemical performance in hydrogen and methane; representative examples are given in Figures 6.14, 6.18 and 6.19 and Tables 6.3 and 6.5. A power output degradation of 1 and 16% after introducing, respectively, 5 and 50 ppm H_2S in H_2 was observed for the cells with $Sr_2MgMoO_{6-\delta}$ (SMM) anode and Pt mesh with a small amount of Pt paste as a current collector [120]. The TECs of SMM ceramics vary in the range 11.7–12.7 in air at 380–1070 K (Table 6.4). Lowering oxygen partial pressure, $p(O_2)$, increases slightly the total conductivity due to formation of n-type charge carriers when molybdenum cations are reduced [120]. The incorporation of La^{3+} cations into $Sr_{2-x}La_xMgMoO_{6-\delta}$ up to $x = 0.8$ improves the cell performance under hydrocarbon fuels, especially in the presence of water vapor [121]. Above 1073–1273 K, $Sr_2MgMoO_{6-\delta}$ was found to chemically interact with CGO, LSGM, and, especially, YSZ electrolytes, thus making it necessary to introduce buffer layers between the electrode and the solid electrolyte [121, 122]. One important comment is that reduction above 1173 K results in decomposition of $Sr_2MgMoO_{6-\delta}$ into a mixture of $n=2$ Ruddlesden–Popper phase with a significantly higher Mo/Mg ratio, and also MgO and Mo; the reduced material was intergrown with the perovskite matrix [123]. XPS and XRD studies on $Sr_2MgMoO_{6-\delta}$ powders, annealed in air and under diluted CH_4 environments, show that the surface is formed mainly by $SrMoO_4$ scheelite and metal carbonates [122]. A number of other degradation phenomena, primarily related to possible molybdenum oxide volatilization during processing and hydration at low temperatures, require further investigations.

Another Mo-containing perovskite, $CaV_{0.5}Mo_{0.5}O_{3-\delta}$, exhibits total conductivity over 520 S cm^{-1} under reducing conditions (10% H_2–90% N_2) at 300–1170 K and was thus proposed for SOFC anodes, though heating in air above 773 K leads to transition of the orthorhombic phase into the tetragonal scheelite phase [124]. This transformation is reversible on further reduction. The TECs of both oxidized and reduced phases are rather similar, 10 and 13×10^{-6} K^{-1}, respectively (Table 6.4). In the $La_{1-x}Sr_xVO_{3-\delta}$ ($x=0.2$–0.5) series, the perovskite phases also exist under reducing conditions only [125]. These vanadates were found chemically compatible with YSZ, at least up to 1573 K. At 1173 K, $La_{0.6}Sr_{0.4}VO_{3-\delta}$–YSZ (50–50 wt%) composite anodes showed the polarization resistances of 0.85 and 1.38 Ω cm^2, and overpotentials of 130 and 200 mV at current densities of 200 mA cm^{-2} in pure H_2 and wet CH_4, respectively (Figures 6.11 and 6.13). The readers interested in detailed information on the alternative anode materials versus conventional compositions may refer to the recent reviews and surveys [5, 8, 10, 70, 72, 77, 78].

6.5
Other Fuel Cell Concepts: Single-Chamber, Micro-, and Symmetrical SOFCs

References [126, 127] present comprehensive reviews on the recent advances in SC-SOFCs (see Chapter 5). In this concept, a mixture of fuel and oxidant is supplied to the anode and cathode, which are placed in one gas chamber and are catalytically active for the oxidation and reduction processes, respectively. The main advantages of this approach are easier fabrication and miniaturization with respect to the conventional SOFC configurations. However, while the electrolyte membrane quality becomes less important and porous membranes may even be beneficial for SC-SOFCs [128], the requirements of the reaction selectivity provided by the electrode materials and the safety of mixing of the oxidizing and reducing gases are crucial. The existing electrode materials provide rather a poor cell efficiency and low fuel utilization, typically less than 10% [126, 127, 129]. Nevertheless, most studies known in the literature focus on attempts to adapt conventional SOFC electrode materials to the SC design (Table 6.8). Note also that apart from the microstructural effects briefly discussed in this section, the performance of SC-SOFCs is considerably affected by the gas molar ratio, gas flow rate, and operating temperature (Figure 6.20 and Table 6.8). In addition, the catalytic

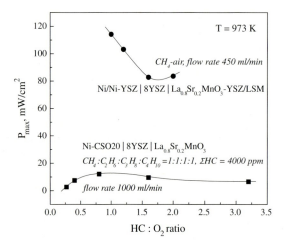

Figure 6.20 Maximum power density of the electrolyte-supported planar single-chamber cells at 973 K as a function of hydrocarbon to oxygen ratio [130, 141]. In all cases, the total gas flow rate was kept constant. Au meshes [130] and Ni and Au meshes [141] were used as current collectors.

oxidation reactions may result in substantial heating of the cells, temperature variations along the cell, and deviations from the initial fuel–air composition [126, 127, 130–132]. For instance, due to the temperature variations, the NiO–YSZ|8YSZ|La$_{0.8}$Sr$_{0.2}$MnO$_{3\pm\delta}$–YSZ cells exhibit a better performance when the small cathodes are located at the inlet and at the CH$_4$/O$_2$ ratio equal to 1 : 1 [130]. Ru–CeO$_{2-\delta}$|Ni–CSO|CSO15 (20 μm)|BSCF–CSO cells with Ag paste and Ag–BSCF current collectors showed P_{max} values of 247 mW cm^{-2} at C$_3$H$_8$/O$_2$ = 4 : 9 and temperature maintained at 853 K without external heating [130].

The SC-SOFC cathodes are not only supposed to be active for oxygen reduction but should also be inert for any reaction with fuels. The major material-related trends remain, in general, similar to those for the conventional SOFC cathodes exposed to atmospheric air [126, 127]. The catalytic activity of La$_{0.8}$Sr$_{0.2}$MnO$_{3\pm\delta}$–8YSZ/La$_{0.8}$Sr$_{0.2}$MnO$_{3\pm\delta}$ cathodes sintered at 1373 K increases on heating and decreases when higher sintering temperatures are used [133]. Above 973 K, the conversion of oxygen species may exceed 30%. However, the half-cells achieve a better performance at 873 K than at 973 K since the catalytic combustion processes become more complex [133]. Despite the problems originating from excessive sintering, from interaction with solid electrolyte, and from a significant activity for fuel oxidation, Sm$_{0.5}$Sr$_{0.5}$CoO$_{3-\delta}$ cathodes display a lower resistance and enable higher power outputs than La$_{0.6}$Sr$_{0.4}$CoO$_{3-\delta}$, LaNi$_{0.6}$Fe$_{0.4}$O$_{3-\delta}$, or lanthanum strontium manganite (LSM) under methane–air gas mixtures [134]. Highest performance of the CSO-based cells with R$_{0.7}$Sr$_{0.3}$Fe$_{0.8}$Co$_{0.2}$O$_{3-\delta}$ (R = La, Pr, Gd) cathodes under CH$_4$–air was achieved for R = Pr [129]. At the same time, LSFC possesses, again, a considerable activity toward electrochemical oxidation of CH$_4$ [135]. BSCF cathodes were reported to have an exceptionally high activity for electrochemical reduction of oxygen and a

Table 6.8 Selected examples of the maximum power density in various planar SC-SOFCs under hydrocarbon–air gas mixtures, reported in 2008–2010.

Anode/current collector	Electrolyte/thickness	Cathode/current collector	Gas mixtures	T (K)[a]	P_{max} (mW cm^{-2})	Reference
Cell type: electrolyte supported						
NiO-8YSZ (45–55 wt %)|NiO/Au mesh	8YSZ/0.5 mm	La$_{0.8}$Sr$_{0.2}$MnO$_{3\pm\delta}$–8YSZ (50–50 wt%)|LSM/Au mesh	CH$_4$/O$_2$ = 1:1	973	114	[130]
NiO–CSO20 /Ni mesh	8YSZ/0.4 mm	La$_{0.8}$Sr$_{0.2}$MnO$_{3\pm\delta}$/Au mesh	ΣHC/O$_2$ = 4:5[b]	973	12	[141]
NiO–CSO17 (60–40 wt%)/Pt paint	CSO17/0.3 mm	La$_{0.7}$Sr$_{0.3}$Fe$_{0.8}$Co$_{0.2}$O$_{3-\delta}$/Ag paint	CH$_4$/O$_2$ = 2:1	873	49	[129]
		Pr$_{0.7}$Sr$_{0.3}$Fe$_{0.8}$Co$_{0.2}$O$_{3-\delta}$/Ag			65	
		Gd$_{0.7}$Sr$_{0.3}$Fe$_{0.8}$Co$_{0.2}$O$_{3-\delta}$/Ag			46	
Ni–CGO20 (90–10 wt %)/the same ink	CGO20-2 wt% Co/0.2 mm	AgO–La$_{0.5}$Sr$_{0.5}$CoO$_{3-\delta}$–CGO20 (10–80–10 wt %)/the same ink	CH$_4$/O$_2$ = 9:10	1013	280	[152]
NiO–PdO–CSO17 (63–7–30 wt %)/porous Pt film	CSO17/0.4 mm	Sm$_{0.5}$Sr$_{0.5}$CoO$_{3-\delta}$/porous Au film	CH$_4$/O$_2$ = 2:1	873	97	[139]
Cell type: anode-supported						
NiO-8YSZ (50–50 wt %)/Ag paste	8YSZ/8 μm	La$_{0.7}$Sr$_{0.3}$MnO$_{3-\delta}$ impregnated with CSO20/Ag paste	CH$_4$/O$_2$ = 3:2	1123	416	[148]
NiO–CSO20/Ag paste	CSO20/20–30 μm	Ba$_{0.5}$Sr$_{0.5}$Co$_{0.8}$Fe$_{0.2}$O$_{3-\delta}$–CSO20 (70–30 wt%)/Ag paste	CH$_4$/O$_2$ = 1:1	873	570	[142]

(Continued)

Table 6.8 (Continued)

Anode/current collector	Electrolyte/thickness	Cathode/current collector	Gas mixtures	T (K)[a]	P_{max} (mW cm^{-2})	Reference
Cell type: cell with one pair of coplanar electrodes						
NiO–8YSZ (55–45 wt %), width 93 µm/Au mesh + paste	8YSZ/0.2 mm, gap size 0.25 mm	$(La_{0.8}Sr_{0.2})_{0.98}MnO_{3\pm\delta}$ –8YSZ (70–30 wt%), width 93 µm/Au mesh + paste	$CH_4/O_2 = 2:1$	973	10.5	[157]

a) The furnace temperature is given. The real cell temperature is supposed to be higher due to exothermic partial oxidation of fuel (see text).
b) CH_4, C_2H_6, C_3H_8, C_4H_{10} – 1000 ppm each, Ar-balance. CSO17 is CeO_2–10 mol % Sm_2O_3.

sufficiently low activity for methane oxidation [136]. The anode-supported Ni–CSO (60–40 wt%)|CSO15 (20 μm)|BSCF–CSO (70–30 wt%) cell with Ag current collectors generated a power density as high as 760 mW cm^{-2} at the nominal furnace temperature of 923 K (cell temperature ~1060 K) when methane, oxygen, and helium flow rates were 87, 80, and 320 ml min^{-1}, respectively [136].

The anodes should be catalytically active and selective for only partial oxidation of fuel and electrochemical oxidation of produced syngas; the optimum theoretical situation, when the direct electrochemical total oxidation of fuel occurs at the SC-SOFC anode, was never achieved in practice. Although carbon deposition is suppressed in fuel–air mixtures, the use of Ni-containing composites may still be problematic due to redox instability of the anode layers and a tendency to form nickel oxide and even volatile hydroxide; these processes decrease the anode conductivity and catalytic activity, and cause microstructural degradation [127, 137]. At temperatures higher than 823 K, the Ni–Ce$_{0.9}$Zr$_{0.1}$O$_{2-\delta}$ catalysts can promote methane conversion up to the level of 90%, with H$_2$ and CO as the main products and without any detectable carbon formation. Addition of Ru catalyst to the Ni–CSO anode improves the cell performance under C$_4$H$_{10}$–air at 473–573 K due to favoring hydrogen formation; the use of Pd has an opposite effect [132]. Nevertheless, the incorporation of Pd in the Ni-containing composites often has a positive impact on the anode performance and Ni reducibility (see Refs [127, 138] and references cited therein).

NiO–PdO–CSO composites showing a high catalytic activity were proposed as the anodes for IT SC-SOFC operating under CH$_4$–air mixtures [139]. The power densities in the cells with CSO electrolyte and Sm$_{0.5}$Sr$_{0.5}$CoO$_{3-\delta}$ cathode were found comparable to those of conventional SOFCs comprising similar materials, though porous Pt and Au films were employed as anode and cathode current collectors, respectively. Notice that the use of metal grids resulted in a lower performance [139]. NiO–(La$_{0.75}$Sr$_{0.25}$)$_{0.9}$Cr$_{0.5}$Mn$_{0.5}$O$_{3-\delta}$ anode layers were tested in Ref. [140]. The Ni–CSO20 anodes appeared active for CH$_4$, C$_2$H$_6$, C$_3$H$_8$, and C$_4$H$_{10}$ oxidation [141]. Prereduction of the anode before the deposition of cathode layer was reported to be more effective than *in situ* reduction by methane–air gas mixtures [142].

The solid electrolyte choice for SC-SOFCs is usually limited to zirconia- and ceria-based materials, although LSGM, BaLaIn$_2$O$_{5.5}$, and proton-conducting BaCe$_{0.8}$Y$_{0.2}$O$_{3-\delta}$, SrCe$_{0.95}$Yb$_{0.05}$O$_{3-\delta}$, and CaZr$_{0.9}$In$_{0.1}$O$_{3-\delta}$ electrolyte-supported cells have also been tested [134, 140, 143–145]. For information on the structure and transport properties of these electrolytes, readers may refer to Chapters 2, 7, 9 and 12 of the first volume. The electrochemical performance of model SC-SOFCs based on YSZ, CSO, and LSGM electrolytes, compared in Refs [134, 146], was found maximum for LSGM at 873–1073 K and for CSO at 623–723 K, with Ni–CSO anode and Sm$_{0.5}$Sr$_{0.5}$CoO$_{3-\delta}$ cathode. If considering the electrolyte membrane microstructure, one interesting approach is associated with the use of porous solid electrolyte, allowing the reaction gases to pass consecutively through the cathode, electrolyte, and anode [147]. The possibility to fabricate a cell stack based on this design has recently been demonstrated [127, 128].

Due to complexity of the catalytic phenomena, the importance of composition, microstructure, and geometry of SC-SOFC current collectors becomes even greater than for the dual-chamber cells. Quite often, Ag current collectors are used both for the single model cells and for SC-SOFC stacks [128, 131, 136, 142, 148, 149]. In order to assess the silver collector stability, microtubular Ni-YSZ|YSZ|LSM cells with Ag paste and current-collecting wires for both electrodes were prepared and tested under CH_4–air at 1023 K and an operating voltage of 0.5 V [150]. The results reveal an increasing porosity in the Ag wires with time, leading finally to rupture, due to both the formation of silver oxide and the volatilization and partial melting. Such degradation processes may also be contributed by the cell overheating, as predicted by two-dimensional axisymmetric numerical modeling [151]. On the other hand, the use of small silver additives in the cathode layers is advantageous [152], as for the dual-chamber arrangement. Pt shows a substantial catalytic activity in the methane–air mixtures, whereas Au is considered essentially inert but interacts with Ni to form a eutectic system above 1073 K [127]. Integrated Pt mesh current collectors were embedded into the electrode layers of the $Sm_{0.5}Sr_{0.5}CoO_{3-\delta}$|CGO10|Ni–CGO20 cells, which indeed showed a high open-circuit voltage (OCV) and a power density of 468 mW cm^{-2} at 873 K under CH_4–air [138]. For microelectrodes, providing an appropriate current collection is associated with serious challenges [127, 153].

The micro-SOFC (μ-SOFC, Chapter 5) concept implies not only the small overall size but also the wide use of microfabrication technologies including those employed in microelectronics [154]. Downscaling of the SC-SOFCs with coplanar electrodes located at the same side of solid electrolyte is considered relatively easy. The efficiency and fuel utilization in this case are, however, lower than those for the symmetric cells [127]. The NiO–8YSZ or NiO–CGO10 anodes and $(La_{0.8}Sr_{0.2})_{0.98}MnO_{3\pm\delta}$–8YSZ or $(La_{0.6}Sr_{0.4})_{0.995}Fe_{0.8}Co_{0.2}O_{3-\delta}$ cathodes applied onto the surface of 8YSZ substrate via direct-write microfabrication were evaluated for single-chamber micro-SOFCs with a coplanar interdigitated (comb-like) electrode configuration [155]. At 973 K under CH_4–air, the best cell performance was obtained for NiO–YSZ and LSM–YSZ electrode pair, regardless of the chemical interaction between the cathode and the anode during sintering. The issues related to preferable shapes, to the geometries and electrochemical properties of electrodes, and to the electrolyte thickness are analyzed in Ref. [156]. One should separately note that the performance of the cells with coplanar microelectrodes significantly depends on the electrode width having both lower and higher limits [157]. Below the former limit, neither a stable difference in the oxygen chemical potential nor OCV can be established; exceeding the higher limit leads to an excessively high cell resistance. For an interelectrode gap of 250 μm, an optimum electrode width was suggested to lie between 550 and 850 μm [157].

The feasibility of using microhotplates, a technology platform initially developed in the field of sensors, was demonstrated for the example of $La_{0.6}Sr_{0.4}Co_{0.2}Fe_{0.8}O_{3-\delta}$ thin-film cathodes deposited by spray pyrolysis [158]. Postdeposition annealing steps in an external furnace and using the integrated heater of the microhotplate showed that despite bending, these microhotplates are suitable for a maximum operation temperature of 1073 K and long-term operation at 973 K. Thin-film μ-SOFCs were fabricated with Pt anode and $La_{0.6}Sr_{0.4}Co_{0.8}Fe_{0.2}O_{3-\delta}$ cathode using sputtering, lithography, and etching [159]. An alternative dual-chamber cell design with 20 μm thick solid electrolyte sintered on the ring of the same material was proposed [160, 161]; at 873 K in wet H_2, the cells with NiO–CGO20 (40–60 wt%) anode, $Sm_{0.5}Sr_{0.5}CoO_{3-\delta}$ cathode, and Pt paste current collectors show P_{max} values of 270 and 290 mW cm^{-2} when using CGO20 and LSGM(10 20) electrolytes, respectively. Another configuration for the coplanar cell was fabricated by applying the Ni–8YSZ and $La_{0.7}Sr_{0.3}MnO_{3-\delta}$ electrodes impregnated with CSO20 in two parallel grooves on YSZ substrate, in order to extend a conductive channel of the oxygen ionic transport and to form a gas separator between the electrodes, thus providing lower ohmic resistance and higher OCV [149]. The functional layers of nanocrystalline $La_{0.6}Sr_{0.4}Co_{0.8}Fe_{0.2}O_{3-\delta}$ and doped zirconia with thicknesses of 50–300 nm were prepared using photolithography, radio frequency (RF) sputtering, liftoff process, and photon-assisted synthesis for potential use in micro-SOFCs [162–164].

The fuel utilization is improved in the microtubular (MT) SC-SOFCs, where gas intermixing between the electrodes is avoided and the gases pass a longer distance over the electrode surface. The 2D distributed finite element model of an MT hollow fiber SOFCs was used in combination with TPB lengths measured by focused ion beam (FIB)-SEM technique to predict the effects of anode microstructure on the distribution of current density and anode activation polarization [165]. As for the planar cells, substantially high thermal gradients were revealed theoretically and experimentally along the microtubes, which may even damage the cell components at the gas inlet where the temperature rise is higher [151, 166]. The anode-supported, mixed reactant MT-SOFC with YSZ electrolyte, Ni–YSZ anode, LSM cathode, and silver ink applied to enhance the electrical conductivity of both electrodes and catalytic activity of the cathode was prepared and tested; at 1023 K, the OCV of 1.05 V and $P_{max} = 122$ mW cm^{-2} were achieved at CH_4/air ratio of 1:4.76 [167]. However, the performance degradation rate was as high as 0.05% per 24 h and current density fluctuations due to redox cycles over the nickel surface were observed [167]. The preparation conditions were adjusted to obtain dense thin-film $CGO10/Zr_{0.89}Sc_{0.1}Ce_{0.01}O_{2-\delta}$ electrolyte double layers by dip coating in aqueous suspensions on a porous microtubular $Nd_{1.95}NiO_{4+\delta}$ cathode substrate, prepared by cold isostatic pressing (CIP) and sintering [168]. The mechanical strength critical for the MT-SOFCs was studied on the example of anode-supported cells consisting of NiO–CGO20 anode, CGO20 electrolyte, and $La_{0.8}Sr_{0.2}Co_{0.6}Fe_{0.4}O_{3-\delta}$–CGO20 cathode, fabricated using extrusion and cofiring techniques [169].

With a very rare exception, the electrochemical properties of individual electrodes (e.g., polarization resistances and overpotentials) are not assessed in the literature reports on SC- and micro-SOFCs due to serious experimental and methodological problems. Instead, the cell voltage and power output are commonly measured and outlet gas mixtures are analyzed, giving some information on the electrode selectivity. The electrochemical activity of $(La_{0.8}Sr_{0.2})_{0.98}MnO_{3\pm\delta}$–YSZ composite cathodes for the oxygen reduction reaction in air, evaluated by the standard three-electrode method, was compared for planar and Ni–YSZ anode-supported closed MT arrangements, with YSZ electrolyte [170]. While the impedance spectra displayed some differences, the total polarization resistances and overpotential values appear similar. Increasing the MT cell sintering temperature from 1373 to 1423 K was found to improve the cathode performance [170]. Anode-supported hollow-fiber SOFC with 12 μm thick 8YSZ electrolyte, Ni–YSZ anode, $(La_{0.8}Sr_{0.2})_{0.95}MnO_{3\pm\delta}$–YSZ cathode, and Ag wires/paste current collectors, fabricated using the phase inversion and vacuum-assisted coating techniques, was reported to provide a maximum power density of 377 mW cm^{-2} at 1073 K under humidified H_2 as a fuel and air as an oxidant [171]. The MT-SOFC with $(La_{0.8}Sr_{0.2})_{0.97}MnO_{3\pm\delta}$–CGO (70–30 wt%) cathode supports fabricated by extrusion, CGO10/10SSZ bilayer electrolyte, NiO–CGO (50–50 wt%) anode, and Ag current collectors shows a power density of 352 mW cm^{-2} at 1023 K, with humidified H_2 and O_2 [172]. A honeycomb design suitable for compact SOFC modules, enabling high volumetric power densities due to a large electrode surface area, was proposed in Ref. [173]. The cathode-supported SOFC prepared via extrusion of an LSM honeycomb monolith and a slurry injection method for the channel surface coating using electrolyte/anode bilayers exhibits power densities of approximately 1200 mW cm^{-3} at 873 K under flowing humidified H_2 fuel and O_2 oxidant. Various concepts of micro-SOFCs are reviewed in detail in Refs [153, 174, 175].

Finally, despite the smaller electronic conductivity and very low oxygen ionic transport (see Chapter 9 of the first volume), the manganite and manganite–chromite-based perovskite phases are attracting much attention due to a moderate thermal expansion, negligible chemical strains, and a good stability under both oxidizing and reducing environments that makes it possible to use these materials simultaneously for cathodes and anodes [20, 68, 104, 111, 113, 119]. The advantages of such "symmetrical" SOFC concept, tested already for various solid electrolytes, include (i) reduced number of steps for the cell fabrication, (ii) lower costs, and (iii) a possibility to reverse the configuration. The latter regime may be important, in particular, for removing carbon and sulfur deposited at the anode surface. The $R_{1-x}A_xCr_{1-y}Mn_yO_{3-\delta}$ family (Section 6.4) has been studied most extensively in this respect. The electrochemical performance of the corresponding cathodes in the IT range is, however, quite poor (Figures 6.10 and 6.21 and Tables 6.4, 6.6 and 6.7). The electrochemical activity of $La_{0.7}A_{0.3}Cr_{0.5}Mn_{0.5}O_{3-\delta}$ (A = Ca, Sr, Ba) for oxygen reduction decreases in a sequence Ca > Sr > Ba, in correlation with the total electrical conductivity in air [113] (Figure 6.21). The performance of $Pr_{0.7}Ca_{0.3}Cr_{1-y}Mn_yO_{3-\delta}$ ($x = 0.2$–0.8) series, where increasing manganese concentration leads to higher conductivity and to lower anode and cathode polarization resistances, is also insufficient [111]. At 1223 K, the symmetrical SOFC with 0.37 mm 8YSZ electrolyte and current collectors formed using Pt ink display power densities of 250 and 160 mW cm^{-2} in humidified H_2 and CH_4 fuels, respectively (Table 6.3). Perovskite-type $La_{0.8}Sr_{0.2}Mn_{0.8}Sc_{0.2}O_{3-\delta}$ possesses a higher conductivity than LSCM under both oxidizing and reducing environments [119]. A symmetric electrolyte-supported cell with $La_{0.8}Sr_{0.2}Mn_{0.8}Sc_{0.2}O_{3-\delta}$ electrodes and 0.3 mm thick scandium-stabilized zirconia electrolyte exhibits,

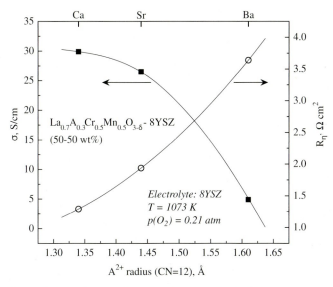

Figure 6.21 Total electrical conductivity of $La_{0.7}A_{0.3}Cr_{0.5}Mn_{0.5}O_{3-\delta}$ ceramics and polarization resistance of porous $La_{0.7}A_{0.3}Cr_{0.5}Mn_{0.5}O_{3-\delta}$–YSZ composite cathodes with Pt mesh current collectors, at 1073 K in air [113].

however, P_{max} values of 310 and 130 mW cm^{-2} at 1173 K in humidified H_2 and CH_4, respectively [119].

6.6
Alternative Fuels: Direct Hydrocarbon and Direct Carbon SOFCs

The direct electrochemical oxidation of dry hydrocarbons on the surface of oxide mixed conductors is quite slow [176]. As a rule, operation of the cells fueled by methane and higher hydrocarbons occurs via the anodic electrochemical oxidation of gaseous products of either reforming (external or internal) or cracking (on metallic catalysts or in the gas phase) [7, 176]. As nickel is catalytically active toward both these processes, large amounts of steam or CO_2 are necessary to avoid the carbon deposition on Ni-containing cermets; alternatively, components with high catalytic activity to the carbon oxidation, such as ceria-based compounds, should be incorporated in the anode layers. At the same time, the endothermic nature of catalytic reforming may cause local thermomechanical stresses and cell failure [176, 177]. Due to the obvious experimental difficulties, information on the microscopic mechanisms of hydrocarbon oxidation in SOFCs is still very scarce; these complex processes require further studies. Special attention should be paid, in particular, to the hermeticity of sealing as air leakages into the fuel chamber should suppress the carbon formation [177]; oxygen-permeable materials, such as silver [178], may often be inappropriate for this purpose. Irrespective of the microscopic mechanisms, a variety of approaches are being tested to overcome problems associated with carbon deposition in the hydrocarbon-fueled SOFCs (see Refs [179–184] and references cited therein). For example, porous disks of CGO-impregnated FeCr were tested as a gas diffusion layer of methane-fueled SOFCs [180]. At 1073 K and ambient pressure, the major reaction in this layer was CO_2 reforming of CH_4 to produce syngas, a mixture of CO and H_2, with subsequent CO oxidation at the Ni–YSZ anode; rather surprisingly, H_2 conversion was reported negligible [180]. An internal reforming layer of $Cu_{1.3}Mn_{1.7}O_4$ spinel, where highly dispersed nano-Cu particles are formed after *in situ* reduction by CH_4, was shown to improve stability of Ni–CSO anode-supported cells that demonstrated $P_{max} = 311$ mW cm^{-2} at 973 K [181]. $La_{0.8}Sr_{0.2}MnO_{3\pm\delta}|8YSZ$ (1 mm)|Pt|Ni–YSZ cell with 0.1 wt% Ir-impregnated CeO_2 catalyst layer coated over the anode was operated under both the gradual internal reforming conditions and pure dry methane at 1173 K, yielding a current density of about 0.1 A cm^{-2} at 0.6 V for 120 h [182]. Despite minor deterioration in the performance, no carbon deposition was observed [182]. Analogously, the incorporation of Rh–alumina catalyst or porous zirconia-based barrier layers made it possible to improve the performance stability of CH_4-fueled Ni–YSZ (50–50 wt%) anodes [177, 178]. The CH_4 conversion over the LSFC–YSZ anode occurs via CH_4 decomposition and electrochemical oxidation of hydrogen and carbon [135]. Various anodes including both Ni–SSZ and Ni–CGO cermets and oxide CGO and $La_{0.85}Sr_{0.15}MnO_{3-\delta}$–CGO composite were employed for rechargeable cells where the solid carbon is supplied by thermal decomposition of hydrocarbon fuel [183, 184].

During the past few years, the direct carbon (DC) SOFCs, in which solid carbonaceous materials (e.g., coal, coke, charcoal, tar, biomass, various solid wastes, etc.) are utilized as fuel, are also receiving an increasing attention [185–200]. In this case, the choice of anode material and configuration (Table 6.9) is determined by methods of delivering solid fuel to the electrochemical reaction zone [185, 188]. In simplest arrangements, the fuel powder may be applied or pressed onto the solid electrolyte surface, thus also serving as an anode [186, 187]. Amorphous carbon materials provide usually higher power densities compared to crystalline graphite [187]. The cells based on fluidized bed (FB) configuration were suggested [188–191] to utilize different gasification approaches; their comparison is found in Ref. [189]. Note that carbon gasified into carbon monoxide by recirculating carbon dioxide, without steam, can be used for the SOFC arrangement coupled to a Boudouard-type FB dry gasifier [188, 191].

Another concept [192] presents a merged SOFC and molten carbonate fuel cell (MCFC) technology with the anodes comprising molten alkaline metal carbonates and/or oxides (e.g., eutectic mixtures of K_2CO_3, Li_2CO_3, and/or Na_2CO_3), which facilitate oxygen transport from the solid oxide electrolyte to the surface of solid fuel [185, 192–197]. Slurries of carbon particles provide electronic transport in the molten carbonate-containing anodes and, hence, act as both an anode component and a fuel. Both the direct oxidation and the Boudouard conversion contribute to the overall reaction [196]. Implementation of this approach requires, however, periodic replacement of the cell components due to their degradation [193, 194]. For instance, YSZ solid electrolyte was found stable in molten Na/K carbonate mixture at 1123 K at least for 24 h, but lithium zirconate is formed in Li/K carbonate at 973 K; ceria seems

Table 6.9 Performance of various DC-SOFCs with planar or tubular zirconia-based electrolyte.

Anode	Electrolyte/thickness	Cathode	Fuel	T (K)	P_{max} (mW cm^{-2})	Reference
Carbon black	$(ZrO_2)(HfO_2)_{0.02}(Y_2O_3)_{0.08}$/0.12 mm	$La_{0.84}Sr_{0.16}MnO_3$	Carbon black	1275	50	[186]
Pt or Pt–YSZ	PSZ[a]/1.3 mm	Pt or Pt–YSZ	Synthetic carbon bed fluidized by He	1073 1173	10 1	[190]
Ni–ceria	YSZ/0.2 mm	LSM		1173	22	[190]
Ni–YSZ	YSZ/10 µm	LSM	Activated carbon bed	1178	220	[189]
Ni–YSZ	YSZ/8–10 µm	Perovskite	Coal bed fluidized by CO_2	1123	450	[191]
Fuel (~40 vol%) in molten mixture of $K_2CO_3 + Li_2CO_3 + Na_2CO_3$	YSZ/0.3–0.8 mm	LSM	Raw PRB[b] coal Biomass (pine saw dust) ATB[c] tar	1223 1223 1173 1173	110 70 40 80	[192]

a) Partially stabilized zirconia.
b) Powder River Basin.
c) Maya atmospheric tower bottom.

quite stable in both cases [194]. A carbon–air semifuel cell/battery concept was developed in Refs [194–197] employing NiO mesh or porous NiO–YSZ anode layers, YSZ electrolyte, and Pt paste or porous LSM–YSZ cathodes; the cosintered tape-cast cells were immersed in $K_2CO_3 + Li_2CO_3$ (38–62 mol%) melt with dispersed super S carbon or pyrolyzed MDF. Liquid tin anode SOFCs with Sn/SnO_2 melt as an anode, YSZ electrolyte, and LSM cathode were also demonstrated to be suitable for continuous long-term operation on JP-8 fuel, and to have prospects of utilizing coal and other fuels [198–200]. For JP-8 with 1400 ppm sulfur, the cell stack achieved a power density of 120 mW cm^{-2} and a fuel utilization of more than 40% [199]. In addition, such an anode stores energy, enabling the cell to deliver power if the fuel flow is interrupted. It is necessary to note, however, that polarization may induce problems caused by the formation of SnO_2, which is poorly soluble in liquid tin and may partially block the solid electrolyte surface [198, 201, 202]. When molten Bi was used, the anode polarization resistance remains low, less than 0.25 Ω cm^2 at 973 K, until essentially all bismuth metal is oxidized to Bi_2O_3 [202]. The volatilization of bismuth oxide and its high reactivity with other materials may, however, hamper applications of the Bi-based anodes.

6.7
Electrode Materials for High-Temperature Fuel Cells with Proton-Conducting Electrolytes

Proton-conducting (PC) solid electrolytes are of potential interest for various electrochemical devices, including hydrogen sensors and pumps, fuel cells, electrolyzers, and hydrogen separation membranes (see Chapters 12 and 13 of the first volume and Chapters 5 and 10 of this book). The use of high-temperature (873–1073 K) proton conductors as alternative electrolyte materials in the fuel cells is expected to provide important advantages, primarily a simplified water management as steam forms at the cathode side [203]. Enhanced sulfur and coking tolerance was reported for $BaCe_{0.7}Zr_{0.1}Y_{0.2-x}Yb_xO_{3-\delta}$ exhibiting a fast diffusion of both protons and oxygen ion vacancies [204]. Among the main issues limiting practical applications of the high-temperature proton conductors, one may note a nonnegligible electronic conduction in many promising materials and a limited chemical stability under CO_2- and SO_x-containing environments. The latter problem is typical for the materials with high barium content, despite a high proton conductivity (Chapter 7 of the first volume). In this respect, recently discovered proton conductors based on rare earth orthoniobates $RNbO_4$ may be advantageous [203, 205, 206]. Since the electrode performance depends on the transport mechanisms in the solid electrolyte and its other properties [7], the trends in behavior of similar electrode materials may differ in the cases when oxygen anion and proton conductors are used. Systematic studies are hence necessary to examine already known electrode materials in contact with proton-conducting electrolytes. Selected data on the electrode performance are presented in Figure 6.22 and Table 6.10.

6.7 Electrode Materials for High-Temperature Fuel Cells with Proton-Conducting Electrolytes

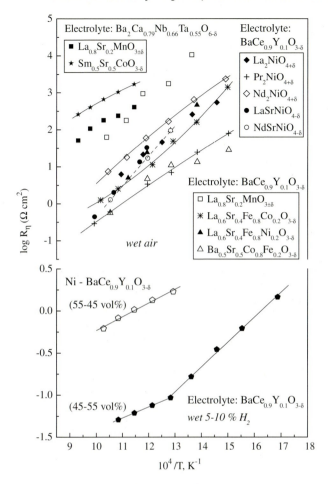

Figure 6.22 Temperature dependences of the cathodic [207, 209] and anodic [211, 212] polarization resistances in the cells with proton-conducting solid electrolytes and Pt mesh [209, 212] or Pt disk [211] current collectors. The gases supplied to the electrodes were humidified under ambient conditions (~3% H_2O), except in cases [207] where ambient air was used.

The double-perovskite $Ba_2Ca_{0.79}Nb_{0.66}Ta_{0.55}O_{6-\delta}$ proton conductor was reported to be stable against chemical reaction with $La_{0.8}Sr_{0.2}MnO_{3-\delta}$ and $Sm_{0.5}Sr_{0.5}CoO_{3-\delta}$ electrodes at 1073–1273 K [207]. The polarization resistance of these and Pt electrodes decreases under wet environments compared to that in air, and is lowest for LSM. The overall performance is much worse than that for the cells based on the oxygen ionic conductors; this fact was explained by the limiting role of protonic transport at the electrolyte and electrode interfaces and/or water effusion through micropores [207]. The cobalt diffusion from electrode into the electrolyte surface may

Table 6.10 Total polarization resistance at 973 K and activation energy for the area-specific electrode conductivity of electrodes in contact with proton-conducting solid electrolytes under open-circuit conditions.

Electrode composition	Current collector	Electrolyte	Atmosphere	$R_\eta^{973\,K}$ ($\Omega\,cm^2$)	T (K)	E_a (kJ mol^{-1})	Reference
Pt	Pt wire	$Ba_2Ca_{0.79}Nb_{0.66}Ta_{0.55}O_{6-\delta}$	Air	78	873–1073	84	[207]
$La_{0.8}Sr_{0.2}MnO_{3\pm\delta}$	Pt wire	$Ba_2Ca_{0.79}Nb_{0.66}Ta_{0.55}O_{6-\delta}$	Air	186.5	873–1073	120	
$La_{0.8}Sr_{0.2}MnO_{3\pm\delta}$	Pt mesh	$BaCe_{0.9}Y_{0.1}O_{3-\delta}$	Wet air	57[a]	730–960	127	[209]
$Sm_{0.5}Sr_{0.5}CoO_{3-\delta}$	Pt wire	$Ba_2Ca_{0.79}Nb_{0.66}Ta_{0.55}O_{6-\delta}$	Air	682.5	873–1073	126	[207]
$La_{0.6}Sr_{0.4}Fe_{0.8}Co_{0.2}O_{3-\delta}$	Pt mesh	$BaCe_{0.9}Y_{0.1}O_{3-\delta}$	Wet air	1[a]	670–980	120	[209]
$La_{0.6}Sr_{0.4}Fe_{0.8}Ni_{0.2}O_{3-\delta}$	Pt mesh	$BaCe_{0.9}Y_{0.1}O_{3-\delta}$	Wet air	1[a]	720–890	142	
$Ba_{0.5}Sr_{0.5}Co_{0.8}Fe_{0.2}O_{3-\delta}$	Pt mesh	$BaCe_{0.9}Y_{0.1}O_{3-\delta}$	Wet air	0.8[a]	670–950	67	
$La_2NiO_{4+\delta}$	Pt mesh	$BaCe_{0.9}Y_{0.1}O_{3-\delta}$	Wet air	3[a]	690–910	104	
$Pr_2NiO_{4+\delta}$	Pt mesh	$BaCe_{0.9}Y_{0.1}O_{3-\delta}$	Wet air	0.4[a]	670–1010	91	
$Nd_2NiO_{4+\delta}$	Pt mesh	$BaCe_{0.9}Y_{0.1}O_{3-\delta}$	Wet air	6[a]	670–960	109	
$LaSrNiO_{4\pm\delta}$	Pt mesh	$BaCe_{0.9}Y_{0.1}O_{3-\delta}$	Wet air	0.9[a]	840–1010	174	
$NdSrNiO_{4\pm\delta}$	Pt mesh	$BaCe_{0.9}Y_{0.1}O_{3-\delta}$	Wet air	0.8[a]	780–950	155	
Pt		$SrCe_{0.95}Yb_{0.05}O_{3-\delta}$	Wet H_2	0.4–1[a]	520–1080	58–84	[216]
Ni–$BaCe_{0.9}Y_{0.1}O_{3-\delta}$ (55–45 vol%)	Pt disk	$BaCe_{0.9}Y_{0.1}O_{3-\delta}$	Wet 5% H_2	0.6	773–973	31	[211]
Ni–$BaCe_{0.9}Y_{0.1}O_{3-\delta}$ (45–55 vol%)	Pt mesh	$BaCe_{0.9}Y_{0.1}O_{3-\delta}$	Wet 10% H_2	0.04[a]	590–780	57	[212]
					780–923	25	

a) Extrapolated or interpolated.

have a positive impact on both electronic and protonic conductivities without stability loss [208]. However, 10 mol% Co doping of $BaCe_{0.5}Zr_{0.4}(Y, Yb)_{0.1-x}Co_xO_{3-\delta}$ lowers its chemical stability in H_2O/H_2 and CO_2. In the series comprising $La_{0.8}Sr_{0.2}MnO_{3\pm\delta}$, $La_{0.6}Sr_{0.4}Fe_{0.8}Co_{0.2}O_{3-\delta}$, $La_{0.6}Sr_{0.4}Fe_{0.8}Ni_{0.2}O_{3-\delta}$, $Ba_{0.5}Sr_{0.5}Co_{0.8}Fe_{0.2}O_{3-\delta}$, $R_2NiO_{4+\delta}$ (R = La, Nd, Pr), and $RSrNiO_{4\pm\delta}$ (R = La, Nd) cathodes in contact with $BaCe_{0.9}Y_{0.1}O_{3-\delta}$ (BCY10) electrolyte under 3% H_2O–air, BSCF and $Pr_2NiO_{4+\delta}$ were identified as most promising despite the lower total conductivity compared to $La_{0.6}Sr_{0.4}Fe_{0.8}M_{0.2}O_{3-\delta}$ and $LaSrNiO_{4\pm\delta}$ [209]. No chemical interaction between these materials and BCY10 was observed at 923 K under humidified ($p(H_2O) = 0.032$ atm) air during 3 weeks [209]. The electrochemical activity of $Sm_{0.5}Sr_{0.5}CoO_{3-\delta}$–$BaCe_{0.8}Sm_{0.2}O_{3-\delta}$, BSCF–$BaCe_{0.8}Sm_{0.2}O_{3-\delta}$, $PrBaCuFeO_{5+\delta}$, $SmBaCuFeO_{5+\delta}$, $SmBaCuCoO_{5+\delta}$, $BaCe_{0.5}Fe_{0.5}O_{3-\delta}$, and $BaCe_{0.5}Bi_{0.5}O_{3-\delta}$ cathodes applied onto the $BaCe_{0.8}Sm_{0.2}O_{3-\delta}$ and $BaCe_{0.7}Zr_{0.1}Y_{0.2}O_{3-\delta}$ electrolytes was also evaluated (see Ref. [210] and references cited therein), although possible influence of Ag paste current collectors should also be examined in this case.

The anode materials for PC-FCs should exhibit mixed protonic and electronic conductivity [203]. As hydrogen dissolves in metallic Ni, nickel-based cermets are often employed for anodes. Typical examples include Ni–BCY10 [211, 212] and Ni–Ba$(Ce,Zr)_{0.8}(Y,M)_{0.2}O_{3-\delta}$ (M = Yb, Zn) [204, 213, 214]. The cermet made of Ni and nanosized BCY10 displays relatively low polarization resistances, around 1 Ω cm^2 at 873 K (Figure 6.22), in contact with BCY10 electrolyte under wet 5% H_2 [211]. Note, however, that this anode material is stable in diluted H_2, while severe degradation occurs after exposure to CO_2 at 973 K. No chemical reaction between $La_{0.995}Ca_{0.005}NbO_{4-\delta}$ and nickel substrate was detected by XPS of samples reduced *in situ* [215]. The electrode morphology issues are discussed in Ref. [216] by the example of Pt layers.

In most cases, the anode-supported PC-SOFC configurations are considered [212–215, 217] as the anodic polarization is typically lower than cathodic, while the moderate levels of the solid electrolyte conductivity make it necessary to use thin electrolyte layers. Typical performances of single PC-SOFCs are compared in Table 6.11. The cells with thin-film $La_{0.995}Sr_{0.005}NbO_{4-\delta}$ electrolyte, cathodes made of $CaTi_{0.9}Fe_{0.1}O_{3-\delta}$, $La_2NiO_{4+\delta}$, and $La_4Ni_3O_{10-\delta}$, and Ni–$La_{0.995}Sr_{0.005}NbO_{4-\delta}$ cermet anodes were produced by a versatile process combining tape casting, spin coating, and screen printing [206]. While formation of La_3NbO_7 and $La(Nb,Ni)O_3$ secondary phases is possible at the $La_2NiO_{4+\delta}/La_{0.995}Sr_{0.005}NbO_{4-\delta}$ interface, the $CaTi_{0.9}Fe_{0.1}O_{3-\delta}$ and $La_4Ni_3O_{10-\delta}$–$La_{0.995}Sr_{0.005}NbO_{4-\delta}$ cathodes showed a good adhesion and no interfacial reaction with the electrolyte. An interesting hybrid system of two different SOFCs, one being based on the oxygen ion conductor and another based on the proton conductor, was proposed in Ref. [218]. Due to an internal reforming of CH_4, the former cell can produce electrical power and exhaust gas containing H_2 and CO, which can be used for proton-conducting cell operation. The remaining CO can further react with H_2O via water–gas shift reaction to produce more H_2 within PC-SOFC, thus reducing carbon deposition and improving the overall system efficiency.

Table 6.11 Performance of various single planar PC-SOFCs under wet H_2.

Anode	Electrolyte/thickness (μm)	Cathode	T (K)	P_{max} (mW cm^{-2})	Reference
Ni–BaCe$_{0.9}$Y$_{0.1}$O$_{3-\delta}$ (45 vol% Ni)	BaCe$_{0.9}$Y$_{0.1}$O$_{3-\delta}$/40	Pr$_2$NiO$_{4+\delta}$	923	130[a]	[212]
Ni–BaCe$_{0.7}$Zr$_{0.1}$(Y,Yb)$_{0.2}$O$_{3-\delta}$ (65 wt% NiO)	BaCe$_{0.7}$Zr$_{0.1}$(Y,Yb)$_{0.2}$O$_{3-\delta}$/10	La$_{0.6}$Sr$_{0.4}$Fe$_{0.8}$Co$_{0.2}$O$_{3-\delta}$ –BaCe$_{0.7}$Zr$_{0.1}$(Y,Yb)$_{0.2}$O$_{3-\delta}$	923	660	[204]
Ni–BaCe$_{0.7}$Zr$_{0.1}$Y$_{0.2}$O$_{3-\delta}$ (60 wt% NiO)	BaCe$_{0.7}$Zr$_{0.1}$Y$_{0.2}$O$_{3-\delta}$/20	PrBa$_{0.5}$Sr$_{0.5}$Co$_2$O$_{5+\delta}$	973	520	[213]
Ni–BaCe$_{0.5}$Zr$_{0.3}$Y$_{0.16}$Zn$_{0.04}$O$_{3-\delta}$ (60 wt% NiO)	BaCe$_{0.5}$Zr$_{0.3}$Y$_{0.16}$Zn$_{0.04}$O$_{3-\delta}$/17	SmBa$_{0.5}$Sr$_{0.5}$Co$_2$O$_{5+\delta}$	973	306	[214]
Ni–Ba$_2$(In$_{0.8}$Ti$_{0.2}$)$_2$O$_{5.2}$ (50 wt% Ni)	Ba$_2$(In$_{0.8}$Ti$_{0.2}$)$_2$O$_{5.2-x}$(OH)$_{2x}$/35	Ba$_{0.5}$Sr$_{0.5}$Co$_{0.8}$Fe$_{0.2}$O$_{3-\delta}$	773	22[b]	[217]

a) Pt mesh current collectors.
b) Au mesh current collectors.

6.8
Electrolyzers, Reactors, and Other Applications Based on Oxygen Ion- and Proton-Conducting Solid Electrolytes

As for SOFCs, perovskite-related materials and Ni-based cermets are widely employed for the electrodes of other electrochemical systems, such as solid oxide electrolysis cells (SOECs). Moreover, the high-temperature steam electrolysis (HTSE) for hydrogen production is often performed by reversing the fuel cell operation regime [219–224]. The overall electrolysis reaction was suggested to be limited by nickel oxidation and steam diffusion in the H_2/H_2O electrode, at least at moderate absolute humidity values (<70 vol%) [219, 225]. As an example, the electrochemical behavior of the cermet electrode composed of NiO–CuO–YSZ–CeO_2 (50–5–40–5 wt%), impregnated with ceria, was studied in contact with YSZ electrode under various environments [226]; near the equilibrium potential, the low-frequency signal in the impedance spectra is ascribed to the gas diffusion limitations of the electrode reaction. Increasing steam partial pressure raises the hydrogen formation rate and, usually, lowers the total electrode polarization. The properties of $(La_{0.75}Sr_{0.25})_{0.95}Cr_{0.5}Mn_{0.5}O_{3-\delta}$ and its composites with YSZ or CGO, discussed above, made it possible to suggest these materials as alternative cathode compositions for YSZ-based SOECs operated under low hydrogen concentrations [227].

Appraising the $LaGaO_3$-based electrolytes for the steam electrolysis applications [230–232] showed, in particular, relatively low cathodic overpotentials of metallic Ni. At 873 K, iron additions enable to increase H_2 formation rate, which is highest at the Ni to Fe weight ratio of 9 : 1 and reaches 180 $\mu mol\, cm^{-2}\, min^{-1}$ for the electrolyte thickness of 0.2 mm [230]. Perovskite-type $Sr_2Fe_{1.5}Mo_{0.5}O_{6-\delta}$ was used for the symmetrical SOEC with LSGM(10 20) solid electrolyte and displayed good long-term stability and a better performance compared to LSM–YSZ|YSZ|Ni–YSZ [232]. Under open-circuit conditions and 60 vol% H_2O in H_2, the polarization resistance of the cell with Pt and Au paste current collectors was as low as $0.26\,\Omega\,cm^2$ at 1173 K.

$La_{0.6}Sr_{0.4}Fe_{0.8}Co_{0.2}O_{3-\delta}$ and $La_{0.8}Sr_{0.2}MnO_{3-\delta}$–YSZ (50–50 vol%) with optimized microstructures were tested as reversible electrodes for oxygen reduction in the fuel cell mode and oxygen evolution in the electrolysis mode in $Zr_{0.81}Sc_{0.18}Ce_{0.01}O_{2-\delta}$-based cells with Ni–8YSZ electrodes under 70% H_2O–6% H_2–N_2 [221]. At 1123 K, the cell with LSFC electrode became apparently reversible with the total polarization resistance of $0.73\,\Omega\,cm^2$, whereas the activation polarization is significant at lower temperatures [221]. In general, the performance of LSFC and BSCF anodes in the electrolysis cells is higher than that of lanthanum–strontium manganites or LSM–YSZ composites [221, 224, 225] (see Figure 6.23 and Table 6.12). As a typical example, at the applied voltage of 1.4 V, the hydrogen production rate in BSCF|YSZ|Ni–YSZ cell is about three times higher than that of LSM|YSZ|Ni–YSZ [225]. BSCF undergoes, however, very fast degradation under the operating conditions [224]. A 3 mol% yttria-stabilized zirconia (3YSZ) electrolyte-supported cell with Ni–CGO and $Nd_2NiO_{4+\delta}$ electrodes was operated reversibly in SOFC/SOEC modes and displayed much higher current densities than those of similar cells with LSM oxygen electrode [222].

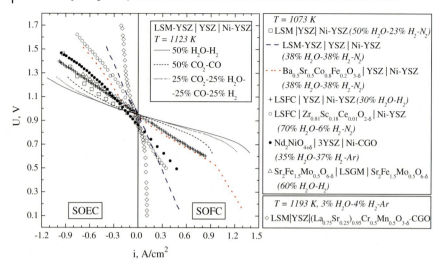

Figure 6.23 Current–voltage curves of various Ni–YSZ electrode-supported [219, 223, 224], electrolyte-supported [221, 222, 227, 232], and porous ferritic steel-supported [228] SOFC/SOEC cells. LSM and LSFC are $La_{0.8}Sr_{0.2}MnO_{3\pm\delta}$ and $La_{0.6}Sr_{0.4}Fe_{0.8}Co_{0.2}O_{3-\delta}$ or unspecified; BSCF is $Ba_{0.5}Sr_{0.5}Co_{0.8}Fe_{0.2}O_{3-\delta}$. The indicated gas mixtures were supplied to the SOEC cathode/SOFC anode, while the oxygen electrodes were operated under air or O_2. Pt mesh [219, 221, 222, 228], Pt wires and paste [224], Au paste and mesh [227], and Pt and Au pastes [232] were used as current collectors.

The polarization resistances in the latter case were slightly lower for the steam electrolysis processes [222]; an opposite behavior was reported for the cells with LSM, LSFC, and BSCF oxygen electrodes [221, 224, 225, 228]. A microtubular SOEC consisting of $La_{0.6}Sr_{0.4}Fe_{0.8}Co_{0.2}O_{3-\delta}$–CGO20 air electrode, CGO20 buffer layer, $Zr_{0.81}Sc_{18}Ce_{0.01}O_{2-\delta}$ electrolyte, and NiO–$Zr_{0.81}Sc_{18}Ce_{0.01}O_{2-\delta}$ supporting steam electrode was also produced and tested [229].

The metal-supported SOECs, consisting of a porous steel support, $La_{0.6}Sr_{0.2}Ca_{0.2}CrO_{3-\delta}$ diffusion barrier layer, Ni–YSZ hydrogen electrode, YSZ electrolyte, and LSFC oxygen electrode, were developed for HTSE [228]. Long-term testing over 2000 h with a steam content of 43% at 1073 K and a current density of $-0.3\,A\,cm^{-2}$ revealed degradation rates of 2.1–3.9% per 1000 h, increasing with time. The polarization resistance tends to increase faster during the electrolysis operation compared to the fuel cell regime [228]. The degradation in the SOEC performance was proved to be essentially caused by impurities absorbed at the cathode; removing these impurities from the inlet gases enables to improve the long-term SOEC durability [223, 233]. The SOECs for CO_2 electrolysis or coelectrolysis of CO_2 and H_2O could be used to recycle CO_2 into sustainable hydrocarbon fuels [223, 233]. The mechanisms of electrode reactions and degradation in such cells were studied in detail by transforming the impedance data to the distribution of relaxation times [223]. The degradation of perovskite-type oxygen (positive) electrodes is related

Table 6.12 Hydrogen production by single SOECs and electrolysis currents at a voltage of 1.3 V.

Anode	Electrolyte/thickness (μm)	Cathode	T (K)	Cathode environment	i (A cm^{-2})	H$_2$ production (ml cm^{-2} h^{-1})	Reference
LSM	YSZ	Ni–YSZ	1073	70% H$_2$O–14% H$_2$–N$_2$	1.06		[219]
LSM–YSZ	YSZ/30	Ni–YSZ (60 wt% NiO)	1123	54% H$_2$O–23% H$_2$–N$_2$	0.46	100	[224]
BSCF, CGO10 interlayer	YSZ/30	Ni–YSZ (60 wt% NiO)	1123	54% H$_2$O–23% H$_2$–N$_2$	0.61	135	
LSFC–CGO20, CGO20 interlayer	Zr$_{0.81}$Sc$_{18}$Ce$_{0.01}$O$_{2-\delta}$/10	Ni–Zr$_{0.81}$Sc$_{18}$Ce$_{0.01}$O$_{2-\delta}$	923	60% H$_2$O–4% H$_2$–Ar	0.57		[229]
LSFC	Zr$_{0.81}$Sc$_{18}$Ce$_{0.01}$O$_{2-\delta}$/155	Ni–YSZ	1073	70% H$_2$O–6% H$_2$–N$_2$	0.35		[221]
			1123		0.68		
LSFC	YSZ/40	Ni–YSZ (50 wt% NiO)	1073	30% H$_2$O–H$_2$	0.77		[228]
Nd$_2$NiO$_{4+\delta}$	3YSZ/90	Ni–CGO	1073	37% H$_2$O–35% H$_2$–Ar	0.64		[222]
			1123		0.87		
La$_{0.4}$Ba$_{0.6}$CoO$_{3-\delta}$	LSGM(10 20)/500	Ni–Fe (90–10 wt%), Ni–CSO20 (10–90 wt%) interlayer	1073	25% H$_2$O–N$_2$	0.31		[231]
Sr$_2$Fe$_{1.5}$Mo$_{0.5}$O$_{6-\delta}$	LSGM(10 20)	Sr$_2$Fe$_{1.5}$Mo$_{0.5}$O$_{6-\delta}$	1173	60% H$_2$O–H$_2$	0.85	380	[232]
LSM	YSZ/250	(La$_{0.75}$Sr$_{0.25}$)$_{0.95}$Cr$_{0.5}$Mn$_{0.5}$O$_{3-\delta}$–CGO	1193	3% H$_2$O–4% H$_2$–Ar	0.15		[227]

LSM: La$_{0.8}$Sr$_{0.2}$MnO$_{3\pm\delta}$ or unknown; LSFC: La$_{0.6}$Sr$_{0.4}$Fe$_{0.8}$Co$_{0.2}$O$_{3-\delta}$ or unknown; BCSF: Ba$_{0.5}$Sr$_{0.5}$Co$_{0.8}$Fe$_{0.2}$O$_{3-\delta}$.

to Cr migration from the other stack components and electrode delamination [234]; the latter mechanism was investigated in Ref. [235].

Numerous efforts were made to experimentally assess and to model various SOEC-based systems (see Refs [236–238] and references cited therein). The SOEC performing at elevated pressures, necessary for the catalyst operation, was proposed to simplify the system [236]. A 16 cm^2 planar Ni–YSZ|YSZ (10 μm)|LSM–YSZ cell was tested at 0.4–10 atm. The SOFC performance increased with total pressure by more than 50%, achieving the power densities of ∼955 mW cm^{-2} at 1023 K using 80% H_2–20% H_2O as fuel; the pressure effect on the SOEC performance was found weaker, despite a pronounced decrease in polarization resistances [236]. The spontaneous electrochemical production of hydrogen and cogeneration of electrical power was achieved in the (H_2O, H_2) Pt|YSZ|Pt (C, CO, CO_2) cells with carbon bed loaded onto the anode [238].

The SOFC-based electrochemical reactors can also be used for syngas production from partial oxidation of methane and other hydrocarbon fuels [178, 180]. For example, the anode-supported $La_{0.8}Sr_{0.2}MnO_{3-\delta}$|8YSZ (12 μm)|Ni–YSZ (50–50 wt %) cells were operated using methane as the fuel and oxygen ion fluxes through the electrolyte membrane close to the molar CH_4 flow rates, when the CH_4/O ratio corresponds to the electrochemical partial oxidation [178]. An 88% CH_4 conversion to syngas and a power density of 936 mW cm^{-2} were initially achieved. Although the conversion decreased during the first 30–40 h of operation due to Ni nanoparticle sintering and resultant lowering of the anode reforming activity, this can be partly suppressed by the incorporation of Rh/Al_2O_3 catalyst [178]. The SOFC-based technologies are also considered feasible for NO removal from industrial flue gases in the stationary sources and electrical power cogeneration [239]. For this goal, the cells were made of Ni–CGO10 (37.5–62.5 wt%) anode, YSZ electrolyte, and $La_{0.58}Sr_{0.4}Fe_{0.8}Co_{0.2}O_{3-\delta}$–CGO10 (67–33 wt%) composite cathodes with small (1.34 wt%) additions of V_2O_5, used in commercial catalysts for selective catalytic reduction processes, or Cu effective for direct electrochemical NO reduction. The electrochemical NO reduction occurred over the cathode with or without the presence of oxygen; P_{max} was greater for Cu catalyst and increased with O_2 concentration [239]. A proton-conducting cell with a thick porous BCY15 anode substrate impregnated with Pt catalyst for dehydrogenation of ethane to ethylene, dense 30 μm thick BCY15 film, and Pt cathode displayed a high (90.5%) selectivity for ethylene at 36.7% ethane conversion, both increasing with current density, and 216 mW cm^{-2} power density at 973 K [240]. The possibility to use Ni–CSO20|8YSZ|$La_{0.8}Sr_{0.2}MnO_{3\pm\delta}$ SC-SOFCs with gas mixtures containing ppm level amounts of CH_4, C_2H_6, C_3H_8, C_4H_{10}, and O_2 as the power generators for exhaust energy recovery was demonstrated in Ref. [141]. Among other applications, one may note catalytic methane sensors [145] based on the SC-SOFC design with high-temperature proton-conducting $SrCe_{0.95}Yb_{0.05}O_{3-\delta}$ or $CaZr_{0.9}In_{0.1}O_{3-\delta}$ and electrodes having different catalytic activities for dry reforming of CH_4. The readers, interested in various applications of oxygen ion- and proton-conducting materials in sensors and relevant electrode materials, may refer to Chapter 13 of the first volume.

6.9
Concluding Remarks

In addition to the electrode and electrolyte composition-related factors, the electrochemical performance of both positive and negative electrodes in solid-state devices is strongly affected by the microstructure and chemical interaction with other cell components, both depending in turn on the material processing conditions and prehistory. As an illustration, the inset in Figure 6.19 and Figure 6.24 display the influence of electrode sintering temperature on the anodic and cathodic polarization. The interdiffusion processes may be suppressed by using protective interlayers or composite and graded electrodes; the latter concepts also enable to minimize thermal expansion mismatch between the cell components. The electrocatalytic activity can be further improved employing infiltration, surface modification, and catalytically active surface additives; very often, the incorporation of nanostructured components in the electrode layers is favorable. In this chapter, it was impossible to address all the fundamental aspects and technological approaches relevant for the electrode developments (see Refs [1–9] and references cited therein). Priority has been given, therefore, to the latest results; information on the well-established materials and technologies is presented in Chapter 5. One should also note that due to an important role of the microstructure- and processing-related factors, literature data on the electrode performance are often contradictory. Some discrepancies can even be found in the information on novel electrode materials reviewed above. As selection of references for this review focused on the past years, further examination and validation of preliminary results will, of course, be performed by numerous research groups working in the field of oxygen ion- and proton-conducting materials, fuel cells, and high-temperature electrolysis of gases.

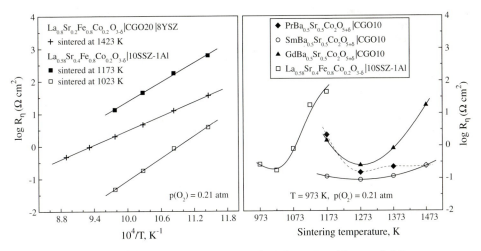

Figure 6.24 Comparison of the polarization resistances of cobaltite [28] and ferrite–cobaltite cathodes [243, 244] sintered on the solid electrolyte surface at various temperatures. 10SSZ-1Al corresponds to the solid electrolyte composition 89 mol% ZrO_2–10 mol% Sc_2O_3–1 mol% Al_2O_3.

References

1 Kharton, V.V. (ed.) (2009) *Solid State Electrochemistry. I. Fundamentals, Materials and their Applications*, Wiley-VCH Verlag GmbH, Weinheim.
2 Srinivasan, S. (2006) *Fuel Cells: From Fundamentals to Applications*, Springer Science + Business Media, New York.
3 Kinoshita, K. (1992) *Electrochemical Oxygen Technology*, Wiley-Interscience, New York.
4 Minh, N.Q. and Takahashi, T. (1995) *Science and Technology of Ceramic Fuel Cells*, Elsevier, Amsterdam.
5 Menzler, N.H., Tietz, F., Uhlenbruck, S., Buchkremer, H.P., and Stöver, D. (2010) *J. Mater. Sci.*, **45**, 3109–3135.
6 Steele, B.C.H. (2001) *J. Mater. Sci.*, **36**, 1053–1068.
7 Tsipis, E.V. and Kharton, V.V. (2008) *J. Solid State Electrochem.*, **12**, 1039–1060.
8 Tsipis, E.V. and Kharton, V.V. (2008) *J. Solid State Electrochem.*, **12**, 1367–1391.
9 Tarancón, A., Burriel, M., Santiso, J., Skinner, S.J., and Kilner, J.A. (2010) *J. Mater. Chem.*, **20**, 3799–3813.
10 Gong, M., Liu, X., Trembly, J., and Johnson, C. (2007) *J. Power Sources*, **168**, 289–298.
11 Baek, S.-W., Bae, J., and Yoo, Y.-S. (2009) *J. Power Sources*, **193**, 431–440.
12 Zhang, G., Dong, X., Liu, Z., Zhou, W., Shao, Z., and Jin, W. (2010) *J. Power Sources*, **195**, 3386–3393.
13 Torres-Garibay, C., Kovar, D., and Manthiram, A. (2009) *J. Power Sources*, **187**, 480–486.
14 Aguadero, A., Pérez-Coll, D., de la Calle, C., Alonso, J.A., Escudero, M.J., and Daza, L. (2009) *J. Power Sources*, **192**, 132–137.
15 Wang, F., Zhou, Q., He, T., Li, G., and Ding, H. (2010) *J. Power Sources*, **195**, 3772–3778.
16 Xia, T., Lin, N., Zhao, H., Huo, L., Wang, J., and Grenier, J.-C. (2009) *J. Power Sources*, **192**, 291–296.
17 Shao, Z.P. and Haile, S.M. (2004) *Nature*, **431**, 170–173.
18 Chen, C.-H., Chang, C.-L., and Hwang, B.-H. (2009) *Mater. Chem. Phys.*, **115**, 478–482.
19 Zuev, A.Yu., Vylkov, A.I., Petrov, A.N., and Tsvetkov, D.S. (2008) *Solid State Ionics*, **179**, 1876–1879.
20 Peña-Martínez, J., Marrero-López, D., Pérez-Coll, D., Ruiz-Morales, J.C., and Núñez, P. (2007) *Electrochim. Acta*, **52**, 2950–2958.
21 Yan, A., Liu, B., Dong, Y., Tian, Z., Wang, D., and Cheng, M. (2008) *Appl. Catal. B*, **80**, 24–31.
22 Arnold, M., Gesing, T.M., Martynczuk, J., and Feldhoff, A. (2008) *Chem. Mater.*, **20**, 5851–5858.
23 Švarcová, S., Wiik, K., Tolchard, J., Bouwmeester, H.J.M., and Grande, T. (2008) *Solid State Ionics*, **178**, 1787–1791.
24 Ovenstone, J., Jung, J.-I., White, J.S., Edwards, D.D., and Misture, S.T. (2008) *J. Solid State Chem.*, **181**, 576–586.
25 Deng, Z.Q., Smit, J.P., Niu, H.J., Evans, G., Li, M.R., Xu, Z.L., Claridge, J.B., and Rosseinsky, M.J. (2009) *Chem. Mater.*, **21**, 5154–5162.
26 Chen, D., Ran, R., Zhang, K., Wang, J., and Shao, Z. (2009) *J. Power Sources*, **188**, 96–105.
27 Tarancón, A., Skinner, S.J., Chater, R.J., Hernández-Ramírez, F., and Kilner, J.A. (2007) *J. Mater. Chem.*, **17**, 3175–3181.
28 Kim, J.H., Cassidy, M., Irvine, J.T.S., and Bae, J. (2009) *J. Electrochem. Soc.*, **156**, B682–B689.
29 Jin, C. and Liu, J. (2009) *J. Alloys Compd.*, **474**, 573–577.
30 Sheptyakov, D.V., Podlesnyak, A., Barilo, S.N., Shiryaev, S.V., Khalyavin, D.D., Chernyshov, D.Yu., and Leonyuk, N.I. (2001) *PSI Sci. Rep.*, **3**, 64.
31 Chmaissem, O., Zheng, H., Huq, A., Stephens, P.W., and Mitchell, J.F. (2008) *J. Solid State Chem.*, **181**, 664–672.
32 Karppinen, M., Yamauchi, H., Fjellvåg, H., and Motohashi, T. (2008) European Patent EP1900706.
33 Tsipis, E.V., Kharton, V.V., and Frade, J.R. (2006) *Solid State Ionics*, **177**, 1823–1826.
34 Tsipis, E.V., Kharton, V.V., Frade, J.R., and Núñez, P. (2005) *J. Solid State Electrochem.*, **9**, 547–557.

35 Tsipis, E.V., Waerenborgh, J.C., Avdeev, M., and Kharton, V.V. (2009) *J. Solid State Chem.*, **182**, 640–643.

36 Vert, V.B., Serra, J.M., and Jordá, J.L. (2010) *Electrochem. Commun.*, **12**, 278–281.

37 Kim, J.-H. and Manthiram, A. (2010) *Chem. Mater.*, **22**, 822–831; Kim, J.-H., Kim, Y.N., Cho, S.M., Wang, H., and Manthiram, A. (2010) *Electrochim. Acta*, **55**, 5312–5317.

38 Nagasawa, K., Daviero-Minaud, S., Preux, N., Rolle, A., Roussel, P., Nakatsugawa, H., and Mentré, O. (2009) *Chem. Mater.*, **21**, 4738–4745.

39 Bahteeva, J.A., Leonidov, I.A., Patrakeev, M.V., Mitberg, E.B., Kozhevnikov, V.L., and Poeppelmeier, K.R. (2004) *J. Solid State Electrochem.*, **8**, 578.

40 Kharton, V.V., Kovalevsky, A.V., Patrakeev, M.V., Tsipis, E.V., Viskup, A.P., Kolotygin, V.A., Yaremchenko, A.A., Shaula, A.L., Kiselev, E.A., and Waerenborgh, J.C. (2008) *Chem. Mater.*, **20**, 6457–6467.

41 Vert, V.B. and Serra, J.M. (2009) *Fuel Cells*, **9**, 663–678.

42 Vidal, K., Rodríguez-Martínez, L.M., Ortega-San-Martin, L., Martínez-Amesti, A., Nó, M.L., Rojo, T., Laresgoiti, A., and Arriortua, M.I. (2009) *ECS Trans.*, **25**, 2427–2434.

43 Vidal, K., Rodríguez-Martínez, L.M., Ortega-San-Martin, L., Martínez-Amesti, A., Nó, M.L., Rojo, T., Laresgoiti, A., and Arriortua, M.I. (2009) *J. Power Sources*, **192**, 175–179.

44 Bidrawn, F., Lee, S., Vohs, J.M., and Gorte, R.J. (2008) *J. Electrochem. Soc.*, **155**, B660–B665.

45 Hou, S., Alonso, J.A., and Goodenough, J.B. (2010) *J. Power Sources*, **195**, 280–284.

46 Jung, W. and Tuller, H.L. (2009) *Solid State Ionics*, **180**, 843–847.

47 Huang, J., Jiang, X., Li, X., and Liu, A. (2009) *J. Electroceram.*, **23**, 67–71.

48 Hjalmarsson, P., Søgaard, M., and Mogensen, M. (2009) *Solid State Ionics*, **180**, 1395–1405.

49 Tsipis, E.V., Kiselev, E.A., Kolotygin, V.A., Waerenborgh, J.C., Cherepanov, V.A., and Kharton, V.V. (2008) *Solid State Ionics*, **179**, 2170–2180.

50 Duval, S.B.C., Graule, T., Holtappels, P., Ouweltjes, J.P., and Rietveld, G. (2009) *Fuel Cells*, **9**, 911–914.

51 Simner, S.P., Bonnett, J.F., Canfield, N.L., Meinhardt, K.D., Sprenkle, V.L., and Stevenson, J.W. (2002) *Electrochem. Solid State Lett.*, **5**, A173–A175.

52 Vogt, U.F., Sfeir, J., Richter, J., Soltmann, C., and Holtappels, P. (2008) *Pure Appl. Chem.*, **80**, 2543–2552.

53 Vogt, U.F., Holtappels, P., Sfeir, J., Richter, J., Duval, S., Wiedenmann, D., and Züttel, A. (2009) *Fuel Cells*, **9**, 899–906.

54 Kammer Hansen, K. (2010) *Mater. Res. Bull.*, **45**, 197–199.

55 Amow, G. and Skinner, S.J. (2006) *J. Solid State Electrochem.*, **10**, 538–546.

56 Amow, G., Davidson, I.J., and Skinner, S.J. (2006) *Solid State Ionics*, **177**, 1205–1210.

57 Kovalevsky, A.V., Kharton, V.V., Yaremchenko, A.A., Pivak, Y.V., Tsipis, E.V., Yakovlev, S.O., Markov, A.A., Naumovich, E.N., and Frade, J.R. (2007) *J. Electroceram.*, **18**, 205–218.

58 Mauvy, F., Lalanne, C., Bassat, J.-M., Grenier, J.-C., Zhao, H., Huo, L., and Stevens, P. (2006) *J. Electrochem. Soc.*, **153**, A1547–A1553.

59 Pérez-Coll, D., Aguadero, A., Escudero, M.J., and Daza, L. (2009) *J. Power Sources*, **192**, 2–13.

60 Aguadero, A., Alonso, J.A., Escudero, M.J., and Daza, L. (2008) *Solid State Ionics*, **179**, 393–400.

61 Kharton, V.V., Tsipis, E.V., Yaremchenko, A.A., and Frade, J.R. (2004) *Solid State Ionics*, **166**, 327–337.

62 Kammer, K. (2009) *Ionics*, **15**, 325–328.

63 Nakayama, S., Kageyama, T., Aono, H., and Sadaoka, Y. (1995) *J. Mater. Chem.*, **5**, 1801.

64 Kharton, V.V., Marques, F.M.B., and Atkinson, A. (2004) *Solid State Ionics*, **174**, 135.

65 Tsipis, E.V., Kharton, V.V., and Frade, J.R. (2007) *Electrochim. Acta*, **52**, 4428–4435.

66 Yaremchenko, A.A., Kharton, V.V., Bannikov, D.O., Znosak, D.V., Frade, J.R., and Cherepanov, V.A. (2009) Solid State Ionics, 180, 878–885.

67 Bonhomme, C., Beaudet-Savignat, S., Chartier, T., Geffroy, P.-M., and Sauvet, A.-L. (2009) J. Eur. Ceram. Soc., 29, 1781–1788.

68 Marrero-López, D., Martín-Sedeño, M.C., Peña-Martínez, J., Ruiz-Morales, J.C., Núñez, P., Aranda, M.A.G., and Ramos-Barrado, J.R. (2010) J. Power Sources, 195, 2496–2506.

69 Jaiswal, A. and Wachsman, E.D. (2009) Ionics, 15, 1–9.

70 Sun, C. and Stimming, U. (2007) J. Power Sources, 171, 247–260.

71 Tietz, F. and Nikolopoulos, P. (2009) Fuel Cells, 9, 867–872.

72 Gorte, R.J. and Vohs, J.M. (2009) Curr. Opin. Colloid Interface Sci., 14, 236–244.

73 Kanjanaboonmalert, T., Tzu, T.W., and Sato, K. (2009) ECS Trans., 16, 23–29.

74 Oishi, N. and Yoo, Y. (2010) Mater. Lett., 64, 876–878.

75 Tsipis, E.V., Kharton, V.V., and Frade, J.R. (2005) J. Eur. Ceram. Soc., 25, 2623–2626.

76 Ouweltjes, J.P., van Tuel, M., Sillessen, M., and Rietveld, G. (2009) Fuel Cells 09, 6, 873–882.

77 Rossmeisl, J. and Bessler, W.G. (2008) Solid State Ionics, 178, 1694–1700.

78 Gross, M.D., Vohs, J.M., and Gorte, R.J. (2007) J. Mater. Chem., 17, 3071–3077.

79 Kiratzis, N.E., Connor, P., and Irvine, J.T.S. (2010) J. Electroceram., 24, 270–287.

80 Nikolla, E., Schwank, J., and Linic, S. (2009) J. Am. Chem. Soc., 131, 2747–2754.

81 Hornés, A., Gamarra, D., Munuera, G., Conesa, J.C., and Martínez-Arias, A. (2007) J. Power Sources, 169, 9–16.

82 Grgicak, C.M., Pakulska, M.M., O'Brien, J.S., and Giorgi, J.B. (2008) J. Power Sources, 183, 26–33.

83 Ju, Y.-W., Eto, H., Inagaki, T., Ida, S., and Ishihara, T. (2010) J. Power Sources, 195, 6294–6300.

84 Park, H.C. and Virkar, A.V. (2009) J. Power Sources, 186, 133–137.

85 Sakai, T., Zhong, H., Eto, H., and Ishihara, T. (2010) J. Solid State Electrochem., 14, 1777–1780.

86 Lu, X.C. and Zhu, J.H. (2007) J. Power Sources, 165, 678–684.

87 Lu, Z.G., Zhu, J.H., Bi, Z.H., and Lu, X.C. (2008) J. Power Sources, 180, 172–175.

88 Lv, H., Yang, D.J., Pan, X.M., Zheng, J.S., Zhang, C.M., Zhou, W., Ma, J.X., and Hu, K.-A. (2009) Mater. Res. Bull., 44, 1244–1248.

89 Wang, S., Hea, Q., and Liu, M. (2009) Electrochim. Acta, 54, 3872–3876.

90 Savaniu, C.-D. and Irvine, J.T.S. (2009) J. Mater. Chem., 19, 8119–8128.

91 Kurokawa, H., Yang, L., Jacobson, C.P., De Jonghe, L.C., and Visco, S.J. (2007) J. Power Sources, 164, 510–518.

92 Ruiz-Morales, J.C., Canalez-Vazquez, J., Marrero-López, D., Irvine, J.T.S., and Núñez, P. (2007) Electrochim. Acta, 52, 7217–7225.

93 Blennow, P., Hansen, K.K., Wallenberg, L.R., and Mogensen, M. (2009) Solid State Ionics, 180, 63–70.

94 Fu, Q.X., Tietz, F., and Stöver, D. (2006) J. Electrochem. Soc., 153, D74–D83.

95 Tao, S. and Irvine, J.T.S. (2004) J. Electrochem. Soc., 151, A252–A259.

96 Zhu, X., Lü, Z., Wei, B., Chen, K., Liu, M., Huang, X., and Su, W. (2009) J. Power Sources, 190, 326–330.

97 Liu, J., Madsen, B., Ji, Z., and Barnett, S. (2002) Electrochem. Solid State Lett., 5, A122–A124.

98 Zha, S., Tsang, P., Cheng, Z., and Liu, M. (2005) J. Solid State Chem., 178, 1844–1850.

99 Tao, S., Irvine, J.T.S., and Kilner, J.A. (2005) Adv. Mater., 17, 1734–1737.

100 Peña-Martínez, J., Marrero-López, D., Ruiz-Morales, J.C., Savaniu, C., Núñez, P., and Irvine, J.T.S. (2006) Chem. Mater., 18, 1001–1006.

101 Jiang, S.P., Chen, X.J., Chan, S.H., Kwok, J.T., and Khor, K.A. (2006) Solid State Ionics, 177, 149–157.

102 Wan, J., Zhu, J.H., and Goodenough, J.B. (2006) Solid State Ionics, 177, 1211–1217.

103 Chen, X.J., Liu, Q.L., Chan, S.H., Brandon, N.P., and Khor, K.A. (2007) J. Electrochem. Soc., 154, B1206–1210.

104 Kharton, V.V., Tsipis, E.V., Marozau, I.P., Viskup, A.P., Frade, J.R., and Irvine, J.T.S. (2007) Solid State Ionics, 178, 101–113.

105 Jiang, S.P., Ye, Y., He, T., and Ho, S.B. (2008) *J. Power Sources*, **185**, 179–182.
106 Chen, X.J., Liu, Q.L., Khor, K.A., and Chan, S.H. (2007) *J. Power Sources*, **165**, 34–40.
107 Kim, G., Lee, S., Shin, J.Y., Corre, G., Irvine, J.T.S., Vohs, J.M., and Gorte, R.J. (2009) *Electrochem. Solid State Lett.*, **12**, B48–B52.
108 van den Bossche, M., Matthews, R., Lichtenberger, A., and McIntosh, S. (2010) *J. Electrochem. Soc.*, **157**, B392–B399.
109 Lu, X.C. and Zhu, J.H. (2007) *Solid State Ionics*, **178**, 1467–1475.
110 van den Bossche, M. and McIntosh, S. (2008) *J. Catal.*, **255**, 313–323.
111 El-Himri, A., Marrero-López, D., Ruiz-Morales, J.C., Peña-Martínez, J., and Núñez, P. (2009) *J. Power Sources*, **188**, 230–237.
112 Bruce, M.K., van den Bossche, M., and McIntosh, S. (2008) *J. Electrochem. Soc.*, **155**, B1202–B1209.
113 Zhang, L., Chen, X., Jiang, S.P., He, H.Q., and Xiang, Y. (2009) *Solid State Ionics*, **180**, 1076–1082.
114 Deleebeeck, L., Fournier, J.L., and Birss, V. (2010) *Solid State Ionics*, **181**, 1229–1237.
115 Lay, E., Gauthier, G., Rosini, S., Savaniu, C., and Irvine, J.T.S. (2008) *Solid State Ionics*, **179**, 1562–1566.
116 Kobsiriphat, W., Madsen, B.D., Wang, Y., Marks, L.D., and Barnett, S.A. (2009) *Solid State Ionics*, **180**, 257–264.
117 Tao, S. and Irvine, J.T.S. (2004) *Chem. Mater.*, **16**, 4116–4121.
118 Haag, J.M., Madsen, B.D., Barnett, S.A., and Poeppelmeier, K.R. (2008) *Electrochem. Solid State Lett.*, **11**, B51–B53.
119 Zheng, Y., Zhang, C., Ran, R., Cai, R., Shao, Z., and Farrusseng, D. (2009) *Acta Mater.*, **57**, 1165–1175.
120 Huang, Y.-H., Dass, R.I., Xing, Z.-L., and Goodenough, J.B. (2006) *Science*, **312**, 254–257; Huang, Y.-H., Dass, R.I., Denyszyn, J.C., and Goodenough, J.B. (2006) *J. Electrochem. Soc.*, **153**, A1266–A1272.
121 Ji, Y., Huang, Y.-H., Ying, J.-R., and Goodenough, J.B. (2007) *Electrochem. Commun.*, **9**, 1881–1885.
122 Marrero-López, D., Peña-Martínez, J., Ruiz-Morales, J.C., Gabás, M., Núñez, P., Aranda, M.A.G., and Ramos-Barrado, J.R. (2010) *Solid State Ionics*, **180**, 1672–1682.
123 Bernuy-Lopez, C., Allix, M., Bridges, C.A., Claridge, J.B., and Rosseinsky, M.J. (2007) *Chem. Mater.*, **19**, 1035–1043.
124 Aguadero, A., de la Calle, C., Alonso, J.A., Pérez-Coll, D., Escudero, M.J., and Daza, L. (2009) *J. Power Sources*, **192**, 78–83.
125 Ge, X.M. and Chan, S.H. (2009) *J. Electrochem. Soc.*, **156**, B386–B391.
126 Yano, M., Tomita, A., Sano, M., and Hibino, T. (2007) *Solid State Ionics*, **177**, 3351–3359.
127 Kuhn, M. and Napporn, T.W. (2010) *Energies*, **3**, 57–134.
128 Riess, I. (2008) *J. Power Sources*, **175**, 325–337.
129 Ruiz de Larramendi, I., Lamas, D.G., Cabezas, M.D., Ruiz de Larramendi, J.I., Walsöe de Reca, N.E., and Rojo, T. (2009) *J. Power Sources*, **193**, 774–778.
130 Morel, B., Roberge, R., Savoie, S., Napporn, T.W., and Meunier, M. (2009) *J. Power Sources*, **186**, 89–95.
131 Shao, Z., Haile, S.M., Ahn, J., Ronney, P.D., Zhan, Z., and Barnett, S.A. (2005) *Nature*, **435**, 795–798.
132 Tomita, A., Hirabayashi, D., Hibino, T., Nagao, N., and Sano, M. (2005) *Electrochem. Solid State Lett.*, **8**, A63–A65.
133 Morel, B., Roberge, R., Savoie, S., Napporn, T.W., and Meunier, M. (2007) *Appl. Catal. A*, **323**, 181–187.
134 Hibino, T., Hashimoto, A., Inoue, T., Tokuno, J., Yoshida, S., and Sano, M. (2000) *J. Electrochem. Soc.*, **147**, 2888–2892.
135 Fisher, J.C., II and Chuang, S.S.C. (2009) *Catal. Commun.*, **10**, 772–776.
136 Shao, Z., Mederos, J., Chueh, W.C., and Haile, S.M. (2006) *J. Power Sources*, **162**, 589–596.
137 Kellogg, I.D., Koylu, U.O., Petrovsky, V., and Dogan, F. (2009) *Int. J. Hydrogen Energy*, **34**, 5138–5143.
138 Buergler, B.E., Siegrist, M.E., and Gauckler, L.J. (2005) *Solid State Ionics*, **176**, 1717–1722.

139 Cabezas, M.D., Lamas, D.G., Bellino, M.G., Fuentes, R.O., Walsöe de Reca, N.E., and Larrondo, S.A. (2009) *Electrochem. Solid State Lett.*, **12**, B34–B37.

140 Asahara, S., Michiba, D., Hibino, M., and Yao, T. (2005) *Electrochem. Solid State Lett.*, **8**, A449–A451.

141 Nagao, M., Yano, M., Okamoto, K., Tomita, A., Uchiyama, Y., Uchiyama, N., and Hibino, T. (2008) *Fuel Cells*, **08**, 322–329.

142 Zhang, C., Zheng, Y., Ran, R., Shao, Z., Jin, W., Xua, N., and Ahn, J. (2008) *J. Power Sources*, **179**, 640–648.

143 Asano, K., Hibino, T., and Iwahara, H. (1995) *J. Electrochem. Soc.*, **142**, 3241.

144 Hibino, T., Asano, K., and Iwahara, H. (1994) *Chem. Lett.*, **3**, 485.

145 van Rij, L.N., Le, J., van Landschoot, R.C., and Schoonman, J. (2001) *J. Mater. Sci.*, **36**, 1069–1076.

146 Hibino, T., Hashimoto, A., Inoue, T., Tokuno, J.I., Yoshida, S.I., and Sano, M. (2000) *Science*, **288**, 2031–2033.

147 Priestnall, M.A., Kotzeva, V.P., Fish, D.J., and Nilsson, E.M. (2002) *J. Power Sources*, **106**, 21–30.

148 Wei., B., Lü, Z., Huang, X., Liu, M., Jia, D., and Su, W. (2009) *Electrochem. Commun.*, **11**, 347–350.

149 Wang, Z., Lü, Z., Wei, B., Huang, X., Chen, K., Liu, M., Pan, W., and Su, W. (2010) *Electrochem. Solid State Lett.*, **13**, B14–B16.

150 Akhtar, N., Decent, S.P., and Kendall, K. (2009) *Int. J. Hydrogen Energy*, **34**, 7807–7810.

151 Akhtar, N., Decent, S.P., and Kendall, K. (2010) *J. Power Sources*, **195**, 7796–7807.

152 Morales, M., Piñol, S., and Segarra, M. (2009) *J. Power Sources*, **194**, 961–966.

153 Kuhn, M., Napporn, T.W., Meunier, M., and Therriault, D. (2008) *J. Electrochem. Soc.*, **155**, B994–B1000.

154 Evans, A., Bieberle-Hütter, A., Galinski, H., Rupp, J.L.M., Ryll, T., Scherrer, B., Tölke, R., and Gauckler, L.J. (2009) *Monatsh Chem.*, **140**, 975–983.

155 Kuhn, M., Napporn, T.W., Meunier, M., and Therriault, D. (2010) *Solid State Ionics*, **181**, 332–337.

156 Fleig, J., Tuller, H.L., and Maier, J. (2004) *Solid State Ionics*, **174**, 261–270.

157 Kuhn, M., Napporn, T.W., Meunier, M., Vengallatore, S., and Therriault, D. (2009) *J. Power Sources*, **194**, 941–949.

158 Beckel, D., Briand, D., Bieberle-Hütter, A., Courbat, J., de Rooij, N.F., and Gauckler, L.J. (2007) *J. Power Sources*, **166**, 143–148.

159 Johnson, A.C., Lai, B.-K., Xiong, H., and Ramanathan, S. (2009) *J. Power Sources*, **186**, 252–260.

160 Joo, J.H. and Choi, G.M. (2009) *Solid State Ionics*, **180**, 839–842.

161 Joo, J.H., Kim, D.Y., and Choi, G.M. (2009) *Electrochem. Solid State Lett.*, **12**, B65–B68.

162 Xiong, H., Lai, B.-K., Johnson, A.C., and Ramanathan, S. (2009) *J. Power Sources*, **193**, 589–592.

163 Tsuchiya, M., Lai, B.-K., Johnson, A.C., and Ramanathan, S. (2010) *J. Power Sources*, **195**, 994–1000.

164 Yan, Y., Rey-Mermet, S., He, Z.B., Deng, G., and Muralt, P. (2009) *Procedia Chem.*, **1**, 1207–1210.

165 Doraswami, U., Shearing, P., Droushiotis, N., Li, K., Brandon, N.P., and Kelsall, G.H. (2009) *Solid State Ionics*. doi: 10.1016/j.ssi.2009.10.013

166 Akhtar, N., Decent, S.P., and Kendall, K. (2010) *J. Power Sources*, **195**, 7818–7824.

167 Akhtar, N., Decent, S.P., Loghin, D., and Kendall, K. (2009) *J. Power Sources*, **193**, 39–48.

168 Luebbe, H., Van herle, J., Hofmann, H., Bowen, P., Aschauer, U., Schuler, A., Snijkers, F., Schindler, H.-J., Vogt, U., and Lalanne, C. (2009) *Solid State Ionics*, **180**, 805–811.

169 Roy, B.R., Sammes, N.M., Suzuki, T., Funahashi, Y., and Awano, M. (2009) *J. Power Sources*, **188**, 220–224.

170 Soderberg, J.N., Sun, L., Sarkar, P., and Birss, V.I. (2009) *J. Electrochem. Soc.*, **156**, B721–B728.

171 Yang, C., Li, W., Zhang, S., Bi, L., Peng, R., Chen, C., and Liu, W. (2009) *J. Power Sources*, **187**, 90–92.

172 Yamaguchi, T., Shimizu, S., Suzuki, T., Fujishiro, Y., and Awano, M. (2009)

Electrochem. Solid State Lett., **12**, B151–B153.

173 Yamaguchi, T., Shimizu, S., Suzuki, T., Fujishiro, Y., and Awano, M. (2009) *J. Am. Ceram. Soc.*, **92**, S107–S111.

174 Lawlor, V., Griesser, S., Buchinger, G., Olabi, A.G., Cordiner, S., and Meissner, D. (2009) *J. Power Sources*, **193**, 387–399.

175 Tuller, H.L., Litzelman, S.J., and Jung, W.C. (2009) *Phys. Chem. Chem. Phys.*, **11**, 3023–3034.

176 Mogensen, M. and Kammer, K. (2003) *Annu. Rev. Mater. Res.*, **33**, 321–331.

177 Pillai, M., Lin, Y., Zhu, H., Kee, R.J., and Barnett, S.A. (2010) *J. Power Sources*, **195**, 271–279.

178 Pillai, M.R., Bierschenk, D.M., and Barnett, S.A. (2008) *Catal. Lett.*, **121**, 19–23.

179 Asamoto, M., Miyake, S., Sugihara, K., and Yahiro, H. (2009) *Electrochem. Commun.*, **11**, 1508–1511.

180 Huang, T.-J. and Huang, M.-C. (2009) *Chem. Eng. J.*, **155**, 333–338.

181 Jin, C., Yang, C., Zhao, F., Coffin, A., and Chen, F. (2010) *Electrochem. Commun.*, **12**, 1450–1452.

182 Klein, J.-M., Hénault, M., Roux, C., Bultel, Y., and Georges, S. (2009) *J. Power Sources*, **193**, 331–337.

183 Ihara, M., Hasegawa, S., Saito, H., and Jin, Y. (2007) *ECS Trans.*, **7**, 1733–1740.

184 Tagawa, Y., Saito, H., and Ihara, M. (2008) *ECS Trans.*, **16**, 287–298.

185 Cao, D.X., Sun, Y., and Wang, G.L. (2007) *J. Power Sources*, **167**, 250–257.

186 Tao, T. (2004) US Patent 6692861.

187 Nürnberger, S., Bußar, R., Desclaux, P., Franke, B., Rzepka, M., and Stimming, U. (2010) *Energy Environ. Sci.*, **3**, 150–153.

188 Gür, T.M. (2010) *J. Electrochem. Soc.*, **157**, B751–B759.

189 Lee, A.C., Mitchell, R.E., and Gür, T.M. (2009) *J. Power Sources*, **194**, 774–785.

190 Li, S., Lee, A.C., Mitchell, R.E., and Gür, T.M. (2008) *Solid State Ionics*, **179**, 1549–1552.

191 Gür, T.M., Homel, M., and Virkar, A.V. (2010) *J. Power Sources*, **195**, 1085–1090.

192 Lipilin, A.S., Balachov, I.I., Dubois, L.H., Sanjurjo, A., McKubre, M.C., Crouch-Baker, S., Hornbostel, M.D., and Tanzclla, F.L.Patent applications WO2005114770-A1, US2006/019132-A1, EP1756895-A1, AU2005246876-A1, CN1969415-A, IN200604273-P4, US2007/0269688-A1, BR200511332-A, JP2007538379-W, ZA200609977-A, RU2361329-C2.

193 Balachov, I. and Wolk, R.Patent application US2010159295-A1.

194 Pointon, K., Lakeman, B., Irvine, J., Bradley, J., and Jain, S. (2006) *J. Power Sources*, **162**, 750–756.

195 Jain, S.L., Lakeman, J.B., Pointon, K.D., and Irvine, J.T.S. (2007) *ECS Trans.*, **7**, 829–836.

196 Jain, S.L., Nabae, Y., Lakeman, B.J., Pointon, K.D., and Irvine, J.T.S. (2008) *Solid State Ionics*, **179**, 1417–1421.

197 Jain, S.L., Lakeman, J.B., Pointon, K.D., Marshall, R., and Irvine, J.T.S. (2009) *Energy Environ. Sci.*, **2**, 687–693.

198 Tao, T.T., Bai, W., and Rackey, S., and Wang, G.International Patent WO/2003/001617.

199 Tao, T.T., McPhee, W.A., Koslowske, M.T., Bateman, L.S., Slaney, M.J., and Bentley, J. (2008) *ECS Trans.*, **12**, 681–690.

200 Tao, T., Bateman, L., Bentley, J., and Slaney, M. (2007) *ECS Trans.*, **5**, 463–472.

201 Bronin, D.I., Karpachev, S.V., and Sal'nikov, V.V. (1982) *Sov. Electrochem.*, **18**, 1299–1301.

202 Jayakumar, A., Lee, S., Hornés, A., Vohs, J.M., and Gorte, R.J. (2010) *J. Electrochem. Soc.*, **157**, B365–B369.

203 Norby, T. (2007) *J. Chem. Eng. Jpn.*, **40**, 1166–1171.

204 Yang, L., Wang, S., Blinn, K., Liu, M., Liu, Z., Cheng, Z., and Liu, M. (2009) *Science*, **326**, 126–129.

205 Haugsrud, R. and Norby, T. (2006) *Nat. Mater.*, **5**, 193–196.

206 Fontaine, M.-L., Larring, Y., Haugsrud, R., Norby, T., Wiik, K., and Bredesen, R. (2009) *J. Power Sources*, **188**, 106–113.

207 Bhella, S.S. and Thangadurai, V. (2009) *J. Electrochem. Soc.*, **156**, B634–B642.

208 Azimova, M.A. and McIntosh, S. (2009) *Solid State Ionics*, **180**, 160–167.
209 Dailly, J., Fourcade, S., Largeteau, A., Mauvy, F., Grenier, J.C., and Marrony, M. (2010) *Electrochim. Acta*, **55**, 5847–5853.
210 Peng, R., Wu, T., Liu, W., Liu, X., and Meng, G. (2010) *J. Mater. Chem.*, **20**, 6218–6225.
211 Chevallier, L., Zunic, M., Esposito, V., Di Bartolomeo, E., and Traversa, E. (2009) *Solid State Ionics*, **180**, 715–720.
212 Taillades, G., Dailly, J., Taillades-Jacquin, M., Mauvy, F., Essouhmi, A., Marrony, M., Lalanne, C., Fourcade, S., Jones, D.J., Grenier, J.-C., and Rozière, J. (2010) *Fuel Cells*, **10**, 166–173.
213 Ding, H. and Xue, X. (2010) *Int. J. Hydrogen Energy*, **35**, 2486–2490.
214 Xu, J., Lu, X., Ding, Y., and Chen, Y. (2009) *J. Alloys Compd.*, **488**, 208–210.
215 Sunding, M.F., Kepaptsoglou, D.M., Diplas, S., Norby, T., and Gunnæs, A.E. (2010) *Surf. Interface Anal.*, **42**, 568–571.
216 Baker, R.T., Salar, R., Potter, A.R., Metcalfe, I.S., and Sahibzada, M. (2009) *J. Power Sources*, **191**, 448–455.
217 Quarez, E., Noirault, S., Le Gal La Salle, A., Stevens, P., and Joubert, O. (2010) *J. Power Sources*, **195**, 4923–4927.
218 Patcharavorachot, Y., Paengjuntuek, W., Assabumrungrat, S., and Arpornwichanop, A. (2010) *Int. J. Hydrogen Energy*, **35**, 4301–4310.
219 Brisse, A., Schefold, J., and Zahid, M. (2008) *Int. J. Hydrogen Energy*, **33**, 5375–5382.
220 Herring, J.S., O'Brien, J.E., Stoots, C.M., Hawkes, G.L., Hartvigsen, J.J., and Shahnam, M. (2007) *Int. J. Hydrogen Energy*, **32**, 440–450.
221 Laguna-Bercero, M.A., Kilner, J.A., and Skinner, S.J. (2010) *Solid State Ionics*. doi: 10.1016/j.ssi.2010.01.003
222 Chauveau, F., Mougin, J., Bassat, J.M., Mauvy, F., and Grenier, J.C. (2010) *J. Power Sources*, **195**, 744–749.
223 Graves, C., Ebbesen, S.D., and Mogensen, M. (2010) *Solid State Ionics*. doi: 10.1016/j.ssi.2010.06.014
224 Kim-Lohsoontorn, P., Brett, D.J.L., Laosiripojana, N., Kim, Y.-M., and Bae, J.-M. (2010) *Int. J. Hydrogen Energy*, **35**, 3958–3966.
225 Bo, Y., Wenqiang, Z., Jingming, X., and Jing, C. (2010) *Int. J. Hydrogen Energy*, **35**, 2829–2835.
226 Osinkin, D.A., Kuzin, B.L., and Bogdanovich, N.M. (2009) *Russ. J. Electrochem.*, **45**, 483–489.
227 Yang, X. and Irvine, J.T.S. (2008) *J. Mater. Chem.*, **18**, 2349–2354.
228 Schiller, G., Ansar, A., Lang, M., and Patz, O. (2009) *J. Appl. Electrochem.*, **39**, 293–301.
229 Wang, Z., Mori, M., and Araki, T. (2010) *Int. J. Hydrogen Energy*, **35**, 4451–4458.
230 Ishihara, T., Jirathiwathanakul, N., and Zhong, H. (2010) *Energy Environ. Sci.*, **3**, 665–672.
231 Ishihara, T. and Kannou, T. (2010) *Solid State Ionics*. doi: 10.1016/j.ssi.2010.06.020
232 Liu, Q., Yang, C., Dong, X., and Chen, F. (2010) *Int. J. Hydrogen Energy.*, **35**, 10039–10044.
233 Ebbesen, S.D., Graves, C., Hauch, A., Jensen, S.H., and Mogensen, M. (2010) *J. Electrochem. Soc.*, **157**, B1419–B1429.
234 Mawdsley, J.R., Carter, J.D., Kropf, A.J., Yildiz, B., and Maroni, V.A. (2009) *Int. J. Hydrogen Energy*, **34**, 4198–4207.
235 Virkar, A.V. (2010) *J. Hydrogen Energy*, **35**, 9527–9543.
236 Jensen, S.H., Sun, X., Ebbesen, S.D., Knibbe, R., and Mogensen, M. (2010) *Int. J. Hydrogen Energy*, **35**, 9544–9549.
237 Koh, J.-H., Yoon, D.-J., and Oh, C.H. (2010) *J. Nucl. Sci. Technol.*, **47**, 599–607.
238 Lee, A.C., Mitchell, R.E., and Gür, T.M. (2010) *Solid State Ionics*. doi: 10.1016/j.ssi.2010.05.034
239 Huang, T.-J. and Chou, C.-L. (2009) *Electrochem. Commun.*, **11**, 477–480.
240 Fu, X.-Z., Luo, J.-L., Sanger, A.R., Xu, Z.-R., and Chuang, K.T. (2010) *Electrochim. Acta*, **55**, 1145–1149.
241 Sakitou, Y., Hirano, A., Hanai, K., Matsumura, T., Imanishi, N., and

Takeda, Y. (2007) *ECS Trans.*, **7**, 1305–1309.
242 Huang, X., Li, T., Lu, Z., Wang, Z., Wei, B., and Su, W. (2010) *J. Phys. Chem. Solids*, **71**, 230–234.
243 Kournoutis, V.C., Tietz, F., and Bebelis, S. (2009) *Fuel Cells*, **9**, 852–860.
244 Lee, S., Song, H.S., Hyun, S.H., Kim, J., and Moon, J. (2009) *J. Power Sources*, **187**, 74–79.

7
Advances in Fabrication, Characterization, Testing, and Diagnosis of High-Performance Electrodes for PEM Fuel Cells

Jinfeng Wu, Wei Dai, Hui Li, and Haijiang Wang

To meet the performance, durability, and cost requirements that will enable the widespread use of proton exchange membrane (PEM) fuel cell technology, researchers have focused a great deal of attention on the fabrication, characterization, testing, and diagnosis of high-performance electrodes. To date, significant efforts have been made to optimize the structure, components, and fabrication methods of PEM fuel cell electrodes, and remarkable advances have been achieved. Concurrently, a wide range of characterization and diagnostic tools have been developed to obtain a fundamental understanding of the electrode process and microstructure and to provide benchmark quality data for modeling research. In this chapter, fundamental and advanced fabrication methods for high-performance electrodes of PEM fuel cells are systematically reviewed. The commonly used techniques for characterizing the physical properties of PEM fuel cell electrodes, such as surface morphology, microstructure, physical characteristics, and composition, are described. Attempts are made to review the various tools for diagnosing electrode behavior and to incorporate the most recent technical advances in PEM fuel cell diagnosis. The capabilities and weaknesses of the different techniques are also discussed.

7.1
Introduction

With their combined advantages of high efficiency, high power density, zero emissions, and modular design, proton exchange membrane (PEM) fuel cells have attracted increasing attention as promising power devices for automotive applications, portable applications, and stationary power generators. Running on sustainable fuels such as hydrogen, this fuel cell technology promises to be one of the few comprehensive, long-term solutions to improve energy efficiency, sustainability, and security, and to reduce greenhouse gases and urban pollution. In the past several decades, numerous industrial developers/manufacturers, universities, and institutes have devoted significant time and resources to the development and commercialization of this technology (see Chapter 5). Currently, PEM fuel cell technology is at the field trial and early commercialization stage, and volume production of PEM fuel cells is

Solid State Electrochemistry II: Electrodes, Interfaces and Ceramic Membranes.
Edited by Vladislav V. Kharton.
© 2011 Wiley-VCH Verlag GmbH & Co. KGaA. Published 2011 by Wiley-VCH Verlag GmbH & Co. KGaA.

expected in a few years. Significant progress has been achieved in recent decades, especially in the areas of increasing volumetric and/or gravimetric specific power density, making material utilization more efficient, and developing more cost-effective fabrication processes. The electrode is the most critical component in a PEM fuel cell, where electrochemical processes, mass transport, and thermal transport take place concurrently. The component materials, structure, and fabrication technologies play important roles in optimizing electrode performance and improving durability. Previous research has shown that the fabrication process has a significant effect on the electrode's microstructure and performance. From the transport perspective, an ideal electrode would allow all active catalytic sites to be accessible to the reactants, protons, and electrons and would at the same time facilitate the effective removal of product water from the electrode. As a result, it is important to apply the knowledge of thermodynamics, fluid dynamics, and electrochemical processes within a fuel cell to develop an electrode fabrication technique that will strike a balance between efficiency and durability, while meeting stringent cost targets.

Fuel cell science and technology cuts across multiple disciplines, including materials science, interfacial science, electrochemistry, transport phenomena, and catalysis. A better understanding of the physical properties and operating parameters influencing the phenomena and function of the entire cell, especially the efficiency-limiting processes, is imperative for developing high-performance electrodes. However, it is always a major challenge to fully understand the electrochemistry, fluid dynamics, and thermodynamic processes within a fuel cell. To achieve this goal, a great deal of effort has been put into characterization, diagnostics, and mathematical modeling. On the one hand, characterization and diagnostic tools can help establish critical process–structure–property–performance relationships between a fuel cell and its components. On the other hand, results obtained from characterization and diagnostics also provide benchmark quality data for fuel cell models, which further improve the prediction, control, and optimization of various transport and electrochemical processes within PEM fuel cells. In addition, the implementation of fundamental characterization and diagnosis can provide ample information about the degradation mechanisms and failure modes of the individual components comprising the electrode. More importantly, characterization and diagnostic tools will enable a better understanding of the structural and compositional changes during aging, and thereby aid in the development of applicable strategies to mitigate performance decay and optimize the electrode fabrication process.

In this chapter, fundamental and advanced fabrication methods for high-performance electrodes of PEM fuel cells are systematically reviewed, with a focus on catalyst layer (CL) preparation. In addition, development trends in fuel cell electrode fabrication are discussed. Common characterization techniques for determining the physical properties of PEM fuel cell electrodes, such as surface morphology and microstructure measurement, physical characterization, and composition analysis, are also addressed. Each characterization technique is analyzed by assessing its merits and weaknesses. For further analysis of relevant methodological and materials science-related aspects, readers are referred to Chapters 4, 5 and 10 of the first volume. Finally, a comprehensive survey of the diagnostic tools presently used in PEM fuel cell research is provided in this chapter. For clarity, the various diagnostic tools are divided into two general categories: electrochemical techniques and chemical/physical methods. The principles, experimental implementation, data processing, advantages, and disadvantages of each diagnostic tool are described in detail.

7.2 Advanced Fabrication Methods for High-Performance Electrodes

A typical electrode in a PEM fuel cell includes a proton exchange membrane, catalyst layers, microporous layers (MPLs), and gas diffusion layers (GDLs), as shown in Figure 7.1. Four transport processes in a PEM fuel cell electrode determine its performance and durability.

1) Proton transport through the membrane and the CL.
2) Electron transport from the CL to the current collector via the MPL and the GDL.
3) Mass transport of reactant gases and water to and from the CL through the MPL and the GDL.
4) Heat transport across the CL, the MPL, and the GDL.

To ensure high performance, compact interfacial contact is required between the different layers to reduce proton and electron conduction resistance. Catalyst loading should be as low as possible while maintaining sufficient performance and durability. Catalyst utilization and the bonding of catalyst particles are significantly affected by the fabrication process. In addition, the MPL and GDL play a critical role in mass transport and water management, so the fabrication of the gas diffusion media is also critical to achieving a high performance of the electrodes.

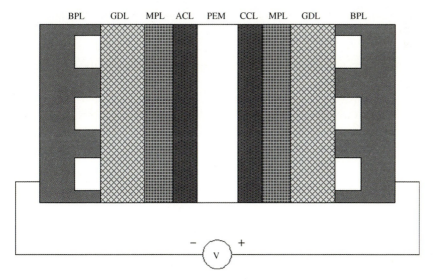

Figure 7.1 Architecture of a single proton exchange membrane fuel cell.

7.2.1
Conventional Hydrophobic PTFE-Bonded Gas Diffusion Electrode

The conventional gas diffusion electrode (GDE) [1–3] uses hydrophobic polytetrafluoroethylene (PTFE, e.g., Teflon®) to bind the catalyst particles onto the surface of a porous GDL such as carbon fiber paper or carbon cloth. The CL is often impregnated with Nafion® ionomer using a spraying or brushing method to improve proton conductivity within the CL. This kind of hydrophobic GDE has dramatically reduced the catalyst loading from 4 to $0.4\,\text{mg cm}^{-2}$ or even lower, compared to a GDE prepared by the very early method of using a paste of carbon-supported catalyst mixed with sulfuric acid. However, the platinum utilization of a hydrophobic GDE is not high, due to low proton transport efficiency in the CL. The general fabrication process of a conventional hydrophobic GDE is as follows [2]:

1) Carbon black and PTFE are thoroughly mixed in a solvent to form an emulsion: carbon ink.
2) The carbon ink is brushed or sprayed onto a wet-proof-treated GDL to form the gas diffusion medium (GDM). The GDM is then dried and sintered in an oven.
3) Pt/C catalyst powder is mixed with solvents by mechanical stirring.
4) The PTFE suspension is then gradually added to the above mixture under ultrasonic mixing until the desired catalyst/PTFE weight ratio is achieved.
5) Bridge builder and peptization agents are added under mechanical stirring. The mixture obtained is catalyst ink or catalyst slurry.
6) The catalyst slurry is then coated on the GDM.
7) The CL is impregnated with Nafion.
8) Finally, the anode GDE, cathode GDE, and pretreated membrane are sandwiched and bonded together by hot pressing to form the membrane electrode assembly (MEA).

7.2.1.1 Catalyst Layer Coating onto Gas Diffusion Layer

Brushing and Screen Printing Methods Brushing and screen printing [4] methods were widely used in the early 1990s to coat the CL onto the GDL. They are simple and cost-effective coating processes involving just a paintbrush or a screen printer. Catalyst ink waste is minimal, especially with the brushing method, but CL uniformity and reproducibility are very hard to control. In addition, these techniques are time consuming because painting, drying, and weighing must be done repeatedly to achieve the desired catalyst loading.

Direct Spray Method Spray painting overcomes many of the problems associated with brush painting and allows a more uniform distribution of catalyst material. It can also be easily scaled up. However, a considerable amount of catalyst is often wasted in the feed lines due to periodic clogging, which can increase the production cost. Kumar et al. [5] prepared their electrode by the spraying technique and achieved a Pt loading of $0.1\,\text{mg cm}^{-2}$, which was found to be comparable in performance to a state-of-the-art electrode with $0.4\,\text{mg cm}^{-2}$ Pt loading. Recently, Oishi et al. [6]

developed a direct solution spray method to deposit the catalyst onto the GDL. A solution of Pt precursor and carbon support was sprayed onto the GDL using an ultrasonic device, and the Pt catalyst precursor was reduced to solid Pt nanoparticles on carbon support, simultaneously.

Inkjet Printing Method Taylor *et al.* [7] used inkjet printing (IJP) to deposit catalyst material on the GDL. Their studies showed that the IJP method can be used to deposit small volumes of water-based catalyst ink solution with picoliter precision, provided the solution properties are compatible with the cartridge design. By optimizing the dispersion of the ink solution, this technique can be used with catalysts supported on different types of carbon black (e.g., XC-72R, Monarch 700, or Black Pearls 2000). Figure 7.2 illustrates how the Pt utilization changes when the loading of the catalyst decreases. Their inkjet-printed MEA with a catalyst loading of 0.020 mg cm^{-2} has shown Pt utilization in excess of 16 000 mW mg^{-1} Pt, which is much higher than the traditional screen-printed MEA (800 mW mg^{-1} Pt). They also prepared MEAs with a gradient distribution of Pt/C catalyst using the standard Johnson Matthey (JM) catalyst, which showed better performance than MEAs with a homogeneous catalyst layer made with JM Pt/C (20% Pt).

7.2.1.2 Nafion Impregnation

To improve proton conductivity within the CL of a PTFE-bonded GDE, the CL is usually impregnated with Nafion ionomer by brushing or spraying after the catalyst coating step. Lee *et al.* [3] investigated the effect of Nafion loading on the performance of a PTFE-bonded GDE with a Pt loading of 0.4 mg cm^{-2}. The results showed a nonlinear relationship between Nafion loading in the CL and electrode performance. The optimal Nafion loadings were 0.5 and 1.9 mg cm^{-2} with air and O$_2$ as the respective oxidants. The difference was due to the mass transport limitation when the

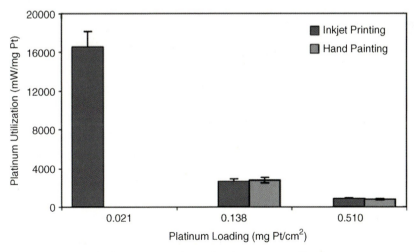

Figure 7.2 Catalyst loading versus Pt utilization by inkjet printing and hand painting [7]. (Reprinted with permission from Elsevier.)

different oxidants were used. Correlation of Nafion loading with activation–polarization characteristics showed an initial quick increase in the electrochemical activity up to a loading of 1.3 mg cm^{-2}, followed by a much slower increase to the maximum at 1.9 mg cm^{-2}. This trend correlated well with the decrease in charge transfer resistance and the increase in electrochemical active surface area. A further loading increase would cause deeper penetration of the Nafion into the CL, decreasing the CL porosity and raising the mass transport limitations.

7.2.1.3 MEA Assembly: Hot Pressing

After the GDEs are prepared, the MEA can be assembled by laminating the anode GDE and the cathode GDE with a pretreated membrane in between, followed by hot pressing of the sandwich. The membrane pretreatment process is as follows:

1) Nafion membrane is first boiled in 3% hydrogen peroxide solution to remove any organic impurities.
2) The membrane is then washed with deionized water and boiled in 1 M sulfuric acid to remove any metallic impurities and to fully convert the membrane to the H$^+$ form.
3) The membrane is finally boiled in distilled water to complete the pretreatment process.

Nakrumpai et al. [8] studied how MEA performance in a PEM fuel cell was affected by varying certain conditions in the hot pressing step: temperature, pressure, and time of compression in the range of 130–150 °C, 50–100 bar, and 1–5 min, respectively. Hot pressing under low pressure at a high temperature and with a long compression time tends to result in better MEA performance. The optimal condition was 64 bar, 137 °C, and 5.5 min of compression time.

7.2.2
Thin-Film Hydrophilic Catalyst Coated Membrane Method

The thin-film hydrophilic catalyst coated membrane (CCM) is a popular fabrication method. In 1993, Wilson [9] patented this breakthrough technology that almost doubled the power density of an electrode having a PTFE-bonded GDE. In this new fabrication method, the hydrophilic perfluorosulfonate ionomer (Nafion) was used to replace the conventional hydrophobic PTFE as the catalyst bonder. The CL was directly coated or transferred (using the decal method) onto the membrane instead of the porous GDM. The CCM technique yielded a thinner catalyst layer than the conventional PTFE-bonded GDE, facilitating proton transport from the catalyst surface to the membrane. The bonding material was the same in the CL and the membrane, which improved the interfacial contact between them and built a better 3D microstructure for gas and proton diffusion than in a PTFE-bonded CL.

The CCM fabrication process according to Wilson's patent [9] is as follows:

1) Catalyst ink preparation:
    a) Combining a 5% solution of solubilized perfluorosulfonate ionomer (such as Nafion) and 20 wt% Pt/C catalyst in a weight ratio of 1 : 3 Nafion/catalyst.

b) Adding water and glycerol in a weight ratio of 1 : 5 : 20 carbon–water–glycerol.

c) Mixing the solution using ultrasound until the catalyst is uniformly dispersed and the mixture is adequately viscous for coating.

2) Coating the catalyst directly onto the membrane:
   a) Ion exchanging the Nafion membrane to the Na^+ form by soaking it in NaOH solution, then rinsing, and let drying.

   b) Applying the carbon–water–glycerol ink to one side of the membrane. Two coats are typically required for adequate catalyst loading. Dry the membrane in a vacuum at a temperature of approximately 160 °C.

   c) Repeating the step (2b) for the other side of the membrane.

   d) Ion exchanging the assembly to the protonated form by lightly boiling the MEA in 0.1 M H_2SO_4 solution and rinsing in deionized water.

3) Decal coating the catalyst layer onto the membrane:
   a) Cleaning a release blank of Teflon film and coating the blank with a thin layer of mold release (e.g., a trifluoroethanol spray). Painting the blank with a layer of ink and baking in an oven at 135 °C until dry. Adding more layers until the desired catalyst loading is achieved.

   b) Forming an assembly of a polymer electrolyte membrane, counter electrode (anode electrode), and the coated blank. Placing the assembly into a conventional hot press and lightly loading the press until it heats to a selected temperature (125 °C for Nafion), and then pressing at 70–90 bar for 90 s.

   c) Cooling the assembly and releasing the blank, leaving the film adhered to the SPE membrane cathode surface.

   d) Placing carbon paper/cloth against the film to produce a MEA.

7.2.2.1 Screen Printing Method

Nafion membrane is usually ion exchanged to the Na^+ form by soaking it in NaOH solution if the catalyst ink is to be directly coated on the Nafion membrane. After the coating step, the CCM has to be boiled in H_2SO_4 solution to the protonated form. Kim et al. [10] and Pettersson et al. [11] used a vacuum table to hold the original Nafion membrane and fabricate the CCM using a screen printing technique without ion exchange steps, which demonstrated repeatable and stable fuel cell performance.

Andrade et al. [12] developed a sieve printing technique using a semiautomatic screen printer (EKRA, Bönnigheim, Germany) to prepare the CCM. The sieve-printed MEA had 39.8% higher power density at 0.6 V compared to MEAs fabricated using spray and hot pressing methods. The sequence of the sieve printing method steps is as follows [12]:

1) Add Nafion solution and solvents (ethylic ester and ultrapure water) into the catalyst, with mechanical stirring and an evaporation process until the paste reaches optimal composition and consistency.

2) Deposit the paste on the sieve, which has been brought into contact with the membrane using a squeegee as the membrane is moved over the sieve.
3) Push the paste into the open area to form a pattern, leaving all the paste that is in the mesh to be deposited on the membrane, and remove the surplus paste using the edge of the squeegee.
4) Treat the CCM with HNO_3 solution (5 wt%) and then rinse the CCM three times with ultrapure water at 80 °C for 1 h to remove any acid traces.

This type of screen printing technique is highly reproducible and significantly more accurate and faster than the spray method. It can be easily scaled up and is very suitable for high-volume, low-cost production.

7.2.2.2 Modified Decal Method

In the conventional decal transfer method, the catalyst layer is directly applied to the decal substrate, which sometimes makes it difficult to realize a high catalyst transfer ratio from the decal to the polymer membrane. If the desired catalyst loading is not transferred to the membrane in a single operation, multiple coats have to be applied to build up the desired loading, which makes the decal method a very slow process. Park et al. [13] modified the conventional decal transfer method to fabricate electrodes for PEM fuel cells. They introduced a breaking layer consisting of carbon powder and PFSA coated on the decal substrate before coating the catalyst layer. Figure 7.3 illustrates the modified decal method. This technique not only ensures the complete transfer of the catalyst but also reduces the process time. In addition, the carbon powder used as the breaking layer material is utilized as a microporous layer in the electrode after fabrication.

7.2.2.3 Inkjet Printing Technology

Towne et al. [14] utilized drop-on-demand inkjet printing technology to fabricate CCMs for PEM fuel cells. Commercial desktop inkjet printers were used to deposit the active catalyst layer directly from ink cartridges onto Nafion membranes in the proton form. Off-the-shelf piezo (Epson) and thermal (HP) inkjet printers were used with only minor modifications. The ink cartridges were opened, cleaned of the original ink, and filled with the catalyst ink using a syringe. The membrane was secured to a cellulose acetate sheet and fed through the printer using the original paper-feed platen.

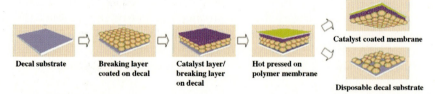

Figure 7.3 Schematic illustration of the modified decal method [13]. (Reprinted with permission from the Electrochemical Society.)

Up to 24 successive layers with varying compositions were printed to build the desired electrode structures [14]. After all layers were printed, the electrode was dried by gently warming it with a heat gun for less than 30 s. The substrate was then turned over and resecured to the acetate sheet, and the printing and drying processes were repeated. The printed electrodes were then processed by a variety of procedures that included hot pressing at various temperatures and water extraction. Hot pressing was intended to improve the integrity and adhesion of the films, while water extraction removed the ethylene glycol ink carrier.

The layers were well adhered, withstood simple tape-peel, bending, and abrasion tests, and did so without any postdeposition hot pressing steps [14]. The elimination of those steps suggested that inkjet-based fabrication or similar processing technologies may provide a route to less expensive large-scale fabrication of CCMs. However, the performance of the inkjet print MEA is not satisfactory. It remains to be determined whether inkjet fabrication offers performance advantages or leads to more efficient utilization of expensive catalyst materials.

7.2.2.4 Direct Spray Coating

Su *et al.* [15] and Leimin *et al.* [16] developed a catalyst-sprayed membrane under irradiation (CSMUI) to prepare ultralow loading CCMs for PEM fuel cells. Catalyst ink was sprayed with an airbrush directly onto the membrane and an infrared light was used simultaneously to evaporate the solvents. Figure 7.4 shows the performances of MEAs with various platinum loadings on cathode and anode. The platinum loadings of the anode and cathode were lowered to 0.04 and 0.12 mg cm^{-2},

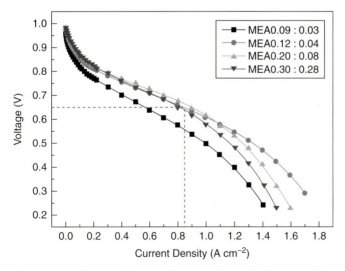

Figure 7.4 Performance of MEAs with various platinum loadings at cathode and anode, evaluated at a cell temperature of 60 °C; humidifier temperatures for hydrogen and air: 65 °C; back pressure of both hydrogen and air: 30 psi [15]. (Reprinted with permission from Elsevier.)

respectively, but still yielded a high performance of 0.7 A cm^{-2} at a cell voltage of 0.7 V. The optimal Nafion content in the catalyst layer was 25 wt%. Scanning electron microscopy (SEM) and electrochemical impedance spectroscopy (EIS) measurements revealed that the CCM had very thin anode and cathode catalyst layers that came in close contact with the membrane, resulting in an MEA with low contact resistance and reduced mass transport limitation. Their results indicated a promising method for preparing MEAs with ultralow Pt loadings but high performance in PEM fuel cells. However, coating efficiency and scalability still need more investigation.

7.2.3
Advanced Fabrication Methods for Low Pt Loading Electrodes

Recently, some novel fabrication methods have been developed that further reduce Pt loading and improve Pt utilization, such as electrophoretic deposition (EPD), pulse electrodeposition and electrospray, sputtering, and sol–gel deposition. These methods can significantly improve the uniform distribution of catalyst particles, reduce catalyst particle size, and provide a more effective catalyst surface area. The Pt loading can be cut to as low as 0.01 mg cm^{-2}. However, these methods still need more investigation to improve scalability and reproducibility for practical mass production.

7.2.3.1 **Electrophoretic Deposition**
Historically, electrophoretic deposition has been a versatile technique for fabricating thin or thick films of both metals and ceramics on conductive substrates. Employing the electrophoretic process to fabricate Pt/C nanocatalysts and Nafion nanoparticles has been demonstrated to effectively reduce catalyst loading and to ensure uniform distribution of the catalyst particles on the electrodes of PEM fuel cells.

Morikawa *et al.* [17] prepared a CCM using EPD. A suspension consisting of ethanol, carbon powders with Pt catalyst, and Nafion polymer was used to obtain a stable dispersed solution. Figure 7.5 shows the schematic of the EPD cell used in this study. The Nafion membrane was set in the middle of the cell. The left compartment contained 0.1 M HClO$_4$ aqueous solution, and the right compartment held a suspension made of Vulcan XC72 carbon powder, Pt catalyst, 5 wt% Nafion polymer solution, and ethanol. Two Pt electrodes were used as cathode and anode to apply a high voltage to the cell.

The EPD preparation process employed by Morikawa *et al.* was as follows [17]:

1) The EPD cell was immersed in an ice bath to maintain the desired temperature over the course of the EPD process.
2) 1000 V was applied to the cell, and the distance between the two electrodes was maintained at roughly 8 cm. The electric field was 125 V cm^{-1}. The duration of the EPD process was changed from 5 to 15 min to control the thickness of the catalyst layer.

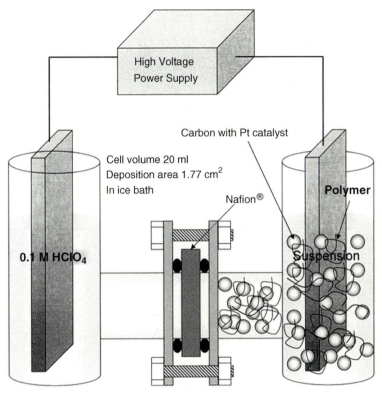

Figure 7.5 Schematic representation of EPD cell [17]. (Reprinted with permission from the Electrochemical Society.)

3) After the catalyst layer was deposited on one side of the Nafion 117 membrane, $HClO_4$ in one side of the compartment and suspension in the other were removed and the membrane was then dried.
4) The side of the compartment that had contained $HClO_4$ was then filled with the suspension, and the other side was filled with 0.1 M $HClO_4$ aqueous solution.
5) The second EPD was carried out to form the catalyst layer on the opposite side of the Nafion membrane, which completed the CCM preparation.

These processes are schematically illustrated in Figure 7.6.

The catalyst layer prepared by the EPD technique had smaller carbon particles and a more uniform structure than that prepared by the hot pressing method. Such a uniform structure yielded a high Pt utilization of 56%.

Louh et al. [18, 19] proposed an EPD process to deposit Pt/C nanopowders onto various GDLs in a better controlled and cost-effective manner. Their method provided high deposition efficiency and a uniform distribution of catalyst and Nafion ionomer on the electrodes of PEM fuel cells. The EPD suspension in their process was prepared by sonicating a mixture of Pt/C nanopowders, Nafion solution, and isopropyl alcohol. The colloidal stability of the EPD suspension with either Pt/C

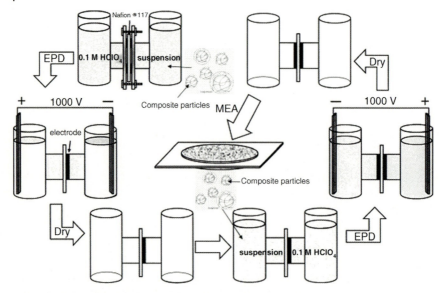

Figure 7.6 EPD process for preparation of MEA [17]. (Reprinted with permission from the Electrochemical Society.)

catalysts or a mixture of Pt/C catalysts and Nafion ionomer was achieved at an optimal pH of ~10, which was controlled using acetic acid and ammonium hydroxide. A nicely distributed deposition of Pt/C nanocatalysts and Nafion ionomer on both hydrophilic and hydrophobic carbon-based electrodes was successfully obtained using a Pt/C concentration of $1.0\,\mathrm{g\,l^{-1}}$, an electrical field of $300\,\mathrm{V\,cm^{-1}}$, and a deposition time of 5 min. Microstructural analysis indicated that the Pt/C nanopowders not only embraced the entire surface of the carbon fibers but also infiltrated the gaps and voids in fiber bundles such that a higher contact area for the same loading of Pt/C nanocatalysts was thus expected. At present, the EPD process can effectively reduce Pt catalyst loading on electrodes in PEM fuel cells better than the conventional methods such as screen printing, brushing, or spraying, and yield a similar level of power output in PEM fuel cells. The process is shown in Figure 7.7.

7.2.3.2 Pulse Electrodeposition

Kim et al. [20, 21] developed a new approach based on pulse electrodeposition to prepare the GDE. The CL performance was optimized by controlling pulse deposition parameters such as peak current density, duty cycle, and total charge density. The peak current density and the pulse duty cycle were found to affect the nucleation rate and decrease the catalyst centric growth. The amount of platinum loading was controlled by the total charge density. The best performance was achieved by preparing an electrode using a peak current density of $400\,\mathrm{mA\,cm^{-2}}$, a duty cycle of 2.9%, and a total charge density of $8\,\mathrm{C\,cm^{-2}}$. In their study, the following fabrication process was employed [20, 21]:

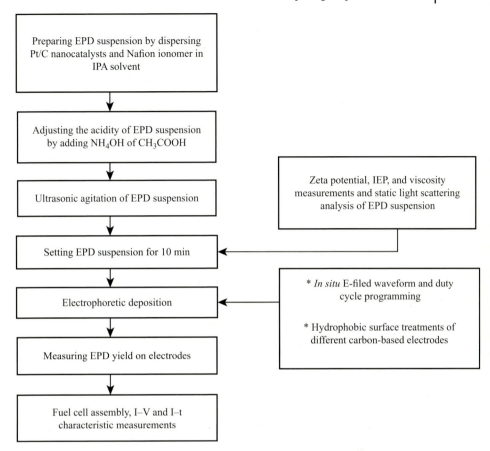

Figure 7.7 Process flow of electrophoretic deposition of Pt/C nanocatalysts and Nafion solution on carbon-based electrodes [19]. (Reprinted with permission from ASME Publication.)

1) Electrodeposition was performed on the carbon black electrode using a Pt plating bath containing $10\,g\,l^{-1}$ H_2PtCl_6 and $60\,g\,l^{-1}$ HCl at room temperature.
2) Platinum gauze was used as a counter electrode.
3) A pulse generator controlled both the pulse wave and the deposition current density. The amount of platinum electrodeposited on the electrode was estimated by the difference in the weights before and after electrodeposition.
4) After electrodeposition, the electrodes were heated at $300\,°C$ in H_2 gas for 2 h.
5) The electrocatalyzed electrode was impregnated with 5 wt% Nafion solution and then dried at $80\,°C$ for 2 h.
6) The impregnated electrodes and the membrane were bonded to form an MEA by hot pressing at $130\,°C$ for 3 min with a pressure of 140 atm.

Later, Lee et al. [22] developed a low Pt loading electrode by galvanostatic pulse electrodeposition of Pt on a Nafion-bonded carbon layer. The deposition process was carried out in a two-electrode cell using an electrochemical instrument (VoltaLab 80, Radiometer Analytical). The following procedures were used [22]:

1) Electrodeposition was carried out on a Nafion-bonded carbon black electrode in a Pt plating bath containing K_2PtC_{14} and NaCl at room temperature.
2) The concentration of electrolyte was kept constant (0.5 M) and the concentration of Pt ranged from 5 to 80 mM.
3) Before Pt deposition, the Nafion-bonded carbon electrode was mounted on an electrode holder coupled with a Pt foil as a current collector. They were assembled with Pt mesh as an anode. The assembled holder was immersed in a Pt bath for 2 min to sufficiently soak the assembly.
4) Immediately after deposition, the electrochemically catalyzed electrode was rinsed thoroughly with ultrapure water to remove any residue of the Pt precursor.
5) After the electrodeposition process, the electrodes were heated at 250 °C in H_2 (10%)/N_2 (90%) gas for 30 min.
6) The protonation process was executed by immersing the electrodeposited layers in 0.1 M H_2SO_4 solution for 30 min at 80 °C. The electrodes were then thoroughly rinsed with a copious amount of ultrapure water.

As shown in Figure 7.8, the performance of the MEA fabricated using electrodeposition (0.025 mg cm^{-2} on the anode and 0.3 mg cm^{-2} on the cathode) is better than that of conventional electrodes (0.3 mg cm^{-2} on both electrodes). A noticeable increase in catalyst utilization is achieved by the deposition of Pt particles taking place only in the three-phase reaction zone.

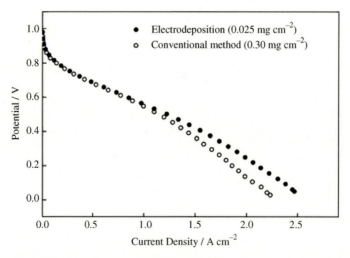

Figure 7.8 The performance of MEA made by electrodeposition and conventional methods [22]. (Reprinted with permission from Elsevier.)

7.2.3.3 Electrospray

Benítez et al. [23] developed a novel electrospray method to prepare electrodes for PEM fuel cells. 3300–4000 V was applied between a capillary tube in which the ink was forced to flow and a carbon cloth substrate. The high electric field generates a mist of highly charged droplets when the solution emerged. During this process, the droplet size was reduced by evaporating the solvent and/or by the so-called Coulomb explosion (droplet subdivision resulting from high charge density). A nitrogen pressurized tank was used to force the catalytic ink through the capillary tube. The catalytic ink was then placed in an ultrasonic bath to maintain its homogeneity. The carbon support was moved by means of an X–Y axis coordinated system controlled by computer software.

Different characterization techniques showed both morphological and structural improvements that contributed to better catalyst utilization than in the electrodes prepared by conventional methods. The MEAs obtained by this electrospray technique also exhibited higher power density than commercial E-TEK MEAs as well as those prepared by impregnation or spray techniques with the same Pt loadings. In addition, this technique is scalable and suitable for high-volume, low-cost production.

7.2.3.4 Plasma Sputtering

Sputtering provides an alternative method of depositing a thin catalyst layer on the GDL and thereby achieving improved performance and ultralow Pt loadings. The sputtering method involves a vacuum evaporation process that removes portions of the target material (e.g., Pt catalyst) and then deposits a thin, resilient film of the target material onto an adjacent substrate (e.g., GDL) [24]. Sputtering technology is a highly commercialized technique for thin film deposition in other industrial applications, such as glass coating. This technique also has potential for use in the large-scale manufacturing of fuel cell electrodes with uniform catalyst layers and low or ultralow Pt loadings.

Hirano et al. [25] deposited a catalyst layer with a low Pt loading (0.1 mg cm^{-2}) onto a GDM using the sputtering deposition technique. The performance of the cathodes thus made was comparable to that of a standard E-TEK electrode. However, such cathodes showed unexpectedly high overpotentials at high current densities, suggesting mass transport limitation issues.

Caillard et al. [26] developed a low-pressure, inductive, high-density plasma sputtering system with a transformer coupled plasma (TCP) antenna to deposit platinum on a GDL (Figure 7.9). Optimal plasma parameters for the Pt concentration profile were deduced assuming linear diffusion in a fractal porous medium. Those optimal parameters provided highly diffusive behavior originating from the plasma's interaction with the substrate.

Huang et al. [27] investigated how two important parameters, input power and sputtering gas pressure, in a radio frequency (RF) magnetron sputtering deposition process, affected the performance of PEM fuel cell electrodes. With a Pt loading of 0.1 mg cm^{-2}, the electrode fabricated at 100 W and 10^{-3} Torr was found to exhibit the best performance, mainly due to its low kinetic (activation) resistance, compared to

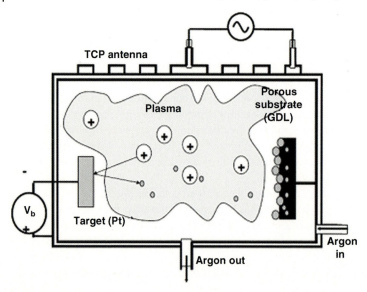

Figure 7.9 Schematic of sputtering in argon TCP plasma [26]. (Reprinted with permission from Elsevier.)

electrodes fabricated under other conditions. In addition, control of the sputtering gas pressure seemed to be more critical for activation loss than control of the RF power.

7.2.3.5 Sol-Gel Method

Khan et al. [28] fabricated novel Pt electrodes with controllable loadings by direct deposition of platinum sols onto a membrane. The sols were prepared with citrate and methanol, and no extra treatment was needed other than evaporation of the solvent. XRD data indicated that the platinum particle size remained similar to that in the original sols. In comparison to electrodes prepared from Pt/C catalysts, the Pt-deposited membrane showed high specific activity and comparable single-cell performance. This electrode preparation method has the advantages of easy preparation and good control over lowering of the platinum loading. The preparation steps were as follows [28]:

1) Pt nanoparticles with an average size of 2–3 nm were prepared by methanol reduction while sodium citrate was used as the stabilizer [29]. In brief, a mixture of 0.1 M $H_2PtCl_{6(aq)}$ and 0.1 M sodium citrate (J.T. Baker, 99.9%) was added to a methanol solution under reflux and stirring at 80 °C. The reaction was stopped by cooling down to room temperature, causing the solution to immediately turn into black Pt sols.
2) Nafion membranes were boiled in deionized water for 30 min and then in 1 M H_2SO_4 solution for 20 min, followed by copious washing with deionized water at 80 °C. The pretreated membranes were stored in deionized water at room temperature.

3) The prepared Pt sols were dispersed evenly on the pretreated Nafion membrane and then the solvent was evaporated at room temperature. The loading was controlled by the amount of Pt sols added for deposition.
4) CCM was hot pressed with a GDM to make an MEA, at a temperature between 30 and 110 °C and a pressure between 1.4 and 9 bar.

7.2.4
Fabrication of Gas Diffusion Media

The gas diffusion media, which in a state-of-the-art PEM fuel cell electrode usually consists of a gas diffusion layer and a microporous layer, is a key component that fulfills several functions: providing access for reactant gases from flow-field channels to the catalyst layers; providing paths for water vapor to be supplied from the humidified reactant gas to the membrane to avoid membrane drying, and paths for liquid water to be removed from the catalyst layer areas to flow-field channels to avoid catalyst layer flooding; conducting electrons and heat; and providing mechanical support to the MEA. Currently, a commercial GDL consists of porous carbon paper or carbon cloth that has both adequate electric conductivity and enough resilience under the corrosive operating conditions of fuel cells. The MPL is a porous layer consisting of carbon black and PTFE, which has been proven to significantly boost PEM fuel cell performance and durability.

7.2.4.1 Fabrication of Gas Diffusion Layer
The GDL used in PEM fuel cells is often wet-proofed by immersing it in a hydrophobic PTFE emulsion. PTFE in the GDL can enhance both gas and water transport when a PEM fuel cell operates under flooding conditions [30]. The porosity of the GDL decreases with increasing PTFE content [31, 32]. In practice, the PTFE concentration used in a GDL is between 0 and ~40 wt%, depending on the application. Lin and Nguyen [30] investigated the effect of GDL thickness on cell performance and found that a thinner GDL was more sensitive to liquid water accumulation and provided less physical support to the MEA than a thicker one.

The typical PTFE wet-proof treatment process is as follows [31]:

1) Cut out the GDL and wash it in acetone to remove dust or other undesirable substances.
2) Dry the washed GDL at 110 °C for 2 h.
3) Dilute 60 wt% PTFE emulsion to make a 5–45 wt% aqueous solutions.
4) Dip the dried GDL entirely in the PTFE solutions for 30 s and dry it again at room temperature for 5 h.
5) Heat-treat the samples in air using several temperature steps (e.g., 80, 110, 290, and 350 °C).

7.2.4.2 Fabrication of Microporous Layer
To some extent, the MPL is more pivotal than the GDL for enhancing PEM fuel cell performance and reliability.

A typical MPL fabrication process is as follows [33]:

1) Mix carbon powder with PTFE-dispersed water, isopropyl alcohol, and glycerol in an ultrasonic bath for 2 h.
2) Spray or print the resulting carbon ink onto one side of a wet-proofed GDL and then dry at 80 °C for 30 min.
3) Heat-treat the GDM sample at 280 °C for 30 min to evaporate all remaining glycerol and then dry at 350 °C for 30 min to uniformly distribute PTFE throughout the MPL.

The size and type of the carbon particles [34–36], the carbon loading [33, 37], and the PTFE content [30, 37–40] used in an MPL may significantly affect PEM fuel cell performance. Park et al. [33, 40] indicated an optimal loading of 20% PTFE with 0.5 mg cm^{-2} carbon, while Qi and Kaufman [37] recommended an optimal loading of 35% PTFE and 2.0 mg cm^{-2} carbon. The MPL thickness was also investigated by Qi and Kaufman [37] and Pasaogullari et al. [41]. The latter concluded that an MPL approximately 50 μm thick would cancel out the positive effects it has on cell performance.

7.3
Characterization of PEM Fuel Cell Electrodes

At present, characterization of PEM fuel cell electrodes mainly concentrates on the surface (morphology and microstructure), physical properties (porosity, permeability, diffusivity, hydrophilicity, conductivity, etc.), and composition analysis.

7.3.1
Surface Morphology Characteristics

Extraordinary advancements have been achieved in characterizing the surface morphology of an electrode, among which microscopy has been a crucial tool in directly visualizing electrode and polymer morphology. The most common imaging techniques used in the analysis of PEM fuel cell materials and components are optical microscopy and electron microscopy. Optical microscopes utilize visible wavelengths of light to obtain an enlarged image of a small object, while electron microscopes use an electronic beam to examine objects on a very fine scale. Microscopy can provide information about not only surface morphology (the shape and size of the particles making up the sample) but also topography (surface features of the sample), composition (the elements and compounds that comprise the sample, and their relative amounts), and crystallographic information (atomic arrangement in a given zone). Two key types of the electron microscopy are SEM and transmission electron microscopy (TEM).

7.3.1.1 Optical Microscopy
The optical microscope, often referred to as the light microscope, uses visible light and a system of lenses to magnify images. A sample is usually mounted on a

motorized stage and illuminated by a diffuse source of light. The image of the sample is projected via a condenser lens system onto an imaging system, such as the eye, a film, or a charge-coupled device (CCD). At very high magnification with transmitted light, point objects are seen as fuzzy disks surrounded by diffraction rings called Airy disks. The resolving power of a microscope is taken as the ability to distinguish between two closely spaced Airy disks. The limit of resolution depends on the wavelength of the illumination source, according to Abbe's theory [42]:

$$d_{rs} = 0.612\lambda/NA_L, \tag{7.1}$$

where d_{rs} is the resolution, λ is the wavelength of light applied, and NA_L is the numerical aperture of the lens. Usually, a wavelength of 550 nm is considered to correspond to green light and consequently the limit of resolution of optical microscopy is about 200 nm. Optical microscopy has long been used to visually analyze the surface or cross section of components used in PEM fuel cells. For instance, using optical microscopy and other diagnostic methods, Tan et al. [43] studied the degradation of four commercially available gasket materials in a fuel cell environment. Their optical micrographs revealed that degradation always starts with surface roughness in the early stages of aging and finally results in cracks over time. Recently, Liu et al. [44] used optical microscopy to measure the decrease in cross-sectional thickness of an MEA degraded by local fuel starvation. Through optical microscopy, Ma et al. [45] observed the adhesive effect that ionomer content in the catalyst layer had on different substrates. The results showed that if ionomer content was below 20 wt%, the catalyst paste adhered poorly to the Nafion membrane.

Although the standard optical microscope is easy to operate, one difficulty is the high-contrast image generated from completely or almost transparent samples, such as proton exchange membranes. To overcome this, fluorescence, dark field, and phase-contrast optical microscopy techniques have been developed, leading to microscopic images with sufficient contrast and high information content [46]. Compared to other types of microscopy for surface imaging, such as SEM and TEM, optical microscopy is restricted by the diffraction limit of visible light to 1000× magnification and 200 nm resolution. One way to improve the lateral resolution of the optical microscope is to use shorter wavelengths of light, such as ultraviolet. Another way is to employ scanning near-field optical microscopy, which has been developed into a powerful surface analytical technique with a spatial resolution of 100 nm or even better [47].

7.3.1.2 Scanning Electron Microscopy

The main difference between optical microscopy and electron microscopy is the substitution of an electron beam and electromagnetic coils for the light source and condenser lens. The operation is based on electron emission from filament to the sample. Emitted electrons are accelerated at a high voltage (up to hundreds of kilovolts). The wavelength is related to the accelerating voltage by the following equation [48]:

$$\lambda = h(2m_e eV)^{-1/2}, \tag{7.2}$$

where h is Planck's constant, V is the accelerating voltage, and m_e and e are the mass and charge of the electron, respectively. The extremely small wavelengths of electrons can push the spatial resolution of electron microscopes to the atomic level. When the electron beam scans across the sample, the interaction between the incident electrons and the sample atoms generates a variety of products, such as secondary electrons, backscattered electrons, X-rays, and Auger electrons. A typical schematic representation of the wealth of information resulting from this interaction is displayed in Figure 7.10.

In SEM, secondary electrons and backscattered electrons are the most common products detected for investigation of a sample's surface morphology. Secondary electrons come from the sample itself, produced as a result of collisions between the beam electrons and weakly bound electrons in the conduction band of the sample. Due to the relatively low energy (several eV) of the secondary electrons, only those formed within the first few nanometers of the sample surface have enough energy to escape and be detected. Therefore, secondary electrons accurately mark the position of the beam and provide a detailed surface morphology with good resolution. The backscattered electrons are derived from the elastic collision between the incident electrons and the sample atomic nucleus, causing the electrons to bounce back with a wide directional change. Elements with higher atomic numbers have more positive

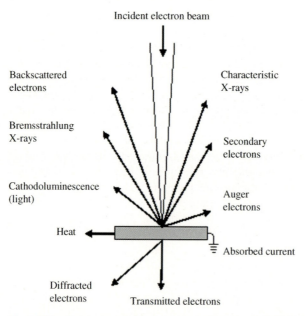

Figure 7.10 Signals generated from the interaction between the high-energy electron beam and the sample in an electron microscope.

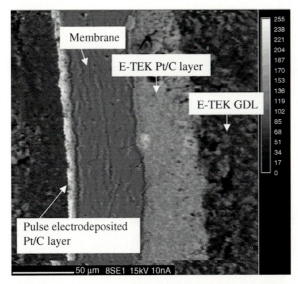

Figure 7.11 Backscattered electron image of the cross section of the membrane and electrode assembly [20]. (Reprinted with permission from the Electrochemical Society.)

charges on the nucleus, and as a result, their backscattered signal will be higher. Therefore, the contrast in a backscattered electron image is sensitive to the atomic number of the imaged material.

SEM can characterize a sample at several nanometers, which makes it suitable to examine the morphology of PEM fuel cell electrodes, such as pore structure in the GDL, and membrane or electrode thickness in cross sections of MEAs. Kim and Popov [20] developed a novel method for preparing PEM fuel cell electrodes based on the pulse electrodeposition technique. The backscattered electrons signal was used to characterize the cross section of an MEA. As shown in Figure 7.11, the thicknesses of the membrane, catalyst layer, and gas diffusion electrode regions are clearly displayed. In this image, the bright portion is associated with the presence of a heavier element, Pt, and the thickness of the Pt electrocatalyst layer is only 5 μm, which is 10 times thinner than a conventional E-TEK catalyst layer. SEM imagery can also be used to quantify morphological changes in PEM fuel cell components before and after degradation tests. Zhang *et al.* [49] studied the effect of open-circuit operation on membrane and catalyst layer degradation in PEM fuel cells. SEM images of the cross-sectioned MEA before and after the degradation test are shown in Figure 7.12. Compared to the "fresh" MEA (Figure 7.12a), the degraded MEA (Figure 7.12b) displays an obvious degree of thinning, with PEM thickness decreasing from 23 to 16 μm at both sides of the reinforced layer. The SEM images show that the PEM thinning was a direct cause of the increased hydrogen crossover that resulted in an unrecoverable decrease in open-circuit voltage and cell performance.

The limitation of this technique is that samples for SEM need to be electrically conducting so that charge built up in the surface from the incident electron beam can

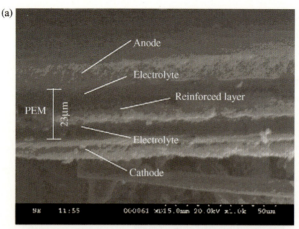

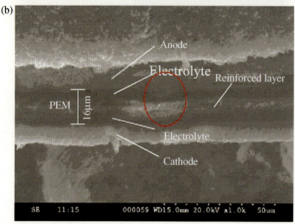

Figure 7.12 SEM images of the MEA (a) before and (b) after OC operation [49]. (Reprinted with permission from Elsevier.)

be conducted away. This problem can be overcome by insulating samples with a covering of thin, conducting coatings, typically of gold or platinum. To image an MEA cross section, preparation is usually carried out by cooling the MEA in liquid nitrogen before fracturing, to avoid deformation [50].

7.3.1.3 Scanning Probe Microscopy

Another option for characterizing the morphology and topography of fuel cell electrodes is scanning probe microscopy (SPM). SPM forms images of the sample surface by scanning an atomically sharp, needlelike probe across the surface. As the probe scans, the probe–surface interaction as a function of position is recorded and variations in the topography of the sample surface can be observed. The resolution of SPM is not limited by diffraction as in electron microscopy, but only by the size of the probe–sample interaction volume, which can be as small as a few picometers. The two main variants of SPM commonly used for morphological measurement are scanning tunneling microscopy (STM) and atomic force microscopy (AFM). For STM, a small electrical current is applied between the probe and the surface of a conducting sample to monitor variations in the surface topography. For AFM, the surface information is detected via the interatomic force between atoms on the probe and those on the surface of either an insulating or a conducting sample.

Ghassemi et al. [51] synthesized several sulfonated–fluorinated poly(arylene ether) multiblock (MB) copolymers with various sulfonation levels for PEM fuel cells. Figure 7.13 compares the representative surface morphology of Nafion 112, BPSH (poly(arylene ether) sulfone random copolymer), and MB copolymer derived from tapping mode AFM (TM-AFM). Images from Nafion 112 and BPSH suggest that the hydrophilic groups aggregate as isolated domains with some local connection of

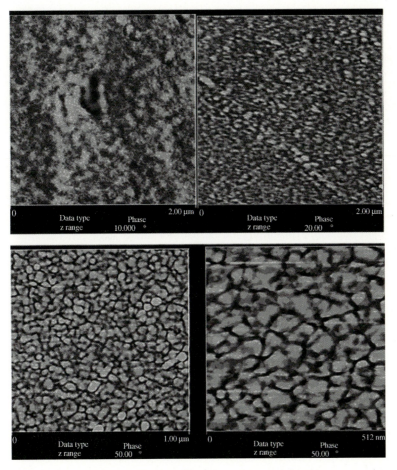

Figure 7.13 AFM tapping phase image for Nafion 112 (upper left), BPSH (upper right), and MB copolymer (bottom images) [51]. (Reprinted with permission from Elsevier.)

hydrophilic domains, as evidenced by the good conductivity of the proton exchange membranes. The images from MB copolymers show a very well-defined phase separation and a distinct morphological architecture compared to random copolymers, which might account for their rather high proton conductivities. Ma et al. [45] employed AFM and Kelvin probe microscopy (KPM) to examine the effect of different ionomer content (20–40 wt%) on the morphology and surface potential mapping of catalyst layers. In their AFM images, grain enlargement, pore reduction, and surface roughness decrease were observed with increased ionomer content in the catalyst layer. The KPM images showed a rise in the surface potential as ionomer content increased, which implies that the protonic conduction network reduced but did not prevent electrical conduction. Inoue et al. [52] investigated the effects of the overall mass of a catalyst layer and the mass ratio of electrolyte in a catalyst layer on PEM fuel cell performance and evaluated the surface roughness of catalyst layers with AFM. The experimental results suggested that the roughness was influenced by the mass ratio of electrolyte, not by the overall mass of a catalyst layer. The structure of the catalyst layer changed significantly at the optimum mass ratio. In the research of Zhang et al. [53], effects of spraying and scraping fabrication methods and hot pressing pretreatment of the anodes on the performance of direct methanol fuel cells were investigated. Compared to the sprayed catalyst layer, the relatively homogeneous and smooth surface of the scraped one was helpful for improving the electrode's mass transport properties and enhancing electrical contact between the smoother catalyst surface and the membrane.

Unlike electron microscopy, which provides a two-dimensional image of a sample, SPM provides a three-dimensional surface profile with even higher resolution. In addition, an electron microscope needs a vacuum environment for proper operation, whereas SPM can work perfectly in ambient air at standard temperature or even in a liquid environment. In general, it is easier to use SPM than electron microscopy because SPM samples require minimal preparation. However, the disadvantages of SPM are that image acquisition is time consuming and the maximum image size is generally small.

7.3.2
Microstructure Analysis

TEM has been popular for gaining insights into PEM fuel cell electrode microstructural properties. In comparison to SEM, TEM focuses on transmitted electron signals, as shown in Figure 7.10. When the electron beam collides with the sample, the electrons transmitted through the sample go on to form a TEM image, while those that are stopped or deflected by heavy atoms in the sample are subtracted from the image. At the bottom of the microscope, the transmitted electrons hit a fluorescent screen, which gives rise to a "shadow image" of the sample with its different parts displayed in varying shades according to their density. Owing to the low wavelength of incident electrons, TEM is capable of imaging a sample at a significantly high resolution, better than 0.2 nm, which makes it suitable for nanoscale characterization in PEM fuel cell research, such as determination of the particle size and distribution. Particle imaging via improved TEM variants, such as high-angle annular dark field

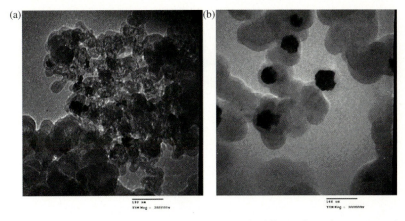

Figure 7.14 TEM images of Pt deposited on carbon under different electrodeposition conditions: (a) 200 mA cm^{-2} of pulse electrodeposition; (b) 10 mA cm^{-2} of direct current electrodeposition [21]. (Reprinted with permission from Elsevier.)

scanning transmission electron microscopy (HAADF-STEM) and high-resolution transmission microscopy (HR-TEM), can characterize the MEA structure at extremely high resolution, below 0.1 nm. Kim *et al.* [21] investigated the effect of peak current density on Pt particle size during the preparation of PEM fuel cell electrodes using pulse electrodeposition. The TEM images presented in Figure 7.14 show that higher peak current density will decrease the nucleation rate, resulting in a smaller Pt particle size on the carbon electrode surface.

However, one major drawback of TEM is the limited penetration of electrons through the sample. Therefore, samples must be extremely thin (approximately 0.1 μm). For the characterization of PEM fuel cell electrodes, this can be achieved by (1) lightly dispersing the powder (as-processed or scraped from an MEA electrode) across a thin (holey/lacy) carbon film or (2) preparing an intact cross section of a three-layer MEA using diamond-knife ultramicrotomy. With the former method, information about particle size and distribution of the catalyst and its support can be obtained at the expense of complete destruction of the electrode structure. The latter method has been used successfully by preparing epoxy-embedded TEM cross sections from thin three-layer MEAs [54, 55]. When the electrode structure is completely embedded with epoxy, imaging the Pt catalyst particles and carbon support within the electrode is straightforward. Unfortunately, the presence of epoxy makes it virtually impossible to identify and characterize the continuous ionomer network that surrounds the catalyst network.

The recently developed ultramicrotomy sample preparation method based on partially embedded electrodes has enabled direct imaging of intact ionomer, carbon/Pt, and pore network surfaces within the MEA [56]. This technique has been employed to characterize differences between catalyst particles before and after degradation, including by determining where inside the MEA structure the growing particles are located. Figure 7.15 presents images of a freshly prepared MEA (a) and

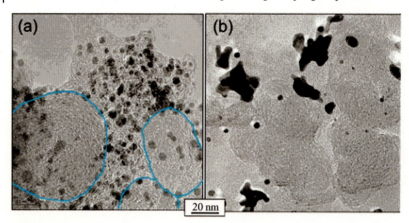

Figure 7.15 (a) TEM of freshly prepared MEA; (b) TEM of MEA after potential cycling at 80–1.2 V for 1500 cycles [57]. (Reprinted with permission from the American Chemical Society.)

an aged MEA (b) that underwent repeated high potential cycling to 1.2 V [57]. In Figure 7.15a, Pt particles of the fresh sample are evident in the ionomer region of the MEA, suggesting that the catalyst particles have separated from the carbon support material. In Figure 7.15b, large particle agglomerates have formed in the ionomer region after aging.

7.3.3
Physical Characteristics

7.3.3.1 Porosimetry
From a mass transport perspective, the proper porosities of the CL and GDL are critical parameters for optimizing PEM fuel cell performance. Due to the electrochemical reactions occurring inside the matrix CL, the reactants must pass through the GDL into the CL, and in the meantime the residual water needs to be removed from the electrodes. Therefore, a porous structure is necessary in both the CL and the GDL. If the thickness and real weight of the CL or GDL are known, the porosity can be easily calculated by dividing the volume of the solid phase by the total volume of the electrode, as expressed in Equation 7.3 [58]:

$$\varepsilon_p = 1 - \frac{W_A}{\varrho_{real} d_{th}}, \tag{7.3}$$

where ε_p is the porosity, W_A is the real weight (g cm^{-2}), ϱ_{real} is the solid-phase density (for carbon-based materials, ϱ_{real} varies between 1.6 and 1.9 g cm^{-3}), and d_{th} is the thickness (either compressed or uncompressed).

Otherwise, the porosity of the electrode needs to be determined by diagnostic methods. So far, several methods have been proposed and implemented to comprehensively characterize porosity, including mercury or gas porosimetry, capillary flow porosimetry, and standard contact porosimetry.

With regard to mercury porosimetry, the applied pressure for injecting mercury into the sample is inversely proportional to the pore radius, in accordance with the Washburn equation, and consequently the pore size distribution can be evaluated. Mercury porosimetry can be used to characterize larger pore sizes (3 nm–300 μm), while gas porosimetry is suited to smaller sizes (0.5–100 nm) [59]. As an example, mercury porosimetry measurements were conducted by Wu et al. [60] to determine changes in the total porosity and pore size distribution of GDL samples before and after degradation. However, one limitation of this method is that the sample has to be cut into small pieces and stacked in the holder, requiring relatively large volumes of sample and thus compromising the measuring accuracy. Another concern is the high pressure required for determining small pore sizes, which may result in collapse of the electrode's pore structure. Jena and Gupta [61] reported using capillary flow porosimetry by monitoring the through-plane and in-plane flow of a gas through dry and wet two-layer porous electrodes. Measurement using this method can yield microstructural information such as largest pore diameter, mean flow pore diameter, cumulative flow percentage, and pore size distribution.

Standard contact porosimetry was recently developed by Volfkovich et al. [62, 63] to characterize porous electrodes. In this method, disks of the sample and two porous standards are first filled with a low-contact angle liquid (e.g., octane or decane) and weighed. Then the sample is sandwiched between the standards and held in compression to attain capillary equilibrium. By heating and/or vacuum treatment or by a flow of dry inert gas through the sandwich, a certain amount of wetting liquid is removed and the liquid in the sample is allowed to reach capillary equilibrium again with the standard. The disks are then disassembled and weighed. This process is repeated until all the liquid is completely evaporated from the sample. Standard contact porosimetry with appropriate standard samples can be used to measure pore sizes in the range of 1 to 3×10^6 nm. The main drawback is that these measurements are time consuming.

7.3.3.2 Permeability and Gas Diffusivity

When the porosity (ε_p) of the electrode is known, the effective diffusivity of the gas phase in this porous medium, D_{eff}, can be calculated according to the Bruggeman correlation

$$D_{eff} = D^0[\varepsilon_p(1-S_{ls})]^\tau, \qquad (7.4)$$

where D^0 is the diffusion coefficient, S_{ls} is the liquid saturation, and τ is the tortuosity. As for the liquid phase, the driving force, capillary pressure (P_c), is related to the porosity and the surface tension (σ_{st}), as shown in Equation 7.5 [64]:

$$P_c = \sigma_{st}\cos(\theta_c)(\varepsilon_p/K_{perm})^{1/2}J(S), \qquad (7.5)$$

where θ_c is the contact angle, K_{perm} represents the absolute permeability, and $J(S)$ is the Leverett function. Based on the fluid flow direction through the porous electrode, the permeability coefficients are normally defined as in-plane permeability (x- and y-directions) and through-plane permeability (z-direction). Experimental determination of permeability has been reported by several research groups using their homemade apparatuses with manometers. The principle of measuring the permeability (k_{perm}) is based on the Darcy formula:

$$v_f = \frac{k_{perm}}{\mu_{fv}} \frac{\Delta P}{l_{th}}, \tag{7.6}$$

where v_f is the fluid velocity, ΔP is the pressure drop, l_{th} is the thickness of the electrode, and μ_{fv} is the fluid viscosity. Then, the permeability of the porous electrode can be calculated by transforming Equation 7.6:

$$k_{perm} = \mu_{fv} v_v \frac{l_{th}}{\Delta P}. \tag{7.7}$$

Williams et al. [65] reported that the through-plane permeability was 0.8–3.1×10^{-11} m^2 for bare carbon paper and that the addition of microporous layers decreased this value by approximately two orders of magnitude. Prasanna et al. [66] reported a permeability of 1–8×10^{-11} m^2 for carbon paper, which decreased significantly as the Teflon loading was increased.

7.3.3.3 Wetting

The hydrophobicity and hydrophilicity of the GDL and CL play complex and critical roles in water management within PEM fuel cells. Adequate hydrophilicity is necessary for better PEM fuel cell performance and extended lifetime. If the amount of water in the membrane is too low, the membrane conductivity will drop, as will the fuel cell performance. However, if an electrode is too hydrophilic, the excess water cannot be removed efficiently; liquid water floods the electrodes, interfering with mass transport of the reactant. Significant effort has been put into investigating water transport and water balance within PEM fuel cells. Research has found that the hydrophobic properties of electrodes can be controlled by the choice of carbon, the ionomer/carbon ratios, the content of the hydrophobic agent, and the pretreatment and fabrication procedures.

The wettability of a PEM fuel cell electrode is normally characterized according to its contact angle to water, which can be divided into two categories: external surface contact angle and internal contact angle. The surface is said to be hydrophilic when its contact angle to water is lower than 90°, and hydrophobic if the contact angle is greater than 90°. Most methods in use today, such as goniometry (the sessile drop method), the capillary meniscus height method, or the Wilhelmy plate gravimetric method [67], aim to determine the external surface contact angle. Goniometry is the most common method of measuring the contact angle of a liquid on a solid surface and involves placing a small liquid droplet on a GDL substrate. The contact angle can be determined by measuring the angle between the tangent of the droplet surface at the contact line and the surface. In this method, the droplet size should be small enough to eliminate the influence of the droplet's weight. For the capillary meniscus height method, the sample is first dipped into water and then an optical technique is used to directly record and measure the capillary meniscus height, as shown in Figure 7.16 [68].

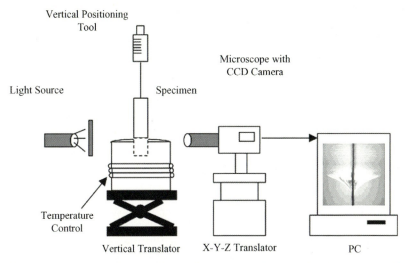

Figure 7.16 Experimental setup of the capillary meniscus height method for measuring the external contact angle [68]. (Reprinted with permission from Elsevier.)

Considering the force balance between gravity and surface tension through a meniscus line, the contact angle (θ_c) between water and the sample has the following relationship with the meniscus height [68]:

$$\sin \theta_c = 1 - \frac{\Delta \rho g h^2}{2\sigma_{st}}, \qquad (7.8)$$

where $\Delta \rho$ is the difference between the densities of water and vapor, g is the gravitational acceleration, h is the meniscus height, and σ_{st} is the liquid–gas surface tension of water. For the Wilhelmy plate method, the sample is dipped into liquid while the force on the sample due to wetting (F_m) is measured via a tensiometer or microbalance. The contact angle between water and the sample can be expressed as [69]

$$\cos \theta_c = \frac{F_m}{2l_w \sigma_{st}}, \qquad (7.9)$$

where l_w is the wetted length of the sample. The advantages of the Wilhelmy plate method include that the angle values obtained represent averages over the sample's entire wetted length and that the temperature of the liquid can be precisely monitored. These methods for determining the external surface contact angle are suitable for materials with smooth surfaces, such as proton exchange membranes, bipolar plates, or CLs prepared on the membrane surface. Brack et al. [70] studied membrane properties by means of goniometry and found that the contact angle increased with dehydration of the membrane. Yu et al. [71] characterized the wetting behavior of a catalyst coated membrane, their results showing that the first attachment of a water droplet at the CL surface presented a relatively high contact angle that then decreased with time.

However, when these methods were applied to characterize porous materials with rough surfaces, such as GDLs, the values of the external contact angle as reported in the literature were between 120° and 140°, or even higher [68, 72]. Since the contact angle to water of pure PTFE is only 108° [73], these large values cannot be explained by the presence of a hydrophobic agent inside the GDL pores but rather by the contribution of GDL surface roughness. For a rough structured surface, these methods do not provide a true measure of the interfacial properties of the material. Instead, microstructural aspects dominate the observed behavior through such phenomena as droplet pinning. To overcome the limitation of these methods for measuring the external surface contact angle of porous electrodes, Parry et al. [74] employed the Washburn method to characterize the internal wetting properties of different GDLs with low PTFE loadings. Capillary rise experiments were performed in a tensiometer by submerging the GDL samples in water. Each sample was held by a metal clamp attached to a microbalance. The mass of liquid water absorbed by the sample was recorded as a function of time. When inertia and gravity forces are negligible, the internal contact angle can be calculated according to the Lucas–Washburn equation [75]

$$\cos \theta_c = \frac{m_L^2}{t_a} \frac{\eta_L}{C_W \varrho_L^2 \gamma_{LV}}, \qquad (7.10)$$

where m_L is the mass of liquid absorbed by the sample in time t_a, C_W is the Washburn constant of the GDL sample, and η_L, ϱ_L, and γ_{LV} are the liquid viscosity, density, and surface tension, respectively. However, the Washburn method is applicable only to hydrophilic materials. Recently, Gurau et al. [67] combined the Washburn method with the Owens–Wendt theory to estimate the internal contact angle to water of hydrophobic GDLs. In their experiments, the Washburn method was first conducted with a set of wetting fluids to find their internal contact angles to the GDL material. The Owens–Wendt theory was then used to extrapolate the data obtained with the Washburn method and estimate the contact angle to water of the GDL material.

7.3.3.4 Conductivity

From the viewpoint of PEM fuel cell performance, high electrical and protonic conductivities are very desirable for optimizing the electrode structure and components. To measure lateral electrical conductivity, a four-probe technique provides more accurate measurements compared to the two-probe method. Both local and large-scale conductivity can be determined, depending on the location and separation of the probes. Although this is not a mainstream characterization method, it is useful as a quality control or development tool [76]. The four-probe technique can yield valuable information on crack formation and membrane deterioration. Measurement of through-plane electrical conductivity can be conducted by sandwiching the electrode sample between two gold-coated copper plates and subsequently compressing the assembly under a certain pressure [60]. A fixed current (I) is applied with a DC power supply and the resulting voltage drop between two gold-coated plates (ΔV)

is measured with a sensitive multimeter. The through-plane electrical resistance (R) can be described by

$$R = \frac{\Delta V}{I} = \frac{\varrho_{el} L_{th}}{A_{el}}, \tag{7.11}$$

where ϱ_{el} is the through-plane electrical resistivity, A_{el} is the area of the sample, and L_{th} is the thickness of the electrode sample under compressive load.

As for the protonic conductivity of the catalyst layer or membrane, the through-plane resistivity can be obtained by AC impedance spectroscopy, which will be further discussed in Section 7.4.1.3. Much attention has been paid to the effects of electrode preparation and fuel cell operating conditions on the protonic conductivity of the membrane and/or CL. However, to the best of our knowledge, there is still no reliable characterization method to determine the protonic conductivity of a CL or membrane separately and accurately.

7.3.4
Composition Analysis

7.3.4.1 Energy Dispersive Spectrometry

When an electron beam interacts with a sample, as shown in Figure 7.10, secondary electrons, backscattered electrons, X-rays, and other signals are generated as a result of collisions between the incident electrons and the electrons within the atoms that make up the sample. These signals carry information about the sample and provide clues to its composition, X-rays and backscattered electrons being most commonly used for investigating a sample's composition. X-ray emission occurs when an electron in a shell of an atom absorbs some energy from an incident electron and ejects to a higher energy level, creating a vacancy (hole) in the original shell. The electron then drops back down to recombine with this vacancy, and the X-ray photon or Auger electron is emitted, equal in energy to the difference between the higher and lower energy levels. The wavelength of the X-ray radiation is generally regarded as the characteristic of the atom. Hence, after being detected and analyzed by an energy dispersive spectrometer (EDS) or wavelength dispersive spectrometer (WDS) attached to the SEM system, a spectrum of the emitted X-ray wavelengths is commonly used for the elemental and compositional analysis. Backscattered electrons are the result of beam electrons being scattered back out of the sample, as mentioned above. The percentage of beam electrons that become backscattered electrons depends on the atomic number, which makes the percentage a useful signal for analyzing the sample composition. An Everhart–Thornley detector or a solid-state detector can be used to collect these backscattered electrons and form an image indicating the compositional information.

As a typical example, Pt dissolution and deposition processes were investigated by Bi et al. [77] under H_2/air and H_2/N_2 potential cycling conditions. In their study, the Pt distributions of (1) a fresh MEA and (2) MEAs after potential cycling were determined by SEM–EDS, as shown in Figure 7.17. No significant amount of Pt was found in the membrane for either the fresh MEA or the H_2/N_2 cycled MEA,

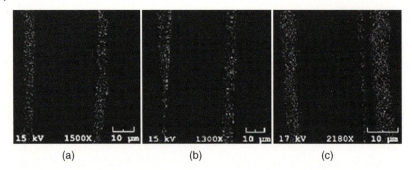

Figure 7.17 Pt distribution maps in the MEAs (left: anode; right: cathode): (a) fresh; (b) H_2/N_2 potential cycled; (c) H_2/air potential cycled [77]. (Reprinted with permission from the Electrochemical Society.)

whereas a clear Pt band formed in the membrane for the H_2/air cycled MEA. The Pt deposition mechanism in the membrane was explored based on SEM–EDS and other diagnostic results. The dissolved Pt species may diffuse into the ionomer phase and subsequently precipitate in the membrane via reduction of Pt ions by the crossover of hydrogen from the anode side, thereby dramatically decreasing membrane stability and conductivity.

7.3.4.2 Thermal Gravimetric Analysis

Thermal analysis comprises a group of techniques in which a physical property of a substance is measured as a function of temperature while the substance is subjected to a controlled temperature program. It generally covers three different experimental techniques in PEM fuel cell research: thermal gravimetric analysis (TGA), differential thermal analysis (DTA), and differential scanning calorimetry (DSC). DSC is a thermoanalytical technique in which the different amounts of heat required to increase the temperature of a sample and of a reference material are measured as a function of temperature. In DTA, the temperature difference between a sample and an inert reference material is measured when both are subjected to identical heat treatments. The main application of DSC and DTA is the determination of phase transitions in various atmospheres, such as exothermic decompositions, that involve energy changes or heat capacity changes.

TGA is performed on samples to determine weight changes in relation to temperature changes and is commonly employed in PEM fuel cell research to characterize the materials composition. For example, TGA diagrams of the proton exchange membrane can provide plenty of information about degradation temperatures, absorbed moisture content, levels of inorganic and organic components, and solvent residues in the membrane. By coupling TGA with gas analysis, Fourier transform infrared spectroscopy (FTIR), and/or mass spectrometry (MS), mass losses and volatile species of decomposition can be determined simultaneously, which will significantly improve the analytic capability of the system. Figure 7.18 shows the TGA–MS data for the decomposition products of Nafion membrane [78].

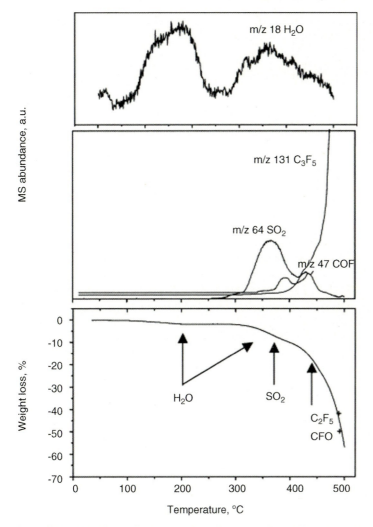

Figure 7.18 TGA–MS results of a relatively dry Nafion membrane [78]. (Reprinted with permission from the American Chemical Society.)

In Figure 7.18, the bottom panel indicates the thermal gravimetric response of Nafion, showing that total decomposition occurs above 400 °C, with minor events starting at ∼300 °C. The top panel indicates the temperature-dependent appearance of species with the mass-to-charge (m/z) ratio equal to 18 amu, which is assigned to water in the gas phase above the sample. Two peaks are observed in which the lower temperature peak is assigned to residual hydration waters in the Nafion polymer, while the higher temperature peak is concomitant with the observation of an $m/z = 64$ amu peak (middle panel) assigned to SO_2. These two signals taken together are a signature for the decomposition of the Nafion sulfonic acid moiety (SO_3H).

The middle panel reports on the temperature response of an $m/z = 47$ peak and an $m/z = 131$ peak, assigned to the Nafion side chain and Nafion backbone decompositions, respectively.

7.4
Testing and Diagnosis of PEM Fuel Cell Electrodes

At present, research efforts in PEM fuel cell diagnostics are mainly concentrated on the following issues [79–81]: (1) mass distribution, especially water distribution over the active electrode, including flooding detection leading to low catalyst utilization, (2) resistance diagnosis and membrane drying detection, which closely relate to membrane conductivity, (3) optimization of electrode structures and components, fuel cell design, and operating conditions, (4) current density distribution in dimensionally large-scale fuel cells, (5) temperature variation resulting from a nonuniform electrochemical reaction and contact resistance in a single cell and different interconnection resistances for a stack, and (6) flow visualization for direct observation of what is occurring within the fuel cell. Due to the complexity of the heat and mass transport processes occurring in fuel cells, there are typically a multitude of parameters to be determined. For all the previous reasons, it is important to examine the operation of PEM fuel cells or stacks with suitable techniques that allow the evaluation of these parameters separately and can determine the influence of each on the overall cell performance. The objective of this section is to provide a brief survey of diagnostic tools presently used in PEM fuel cell testing and research. For clarity, in this section the various testing and diagnostic tools are divided into two categories: electrochemical techniques and chemical/physical methods.

7.4.1
Electrochemical Techniques

7.4.1.1 Polarization Curves
Analyzing a plot of cell potential against current density under a set of constant operating conditions, known as a polarization curve, is the standard electrochemical approach for characterizing the performance of fuel cells (both single cells and stacks) [58, 82, 83]. While a steady-state polarization curve can be recorded in the potentiostatic or galvanostatic regimes, non-steady-state polarization is analyzed using a rapid current sweep [84]. By measuring polarization curves, certain parameters such as the effects of the composition, flow rate, temperature, and relative humidity of the reactant gases on cell performance can be characterized and compared systematically. The ideal polarization curve for a single hydrogen/air fuel cell has three major regions, which are shown in Figure 7.19 [58]. At low current densities (the region of activation polarization), the cell potential drops sharply, the majority of these losses being due to the sluggish kinetics of the oxygen reduction reaction (ORR) [58]. At intermediate current densities (the region of ohmic polarization), the voltage loss caused by ohmic resistance becomes significant and results

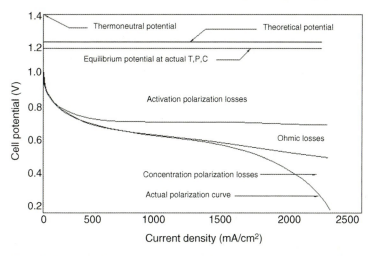

Figure 7.19 An ideal polarization curve with the corresponding regions and overpotentials [79].

mainly from resistance to the flow of ions in the electrolyte and resistance to the flow of electrons through the electrode [85]. In this region, the cell potential decreases nearly linearly with current density, while the activation overpotential reaches a relatively constant value [58]. At high current densities (the region of concentration polarization), mass transport effects dominate due to the transport limit of the reactant gas through the pore structure of the GDL and CL, and cell performance drops drastically [86]. From Figure 7.19, it can also be seen that the difference between the theoretical cell potential (1.23 V) and the thermoneutral voltage (1.4 V) represents the energy loss under the reversible condition (the reversible loss) [83]. Very often, polarization curves are converted to power density versus current density plots by multiplying the potential by the current density at each point of the curve.

Polarization curves provide information on the performance of the cell or stack as a whole. However, these measurements fail to produce much information about the performance of individual components within the cell and cannot be performed during normal operation of a fuel cell, requiring significant amounts of time to finish. In addition, they fail to differentiate different mechanisms from each other; for example, both flooding and drying inside a fuel cell cannot be distinguished in a single polarization curve [58]. Resolving time-dependent processes occurring in the fuel cell and the stack is another important problem [59]. For the latter purpose, current interruption, electrochemical impedance spectroscopy measurements, and other electrochemical approaches are preferable.

7.4.1.2 Current Interrupt

In general, the current interrupt method is used to measure the ohmic losses in a PEM fuel cell. The principle is that the ohmic losses vanish much more quickly than the electrochemical overpotentials when the current is interrupted [87]. As shown schematically in Figure 7.20, a typical current interrupt result is obtained by

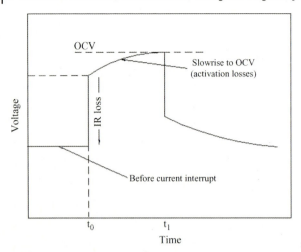

Figure 7.20 Ideal voltage transient in a PEM fuel cell after current interruption. The cell is operated at a fixed current. At $t = t_0$, the current is interrupted and the ohmic losses vanish almost immediately. After the current interruption, overpotentials start to decay and the voltage rises exponentially toward the open-circuit voltage. At $t = t_1$, the current is again switched on [88]. (Reprinted with permission from Elsevier.)

recording the voltage transient upon interruption of the current after the fuel cell has been operated at a constant current. The ohmic losses disappear almost immediately, whereas the electrochemical (or activation) overpotentials decline to the open-circuit value at a considerably slower rate. Therefore, rapid acquisition of the transient data is of vital importance to adequately differentiate the ohmic and activation losses [88].

Using the current interrupt method, Noponen et al. [89] recognized the effect of operating conditions on ohmic resistances in a free-breathing PEM fuel cell. The same method was employed by Mennola et al. [88] to determine the ohmic resistances in the individual cells of a PEM fuel cell stack. This was achieved by producing voltage transients and monitoring them with a digital oscilloscope connected in parallel with the individual cell. Their results showed good agreement between the ohmic losses in the entire stack and the sum of the ohmic losses in each individual cell.

Compared to other methods, such as impedance spectroscopy, the current interrupt method has the advantage of relatively straightforward data analysis. However, one of the weaknesses of this method is that the information obtained for a single cell or stack is limited. Another issue is the difficulty in determining the exact point at which the voltage jumps instantaneously.

7.4.1.3 Electrochemical Impedance Spectroscopy

In contrast to linear sweep and potential step methods where the system is far from equilibrium, electrochemical impedance spectroscopy applies a small AC voltage or current perturbation/signal (of known amplitude and frequency) to the cell, and the amplitude and phase of the resulting signal are measured as a function of frequency.

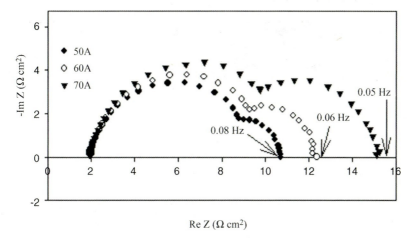

Figure 7.21 Typical impedance spectra of a PEM fuel cell. The spectra were obtained at 30 °C using a Ballard Mark V six-cell stack with an active area of 280 cm^2 [100]. (Reprinted with permission from Elsevier.)

Basically, impedance is a measure of the ability of a system to impede the flow of electrical current; thus, EIS is a powerful technique that can resolve various sources of polarization loss in a short time, and it has been widely applied to PEM fuel cells. Common uses for EIS analysis in PEM fuel cells are to study the ORR [90, 91], to characterize transport (diffusion) losses [92–95], to evaluate ohmic resistance and electrode properties such as charge transfer resistance and double-layer capacitance [96], and to evaluate and optimize the MEA [97–99].

Impedance spectra are conventionally plotted in both Bode and Nyquist forms. In a Bode plot, the amplitude and phase of the impedance are plotted as a function of frequency, while in a Nyquist plot the imaginary part of the impedance is plotted against the real part at each frequency. Figure 7.21 shows a typical impedance spectra for PEM fuel cells in the Nyquist form with two arcs, where the frequency increases from the right to the left [100]. The high-frequency arc reflects the combination of the double-layer capacitance in the CLs, the effective charge transfer resistance, and the ohmic resistance, through which the latter can be directly compared to the data obtained from current interrupt measurements. The low-frequency arc reflects the impedance due to mass transport limitations [101, 102].

Due to the fast hydrogen reduction reaction (HRR), the impedance spectra of the PEM fuel cells are often nearly equal to the cathode impedance. In these cases, the fuel cell impedance under H_2/O_2 gradients can be used to study cathode behavior.

Many researchers have studied the charge transfer of the cathode CL, which is represented by the high-frequency arc. The ohmic resistance, including the membrane resistance together with the contact resistances and resistances of the GDL and bipolar plates, is given by the real part of the high-frequency end of the high-frequency arc [90, 103]. It is often assumed that any change in this value during operation is caused by a change in membrane ionic resistivity due to membrane

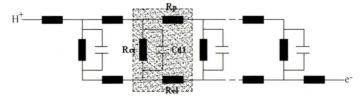

Figure 7.22 Transmission line equivalent circuit describing the impedance behavior of the CL [93]. (Reprinted with permission from Elsevier.)

hydration; the high-frequency resistance is therefore a measure of membrane water content.

While the high-frequency intercept corresponds to the ohmic resistance, the diameter of the arc relates to the charge transfer resistance of the CLs at high frequencies. The standard approaches to analyze the spectra using equivalent circuit models include the nonlinear least-squares (NLSQ) procedure [104] and the transmission line model [93]. Consequently, the effects due to charge transfer, air diffusion through the pores of the CL, and diffusion in the ionomer layer surrounding the catalyst particles can be separated.

Figure 7.22 illustrates the equivalent circuit of the transmission line model [93], which consists of two parallel resistive elements: one for electron transport through the conducting carbon particles (R_{el}) and the other for proton transport in the CL (R_p). The resistive elements are connected by double-layer capacitances (C_{dl}) in parallel to the charge transfer resistance (R_{ct}). The transmission line model was also employed by Makharia et al. [105] to mimic the CLs in H_2/N_2 and H_2/O_2 fueled PEM fuel cells. Parameters such as cell ohmic resistance, CL electrolyte resistance, and double-layer capacitance were extracted. More recently, Lefebvre et al. [94] and Jia et al. [106] used the transmission line model to simulate the impedance behavior of PEM fuel cell CLs under a N_2 atmosphere at the cathode side and, ultimately, the ionomer loading in the CL [107] was optimized. The EIS was further used to study the capacitance and ion transport properties of CLs under H_2/O_2 conditions [108].

In many cases, the appearance of the low-frequency arc, illustrated in Figure 7.21, can be attributed to the limitations on oxygen diffusion through nitrogen in the pores of the electrode [109, 110]. Evidence to support this conclusion comes from the absence of a low-frequency arc for operation with pure oxygen and an increase in the arc radius when operated under air or with increasing GDL thickness [111, 112].

CO poisoning is a significant issue for PEM fuel cells, lowering performance due to the deactivation of the Pt anode catalyst. Some researchers have employed the EIS technique to study CO poisoning of the anode [113–116]. However, two major problems exist. One complex issue is the separation of the anode and cathode impedances [113]; another problem is that the poisoning causes a change in the state of the fuel cell, which is reflected in the recorded impedance spectra [114].

In summary, EIS is a powerful technique for fuel cell studies. This dynamic method can yield more information than steady-state experiments and can provide diagnostic criteria for evaluating PEM fuel cell performance. The main advantage of EIS as a diagnostic tool for evaluating fuel cell behavior is its ability to resolve, in the frequency domain, the individual contributions of the various factors determining the overall PEM fuel cell power losses: ohmic, kinetic, and mass transport. Such a separation provides useful information for both optimization of the fuel cell design and selection of the most appropriate operating conditions.

7.4.1.4 Other Electrochemical Methods

Cyclic Voltammetry Cyclic voltammetry (CV) is a commonly used *in situ* approach for fuel cell research, especially to assess catalyst activity. The *in situ* CV technique has proven to be quite valuable for ascertaining the electrochemical surface area (ECA) of gas diffusion electrodes [117–120]. In this technique, the potential of a system is swept back and forth between two voltage limits while the current is recorded. The voltage sweep is normally linear with time and the plot of the current versus voltage is called a cyclic voltammogram. In a fuel cell, H_2 is fed to one side of the fuel cell, acting as both counter electrode and reference electrode, to function as a dynamic hydrogen electrode (DHE). The other side is flushed with inert gas (N_2 or Ar) and connected to the working electrode. Voltammetric studies are performed using a potentiostat/galvanostat, and the cyclic voltammograms can be recorded at different sweep rates. The ECA of the electrode is estimated based on the relationship between the surface area and the H_2 adsorption charge on the electrode, as determined from the CV measurement. The H_2 adsorption charge on a smooth Pt electrode has been measured to be approximately $210\,\mu C\,cm^{-2}$ of Pt loading in the CL. The ECA is then evaluated as [121]

$$\text{ECA}\,(cm^2\,Pt\,g^{-1}\,Pt) \frac{\text{charge}\,(\mu C\,cm^{-2})}{210\,(\mu C\,cm^{-2}\,Pt) \times \text{catalyst loading}\,(g\,Pt\,cm^{-2})}. \quad (7.12)$$

The disadvantage of this technique for assessing supported electrocatalysts is that the carbon features mask the H_2 adsorption and desorption characteristics, for example, the double-layer charging and redox behavior of surface active groups on carbon. In Figure 7.23, a typical fuel cell CV can be observed. The two peaks identified represent the hydrogen adsorption and desorption peaks on the Pt catalyst surface [122]. The forward and reverse double-layer current density $I_{dl\,charging}$, which is offset from the zero current value by the crossover current density $i_{crossover}$, is also shown.

CO Stripping Voltammetry CO stripping voltammetry is another common technique for determining the ECA of electrodes through the oxidation of adsorbed CO at room temperature, operating under the same principle as CV [123, 124]. One side of the fuel cell is supplied with CO plus inert gas or humidified high-purity inert gas (Ar or N_2) and connected to the working electrode, while humidified H_2 is fed to the other side, serving as the DHE [124]. During the CO adsorption process, CO plus inert gas is supplied to the anode at a certain flow rate, while the electrode potential is kept at about 0.1 V (versus DHE). Then, the gas is switched to high-purity Ar for a long time

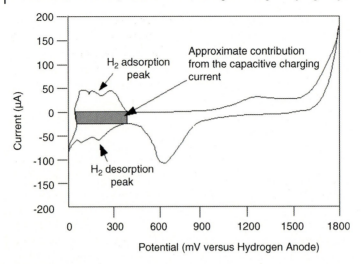

Figure 7.23 Schematic of a fuel cell CV. The two peaks identified represent the hydrogen adsorption and desorption peaks on the platinum fuel cell catalyst surface. The forward and reverse double-layer charging current density $I_{dl\ charging}$ and the crossover current density $i_{crossover}$ are also shown [79].

to remove any CO from the gas phase. To record the CO stripping voltammogram, the potential is scanned from the adsorption value to nearly 0.9 V. When assessing ECA based on CO stripping voltammetry, one can also employ Equation 7.12 by utilizing a value of $424\ \mu C\ cm^{-2}$ for polycrystalline Pt [125]. The CO stripping peak charge can provide information about the active surface sites of the CL. Song et al. [126] used this technique to investigate the effect of different electrode fabrication procedures on the structural properties of MEAs. It has also been found that exposing CO to platinum and the subsequent removal of that CO by electrochemical stripping is an excellent method of cleaning and activating Pt [123].

Linear Sweep Voltammetry Crossover of hydrogen and oxygen through the membrane is considered to be one of the most important phenomena in PEM fuel cells and leads to fuel inefficiency [127]. Linear sweep voltammetry (LSV) experiments are always conducted at room temperature to evaluate and monitor fuel crossover and to check for the presence of electrical shorts [128, 129]. The experimental procedure is similar to the CV technique, the principal difference being the irreversibility of the scan. Humidified H_2 and N_2 are supplied to the anode and cathode sides of the fuel cell, respectively. The scan potential ranges from 0 to 0.8 V, with higher voltages avoided to prevent Pt oxidation [128]. The experimental procedure involves controlling the potential of the fuel cell cathode (working electrode) and monitoring any electrochemical activity that occurs in the form of a current. Since N_2 gas is the only substance introduced into the cathode chamber, any current generated in the given potential range is solely attributable to the electrochemical oxidation of H_2 gas that crosses over from the anode side through the membrane. The crossover current typically increases with the scan potential and rapidly reaches a limiting value when the potential grows to around 300 mV [129]. At this value, all crossover H_2 is instantaneously oxidized due to the high overpotential applied. Based on the limiting current, one can ultimately calculate the flux of H_2 gas using Faraday's law [129]. Using this diagnostic method, Kocha et al. [127] examined the effects of various operating temperatures, gas pressures, and relative humidity on hydrogen crossover.

7.4.2
Physical and Chemical Methods

7.4.2.1 Species Distribution Mapping

Pressure Drop Measurements Pressure drop can be an important design parameter and diagnostic tool, especially at the cathode where water is produced. The cause of pressure drop is friction between the reactant gases and the flow-field passages and/or the GDL. A pump or blower should provide any required increase in pressure, resulting in an increase in parasitic power (load) loss. However, a higher pressure drop also results in more effective removal of excess liquid from the fuel cell. Therefore, the pressure drop must be carefully considered. During flooding, too much liquid water in the flow channels increases gas flow resistance, which impedes reactant gas transport and leads to performance losses related to mass transport. Pressure drop on the cathode side increases with cell flooding, while it remains unchanged with cell drying, thus clearly distinguishing between the two phenomena [130]. An increase in pressure drop, particularly on the cathode side of PEM fuel cell, is a reliable indicator of PEM fuel cell flooding. If the pressure drop exceeds the determined threshold, corrective measures are to be automatically initiated, such as dehumidifying the gases, increasing the gas mass flow rate, reducing the gas pressure, and/or reducing the current drain [131].

Gas Composition Analysis The species distribution within a PEM fuel cell is critical to fully characterize the local performance and accurately quantify the various modes of water transport. The most commonly used analytical technique for measuring the gas composition is gas chromatography (GC). Mench et al. [132] demonstrated the measurement of water vapor, hydrogen, and oxygen concentration distributions at steady state. A micro gas chromatograph and eight different sampling ports at various locations along the anode and cathode flow paths of a specially designed fuel cell were used. While gas chromatography provides a high accuracy (errors less than 2 mol%), this method is still essentially limited to analyzing steady-state species distribution due to significant data acquisition times. In general, the weakness of utilizing GC to measure species distribution is that only discrete data at desired positions can be determined.

Mass spectrometry is another analytical technique capable of determining the gas composition in an operating fuel cell. Partridge et al. [133] utilized spatially resolved capillary inlet mass spectrometry (SpaciMS) to successfully conduct measurements (with a temporal resolution of 104 ms) at realistic humidity levels, despite the concern that liquid water could block the capillaries used for gas sampling throughout the active area of the fuel cell. They analyzed the effect of load switching on the species concentrations and observed concentration gradients and nonuniformities.

Neutron Imaging Liquid water plays a key role in PEM fuel cells because its presence is closely linked to the functionality of the main components of the unit cell. To date, several universities and institutes have explored the neutron imaging method as an experimental tool to study water management in fuel cells. Because the neutron incoherent scattering length for hydrogen is almost two orders of magnitude larger than the length for almost all other elements, the neutrons are ideal for studying hydrogen-containing compounds such as water [134]. Equipped with good quality beam lines to probe the cell and a high-sensitivity scintillator/CCD camera as the detector system to record the images, the neutron imaging technique shows the potential for discerning two-phase flows within the fuel cell in real time or steady state [135–137]. Figure 7.24 shows a typical setup for neutron imaging of a fuel cell [135], in which neutrons are converted to light using a scintillator screen and the light is focused on the CCD chip.

Neutron imaging with sufficient spatial and temporal resolution has proven to be a powerful technique for obtaining qualitative and quantitative information on liquid water content and distribution. However, the need for a neutron source with a high fluence rate may hamper wide application of this method. In addition, cell rotation is imperative for present tomographic imaging methods to gain three-dimensional information on an operating PEM fuel cell [137].

Magnetic Resonance Imaging Magnetic resonance imaging (MRI), based on the nuclear magnetic resonance phenomenon, is an imaging technique using gradient radio frequency pulses in a strong magnetic field. The basic principle of MRI is that certain atomic nuclei within an object, if placed in a magnetic field, can be stimulated

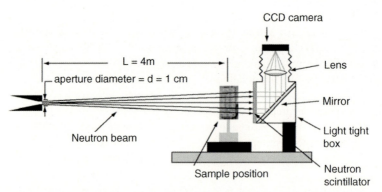

Figure 7.24 Typical experimental setup for neutron imaging of a fuel cell [135]. (Reprinted with permission from Elsevier.)

by the correct radio frequency pulses. After this stimulation, the nuclei relax while energy is induced into a receiver antenna to further obtain a viewable image. The nondestructive nature of MRI enables one to obtain unique *in situ* information from a multitude of systems [138]. In the case of PEM fuel cells, the hydrogen atom is always observed in MRI experiments due to the excellent correlation between its signal intensity and water content in the membrane [139]. MRI experiments by Tsushima and coworkers demonstrated the effects of operating conditions [140], membrane thickness [141], and liquid water supply [142] on water distribution within polymer electrolyte membranes in fuel cells. The weakness of the MRI technique lies mainly in the requirement that the materials be nonmagnetic. For this reason, the water content in the CL and GDL, whether these are made from nonwoven carbon paper or from woven carbon cloth, will be difficult to visualize with MRI [143].

7.4.2.2 Temperature Distribution Mapping

Heat generation always accompanies the operation of a fuel cell, due to thermodynamic irreversibility in the transport and electrochemical processes, fuel crossover (residual diffusion through the solid electrolyte membrane), and electrical heat from interconnection resistances. Spatial temperature variation can occur if any of these heat generating processes occur preferentially in different parts of the fuel cell stack.

IR Transparent Fuel Cells In recent years, a number of studies on temperature distribution along the active area of a fuel cell have used infrared cameras. Hakenjos *et al.* [144] designed a cell for the combined measurement of current and temperature distribution; for temperature distribution, an IR transparent window made out of zinc selenide was located at the cathode side. Sailler *et al.* [145] conducted combined measurements of current density and temperature distribution. The current density was measured with magnetic sensors, while the temperature distribution was measured with an array of nine thermocouples and an IR camera. The current density and temperature distributions highlighted heterogeneous distribution influenced by cell geometry and collector locations.

Embedded Sensors The most commonly utilized embedded sensor for temperature distribution mapping is the thermocouple. Wilkinson *et al.* [146] developed a simple, *in situ*, noninvasive method of measuring the temperature distribution of a fuel cell with microthermocouples. In this study, the thermocouples were located in the landing area of the flow-field plates (in contact with the GDL of the MEA). The temperature data taken at different locations along the flow channel were then used to find each temperature slope, which in turn was related to the local current density by modeling. Fiber optic sensors are an alternative to thermocouples as embedded temperature distribution mapping sensors. Recently, McIntyre *et al.* [147] developed two distinct fiber optic temperature probe technologies for fuel cell applications (free space probes and optical fiber probes). Both sensor technologies showed similar trends in fuel cell temperature and were also used to study transient conditions.

7.4.2.3 Current Distribution Mapping

In PEM fuel cells, uniformity of the current density across the entire active area is critical for optimizing cell performance. A nonuniform current density can have drastic effects on different parameters, such as reducing reactant and catalyst utilization along the active area, decreasing the total efficiency and lifetime, and/or causing fuel cell failure. Thus, determination of the current density distribution is vital for designing PEM fuel cells that can achieve higher performance and longer life [148]. A number of methods for measuring current distribution in PEM fuel cells have been demonstrated.

Partial MEA The partial MEA approach involves the use of several MEAs, each with different sections of the catalyzed active area covered, thus reducing the total active area of each MEA [148], as shown in Figure 7.25. Appointed sectional performance can be achieved by subtracting one steady-state polarization curve from another one. The main advantage of this method is that it is relatively simple to implement with respect to other techniques. In addition, this technology allows the cell performance to be analyzed in a steady state, with low spatial resolution. Although the resolution can be improved by increasing the number of segments or portions to be tested, significant errors arise due to inherent variations in electrical, transport, and kinetic properties between different MEAs.

Segmented Cells A number of research groups have presented segmented cell approaches and combined them with electrochemical methods, for example, EIS. These diagnostic approaches provide direct information not only on the current distribution of the cell but also on other phenomena that are occurring inside the cell under various operating conditions. To date, a parallel effort has been made to achieve sufficient spatial and temporal resolution by designing segmented flow-field plates. The basic concept of the segmented cell (or segmented flow-field plate) approach is to divide the anode or/and cathode plate into conductive segments that are electrically isolated from each other.

Subcells Stumper et al. [148] presented the subcell approach to measure localized currents and localized electrochemical activity in a fuel cell. In this method, a number of subcells were situated in different locations along the cell's active area and each subcell was electrically isolated from the others and from the main cell. Separate load banks controlled each subcell. Figure 7.26 shows the subcells in both the cathode and anode flow-field plates (the MEA also had such subcells). The current–voltage characteristics for the subcells, when compared to those of the main cell, are indicative of local fuel cell performance [148]. Although this method provides good understanding of the current distribution along a flow field, the manufacturing of the modified flow-field plates and MEA makes it very complex and difficult. Recently, Mench et al. [149], Yang et al. [150], and Rajalakshmi et al. [151] also used the subcell approach to determine the current distributions under various operating conditions in PEM fuel cells with a single serpentine flow field.

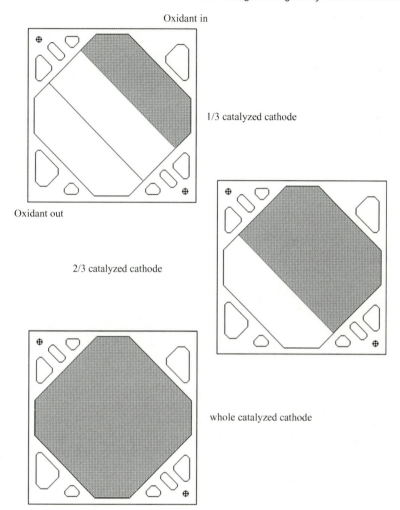

Figure 7.25 Sketch of MEA design showing active cathode area for the partial MEA method [148]. (Reprinted with permission from Elsevier.)

Segmented Plates In order to have spatially resolved performance data, Stumper *et al.* [148] used a passive resistor network made from resin-isolated graphite blocks located between the flow-field plate and the current collector. The potential drops across these blocks were monitored to establish the current flowing through them, using Ohm's law. By scanning the entirety of the graphite blocks, the electrode's current distribution was mapped. One of the main advantages to this approach is that time-dependent phenomena can be monitored in real time, enabling researchers to observe sudden changes after certain parameters have been modified. But a key issue about this configuration arises from the fact that the graphite blocks are

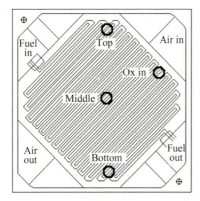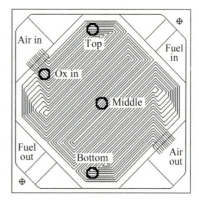

Figure 7.26 Schematic diagram of subcells method, showing fuel and oxidant flow fields separately [148]. (Reprinted with permission from Elsevier.)

not part of the flow-field plate; thus, with these passive resistors, there is the possibility of low spatial resolution due to the lateral in-plane current through the flow-field plate. The setup of these resistors also makes this approach quite complex and tedious. In another study by Wieser et al. [152, 153], a magnetic loop array was employed to determine the currents generated by segmented fuel cells. A flow-field plate divided into 40 electrically isolated segments was used to avoid lateral currents between adjacent segments. The current sensors that consisted of Hall sensors fixed in the air gaps of annular soft magnetic ferrites were placed around a gudgeon on each flow-field segment. The advantage is that a combination of high spatial and time resolution can be achieved and, in principle, integration into fuel cell stacks is possible. However, installation of the Hall sensors into the flow-field plate makes maintenance difficult and the investigation of different flow-field structures expensive.

Printed Circuit Board Cleghorn et al. [154] performed pioneering work using printed circuit board (PCB) technology to create a segmented anode current collector and anode flow field to measure current distribution in PEM fuel cells. In this study, the gas diffusion backing and the catalyst layer in the anode side were also segmented and each segment had an active area of $4.4\,cm^2$. The segmented fuel cell was designed using PCB technology. Through the use of current distribution, they were able to investigate different flow-field designs and optimize utilization of the active electrode area with the best humidification conditions and reactant stoichiometry. It is important to note that some of the mentioned segmented cell approaches have also been used to develop other diagnostic tools. For example, Brett et al. [155] used a segmented cell [156] to demonstrate localized EIS response over a frequency range of 0.1 Hz–10 kHz as a function of position in a PEM fuel cell. Their results proved that integral EIS measurement alone was not sufficient to properly characterize the operation of a fuel cell.

7.5
Final Comments

To ensure that the appropriate electrochemical reactions take place in the three-phase areas of a fuel cell catalyst layer, the proton conductor, electron conductor, and reactants must meet at the active catalyst sites. Passages for protons, electrons, gases, and water transport need to coexist in the porous electrode. During the development of PEM fuel cells, various types of electrodes have been designed and, accordingly, many techniques have also been proposed and implemented to fabricate these electrodes. In the past several decades, significant achievements have been made in electrode performance improvement and noble catalyst loading reduction. This chapter has reviewed various fabrication technologies for high-performance PEM fuel cell electrodes. There is no doubt that a high-performance electrode is primarily shaped by its preparation procedure. However, to fabricate durable and cost-effective electrodes for PEM fuel cell commercialization, we still need a breakthrough in materials exploration, such as non-noble catalysts and high-temperature membrane materials.

The optimized physical features of an electrode usually include surface morphology, microstructural properties, physical characteristics, and composition. Very often, it becomes important to know as much as possible about the microstructure of an electrode to determine how to improve its efficiency in carrying out the relevant electrochemical reaction. In this chapter, the commonly used techniques for characterizing PEM fuel cell electrodes have been addressed. Our discussion has focused on the merits and limitations of the techniques applied to PEM fuel cell research.

This chapter has also covered various diagnostic techniques currently employed in PEM fuel cell research. All the diagnostic approaches explained here were divided into two main categories: (1) electrochemical techniques and (2) chemical and physical methods. Among the various electrochemical techniques, the polarization curve has emerged as the simplest way of characterizing a fuel cell, while EIS is considered a powerful technique for determining electrode parameters as functions of structure and composition. With regard to chemical and physical methods, advancements in relevant equipment and materials are playing an important role in the application of the techniques to PEM fuel cells.

As shown in this chapter, test equipment integrated with several diagnostic techniques is a preferable strategy for providing deeper insights into the mechanisms that cause performance losses and spatial nonuniform distribution. Consequently, new information gathered with these diagnostic tools will either strongly support the development of empirical models or validate theoretical models predicting performance as a function of operating conditions and fuel cell characteristic properties.

Acknowledgments

The authors gratefully acknowledge the NRC-MOST Joint Research Program, BCICs ICSD Program, and the NRC-Helmholtz Joint Research Program for their financial support.

References

1 Ticianelli, E.A., Derouin, C.R., Redondo, A., and Srinivasan, S. (1988) *J. Electrochem. Soc.*, **135**, 2209–2214.
2 Chun, Y.G., Kim, C.S., Peck, D.H., and Shin, D.R. (1998) *J. Power Sources*, **71**, 174–178.
3 Lee, S.J., Mukerjee, S., McBreen, J., Rho, Y.W., Kho, Y.T., and Lee, T.H. (1998) *Electrochim. Acta*, **43**, 3693–3701.
4 Passalacqua, E., Lufrano, F., Squadrito, G., Patti, A., and Giorgi, L. (1998) *Electrochim. Acta*, **43**, 3665–3673.
5 Kumar, G.S., Raja, M., and Parthasarathy, S. (1995) *Electrochim. Acta*, **40**, 285–290.
6 Oishi, K. and Savadogo, O. (2008) *J. New Mater. Electrochem. Syst.*, **11**, 221–227.
7 Taylor, A.D., Kim, E.Y., Humes, V.P., Kizuka, J., and Thompson, L.T. (2007) *J. Power Sources*, **171**, 101–106.
8 Nakrumpai, B., Pruksathorn, K., and Piumsomboon, P. (2006) *Korean J. Chem. Eng.*, **23**, 570–575.
9 Wilson, M.S. (1993) Membrane catalyst layer for fuel cells. US Patent 5,234,777.
10 Kim, C.S., Chun, Y.G., Peck, D.H., and Shin, D.R. (1998) *Int. J. Hydrogen Energy*, **23**, 1045–1048.
11 Pettersson, J., Ramsey, B., and Harrison, D.J. (2006) *Electron. Lett.*, **42**, 1444–1446.
12 Andrade, A.B., Bejarano, M.L.M., Cunha, E.F., Robalinho, E., and Linardi, M. (2009) *J. Fuel Cell Sci. Technol.*, **6**, 021305-1–021305-3.
13 Park, H.S., Cho, Y.H., Park, I.S., Jung, N., Ahn, M., and Sung, Y.E. (2008) *J. Electrochem. Soc.*, **155**, B455–B460.
14 Towne, S., Viswanathan, V., Holbery, J., and Rieke, P. (2007) *J. Power Sources*, **171**, 575–584.
15 Su, H.N., Liao, S.J., Shu, T., and Gao, H.L. (2010) *J. Power Sources*, **195**, 756–761.
16 Leimin, X., Shijun, L., Lijun, Y., and Zhenxing, L. (2009) *Fuel Cells*, **9**, 101–105.
17 Morikawa, H., Tsuihiji, N., Mitsui, T., and Kanamura, K. (2004) *J. Electrochem. Soc.*, **151**, A1733–A1737.
18 Louh, R.F., Huang, H., and Tsai, F. (2005) Proceedings of the 1st European Fuel Cell Technology and Applications Conference 2005: Book of Abstracts, vol. 2005, p. 206.
19 Louh, R.F., Huang, H., and Tsai, F. (2007) *J. Fuel Cell Sci. Technol.*, **4**, 72–78.
20 Kim, H.S. and Popov, B.N. (2004) *Electrochem. Solid State Lett.*, **7**, A71–A74.
21 Kim, H.S., Subramanian, N.P., and Popov, B.N. (2004) *J. Power Sources*, **138**, 14–24.
22 Lee, J., Seo, J., Han, K.K., and Kim, H. (2006) *J. Power Sources*, **163**, 349–356.
23 Benítez, R., Soler, J., and Daza, L. (2005) *J. Power Sources*, **151**, 108–113.
24 Sasikumar, G., Ihm, J.W., and Ryu, H. (2004) *J. Power Sources*, **132**, 11–17.
25 Hirano, S., Kim, J., and Srinivasan, S. (1997) *Electrochim. Acta*, **42**, 1587–1593.
26 Caillard, A., Brault, P., Mathias, J., Charles, C., Boswell, R.W., and Sauvage, T. (2005) *Surf. Coat. Technol.*, **200**, 391–394.
27 Huang, K.L., Lai, Y.C., and Tsai, C.H. (2006) *J. Power Sources*, **156**, 224–231.
28 Khan, M.R. and Lin, S.D. (2006) *J. Power Sources*, **162**, 186–191.
29 Lin, C.S., Khan, M.R., and Lin, S.D. (2006) *J. Colloid Interface Sci.*, **299**, 678–685.
30 Lin, G. and Nguyen, T.V. (2005) *J. Electrochem. Soc.*, **152**, A1942–A1948.
31 Park, G.G., Sohn, Y.J., Yang, T.H., Yoon, Y.G., Lee, W.Y., and Kim, C.S. (2004) *J. Power Sources*, **131**, 182–187.
32 Giorgi, L., Antolini, E., and Pozio, A. (1998) *Electrochim. Acta*, **43**, 3675–3680.
33 Park, S., Lee, J.W., and Popov, B.N. (2006) *J. Power Sources*, **163**, 357–363.
34 Chen, J., Matsuura, T., and Hori, M. (2004) *J. Power Sources*, **131**, 155–161.
35 Passalacqua, E., Squadrito, G., Lufrano, F., Patti, A., and Giorgi, L. (2001) *J. Appl. Electrochem.*, **31**, 449–454.
36 Wang, X.L., Zhang, H.M., Zhang, J.L., Xu, H.F., Tian, Z.Q., Chen, J.,

Zhong, H.X., Liang, Y.M., and Yi, B.L. (2006) *Electrochim. Acta*, **51**, 4909–4915.
37. Qi, Z. and Kaufman, A. (2002) *J. Power Sources*, **109**, 38–46.
38. Antolini, E., Pozio, A., Giorgi, L., and Passalacqua, E. (1998) *J. Mater. Sci.*, **33**, 1837–1843.
39. Antolini, E., Passos, R.R., and Ticianelli, E.A. (2002) *J. Appl. Electrochem.*, **32**, 383–388.
40. Park, S., Lee, J.W., and Popov, B.N. (2008) *J. Power Sources*, **177**, 457–463.
41. Pasaogullari, U. and Wang, C.Y. (2004) *Electrochim. Acta*, **49**, 4359–4369.
42. Abbe, E. (1873) *Arch. Microsk.*, **9**, 413–468.
43. Tan, J., Chao, Y.J., Van Zee, J.W., and Lee, W.K. (2007) *Mater. Sci. Eng. A*, **445–446**, 669–675.
44. Liu, Z.Y., Brady, B.K., Carter, R.N., Litteer, B., Budinski, M., Hyun, J.K., and Muller, D.A. (2008) *J. Electrochem. Soc.*, **125**, B979–B984.
45. Ma, S., Solterbeck, C.H., Odgaard, M., and Skou, E. (2009) *Appl. Phys. A*, **96**, 581–589.
46. Murphy, D.B. (2001) *Fundamentals of Light Microscopy and Electronic Imaging*, Wiley-VCH Verlag GmbH, Weinheim.
47. Rasmussen, A. and Deckert, V. (2005) *Anal. Bioanal. Chem.*, **381**, 165–172.
48. Spence, J.C.H. (2003) *High-Resolution Electron Microscopy*, 3rd edn, Oxford University Press, New York.
49. Zhang, S.S., Yuan, X.Z., Hin, J.N.C., Wang, H.J., Wu, J.F., Friedrich, K.A., and Schulze, M. (2010) *J. Power Sources*, **195**, 1142–1148.
50. Wasmus, S. and Küver, A. (1999) *J. Electroanal. Chem.*, **461**, 14–31.
51. Ghassemi, H., McGrath, J.E., and Zawodzinski, T.A., Jr. (2006) *Polymer*, **47**, 4132–4139.
52. Inoue, H., Daiguji, H., and Hihara, E. (2004) *JSME Int. J. B*, **47**, 228–234.
53. Zhang, J., Yin, G., Wang, Z., and Shao, Y. (2006) *J. Power Sources*, **160**, 1035–1040.
54. Blom, D.A. and Dunlap, J.R. (2003) *Microsc. Microanal.*, **9** (Suppl. 2), 802–803.
55. Reeves, K.S., More, K.L., Walker, L.R., and Xie, J. (2004) *Microsc. Microanal.*, **10** (Suppl. 2), 1368–1369.
56. More, K.L. and Reeves, K.S. (2005) *Microsc. Microanal.*, **11** (Suppl. 2), 2104–2105.
57. Borup, R., Meyers, J., Pivovar, B., Kim, Y.S., Mukundan, R., Garland, N., Myers, D., Wilson, M., Garzon, F., Wood, D., Zelenay, P., More, K., Stroh, K., Zawodzinski, T., Boncella, J., McGrath, J.E., Inaba, M., Miyatake, K., Hori, M., Ota, K., Ogumi, Z., Miyata, S., Nishikata, A., Siroma, Z., Uchimoto, Y., Yasuda, K., Kimijima, K., and Iwashita, N. (2007) *Chem. Rev.*, **107**, 3904–3951.
58. Barbir, F. (2005) *PEM Fuel Cells: Theory and Practice*, Elsevier Academic Press, Burlington, MA.
59. Hinds, G. (2004) Performance and durability of PEM fuel cells: a review. NPL report DEPC-MPE 002, National Physics Laboratory, Teddington, UK.
60. Wu, J.F., Martin, J.J., Orfino, F.P., Wang, H.J., Legzdins, C., Yuan, X.Z., and Sun, C. (2010) *J. Power Sources*, **195**, 1888–1894.
61. Jena, A. and Gupta, K. (2001) *J. Power Sources*, **96**, 214–219.
62. Volfkovich, Y.M., Bagotzky, V.S., Sosenkin, V.E., and Blinov, I.A. (2001) *Colloids Surf. A*, **187–188**, 349–365.
63. Volfkovich, Y.M. and Bagotzky, V.S. (1994) *J. Power Sources*, **48**, 327–338.
64. Wang, C.Y. and Cheng, P. (1997) *Adv. Heat Transfer*, **30**, 93–196.
65. Williams, M.V., Begg, E., Bonville, L., and Kunz, H.R. (2004) *J. Electrochem. Soc.*, **151**, A1173–A1180.
66. Prasanna, M., Ha, H.Y., Cho, E.A., Hong, S.A., and Oh, I.H. (2004) *J. Power Sources*, **131**, 147–154.
67. Gurau, V., Bluemle, M.J., De Castro, E.S., Tsou, Y.M., Mann, J.A., Jr., and Zawodzinski, T.A., Jr. (2006) *J. Power Sources*, **160**, 1156–1162.
68. Lim, C. and Wang, C.Y. (2004) *Electrochim. Acta*, **49**, 4149–4156.
69. Holmberg, K. (2002) *Handbook of Applied Surface and Colloid Chemistry*, John Wiley & Sons, Ltd, Chichester.
70. Brack, H.P., Slaski, M., Gubler, L., Scherer, G.G., Alkan, S., and Wokaun, A. (2004) *Fuel Cells*, **4**, 141–146.

71 Yu, H.M., Ziegler, C., Oszcipok, M., Zobel, M., and Hebling, C. (2006) *Electrochim. Acta*, **51**, 1199–1207.

72 Mathias, M., Roth, J., Fleming, J., and Lehnert, W. (2003) *Handbook of Fuel Cells: Fundamentals, Technology and Applications*, vol. 3 (eds W. Vielstich, H.A., Lamm, and A. Gasteiger), John Wiley & Sons, Ltd, Chichester, pp. 1–21.

73 Owens, D.K. and Wendt, R.C. (1969) *J. Appl. Polym. Sci.*, **13**, 1741–1747.

74 Parry, V., Appert, E., and Joud, J.C. (2010) *Appl. Surf. Sci.*, **256**, 2474–2478.

75 Washburn, E.W. (1921) *Phys. Rev.*, **17**, 273–283.

76 Srinivasan, S. and Kirby, B. (2006) *Fuel Cells: From Fundamentals to Applications* (ed. S. Srinivasan), Springer, New York, pp. 441–573.

77 Bi, W., Gray, G.E., and Fuller, T.F. (2007) *Electrochem. Solid State Lett.*, **10**, B101–B104.

78 Adjemian, K.T., Dominey, R., Krishnan, L., Ota, H., Majsztrik, P., Zhang, T., Mann, J., Kirby, B., Gatto, L., Velo-Simpson, M., Leahy, J., Srinivasan, S., Benziger, J.B., and Bocarsly, A.B. (2006) *Chem. Mater.*, **18**, 2238–2248.

79 Wu, J.F., Yuan, X.Z., Wang, H.J., Blanco, M., Martin, J.J., and Zhang, J.J. (2008) *Int. J. Hydrogen Energy*, **33**, 1735–1746.

80 Wu, J.F., Yuan, X.Z., Wang, H.J., Blanco, M., Martin, J.J., and Zhang, J.J. (2008) *Int. J. Hydrogen Energy*, **33**, 1747–1757.

81 Wang, C.Y. (2004) *Chem. Rev.*, **104**, 4727–4766.

82 Barbir, F. (2006) *Fuel Cell Technology: Reaching Towards Commercialization* (ed. N. Sammes), Springer, London, pp. 27–31.

83 Li F X. (2006) *Principle of Fuel Cells*, Taylor & Francis, New York.

84 Lim, C.Y. and Haas, H.R. (2006) A diagnostic method for an electrochemical fuel cell and fuel cell components. International Patent No. WO2006,029,254.

85 Hirschenhofer, J.H., Stauffer, D.B., Engleman, R.R., and Klett, M.G. (1998) *Fuel Cell Handbook*, 4th edn, Parsons Corporation, Reading, PA.

86 Ju, H. and Wang, C.Y. (2004) *J. Electrochem. Soc.*, **151**, A1954–A1960.

87 Wruck, W.J., Machado, R.M., and Chapman, T.W. (1987) *J. Electrochem. Soc.*, **134**, 539–546.

88 Mennola, T., Mikkola, M., and Noponen, M. (2002) *J. Power Sources*, **112**, 261–272.

89 Noponen, M., Hottinen, T., Mennola, T., Mikkola, M., and Lund, P. (2002) *J. Appl. Electrochem.*, **32**, 1081–1089.

90 Parthasarathy, A., Dave, B., Sirinivasan, S., and Appleby, A.J. (1992) *J. Electrochem. Soc.*, **139**, 1634–1641.

91 Antoine, O., Bultel, Y., and Durand, R. (2001) *J. Electroanal. Chem.*, **499**, 85–94.

92 Springer, T.E., Zawodzinski, T.A., Wilson, M.S., and Gottesfeld, S. (1996) *J. Electrochem. Soc.*, **143**, 587–599.

93 Eikerling, M. and Kornyshev, A.A. (1999) *J. Electroanal. Chem.*, **475**, 107–123.

94 Lefebvre, M.C., Martin, R.B., and Pickup, P.G. (1999) *Electrochem. Solid State Lett.*, **2**, 259–261.

95 Saab, P., Garzon, F.H., and Zawodzinski, T.A. (2003) *J. Electrochem. Soc.*, **150**, A214–A218.

96 Cooper, K.R., Ramani, V., Fenton, J.M., and Kunz, H.R. (2005) *Experimental Methods and Data Analyses for Polymer Electrolyte Fuel Cells*, Scribner Associates Inc., Southern Pines, NC.

97 Romero-Castanon, T., Arriaga, L.G., and Cano-Castillo, U. (2003) *J. Power Sources*, **118**, 179–182.

98 Song, J.M., Cha, S.Y., and Lee, W.M. (2001) *J. Power Sources*, **94**, 78–84.

99 Wagner, N. (2002) *J. Appl. Electrochem.*, **32**, 859–863.

100 Yuan, X.Z., Sun, J.C., Blanco, M., Wang, H.J., Zhang, J.J., and Wilkinson, D.P. (2006) *J. Power Sources*, **161**, 920–928.

101 Ivers-Tiffée, E. and Weber, A. (2003) *Handbook of Fuel Cells: Fundamentals, Technology and Applications*, vol. 2 (eds W. Vielstich, H.A., Gasteiger, and A. Lamm), John Wiley & Sons, Ltd, New York, pp. 220–235.

102 Perry, M.L., Newman, J., and Cairns, E.J. (1998) *J. Electrochem. Soc.*, **145**, 5–15.

103 Kurzweil, K. and Fischle, H.J. (2004) *J. Power Sources*, **127**, 331–340.
104 Ciureanu, M. and Roberge, R. (2001) *J. Phys. Chem. B*, **105**, 3531–3539.
105 Makharia, R., Mathias, M.F., and Baker, D.R. (2005) *J. Electrochem. Soc.*, **152**, A970–A977.
106 Jia, N., Martin, R.B., Qi, Z., Lefebvre, M.C., and Pickup, P.G. (2001) *Electrochim. Acta*, **46**, 2863–2869.
107 Li, G. and Pickup, P.G. (2003) *J. Electrochem. Soc.*, **150**, C745–C752.
108 Easton, E.B. and Pickup, P.G. (2005) *Electrochim. Acta*, **50**, 2469–2474.
109 Paganin, V.A., Olivaira, C.L.F., Ticianelli, E.A., Springer, T.E., and Gonzalez, E.R. (1998) *Electrochim. Acta*, **43**, 3761–3766.
110 Mueller, J.T. and Urban, P.M. (1998) *J. Power Sources*, **75**, 139–143.
111 Wilson, M.S., Springer, T.E., Davey, J.R., and Gottesfeld, S. (1995) Proceedings for the First International Symposium on Proton Conducting Membrane Fuel Cells, Pennington, NJ, pp. 115–126.
112 Freire, T.J.P. and Gonzalez, E.R. (2001) *J. Electroanal. Chem.*, **503**, 57–68.
113 Mazurek, M., Benker, N., Roth, C., Buhrmester, T., and Fuess, H. (2006) *Fuel Cells*, **1**, 16–20.
114 Wagner, N. and Gülzow, E. (2004) *J. Power Sources*, **127**, 341–347.
115 Schiller, C.A., Richter, F., Gülzow, E., and Wagner, N. (2001) *Phys. Chem. Chem. Phys.*, **3**, 2113–2116.
116 Schiller, C.A., Richter, F., Gülzow, E., and Wagner, N. (2001) *Phys. Chem. Chem. Phys.*, **3**, 374–378.
117 Ticianelli, E.A., Derouin, C.R., and Srinivasan, S. (1988) *J. Electroanal. Chem.*, **251**, 275–295.
118 Koponen, U., Kumpulainen, H., Bergelin, M., Keskinen, J., Peltonen, T., Valkiainen, M., and Wasberg, M. (2003) *J. Power Sources*, **118**, 325–333.
119 Wang, M., Woo, K., Lou, T., Lou, Y., Zhai, Y., and Kim, D. (2005) *Int. J. Hydrogen Energy*, **30**, 381–384.
120 Tamizhmani, G. and Capuano, G.A. (1994) *J. Electrochem. Soc.*, **141**, 968–975.
121 Ralph, T.R., Hards, G.A., Keating, J.E., Campbell, S.A., Wilkinson, D.P., David, M., St-Pierre, J., and Johnson, M.C. (1997) *J. Electrochem. Soc.*, **144**, 3845–3857.
122 O'hayre, R., Cha, S.W., Colella, W., and Brinz, F.B. (2006) *Fuel Cell Fundamentals*, John Wiley & Sons, Inc., New York.
123 Brett, D.J.L., Atkins, S., Brandon, N.P., Vesovic, V., Vasileiadis, N., and Kucernak, A.R. (2004) *J. Power Sources*, **133**, 205–213.
124 Salgado, J.R.C., Antolini, E., and Gonzalez, E.R. (2004) *J. Electrochem. Soc.*, **151**, A2143–A2149.
125 Pozio, A., Francesco, M.D., Cemmi, A., Cardellini, F., and Giorgi, L. (2002) *J. Power Sources*, **105**, 13–19.
126 Song, S.Q., Liang, Z.X., Zhou, W.J., Sun, G.Q., Xin, Q., Stergiopoulos, V., and Tsiakaras, P. (2005) *J. Power Sources*, **145**, 495–501.
127 Kocha, S.S., Yang, J.D., and Yi, J.S. (2005) *AIChE J.*, **52**, 1916–1925.
128 Ramani, V., Kunz, H.R., and Fenton, J.M. (2004) *J. Membr. Sci.*, **232**, 31–44.
129 Ramani, V., Kunz, H.R., and Fenton, J.M. (2005) *J. Power Sources*, **152**, 182–188.
130 Görgün, H., Arcak, M., and Barbir, F. (2006) *J. Power Sources*, **157**, 389–394.
131 Bosco, A.D. and Fronk, M.H. (2000) Fuel cell flooding detection and correction. US Patent 6,103,409.
132 Mench, M.M., Dong, Q.L., and Wang, C.Y. (2003) *J. Power Sources*, **124**, 90–98.
133 Partridge, W.P., Toops, T.J., Green, J.B., and Armstrong, T.R. (2006) *J. Power Sources*, **160**, 454–461.
134 El-Abd, A.E.-G., Czachor, A., Milczarek, J.J., and Pogorzelski, J. (2005) *IEEE Trans. Nucl. Sci.*, **52**, 299–304.
135 Satija, R., Jacobson, D.L., Arif, M., and Werner, S.A. (2004) *J. Power Sources*, **129**, 238–245.
136 Pekula, N., Heller, K., Chuang, P.A., Turhan, A., Mench, M.M., Brenizer, J.S., and Ünlü, K. (2005) *Nucl. Instrum. Methods Phys. Res. A*, **542**, 134–141.
137 Hickner, M.A., Siegel, N.P., Chen, K.S., Mcbrayer, D.N., Hussey, D.S., Jacobson,

D.L., and Arif, M. (2006) *J. Electrochem. Soc.*, **153**, A902–A908.

138 Callaghan, P.T. (1991) *Principles of Nuclear Magnetic Resonance Microscopy*, Oxford University Press, New York.

139 Bunce, N.J., Sondheimer, S.J., and Fyfe, C.A. (1986) *Macromolecules*, **19**, 333–339.

140 Tsushima, S., Teranishi, K., and Hirai, S. (2005) *Energy*, **30**, 235–245.

141 Teranishi, K., Tsushima, S., and Hirai, S. (2005) *Electrochem. Solid State Lett.*, **8**, A281–A284.

142 Tsushima, S., Teranishi, K., Nishida, K., and Hirai, S. (2005) *Magn. Reson. Imaging*, **23**, 255–258.

143 Feindel, K.W., Bergens, S.H., and Wasylishen, R.E. (2006) *Chem. Phys. Chem.*, **7**, 67–75.

144 Hakenjos, A., Muenter, H., Wittstadt, U., and Hebling, C. (2004) *J. Power Sources*, **131**, 213–216.

145 Sailler, S., Rosini, S., Chaib, M.A., Voyant, J.Y., Bultel, Y., Druart, F., and Ozil, P. (2007) *J. Appl. Electrochem.*, **37**, 161–171.

146 Wilkinson, M., Blanco, M., Gu, E., Martin, J.J., Wilkinson, D.P., Zhang, J.J., and Wang, H.J. (2006) *Electrochem. Solid State Lett.*, **9**, A507–A511.

147 McIntyre, J., Allison, S.W., Maxey, L.C., Partridge, W.P., Cates, M.R., Lenarduzzi, R., Britton, C.L., Jr., Garvey, D., and Plant, T.K. (2005). DOE Hydrogen Program FY 2005 Progress Report, US Department of Energy, p. 989.

148 Stumper, J., Campbell, S.A., Wilkinson, D.P., Johnson, M.C., and Davis, M. (1998) *Electrochim. Acta*, **43**, 3773–3783.

149 Mench, M.M. and Wang, C.Y. (2003) *J. Electrochem. Soc.*, **150**, A1052–A1059.

150 Yang, X.G., Burke, N., Wang, C.Y., Tajiri, K., and Shinohara, K. (2005) *J. Electrochem. Soc.*, **152**, A759–A766.

151 Rajalakshmi, N., Raja, M., and Dhathathreyan, K.S. (2002) *J. Power Sources*, **112**, 331–336.

152 Wieser, C., Helmbold, A., and Gülzow, E. (2000) *J. Appl. Electrochem.*, **30**, 803–807.

153 Wieser, C. and Helmbold, A. (1999) Method for determining materials conversion in electrochemical reactions and electrochemical unit. International Patent No. WO9,926,305.

154 Cleghorn, S.J.C., Derouin, C.R., Wilson, M.S., and Gottesfeld, S. (1998) *J. Appl. Electrochem.*, **28**, 663–672.

155 Brett, D.J.L., Atkins, S., Brandon, N.P., Vesovic, V., Vasileiadis, N., and Kucernak, A. (2003) *Electrochem. Solid State Lett.*, **6**, A63–A66.

156 Brett, D.J.L., Atkins, S., Brandon, N.P., Velisa, V., Vasileiadis, N., and Kucernak, A.R. (2001) *Electrochem. Commun.*, **3**, 628–632.

8
Nanostructured Electrodes for Lithium Ion Batteries
Ricardo Alcántara, Pedro Lavela, Carlos Pérez-Vicente, and José L. Tirado

Nanoscience and nanotechnology are making a strong impact on the development of materials for energy storage, particularly for lithium ion batteries. Net improvements in total capacity, rate performance, and electrode stability are some of the advantages of using nanomaterials. Also, a combination of electrical and interfacial effects leading to non-Faradaic capacities together with new reaction mechanisms such as displacement or conversion reactions has given breakthrough results for both electrodes of Li-ion batteries. While layered-, spinel-, and olivine-related solids commonly used as cathode materials improve their capacity and rate performance by nanodispersion, new materials in the fluoride or silicate world emerge as potential candidates to replace the currently applied materials. For the anode, nanodispersed forms of tin and, more recently, nanodispersed forms of silicon provide veritable alternatives to the conventional carbon-based materials.

8.1
Introduction

One of the greatest challenges for the scientific community is to satisfactorily address the worldwide energy needs while minimizing global warming emissions. Particularly, the dependence on nearly depleted fossil energy sources justifies the urgent necessity to shift toward renewable and less polluting energy sources. For properly using intermittent solar or wind energies, load leveling and distribution networks require advanced systems to store excess electricity. For less polluting electric vehicles (EVs) and hybrid electric vehicles (HEVs), the requirements are even more restrictive. Thus, it is imperative to find new high-capacity and high-power storage systems that are cheaper and environmentally friendlier than the existing ones. No doubt, the most promising rechargeable systems are lithium ion batteries. Even though they are extensively used at present, they still suffer from a deficiency of optimal materials for the electrodes and the electrolyte. Another important limitation of this type of batteries is their high cost in large-scale production that at present does not permit lithium ion systems to satisfy the continuously increasing global demands.

Solid State Electrochemistry II: Electrodes, Interfaces and Ceramic Membranes.
Edited by Vladislav V. Kharton.
© 2011 Wiley-VCH Verlag GmbH & Co. KGaA. Published 2011 by Wiley-VCH Verlag GmbH & Co. KGaA.

Although nanomaterials promptly emerged in many fields such as catalysis or electronics, they arrived much more recently in the field of energy storage. A brief analysis of fundamental aspects relevant for the use of nanomaterials in solid-state electrochemical devices and important examples can be found in the first volume of this handbook (Chapters 1 and 4). Selected methodological and structural aspects important for understanding ion insertion and diffusion processes are described in Chapters 5 and 7 of the first volume. In this chapter, we describe how current research trends depart from the usual strategy in order to develop the foundations for the future generation of energy storage systems by using different concepts imported from nanoscience and nanotechnology. Recent results have demonstrated the viability of extending reversible reaction with lithium to new electrode materials in the nanoscale, which are different from previously known materials in terms of

- a huge increase in capacity due to a favored accessibility of the reaction to the complete mass of active material;
- new reaction pathways unsuitable in the absence of nanodispersion, which may increase the number of exchangeable electrons per unit formula;
- excellent rate performance by reducing the diffusion lengths combined, in some cases, with extra capacitive contributions;
- a wide range of working voltages due to the involvement of the surface energy in addition to the reaction energy;
- increased safety due to replacing the low-potential graphite intercalation compounds (GICs) by new anode materials;
- less polluting systems, resulting from the replacement of cobalt by other environment-friendly elements; this in turn may reduce production costs.

In summary, recent discoveries have demonstrated the exciting possibilities of using nanomaterials and nanostructures for both the anode and the cathode of the next-generation lithium ion batteries, leading to a revolutionary form of energy storage and conversion. The rest of the chapter contains different convincing demonstrations in this direction.

8.2
Positive Electrodes: Nanoparticles, Nanoarchitectures, and Coatings

8.2.1
Layered Oxides: LiMO$_2$

The suitability of LiCoO$_2$ as electrode material was proposed two decades ago and one decade before the start of the commercial use of lithium ion batteries [1]. Since 1991, lithium cobaltate with a layered structure has been the most commonly used positive electrode for lithium ion batteries, mainly due to its high energy density and easy preparation [2, 3]. Due to its cost and toxicity, the replacement of cobalt would be recommendable [4]. Otherwise, the cobalt recuperation contributes to the economic viability of recycling of the lithium ion batteries. The electron paramagnetic resonance (EPR) data on LiNiO$_2$ suggest migration of nickel after cycling, yielding poor stability [5]. To increase the safety, LiCoO$_2$ particles of 20 μm were specifically prepared and used in the batteries of Sony Corporation [2].

The preparation of LiCoO$_2$ [6] and LiNiO$_2$ [7] nanoparticles by thermal decomposition of metal-organic precursors was reported early. The decrease in the particle size down to the nanometric domain can be advantageous in minimizing the lithium diffusion pathway and increasing the charge rate. However, small particles with large surface area can enhance the electrolyte decomposition during the charge process, and this is not desirable from the point of view of safe batteries. Nanostructured LiCoO$_2$ fibers were prepared by a sol–gel-related electrospinning technique. Very large capacities in the first cycles but strong capacity fade were observed upon cycling [8]. A "desert rose" form of LiCoO$_2$ with nanostructure and with high electrochemical capacity at high rates has been reported [9]. Electrochemical measurements and theoretical analyses on nanocrystalline LiCoO$_2$ revealed that particle size below 15 nm is not favorable for most applications. An excellent high-rate capability was observed in nanocrystalline LiCoO$_2$ with a particle size of 17 nm [10].

Surface coating with nanometric layers of a protective material is a parallel way to improve electrode stability in layered oxides. Mladenov et al. [11] studied the Mg doping and MgO surface modification of LiCoO$_2$. Later, Liu et al. [12] reported LiCoO$_2$ coating with Al$_2$O$_3$. More recently, ZrO$_2$ and AlPO$_4$ coatings at the nanoscale of layered compounds could help to maintain the electrode integrity upon cycling [13]. Also, Cho et al. [14] and later Chen and Dahn [15] studied the effect of a ZrO$_2$ coating on lithium cobaltate. A model based on the mechanical stresses was proposed in the former study. Coating of LiNi$_y$Co$_{1-y}$O$_2$ with MgO has also been reported [16].

8.2.2
Spinel Compounds

8.2.2.1 Materials for 4 V Electrodes
Manganese oxides are suitable hosts for lithium insertion/extraction reactions [17]. LiMn$_2$O$_4$ is an attractive positive electrode due to several advantages, such as low cost, low toxicity, and high discharge capacity [18, 19]. However, the use of LiMn$_2$O$_4$ in lithium ion batteries is restricted by the Jahn–Teller effect on manganese ions. Thus, capacity rapidly fades during electrode cycling. A dynamic Jahn–Teller effect is induced by the presence of Mn^{3+} and the manganese environment is affected by a tetragonal distortion. The continuous interconversion between Mn^{3+} and Mn^{4+} redox species induces lattice stresses and eventually structural degradation [20]. Several solutions have been proposed to overcome this problem including the partial substitution of Mn^{3+} ions by other transition metal ions. In addition, batteries with this spinel are particularly unstable at high temperature due to surface reactions. Nevertheless, nanoarchitectured materials have also been suggested for LiMn$_2$O$_4$ both to improve rate performance and to better use the theoretical capacity. The template method of synthesis of nanostructured materials is a simple and straight-

forward procedure that has provided interesting results. One of the most versatile templates are porous alumina membranes that can be prepared electrochemically from aluminum metal and are also commercially available. The pores in these membranes are arranged in a regular hexagonal lattice. Pore densities as high as 10^{11} pores cm^{-2} can be achieved [21]. After chemical reactions inside the porous volume, leading to the desired product, the template can be easily dissolved in alkali, thanks to the amphoteric nature of alumina. $LiMn_2O_4$ nanotubes with 200 nm of outer diameter were prepared using a nanoporous alumina membrane as a template and showed good electrochemical behavior in the first cycles [22]. Another example is provided by mesoporous $LiMn_2O_4$, which recently showed high-rate capability and excellent cycling [23].

Zheng et al. [24] studied $Li_{1.03}Mn_{1.97}O_4$ coated with a thin layer of SiO_2 obtained by a sol-gel method. Cho et al. [25] studied $LiMn_2O_4$ coated with Co_3O_4 where the formation of a solid solution $Li_zMn_{2-x}Co_xO_4$ thin-film phase was reported.

8.2.2.2 Materials for Higher Voltages

Most studies on lithium ion batteries have used the so-called 4 V cathodes based on $LiCoO_2$, $LiNiO_2$, and $LiMn_2O_4$. In spinel oxides, several quaternary compositions with a general stoichiometry $LiM_xMn_{2-x}O_4$ (M = Cr, Fe, and Co) [26] and $LiNi_{0.5}Mn_{1.5}O_4$ [27] have been assayed. In all cases, the Mn^{4+}/Mn^{3+} redox couple is substituted by M^{4+}/M^{3+} or Ni^{4+}/Ni^{2+}. Thus, Mn^{3+} is avoided and subsequently a notorious increase in the working voltage is detected from 4 to 4.8 V. Among these, special attention has been paid to $LiNi_{0.5}Mn_{1.5}O_4$ because its discharge capacity is delivered in a unique voltage interval at around 4.7 V, whereas other materials' capacity is shared by two plateaus at quite different voltages (about 4.0 and 5 V).

The conventional solid-state synthesis is an easy route to prepare a large amount of the product. Nevertheless, most of the reported cases needed multiple steps, regrinding procedures, or high temperatures to eventually obtain the pure phase. Problems as oxygen release at high temperatures ought to be solved by careful cooling or reheating [27]. However, the rate capability of these high-voltage cathode materials has to be improved to meet high power demands of lithium ion batteries in new applications. For this purpose, the most common strategy is the control of the microstructure [28] or the diminution of the diffusion length to enhance lithium ion insertion [29].

Several methods of synthesis have been proposed to prepare nanostructured $LiNi_{0.5}Mn_{1.5}O_4$ including a composite carbonate process [30], soft combustion [31], or a sucrose-aided combustion method [32]. Kunduraci and Amatucci [33] have reported a successful combination of these strategies. Besides cubic spinel structure/space group $Fd3m$, $LiNi_{0.5}Mn_{1.5}O_4$ synthesized under O_2 atmosphere is characterized by an ordered $P4_332$ structure [34]. Transition metal ions are ordered on the 4b and 12d octahedral sites when Mn^{3+} content decreases. Kunduraci et al. have demonstrated a correlation between the decrease in Mn^{3+} ions and the electronic conductivity. A higher electronic conduction of disordered ($Fd3m$) spinel phases than the ordered ($P4_332$) spinel was observed [35]. In fact, a better electrochemical behavior and structure reversibility for $Fd3m$ $LiNi_{0.5}Mn_{1.5}O_4$ has been reported than for $P4_332$ $LiNi_{0.5}Mn_{1.5}O_4$ [36]. The synthesis of nanostructured 4.7 V $Li_xMn_{1.5}Ni_{0.5}O_4$ spinels provided rate capabilities exceeding 80% utilization at 6C.

Nevertheless, nanostructured materials exhibited poorer cycle life than microstructured counterparts [33]. Effectively, even if the rate capabilities are enhanced in nanostructured electrode materials, cathode materials constituted by nanocrystallites often favor parallel reactions with the electrolyte. It leads to a poorer extent of useful cycling and safety issues.

Coating was also successfully applied to improve the stability of HV electrodes in contact with the electrolyte. Thus, different authors [37–40] have reported ZnO-coated $LiNi_{0.5}Mn_{1.5}O_4$ and its electrochemical performance. A partial incorporation of zinc into the spinel lattice was suggested [40].

8.2.3
Olivine Phosphates

The olivine-related $LiFePO_4$ [41], a cathode material used in commercial Li-ion products for power tools and HEV [42] and an environment-friendly and low-cost material, owes its success to nanoscience. In general, $LiMPO_4$, olivine-related phosphates, show poor electrochemical performances due to their low intrinsic electronic conductivity. Lithium deinsertion/insertion from/into $LiFePO_4$ takes place through a two-phase reaction mechanism between $Li_{1-\delta}FePO_4$ and $Li_\delta FePO_4$. Recent results indicate that for nanostructured materials the growth reaction is considerably faster than its nucleation (described by the authors as "domino-cascade model"), explained by the existence of structural constraints occurring just at the reaction interface [43]. Anyway, by modifying the particle size and ion ordering (the presence of defects and cation vacancies) in the material, the two-phase mechanism described at room temperature can be modified resulting in a single-phase mechanism [44].

The improvement in the rate capacity of doped nanostructured $LiFePO_4$ (obtained by ball milling and heating under Ar atmosphere) was first reported by Chung et al., with a reversible capacity of about 120 mAh g^{-1} at 1C rate [45]. The synthesis by microwave heating with multiwall carbon nanotubes (MWCNTs) is also quite effective to obtain MWCNT-coated nanoparticles of $LiFePO_4$, although the electrochemical test showed a relatively low capacity of 145 mAh g^{-1} at C/2 rate [46]. The polyol method also allows to obtain nanoscaled $LiFePO_4$ with high crystallinity. This material showed an initial capacity of 166 mAh g^{-1} at 0.1 mA cm^{-2} (about 98% of the theoretical one), with an excellent capacity retention (163 mAh g^{-1} from the 2nd to 50th cycle) and good rate performance at high current rates (58 and 47% of capacity retention at 30C and 60C rate, respectively) [47]. Coating with ZnO nanofilms was also proposed to avoid dissolution of the phosphate on cycling [48]. A net improvement is observed by this approach (Figure 8.1).

An interesting and new approach is the synthesis of nanowire and hollow $LiFePO_4$ using hard templates. Both nanowire and hollow $LiFePO_4$ cathodes show excellent capacity, about 155 and 162 mAh g^{-1}, respectively, at 1C rate, without capacity fading

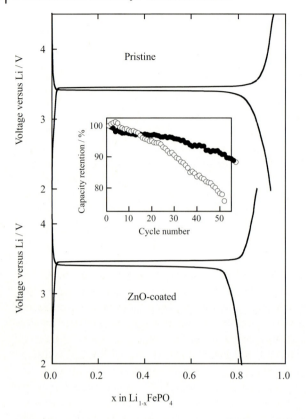

Figure 8.1 First galvanostatic charge and discharge cycle (solid lines) and capacity retention at C rate of lithium test cells using pristine (open circles) and ZnO-coated (filled circles) LiFePO$_4$ samples as active material for the positive electrode. ZnO coating was achieved by immersion in a zinc acetate bath followed by heat treatment at 400 °C. The coated material displays better capacity retention under the experimental conditions used in the study.

after 80 cycles. Even at a 15 C rate, the capacity values are about 137 and 153 mAh g^{-1}, close to the 90% capacity retention of the initial capacity at C/5 [49].

Kang and Ceder recently claimed an ultrafast nanostructured form of LiFePO$_4$ obtained by solid-state reaction (about 100 mAh g^{-1} nearly with negligible capacity fade after 50 cycles at 60C) [50]. However, this study has generated some controversy [51].

Nanostructured LiCoPO$_4$/C composite can be obtained by the microwave heating method, yielding particles with about 150 nm size, coated with a 10 nm thick uniform amorphous carbon film. Although electrochemical measurements indicate that carbon-coating film increases the electrical conductivity and lithium ion diffusion coefficient, exhibiting an initial capacity of 137 mAh g^{-1} at C rate and 144 at C/10, it rapidly decreases to below 100 mAh g^{-1} after 10 cycles [52].

An interesting nanostructured form is obtained by a combined microwave solvothermal approach that requires a short reaction time (5–15 min). The result is a monodispersed, single-crystalline LiMPO$_4$ (Mn, Fe, Co, and Ni) with nanothumb-like shapes. An additional treatment with MWCNT yields the final nanothumbed composite material. LiFePO$_4$/MWCNT shows high capacity (more than 150 mAh g^{-1} at 1C rate) with excellent rate capability. For M = Co, the capacity decreases down to 90 mAh g^{-1}. Although the nanonetworking of LiMnPO$_4$/MWCNT increases the capacity values and cyclability compared to LiMnPO$_4$, it exhibits poor electrochemical performance with a low capacity even at C/10 rate, about 45 mAh g^{-1} [53].

The olivine-related LiMnPO$_4$ material found almost no practical application as electrode material due to the low intrinsic conductivity (about 10^{-14} S cm^{-1}; compare, for example, with LiFePO$_4$, 10^{-9} S cm^{-1}) [54], resulting in a low electrochemical activity. It shows a high polarization in charge and discharge branches, mainly caused by the low ionic/electronic transport within particles. The reversible capacity was assigned to the reversible reaction Mn^{II}/Mn^{III}, about 140 mAh g^{-1} at a low rate as C/15 [55], while the theoretical capacity is about 170 mAh g^{-1}. Kwon et al. showed that the particle size plays an important role in their electrochemical performances. Improvement in ion transport was achieved by reducing the particle size, while enhancement of the electron transport was obtained by carbon coating [56]. Thus, they reached about 130 mAh g^{-1} at C/10 rate for 140 nm particle size, while increasing the particle size resulted in a progressive decrease in the capacity down to 60 mAh g^{-1} for 830 nm particle size.

Thin films of LiMPO$_4$ (M = Mn and Fe) and LiFe$_{0.6}$Mn$_{0.4}$PO$_4$ can be obtained by electrostatic spray deposition combined with the sol-gel method, having a uniform surface morphology with average grain size less than 100 nm. The electrochemical performance was similar to powder electrode reported previously, although LiFe$_{0.6}$Mn$_{0.4}$PO$_4$ film electrode exhibited a larger capacity than the pure LiFePO$_4$ and LiMnPO$_4$ film electrodes [57]. Anyway, the capacity values on discharge (about 50–60 mAh g^{-1}) are considerably lower than those reported for the carbon-containing LiFe$_y$Mn$_{1-y}$PO$_4$ compounds reported by Li et al. (about 165 mAh g^{-1} for y = 0.75) [58].

The use of the polyol method in the synthesis of LiMnPO$_4$ also yields a nanostructured material, although the specific conditions allow to control the final particle size: about 40–50 nm after 12 h of treatment and about 150 nm after 24 h. An additional heat treatment at 500 °C makes the particle size bigger and the distribution of the particle size broadened. As the particle size increases, the reversible capacity decreases from about 115 mAh g^{-1} (50 nm) down to about 100 mAh g^{-1} (150 nm) and about 80 mAh g^{-1} for the heated sample at about C/5 rate, revealing the importance of the particle size on the electrochemical performances [59].

If LiMnPO$_4$ powder obtained by the polyol method is ball-milled with 20 wt% carbon black, this results in approximately 30 nm particles of carbon-coated LiMnPO$_4$ (about 15 nm). The reversible capacity strongly increases up to 130 mAh g^{-1} at C/5 rate and 100 mAh g^{-1} at 1C rate after 100 cycles. This last capacity is also obtained when cycled at 5C at 50 °C [60]. Nanostructured carbon-coated LiMnPO$_4$ with platelet morphology also obtained by the polyol method showed good electrochemical performances, with approximately 110 mAh g^{-1} at 1C rate at room temperature and about 140 mAh g^{-1} at 50 °C [61].

These results clearly show that nanostructured carbon-coated LiMPO$_4$ can overcome the problems associated with ion (reducing the distance for Li$^+$ transport) and electronic (increasing the electronic contact between particles) transport, thus confirming their potential as a cathode material for Li-ion batteries.

8.2.4
Silicates

Probably as a consequence of the successful use of lithium iron phosphate with an olivine structure in commercial lithium ion batteries, the search for new cathode materials has been extended to the silicate world. An early attempt was made by Patoux and Masquelier [62]. In this work, Li$_2$TiSiO$_5$ among other Ti compounds was suggested as potential positive electrode in lithium batteries. However, it was found that the electrochemical performance of Li$_2$TiSiO$_5$ was poor. Until now, the best results have been attained for Li$_2$TSiO$_4$ (T = Mn, Fe, Co). However, these solids display almost insulating properties, which jeopardize their potential applicability. According to Arroyo-de Dompablo et al. [63], the origin of the limited conductivity relies on the presence of half-closed shell transition metal ions (Mn^{2+} in the discharged state, Fe^{3+} in the charged state). As in LiFePO$_4$, the conductivity problem of the silicate materials has been circumvented by the use of nanotextured materials, in which nanoparticles of the silicate and an electronic conducting material such as carbon are intimately mixed at the nanoscale. Further improvement has also been proposed by the preparation of silicate materials in a nanoparticulate or nanotextured form in intimate contact both with an electron-conducting phase and with an ion-conducting phase [64].

Li$_2$FeSiO$_4$ possesses a lithium superionic conductor (LISICON)-related structure (Figure 8.2, see also Chapter 7 of the first volume). The structure was recently revised and monoclinic parameters were introduced [65]. Concerning the synthesis of Li$_2$FeSiO$_4$ nanoparticles and carbon-containing nanocomposites, hydrothermal-assisted and sol-gel methods have been found particularly useful due to the molecular level mixing of the starting reagents. The addition of sources of carbon (e.g., sucrose) can also be used in conjunction with these methods [66].

Nytén et al. [67] first reported the electrochemical properties of Li$_2$FeSiO$_4$ at 60 °C. An initial discharge capacity close to the theoretical value decreased after a few cycles to about 140 mAh g^{-1}. A reaction involving phase transitions to more stable structures after the initial cycle was suggested. At room temperature, Dominko et al. [68] described a reversible lithium extraction reinsertion of 0.6 Li per formula at room temperature, in the 2.0–4.2 V range. The ideal half reaction in that voltage interval could reach a maximum oxidation of iron to the +3 oxidation state according to

$$Li_2FeSiO_4 \rightarrow LiFeSiO_4 + Li^+ + e^- \tag{8.1}$$

However, the door was left open for further possible extraction at higher voltages provided that a sufficiently stable electrolyte is found. A large discharge capacity of about 160 mAh g^{-1} at C/16 rate and superior charge and discharge capabilities under

8.2 Positive Electrodes: Nanoparticles, Nanoarchitectures, and Coatings | 391

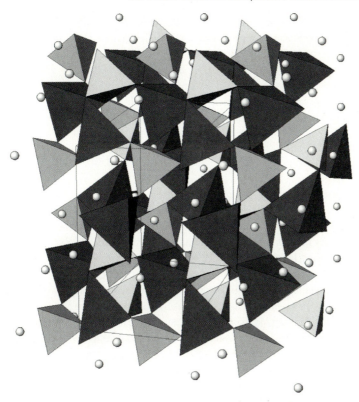

Figure 8.2 Monoclinic structure of Li_2FeSiO_4. SiO_4: light tetrahedra; FeO_4: dark tetrahedra; Li: spheres.

high rate conditions were found by carbon coating [66]. By using nanoparticulate materials, a discharge capacity of 100 mAh g^{-1} at a current density of 2C has been reported [69].

Similar behavior has been found for the isostructural Li_2MnSiO_4. However, formation of an amorphous phase shell during the first oxidation of Li_2MnSiO_4 has been recently suggested from *ex situ* X-ray diffraction data [70]. Similarly, DFT calculations suggested that both lithium atoms were extracted from the crystal cell unit (with a change in the oxidation state of manganese from Mn^{2+} to Mn^{4+}) simultaneously, before the extraction of lithium from neighbor crystal cell unit started to proceed. Solid solutions of the type $Li_2Mn_xFe_{1-x}SiO_4$ have also been tested [64].

Recently, a related cobalt silicate was studied as lithium intercalation compound. Pristine Li_2CoSiO_4 prepared by a solution route shows poor electrochemical performance. After coating with carbon by ball milling, a reversible lithium extraction of 0.46 Li per unit formula at an average voltage of about 4.1 V versus lithium was reported for Li_2CoSiO_4 [71]. So far, the electrochemical behavior of three polymorphs of the Li_2CoSiO_4, β_I, β_{II}, and γ_0, has been investigated as positive electrodes in rechargeable lithium batteries [72].

Finally, a vanadium-containing silicate has also been tested. Li_2VOSiO_4 crystallizes in a tetragonal structure (space group $P4/nmm$), in which VO_5 square pyramids with shared SiO_4 tetrahedra form layers. Li ions are accommodated between the layers. A reversible capacity of 0.6 Li per unit formula at an average voltage of 3.6 V was reported, which was taken as an unenviable limit to its practical application [73].

In summary, much work is still needed to reach suitable electrode materials for advanced Li-ion batteries in the silicate world.

8.2.5
Transition Metal Fluorides

Conversion electrode materials – particularly oxides and oxycompounds – for the negative electrode of Li-ion batteries will be discussed in Section 8.3.3. However, conversion electrodes are not exclusive to metal oxides. In fact, metal halides have also been successfully used as active electrode material of primary cells. The Li/AgCl was proposed in the early 1960s by NASA researchers as a high energy density cell [74], and the $Li/NiCl_2$ cell is close to be a commercial product by the MES-DEA company [75]. More recently, several authors have examined the possibilities of using metal fluoride electrodes as a logical continuation of the study of the reversible oxide conversion electrodes. As expected from the different Gibbs free energy change in the conversion reactions [76], working voltages make metal fluorides more appropriate for the cathodes of Li-ion batteries.

However, metal fluorides are electronic insulators and cannot be used as conventional electrode materials. The introduction of carbon metal fluoride nanocomposites (CMFNCs) made it possible to carry out the reversible conversion process according to

$$MF_x + xLi^+ + xe^- = M + xLiF \tag{8.2}$$

which occurs at high voltages (4.5–2.5) V for FeF_3, as expected from the free energy of formation of FeF_3. The theoretical capacity (237 mAh g^{-1}) could be mostly recovered with a total CMFNC-specific capacity close to 200 mAh g^{-1} [77].

Four parallel effects have been suggested to contribute to the enhanced activity: (i) an increased surface activity of the metal fluoride resulting from defects created by the profound reduction in crystallite size, (ii) the electrical connections between the fluoride grains by the highly conducting carbon nanoparticles, (iii) the short diffusion pathways for lithium, and (iv) the possibility of electron tunneling.

Li et al. [78] have shown good cycling properties of TiF_3 and revealed that extraction of lithium from LiF/Ti amorphous mixtures results in the formation of nanocrystalline TiF_3. Liao et al. [79], using combinatorial sputtering tables, were also able to prepare electroactive $[LiF]_{1-x}Fe_x$ films in nanocomposite libraries. The experimental capacities showed a peak around $x = 0.25$ (F/Fe = 3), suggesting the formation of FeF_3 upon charge. For lower F/Fe ratios, the capacity was higher than the theoretical values, which was interpreted in terms of surface storage.

A recent improvement resulted from the use of mixed conducting matrices, such as MoO_3, which showed the best performance for electroactive CuF_2. Compared to pure CuF_2 or nanocomposites fabricated with carbon matrices, the oxide matrices induced a distinct crystallographic change to the CuF_2 monoclinic crystallites. Near-theoretical capacity (450 mAh g^{-1} for the composite or 525 mAh g^{-1} for the pure compound) was reported for CuF_2. [80].

Finally, bismuth fluoride behaves similarly to FeF_3. According to Bervas et al. [81], both lithiation and delithiation reactions are two-phase processes.

The reversible electroactivity can be extended to oxyfluorides. Particularly, bismuth and more recently FeOF have been assayed. For BiO_xF_{3-2x} compounds, Bervas et al. [82] found that addition of small amounts of oxygen to the fluoride structure increased the electronic conductivity, which facilitates a better utilization of the theoretical capacity of the material. On decreasing the voltage of lithium test cells, BiOF was shown to react with lithium by two successive two-phase reactions, which may be written as follows:

$$3\, BiOF + 3\, Li^+ + 3\, e^- = Bi_2O_3 + 3\, LiF + Bi \tag{8.3}$$

$$Bi_2O_3 + 6\, Li^+ + 6\, e^- = 2\, Bi + 3\, Li_2O \tag{8.4}$$

The initial BiOF compound was found to reform upon oxidation. A similar mechanism was also suggested for $BiO_{0.5}F_2$.

For iron oxyfluoride, recently Pereira et al. [83] showed that the introduction of oxygen into the fluoride sublattice increases the average output voltage and has a beneficial impact on cycling stability and reversibility of the conversion process.

8.3
Negative Electrodes

8.3.1
Intercalation Nanomaterials

8.3.1.1 Graphene, Fullerene, and Carbon Nanotubes

Since the first reports on the isolation of graphene (a sheet of carbon atoms arranged in a honeycomb pattern) [84, 85], the special conduction properties of this material – "the ultimate incarnation of the surface [86]" – have been unfolded. Temperature-independent carrier mobilities, ambipolar carrier conduction, and changes in bandgap with ribbon width, doping [87, 88], and compound formation [89] are some examples. Also, different preparation methods have been developed, from the initial mechanical exfoliation of graphite leading both to multilayer graphenes and to single-layer graphene. Particularly, chemical vapor deposition (CVD) growth on a substrate was shown to be useful. Related to the chemical exfoliation of graphite, it was recently shown that MWCNTs can be opened longitudinally by intercalation of lithium and ammonia followed by exfoliation [90].

In connection with the reactivity of graphene with lithium, at least doubling the capacity of graphite is expectable by simply accommodating lithium in both external

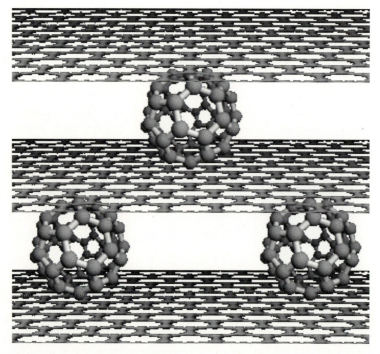

Figure 8.3 Artist's impression of graphene layers pillared by fullerene molecules, thus enhancing the surface exposed to lithium attachment.

surfaces. Moreover, the possible arrangement of single-layer sheets in a "falling cards" model [91] could facilitate the effective use of both sides. An alternative procedure was recently suggested [92]. A good control over the intergraphene sheet distance was achieved by using large molecular "pillars" such as carbon nanotubes or fullerenes (Figure 8.3). Both molecules caused an increase in the average d-spacing of about 0.40 nm, which resulted in capacity values that doubled the theoretical value of graphite.

The solid-state structure of C_{60} fullerene apparently offers many possibilities for lithium insertion. Exohedral lithium could be found in the different interstices between C_{60} molecules in a "virtual fcc packing" (virtual, as the fullerene molecules are not rigid spheres but the well-known I_h polyhedron [93]). In a simplified $Fm3m$ space group with the C_{60} molecules in 4a sites, different stoichiometries could be derived by filling the 4b (LiC_{60}), 8c (Li_2C_{60}), 24e (Li_6C_{60}), and 32f (Li_8C_{60}) sites, or a combination of all of them with possible formation of lithium clusters (up to $Li_{17}C_{60}$). Besides, endohedral clusters could also be formed. In fact, the closed micropores found in other forms of electroactive carbon have certain similarities to fullerenes, [94] and fullerenes have been considered to be an idealized model of nanopores in carbon materials, as the presence of sp_3-like bonds introduces curvature into the sp_2-bonded graphitic planes [95].

Table 8.1 Electrochemical performance of different forms of carbon nanotubes in lithium test cells.

CNT	Capacity (mAh g^{-1})	Efficiency (%)	Reference
SWCNT	460	27.7	[98]
MWCNT			
Graphitic	282	91.5	[99]
Nongraphitic	640	65.3	
Bamboo-CNT	460	—	[100]
MWCNT (B-doped)	180	—	[101]

The chemical lithiation of fullerene was reported by using lithium azide as the source of lithium [96], according to

$$C_{60} + x\,LiN_3 \rightarrow Li_xC_{60} + (3x/2)\,N_2 \text{ (with } 0 \leq x \leq 14 \text{ in fcc structure)} \quad (8.5)$$

Apparently, these were promising results; especially if one takes into account that in order to reach the theoretical capacity of graphite, only the $Li_{10}C_{60}$ stoichiometry is needed. However, subsequent assays of electrochemical insertion were deceptive. On the one hand, fullerene was found to have little stability and high solubility in common organic solvents used for the electrolyte in Li-ion battery technology. Moreover, the C_{60} unit was found to be impenetrable to lithium, in contrast to other species, which also limits the possibilities of extending insertion capability [97].

Carbon nanotubes were considered potential candidates in Li-ion technology (Table 8.1) [98–102]. These studies showed that both single-walled (SW) and multi-walled CNTs have noncompetitive performances. The mechanism of lithium intercalation into carbon nanofibers and nanotubes has been previously explored [103–106]. The ^{7}Li nuclear magnetic resonance (NMR) spectra of lithiated nongraphitizable carbon fibers showed the occurrence of lithium species with metallic character. In addition, it was shown that lithium is stored in nongraphitic carbon fibers in at least two forms: lithium as mainly ionic character at about 0 ppm and inserted as lithium with less ionic character at about +5 ppm. The relative extension of the two NMR peaks corresponding to each type of lithium changes with the origin of the carbon fibers and the prehistory of the samples.

On the other hand, an anomalous behavior was found for MWCNTs from the voltage versus capacity curves recorded at different rates. In the voltage range from 0 to 2.2 V, the capacity of the Li/MWNTs cells observed using larger current densities is higher than that observed for smaller current densities. The slow diffusion of lithium ions is not sufficient to account for this behavior. The existence of profound potential wells was shown to contribute to the unexpected rate dependence by modifying the voltage values in the galvanostatic experiments. [107]. Moreover, inner "spring-like" nanotubes were suggested to duplicate the global insertion capacity [108]. More recently, carbon nanosprings [109] exhibited good coulombic efficiency on prolonged cycling and capacities around 160 mAh g^{-1} at a current density as high as 3 A g^{-1}.

8.3.1.2 Titanium Oxides

Among the numerous transition metal oxides that provide suitable host lattices for reversible lithium insertion reactions, titanium dioxides are interesting candidates. In fact, titanium is abundant in nature, and titania is chemically stable, nontoxic, environment-friendly "White Knight," and an inexpensive option. The different TiO_2 polymorphs have been assayed so far as active electrode material. In early reports, rutile, hollandite, and brookite were considered as having a limited lithium intercalation capacity at room temperature compared to anatase and the synthetic polymorph TiO_2 (B) [110–112]. However, nanodispersion and nanostructuring recently allowed a dramatic improvement in the performance of different forms of titania, as discussed below.

The reaction of anatase with lithium takes place at about 1.75 V versus Li, reaching a maximum $Li_{0.5}TiO_2$ stoichiometry by a biphasic mechanism [113]. The performance is significantly improved by using nanostructured anatase electrodes (NAE), which enhances the lithium insertion rate while decreasing the diffusion lengths [114]. Thus, NAE-based anodes reduce the overall cell voltage, but provide cells with enhanced safety, good capacity retention on cycling, and low self-discharge. The preparation of nanodisperse, nanoporous, or nanostructured anatase has deserved the interest of many researchers due to its applicability as photoanode in regenerative photoelectrochemical cells. The protolysis of titanium isopropoxide is a straightforward procedure [115]. For nanosized anatase, a solid solution domain was observed, prior to the classical biphasic transition [116].

On the other hand, self-organized TiO_2 nanotube layers prepared by electrochemical techniques have been the subject of recent investigations. In 1979, Kelly [117] reported the preparation of porous nanostructured titania by anodization of Ti foils in fluoride-containing medium. Self-organized TiO_2 nanotubes (nt-TiO_2) have been proposed as an alternative electrode for lithium ion batteries [118–120]. The nt-TiO_2 layers can be prepared by a simple anodization procedure providing subsequent good electrical contacts between the active materials and the current collector. The powder-free fabrication method allows to prepare the active material directly on the current collector, without using additives to improve the mechanical (polymer binders) or conductive (carbon black) properties. The capacity is substantially higher than that of the compact thin films of titania prepared on similar substrates. Moreover, nt-TiO_2 layers were supported on Si substrate so that the resulting flat surface can be used as an alternative electrode for rechargeable on-chip 2D Li-ion microbatteries.

Concerning $TiO_2(B)$, a significant advancement emerged from the use of this solid in the form of nanowires [121, 122]. A working voltage of about 1.6 V versus Li, a maximum capacity of 305 mAh g^{-1}, excellent cyclability, and a better rate performance than $TiO_2(B)$ or anatase particles having a size similar to the nanorod diameter were reported.

Finally, rutile nanoparticles have emerged as a novel material of interest. Lithium insertion into bulk rutile proceeds at about 0.4 V below the insertion potential into anatase, but lower capacities than anatase were also obtained at room temperature. In 1996, the use of rutile nanoparticles provided examples of a net improvement in this picture [123, 124]. The appearance of two solid solution domains, followed by the formation of electroactive rock salt-type $LiTiO_2$, was the mechanism found for nanoparticles of rutile [125]. More recently, the insertion of lithium into rutile nanorods was shown to cause two consecutive structural phase transformations to lithium titanate phases with spinel and rock salt structures at $x = 0.46$ and 0.88, respectively [126].

Concerning the spinel $Li_4Ti_5O_{12}$, with a $(Li)_{8a}[Li_{1/3}Ti_{5/3}]_{16d}O_4$ cation distribution in the $Fd\bar{3}m$ space group, Ohzuku et al. [127] reported in 1995 its possible use as a lithium insertion material. After lithium insertion up to 1 Li per formula, they proposed that the cations are redistributed using the 16c sites that are empty in the normal spinel, leading to $[Li_2]_{16c}[Li_{1/3}Ti_{5/3}]_{16d}O_4$. This process was found to induce little changes in the unit cell parameter. Such a zero-strain property allowed an enhanced stability and little damage to the solid on cycling. The electrochemical curves show a plateau at about 1.5 V and capacity values close to 150 mAh g^{-1} are preserved during 100 cycles. Recently, Kavan and Grätzel [128] have emphasized the role of the synthesis procedure in the electrochemical behavior of $Li_4Ti_5O_{12}$. Thus, thin-film electrodes obtained from nanocrystalline $Li_4Ti_5O_{12}$, which was prepared by a sol-gel route employing lithium ethoxide and Ti(IV) alkoxides, showed an excellent behavior.

8.3.2
Intermetallic Compounds

Intermetallic compounds (compounds containing exclusively two or more kinds of metal or metalloid elements) and alloys are important functional materials, have a wide range of applications, and are very interesting from a fundamental point of view [129, 130]. It is known that nanoparticles may show modified properties – such as reactivity – compared to the bulk phase. In addition, structures not observed using traditional bulk syntheses could be stabilized as nanocrystals using low-temperature synthetic routes such as the polyol process [131].

The electrochemical behavior of an intermetallic phase that constitutes the electrode active material in batteries can be influenced by particle size [132]. In crystalline materials, new intermetallic phases are formed upon lithium insertion leading to inhomogeneous volume expansions in the two-phase regions that can cause cracking and pulverization of the electrode. It is believed that starting with small grain sizes helps to keep the tin regions in the anode small. The grain volume changes associated with the reaction with lithium can be buffered by using nanoparticles, and it leads to improved electrochemical behavior upon prolonged cycling without fracturing. Small particle sizes and high surface areas can improve the electrochemical performance at high rates [133]. Since the small particles have a natural tendency to agglomerate in order to reduce their surface energy, the main challenge is to maintain the small particle size upon cycling.

8.3.2.1 Tin and Tin Composites
Tin-based electrode materials are present in the second generation of commercial Li-ion batteries. Early reports revealed the possibilities of forming Li–Sn intermetallics, which provide theoretical capacities close to 990 mAh g^{-1} [134, 135]. However, it was

soon realized that reversible alloying–dealloying processes are accompanied by colossal changes in the unit cell volume that can produce strain and destroy the electrode (a detailed illustration of this phenomenon is reported in Ref. [136]). In 1997, an alternative material was patented by Fuji: tin composite oxides in which tin oxides were the starting material [137]. More recently, commercialization of Nexelion™ product by Sony [138] gave a new impetus to tin-based electrodes, with the special feature of being noncrystalline.

At present, two main strategies are envisaged to stabilize the Sn-based electrodes and to improve their electrochemical performance: the use of nanoparticles and the incorporation of other elements. The use of small particle size can have other advantages such as to reduce the length of the lithium diffusion path. Porous Sn electrodes can also be advantageous due to high lithium diffusion coefficient and small volume change [139]. To preserve nanoparticles from coalescence and growth, an inert matrix can be used. The concept of using an inert matrix as a battery anode was proposed by Huggins [140]. Infiltration of organometallic compounds into an organic gel (e.g., resorcinol formaldehyde), followed by calcinations under argon, also yields metal–carbon nanostructured nanocomposites [141]. Glucose has been used to encapsulate tin nanoparticles within amorphous carbon [142]. Surfactants and micelles can be used to obtain carbon-encapsulated intermetallic compounds [143].

The best electrochemical results are usually obtained with two-phase metallic matrices allowing the expansion of the more reactive phase material embedded in the still unreacted and ductile phase material [144]. In fact, Nexelion electrodes use highly uniform amorphous tin–cobalt electrodes (Co/Sn atomic ratio = 1 : 1.1) of 5 nm diameter, graphite, and around 5% of titanium. The role of titanium is probably to stabilize the tin–cobalt system in the amorphous state [145].

In Figure 8.4, the capacity values of binary tin compounds with micro- and nanosizes are compared. The data shown are from our experimental results and also from the literature. Several compounds exhibit great capacity to react with Li, irrespective of the particle size. In contrast, others only react appreciably when tested in the form of nanoparticles. The cobalt–tin system is particularly interesting [146–148]. The electrochemical behavior differs greatly from one phase to another, and it is also strongly influenced by particle size (Figure 8.4a). The $CoSn_2$ phase with microsize particles shows great ability to react reversibly with lithium [149]. In contrast, CoSn [150] and Co_3Sn_2 [151, 152] show only significant capacities when nanometric particles are used. Moreover, stabilization of Co–Sn electrodes may involve the use of unknown phases [153, 154]. Large particles of the Co-rich compounds exhibit low capacity to react with lithium due to a slow diffusion of lithium that results in electrode overpotential. In nanoparticulate materials, the lithium diffusion path length is shorter and the cobalt content is less critical.

$FeSn_2$ might yield a maximum capacity of 802 mAh g^{-1} according to the following overall reaction:

$$5\,FeSn_2 + 44\,Li^+ + 44\,e^- = 2\,Li_{22}Sn_5 + 5\,Fe \qquad (8.6)$$

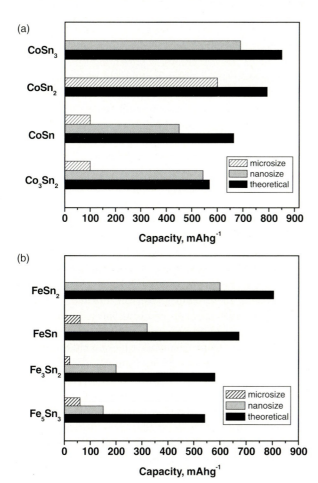

Figure 8.4 Capacities for (a) CoSn$_x$ and (b) FeSn$_x$ intermetallic electrodes. The theoretical capacity values calculated from the respective compositions and the experimental capacity values delivered by micrometric and nanometric particle size are shown.

Mao et al. studied the electrochemical behavior of FeSn$_2$ obtained by arc melting followed by mechanical alloying [155]. It is worth noting that isostructural CoSn$_2$ and FeSn$_2$ (I4/mcm space group) exhibit great capacity, irrespective of particle size. This feature suggests that structure plays an important role in making easier the insertion of lithium into these intermetallics. Several FeSn$_x$ phases with higher iron content, such as FeSn, Fe$_3$Sn$_2$, and Fe$_5$Sn$_3$, tend to form an electrochemically inactive "skin" of Fe over the surface of the active particles that prevents the full reaction with lithium [156] and the resulting capacities are lower than the theoretical values (Figure 8.4b). Nanosized FeSn$_2$ exhibits 500 mAh g^{-1} of reversible capacity [157]. In order to improve the electrochemical behavior, a composite electrode obtained by ball milling of Fe, Sn, and C was also reported [158].

Nanocarbons are a suitable option to support tin-based nanoparticles [159]. Nanoparticles of CoSn and Co_3Sn_2 deposited on carbon nanotubes have been prepared and the influence of the carbon nanotubes in preventing agglomeration of intermetallic particles was proposed [160]. Addition of organic polymers to tin-based electrodes has been envisaged as a promising option [161]. The pyrolyzed polymer can form a matrix that improves the electrochemical behavior due to the formation of a stable solid/electrolyte interface and the ability to avoid nanoparticle aggregation upon cycling.

8.3.2.2 Silicon

Pioneering reports on the reversible electrochemical formation of Li–Si intermetallics started in the mid-1970s [162–165]. In these studies, high-temperature conditions (above 400 °C) in molten salt electrolyte cells were needed, to record the potential plateaus below 0.5 V versus Li. Also, the definition of multiphase regions between the intermetallic phases $Li_{12}Si_7$, Li_7Si_3, $Li_{13}Si_4$, and $Li_{22}Si_5$ was achieved. These stoichiometries reveal a maximum theoretical capacity of 4200 mAh g^{-1}. Almost 20 years later, the first electrochemical experiments on pure phases of the lithium–silicon system carried out at room temperature with organic electrolyte solvents were reported [166]. Reversible capacities of up to 500 mA h g^{-1} were observed. This value could be increased by using a metal silicide (e.g., $CrSi_2$) as starting material. Different experiments demonstrated that the initial lithiation of metal disilicides, silicon boride, and silicon monoxide results in the formation of an amorphous lithium (a-Si)-containing product from which lithium can be subsequently extracted and inserted. The resulting Li_xSi material is equivalent to the products of electrochemical lithiation of amorphous silicon [167]. An in situ XRD study of the reaction of lithium with amorphous silicon films demonstrated that the film thickness plays an important role in determining if the sample crystallizes during lithiation. Thus, below 0.03 V, thick films of a-Si rapidly crystallize to form a crystalline $Li_{15}Si_4$ phase, with a theoretical capacity of 3579 mAh g^{-1}. However, the a-Si phase reforms on delithiation [168].

On the other hand, the reversible lithium insertion in carbons containing nanodispersed silicon was reported as early as 1995. Compared to pure carbons, composite materials significantly increased the reversible capacity in electrochemical lithium cells [169]. More recently, Lee and Lee reported on a good cycling performance of carbon-coated nano-Si dispersed oxides/graphite composites [170]. These results stress the importance of particle size and microstructure to improve the performance of silicon-based electrodes. It should be noted that the volume changes associated with the formation of Li–Sn intermetallics have the same undesirable effects on electrode performance as those described for tin-based materials. Uono et al. [171] have proposed the optimized structure of the Si/carbon/graphite composites to improve the cycling performance, based on the analysis of deteriorated electrodes after cycling. Combinatorial methods have also contributed to a deeper understanding of $Si_{1-x}C_x$ electrode materials [172].

A particularly relevant improvement has been recently achieved in an impactful example of how nanoscience and nanotechnology are bringing a real revolution in the battery field. Silicon can be prepared as nanowire on a conducting substrate. Apart from reducing the lithium diffusion path length, silicon nanowires can accommodate large strains without electrochemical grinding and keep good electrical contact with the substrate [173]. Thus, the theoretical charge capacity for silicon anodes was achieved. On prolonged cycling, almost 75% of this maximum was maintained and a good rate performance was observed. The possible use of other elongated Si nanoparticles such as bamboo silicon nanotubes (b-SiNT) [174] or the different Si wires, flamingos, forests, droplets, mushrooms, and so on obtained by vapor–liquid–solid (VLS) growth, with In catalyst, and double magnetron sputtering, by Abraham, Thackeray, and coworkers could open new possibilities to these phenomena.

In summary, nanostructured silicon could offer the highest capacities. However, several difficult problems are still there to be resolved such as huge irreversible capacity and not easily scalable preparative routes.

8.3.3
Nanomaterials Obtained *In Situ*

The electrochemical reactions leading to the formation of ultrafine particles of the reaction products inside the electrochemical cell can be considered as "electrochemical *in situ* nanodispersion." In these reactions, the properties of the electrode are significantly modified from the first discharge to further cycles. Thus, new phases may be involved in the electrochemical reactions, surface energy effects become notorious, and the contributions of surface storage and double-layer effects appear. Among these electrochemical reactions, displacement (conversion) reactions are a hot topic in Li-ion battery materials.

Different examples of conversion reactions involve not only transition metal oxides [175] but also sulfides [176], nitrides [177], phosphides [178], antimonides [179], fluorides (see Section 8.2.5), and, more recently, oxysalts [180]. Several examples of oxygen-containing compounds for the anode of Li-ion cells are discussed below.

8.3.3.1 Binary Oxides for Conversion Electrodes

The use of transition metal oxides as conversion cathodes for primary lithium batteries is well documented. Among them, CuO was considered particularly useful because of its high Faradaic capacity per unit volume [181]. In the early 1980s, several authors focused their efforts on untangling the lithium reaction mechanism in Fe_3O_4, Fe_2O_3 [182], Co_3O_4 [183], and CuO [184]. They detected the following displacement reactions on deep lithiation:

$$Fe_3O_4 + 8\,Li^+ + 8\,e^- \rightarrow 3\,Fe + 4\,Li_2O \qquad (8.7)$$

$$CoO + 2\,Li^+ + 2\,e^- \rightarrow Co + Li_2O \qquad (8.8)$$

The exoergic nature of the reaction is evidenced by the evaluation of the free energies for the oxide formation ($Li_2O = -561.2\,kJ\,mol^{-1}$; $\Delta G_f^0(CoO) = -214.2\,kJ\,mol^{-1}$) and the reversibility was experimentally demonstrated [182]. However, the restricted performance of the LiAl/LiCl,KCl/CoO cell was attributed to such factors as the instability of cobalt oxide in the melted electrolyte and Li_2O and Co dispersion into the electrolyte. For this reason, further studies on these materials mainly focused on the structural modification induced by phase transitions and/or insertion processes occurring in the first stages of the electrochemical reaction. The successful implementation of the Li-ion concept eventually led to the withdrawal of the use of these oxides as cathodes. Li-ion batteries required higher voltage electrodes including lithium in their composition.

Tarascon and coworkers [175] recently claimed the ability of transition metal oxides (MO, where M is Co, Ni, Cu, or Fe) to deliver capacities as high as 700 mAh g^{-1}, with 100% capacity retention for up to 100 cycles and high recharging rates. The advances achieved in the research of new 5 V cathode materials (see Section 8.2.2.2) and the unsolved limitations existing in carbonaceous materials and lithium alloy-based anodes allowed to consider this transition metal compound as a promising material as anode in lithium ion batteries.

The first discharge of a lithium test cell is characterized by two different regions (Figure 8.5). A voltage plateau around 1 V extends to charge values that fairly match with the electron consumption expected to the complete metal reduction and a steady voltage decrease to 0 V. The precise voltage plateau depends on the transition metal

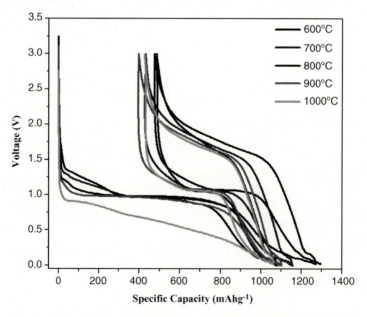

Figure 8.5 Charge and discharge cycles recorded by the galvanostatic method. The displayed electrochemical curves correspond to the first and second discharge and first charge. Two electrode cells were assembled with a lithium metal anode and $CuFe_2O_4$ samples prepared at several temperatures as working electrodes. The cells were cycled at a 1C rate.

composing the oxide. The electrochemical metal reduction to the metallic state has been undoubtedly evidenced by several techniques such as X-ray absorption [185, 186], Mössbauer spectroscopy [187–189], and X-ray photoemission [190, 191]. The final product of this reduction is a composite matrix made of metallic nanograins (M^0) dispersed into Li_2O [192]. The nanometric size of the metallic particles involves very short diffusion length for Li^+ and the composite can be considered a near-liquid phase. It facilitates the species transport and is responsible for the unexpected reversibility of the electrochemical reaction. The voltage plateau during the first discharge is lower than that of the theoretical predictions. Such deviation is of kinetic nature and represents the important overpotential needed to initiate and continue with the decomposition reaction.

The importance of the second region in the first discharge is essential for several reasons. It provides an extra reversible capacity to the cell. On an extended cycling, the charge capacity related to this effect seems to be less affected, while a progressive shortening of the metal reduction plateau is observed [193]. The origin of this extra capacity has been subjected to various interpretations. Tarascon and coworkers claimed the reversible formation/dissolution of polymeric species at low potential in alkyl carbonate solution. It is reflected in the pseudocapacitive behavior and promoted by the highly reactive pristine metallic nanograins formed upon reduction [194]. Alternatively, Maier and coworkers found that formation of the polymer film was not constrained to this region in RuO_2 [195]. An electron microscopy study on this composition revealed that even if the polymer is formed mainly in the sloped region on discharge, it decomposes beyond this region on charging. Thus, the authors proposed an interfacial charge storage mechanism. An excess of charge (lithium and electrons) can be stored in the boundary regions between metal and Li_2O nanoparticles through a charge separation until the cell potential reaches zero [196]. According to these authors, this mechanism is a hybrid between a supercapacitor and a battery electrode offering a compromise between kinetic rate and capacity.

On charging, the polymer film is decomposed and the oxidized transition metal compound is partially recovered. Thus, spectroscopic techniques revealed the oxidation of the transition metal atoms but X-ray diffraction patterns showed poor crystalline materials [188]. The second and subsequent discharges show a different profile. The flat plateau becomes a continuous voltage decrease. Apart from surface energy contributions, the small particle size and inhomogeneous lithium distribution of the amorphized material provide energetically distinct environments for further cycling.

In spite of the good performance of these electrode materials, several drawbacks, such as charge curve polarization and initial irreversibility, must be overcome to produce commercial prototypes. Different approaches have been studied to solve these problems. In this sense, several preparation routes have been assayed to produce controlled size transition metal oxides such as the use of polyols [197], reverse micelles [198], thin films [199], and mesoporous silica templating [200]. A detailed study of the Li-driven conversion reaction by electron microscopy revealed the continuous progression of the electrochemical reaction from the surface toward the core of the particles [201]. Considering this fact, it could be inferred that samples with low particles and high surface areas should offer a better kinetic response and consequently a more performing behavior. However, the study of monodisperse Cu_2O samples prepared in the submicron and micron ranges showed better capacity retention for the latter sample [202]. The authors point out the enhanced solvent degradation promoted by the increase in the metal–electrolyte interphase as mainly responsible for the negative results. It would create an electronic isolating layer on the particle surface that eventually led to the lost contact between nanograins.

8.3.3.2 Multinary Oxides

In order to improve the performance of conversion oxide electrodes, the search for possible synergistic effects resulting from mixed transition metal stoichiometries has been undertaken. In addition, some drawbacks of binary oxides NiO, CoO, and Co_3O_4 related to their use, such as cost, toxicity, and high working voltage, can be overcome by using multinary oxides containing manganese or iron. The use of manganese considerably decreases these disadvantages offering a low working voltage for the anode material. Previous reports on the solid solution of manganese oxides and cobalt [203] or nickel [204] oxides have evidenced the benefits of ternary oxides in achieving a compromise solution between capacity and cell voltage.

Numerous ternary spinel oxides contain iron in their composition. This fact allows to apply the ^{57}Fe Mössbauer spectroscopy as a tool to unfold the lithium conversion reaction mechanism. $CoFe_2O_4$ prepared by a sol-gel route was able to sustain capacity values over 700 mAh g^{-1} after 75 cycles. This cobalt ferrite evidenced the presence of Fe^{2+} ions as an intermediate of the iron reduction and oxidation. Also, the high disorder of the charged electrode hinders the recovery of the ferrimagnetic character [205]. Instead, Mössbauer spectroscopy did not reveal the presence of Fe^{2+} upon $CuFe_2O_4$ electrochemical reduction [206]. This study revealed the importance of particle morphology in performing a reversible conversion reaction in these oxides. The macroporous texture produced by the interconnection of nanoparticles ensures a good electronic conductivity and contacts between the oxide surface and the electrolyte solution, favoring the Li storage capacity (Figure 8.5). After a number of discharges of $CoFe_2O_4$, the transition metal atoms are less effectively reduced and the population of bulk atoms eventually disappears explaining the capacity fading observed during the first cycles [193, 207].

8.3.3.3 Transition Metal Oxysalts

A convenient synthetic route of transition metal oxide nanoparticles is the thermal decomposition of carbon-containing oxysalts, such as carbonates, oxalates, and so on, which give gaseous carbon oxides as main by-products at moderate temperatures. Also, these oxysalts can be prepared in the form of nanoparticles by nonconventional routes, such as the reverse micelles procedure. Surprisingly, the transition metal oxysalts themselves may allow novel potential alternatives of application in the field of advanced lithium ion batteries. The expected reaction mechanism when oxysalts are used as anode material in lithium batteries is somewhat similar to that of oxides, that is, a conversion reaction mechanism, as discussed in previous sections, but accompanied with the formation of lithium oxysalts instead of lithium oxide:

$$MC_xO_y + 2\,Li^+ + 2\,e^- \rightarrow Li_2C_xO_y + M \tag{8.9}$$

where $(C_xO_y)^{2-}$ is the anion of the oxysalt and M^{2+} is the transition metal cation. Here, we examine the possibility of extending the range of new anode materials to submicrometric and nanometric particles of transition metal oxysalts.

The synthesis of manganese carbonate by the reverse micelle method [208] yields monodisperse submicron, well-crystallized particles with uniform size and rhombic shape habit, with calcite-type rhombohedral structure. Concerning the electrochemical behavior, although the initial reversible capacity is lower for the carbonate compared to that of MnO oxide, the capacity retention is significantly better for $MnCO_3$. Thus, after 10 cycles, the reversible capacity of carbonate is clearly better than that of oxide (about 550 mAh g^{-1} versus about 450 after 15 cycles). The behavior of MnO electrodes was previously described by Tarascon and coworkers [192] who found that bulk crystalline MnO shows poorer ability to electrochemically intercalate lithium compared to other first-row transition metal oxides. The discharge characteristics on prolonged cycling also resemble that of the Mn_3O_4 nanofibers recently described by Fan and Whittingham [209]. These capacity values are improved by the use of an oxysalt, such as manganese carbonate. Compared to other displacement reactions, so far mostly restricted to metal oxides and fluorides, the use of a low molecular weight salt such as $MnCO_3$ does not penalize the capacity while giving extra stability due to the formation of lithium carbonate as the main side product.

The synthesis of transition metal oxalates by precipitation techniques commonly yields dehydrated compounds that crystallize into two allotropic forms: monoclinic (*C2/c*) and orthorhombic (*Cccm*) [210]. Both commercial and micrometric synthesized oxalates show the monoclinic structure. In contrast, the oxalates synthesized by the reverse micelles procedure showed the orthorhombic structure. There are also significant differences in particle size and morphology. While the first show irregularly shaped particles of several micrometers, the products obtained using reverse micelles show nanoribbons about 30 nm wide. These results emphasize the effectiveness of the reverse micelles procedure to obtain nanoparticles with controlled size and shape. Heating the samples results in a first weight loss corresponding to two water molecules (below 200 °C), followed by decomposition of the oxalate anion to yield oxides. Stopping the heating at 200 °C allows to obtaining dehydrated oxalate, keeping the integrity of the oxalate groups as well as particle size and nanoribbon morphology (Figure 8.6).

The electrochemical reaction of oxalates with lithium took place with a first extended plateau at about 1 V versus Li leading to an X-ray amorphous product. At the end of the first discharge, reduction of iron and cobalt was confirmed by ^{57}Fe Mössbauer spectroscopy [211] and X-ray absorption near edge structure (XANES) at Co-K edge [212], respectively, while Fourier transform infrared (FTIR) spectroscopy confirms the integrity of the oxalate anions on cycling and the formation of lithium

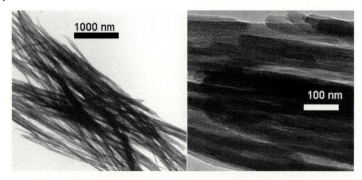

Figure 8.6 TEM images at low and high magnification of cobalt oxalate, showing the nanoribbon morphology.

oxalate as well, in agreement with Equation 8.9. Anyway, the charge–discharge branches display a significant hysteresis phenomenon, a feature that is commonly observed, and limits the use of conversion electrodes.

In addition, it is worth noting that the experimental capacity exceeds the theoretical value of the pure conversion reaction, according to Equation 8.9. The study carried out in order to evaluate the possible contributions of Faradaic and non-Faradaic (capacitance) processes to the global observed values, recently described by Brezesinski et al. for TiO_2 electrodes [213], indicates that the first discharge is mostly Faradaic and irreversible and overpasses the theoretical capacity, a process that mainly arises from irreversible reactions with the electrolyte. This large irreversible capacity is commonly found in conversion electrodes (see previous sections). On further cycles, the Faradaic capacity gets closer to the theoretical capacity of the proposed conversion reaction, while the extra capacity comes from non-Faradaic origin.

Compared to non-nanostructured products, a higher capacity on extended cycling is observed for ribbon-like nanoparticles obtained by the reverse micelles synthesis, even at higher rate (Figure 8.7). Moreover, mixed iron, cobalt oxalates seems to have a synergic effect compared to nonmixed oxalates, with capacity values close to 700 mAh g^{-1} after more than 70 cycles [214].

8.4
Concluding Remarks

The active materials of both electrodes in present and future generations of lithium ion batteries are strongly affected by the recent scientific developments at the nanoscale.

Layered-, spinel-, and olivine-related solids commonly used as cathode materials improve their capacity and rate performance by nanodispersion. Moreover, the reaction mechanism may change between single-phase and multiphase or between crystalline and amorphous by changing from micro- to nanoparticles. Nanometric

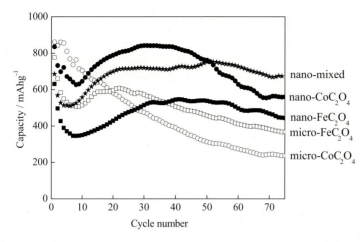

Figure 8.7 Plot of discharge capacity versus cycle number upon glavanostatic cycling at 2C of lithium cells using anhydrous Fe, Co, and mixed $Fe_{0.75}Co_{0.25}$ oxalates as active material of the positive electrode. The highest capacities were achieved for CoC_2O_4 during the first cycles, while the mixed oxalate showed better capacity retention on more prolonged cycling.

films of oxides or carbon can both protect these solids from solubilization during cell utilization and circumvent their limitations in electronic conductivity. Finally, new materials in the fluoride or silicate families emerge as potential candidates to replace the currently applied materials. Their use is unavoidably connected to nanodispersion and nanostructuring, in interconnected composites that may include electronic and ionic conducting phases.

The replacement of carbon-based negative electrodes is still a difficult task. New insertion, alloying, and conversion electrode materials have emerged thanks to nanoscience and have also given rise to commercial products. Particularly, tin-based materials have demonstrated their usefulness and experienced dramatic improvements, which make them a hot topic in the present state of the art. A tough competition may result from ultrahigh-capacity silicon nanoforms provided the problems associated with irreversible capacity and scaling up of expensive processes are solved. Metal oxides for the conversion electrodes are still an incipient area. Unexpected conversion solids such as transition metal oxysalts have emerged recently. The possibilities of using combined double-layer capacitance with insertion and/or conversion reactions provide new directions in the borderline between lithium batteries and supercapacitors.

References

1 Mizushima, K., Jones, P.C., Wiseman, P.J., and Goodenough, J.B. (1980) *Mater. Res. Bull.*, **15**, 783.
2 Nishi, Y. (2001) *Chem. Record*, **1**, 406.
3 Alcántara, R., Lavela, P., Tirado, J.L., Zhecheva, E., and Stoyanova, R. (1999) *J. Solid State Electrochem.*, **3**, 121.

4 Alcántara, R., Lavela, P., Relaño, P.L., Tirado, J.L., Zhecheva, E., and Stoyanova, R. (1998) *Inorg. Chem.*, **37**, 264.

5 Alcántara, R., Lavela, P., Tirado, J.L., Stoyanova, R., Kuzmanova, E., and Zhecheva, E. (1997) *Chem. Mater.*, **9**, 2145.

6 Zhecheva, E., Stoyanova, R., Gorova, M., Alcántara, R., Morales, J., and Tirado, J.L. (1996) *Chem. Mater.*, **8**, 1429.

7 Alcántara, R., Lavela, P., Tirado, J.L., Stoyanova, R., Kuzmanova, E., and Zhecheva, E. (1997) *Chem. Mater.*, **9**, 2145.

8 Gu, Y.X., Chen, D.R., and Jiao, M.L. (2005) *J. Phys. Chem. B*, **109**, 17091.

9 Chen, H. and Grey, C.P. (2008) *Adv. Mater.*, **20**, 2206.

10 Okubo, M., Hosono, E., Kim, J., Enomoto, M., Kojima, N., Kudo, T., Zhou, H.S., and Honma, I. (2007) *J. Am. Chem. Soc.*, **129**, 7444.

11 Mladenov, M., Stoyanova, R., Zhecheva, E., and Vassilev, S. (2001) *Electrochem. Commun.*, **3**, 410.

12 Liu, L., Wang, Z., Li, H., Chen, L., and Huang, X. (2002) *Solid State Ionics*, **152–153**, 341.

13 Lee, H., Kim, M.G., and Cho, J. (2007) *Electrochem. Commun.*, **9**, 149.

14 Cho, J., Kim, Y.J., Kim, T.J., and Park, B. (2001) *Angew. Chem., Int. Ed.*, **40**, 3367.

15 Chen, Z. and Dahn, J.R. (2002) *Electrochem. Solid State Lett.*, **5**, A213.

16 Kweon, H.J., Kim, S.J., and Park, D.G. (2000) *J. Power Sources*, **88**, 255.

17 Goodenough, J.B., Thackeray, M.M., David, W.I.F., and Bruce, P.G. (1984) *Rev. Chim. Minér.*, **21**, 435.

18 Xia, Y., Zhou, Y., and Yoshio, M. (1997) *J. Electrochem. Soc.*, **144**, 2593.

19 Amatucci, G.G., Schmutz, C.N., Blyr, A., Sigala, C., Gozdz, A.S., Larcher, D., and Tarascon, J.M. (1997) *J. Power Sources*, **69**, 11.

20 Gummow, R.J. and Thackeray, M.M. (1994) *J. Electrochem. Soc.*, **141**, 1178.

21 AlMawiawi, D., Coombs, N., and Moskovits, M. (1991) *J. Appl. Phys.*, **70**, 4421.

22 Nishizawa, M., Mukai, N., Kuwabata, S., Martin, C.R., and Yoneyama, H. (1997) *J. Electrochem. Soc.*, **144**, 1923.

23 Luo, J., Wang, Y., Xiong, H., and Xia, Y. (2007) *Chem. Mater.*, **19**, 4791.

24 Zheng, Z., Tang, Z., Zhang, Z., Sheng, W., and Lin, Y. (2002) *Solid State Ionics*, **148**, 317.

25 Cho, J., Kim, T.-J., Kim, Y.J., and Park, B. (2001) *Chem. Commun.*, **16**, 1074.

26 Kawai, H., Nagata, M., Tukamoto, H., and West, A.R. (1999) *J. Power Sources*, **81–82**, 67.

27 Zhong, Q., Bonakdarpour, A., Zhang, M., Gao, Y., and Dahn, J.R. (1997) *J. Electrochem. Soc.*, **144**, 205.

28 Alcántara, R., Jaraba, M., Lavela, P., Tirado, J.L., Biensan, P., De Guibert, A., Jordy, C., and Peres, J.P. (2003) *Chem. Mater.*, **15**, 2376.

29 Kovacheva, D., Markovsky, B., Salitra, G., Talyosef, Y., Gorova, M., Levi, E., Riboch, M., Kim, H.J., and Aurbach, D. (2005) *Electrochim. Acta*, **50**, 5553.

30 Lee, Y.S., Sun, Y.K., Ota, S., Miyashita, T., and Yoshio, M. (2002) *Electrochem. Commun.*, **4**, 989.

31 Zhao, Z., Ma, J., Tian, H., Xie, L., Zhou, J., Wu, P., Wang, Y., Tao, J., and Zhu, X. (2005) *J. Am. Ceram. Soc.*, **88**, 3549.

32 Amarilla, J.M., Rojas, R.M., Pico, F., Pascual, L., Petrov, K., Kovacheva, D., Lazarraga, M.G., Lejona, I., and Rojo, J.M. (2007) *J. Power Sources*, **174**, 1212.

33 Kunduraci, M. and Amatucci, G.G. (2006) *J. Electrochem. Soc.*, **153**, A1345.

34 Idemoto, Y., Narai, H., and Koura, N. (2003) *J. Power Sources*, **119–121**, 125.

35 Kunduraci, M., Al-Sharab, J.F., and Amatucci, G.G. (2006) *Chem. Mater.*, **18**, 3585.

36 Kim, J.H., Myung, S.T., Yoon, C.S., Kang, S.G., and Sun, Y.K. (2004) *Chem. Mater.*, **16**, 906.

37 Sun, Y.K., Lee, Y.-S., Yoshio, M., and Amine, K. (2002) *Electrochem. Solid State Lett.*, **5**, A99.

38 Sun, Y.K., Hong, K.-J., Prakash, J., and Amine, K. (2002) *Electrochem. Commun.*, **4**, 344.

39 Sun, Y.K., Yoon, C.S., and Oh, I.H. (2003) *Electrochim. Acta*, **48**, 503.

40 Alcántara, R., Jaraba, M., Lavela, P., and Tirado, J.L. (2004) *J. Electroanal. Chem.*, **566**, 187.

41 Padhi, A.K., Nanjundaswamy, K.S., and Goodenough, J.B. (1997) *J. Electrochem. Soc.*, **144**, 1188.
42 http://www.a123systems.com/applications/hybrid-electric (accessed July 20, 2009).
43 Delmas, C., Maccario, M., Croguennec, L., Le Cras, F., and Weill, F. (2008) *Nat. Mater.*, **7**, 665.
44 Gibot, P., Casas-Cabanas, M., Laffont, L., Levasseur, S., Carlach, P., Hamelet, S., Tarascon, J.M., and Masquelier, C. (2008) *Nat. Mater.*, **7**, 741.
45 Chung, S.U., Bloking, J.T., and Chiang, Y.M. (2002) *Nat. Mater.*, **11**, 23.
46 Wang, L., Huang, Y., Jiang, R., and Jiaz, D. (2007) *J. Electrochem. Soc.*, **154**, A1015.
47 Kim, D.H. and Kim, J. (2006) *Electrochem. Solid State Lett.*, **9**, A439.
48 León, B., Pérez Vicente, C., Tirado, J.L., Biensan, Ph., and Tessier, C. (2008) *J. Electrochem. Soc.*, **155**, A211.
49 Lim, S., Yoon, C.S., and Cho, J. (2008) *Chem. Mater.*, **20**, 4560.
50 Kang, B. and Ceder, G. (2009) *Nature*, **458**, 190.
51 Zaghib, K., Goodenough, J.B., Mauger, A., and Julien, C. (2009) *J. Power Sources*, **194**, 1021.
52 Li, H.H., Jin, J., Wei, J.P., Zhou, Z., and Yan, J. (2009) *Electrochem. Commun.*, **11**, 95.
53 Vadivel Murugan, A., Muraliganth, T., Ferreira, P.J., and Manthiram, A. (2009) *Inorg. Chem.*, **48**, 946.
54 Delacourt, C., Laffont, L., Bouchet, R., Wurm, C., Leriche, J.B., Morcrette, M., Tarascon, J.M., and Masquelier, C. (2005) *J. Electrochem. Soc.*, **152**, A913.
55 Li, G.H., Azuma, H., and Tohda, M. (2002) *Electrochem. Solid State Lett.*, **5**, A135.
56 Kwon, N.H., Drezen, T., Exnar, I., Teerlinck, I., Isono, M., and Graetzel, M. (2006) *Electrochem. Solid State Lett.*, **9**, A277.
57 Ma, J. and Qin, Q.Z. (2005) *J. Power Sources*, **148**, 66.
58 Li, G., Azuma, H., and Tohda, M. (2002) *J. Electrochem. Soc.*, **149**, A743.
59 Kim, T.R., Kim, D.H., Ryu, H.W., Moon, J.H., Lee, J.H., Boo, S., and Kim, J. (2007) *J. Phys. Chem. Solids*, **68**, 1203.
60 Martha, S.K., Markovsky, B., Grinblat, J., Gofer, Y., Haik, O., Zinigrad, E., Aurbach, D., Drezen, T., Wang, D., Deghenghi, G., and Exnar, I. (2009) *J. Electrochem. Soc.*, **156**, A541.
61 Wang, D., Buqa, H., Crouzet, M., Deghenghi, G., Drezen, T., Exnar, I., Kwon, N.H., Miners, J.H., Poletto, L., and Grätzel, M. (2009) *J. Power Sources*, **189**, 624.
62 Patoux, S. and Masquelier, C. (2002) *Chem. Mater.*, **14**, 5057.
63 Arroyo-de Dompablo, M.E., Armand, M., Tarascon, J.M., and Amador, U. (2006) *Electrochem. Commun.*, **8**, 1292.
64 Dominko, R. (2008) *J. Power Sources*, **184**, 462.
65 Nishimura, S.-I., Hayase, S., Kanno, R., Yashima, M., Nakayama, N., and Yamada, A. (2008) *J. Am. Chem. Soc.*, **130**, 13212.
66 Gong, Z.L., Li, Y.X., He, G.N., Li, J., and Yang, Y. (2008) *Electrochem. Solid State Lett.*, **11**, A60.
67 Nytén, A., Abouimrane, A., Armand, M., Gustafsson, T., and Thomas, J.O. (2005) *Electrochem. Commun.*, **7**, 156.
68 Dominko, R., Bele, M., Gaberscek, M., Meden, A., Remskar, M., and Jamnik, J. (2006) *Electrochem. Commun.*, **8**, 217.
69 Zhang, S., Deng, C., and Yang, S. (2009) *Electrochem. Solid State Lett.*, **12**, A136.
70 Kokalj, A., Dominko, R., Gaberscek, M., Bele, M., Mali, G., Remskar, M., and Jamnik, J. (2007) *Chem. Mater.*, **19**, 3633.
71 Gong, Z.L., Li, Y.X., and Yang, Y. (2007) *J. Power Sources*, **174**, 524.
72 Lyness, C., Delobel, B., Armstrong, A.R., and Bruce, P.G. (2007) *Chem. Commun.*, 4890.
73 Prakash, A.S., Rozier, P., Dupont, L., Vezin, H., Sauvage, F., and Tarascon, J.-M. (2006) *Chem. Mater.*, **18**, 407.
74 Chilton, J.E., Jr., Conner, W.J., and Holsinger, R.W. (1964) Scientific and Technical Aerospace Report 2, p. 2451.
75 Resmini, F. and Ohlson, J. (2009) EVS24 International Battery, Hybrid and Fuel Cell Electric Vehicle Symposium, p. 1.
76 Bruce, P.G., Scrosati, B., and Tarascon, J.M. (2008) *Angew. Chem., Int. Ed.*, **47**, 2930.
77 Badway, F., Pereira, N., Cosandey, F., and Amatucci, G.G. (2003) *J. Electrochem. Soc.*, **150**, A1209.

78 Li, H., Balaya, P., and Maier, J. (2004) *J. Electrochem. Soc.*, **151**, A1878.
79 Liao, P., MacDonald, B.L., Dunlap, R.A., and Dahn, J.R. (2008) *Chem. Mater.*, **20**, 454.
80 Badway, F., Mansour, A.N., Pereira, N., Al-Sharab, F.J., Cosandey, F., Plitz, I., and Amatucci, G.G. (2007) *Chem. Mater.*, **19**, 4129.
81 Bervas, M., Mansour, A.N., Yoon, W.-S., Al-Sharab, J.F., Badway, F., Cosandey, F., Klein, L.C., and Amatucci, G.G. (2006) *J. Electrochem. Soc.*, **153**, A799.
82 Bervas, M., Klein, L.C., and Amatucci, G.G. (2006) *J. Electrochem. Soc.*, **153**, A159.
83 Pereira, N., Badway, F., Wartelsky, M., Gunn, S., and Amatucci, G.G. (2009) *J. Electrochem. Soc.*, **156**, A407.
84 Novoselov, K.S., Geim, A.K., Morozov, S.V., Jiang, D., Katsnelson, M.I., Grigorieva, I.V., Dubonos, S.V., and Firsov, A.A. (2005) *Nature*, **438**, 197.
85 Zhang, Y., Tan, Y.W., Stormer, H.L., and Kim, P. (2005) *Nature*, **438**, 201.
86 Geim, A.K. (2009) *Science*, **324**, 1530.
87 Chen, W., Chen, S., Qi, D.C., Gao, X.Y., and Wee, A.T.S. (2007) *J. Am. Chem. Soc.*, **129**, 10418.
88 Wei, D., Liu, Y., Wang, Y., Zhang, H., Huang, L., and Yu, G. (2009) *Nano Lett.*, **9**, 1752.
89 Elias, D.C., Nair, R.R., Mohiuddin, T.M.G., Morozov, S.V., Blake, P., Halsall, M.P., Ferrari, A.C., Boukhvalov, D.W., Katsnelson, M.I., Geim, A.K., and Novoselov, K.S. (2009) *Science*, **323**, 610.
90 Cano-Márquez, A.G., Rodríguez-Macías, F.J., Campos-Delgado, J., Espinosa-González, C.G., Tristán-López, F., Ramírez-González, D., Cullen, D.A., Smith, D.J., Terrones, M., and Vega-Cantu, Y.I. (2009) *Nano Lett.*, **9**, 1527.
91 Xue, J.S. and Dahn, J.R. (1995) *J. Electrochem. Soc.*, **142**, 3668.
92 Yoo, E.J., Kim, J., Hosono, E., Zhou, H.S., Kudo, T., and Honma, I. (2008) *Nano Lett.*, **8**, 2277.
93 Kroto, H.W., Heath, J.R., O'Brien, S.C., Curl, R.F., and Smalley, R.E. (1985) *Nature*, **318**, 162.
94 Alcántara, R., Ortiz, G.F., Lavela, P., Tirado, J.L., Stoyanova, R., and Zhecheva, E. (2006) *Chem. Mater.*, **18**, 2293.
95 Amaratunga, G.A.J., Chowalla, M., Kiely, C.J., Alexandrou, I., Aharonov, R., and Devenish, R.M. (1996) *Nature*, **383**, 321.
96 Yasukawa, M. and Yamanaka, S. (2001) *Chem. Phys. Lett.*, **341**, 467.
97 Buiel, et al. (1998) *J. Electrochem. Soc.*, **145**, 2252.
98 Claye, A., Fischer, J.E., Huffman, C.B., Rinzler, A.G., and Smalley, R.E. (2000) *J. Electrochem. Soc.*, **147**, 2845.
99 Wu, G.T., Wang, C.S., Zhang, X.B., Yang, H.S., Qi, Z.F., He, P.M., and Li, W.Z. (1999) *J. Electrochem. Soc.*, **146**, 1696.
100 Wang, Q., Li, H., Chen, L., Huang, X., Zhong, D., and Wang, E. (2003) *J. Electrochem. Soc.*, **150**, A1281.
101 Mukhopadhyay, I., Hoshino, N., Kawasaki, S., Okino, F., Hsu, W.K., and Touhara, H. (2002) *J. Electrochem. Soc.*, **149**, A39.
102 Maurin, G., Henn, F., Simon, B., Colomer, J.F., and Nagy, J.B. (2001) *Nano Lett.*, **1**, 75.
103 Eom, J.Y., Kim, D.Y., and Kwon, H.S. (2006) *J. Power Sources*, **157**, 507.
104 Duclaux, L. (2002) *Carbon*, **40**, 1751.
105 Tatsumi, K., Conard, J., Nakahara, M., Menu, S., Lauginie, P., Sawada, Y., and Ogumi, Z. (1997) *Chem. Commun.*, 687.
106 Alcántara, R., Lavela, P., Ortiz, G.F., Tirado, J.L., Zhecheva, E., and Stoyanova, R. (2007) *J. Electrochem. Soc.*, **154**, A964.
107 Wang, Q., Chen, L., and Huang, X. (2002) *Electrochem. Solid State Lett.*, **5**, A188.
108 Che, G., Lakshmi, B.B., Fisher, E.R., and Martin, C.R. (1998) *Nature*, **393**, 346.
109 Wu, X.L., Liu, Q., Guo, Y.G., and Song, W.G. (2009) *Electrochem. Commun.*, **11**, 1468.
110 Kavan, L., Graetzel, M., Gilbert, S.E., Klemenz, C., and Scheel, H.J. (1996) *J. Am. Chem. Soc.*, **118**, 6716.
111 Nuspl, G., Yoshizawa, K., and Yamabe, T. (1997) *J. Mater. Chem.*, **7**, 2529.
112 Noailles, L.D., Johnson, C.S., Vaughey, J.T., and Thackeray, M.M. (1999) *J. Power Sources*, **82**, 259.
113 van de Krol, R., Goznes, A., and Meulenkamp, E.A. (1999) *J. Electrochem. Soc.*, **146**, 3150.

114 Huang, S., Kavan, L., Exnar, I., and Grätzel, M. (1995) *J. Electrochem. Soc.*, **142**, L142.
115 O'Regan, B. and Grätzel, M. (1991) *Nature*, **533**, 737.
116 Sudant, G., Baudrin, E., Larcher, D., and Tarascon, J.M. (2005) *J. Mater. Chem.*, **15**, 1263.
117 Kelly, J.J. (1979) *Electrochim. Acta*, **24**, 1273.
118 Ortiz, G.F., Hanzu, I., Djenizian, T., Lavela, P., Tirado, J.L., and Knauth, P. (2009) *Chem. Mater.*, **21**, 63.
119 Ortiz, G.F., Hanzu, I., Knauth, P., Lavela, P., Tirado, J.L., and Djenizian, T. (2009) *Electrochim. Acta*, **54**, 4262.
120 Ortiz, G.F., Hanzu, I., Knauth, P., Lavela, P., Tirado, J.L., and Djenizian, T. (2009) *Electrochem. Solid State Lett.*, **12**, A186.
121 Armstrong, A.R., Armstrong, G., Canales, J., and Bruce, P.G. (2004) *Angew. Chem., Int. Ed.*, **43**, 2286.
122 Armstrong, A.R., Armstrong, G., Canales, J., García, R., and Bruce, P.G. (2005) *Adv. Mater.*, **17**, 862.
123 Reddy, M.Anji, Kishore, M.Satya, Pralong, V., Caignaert, V., Varadaraju, U.V., and Raveau, B. (2006) *Electrochem. Commun.*, **8**, 1299.
124 Hu, Y.S., Kienle, L., Guo, Y.G., and Maier, J. (2006) *Advan. Mater.*, **18**, 1421.
125 Baudrin, E., Cassaignon, S., Koelsch, M., Jolivet, J.P., Dupont, L., and Tarascon, J.M. (2007) *Electrochem. Commun.*, **9**, 337.
126 Vijayakumar, M., Kerisit, S., Wang, C., Nie, Z., Rosso, K.M., Yang, Z., Graff, G., Liu, J., and Hu, J. (2009) *J. Phys. Chem.*, **113** (46), 20108–20116.
127 Ohzuku, T., Ueda, A., and Yamamoto, N. (1995) *J. Electrochem. Soc.*, **142**, 1431.
128 Kavan, L. and Grätzel, M. (2002) *Electrochem. Solid State Lett.*, **5**, A39.
129 Cahn, R.W. (2001) *Contemp. Phys.*, **42**, 365.
130 Chou, N.H. and Schaak, R.E. (2007) *J. Am. Chem. Soc.*, **129**, 7339.
131 Cable, R.E. and Schaak, R.E. (2005) *Chem. Mater.*, **17**, 6835.
132 Winter, M., Besenhard, J.O., Spahr, M.E., and Novák, P. (1998) *Adv. Mater.*, **10**, 725.
133 Lee, H. and Cho, J. (2007) *Nano Lett.*, **7**, 2638.
134 Foster, M.S., Crouthamel, C.E., and Wood, S.E. (1966) *J. Phys. Chem.*, **70**, 3042.
135 Wen, C.J. and Huggins, R.A. (1981) *J. Electrochem. Soc.*, **128**, 1181.
136 Beaulieu, L.Y., Eberman, K.W., Turner, R.L., Krause, L.J., and Dahn, J.R. (2001) *Electrochem. Solid State Lett.*, **4**, A137.
137 Idota, Y., Kubota, T., Matsufuji, A., Maekawa, Y., and Miyasaka, T. (1997) *Science*, **276**, 1395.
138 http://www.sony.net/SonyInfo/News/Press/200502/05-006E/index.html.
139 Hosono, E., Matsuda, H., Honma, I., Ichihara, M., and Zhou, H. (2007) *J. Electrochem. Soc.*, **154**, A146.
140 Wang, Y., Raistrick, I.D., and Huggins, R.A. (1986) *J. Electrochem. Soc.*, **133**, 457.
141 Derrien, G., Hassoun, J., Panero, S., and Scrosati, B. (2007) *Adv. Mater.*, **19**, 2336.
142 Noh, M., Kwon, Y., Lee, H., Cho, J., Kim, Y., and Kim, M.G. (2005) *Chem. Mater.*, **17**, 1926.
143 Wang, K., He, X., Wang, L., Ren, J., Jiang, C., and Wang, C. (2006) *J. Electrochem. Soc.*, **153**, A1859.
144 Yang, J., Winter, M., and Besenhard, J.O. (1996) *Solid State Ionics*, **90**, 281.
145 Fan, Q., Chupas, P.J., and Whittingham, M.S. (2007) *Electrochem. Solid State Lett.*, **10**, A274.
146 Kim, H. and Cho, J. (2007) *Electrochim. Acta*, **52**, 4197.
147 Hassoun, J., Ochal, P., Panero, S., Mulas, G., Bonatto Minella, C., and Scrosati, B. (2008) *J. Power Sources*, **180**, 568.
148 Todd, A.D., Dunlap, R.A., and Dahn, J.R. (2007) *J. Alloys Compd.*, **443**, 114.
149 Ionica-Bousquet, C.M., Lippens, P.E., Aldon, L., Olivier-Fourcade, J., and Jumas, J.C. (2006) *Chem. Mater.*, **18**, 6442.
150 Alcántara, R., Rodríguez, I., and Tirado, J.L. (2008) *ChemPhysChem*, **9**, 1171.
151 Xie, J., Zhao, X.B., Cao, G.S., and Tu, J.P. (2007) *J. Power Sources*, **164**, 386.
152 Alcántara, R., Ortiz, G., Rodríguez, I., and Tirado, J.L. (2009) *J. Power Sources*, **189**, 309.
153 Dahn, J.R., Mar, R.E., and Abouzeid, A. (2006) *J. Electrochem. Soc.*, **153**, A361.
154 Ortiz, G.F., Alcántara, R., Rodríguez, I., and Tirado, J.L. (2007) *J. Electroanal. Chem.*, **605**, 98.

155 Mao, O., Dunlap, R.D., and Dahn, J.R. (1999) *J. Electrochem. Soc.*, **146**, 405.
156 Mao, O. and Dahn, J.R. (1999) *J. Electrochem. Soc.*, **146**, 414.
157 Zhang, C.Q., Tu, J.P., Huang, X.H., Yuan, Y.F., Wang, S.F., and Mao, F. (2008) *J. Alloys Compd.*, **457**, 81.
158 Mao, O., Turner, R.L., Courtney, I.A., Fredericksen, B.D., Buckett, M.I., Krause, L.J., and Dahn, J.R. (1999) *Electrochem. Solid State Lett.*, **2**, 3.
159 Ortiz, G.F., Alcántara, R., Lavela, P., and Tirado, J.L. (2005) *J. Electrochem. Soc.*, **152**, A1797.
160 Huang, L., Cai, J.S., He, Y., Ke, F.S., and Sun, S.G. (2009) *Electrochem. Commun.*, **11**, 950.
161 Nacimiento, F., Alcántara, R., and Tirado, J.L. (2009) *J. Alloys Compd.* doi: 10.1016/j.jallcom.2009.097.
162 Sharma, R.A. and Seefurth, R.N. (1976) *J. Electrochem. Soc.*, **123**, 1763.
163 Seefurth, R.N. and Sharma and, R.A. (1977) *J. Electrochem. Soc.*, **124**, 1207.
164 Boukamp, B.A., Lesh, G.C., and Huggins, R.A. (1981) *J. Electrochem. Soc.*, **128**, 725.
165 Wen, C.J. and Huggins, R.A. (1981) *J. Solid State Chem.*, **37**, 271–278.
166 Weydanz, W.J., Wohlfahrt-Mehrens, M., and Huggins, R.A. (1999) *J. Power Sources*, **81–82**, 237.
167 Netz, A., Huggins, R.A., and Weppner, W. (2003) *J. Power Sources*, **119–121**, 95.
168 Hatchard, T.D. and Dahn, J.R. (2004) *J. Electrochem. Soc.*, **151**, A838.
169 Wilson, A.M. and Dahn, J.R. (1995) *J. Electrochem. Soc.*, **142**, 326.
170 Lee, H.Y. and Lee, S.M. (2004) *Electrochem. Commun.*, **6**, 465.
171 Uono, H., Kim, B.C., Fuse, T., Ue, M., and Yamaki, J.I. (2006) *J. Electrochem. Soc.*, **153**, A1708.
172 Timmons, A., Todd, A.D.W., Mead, S.D., Carey, G.H., Sanderson, R.J., Mar, R.E., and Dahn, J.R. (2007) *J. Electrochem. Soc.*, **154**, A865.
173 Chan, C.K., Peng, H., Liu, G., McIlwrath, K., Zhang, X.F., Huggins, R.A., and Cui, Y. (2008) *Nat. Nanotechnol.*, **3**, 31.
174 Li, C., Liu, Z., Gu, C., Xu, X., and Yang, Y. (2006) *Adv. Mater.*, **18**, 228.
175 Poizot, P., Laruelle, S., Grugeon, S., Dupont, L., and Tarascon, J.M. (2000) *Nature*, **407**, 496.
176 Shao-Horn, Y., Osmialowski, S., and Horn, Q.C. (2002) *J. Electrochem. Soc.*, **149**, A1547.
177 Pereira, N., Dupont, L., Tarascon, J.M., Klein, L.C., and Amatucci, G.G. (2003) *J. Electrochem. Soc.*, **150**, A1273.
178 Alcántara, R., Tirado, J.L., Jumas, J.C., Monconduit, L., and Olivier-Fourcade, J. (2002) *J. Power Sources*, **109**, 308.
179 Alcántara, R., Fernández-Madrigal, F.J., Lavela, P., Tirado, J.L., Jumas, J.C., and Olivier-Fourcade, J. (1999) *J. Mater. Chem.*, **9**, 2517.
180 Aragón, M.J., León, B., Pérez Vicente, C., and Tirado, J.L. (2009) *J. Power Sources*, **189**, 823.
181 Iijima, T., Toyoguchi, Y., Nishimura, J., and Ogawa, H. (1980) *J. Power Sources*, **5**, 99.
182 Thackeray, M.M., Baker, S.D., and Coetzer, J. (1982) *Mater. Res. Bull.*, **17**, 405.
183 Thackeray, M.M., David, W.I.F., and Goodenough, J.B. (1982) *Mater. Res. Bull.*, **17**, 785.
184 Novak, P. (1985) *Electrochim. Acta*, **30**, 1687.
185 Choi, H.C., Lee, S.Y., Kim, S.B., Kim, M.G., Lee, M.K., Shin, H.J., and Lee, J.S. (2002) *J. Phys. Chem. B*, **106**, 9252.
186 Chadwick, A.V., Savin, S.L., Fiddy, S., Alcantara, R., Lisbona, D.F., Lavela, P. et al. (2007) *J. Phys. Chem. C*, **111**, 4636.
187 Larcher, D., Bonnin, D., Cortes, R., Rivals, I., Personnaz, L., and Tarascon, J.M. (2003) *J. Electrochem. Soc.*, **150**, A1643.
188 Alcántara, R., Jaraba, M., Lavela, P., and Tirado, J.L. (2003) *Electrochem. Commun.*, **5**, 16.
189 Obrovac, M.N. and Dunlap, R.A.R.J. (2001) *J. Electrochem. Soc.*, **148**, A576.
190 Dedryvere, R., Laruelle, S., Grugeon, S., Poizot, P., Gonbeau, D., and Tarascon, J.M. (2004) *Chem. Mater.*, **16**, 1056.
191 Thissen, A., Jaegermann, W., Alcantara, R., Lavela, P., Ortiz, G.F., and Tirado, J.L. (2005) *J. Electroanal. Chem.*, **584**, 147.
192 Poizot, P., Laruelle, S., Grugeon, S., and Tarascon, J.M. (2002) *J. Electrochem. Soc.*, **149**, A1212.

193 Lavela, P., Tirado, J.L., Womes, M., and Jumas, J.C. (2009) *J. Electrochem. Soc.*, **156**, A589.
194 Laruelle, S., Grugeon, S., Poizot, P., Dollé, M., Dupont, L., and Tarascon, J.M. (2002) *J. Electrochem. Soc.*, **149**, A627.
195 Balaya, P., Bhattacharyya, A.J., Jamnik, J., Zhukovskii, Y.F., Kotomin, E.A., and Maier, J. (2006) *J. Power Sources*, **159**, 171.
196 Jamnik, J. and Maier, J. (2003) *Phys. Chem. Chem. Phys.*, **5**, 5215.
197 Larcher, D., Sudant, G., Patrice, R., and Tarascon, J.M. (2003) *Chem. Mater.*, **15**, 3543.
198 Vidal-Abarca, P., Lavela, P., and Tirado, J.L. (2008) *Electrochem. Solid State Lett.*, **11**, A198.
199 Pralong, V., Leriche, J.B., Beaudoin, B., Naudin, E., Morcrette, M., and Tarascon, J.M. (2004) *Solid State Ionics*, **166**, 295.
200 Jiao, F., Harrison, A., Jumas, J.C., Chadwick, A.V., Kockelmann, W., and Bruce, P.G. (2006) *J. Am. Chem. Soc.*, **128**, 5468.
201 Débart, A., Dupont, L., Poizot, P., Leriche, J.B., and Tarascon, J.M. (2001) *J. Electrochem. Soc.*, **148**, A1266.
202 Grugeon, S., Laruelle, S., Herrera-Urbina, R., Dupont, L., Poizot, P., and Tarascon, J.M. (2001) *J. Electrochem. Soc.*, **148**, A285.
203 Lavela, P., Tirado, J.L., and Vidal-Abarca, C. (2007) *Electrochim. Acta*, **52**, 7986.
204 Liu, X., Yasuda, H., and Yamachi, M. (2005) *J. Power Sources*, **146**, 510.
205 Lavela, P. and Tirado, J.L. (2007) *J. Power Sources*, **172**, 379.
206 Bomio, M., Lavela, P., and Tirado, J.L. (2007) *ChemPhysChem*, **8**, 1999.
207 Lavela, P., Ortiz, G.F., Tirado, J.L., Zhecheva, E., Stoyanova, R., and Ivanova, S. (2007) *J. Phys. Chem. C*, **111**, 14238.
208 Aragón, M.J., Pérez-Vicente, C., and Tirado, J.L. (2007) *Electrochem. Commun.*, **9**, 1744.
209 Fan, Q. and Whittingham, M.S. (2007) *Electrochem. Solid State Lett.*, **10**, A48.
210 Deyrieux, R., Berro, C., and Peneloux, A. (1973) *Bull. Soc. Chim. Fr.*, **12**, 25.
211 Aragón, M.J., León, B., Pérez Vicente, C., and Tirado, J.L. (2008) *Inorg. Chem.*, **47**, 10336.
212 Aragón, M.J., León, B., Pérez Vicente, C., Tirado, J.L., Chadwick, A.V., Berko, A., and Beh, S.Y. (2009) *Chem. Mater.*, **21**, 1834.
213 Brezesinski, T., Wang, J., Polleux, J., Dunn, B., and Tolbert, S.H. (2009) *J. Am. Chem. Soc.*, **131**, 1802.
214 Serrano, T., Aragón, M.J., Leon, B., Pérez Vicente, C., and Tirado, J.L. (2009) Congreso Internacional de Química Industrial 2009, Monterrey, NL, Mexico.

9
Materials Science Aspects Relevant for High-Temperature Electrochemistry

Annika Eriksson, Mari-Ann Einarsrud, and Tor Grande

High-temperature electrochemistry has been driven by the progress in solid-state ionics and by the progress in ceramic processing of membrane reactors and fuel cells. An overview of oxide powder fabrication, ceramic processing of dense films on porous substrates, and sintering of the materials is given here. Further progress calls for an improved understanding of the underlying mechanisms for degradation of such devices. A review of cation diffusion in oxygen ion conductors, which control the stability of electrochemical devices, is therefore given. Moreover, stresses induced by mismatch in thermal expansion between materials and by chemical expansion are discussed in relation to the mechanical properties of relevant materials. Finally, the thermodynamic stability of relevant materials is discussed.

9.1
Introduction

High-temperature solid-state electrochemistry is one of the key elements for the development of green technologies. Among the most important solid-state electrochemical devices are solid oxide fuel cells (SOFCs) and dense ceramic membranes for gas separation (see Chapter 12 of the first volume and Chapters 5 and 10 of this book). There has been a tremendous development in the ionic conductivity of oxide materials, accompanied by the advances in ceramic processing, and the performance of fuel cells and membranes meets the targets for commercialization. Despite promising achievements, there are still issues related to degradation of performance and to lifetime of the devices, which are associated with the thermal and chemical stability of the materials.

In this chapter, we have focused on some of the most relevant aspects of the materials science related to high-temperature solid-state electrochemistry. In Section 9.2, the ceramic processing science is reviewed with focus on powder fabrication, ceramic processing of dense films on porous substrates, and, finally, sintering of materials. Although the primary attention is given to SOFCs and ceramic membranes, the experimental and methodological aspects discussed in this section are important for all areas of solid-state electrochemistry where ceramic materials are used, such as

Solid State Electrochemistry II: Electrodes, Interfaces and Ceramic Membranes.
Edited by Vladislav V. Kharton.
© 2011 Wiley-VCH Verlag GmbH & Co. KGaA. Published 2011 by Wiley-VCH Verlag GmbH & Co. KGaA.

batteries, sensors, and electrolyzers. Section 9.3 is concerned with cation diffusion in oxygen ion conductors, which is vital for the understanding of the long-term stability of electrochemical high-temperature devices. Section 9.4 is devoted to the thermomechanical stability of materials with particular focus on stresses induced by mismatch in thermal expansion between materials and chemical expansion. The mechanical properties of SOFC and ceramic membrane materials are also reviewed in this section. The final part, Section 9.5, is focused on the thermodynamic stability of relevant materials.

9.2
Powder Preparation, Forming Processes, and Sintering Phenomena

One of the main challenges in the development of SOFCs and dense membrane systems is to prepare materials with the desired density and micro(nano)structure fabricated into systems with the desired geometry. A dense, nonpermeable material is required for the membranes and electrolytes for the SOFC. The electrodes must be porous and permeable to gas transport, but with sufficient mechanical strength. To be able to control the density and microstructure of the ceramic materials, the whole preparation process starting from the powders to the final fuel cell system or membrane reactor has to be well designed. We therefore start here to look at the powder processing before we consider the shaping and densification. There are many different materials of interest considering membranes and electrodes for SOFCs and this chapter cannot cover all of them. The examples described in this chapter have therefore been selected to be perovskites based on $LaCoO_{3-\delta}$ [1–4], $(La,Sr)(Co,Fe)O_{3-\delta}$ (LSCF) [5–15], $(Ba,Sr)(Co,Fe)O_{3-\delta}$ (BSCF) [16], $(La,Sr)MnO_{3\pm\delta}$ (LSM) [17–19], and $LaGaO_3$ [20, 21] as well as $La_2NiO_{4+\delta}$ with K_2NiF_4-type structure [22–25] as oxygen-permeable membranes and/or cathodes for SOFCs, as well as the state-of-the-art proton conductors, including $BaZrO_3$ [26–31], $SrCeO_3$ [32], and $LaNbO_4$ [33, 34]. The traditional fluorite-structure solid electrolytes based on ZrO_2 [35] and CeO_2 [36, 37] will be only briefly covered. Description of the crystal structures, ionic transport properties, and selected applications of these materials can be found in Chapters 2, 7, 9, 12 and 13 of the first volume and Chapters 6 and 11 of this book.

9.2.1
Powder Processing and Forming Techniques

To achieve the target properties of the ceramic components (e.g., electrodes or membranes), the following properties of the ceramic powders should be considered: particle size, particle size distribution, particle morphology, degree of agglomeration, phase purity, and chemical purity. It is important to select a powder with high reactivity for fabrication of dense materials and the powder should therefore have a small particle size and a low degree of agglomeration. To prepare porous structures, other powder characteristics are normally desirable.

Ceramic powders of complex oxides are traditionally prepared by the solid-state reaction method where a mixture of binary oxides is subsequently exposed to firing and grinding until a phase pure material is obtained. However, this preparation method normally gives coarse agglomerated powders with low sintering reactivity. Therefore, several wet chemistry-based methods have been developed where the precursor is a homogeneous stoichiometric solution of the constituting cations of the oxide. The main advantage with these methods is the possibility to maintain the homogeneous mixing of the cations. Control of the chemical composition by these methods is, however, demanding, especially for substituted materials. Even a small deviation from stoichiometry might be detrimental for the phase composition, sinterability, and final microstructure. For the membrane and electrode materials, the key methods include coprecipitation, sol-gel-based techniques, combustion synthesis such as glycine nitrate, and spray pyrolysis, among others. One prerequisite for several of these methods is to have (water) soluble precursors that can be formed into a stable solution together with the other cations. Therefore, the use of different complexing agents is normally necessary like in the citric acid and the glycine nitrate routes. An overview of a selection of methods and their advantages and disadvantages are given in Table 9.1.

For the application in fuel cells and membrane systems, the ceramic powder has to be shaped into films of thickness varying from around 1 μm up to millimeter size and the layers have often to be combined. The membranes and electrolytes must be dense without any defects or pinholes and should preferably be quite thin (down to a few microns). To ensure sufficient mechanical stability, they have to be supported by, for example, the electrode (about 200–1500 μm thick). The electrodes must be made porous with a controlled pore size (or a controlled gradient in pore size). There are many methods described in the literature for the fabrication of such supported membrane structures [49, 50], but the recent developments have mainly been focused on methods based on wet ceramic technology due to cost efficiency and ease of upscaling, and these methods will be reviewed here. Planar and tubular geometries of the SOFCs or membranes require different fabrication techniques. Considering the planar design first, the most common configuration is anode-supported structures. The anode supports are almost always shaped by tape casting. Powders are dispersed in a solvent and plasticizers and binders are added. To increase the porosity of the anode, pyrolyzable pore formers such as carbon or starch are normally added. The slip is cast on a supporting tape and thickness is controlled by the height of the doctor blade. After careful drying, binder/pore former removal, and firing, the porous anode is obtained. Efforts have also been made to develop environment-friendly aqueous slips [51]. On top of the porous anode, an active anode might be deposited by similar methods as the membrane/electrolyte. This includes methods dependent on a stable slip or dispersion of the ceramic powder, such as vacuum slip casting, spin coating, dip coating, and spray coating. In vacuum slip casting, the porous substrate is used as a mold for building up the next layer. In spin coating, the slip is spun on a rotating substrate, while in dip coating the substrate is dipped into the slip and withdrawn

Table 9.1 Overview of selected methods used for the synthesis of oxide powders for high-temperature electrochemical applications.

Synthesis method	Advantages	Disadvantages	Examples	References
Solid-state method	Simple method Possible route for compounds containing elements with limited number of water-soluble precursors	Coarse agglomerated powder obtained	$BaZrO_3$ $LaCrO_3$ $SrCeO_3$-based materials	[30, 38, 39] [40] [41–44]
Coprecipitation	Easy upscaling	Homogeneity difficult to achieve Calcination necessary	$La_2NiO_{4+\delta}$ $BaZrO_3$ $(La,Sr)CoO_3$	[24] [31, 45] [1, 2]
Sol-gel-based methods	High degree of homogeneity Fine particle size obtained	Calcination necessary	$La_{0.6}Sr_{0.4}CoO_3$ $La(Sr)Fe(Co)O_3$ La_2NiO_4 $BaZrO_3$-based materials	[14, 15] [6, 9, 11] [23, 25] [28]
Combustion method including glycine nitrate	Fine particle size obtained Crystalline particles High degree of homogeneity	Difficulty in upscaling due to violent reaction	CeO_2 $BaZrO_3$ LSGM $(La,Sr)(Ga,Fe)O_3$ YSZ	[37] [26, 31, 46] [21] [10] [47]
Spray pyrolysis	Easy upscaling Homogeneous powder obtained No or simple calcination	Milling necessary	$LaNbO_4$ $BaZrO_3$-based materials $SrCeO_3$-based materials LSGM	[33] [27, 29, 30, 48] [32] [20]

with a given rate. The simplest of these routes is probably spray coating where the slip is sprayed through a two-phase or ultrasound nozzle onto the substrate. An alternative and also simple route for the preparation of the electrolyte is screen printing where the ceramic powder is mixed with a binder in approximately 50:50 ratio and this ink is pressed onto the substrate through a mesh screen. The critical step of removing the binder system is performed before the sintering step is done. A summary of the different green body shaping methods for a planar geometry is summarized in Table 9.2 together with the approximate thicknesses obtained for each of the methods.

Table 9.2 Overview of common wet ceramic techniques used for fabrication of ceramic membrane systems.

Design	Part of system	Fabrication method	Thickness (μm)	References
Planar	Support (anode)	Warm pressing	1500	[49, 54]
		Tape casting	200–1700	[50, 51, 55, 56]
	Active anode	Vacuum slip casting	5–15	[49]
	Membrane or electrolyte	Vacuum slip casting	5–30	[49]
		Spin coating	4–26	[55, 56]
		Dip coating	10–30	[51, 57]
		Spray coating	5–40	[51, 58]
		Screen printing	50	[54]
	Electrode (cathode)	Spray coating	<1000	[49, 51, 58]
		Screen printing	20–50	[54, 55]
Tubular	Electrode (cathode)	Extrusion	~2000	[50, 52, 59, 60]
		Isostatic pressing	400	
	Membrane or electrolyte	Spray coating	400	[52, 61]
		Electrophoresis	50	
	Electrode (anode)	Dip coating	10	[52]

Several of the most common methods to shape planar membrane structures, such as tape casting and screen printing, cannot be simply used for the tubular geometry where both cathode- and anode-supported structures are reported. The support can be shaped by extrusion where the ceramic powder is mixed into a plastic mass by the addition of a binder system (thermoset or thermoplastic polymer in addition to solvent, plasticizers, and lubricants). The tube-shaped geometry is formed by forcing the plastic mass through a designed die. The most critical step is the binder burnout in addition to avoiding defects such as tearing during the extrusion itself. The tube-shaped support can also be formed by isostatic pressing with an inner steel mandrel to shape the tube [52] or wet chemical techniques such as electrophoresis can be used [53]. A summary of the most applied fabrication methods for tubular geometry is also included in Table 9.2.

There is a challenge to prepare very thin dense membranes (<~2 μm) without any pinholes by using wet ceramic technology. The underlying substrate has to be very smooth and exhibits a small pore size that does not introduce any defects in the membrane. Moreover, clean room to avoid defects due to dust particles is also necessary. For research applications, several other methods such as atomic layer deposition (ALD) and metal organic chemical vapor deposition (MOCVD) are also recommended [62–65].

The ceramic processes described above to prepare the layered structures all contain several steps and for some of them even several heating steps are necessary. Therefore, technology that can simplify the overall fabrication process is desired.

Learning from the polymer membrane technology to prepare asymmetric ceramic hollow fibers for oxygen permeation is therefore interesting [66, 67]. These hollow fibers are prepared in one step using an immersion-induced phase inversion technique. Commercially available ceramic powders are used in addition to a binder, a solvent, and an additive.

There is also a large focus on the development of miniaturized power sources, in particular, micro-SOFCs (so-called μ-SOFCs), for electronic devices. μ-SOFCs are normally fabricated by microelectromechanical system (MEMS) technology and were thoroughly described in a recent review by Evans et al. [53] and are therefore not covered here. We would, however, like to point out the fact that since the dimensions of such μ-SOFCs are much smaller than those of conventional cells, the reactivity and stabilities might be different from their more bulk counterparts due to the small diffusion distances.

9.2.2
Densification, Grain Growth, and Pore Coalescence

After green body formation, the ceramic body is subjected to a heating process to facilitate mass transport and densification. In this section, we will focus on sintering phenomena in general, while sintering studies of specific materials and membrane systems will be covered in Section 9.2.3.

The macroscopic driving force for the mass transport during sintering is reduction in surface area and the mass transport during solid-state sintering can take place through the following mechanisms: (i) evaporation–condensation, (ii) surface diffusion, (iii) lattice diffusion from the surface, (iv) lattice diffusion from the grain boundaries, and (v) grain boundary diffusion. These mass transport mechanisms are illustrated in Figure 9.1.

Only mechanisms (iv) and (v) provide densification and these have therefore to be encouraged to occur to achieve a high-density membrane. The other three

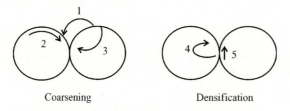

Coarsening   Densification

(1) Evaporation – condensation
(2) Surface diffusion
(3) Lattice diffusion from surface
(4) Lattice diffusion from grain boundary
(5) Grain boundary diffusion

Figure 9.1 Illustration of mass transport mechanisms occurring during solid-state sintering of two ceramic particles (see text).

mechanisms provide coarsening and should be discouraged to obtain high density. However, these mechanisms should be encouraged when porous support materials with high strength should be prepared. Different sintering parameters such as temperature, heating program, particle size, pressure, atmosphere, and doping can be used to encourage/discourage the different mass transport mechanisms and these should be considered differently for the dense membranes and the porous supports. A schematic drawing of the different steps of the sintering process to prepare dense ceramic materials and porous structures is provided in Figure 9.2.

The sintering cycle is usually divided into initial, intermediate, and final stages. The initial stage provides little shrinkage, but a coarsening of the particles due to evaporation and condensation (if there are volatile species) as well as surface diffusion occurs, and necks between the particles are formed. To obtain a highly dense material, these mass transport mechanisms should be discouraged (i.e., rapid heating) at this stage as they reduce the driving forces for the following densification. The intermediate stage gives a high degree of densification, but the pores are still

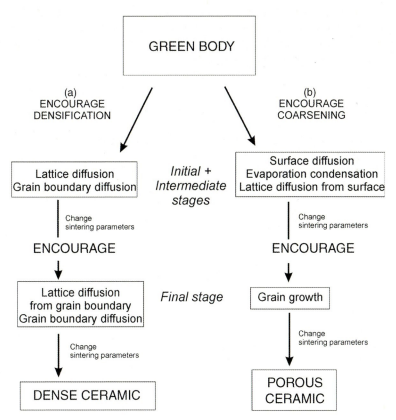

Figure 9.2 Schematic drawing of mass transport mechanisms that should be encouraged to obtain (a) dense ceramic body and (b) porous structure.

open. The densification is caused by mass transport via lattice diffusion from grain boundaries and grain boundary diffusion that should be encouraged to obtain high density. During the final stage, the pores are no longer large enough to prevent grain growth and this is the major process going on in addition to the final densification. With the increasing grain size, the densification rate decreases as the distance to the grain boundaries increases. Grain growth also gives pore coalescence where smaller pores are merged together into larger ones that also reduce the densification rate. Grain growth is therefore normally discouraged to achieve high density and additives are normally necessary to avoid this. On the other hand, to achieve a highly porous ceramic body (porosity $>\sim 20\%$), mass transport mechanisms providing coarsening have to be encouraged. However, normally it is not possible to prevent densification to this large extent and the application of pyrolyzable pore fillers might be necessary [68, 69]. During heating, the pore fillers are burned off leaving large voids that are not densified during the subsequent firing to higher temperatures. As long as large voids are formed in the ceramic body, there is no driving force for them to disappear.

Recommended heating program for solid-state sintering can be elucidated from the rate versus temperature diagram given in Figure 9.3a. The rate of coarsening, densification, and grain growth is presented as a function of inverse temperature. At low temperatures, coarsening is the major process and hence the heating up to the sintering temperature should be fast. When entering the temperature range where densification is the major process, the heating rate should be decreased. However, to reach a high density, the temperature has to be increased to the upper limit for densification as densification normally is a slow process. Since pores present in the sintering body will retard the rate for grain growth, it is also possible to increase the temperature into the grain growth region to achieve a significantly high density. The heating program illustrated in Figure 9.3b is therefore recommended. For solid-state sintering, the densification temperature should be in the range of $(2/3)T_m$, where T_m is the melting temperature in Kelvin. For membrane systems of interest here the major fabrication cost lies in sintering, and cofiring methods are therefore developed to avoid several firing steps.

The use of additives to, for example, reduce grain growth or the presence of nonstoichiometry may give secondary phases that can influence the densification process. If the secondary phase melts during the sintering process, the liquid-phase sintering is introduced and mass transport is enhanced by a dissolution/reprecipitation mechanism [3]. The liquid-phase sintering is much faster than the solid-state sintering and exaggerated grain growth may occur giving an undesirable microstructure.

In addition to the conventional sintering, other methods have also been used to reduce the sintering temperature and time necessary to obtain materials with a much smaller grain size. These methods include two-step sintering, hot pressing (HP) and spark plasma sintering (SPS). Using the two-step sintering approach, it has been demonstrated that the densification and grain growth can be separated in temperature in order to obtain a small degree of grain growth [70]. During hot pressing, additional mass transport mechanisms occur and the sintering temperature can be reduced by 200–300 °C and fine-grained microstructures are

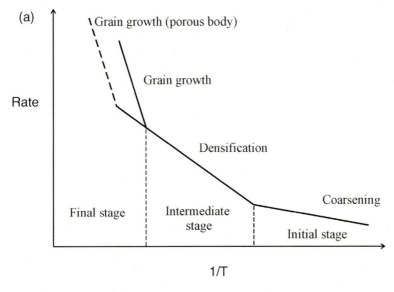
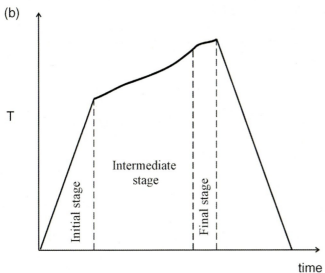

Figure 9.3 Schematic drawing showing the main processes taking place during the three stages of solid-state sintering. (a) Rates of coarsening, densification, and grain growth as a function of inverse temperature. (b) Recommended temperature program for solid-state sintering. Modified from Ref. [228].

obtainable [29]. By using an additional spark, the densification rate can be further increased and densification within minutes at a rather low temperature is possible [71]. The applicability of these methods to produce fuel cells or membranes is, however, questionable.

9.2.3
Sintering of Oxide Electrolytes and Ceramic Membrane Materials

A high density without defects or flaws is required for oxide ceramics to be used as SOFC electrolytes and membranes. A homogeneous microstructure and controlled distribution of any secondary phase present are therefore required. In the following, a brief review of the densification of selected electrolyte/membrane/electrode materials is given with the focus on how to achieve a well-designed microstructure.

Of the fluorite-type electrolytes, a comprehensive work on the ionic conductivity dependence on the grain size has been performed on ceria-based materials and a large effort has been made to prepare nanocrystalline materials [72, 73]; relevant references can be found in Chapter 4 of the first volume. Preparation of polycrystalline materials with nanosized grains is, however, challenging since grain growth is accompanying the last stage of the sintering process. Therefore, to prepare CeO_2-based materials with a very small grain size, hot pressing or SPS has to be used. Dense CeO_{2-x} with grain size of $\sim$10 nm [74] was prepared by hot pressing at a very high pressure of 1.1 GPa at 600 °C and grain size of 160 nm has been obtained by SPS [75].

Perovskite membrane and electrode materials, where A-sites in the ABO_3 perovskite lattice are occupied with rare earth and alkaline earth cations and B-sites contain transition metal cations, are complex materials and the sintering has been studied for a variety of compositional families. Since $LaMnO_{3+\delta}$ is an excess oxygen perovskite with cation vacancies, the sintering is challenging as the oxygen excess and hence also the cation vacancy concentration decreases with increasing temperature. Enhanced sintering rate has been observed for La-deficient material showing that increased concentration of La vacancies increases the densification rate and hence diffusion of La was supposed to be the rate-limiting step [76, 77]. On substitution with Sr in LSM, the oxygen content is lowered and thereby the A-vacancy concentration decreases. The sintering temperature is therefore increased for A-site substituted materials. Also for Sr-substituted materials, A-site cation deficiency promotes densification [78]. Recently, McCarthy et al. [19] showed that the intermediate stage sintering of LSM was enhanced by subjecting the sample to alternating flows of air and nitrogen. A mechanism including the formation of transient cation vacancy gradients leading to higher cation mobilities was proposed. The densification and grain growth on LSM have also been studied under cathodic polarization conditions that have been shown to hinder both the densification and grain growth rates [18].

Sagdahl et al. [12] sintered both Fe-excess and Fe-deficient $LaFeO_3$ and showed that secondary phases played a major role in the densification rate and microstructure. Fe excess led to the formation of a liquid phase above 1430 °C resulting in exaggerated grain growth and swelling. For Fe deficiency, the presence of La_2O_3 inhibited densification and grain growth. Dense $LaFeO_3$ could only be obtained in a narrow composition and temperature range that is quite representative for other ceramics of this type. By substitution with Sr on A-site in cation nonstoichiometric $LaFeO_3$, Sagdahl et al. [13] showed that the presence of secondary phases increased the densification temperature. Swelling at higher sintering temperatures was evident in all the compositions and this was explained by a temporary overpressure of oxygen in closed pores due to the thermal reduction of Fe during heating. The swelling was extremely rapid in the samples where a liquid phase was formed at high temperature due to viscous flow. The densification behavior of such ferrite-based materials can also be changed by the addition of sintering aids. For example, Kharton et al. [79] found that moderate additions of alumina improved the sinterability by the segregation of alumina-rich phases and the formation of A-site deficient perovskite.

The sintering of $LaCoO_3$-based materials has been studied by several groups [3, 80]. It is clear that to reduce the sintering temperature fine powders from wet chemical synthesis routes are necessary. By studying stoichiometric, Co-excess and Co-deficient samples of $La_{1-x}M_xCoO_{3-\delta}$ (M = Sr, Ca), Kleveland et al. [3] showed that the sintering temperature was decreased by Sr or Ca substitution and that dense materials were obtained at 1200 °C for stoichiometric materials. Considerable grain growth was observed at higher temperatures. The presence of other crystalline phases and a liquid in nonstoichiometric $La_{0.8}Ca_{0.2}CoO_{3-\delta}$ significantly influenced the densification rate and the microstructure of the materials. The presence of the liquid caused exaggerated grain growth.

It is clear from these studies of the perovskite-based materials that secondary phases, either added as sintering aids or formed by deviations from cation stoichiometry, play a very important role during sintering. The secondary phases are often formed on purpose, but improper control of composition has also in many cases resulted in unwanted effects. A very careful control of composition, sintering temperature, and amount of sintering aids is necessary to avoid undesired effects such as abnormal grain growth, evaporation, and swelling. It is also clear that for some of these materials, the temperature window to obtain a dense material with the desired microstructure is very narrow.

Relatively little focus has been put on fundamental studies of densification and the development of the microstructure during sintering of state-of-the-art proton-conducting oxides. Generally sintering temperatures in the range 1500–1650 °C are needed for preparing dense (>95%) $SrCeO_3$ and $SrCe_{2.95}Yb_{0.05}O_{3-\delta}$ materials by the solid-state ceramic method [41–44]. Sintering temperatures have been reduced to 1300–1400 °C by using powders prepared by wet chemical synthesis [32, 81]. Problems related to carbonate formation in these Sr-containing materials (see Section 9.5.2) make it necessary to calcine the powders in a CO_2-free atmosphere prior to sintering. A simple approach to $SrCeO_3$-based materials (Nd- and Y-substituted) has been shown by Liou and Yang [82] who fired a pressed green body of the constituting oxides and carbonates and formed highly dense materials. B_2O_3 was necessary as sintering aid to reduce the sintering temperature.

$BaZrO_3$-based ceramic powders are normally formed by solid-state reaction giving coarse powders with sintering temperatures above 1700 °C [38, 39]. At these high temperatures, $BaZrO_3$ may decompose according to Equation 9.1 [83] giving non-stoichiometric materials.

$$BaZrO_3(s) = BaO(g) + ZrO_2(s) \qquad (9.1)$$

Excess BaO has therefore been added during sintering to compensate for the evaporation [46]. Additives such as oxides of Zn, Ga, Ti, Mg, In, and Al have been used to lower the sintering temperature, but it has proven difficult to find an additive that enhances sintering without a negative effect on the conductivity [84]. Finer powders of $BaZrO_3$-based materials have been prepared by different methods such as coprecipitation [45], sol-gel techniques [28], spray pyrolysis [27, 29, 30, 48], and glycine nitrate method [26, 31, 46]. These finer powders possess higher sintering reactivity. By hot pressing of fine powder prepared by spray pyrolysis, Dahl et al. prepared dense $BaZrO_3$ at 1500 °C and $BaZr_{0.9}Yb_{0.1}O_{2.95}$ at 1550 °C [29]. Dense (>95%) ceramics of undoped and Y-substituted $BaZrO_3$ have been obtained by SPS [85].

9.3
Cation Diffusion

9.3.1
Theoretical Aspects of Cation Diffusion

Despite the potential use of oxide materials in high-temperature electrochemical devices, the fundamental aspects of the mobility of all other ions except oxygen anions have not received much attention. As the computational methods have become more powerful tools, the theoretical insight into cation migration has emerged accompanied by an increasing body of experimental data on the topic.

If we consider the two most important structures related to this field, the perovskite with the formula ABO_3 and fluorite crystal structure with the molecular formula AO_2, the possible pathways for cation migration can be understood from the crystal lattice, depictured in Figure 9.4, and the particular point defect equilibrium encountered for the two types of crystal lattices.

The fluorite structure (Figure 9.4a) consists of a cubic closed array of cations in which all the tetrahedral sites are occupied by anions. The coordination numbers of the cations and anions are 8 and 4, respectively. The perovskite structure on the other hand contains two cation sublattices (Figure 9.4b). It can be described by a network of corner-sharing BO_6 octahedra with the large A-cation located in the middle of eight of the octahedra. An alternative description is a ccp lattice of AO_3 layers with the B-cation located in octahedral holes surrounded by only oxygen.

The fluorite lattice, and particularly the perovskite lattice, is characterized by a high packing density and interstitial cations are energetically not favorable. Theoretical considerations of cation migration have therefore been based on hopping of cation vacancies. Relatively high activation energies for cation migration have been found for $LaGaO_3$ [86] and $LaMnO_3$ [87]. The highest activation energies in the range 6.7–14 eV have been found for B-site migration. For A-site vacancy migration, the activation energy is in the range 2.8–4.7 eV. A-site vacancy migration is therefore favorable based on static energy considerations. The most favorable pathway was found for A-site migration along $\langle 100 \rangle$ direction (see Figure 9.4b). For B-site

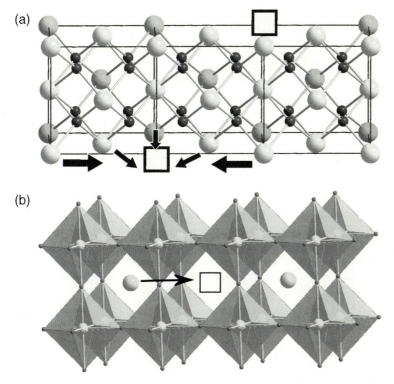

Figure 9.4 Crystal structures of AO_2 fluorite (a) and ABO_3 perovskite (b) showing the most energetically favorable pathways for the A-site vacancy migration. The vacancies and A-site cations are shown as empty squares and large gray spheres, respectively. The octahedra in the perovskite structure are formed by B-site cations and oxygen anions.

migration, the pathway is more complicating and has curved path onto the (001) plane (Figure 9.5).

Oxides with fluorite structure have not been studied in much detail, but a molecular dynamics (MD) study has shown activation energies (4.4 eV) for cation migration in very good accordance with experimental values (4.3–5.3 eV) [88].

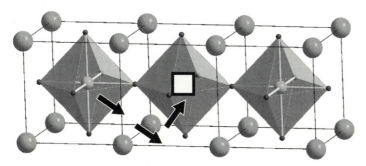

Figure 9.5 B-site cation migration path in the perovskite structure.

Vacant sites in the lattice have to be created in order to facilitate cation motion, and the relevant point defect equilibria for fluorite and perovskite materials are summarized below.

For the fluorite structure, the two dominating point defect equilibria are the Schottky-type (Equation 9.2) and Frenkel-type (Equation 9.3) disorders:

$$\text{nill} = V_A'' + 2V_O^{\bullet\bullet} \tag{9.2}$$

$$O_O = O_i'' + V_O^{\bullet\bullet} \tag{9.3}$$

Note that the Kröger–Vink notation used in these equations and the principles of point defect equilibria in solids were explained in Chapter 3 of the first volume. Cation vacancies are created by the Schottky defect equilibrium, but since the oxygen vacancy concentration is influenced by the Frenkel disorder (Equation 9.3), the cation vacancy concentration is also influenced by this defect equilibrium. In electrolytes based on the fluorite structure, such as yttria-stabilized zirconia (YSZ) and gadolinia-substituted ceria (GSC), the chemical composition is optimized in order to enhance the oxygen anion mobility. The defect equilibrium due to Y_2O_3 substitution in ZrO_2 can be described as

$$\tfrac{1}{2}Y_2O_3 = Y_{Zr}' + \tfrac{3}{2}O_O^X + \tfrac{1}{2}V_O^{\bullet\bullet} \tag{9.4}$$

The high anion vacancy concentration shifts the cation vacancy concentration to very low values and cation diffusion in fluorite electrolytes is therefore typically on the order of 10^{-20} cm^2 s^{-1} at 950 °C, which is $>10^{-13}$ order of magnitude lower than the diffusion coefficient of oxygen anions [89]. The MD simulations showed that the migration path goes through vacancies and not via interstitials [88]. The activation energies are in good accord with experimental values and are mainly governed by the migration enthalpy. The migrating pathway has been suggested to incorporate associated vacancy pairs. To our knowledge, no theoretical consideration of cation diffusion in ceria has been reported.

For perovskites, Frenkel disorder is not likely and Schottky defects are the key defects. The Schottky point defects in a perovskite, ABO_3, can be expressed as

$$\text{nill} = V_A''' + V_B''' + 3V_O^{\bullet\bullet} \tag{9.5}$$

The perovskite lattice can accommodate a whole range of cations with different charge. The possibility of mixed valence state of metal ions does therefore often occur and the point defect equilibrium associated with oxidation may give an explanation for the oxygen partial pressure dependence of cation diffusion:

$$\tfrac{3}{2}O_2(g) = V_A''' + V_B''' + 6h^{\bullet} + 3O_O \tag{9.6}$$

Correspondingly, thermal reduction of the perovskite may control (depress) the defect concentration by enhancing the oxygen vacancy concentration:

$$O_O = V_O^{\bullet\bullet} + 2e' + \tfrac{1}{2}O_2(g) \tag{9.7}$$

Theoretical simulation of cation migration has been performed for several perovskite materials [86, 87]. Generally the activation energy found by computational studies is significantly higher than the values observed experimentally, as discussed in Section 9.3.4. The discrepancy has been explained by involving association of point defects and defect clusters [90, 91].

The diffusion of cations is an activated process and the diffusion constant can be expressed as

$$D = D_0 \exp(-E_A/RT), \quad (9.8)$$

where E_A is the activation energy for diffusion and the preexponential factor D_0 can be expressed as

$$D_0 \approx \gamma_D a_0^2 v_f \exp(S_D/k_B), \quad (9.9)$$

where γ_D is a geometrical factor, a_0 is the jump distance between the lattice sites, v_f is the vibrational frequency, S_D is an entropy term associated with the diffusion, and k_B is the Boltzmann constant.

For a vacancy diffusion mechanism, the self-diffusion coefficient D is proportional to the vacancy concentration fraction in the relevant lattice [V] and the vacancy concentration diffusion coefficient D_V:

$$D \approx D_V[V]. \quad (9.10)$$

The apparent activation energy for diffusion can be divided into two terms – the migration energy E_m and the vacancy formation enthalpy ΔH_V:

$$E_A = E_m + \Delta H_V. \quad (9.11)$$

Since cation migration is often slow, there is considerable chance to *freeze in* a particular vacancy concentration during crystal growth or sintering of bulk materials especially during cooling. The determination of the contribution from migration energy and vacancy formation may therefore be difficult if slow relaxation of the equilibrium concentration takes place during the time frame of an experiment.

9.3.2
Grain Boundary and Bulk Diffusion

In single crystals, diffusion of cations is dominated by bulk, but significantly enhanced diffusion along dislocations has been reported for zirconia [92]. In polycrystalline materials, the grain boundaries open up for two diffusion mechanisms, bulk and grain boundary (gb) diffusion. The effective diffusion coefficient D_{eff} can be given by the Hart's equation [93]

$$D_{\text{eff}} = r_{\text{gb}} D_{\text{gb}} + (1 - r_{\text{gb}}) D_{\text{bulk}}, \quad (9.12)$$

where D_{gb} and D_{bulk} are the grain boundary and bulk diffusion coefficients, respectively. The volume fraction of grain boundaries, r_{gb}, can be related to the grain diameter d_g and the grain boundary width w_{gb}. ($r_{gb} \approx 3w_{gb}/d_g$). In cases where $r_{gb} \ll 1$, Equation 9.12 simplifies to $D_{eff} \approx D_{bulk} + r_{gb}D_{gb}$. Harrison [94] provided the analysis of the problem using a model of parallel grain boundaries. This model distinguishes three limiting cases of diffusion kinetics. The so-called type A kinetics is found when the penetration depth due to bulk diffusion is large enough, that is, $\sqrt{D_{bulk}t} > d_g$, where t is time, and the diffusion profiles appear to follow Fick's law for a homogeneous system. In the opposite case, $\sqrt{D_{bulk}t} < w_{gb}$, diffusion only along grain boundaries occur (type C kinetics). Finally, in the intermediate case (type B kinetics), fast diffusion along the grain boundaries is accompanied by bulk diffusion ($w_{gb} < \sqrt{D_{bulk}t} < d_g/2$). Harrison's model is important for the analysis of tracer or impurity diffusion experiments as discussed below.

For both perovskite and fluorite type of oxide materials, enhanced cation diffusion along grain boundaries relative to bulk has been reported. For YSZ, D_{gb}/D_{bulk} ratio is on the order of 10^5–10^6 [95, 96]. For perovskite materials, the corresponding values can be found in Refs [97–102].

There is increasing interest in "nanoionics" [103] where bulk materials possess grains in the nanorange, and the grain boundary volume becomes much more pronounced than in traditional bulk polycrystals. Grain boundary diffusion of Zr in nanocrystalline YSZ has been reported to be 10^7 times higher than bulk diffusion [95]. The long-term stability of nanocrystalline YSZ relative to microcrystalline YSZ should therefore be addressed.

9.3.3
Experimental Methods for Determination of Cation Diffusion

The mobility of cations is relatively low compared to the mobility of other charge carriers and the cation mobility cannot be studied by indirect methods based on flux or conductivity measurements. There are in principle four different methods to study cation diffusion in solids: (i) tracer or impurity diffusion, (ii) interdiffusion, (iii) diffusion-controlled solid-state reaction between two binary oxides, and finally (iv) creep measurement.

Tracer or impurity diffusion is a very powerful tool to investigate the diffusion of cations in oxide materials [104]. A tracer of component M, denoted M*, is chemically identical to M and has a low concentration. The diffusion coefficient of the tracer, D_M^*, can be found by an analysis of the tracer diffusion profile in chemically homogeneous solids. Both radioactive and stable isotopes of elements can be used as tracers, and alternatively an impurity element X_M with very similar chemical properties can be used if no isotopes are available. Radiotracers are analyzed by means of their emitted radiation, while the concentration profile of stable isotopes or impurities is most widely accomplished by ion beam analysis techniques. Here, secondary ion mass spectroscopy (SIMS) is most widely applicable and has been used to study diffusion in ceramic materials [104]. First, a thin film of the tracer is deposited on a smooth and polished surface of a single crystal or polycrystal of the material. The material is then annealed at a specific temperature for a certain time and quenched at ambient temperature (see Figure 9.6a). The obtained concentration profile of the tracer is

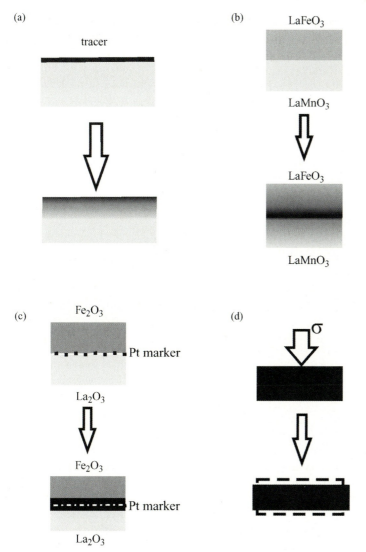

Figure 9.6 Schematic representation of the main methods for experimental investigation of cation diffusion in solids. (a) Tracer or impurity diffusion where a thin film of the tracer is covering the sample. (b) Interdiffusion between two materials with the same crystal structure, using LaFeO$_3$ and LaMnO$_3$ as an example. (c) Diffusion couples between Fe$_2$O$_3$ and La$_2$O$_3$ resulting in the growth of a LaFeO$_3$ layer, where a Pt marker between the La$_2$O$_3$ and Fe$_2$O$_3$ is used to identify the dominating diffusion species. (d) Compressive creep testing where the strain is monitored as a function of stress.

then analyzed using appropriate solutions of the diffusion equation. For pure bulk diffusion from a thin film source [105], the concentration profile can be fitted to the equation

$$\frac{\Delta c(x, t,)}{c(0, t)} = \exp\left(\frac{-x^2}{4D_{\text{bulk}}t}\right), \tag{9.13}$$

where $\Delta c(x, t,)/c(0, t)$ is the normalized concentration versus the distance x from the surface. In cases where both bulk and grain boundary diffusion take place, and the bulk diffusion can be separated from the grain boundary diffusion (type B kinetics), the grain boundary tail can be fitted by using

$$s_f w_{gb} D_{gb} = 0.3292 \left(\frac{D_{\text{bulk}}}{t}\right)^{1/2} \left(\frac{\partial \log c}{\partial x^{6/5}}\right)^{-5/3}, \tag{9.14}$$

where s_f is a segregation factor. D_{gb} can then be found by plotting the normalized intensity $\log (I/I_0)$ versus $x^{6/5}$ [106].

Interdiffusion experiments are based on annealing of diffusion couples of two materials, for example, $LaMnO_3$ and $LaFeO_3$, both having perovskite crystal structures (see Figure 9.6b). After annealing the diffusion couple for a certain time at constant temperature, the concentration profile of the two cations can be found by, for example, electron microprobe analysis [107]. The expression for the bulk diffusion concentration of the diffusing cation is obtained by solving Fick's second law (assuming a constant source and constant diffusion coefficient) and is given by

$$c(x) = \frac{c_1 - c_2}{2} \operatorname{erfc}\left(\frac{x}{2\sqrt{D_{\text{bulk}}t}}\right) + c_2, \tag{9.15}$$

where c_1 and c_2 are constants apparent from the experimental concentration profile and erfc is the complementary error function [105]. With type B kinetics, the grain boundary diffusion can also be determined in interdiffusion experiments using the same equation as used for tracer diffusion profiles.

The third method used to explore cation diffusion is *diffusion-controlled solid-state reaction between two binary oxides*. In this experiment, the solid-state reaction between two solids is examined by annealing diffusion couples of the two solids at constant temperature and time. The compound formed during the thermal treatment and the thickness of the growing layer is monitored as a function of time at a given condition in temperature and partial pressure of oxygen. An example is the diffusion couple between La_2O_3 and Fe_2O_3, where $LaFeO_3$ is formed. The principle is illustrated in Figure 9.6c. A Pt inert marker at the initial interface can be used to determine the mobile cation in the growth process. If the thickness x of the new compound follows the parabolic rate law ($x^2 = k_p t$, k_p = rate constant), the kinetics of the growth process is controlled by diffusion, and the chemical diffusion constant of the most mobile cation can be determined. In the La_2O_3/Fe_2O_3 couple (Figure 9.6c), where $LaFeO_3$ is formed, diffusion of Fe^{3+} is dominating [108]. The parabolic rate constant can be related to the diffusion coefficient by

$$k_p = -\frac{V_m}{RT} \int_{\mu_{Fe(La_2O_3)}}^{\mu_{Fe(La_2O_3)}} D_{Fe} c_{Fe} \, d\mu_{Fe}, \qquad (9.16)$$

where V_m is the molar volume of reaction product formed by transport of Fe^{3+}, c_{Fe} is the molar concentration of Fe^{3+}, and $\mu_{Fe(Fe_2O_3)}$ and $\mu_{Fe(La_2O_3)}$ are the chemical potentials of Fe cations at the two interfaces.

Finally, *creep measurements* are based on the recording of the macroscopic plastic strain of a sample under a constant load (see Figure 9.6d). The strain rate can be related to a set of parameters given by

$$\dot{\varepsilon} = A_c \left(\frac{1}{d_g}\right)^p (p_{O_2})^m \sigma^n \exp\left(-\frac{Q}{RT}\right), \qquad (9.17)$$

where $\dot{\varepsilon}$ is the steady-state creep rate, A_c is a constant, d_g is the grain size, p is the inverse grain size exponent, p_{O_2} is the partial pressure of oxygen, m is the oxygen partial pressure exponent, σ is the applied stress, n is the stress exponent, and Q is the activation energy. Since the strain rate depends on factors related to both the crystal structure (point defects) and microstructure (grain size), the interpretation of creep data can be quite challenging. However, it is a quite powerful tool to investigate the effect of oxygen partial pressure.

9.3.4
Cation Diffusion in Perovskite and Fluorite Oxide Materials

The available data for cation diffusion in oxide materials with the fluorite structure are displayed in Figure 9.7. The data show that the cation mobility is significantly decoupled from the oxygen ion mobility ($D_O/D_{cations} \sim 10^{13}$ at 950 °C), and YSZ materials are therefore very good candidate electrolytes for fuel cells, oxygen pumps, or sensors. The low diffusion constants also explain the relatively high sintering temperature of YSZ, which is usually above 1350 °C. At this temperature, the bulk mobility is typically 10^{-15}–10^{-16} cm^2 s^{-1}, while the grain boundary diffusion is $\sim 10^5$–10^7 times higher. This suggests that the sintering is dominated by grain boundary diffusion (the average diffusion lengths during 1 h sintering by bulk and grain boundary diffusion are 0.02 and 6 μm, respectively). Unfortunately, experimental data for ceria-based electrolytes are not available. Creep data of ceria have, however, been reported, and the creep resistance in ceria is higher than that for zirconia [109]. The cation diffusion in ceria is therefore anticipated to be comparable to zirconia materials.

The corresponding data for cation diffusion in oxide materials with perovskite structure are displayed in Figure 9.8. The data show that the cation mobility is significantly higher than that in fluorite materials. The operation temperature of devices based on perovskites should therefore be lower to hinder long-term degradation

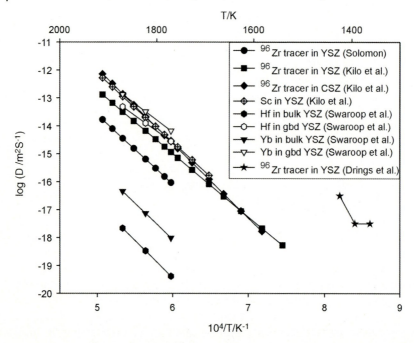

Figure 9.7 Bulk and grain boundary diffusion coefficients of zirconia-based materials [92, 95, 96].

processes due to cation transport. The decoupling between cations and oxygen anions in perovskite materials is on the same order of magnitude for LSGM ($D_O/D_{cations} \sim 10^{12}$ at 1000 °C) compared to YSZ, while it seems to be significantly lower for LaCoO$_3$-based materials ($D_O/D_{cations} \sim 10^9$–$10^{10}$ at 1000 °C). The typical cathode materials are therefore less thermally stable with respect to the decoupling. The activation energy found for most of the materials shown in Figure 9.8 is lower than that found by theoretical considerations and there is no consensus concerning the diffusion mechanism in perovskite materials.

9.3.5
Kinetic Demixing and Decomposition

In any high-temperature electrochemical device, the components are placed under electric field and/or under gradients of thermodynamic potentials such as temperature, pressure, and chemical potential. Without a gradient, the device would not work and transport of ions would not occur. The design of such a device is based on certain criteria, including the transport numbers of the charge carriers. For an electrolyte, the transport number of the mobile potential-determining ions is in the ideal case one, while for gas separation membranes the transport of one specific type of ions is compensated by electronic charge carriers. However, the diffusion coefficients of the residual ions, in the present context cations, are not zero, and the potential gradient acts as a thermodynamic force that will induce mass fluxes. These fluxes

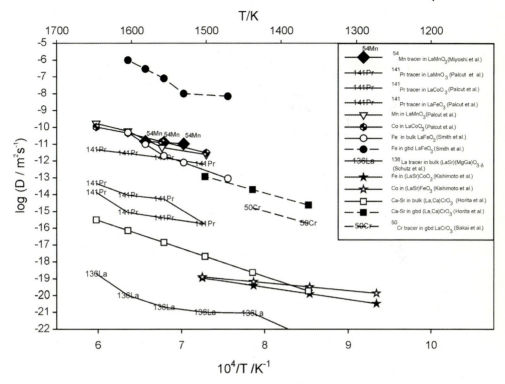

Figure 9.8 Bulk and grain boundary diffusion coefficients of oxide materials with perovskite structure. LSC and LSF are $(La,Sr)CoO_{3-\delta}$ and $(La,Sr)FeO_{3-\delta}$, respectively. [90, 91, 97, 98, 99, 100, 108, 239, 240].

may lead to three basic degradation phenomena of the materials [110, 111]: (i) kinetic demixing, (ii) kinetic decomposition, and (iii) morphological instability.

Kinetic demixing of a chemical homogeneous multicomponent material results in a chemical inhomogeneous material due to the difference in mobility of the ions. We will illustrate this phenomenon by looking at an oxygen-permeable membrane with composition $La_{1-x}Sr_xCoO_{3-\delta}$. At constant temperature and pressure, this oxide has four thermodynamic degrees of freedom, namely, the chemical potentials of the four elements. If the membrane is exposed to a gradient in the chemical potential of oxygen, $\mu_{O_2} = \mu_{O_2}^0 + RT \ln(p_{O_2}/p_{O_2}^0)$, gradients in the chemical potentials of the three metallic elements La, Sr, and Co are induced as consequence of the Gibbs–Duhem relationship

$$x_{La}\mu_{La} + x_{Sr}\mu_{Sr} + x_{Co}\mu_{Co} + x_O\mu_O = 0, \tag{9.18}$$

where x_i is the molar fraction of *i*th species. For the analysis, an incremental gradient and quasiequilibrium across the membrane can be assumed; these are always observed for any steady state and under moderate deviations from stationary conditions (see Chapter 3 of the first volume). The situation is illustrated in Figure 9.9. The potential gradients represent a thermodynamic driving force for the flux of

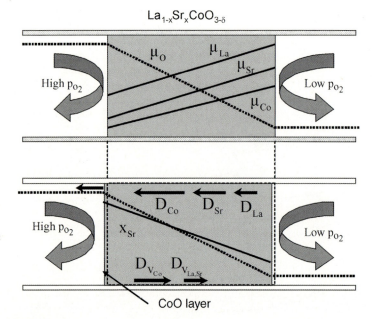

Figure 9.9 One $La_{1-x}Sr_xCoO_{3-\delta}$ membrane in a gradient of the chemical potential of oxygen. A flux of oxygen from the high to the low p_{O_2} side creates a gradient in the chemical potential of the cations and a driving force for the flux of cations to the high p_{O_2} side. Both a gradient in the composition due to kinetic demixing represented by x_{Sr} and kinetic decomposition represented by a layer of CoO at the feed side may occur if the cations are sufficiently mobile.

mobile species across the membrane. If we consider that the cations move by means of cation vacancies, chemical reactions with the atmosphere will take place when the cations and cation vacancies arrive at the surface. At the feed side of the membrane, oxidation takes place and oxides are formed, while at the permeate side the opposite reactions take place.

The formal treatment of demixing in an oxygen potential gradient of a ternary oxide with the perovskite structure has so far not been reported. The problem is not trivial since the oxide consists of three different sublattices, one for oxygen and two for cations. The situation may even become more complicated if additional chemical reactions, such as interaction with CO_2 (see Section 9.5.2), occur at either membrane side. Several reports have, however, given evidence for chemical demixing and chemical decomposition in oxygen-permeable membranes [67, 112, 113]. Accumulation of Co or Fe oxides at the feed side of LSCF membranes has been reported in several of these reports, reflecting the fast diffusion of B-cations in these materials. Lein [114] reported that this could be changed by changing the overall cation stoichiometry, which indicates that grain boundary diffusion of cations may also play a role.

Kinetic decomposition of a membrane material may also occur despite the fact that the material is thermodynamically stable at the operation conditions [111]. This has been reported for an oxygen-permeable membrane of $La_{1-x}Sr_xFeO_{3-\delta}$ where a layer of the phase $SrFe_{12}O_{19}$ was formed at the feed side [112, 114].

Finally, *morphological instabilities* may also occur due to cation motion. The morphological stabilities of interfaces belong to a wide class of self-organization or pattern formation problems arising in biology, physics, chemistry, and geology [111]. The evolution of morphology is governed by the nonequilibrium conditions.

Although the fundamentals of demixing and decomposition are far from being understood, the above reports give clear evidence that the cation mobility in mixed conducting materials is a matter of concern and may limit the lifetime of the membranes. Fundamental studies of these phenomena are therefore essential.

9.4
Thermomechanical Stability

9.4.1
Thermal Expansion of Oxide Electrolyte and Mixed Conducting Ceramics

Early in the field of SOFC, it was recognized that any mismatch in the thermal expansion between the different fuel cell materials will lead to mechanical detachment (delamination) and cracking. The thermal expansion is therefore one of the most important material parameters to be considered. The volume thermal expansion is defined in thermodynamics as $\alpha_V = 1/V(\partial V/\partial T)_P$, where V is the volume and P is the total pressure. In engineering, the linear thermal expansion coefficient (TEC) α_L is more appropriate to use, which is defined as

$$\alpha_L = \frac{1}{L_o}\left(\frac{\partial L}{\partial T}\right) \approx \frac{\Delta L}{L_o \Delta T}, \tag{9.19}$$

where L_o is the original length of the sample. If the volume thermal expansion is known, linear TEC can be estimated by $\alpha_L \approx \alpha_V/3$ for an isotropic material. TEC for a whole range of materials is summarized in Table 9.3. Of particular importance for the replacement of $LaCrO_3$-based interconnects with metallic materials in intermediate temperature SOFC is to find materials with compatible TEC with the most promising metallic interconnects. Some of the most promising electrode materials based on $LaCoO_3$ are unfortunately not suitable due to the high TEC values.

Accurate calculation of thermomechanical stress due to mismatch in TECs depends on the fuel cell or membrane reactor design and is not straightforward to perform. A simple estimation of the thermomechanical in-plane stress of a planar membrane on an inert support, which is stiff and puts constraint on the film, is

$$\sigma_m = \frac{E_{YM}}{1-\nu_P}(\alpha_s - \alpha_m)\Delta T, \tag{9.20}$$

where E_{YM} and ν_P are the Young's modulus and the Poisson ratio of the film, and α_s and α_m are the thermal expansion coefficients of the substrate and the film, respectively. An estimation of the thermomechanical stress is shown in Figure 9.10. The stress can easily exceed the fracture strength of relevant materials (Table 9.5),

Table 9.3 Linear thermal expansion coefficients of selected solid electrolytes and mixed conducting oxide materials.

Material	α ($\times 10^6$ K^{-1})	Temperature range (°C)	References
Oxygen ionic and protonic conductors			
YSZ (4.5% Y^{3+})	10.9, 9.9	50–1200, 600	[115–117]
Ce$_{1-x}$Gd$_x$O$_{2-x/2}$ ($x = 0.0$–0.4)	10.5–12.0, 12.1–13.5	77–327, 427–1000	[118–120]
Ce$_{1-x}$Ca$_x$O$_{2-x/2}$ ($x = 0.1$–0.2)	12.8–13.6	50–1000	[120]
Ce$_{1-x}$Sm$_x$O$_{2-x/2}$ ($x = 0.0$, 0.2)	11.8, 11.4	25–900	[121]
La$_{0.9}$Sr$_{0.1}$Ga$_{0.8}$Mg$_{0.2}$O$_{2.85}$ (LSGM)	11.01	25–800	[117]
BaZrO$_3$	7.13	25–727	[122, 123]
SrCeO$_3$	11.1	25–727	[123, 124]
BaCeO$_3$	11.2	25–727	[123]
LaNbO$_4$	14, 8.4	RT–504, 504–1000	[125]
Mixed ionic–electronic conductors			
La$_{1-x}$Sr$_x$MnO$_3$ ($0 < x < 0.3$)	11.3–13.7	RT–900	[126]
La$_{0.3}$Sr$_{0.7}$CoO$_{3-\delta}$	19.6, 28.8	20–367, 367–747	[127]
La$_{0.5}$Sr$_{0.5}$Co$_{1-x}$Fe$_x$O$_{3-\delta}$ ($0 \leq x \leq 0.5$)	16.9–17.9	20–400	[128]
	28.6–30.7	700–1000	
La$_{0.5}$Sr$_{0.5}$FeO$_{3-\delta}$	16.9, 18.4	20–400, 700–1000	[128]
La$_{0.8}$Sr$_{0.2}$Fe$_{1-x}$Co$_x$O$_{3-\delta}$ ($0.9 \leq x \leq 0.1$)	14.5–20.7	100–900	[129]
(Ln$_{0.6}$Sr$_{0.4}$)$_{0.99}$Fe$_{0.8}$Co$_{0.2}$O$_{3-\delta}$ (Ln = La, Sm, Pr, Gd)	19.5, 19.7, 19.9, 18.1	100–1000	[130]
Sr$_{0.75}$Y$_{0.25}$Co$_{0.5}$Mn$_{0.5}$O$_{3-\delta}$	13.3, 19.6	200–600, 600–800	[131]
La$_{0.6}$Sr$_{0.4}$Fe$_{1-x}$Ga$_x$O$_{3-\delta}$ ($0.2 \leq x \leq 0.4$)	9.8–11.7	20–600	[132]
	19.2–20.8	850–1050	
SrFe$_{0.5}$Co$_{0.5}$O$_{3-\delta}$	18, 33	27–477, 527–827	[133]
La$_2$NiO$_{4+\delta}$	13, 11.9	25–1000	[134–136]
LaCrO$_3$	7.7, 8.6–9.4	25–1000	[137, 138]
LaCr$_{1-x}$Co$_x$O$_3$ ($0.1 \leq x \leq 0.2$)	9.3–10.7	25–1000	[137]
LaCoO$_{3-\delta}$	22.3, 20–22	25–827	[137, 139]
LaFeO$_{3-\delta}$	10.9–11.6	200–420, 480–920	[140]
(Ba$_{0.5}$Sr$_{0.5}$)$_{1-x}$Sm$_x$Co$_{0.8}$Fe$_{0.2}$O$_{3-\delta}$ ($0.05 \leq x \leq 0.2$)	19.5–20.1	30–800	[141]
La$_{0.8}$Sr$_{0.2}$Mn$_{1-x}$M$_x$O$_{3+\delta}$ (M = Mg, Al, Ti, Mn, Fe, Co, Ni, $0 < x < 0.1$)	11.3–12.8	25–1000	[142]
Interconnects			
La$_{0.7}$Ca$_{0.3}$Cr$_{0.97}$O$_{3-\delta}$ (LCC97)	10.2	25–1000	[143]
LCC97/NiO	12.8	25–1000	[143]
LCC97/YSZ	11.6	25–1000	[143]
LCC97/Sm$_{0.8}$Ce$_{0.2}$O$_3$	13.7	25–1000	[143]
LCC97/La$_{0.8}$Sr$_{0.2}$MnO$_{3-\delta}$	11.9	25–1000	[143]
Fe–Cr alloy (CrFe$_5$Y$_2$O$_3$)	11.3–12.0	25–1000	[144]
Ferritic steel (X$_{10}$CrAl$_{18}$)	13.9	25–1000	[144]
Fe–Cr alloy (Crofer 22APU)	12.7	25–1000	[144]
Fe–Cr–Ni alloy (253MA)	19.5	25–1000	[144]

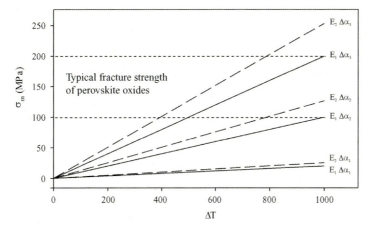

Figure 9.10 Estimated thermomechanical stress due to mismatch in TECs between a stiff substrate and a film according to Equation 9.20. E_i and $\Delta\alpha_i$ are the Young's modulus (E_{YM}) and the difference in linear TECs ($\Delta\alpha_L$) in the particular cases under consideration, respectively. $E_1 = 120\,\text{GPa}$, $E_2 = 190\,\text{GPa}$, $\Delta\alpha_1 = 1\times 10^{-6}\,\text{K}^{-1}$, $\Delta\alpha_2 = 5\times 10^{-6}\,\text{K}^{-1}$, and $\Delta\alpha_3 = 1\times 10^{-5}\,\text{K}^{-1}$.

which is typically between 100 and 200 MPa if the mismatch in TECs, $\Delta\alpha_L$, is larger than $\sim 5\times 10^{-6}\,\text{K}^{-1}$. Note than when $\Delta\alpha_L$ is positive (the substrate has the highest α_L), the film will be in compression and can withstand stresses about five times higher than the fracture strength. On the other hand, when $\Delta\alpha_L$ is negative (the film has the highest α_L), the film will be in tension and fracture might occur at lower values than those indicated in Figure 9.10.

9.4.2
Chemical Expansion

The possibility to change the valance state of the transition metal ions results in the so-called chemical expansion. This effect was probably first observed for alkaline earth substituted $LaCrO_3$ interconnects used in SOFC. The lattice expansion caused by chemical reduction can be characterized by a linear strain ε_c:

$$\varepsilon_c = \frac{L-L_0}{L_0}, \tag{9.21}$$

where L_0 and L are the initial and final lengths of the specimen. The reduction takes place at constant pressure with a change in the partial pressure of oxygen. A chemical xpansion coefficient can be defined if the change in chemical composition (increased oxygen deficiency) is known for the process:

$$\alpha_{V,c} = \frac{1}{V}\left(\frac{\partial V}{\partial x_o}\right)_{T,P}, \tag{9.22}$$

Table 9.4 Chemical expansion of selected oxide materials containing variable valence cations.

Material	Chemical expansion $\varepsilon_c/\Delta\delta$	Temperature range (°C)	References
$La_2NiO_{4+\delta}$	0.01	700–950	[136]
$La_{0.5}Sr_{0.5}FeO_{3-\delta}$	0.059	800	[140]
$La_{0.5}Sr_{0.5}Co_{1-x}Fe_xO_{3-\delta}$ ($x = 0.0, 0.5$)	0.035, 0.039	800–1000	[128, 146]
$La_{0.7}Ca_{0.3}Cr_{1-x}Al_xO_{3-\delta}$ ($0.0 \leq x \leq 0.1$)	0.036–0.034	1000	[147]
$La_{0.6}Sr_{0.4}Fe_{0.2}Co_{0.8}O_{3-\delta}$	0.022	800	[148]
$La_{0.8}Sr_{0.2}Cr_{0.97}V_{0.03}O_{3-\delta}$	0.030	1000	[149]
$La_{0.3}Sr_{0.7}Fe_{1-x}Ga_xO_{3-\delta}$ ($x = 0, 0.4$)	0.032, 0.44	875	[150]
$LaMnO_{3+\delta}$	0.024	700	[151]
$La_{0.8}Sr_{0.2}CrO_{3-\delta}$	0.024	1000	[149]
$Ce_{1-x}Gd_xO_{1.95-\delta}$ ($x = 0.1$–0.2)	0.050, 0.66	1000	[152]
$Ba_{0.5}Sr_{0.5}Co_{0.8}Fe_{0.2}O_{3-\delta}$	0.026	600–900	[153]

where x_o is the oxygen content. The linear chemical expansion ($\alpha_c \approx \alpha_{V,c}/3$ for isotropic materials) can be calculated by the simple expression $\varepsilon_c/\Delta\delta$, where $\Delta\delta$ is the change in the oxygen deficiency due to the change in the partial pressure of oxygen.

Chemical expansion data for a range of materials are summarized in Table 9.4. The chemical expansion of ceria-based materials is considerably larger than that for perovskite-based materials. $La_2NiO_{4+\delta}$ is of particular interest due to its low chemical expansion compared to the perovskite materials. Attempts to correlate the observed chemical expansion with the change in the ionic radius of the transition metal ion has not fully explained the observed chemical expansion possibly due to an additional volumetric effect that depends on the oxygen vacancy concentration [145].

Chemically induced stress due to gradients in the chemical potential of oxygen can be induced due to chemical expansion. The maximum stress in a membrane film σ_{mc} can be estimated by

$$\sigma_{mc} = \frac{E_{YM}}{1-\nu_P} \alpha_c \Delta\delta, \qquad (9.23)$$

where $\Delta\delta$ corresponds to the change in the oxygen deficiency due to the difference in chemical potential of oxygen across the membrane. The maximum chemical stress, calculated by Equation 9.23 is shown as a function of temperature and partial pressure in Figure 9.11. Stress on the order of 100 MPa is easily induced if chemical expansion is induced.

9.4.3
Mechanical Properties

The mechanical properties of ceramic materials are one of the key topics in ceramic science and engineering and have been covered extensively in several textbooks [154, 155]. The mechanical strength of ceramic materials can be understood from Griffith's postulate

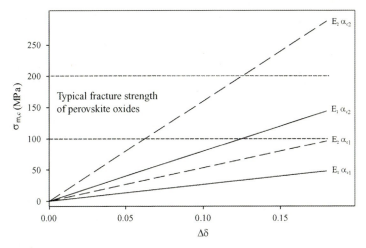

Figure 9.11 Estimated maximum chemical-induced stress due to chemical expansion according to Equation 9.23. $E_1 = 100$ GPa, $E_2 = 200$ GPa, $\alpha_{c1} = 0.02$, and $\alpha_{c2} = 0.06$.

$$\sigma_f = A_g \left(\frac{E_{YM} \gamma_f}{c_{cf}} \right)^{1/2}, \tag{9.24}$$

where A_g is a geometrical constant, γ_f is the fracture energy, and c_{cf} is the critical flaw size. The equation illustrates the strong dependence of the macrostructure of the materials (flaw size) and the significantly lower fracture strength, which is typically 1/100 of the theoretical strength. Moreover, a high Young's modulus not only gives higher strength but also results in higher stress induced by temperature or chemical changes (Sections 9.4.1 and 9.4.2). Fracture strength of materials can be measured by several different methods, and the measured strength depends on the technique [154]. The reason for this can be understood from the distribution of flaws in the material and the stress distribution in the test specimen. The four-point bending test is the most common test for measurement of fracture strength of fuel cell and membrane materials. Fracture strength and Young's modulus of some relevant materials are summarized in Table 9.5. For most materials, the strength reported is relatively moderate compared to the strength of engineering materials such as alumina or transformation toughened zirconia, but this does not necessarily reflect poor mechanical properties but rather the relatively large flaws observed by fractography or stresses induced by *frozen-in* gradients of the oxygen content [156, 157]. Improved ceramic processing of the materials should therefore give an improved strength.

Another important material property with respect to mechanical performance is the fracture toughness. The fracture toughness is a measure of resistance toward crack propagation in a material. Stresses are often induced in fuel cells or membranes due to thermal cycling and chemically induced stresses and crack propagation from surface flaws or interfaces are a matter of concern. The fracture toughness K_{IC} is defined by

$$K_{IC} = \sigma Y c_{cf}^{1/2}, \tag{9.25}$$

Table 9.5 Mechanical properties of selected oxide materials.

Material	Fracture strength at 25°C (MPa)	Young's modulus (GPa)	Fracture toughness (MPa m$^{1/2}$)	References
$LaFeO_3$	202	213	2.5	[158]
$LaCoO_3$	53	83	1.3	[156, 159, 160]
$La_{0.5}Sr_{0.5}Co_{1-x}Fe_xO_{3-\delta}$ (x = 0, 0.5)	138, 121	135, 131	1.5, 1.3	[157]
$LaCr_{0.9}Mg_{0.1}O_3$	140	—	2.8	[161]
$La_{0.8}Ca_{0.2}CrO_3$	76–123	104–187	—	[162]
$Ce_{1-x}Gd_xO_{1.95-\delta}$ (x = 0.1–0.2)	120–138	200–187	—	[163]
YSZ	214	190	1.61–1.79	[163, 164]
NiO–YSZ	180	161	—	[163]
$SrCeO_3$:5% Yb				[165, 166]
Nonprotonated	175–177	145	2.08	
Protonated	194	—	—	
$SrZrO_3$:5% Yb	31–66	131	1.54	[166]
$La_{1-x}Sr_xGa_{0.8}Mg_{0.2}O_{3-x/2}$ (x = 0.1–0.2)	150	—	2.0–2.4	[78]
	121–126	175	1.22–1.31	[167]

where Y is a parameter determined by the crack configuration and loading geometry. Fracture toughness for some relevant materials is also given in Table 9.5. Most of the materials suffer due to a low fracture toughness, and possible toughening mechanisms of noncubic perovskite materials due to ferroelasticity have been suggested [156, 168]. Corresponding transformation toughening of tetragonal zirconia is also well known [169, 170], but this is not relevant for cubic stabilized zirconia.

9.4.4
Degradation Due to Fracture

Degradation of the materials may occur both during fabrication and under operation conditions. Fracture during fabrication may occur due to thermal shock if too fast cooling rates are applied. The thermal stress

$$\sigma_{th} = E_{YM}\alpha_L \Delta T/(1-\nu_P) \tag{9.26}$$

is a matter of concern particularly for materials with high thermal expansion and high Young's modulus. The chemical expansion of mixed conductors may also cause significant problems of fracture during cooling since reoxidation of the transition metal may take place. However, the kinetics of the oxidation slows down as the temperature is lowered, and a gradient in the oxygen content results. This causes a frozen-in tensile stress at the surface due to the volume contraction accommodating the oxidation process. Surface fracture has also been reported for materials with substantial chemical expansion [157].

Similar gradients in oxygen content are also present in oxygen-permeable membranes on reduction and oxidation. However, to our knowledge, there have not been reported severe problems with fracture due to the chemical potential gradient under operation, which in some way is surprising since the stress may become larger than the typical fracture strength of membrane materials. To a definite extent, this is associated with the complexity and substantial costs of the membrane tests, which require a careful preliminary optimization of the membrane operation conditions. There are, however, reports on fracture of materials that either decompose due to thermodynamic instability or go through a first-order phase transition at reducing conditions (see Section 9.5). In laboratory membrane reactors, catastrophic failure may have been avoided due to stress relaxation by high-temperature creep as discussed below. For $LaCrO_3$-based interconnects, which have a significantly higher creep resistance, fracture can be related to the chemically induced stress due to the gradient in chemical potential of oxygen, as discussed in Section 9.4.2.

Degradation of SOFC due to thermal cycling and reduction/reoxidation cycles is a very important limitation for the lifetime of SOFC stacks. Advanced methods and modeling have been developed, but more efforts are needed to fully understand the thermomechanical changes in SOFCs [171]. Fracture occurs due to thermomechanical incompatibility, but microstructural changes are also known to induce stresses and possibly failure.

9.4.5
Chemical Compatibility of Materials

The long-term stability of SOFCs implies that the materials with common interfaces are not reacting over time or that they are chemically compatible. Yokokawa has provided an excellent review of materials compatibility with emphasis on thermodynamics (explaining the chemical compatibility) [172]. The two main phenomena that may occur at an interface between two materials are the formation of a new phase at the interface and interdiffusion across the interface. These are illustrated in Figure 9.12. The formation of new phases can be understood in terms of a simple

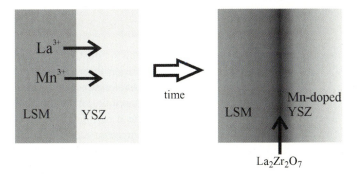

Figure 9.12 Illustration of interdiffusion of Mn and La and formation of $La_2Zr_2O_7$ at the interface between YSZ electrolyte and LSM electrode that may occur due to chemical incompatible interfaces.

chemical equilibrium. One of the most studied systems in this context is the interface between YSZ and LSM. The chemical reaction taking place at this interface can be understood from the following reaction:

$$\text{LaMnO}_3 + x\left\{\text{ZrO}_2 + \tfrac{3}{2}\text{O}_2(\text{g})\right\} = (x/2)\text{La}_2\text{Zr}_2\text{O}_7 + \text{La}_{1-x}\text{MnO}_3 \qquad (9.27)$$

where "g" corresponds to the gaseous phase and the influence of Sr in LSM and Y in YSZ is neglected to simplify the reaction. The driving force for the chemical reaction is oxidation from Mn(III) to Mn(IV). A close to stable YSZ–LSM interface is obtained by Sr substitution of LaMnO$_3$, since this increases the Mn(IV) content. At too high substitution level, the interface becomes unstable due to formation of SrZrO$_3$. Yokokawa has provided an excellent thermodynamic model for this system and a predominance diagram (chemical potential diagrams), showing the coexistence of La$_{1-x}$MnO$_3$ and YSZ [172].

In cases where secondary phases are not formed at the interface, interdiffusion will occur until there is no gradient in the chemical potential of any of the components in the two materials. If the solid solubility is extensive, the properties of the materials will be affected by interdiffusion. For example, extensive diffusion of Mn into YSZ results in a higher electrical conductivity of YSZ.

There are not many systems except the LSM–YSZ where the chemical compatibility of a SOFC component has been investigated in such a detail. The chemical compatibility issues between electrode and electrolyte materials have been reviewed in Refs [173, 174].

Recently, proton conductivity in LaNbO$_4$ has been reported [175, 176]. Other proton conductive electrolytes are based on SrCeO$_3$ and BaZrO$_3$. Attempts to identify oxide cathode materials for SOFCs based on SrCeO$_3$, BaZrO$_3$, and LaNbO$_4$ electrolytes have been performed by Tolchard [122, 124, 177]. In these works, material compatibility has been just a guideline.

9.4.6
High-Temperature Creep

At ambient temperature, ceramic materials respond elastically to mechanical stress according to Hooke's law

$$\sigma = E_{\text{YM}}\varepsilon, \qquad (9.28)$$

where ε is the induced strain. In this context, it is interesting to note that noncubic perovskite materials have been reported to deform nonelastically due to ferroelastic domain switching [156, 168]. At elevated temperatures, mechanical stress may lead to nonelastic or plastic deformation over time due to the onset of thermally activated mass transport. The term "creep" is usually used to refer to plastic deformation at constant stress as a function of time and temperature. Creep may represent a possible mechanism for relaxation of induced stresses, but a high creep resistance is an advantage to secure geometrical stability of a device. For example, in membranes operated with substantial gradients in pressure, creep would induce deformation due to the stress induced by the pressure gradient.

Steady-state creep can be described by the following equation:

$$\dot{\varepsilon} = A_c \sigma^n D_{\text{eff}}, \tag{9.29}$$

where D_{eff} is the effective diffusion constant (see Equations 9.12 and 9.17). If the deformation occurs via a mechanism controlled by diffusion of point defects, for example, Nabarro-Herring (bulk) or Coble creep (grain boundary diffusion), Equation 9.29 can be modified as

$$\dot{\varepsilon} = \frac{B_c \sigma \Omega}{d_g^2 k_B T} \left(D_{\text{bulk}} + D_{\text{gb}} \frac{\pi w_{\text{gb}}}{d_g} \right), \tag{9.30}$$

where B_c is a constant and Ω is the atomic volume of the rate-controlling diffusing species [178]. The possibility to have two diffusion pathways for the rate-controlling species is a major challenge for the materials containing several different cation species (diffusion of cations is rate limiting for the mass transport in the oxide materials). The relationship between creep and point defects associated with changes in partial pressure of oxygen can be described by Equation 9.17.

Creep data for some relevant materials are shown in Figure 9.13. The activation energy, creep exponent, inverse grain size exponent, and oxygen partial pressure exponent of some of the materials are given in Table 9.6. The creep resistance of fluorite-type materials is significantly higher than that for perovskite materials. Among the perovskite materials, the electrolyte material LSGM has a higher creep resistance than the other perovskite materials with mixed electronic and ionic conductivity. Yi et al. have reported a correlation between the creep rate and the oxygen ion conductivity [179]. The creep rate of some perovskite materials is very

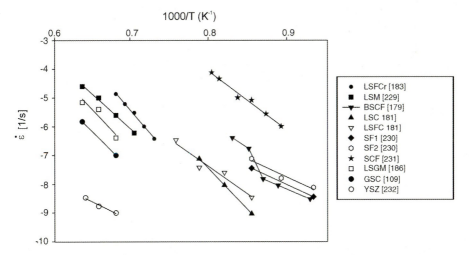

Figure 9.13 Creep rate $\dot{\varepsilon}$, at a stress of 10 MPa, of some relevant perovskite and fluorite materials: $(La,Sr)(Fe,Cr)O_{3-\delta}$ (LSFCr) [183], LSM [229], BSCF [179], LSC, LSFC [181], $SrFeO_{3-\delta}$ (SF1-2) [230], $SrCo_{0.8}Fe_{0.2}O_{3-\delta}$ (SCF) [231], LSGM [186], $Ce_{0.9}Gd_{0.1}O_{1.95}$ (GSC) [109], and YSZ [232].

Table 9.6 Creep parameters[a] for perovskite oxides under low stresses.

Material	Stress (MPa)	T (K)	n	p	m	Q (kJ mol^{-1})	Reference
$SrCo_{0.8}Fe_{0.2}O_{3-\delta}$	10–20	1123–1248	1.0–1.5	1	—	≥275	[180]
$Ba_{0.5}Sr_{0.5}Co_{0.8}Fe_{0.2}O_{3-\delta}$	5–20	1078–1208	0.76	2.5	−0.63	≥250	[179]
$La_{0.5}Sr_{0.5}CoO_{3-\delta}$	5–28	1173–1273	1.2	1.3	−0.46	619	[181]
$La_{1-x}Sr_xMnO_3$ ($x = 0.1–0.25$)	5–30	1423–1573	0.9–0.11	—	—	460–490	[182]
$La_{0.2}Sr_{0.8}Fe_{0.8}Cr_{0.2}O_{3-\delta}$	10–30	1646–1746	1.4	2.3	0.04 (>10^{-4} Pa) 0.5 (<10^{-4} Pa)	566	[183, 184]
$La_{0.8}Sr_{0.2}Ga_{0.85}Mg_{0.15}O_{2.825}$	5–20	1473–1573	1.2	1.7	0	521	[185, 186]

a) n is the strain exponent, p is the inverse grain size exponent, m is the oxygen partial pressure exponent, and Q is the activation energy.

high, implying that even small stresses will induce substantial deformations. This is particularly true for the perovskites with only alkaline earth on the A-site in the crystal structure.

As reflected in the slope of the creep data in Figure 9.13 (reflecting the activation energy for creep) and the data summarized in Table 9.6, the rate-controlling mass transport mechanism for the creep observed for different materials is not straightforward to determine. Several possible mechanisms have been reported for various materials, such as diffusion accommodated grain boundary creep mechanism [187, 188]. It is also important to note that the oxygen partial pressure exponent m (see Equation 9.17) is negative for most of the materials. This means that creep is enhanced at reducing environment. Referring to the Schottky equilibrium (Equation 9.5), the cation vacancy concentration should not be enhanced by reducing the partial pressure of oxygen. The increasing mass transport under reducing conditions cannot reflect a change in the bulk diffusion of cations. More sophisticated models for the creep are therefore needed.

9.5
Thermodynamic Stability of Materials

9.5.1
Phase Decomposition and Solid-State Transformation

The chemical composition of most solid electrolytes or mixed conductors is always to be optimized in order to increase the concentration of ionic charge carriers, a prerequisite for high ionic conductivity. For the materials with vacancy diffusion mechanism, the high degree of oxygen deficiency gives substantial disorder at elevated temperatures resulting in crystal structures with usually high symmetry, such as the cubic perovskite or fluorite. The incorporation of oxygen vacancies by chemical substitution has a certain limit since the thermodynamic stability of the lattice is challenged at high nonstoichiometry levels. If the stability is considered at isothermal conditions, this thermodynamic instability is in some cases recognized by the formation of new ordered crystal structures, in which the oxygen vacancies become less mobile. An example is the formation of $Ca_6Zr_{19}O_{44}$ in calcia-stabilized zirconia (CSZ) [189]. Other relevant examples can be found in Chapter 2 of the first volume. The instability does not necessarily result in such long-range order, but short-range order in the form of point defect clusters that give rise to substantial reduction of the oxygen vacancy mobility. Point defect clusters are well known for zirconia- and ceria-based electrolytes [89, 190–195]. The thermodynamic instability at high substitution level may also be recognized by simply formation of secondary phases as in the case of $La_{1-y}Sr_yGa_{1-x}Mg_xO_{3-\delta}$ (LSGM), where the solubility of SrO and MgO, which increases the oxygen vacancy concentration, is thermodynamically limited [196].

If we now consider the thermodynamic stability of the materials with respect to changes in temperature, there are distinct temperature limits for the stability of the materials. Intuitively, there is always a high-temperature limit where the material decomposes or melts congruently or incongruently, but this is usually only a matter of concern during fabrication of the device (see Section 9.2). More interesting is the low-temperature limit, where the highly conducting disordered crystal structure is transformed to a low-temperature ordered crystal structure. With respect to the high oxygen vacancy mobility, any ordering process will reduce the oxygen vacancy mobility and is thus not desired. For mixed conducting perovskite materials, the most well-known phase transition to an ordered crystal structure is the transition from the high-temperature disordered cubic perovskite structure to the low-temperature ordered brownmillerite crystal structure, illustrated in Figure 9.14a and b. Classical examples are the oxygen-deficient phases of $SrFeO_{3-\delta}$ [197], $SrCoO_{3-\delta}$ [198, 199], $SrMnO_{3-\delta}$ [200, 201], and $CaMnO_{3-\delta}$ [200, 202]. For some of these examples, the ordered crystal structure is an oxygen-deficient perovskite that can be related to brownmillerite structure. Similar types of ordering in cubic perovskite materials may occur for compositions corresponding to $A_nB_nO_{3n-1}$. Reduced oxygen permeability due to formation of such ordered crystal structures has been reported for heavily alkali earth substituted $LaFeO_{3-\delta}$ [203–206].

The formation of hexagonal perovskite-like structures is also known to limit the low-temperature stability of some oxygen-deficient cubic perovskite materials. The hexagonal perovskite-like structure [16] shown in Figure 9.14c is favored for compounds that possess a high Goldschmidt tolerance factor. Perovskites with large A-cations such as Sr and Ba are therefore candidate materials for the formation of crystal structures with hexagonal perovskite structure or related structures. The most

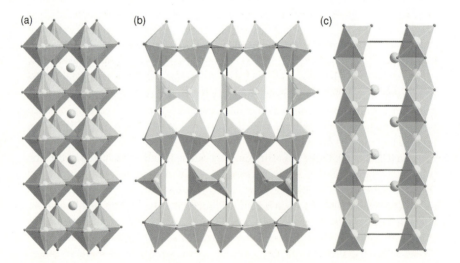

Figure 9.14 Perovskite-like phases as a function of temperature and oxygen vacancy concentration: ordered or disordered cubic phase (a), ordered brownmillerite phase (b), and hexagonal phase (c). The polyhedra (octahedral or tetrahedral) are formed by B-site cations and oxygen anions. The A-site cations are shown as large spheres.

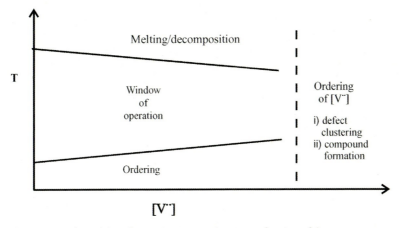

Figure 9.15 The stability of an oxygen ion conductor as a function of the oxygen vacancy concentration (composition) and temperature.

well-known example is the formation of $Sr_6Co_5O_{15}$ during cooling of $SrCoO_{3-\delta}$ [198] and also cubic $SrMnO_{3-\delta}$ [200, 201], which is hexagonal below 1400 °C. In Figure 9.15, the concentration of oxygen vacancies [$VO^{\bullet\bullet}$] is linked to the stability of ordered/disordered $ABO_{3-\delta}$ perovskite phases as a function of temperature.

Finally, the thermodynamic stability of the materials also has a finite limit with respect to changes in the partial pressure of oxygen (chemical potential). Here, two possible scenarios may occur. First, reduced oxygen partial pressure will cause the oxygen vacancy concentration to increase, which may shift the material into the stability field of a phase with ordered oxygen vacancies (similar to what may occur with chemical substitution of a metal oxide with lower oxidation state). Second, the stability of perovskite materials may also be related to the valence state of the transition metal ions and the oxygen vacancy concentration, which is related to their valence state. We may use the decomposition of $La_{0.6}Sr_{0.4}MnO_{3-\delta}$ as an example [207, 208]:

$$2La_{0.6}Sr_{0.4}MnO_{3-\delta} \rightarrow (La_{0.6}Sr_{0.4})_2MnO_4 + MnO + \{(1-2\delta)/2\}O_2(g) \quad (9.31)$$

Here, it is the stability of Mn^{3+} relative to Mn^{2+} that determines the conditions for the decomposition. Reduction of the valence state of Mn causes the oxygen vacancy concentration (δ) to increase and the material becomes unstable with respect to the new compounds (reaction products). In most cases (perovskites with only alkali earth on A-site are exceptions), the materials become unstable when $\delta > 0.5$. Moreover, it has been shown that the stability of the perovskite materials $La_{1-x}A_xBO_{3-\delta}$ (A = Sr, Ca, etc.) increases in the order Cu, Ni, Co, Fe, Mn, and Ti/Cr on the B-site [209]. Perovskite materials based on Cu, Ni, Co, and possibly also Fe are not stable under reducing conditions (i.e., in synthesis gas mixtures of H_2 and CO). The high oxygen flux in perovskite materials with Co and Fe on B-site has given motivation to increase the stability of the materials under reducing conditions by chemical substitution. Here, it is important to note that ternary oxides cannot be stabilized substantially by doping [210].

For fluorite-type materials, the thermodynamic stability of YSZ is superior since the constituent oxides are among the most stable oxides. Ceria-based materials, on the other hand, do not have the same stability due to the possibility of reducing ceria from $+4$ to $+3$. In most cases, the reduction does not cause decomposition of the material, but the electronic conductivity becomes dominant and Ce^{3+}/Ce^{4+} and oxygen vacancy ordering may occur [195, 211, 212]; see also Chapter 9 of the first volume.

9.5.2
Reactions with Gaseous Species

The reaction with gas mixtures containing reducing gases such as H_2 or CO may cause decomposition of the material to oxides with lower oxidation state as illustrated by reaction (9.31) or even to metals (Co, Ni, and Cu) if the partial pressure of oxygen in the reducing atmosphere is below the corresponding phase boundary. There is a substantial amount of experimental evidence for the stability limit for a whole range of materials; see, for example, the excellent overview of $LaCoO_3$-based materials [213].

Reactive gaseous species (CO_2, SO_2, and H_2O) may also be found under oxidizing conditions. The combination of acidic gases and basic oxides may lead to degradation and instability of the solid electrolytes, electrodes, and membrane materials. The reaction with both water vapor and CO_2 may be detrimental for the performance of both fuel cells and oxygen or hydrogen separation membranes. To illustrate the reactions, we have used $SrCeO_3$ as a model material. The possible reactions with acidic gases are, for example [214],

$$SrCeO_3(s) + CO_2(g) = SrCO_3(s) + CeO_2(s) \tag{9.32}$$

$$SrCeO_3(s) + SO_2(g) + \tfrac{1}{2}O_2(g) = SrSO_4(s) + CeO_2(s) \tag{9.33}$$

and the corresponding reaction with water vapor [215] is

$$SrCeO_3(s) + 2H_2O(g) = Sr(OH)_2 \cdot H_2O(s) + CeO_2(s) \tag{9.34}$$

Note that, as accepted in thermodynamics, the symbol "s" in these equations corresponds to the solid phases. The extent of reactions (9.32)–(9.34) is closely related to the Gibbs energy of formation of the perovskite relative to the binary constituent oxides and the basicity of the components. Let us first discuss the Gibbs energy of formation of the perovskite relative to their binary constituent oxides. The formation energy is the Gibbs energy for the following reaction using $SrCeO_3$ as an example:

$$SrO(s) + CeO_2(s) = SrCeO_3(s) \tag{9.35}$$

To a first approximation, the Gibbs energy can be assumed to be equal to the enthalpy of the same reaction as the entropy term should be low since there are only solid

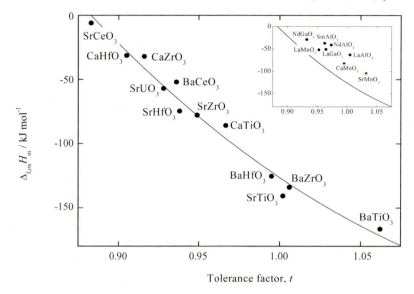

Figure 9.16 Enthalpy of formation of selected perovskite-type oxides as a function of the tolerance factor. The main figure shows data for ABO$_3$ where A is a group 2 element and B is a d- or f-element that readily takes a tetravalent state [233, 234]. Inset shows enthalpies of formation of perovskite-type oxides where A is a trivalent lanthanide metal [235] or a divalent alkaline earth metal [210], whereas B is a late transition metal atom.

products and reactants. It has been shown by Yokokawa et al. [216] that the enthalpy of formation of perovskite materials can be related to the Goldschmidt tolerance factor (t). This is shown in Figure 9.16 for a range of different materials. The higher the tolerance factor, the more stable the perovskite materials relative to their binary constituent oxides.

The reaction with water and the acidic gases can be further related to the basicity of the constituent oxides. Here, one may use the factor $r^{1/2}/Z^*$, where r is the ionic radius of the cation in the constituent oxide and Z^* is the effective charge of the cation, to express the basicity. The enthalpy of decomposition ($\Delta H^0_{298,\,\text{decomposition}}$) of hydroxides and carbonates of the binary oxides can be related to this factor according to Stern [217]. The decomposition reactions using SrO as an example are

$$SrCO_3(s) = SrO(s) + CO_2(g) \tag{9.36}$$

$$Sr(OH)_2(s) = SrO(s) + H_2O(g) \tag{9.37}$$

The enthalpy of decomposition of hydroxides and carbonates follows a linear trend with the size of the effective ionic radius ($r^{1/2}$) [218] and effective charge $1/Z^*$ [219], as shown in Figure 9.17. Thus, in order to obtain stable materials in H$_2$O- and CO$_2$-containing atmospheres, one should have a high Goldschmidt tolerance factor and reduce or limit the concentration of basic oxides such as the alkali or alkaline earth metal oxides. Note that more complex correlations may be observed for heavily

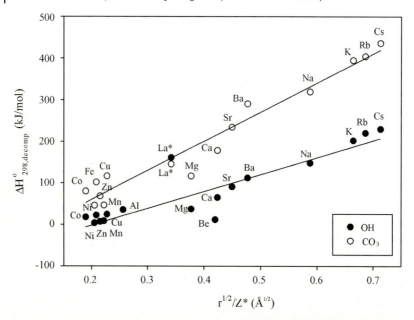

Figure 9.17 Enthalpy of decomposition of hydroxides and carbonates [236]. La* involves decomposition of hydroxide-oxide (LaOOH) and dioxycarbonate (La$_2$O$_2$CO$_3$) [237, 238].

doped multicomponent materials, where the oxygen coordination of the cations and, therefore, tolerance factor cannot be a priori defined and vacancy ordering may occur.

Finally, we will briefly describe possible reactions leading to formation of stable silicate phases. It is because of the awareness in the solid-state community that silica-rich phases have a poor influence on the performance of fuel cells and membranes. Silica-containing phases are used as sealing materials and can cause silica poisoning on the surface of the electrodes and membranes. It is therefore still an open question if silica-based sealants can be used in SOFCs. Less known is the volatility of silicon or silicon oxide in reducing or humid atmospheres, which enable transport of silicon through the vapor phase. The vaporization of SiO$_2$ in reducing condition can be illustrated by the reaction with H$_2$ [220]:

$$SiO_2(s) + H_2(g) = SiO(g) + H_2O(g) \tag{9.38}$$

SiO(g) is a strong reducing agent and will easily condense and react. The reaction with SrCoO$_{3-\delta}$ is given as an example:

$$2SrCoO_{3-\delta}(s) + SiO(g) = Sr_2SiO_4(s) + CoO(s) + (1-\delta)O_2(g) \tag{9.39}$$

Formation of orthosilicates such as Sr$_2$SiO$_4$(s) has been shown by Faaland et al. [221].

The volatilization of SiO$_2$ due to reaction with water vapor is quite complex and leads to the formation of gaseous hydroxides and oxyhydroxides [222]:

$$SiO_2(s) + \tfrac{1}{2}H_2O(g) = SiO(OH)(g) + \tfrac{1}{4}O_2(g) \tag{9.40}$$

$$SiO_2(s) + H_2O(g) = SiO(OH)_2(g) \tag{9.41}$$

$$SiO_2(s) + 2H_2O(g) = Si(OH)_4(g) \tag{9.42}$$

$$2SiO_2(s) + 3H_2O(g) = Si_2O(OH)_6(g) \tag{9.43}$$

A possible source of silicon does not need to be oxides but metallic materials with small levels of silicon, which will oxidize even under reducing conditions [223].

9.5.3
Volatilization of Components

Oxidation and degradation processes in the presence of humid atmospheres are other examples of volatilization of oxide materials. Both basic oxides such as SrO and more acidic oxides such as MnO_2 may become volatile in the presence of humid gases. The formation of volatile hydroxides in the presence of water vapor is illustrated in the following reactions [224]:

$$SrO(s) + \tfrac{1}{2}H_2O(g) = SrOH(g) + \tfrac{1}{4}O_2(g) \tag{9.44}$$

$$SrO(s) + H_2O(g) = Sr(OH)_2(g) \tag{9.45}$$

$$MnO_2(s) + \tfrac{1}{2}H_2O(g) = MnOH(g) + \tfrac{3}{4}O_2(g) \tag{9.46}$$

$$MnO_2(s) + H_2O(g) = Mn(OH)_2(g) + \tfrac{1}{2}O_2(g) \tag{9.47}$$

The vapor pressure of these volatile species is naturally lower if the enthalpy of formation of the more complex oxides is more exothermic.

Contamination and loss of performance due to reactions with chromium is a well-known effect in SOFCs [50, 171, 225]. The dominating reactions for the volatility of chromium are [226]

$$\tfrac{1}{2}Cr_2O_3(s) + \tfrac{1}{2}H_2O = CrOH(g) + \tfrac{1}{2}O_2(g) \tag{9.48}$$

$$\tfrac{1}{2}Cr_2O_3(s) + H_2O(g) + \tfrac{3}{4}O_2(g) = CrO_2(OH_2)(g) \tag{9.49}$$

Finally, the chemical stability of certain metal oxides at high temperatures or reducing conditions has to be taken under consideration in preparative and application techniques. For example, metal oxides with known low melting point, such as thallium (Tl_2O_3, mp = 717 °C), lead (PbO) (mp = 888 °C), bismuth (Bi_2O_3, mp = 817 °C), and also barium oxides [227], are volatile. The volatilization can be detained by integration into an oxide system such as zirconate (see Figure 9.17), which exhibits higher lattice energies and decomposition temperatures.

References

1 Ge, L., Shao, Z., Zhang, K., Ran, R., Diniz da Costa, J.C., and Liu, S. (2009) Evaluation of mixed-conducting lanthanum–strontium–cobaltite ceramic membrane for oxygen separation. *AIChE J.*, **55** (10), 2603–2613.

2 Hong, L., Chen, X., and Cao, Z. (2001) Preparation of a perovskite $La_{0.2}Sr_{0.8}CoO_{3-x}$ membrane on a porous MgO substrate. *J. Eur. Ceram. Soc.*, **21**, 2207–2215.

3 Kleveland, K., Einarsrud, M.-A., and Grande, T. (2000) Sintering of $LaCoO_3$-based ceramics. *J. Eur. Ceram. Soc.*, **20**, 185–193.

4 Tao, Y., Shao, J., Wang, J., and Wang, W.G. (2008) Synthesis and properties of $La_{0.6}Sr_{0.4}CoO_{3-\delta}$. *J. Power Sources*, **185**, 609–614.

5 Büchler, O., Serra, J.M., Meulenberg, W.A., Sebold, D., and Buchkremer, H.P. (2007) Preparation and properties of thin $La_{1-x}Sr_xCo_{1-y}Fe_yO_{3-\delta}$ perovskitic membranes supported on tailored ceramic substrates. *Solid State Ionics*, **178**, 91–99.

6 Chick, L.A., Maupin, G.D., Graff, G.L., Pederson, L.R., McCready, D.E., and Bates, J.L. (1992) Redox effects in self-sustaining combustion synthesis of oxide ceramic powders. *Mater. Res. Soc. Symp. Proc.*, **249**, 159–164.

7 Fossdal, A., Sagdahl, L.T., Einarsrud, M.-A., Wiik, K., Larsen, P.H., Poulsen, F.W., and Grande, T. (2001) Phase equilibria and microstructure in $Sr_4Fe_{6-x}Co_xO_{13}$, $0 < x < 4$, mixed conductors. *Solid State Ionics*, **143**, 367–377.

8 Kleveland, K., Grande, T., and Einarsrud, M.-A. (2000) Sintering behavior, microstructure and phase composition of $Sr(Fe, Co)O_{3-\delta}$ ceramics. *J. Am. Ceram. Soc.*, **83**, 3158–3164.

9 Li, S., Jin, W., Xu, N., and Shi, J. (1999) Synthesis and oxygen permeation properties of $La_{0.2}Sr_{0.8}Co_{0.2}Fe_{0.8}O_{3-\delta}$ membranes. *Solid State Ionics*, **124**, 161–170.

10 Ming, Q., Nerseyan, M.D., Wagner, A., Richie, J., Richardson, J.T., Luss, D., Jacobson, A.J., and Yang, Y.L. (1999) Combustion synthesis and characterization of Sr and Ga doped $LaFeO_3$. *Solid State Ionics*, **122**, 113–121.

11 Murata, K., Fukui, T., Abe, H., Naito, M., and Nogi, K. (2005) Morphology control of $La(Sr)Fe(Co)O_3$ cathodes for IT-SOFCs. *J. Power Sources*, **145**, 257–261.

12 Sagdahl, L.T., Grande, T., and Einarsrud, M.-A. (2000) Sintering of $LaFeO_3$ ceramics. *J. Am. Ceram. Soc.*, **83**, 2318–2320.

13 Sagdahl, L.T., Grande, T., and Einarsrud, M.-A. (2006) Sintering behaviour of $La_{1-x}Sr_xFeO_{3-\delta}$ mixed conductors. *J. Eur. Ceram. Soc.*, **26**, 3665–3673.

14 Sunstrom, J.E., IV, Ramanujachary, K.V., Greenblatt, M., and Croft, M. (1998) The synthesis and properties of the chemically oxidized perovskite $La_{1-x}Sr_xCoO_{3-\delta}$ ($0.5 \leq x \leq 0.9$). *J. Solid State Chem.*, **139**, 388–397.

15 Tao, Y., Shao, J., Wang, J., and Wang, W.G. (2008) Synthesis and properties of $La_{0.6}Sr_{0.4}CoO_{3-\delta}$ nanopowder. *J. Power Sources*, **185**, 609–614.

16 Svarcová, S., Wiik, K., Tolchard, J.R., Bouwmeester, H.J.M., and Grande, T. (2008) Structural instability of cubic perovskite $Ba_xSr_{1-x}Co_{1-y}Fe_yO_{3-\delta}$. *Solid State Ionics*, **178** (35–36), 1787–1791.

17 Faaland, S., Knudsen, K., Einarsrud, M.-A., Rørmark, L., Høier, R., and Grande, T. (1998) Structure, stoichiometry and phase purity of calcium-doped lanthanum manganite powders. *J. Solid State Chem.*, **140**, 320–330.

18 Jiang, S.P. and Wang, W. (2005) Sintering and grain growth of (La, Sr)MnO_3 electrodes of solid oxide fuel cells under polarization. *Solid State Ionics*, **176**, 1185–1191.

19 McCarthy, B.P., Pederson, L.R., Williford, R.E., and Zhou, X.-D. (2009) Low-temperature densification of lanthanum strontium manganite ($La_{1-x}Sr_xMnO_{3+\delta}$), $x = 0.0$–0.20. *J. Am. Ceram. Soc.*, **92** (8), 1672–1678.

20 Djurado, E. and Labeau, M. (1998) Second phases in doped lanthanum

gallate perovskites. *J. Eur. Ceram. Soc.*, **18**, 1397–1404.

21. Stevenson, J.W., Armstrong, T.R., Pederson, L.R., Li, J., Lewinsohn, C.A., and Baskaran, S. (1998) Effect of A-site cation nonstoichiometry on the properties of doped lanthanum gallate. *Solid State Ionics*, **113–115**, 571–583.

22. Amow, G. and Skinner, S.J. (2006) Recent developments in Ruddlesden–Popper nickelate systems for solid oxide fuel cell cathodes. *J. Solid State Electrochem.*, **10**, 538–546.

23. Guo, C., Zhang, X., Zhang, J., and Wang, Y. (2007) Preparation of La_2NiO_4 catalyst and catalytic performance for partial oxidation of methane. *J. Mol. Catal. A*, **269**, 254–259.

24. Huang, D.-P., Xu, Q., Zhang, F., Chen, W., Liu, H.-X., and Zhou, J. (2006) Synthesis and electrical conductivity of $La_2NiO_{4+\delta}$ derived from a polyaminocarboxylate complex precursor. *Mater. Lett.*, **60**, 1892–1895.

25. Li, C., Hu, T., Chen, Y., Jin, J., and Yang, N. (2003) Preparation and characterization of supported dense oxygen permeating membrane of mixed conductor $La_2NiO_{4+\delta}$. *J. Membr. Sci.*, **226**, 1–7.

26. Babilo, P., Uda, T., and Haile, S.M. (2007) Processing of yttrium-doped barium zirconate for high proton conductivity. *J. Mater. Res.*, **22** (5), 1322–1330.

27. Bućko, M.M. and Oblakowski, J. (2007) Preparation of $BaZrO_3$ nanopowders by spray pyrolysis method. *J. Eur. Ceram. Soc.*, **27**, 3625–3628.

28. Cervera, R.B., Oyama, Y., and Yamaguchi, S. (2007) Low temperature synthesis of nanocrystalline proton conducting $BaZr_{0.8}Y_{0.2}O_{3-\delta}$ by sol-gel method. *Solid State Ionics*, **178**, 569–574.

29. Dahl, P.I., Lein, H.L., Yu, Y., Grande, T., Einarsrud, M.-A., Kjølseth, C., Norby, T., and Haugsrud, R. (2011) Microstructural characterization and electrical properties of spray pyrolyzed conventionally sintered or hot-pressed $BaZrO_3$ and $BaZr_{0.9}Y_{0.1}O_{2.95}$. *Solid State Ionics*, **182**, 32–40.

30. Iguchi, F., Yamada, T., Sata, N., Tsurui, T., and Yugami, H. (2006) The influence of grain structures on the electrical conductivity of a $BaZr_{0.95}Y_{0.05}O_3$ proton conductor. *Solid State Ionics*, **177**, 2381–2384.

31. Magrez, A. and Schober, T. (2004) Preparation, sintering, and water incorporation of proton conducting $Ba_{0.99}Zr_{0.8}Y_{0.2}O_{3-\delta}$: comparison between three different synthesis techniques. *Solid State Ionics*, **175**, 585–588.

32. Dahl, P.I., Haugsrud, R., Lein, H.L., Grande, T., Norby, T., and Einarsrud, M.-A. (2007) Synthesis, densification and electrical properties of strontium cerate ceramics. *J. Eur. Ceram. Soc.*, **27**, 4461–4471.

33. Mokkelbost, T., Andersen, Ø., Strøm, R.A., Wiik, K., Grande, T., and Einarsrud, M.-A. (2007) High temperature proton conducting $LaNbO_4$-based materials. Powder synthesis by spray pyrolysis. *J. Am. Ceram. Soc.*, **90** (11), 3395–3400.

34. Mokkelbost, T., Kaus, I., Haugsrud, R., Norby, T., Grande, T., and Einarsrud, M.-A. (2008) High temperature proton conducting $LaNbO_4$-based materials. Part II. Sintering properties and solubility of alkaline earth oxides. *J. Am. Ceram. Soc.*, **91** (3), 879–886.

35. Subbarao, E.C. (1981) *Zirconia: An Overview*, American Ceramic Society, Cleveland, OH.

36. Hartmanova, M., Gmucova, K., Jergel, M., Thurzo, I., Kundracik, F., and Brunel, M. (1999) Structural and electrical properties of double-layer ceria/yttria stabilized zirconia deposited on silicon substrate. *Solid State Ionics*, **119** (1–4), 85–90.

37. Mokkelbost, T., Kaus, I., Grande, T., and Einarsrud, M.-A. (2004) Combustion synthesis and characterization of nanocrystalline CeO_2-based powders. *Chem. Mater.*, **16**, 5489–5494.

38. Katahira, K., Kohchi, Y., Shimura, T., and Iwahara, H. (2000) Protonic conduction in Zr-substituted $BaCeO_3$. *Solid State Ionics*, **138**, 91–98.

39. Kreuer, K.D., Adams, S., Münch, W., Fuchs, A., Klock, U., and Maier, J. (2001) Proton conducting alkaline earth

40 Minh, N.Q. (1993) Ceramic fuel cells. J. Am. Ceram. Soc., 76 (3), 563–588.

41 Iwahara, H., Yajima, T., Hibino, T., Ozaki, K., and Suzuki, H. (1993) Protonic conduction in calcium, strontium and barium zirconates. Solid State Ionics, 61, 65–69.

42 Kosacki, I. and Tuller, H.L. (1995) Mixed conductivity in $SrCe_{0.95}Yb_{0.05}O_3$ protonic conductors. Solid State Ionics, 81, 223–229.

43 Matsumoto, H., Shimura, T., Iwahara, H., Higuchi, T., Yashiro, K., Kaimai, A., Kawada, T., and Mizusaki, J. (2006) Hydrogen separation using proton-conducting perovskites. J. Alloys Compd., 408, 456–462.

44 Matsunami, N., Yajima, T., and Iwahara, H. (1992) Permeation of implanted deuterium through $SrCeO_3$ (5% Yb). Nucl. Instrum. Methods Phys. Res. B, 65, 278–281.

45 Boschini, F., Robertz, B., Rulmont, A., and Cloots, C. (2003) Preparation of nanosized barium zirconate powder by thermal decomposition of urea in an aqueous solution containing barium and zirconium, and by calcination of the precipitate. J. Eur. Ceram. Soc., 23, 3035–3042.

46 Babilo, P., Uda, T., and Haile, S.M. (2007) Processing of yttrium-doped barium zirconate for high proton conductivity. J. Mater. Res., 22 (5), 1322–1330.

47 Kaus, I., Dahl, P.I., Mastin, J., Grande, T., and Einarsrud, M.-A. (2006) Synthesis and characterization of nanocrystalline YSZ powder by smoldering combustion synthesis. J. Nanomater., 2006, 41–47.

48 Stuart, P.A., Unno, T., Ayres-Rocha, R., Djuado, E., and Skinner, S.J. (2009) The synthesis and sintering behaviour of $BaZr_{0.9}Y_{0.1}O_{3-\delta}$ powders prepared by spray pyrolysis. J. Eur. Ceram. Soc., 29, 697–702.

49 Stöver, D., Buchkremer, H.P., and Uhlenbruck, S. (2004) Processing and properties of the ceramic conductive multilayer device solid oxide fuel cell (SOFC). Ceram. Int., 30, 1107–1113.

50 Tietz, F., Buchkremer, H.P., and Stöver, D. (2002) Components manufacturing for solid oxide fuel cells. Solid State Ionics, 152–153, 373–381.

51 Lein, H.L., Tezuka, T., Grande, T., and Einarsrud, M.-A. (2008) Asymmetric proton conducting oxide membranes and fuel cells prepared by aqueous tape casting. Solid State Ionics, 179 (21–26), 1146–1150.

52 Campana, R., Merino, R.I., Larrea, A., Villarreal, I., and Orera, V.M. (2009) Fabrication, electrochemical characterization and thermal cycling of anode supported microtubular solid oxide fuel cells. J. Power Sources, 192 (1), 120–125.

53 Evans, A., Bieberle-Hütter, A., Rupp, J.L.M., and Gauckler, L.J. (2009) Review on microfabricated micro-solid oxide fuel cell membranes. J. Power Sources, 194, 119–129.

54 Lin, B., Zhang, X., Wang, S., Zhao, L., Fang, D., and Meng, G. (2008) Stable proton-conducting Ca-doped $LaNbO_4$ thin electrolyte-based protonic ceramic membrane fuel cells by in situ screen printing. J. Alloys Compd., 478 (1–2), 355–357.

55 Fontaine, M.-L., Larring, Y., Haugsrud, R., Norby, T., Wiik, K., and Bredesen, R. (2009) Novel high temperature proton conducting fuel cells: production of $La_{0.995}Sr_{0.005}NbO_{4-\delta}$ electrolyte thin films and compatible cathode architectures. J. Power Sources, 188 (1), 106–113.

56 Fontaine, M.-L., Larring, Y., Smith, J.B., Ræder, H., Andersen, Ø.S., Einarsrud, M.-A., Wiik, K., and Bredesen, R. (2009) Shaping of advanced asymmetric structures of proton conducting ceramic materials for SOFC and membrane-based process applications. J. Eur. Ceram. Soc., 29, 931–935.

57 Zhang, Y., Gao, J., Peng, D., Guangyao, M., and Liu, X. (2004) Dip-coating thin yttria-stabilized zirconia films for solid oxide fuel cell applications. Ceram. Int., 30 (6), 1049–1053.

58 Schüller, E., Vaßen, R., and Stöver, D. (2002) Thin electrolyte layers for SOFC via wet powder spraying (WPS). Adv. Eng. Mater., 4 (9), 659–662.

59 Trunec, M., Cihlar, J., Diethelm, S., and Van herle, J. (2006) Tubular $La_{0.7}Ca_{0.3}Fe_{0.85}Co_{0.15}O_{3-\delta}$ perovskite membranes. Part I. Preparation and properties. *J. Am. Ceram. Soc.*, **89** (3), 949–954.

60 Wang, H., Cong, Y., and Yang, W. (2003) Investigation on the partial oxidation of methane to syngas in a tubular $Ba_{0.5}Sr_{0.5}Co_{0.8}Fe_{0.2}O_{3-\delta}$ membrane reactor. *Catal. Today*, **82** (1–4), 157–166.

61 Falk, G., Bohm, N., Delaporte, P.G., Clasen, R., and Kuhn, S. (2008) *Electrophoretic Deposition and Sintering of Tubular Anode Supported Gadolinium Doped Ceria Solid Oxide Fuel Cell*, Electrochemical Society, Inc., Daytona Beach, FL.

62 Abrutis, A., Plausinaitiene, V., Kubilius, V., Teiserskis, A., Saltyte, Z., Butkute, R., and Senateur, J.P. (2002) Magnetoresistant $La_{1-x}Sr_xMnO_3$ films by pulsed injection metal organic chemical vapor deposition: effect of deposition conditions, substrate material and film thickness. *Thin Solid Films*, **413** (1–2), 32–40.

63 Lie, M., Nilsen, O., Fjellvåg, H., and Kjekshus, A. (2009) Growth of $La_{1-x}Sr_xFeO_3$ thin films by atomic layer deposition. *Dalton Trans.* (3), 481–489.

64 Nilsen, O., Fjellvåg, H., and Kjekshus, A. (2003) Growth of manganese oxide thin films by atomic layer deposition. *Thin Solid Films*, **444** (1–2), 44–51.

65 Xia, C.F., Ward, T.L., Atanasova, P., and Schwartz, R.W. (1998) Metal-organic chemical vapor deposition of Sr–Co–Fe–O films on porous substrates. *J. Mater. Res.*, **13** (1), 173–179.

66 Li, K., Tan, X., and Liu, Y. (2006) Single-step fabrication of ceramic hollow fibers for oxygen permeation. *J. Membr. Sci.*, **272**, 1–5.

67 Wang, B., Zydorczak, B., Wu, Z.-T., and Li, K. (2009) Stabilities of $La_{0.6}Sr_{0.4}Co_{0.2}Fe_{0.8}O_{3-\delta}$ oxygen separation membranes: effects of kinetic demixing/decomposition and impurity segregation. *J. Membr. Sci.*, **344**, 101–106.

68 Clemmer, R.M.C. and Corbin, S.F. (2004) Influence of porous composite microstructure on the processing and properties of solid oxide fuel cell anodes. *Solid State Ionics*, **166**, 251–259.

69 Tang, F., Fudouzi, H., Uchikoshi, T., and Sakka, Y. (2004) Preparation of porous materials with controlled pore size and porosity. *J. Eur. Ceram. Soc.*, **24**, 341–344.

70 Chen, I.W. and Wang, X.H. (2000) Sintering dense nanocrystalline ceramics without final-stage grain growth. *Nature*, **404** (6774), 168–171.

71 Nygren, M. and Shen, Z.J. (2003) On the preparation of bio-, nano- and structural ceramics and composites by spark plasma sintering. *Solid State Sci.*, **5** (1), 125–131.

72 Toshio, S., Kosacki, I., and Anderson, H.U. (2002) Defect and mixed conductivity in nanocrystalline doped cerium oxide. *J. Am. Ceram. Soc.*, **85** (6), 1492–1498.

73 Tuller, H.L. (2000) Ionic conduction in nanocrystalline materials. *Solid State Ionics*, **131** (1–2), 143–157.

74 Chiang, Y.M., Lavik, E.B., Kosacki, I., Tuller, H.L., and Ying, J.Y. (1996) Defect and transport properties of nanocrystalline CeO_{2-x}. *Appl. Phys. Lett.*, **69** (2), 185–187.

75 Mokkelbost, T. (2006) Synthesis and characterization of CeO_2- and $LaNbO_4$-based ionic conductors. PhD thesis, Department of Materials Science and Engineering, NTNU, Trondheim, Norway.

76 Meixner, D.L. and Cutler, R.A. (2002) Sintering and mechanical characteristics of lanthanum strontium manganite. *Solid State Ionics*, **146** (3–4), 273–284.

77 van Roosmalen, J.A.M., Cordfunke, E.H.P., and Huijsmans, J.-P.-P. (1993) Sinter behaviour of (La,Sr)MnO_3. *Solid State Ionics*, **66**, 285–293.

78 Stevenson, J.W., Hallman, P.F., Armstrong, T.R., and Chick, L. (1995) Sintering behavior of doped lanthanum and yttrium manganite. *J. Am. Ceram. Soc.*, **78**, 507–512.

79 Kharton, V.V., Shaula, A.L., Snijkers, F.M.M., Cooymans, J.F.C., Luyten, J.J., Marozau, I.P., Viskup, A.P., Marques, F.M.B., and Frade, J.R. (2006) Oxygen transport in ferrite-based ceramic

membranes: effects of alumina sintering aid. *J. Eur. Ceram. Soc.*, **26**, 3695–3704.

80 Kharton, V.V., Figueiredo, F.M., Kovalevsky, A.V., Viskup, A.P., Naumovich, E.N., Yaremchenko, A.A., Bashmakov, I.A., and Marques, F.M.B. (2001) Processing, microstructure and properties of LaCoO$_{3-\delta}$ ceramics. *J. Eur. Ceram. Soc.*, **21**, 2301–2309.

81 de Vires, K.J. (1997) Electrical and mechanical properties of proton conducting SrCe$_{0.95}$Yb$_{0.05}$O$_{3-\alpha}$. *Solid State Ionics*, **100**, 193–200.

82 Liou, Y.-C. and Yang, S.-L. (2008) A simple and effective process for Sr$_{0.995}$Ce$_{0.95}$Y$_{0.05}$O$_{3-\delta}$ and BaCe$_{0.9}$Nd$_{0.1}$O$_{3-\delta}$ solid electrolyte ceramics. *J. Power Sources*, **179**, 553–559.

83 Matsui, T. (1995) Thermodynamic properties of ternary barium oxides. *Thermochim. Acta*, **253**, 155–165.

84 Duval, S.B.C., Holtappels, P., Stimming, U., and Gaule, T. (2008) Effect of minor element addition on the electrical properties of BaZr$_{0.9}$Y$_{0.1}$O$_{3-\delta}$. *Solid State Ionics*, **179** (21–26), 1112–1115.

85 Anselmi-Tamburini, U., Buscaglia, M.T., Viviani, M., Bassoli, M., Bottini, C., Buscaglia, V., Nanni, P., and Munis, Z.A. (2006) Solid-state synthesis and spark plasma sintering of submicron BaY$_x$Zr$_{1-x}$O$_{3-(x/2)}$ ($x = 0$, 0.08 and 0.16) ceramics. *J. Eur. Ceram. Soc.*, **26**, 2313–2318.

86 De Souza, R.A. and Maier, J. (2003) A computational study of cation defects in LaGaO$_3$. *Phys. Chem. Chem. Phys.*, **5**, 740–748.

87 De Souza, R.A., Saiful Islam, M., and Ivers-Tifflée, E. (1999) Formation and migration of cation defects in the perovskite oxide LaMnO$_3$. *J. Mater. Chem.*, **9**, 1621–1627.

88 Kilo, M., Taylor, M.A., Argirusis, C., Borchardt, G., Jackson, R.A., Schulz, O., Martin, M., and Weller, M. (2004) Modeling of cation diffusion in oxygen ion conductors using molecular dynamics. *Solid State Ionics*, **175**, 823–827.

89 Kilner, J.A. (2000) Fast transport in acceptor doped oxides. *Solid State Ionics*, **129**, 13–23.

90 Miyoshi, S. and Martin, M. (2009) B-site cation diffusivity of Mn and Cr in perovskite-type LaMnO$_3$ with cation-deficit nonstoichiometry. *Phys. Chem. Chem. Phys.*, **11**, 3063–3070.

91 Schulz, O., Martin, M., Argirusis, C., and Borchardt, G. (2003) Cation tracer diffusion of ^{138}La, ^{84}Sr and ^{25}Mg in polycrystalline La$_{0.9}$Sr$_{0.1}$Ga$_{0.9}$Mg$_{0.1}$O$_{2.9}$. *Phys. Chem. Chem. Phys.*, **5**, 2308–2313.

92 Kilo, M., Taylor, M.A., Argirusis, C., Borchardt, G., Lesage, B., Weber, S., Scherrer, S., Scherrer, H., Schroeder, M., and Martin, M. (2003) Cation self-diffusion of ^{44}Ca, ^{88}Y, and ^{96}Zr in single crystalline calcia- and yttria-doped zirconia. *J. Appl. Phys.*, **94** (12), 7547–7552.

93 Hart, E.W. (1957) On the role of dislocations in bulk diffusion. *Acta Metall.*, **5** (10), 597.

94 Harrison, L.G. (1961) Influence of dislocations on diffusion kinetics in solids with particular reference to the alkali halides. *Trans. Faraday Soc.*, **57**, 1191–1199.

95 Drings, H., Brossman, U., Carstanjen, H.-D., Székfalvi-Nagy, Á., Noll, C., and Schaefer, H.-E. (2009) Enhanced ^{95}Zr diffusion in grain boundaries of nano-crystalline ZrO$_2$·9.5 mol% Y$_2$O$_3$. *Physica Status Solidi a*, **206** (1), 54–58.

96 Swaroop, S., Kilo, M., Argirusis, C., Borchardt, G., and Chokshi, A.H. (2005) Lattice and grain boundary diffusion of cations in 3YTZ analyzed using SIMS. *Acta Mater.*, **53**, 4975–4985.

97 Horita, T., Ishikawa, M., Yamaji, K., Sakai, N., Yokokawa, H., and Dokiya, M. (1999) Calcium tracer diffusion in (La,Ca)CrO$_3$ by SIMS. *Solid State Ionics*, **124**, 301–307.

98 Palcut, M., Christensen, J.S., Wiik, K., and Grande, T. (2008) Impurity diffusion of ^{141}Pr in LaMnO$_3$, LaCoO$_3$ and LaFeO$_3$ materials. *Phys. Chem. Chem. Phys.*, **10**, 6544–6552.

99 Palcut, M., Wiik, K., and Grande, T. (2007) Cation self-diffusion and nonstoichiometry of lanthanum manganite studied by diffusion couple measurements. *J. Phys. Chem. C*, **111**, 813–822.

100 Palcut, M., Wiik, K., and Grande, T. (2007) Cation self-diffusion in LaCoO$_3$ and La$_2$CoO$_4$ studied by diffusion couple experiments. *J. Phys. Chem. B*, **111**, 2299–2308.

101 Wærnhus, I., Grande, T., and Wiik, K. (2005). Electronic properties of polycrystalline LaFeO$_3$. Part II. Defect modelling including Schottky defects. *Solid State Ionics*, **176**, 35–36.

102 Wærnhus, I., Vullum, P.E., Holmestad, R., Grande, T., and Wiik, K. (2005) Electronic properties of polycrystalline LaFeO$_3$. Part I. Experimental results and the qualitative role of Schottky defects. *Solid State Ionics*, **176**, 2783–2790

103 Maier, J. (2005) Nanoionics: ion transport and electrochemical storage in confined systems. *Nat. Mater.*, **4** (11), 805–815.

104 De Souza, R.A. and Martin, M. (2004) Secondary ion mass spectroscopy: a powerful tool for diffusion studies in solids. *Arch. Metall. Mater.*, **49** (2), 431–446.

105 Crank, J. (1956) *Mathematics of Diffusion*, Oxford University Press, London.

106 Le Claire, A.D. (1963) The analysis of grain boundary diffusion measurements. *Br. J. Appl. Phys.*, **14**, 351–356.

107 Palcut, M. (2007) Cation diffusion in LaMnO$_3$, LaCoO$_3$ and LaFeO$_3$ materials. PhD thesis, Department of Materials Science and Engineering, NTNU, Trondheim, Norway.

108 Smith, J.B. and Norby, T. (2006) Cation self-diffusion in LaFeO$_3$ measured by the solid state reaction method. *Solid State Ionics*, **177**, 639–646.

109 Routbort, J., Goretta, K.C., de Arellano-López, A.R., and Wolfenstine, J. (1998) Creep of Ce$_{0.9}$Gd$_{0.1}$O$_{1.95}$. *Scr. Mater.*, **38** (2), 315–320.

110 Martin, M. (2000) Electrotransport and demixing in oxides. *Solid State Ionics*, **136–137**, 331–337.

111 Martin, M. (2003) Materials in thermodynamic potential gradients. *Pure Appl. Chem.*, **75** (7), 889–903.

112 Lein, H.L., Wiik, K., and Grande, T. (2006) Kinetic demixing and decomposition of oxygen permeable membranes. *Solid State Ionics*, **177**, 1587–1590.

113 van Doorn, R.H.E., Bouwmeester, H.J.M., and Burggraaf, A.J. (1998) Kinetic decomposition of La$_{0.3}$Sr$_{0.7}$SrO$_{3-\delta}$, $x = 0$ and 0.1, measured by SIMS. *Solid State Ionics*, **111**, 263–272.

114 Lein, H.L. (2005) Mechanical properties and phase stability of oxygen permeable membranes La$_{0.5}$Sr$_{0.5}$Fe$_{1-x}$Co$_x$O$_{3-\delta}$. PhD thesis, Department of Materials Science and Engineering, NTNU, Trondheim, Norway.

115 Hayashi, H., Saitou, T., Maruyama, N., Inaba, H., Kawamura, K., and Mori, M. (2005) Thermal expansion coefficient of yttria stabilized zirconia for various yttria contents. *Solid State Ionics*, **176** (5–6), 613–619.

116 Liu, Z.-G., Ouyang, J.-H., and Zhou, Y. (2009) Influence of gadolinia on thermal expansion property of ZrO$_2$–4.5 mol% Y$_2$O$_3$ ceramics. *J. Alloys Compd.*, **473** (1–2), L17–L19.

117 Sammes, N.M. and Du, Y. (2007) Fabrication and characterization of tubular solid oxide fuel cells. *Int. J. Appl. Ceram. Technol.*, **4** (2), 89–102.

118 Hayashi, H., Kanoh, M., Quan, C.J., Inaba, H., Wang, S.R., Dokiya, M., and Tagawa, H. (2000) Thermal expansion of Gd-doped ceria and reduced ceria. *Solid State Ionics*, **132** (3–4), 227–233.

119 Körner, R., Ricken, M., Nölting, J., and Riess, I. (1989) Phase transformations in reduced ceria: determination by thermal expansion measurements. *J. Solid State Chem.*, **78** (1), 136–147.

120 Mogensen, M., Lindegaard, T., Hansen, U.R., and Mogensen, G. (1994) Physical properties of mixed conductor solid oxide fuel cell anodes of doped CeO$_2$. *J. Electrochem. Soc.*, **141** (8), 2122–2128.

121 Sameshima, S., Ichikawa, T., Kawaminami, M., and Hirata, Y. (1999) Thermal and mechanical properties of rare earth-doped ceria ceramics. *Mater. Chem. Phys.*, **61** (1), 31–35.

122 Tolchard, J.R. and Grande, T. (2007) Chemical compatibility of candidate oxide cathodes for BaZrO$_3$ electrolytes. *Solid State Ionics*, **178** (7–10), 593–599.

123 Yamanaka, S., Kurosaki, K., Maekawa, T., Matsuda, T., Kobayashi, S., and Masayoshi, U.J. (2005) Thermochemical and thermophysical properties of

alkaline-earth perovskites. *Nucl. Mater.*, **344** (1–3), 61–66.

124 Tolchard, J.R. and Grande, T. (2007) Physicochemical compatibility of SrCeO$_3$ with potential SOFC cathodes. *J. Solid State Chem.*, **180** (10), 2808–2815.

125 Mokkelbost, T., Lein, H.L., Vullum, P.-E., Holmestad, R., Grande, T., and Einarsrud, M.-A. (2009) Thermal and mechanical properties of LaNbO$_4$-based ceramics. *Ceram. Int.*, **35** (7), 2877–2883.

126 Aruna, S.T., Muthuraman, M., and Patil, K.C. (1997) Combustion synthesis and properties of strontium substituted lanthanum manganites La$_{1-x}$Sr$_x$MnO$_3$ ($0 \leq x \leq 0.3$). *J. Mater. Chem.*, **7** (12), 2499–2503.

127 Kharton, V.V., Yaremchenko, A.A., Kovalevsky, A.V., Viskup, A.P., Naumovich, E.N., and Kerko, P.F. (1999) Perovskite-type oxides for high-temperature oxygen separation membranes. *J. Membr. Sci.*, **163**, 307–317.

128 Lein, H.L., Wiik, K., and Grande, T. (2006) Thermal and chemical expansion of mixed conducting La$_{0.5}$Sr$_{0.5}$Fe$_{1-x}$Co$_x$O$_{3-\delta}$ materials. *Solid State Ionics*, **177** (19–25), 1795–1798.

129 Tai, L.-W., Nasrallah, M.M., Anderson, H.U., Sparlin, D.M., and Sehlin, S.R. (1995) Structure and electrical properties of La$_{1-x}$Sr$_x$Co$_{1-y}$Fe$_y$O$_3$. Part 1. The system La$_{0.8}$Sr$_{0.2}$Co$_{1-y}$Fe$_y$O$_3$. *Solid State Ionics*, **76**, 259–271.

130 Kammer, K. (2006) Studies of Fe–Co based perovskite cathodes with different A-site cations. *Solid State Ionics*, **177** (11–12), 1047–1051.

131 Burmistrov, I., Drozhzhin, O.A., Istomin, S.Y., Sinitsyn, V.V., Antipov, E.V., and Bredikhin, S.I. (2009) Sr$_{0.75}$Y$_{0.25}$Co$_{0.5}$Mn$_{0.5}$O$_{3-y}$ perovskite cathode for solid oxide fuel cells. *J. Electrochem. Soc.*, **156** (10), B1212–B1217.

132 Juste, E., Julian, A., Etchegoyen, G., Geffroy, P.M., Chartier, T., Richet, N., and Del Gallo, P. (2008) Oxygen permeation, thermal and chemical expansion of (La,Sr)(Fe,Ga)O$_{3-\delta}$ perovskite membranes. *J. Membr. Sci.*, **319** (1–2), 185–191.

133 Kharton, V.V., Tikhonovich, V.N., Li, S.B., Naumovich, E.N., Kovalevsky, A.V., Viskup, A.P., and Yaremchenko, A.A. (1998) Ceramic microstructure and oxygen permeability of SrCo(Fe,M)O$_{3-\delta}$ (M=Cu or Cr) perovskite membranes. *J. Electrochem. Soc.*, **145**, 1363–1373.

134 Al Daroukh, M., Vashook, V.V., Ullmann, H., Tietz, F., and Arual Raj, I. (2003) Oxides of the AMO$_3$ and A$_2$MO$_4$-type: structural stability, electrical conductivity and thermal expansion. *Solid State Ionics*, **158** (1–2), 141–150.

135 Boehm, E., Bassat, J.M., Steil, M.C., Dordor, P., Mauvy, F., and Grenier, J.C. (2003) Oxygen transport properties of La$_2$Ni$_{1-x}$Cu$_x$O$_{4+\delta}$ mixed conducting oxides. *Solid State Sci.*, **5** (7), 973–981.

136 Kharton, V.V., Kovalevsky, A.V., Avdeev, M., Tsipis, E.V., Patrakeev, M.V., Yaremchenko, A.A., Naumovich, E.N., and Frade, J.R. (2007) Chemically induced expansion of La$_2$NiO$_{4+\delta}$-based materials. *Chem. Mater.*, **19** (8), 2027–2033.

137 Gilbu, B., Fjellvåg, H., and Kjekshus, A. (1994) Properties of LaCo$_{1-t}$Co$_t$O$_3$. I. Solid solubility, thermal expansion and structural transition. *Acta Chem. Scand.*, **48**, 37–45.

138 Sakai, N., Yokokawa, H., Horita, T., and Yamaji, K. (2004) Lanthanum chromite-based interconnects as key materials for SOFC stack development. *Int. J. Appl. Ceram. Technol.*, **1** (1), 23–30.

139 Kharton, V.V., Naumovich, E.N., Kovalevsky, A.V., Viskup, A.P., Figueiredo, F.M., Bashmakov, I.A., and Marques, F.M.B. (2000) Mixed electronic and ionic conductivity of LaCo(M)O$_3$ (M=Ga, Cr, Fe or Ni). IV. Effect of preparation method on oxygen transport in LaCoO$_{3-\delta}$. *Solid State Ionics*, **138** (1–2), 135–148.

140 Fossdal, A., Menon, M., Værnhus, I., Wiik, K., Einarsrud, M.-A., and Grande, T. (2004) Crystal structure and thermal expansion of La$_{1-x}$Sr$_x$FeO$_{3-\delta}$ materials. *J. Am. Ceram. Soc.*, **87** (10), 1952–1958.

141 Li, S., Lü, Z., Huang, X., Wei, B., and Su, W. (2007) Electrical and thermal properties of (Ba$_{0.5}$Sr$_{0.5}$)$_{1-x}$Sm$_x$Co$_{0.8}$Fe$_{0.2}$O$_{3-\delta}$

142 Mori, M. (2004) Effect of B-site doping on thermal cycle shrinkage for $La_{0.8}Sr_{0.2}Mn_{1-x}M_xO_{3+\delta}$ perovskites (M=Mg, Al, Ti, Mn, Fe, Co, Ni; $0 \leq x \leq 0.1$). *Solid State Ionics*, **174** (1–4), 1–8.

143 Wang, S., Lin, B., Chen, Y., Liu, X., and Meng, G. (2009) Evaluation of simple, easily sintered $La_{0.7}Ca_{0.3}Cr_{0.97}O_{3-\delta}$ perovskite oxide as novel interconnect material for solid oxide fuel cells. *J. Alloys Compd.*, **479** (1–2), 764–768.

144 Tietz, F. (1999) Thermal expansion of SOFC materials. *Ionics*, **5**, 129–139.

145 Chen, X., Yu, J., and Adler, S.B. (2005) Thermal and chemical expansion of Sr-doped lanthanum cobalt oxide ($La_{1-x}Sr_xCoO_{3-\delta}$). *Chem. Mater.*, **17** (17), 4537–4546.

146 Adler, S.B. (2001) Chemical expansivity of electrochemical ceramics. *J. Am. Ceram. Soc.*, **84** (9), 2117–2119.

147 Armstrong, T.R., Stevenson, J.W., Pederson, L.R., and Raney, P.E. (1996) Dimensional instability of doped lanthanum chromite. *J. Electrochem. Soc.*, **143** (9), 2919–2925.

148 Wang, S., Katsuki, M., Dokiya, M., and Hashimoto, T. (2003) High temperature properties of $La_{0.6}Sr_{0.4}Co_{0.8}Fe_{0.2}O_{3-\delta}$ phase structure and electrical conductivity. *Solid State Ionics*, **159**, 71–78.

149 Zuev, A., Singheiser, L., and Hilpert, K. (2002) Defect structure and isothermal expansion of A-site and B-site substituted lanthanum chromites. *Solid State Ionics*, **147**, 1–11.

150 Kharton, V.V., Yaremchenko, A.A., Patrakeev, M.V., Naumovich, E.N., and Marques, F.M.B. (2003) Thermal and chemical induced expansion of $La_{0.3}Sr_{0.7}(Fe,Ga)O_{3-\delta}$ ceramics. *J. Eur. Ceram. Soc.*, **23**, 1417–1426.

151 Miyoshi, S., Hong, J.-O., Yashiro, K., Kaimai, A., Nigara, Y., Kawamura, K., Kawada, T., and Mizusaki, J. (2003) Lattice expansion upon reduction of perovskite-type $LaMnO_3$ with oxygen-deficit nonstoichiometry. *Solid State Ionics*, **161**, 209–217.

152 Atkinson, A. and Ramos, T.M.G.M. (2000) Chemically-induced stresses in ceramic oxygen ion-conducting membranes. *Solid State Ionics*, **129**, 259–269.

153 McIntosh, S., Vente, J.F., Haije, W.G., Blank, D.H.A., and Bouwmeester, H.J.M. (2006) Oxygen stoichiometry and chemical expansion of $Ba_{0.5}Sr_{0.5}Co_{0.8}Fe_{0.2}O_{3-\delta}$ measured by *in situ* neutron diffraction. *Chem. Mater.*, **18** (8), 2187–2193.

154 Cranmer, D.C. and Richerson, D.W. (1998) *Mechanical Testing Methodology for Ceramic Design and Reliability*, Marcel Dekker, Inc., New York.

155 Wachtman, J.B. (1996) *Mechanical Properties of Ceramics*, John Wiley & Sons, Inc., New York.

156 Kleveland, K., Orlovskaya, N., Grande, T., Moe, A.M.M., Einarsrud, M.-A., Breder, K., and Gogotsi, G. (2001) Ferroelastic behavior of $LaCo_3$-based ceramics. *J. Am. Ceram. Soc.*, **84**, 2029–2033.

157 Lein, H.L., Skottun Andersen, Ø., Vullum, P.E., Lara-Curzio, E., Holmestad, R., Einarsrud, M.-A., and Grande, T. (2006) Mechanical properties of mixed conducting $La_{0.5}Sr_{0.5}Fe_{1-x}Co_xO_{3-\delta}$ ($0 \leq x \leq 1$) materials. *J. Solid State Electrochem.*, **10**, 635–642.

158 Fossdal, A., Einarsrud, M.-A., and Grande, T. (2005) Mechanical properties of $LaFeO_3$ ceramics. *J. Eur. Ceram. Soc.*, **25**, 927–933.

159 Orlovskaya, N., Kleveland, K., Grande, T., and Einarsrud, M.-A. (2000) Mechanical properties of $LaCoO_3$ based ceramics. *J. Eur. Ceram. Soc.*, **20**, 51–56.

160 Seluk, A. and Atkinson, A. (1997) Elastic properties of ceramic oxides used in solid oxide fuel cells (SOFCs). *J. Eur. Ceram. Soc.*, **17**, 1523–1532.

161 Montross, C.S., Yokokawa, H., Dokiya, M., and Bekessy, L. (1995) Mechanical properties of magnesia-doped lanthanum chromite versus temperature. *J. Am. Ceram. Soc.*, **78**, 1869–1872.

162 Krogh, B., Brustad, M., Dahle, M., Eilertsen, J.L., and Ødegård, R. (1997) Increased mechanical strength of $La_{0.8}Ca_{0.2}CrO_3$ by improvement of powder properties and sintering conditions. *Proc. Electrochem. Soc.*, **97**, 40.

163 Atkinson, A. and Selçuk, A. (2000) Mechanical behaviour of ceramic oxygen ion-conducting membranes. *Solid State Ionics*, **134** (1–2), 59–66.

164 Vassen, R., Cao, X., Tietz, F., Basu, D., and Stöver, D. (2000) Zirconates as new materials for thermal barrier coatings. *J. Am. Ceram. Soc.*, **83** (8), 2023–2028.

165 de Vries, K.J. (1997) Electrical and mechanical properties of proton conducting $SrCe_{0.95}Yb_{0.05}O_{3-\alpha}$. *Solid State Ionics*, **100** (3–4), 193–200.

166 Hassan, D., Janes, S., and Clasen, R. (2003) Proton-conducting ceramics as electrode/electrolyte materials for SOFC's. Part I. Preparation, mechanical and thermal properties of sintered bodies. *J. Eur. Ceram. Soc.*, **23** (2), 221–228.

167 Pathak, S., Steinmetz, D., Kuebler, J., Payzant, E.A., and Orlovskaya, N. (2009) Mechanical behavior of $La_{0.8}Sr_{0.2}Ga_{0.8}Mg_{0.2}O_3$ perovskites. *Ceram. Int.*, **35**, 1235–1241.

168 Faaland, S., Grande, T., Einarsrud, M.A., Vullum, P.E., and Holmestad, R. (2005) Stress–strain behavior during compression of polycrystalline $La_{1-x}Ca_xCoO_3$ ceramics. *J. Am. Ceram. Soc.*, **88** (3), 726–730.

169 Cain, M.G. and Lewis, M.H. (1990) Evidence of ferroelasticity in Y-tetragonal zirconia polycrystals. *Mater. Lett.*, **9** (9), 309–312.

170 Foitzik, A., Stadtwaldklenke, M., and Ruhle, M. (1993) Ferroelasticity of t-ZrO_2. *Z. Metallkd.*, **84** (6), 397–404.

171 Yokokawa, H., Tu, H., Iwanschitz, B., and Mai, A. (2008) Fundamental mechanisms limiting solid oxide fuel cell durability. *J. Power Sources*, **182** (2), 400–412.

172 Yokokawa, H. (2003) Understanding materials compatibility. *Annu. Rev. Mater. Res.*, **33**, 581–610.

173 Tsipis, E.V. and Kharton, V.V. (2008) Electrode materials and reaction mechanisms in solid oxide fuel cells: a brief review. *J. Solid State Electrochem.*, **12** (11), 1367–1391.

174 Tsipis, E.V. and Kharton, V.V. (2008) Electrode materials and reaction mechanisms in solid oxide fuel cells: a brief review. *J. Solid State Electrochem.*, **12** (9), 1039–1060.

175 Haugsrud, R. and Norby, T. (2006) High-temperature proton conductivity in acceptor-doped $LaNbO_4$. *Solid State Ionics*, **177** (13–14), 1129–1135.

176 Haugsrud, R. and Norby, T. (2006) Proton conduction in rare-earth *ortho*-niobates and *ortho*-tantalates. *Nat. Mater.*, **5** (3), 193–196.

177 Tolchard, J.R., Lein, H.L., and Grande, T. (2009) Chemical compatibility of proton conducting $LaNbO_4$ electrolyte with potential oxide cathodes. *J. Eur. Ceram. Soc.*, **29** (13), 2823–2830.

178 Kingery, W.D., Bowen, H.K., and Uhlmann, D.R. (1976) *Introduction to Ceramics*, 2nd edn, Wiley Series on the Science and Technology of Materials, John Wiley & Sons, Inc., New York.

179 Yi, J.X., Lein, H.L., Grande, T., Yakovlev, S., and Bouwmeester, H.J.M. (2009) High-temperature compressive creep behaviour of perovskite-type oxide $Ba_{0.5}Sr_{0.5}Co_{0.8}Fe_{0.2}O_{3-\delta}$. *Solid State Ionics*, **180**, 1564–1568.

180 Majkic, G., Wheeler, L.T., and Salama, K. (2000) Creep of polycrystalline $SrCo_{0.8}Fe_{0.2}O_{3-\delta}$. *Acta Mater.*, **48**, 1907–1917.

181 Lein, H.L., Wiik, K., Einarsrud, M.-A., and Grande, T. (2006) High-temperature creep behaviour of mixed conducting $La_{0.5}Sr_{0.5}Fe_{1-x}Co_xO_{3-\delta}$ ($0.5 \leq x \leq 1$) materials. *J. Am. Ceram. Soc.*, **89** (9), 2895–2898.

182 Wolfenstine, J., Goretta, K.C., Cook, R.E., and Routbort, J.L. (1996) Use of diffusional creep to investigate mass transport in (La,Sr)MnO_3. *Solid State Ionics*, **92**, 75–83.

183 Majkic, G., Wheeler, L.T., and Salama, K. (2002) High-temperature deformation of $La_{0.2}Sr_{0.8}Fe_{0.8}Cr_{0.2}O_{3-\delta}$-mixed ionic-electronic conductor. *Solid State Ionics*, **146**, 393–404.

184 Majkic, G., Wheeler, L.T., and Salama, K. (2003) Stress-induced diffusion and defect chemistry of $La_{0.2}Sr_{0.8}Fe_{0.8}Cr_{0.2}O_{3-\delta}$. Part 1. Creep in controlled-oxygen atmosphere. *Solid State Ionics*, **164**, 137–148.

185 Wolfenstine, J. (1999) Rate-controlling species for creep of the solid state

electrolyte: doped lanthanum gallate. *Solid State Ionics*, **126**, 293–298.

186 Wolfenstine, J., Huang, P., and Petric, A. (2000) High-temperature mechanical behaviour of the solid-state electrolyte: $La_{0.8}Sr_{0.2}Ga_{0.85}Mg_{0.15}O_{2.825}$. *J. Electrochem. Soc.*, **147** (5), 1668–1670.

187 Majkic, G., Mironova, M., and Salama, K. (2001) A transmission electron microscopy study of polycrystalline $SrCo_{0.8}Fe_{0.2}O_{3-\delta}$ creep in the diffusion-to-power law transition regime. *Philos. Mag. A*, **81** (11), 2675–2688.

188 Majkic, G., Mironova, M., Wheeler, L.T., and Salama, K. (2004) Stress-induced diffusion and defect chemistry of $La_{0.2}Sr_{0.8}Fe_{0.8}Cr_{0.2}O_{3-\delta}$. 2. Structural, elemental and chemical analysis. *Solid State Ionics*, **167**, 243–254.

189 Stubican, V.S., Hellmann, J.R., and Ray, S.P. (1982) Defects and ordering in zirconia crystalline solutions. *Mater. Sci. Monogr.*, **10**, 257–61.

190 Hellmann, J.R. and Stubican, V.S. (1983) Stable and metastable phase relations in the system ZrO_2–CaO. *J. Am. Ceram. Soc.*, **66** (4), 260–264.

191 Kilner, J.A. and Brook, R.J. (1982) A study of oxygen ion conductivity in doped non-stoichiometric oxides. *Solid State Ionics*, **6**, 237–252.

192 Sakib Khan, M., Saiful Islam, M., and Bates, D.R. (1998) Cation doping and oxygen diffusion in zirconia: a combined atomistic simulation and molecular dynamics study. *J. Mater. Chem.*, **8** (10), 2299–2307.

193 Wang, D.Y., Park, D.S., Griffith, J., and Nowick, A.S. (1981) Oxygen-ion conductivity and defect interactions in yttria-doped ceria. *Solid State Ionics*, **2** (2), 95–105.

194 Zhang, T.S., Ma, J., Huang, H.T., Hing, P., Xia, Z.T., Chan, S.H., and Kilner, J.A. (2003) Effects of dopant concentration and aging on the electrical properties of Y-doped ceria electrolytes. *Solid State Sci.*, **5** (11–12), 1505–1511.

195 Zhang, T.S., Ma, J., Kong, L.B., Chan, S.H., and Kilner, J.A. (2004) Aging behavior and ionic conductivity of ceria-based ceramics: a comparative study. *Solid State Ionics*, **170** (3–4), 209–217.

196 Huang, K., Tichy, R.S., and Goodenough, J.B. (1998) Superior perovskite oxide-ion conductor; strontium- and magnesium-doped $LaGaO_3$. I. Phase relationships and electrical properties. *J. Am. Ceram. Soc.*, **81** (10), 2565–2575.

197 Tofield, B.C., Greaves, C., and Fender, B.E.F. (1975) The $SrFeO_{2.5}$–$SrFeO_{3.0}$ system. Evidence of a new phase $Sr_4Fe_4O_{11}$ ($SrFeO_{2.75}$). *Mater. Res. Bull.*, **10** (7), 737–746.

198 Harrison, W.T.A., Hegwood, S.L., and Jacobson, A.J. (1995) A powder neutron diffraction determination of the structure of $Sr_6Co_5O_{15}$, formerly described as the low-temperature hexagonal form of $SrCoO_{3-x}$. *J. Chem. Soc. Chem. Commun.*, 1953–1954.

199 Rodriguez, J., Gonzalez-Calbet, J.M., Grenier, J.C., Pannetier, J., and Anne, M. (1987) Phase transitions in $Sr_2Co_2O_5$: a neutron thermodiffractometry study. *Solid State Commun.*, **62** (4), 231–234.

200 Rørmark, L., Mørch, A.B., Wiik, K., Stølen, S., and Grande, T. (2001) Enthalpies of oxidation of $CaMnO_{3-\delta}$, $CaMnO4-\delta$ and $SrMnO3-\delta$: deduced redox properties. *Chem. Mater.*, **13**, 4005–4013.

201 Rørmark, L., Wiik, K., Stølen, S., and Grande, T. (2002) Oxygen stoichiometry and structural properties of $La_{1-x}A_xMnO_{3+\delta}$ (A=Ca or Sr and $0 \leq x \leq 1$). *J. Mater. Chem.*, **12**, 1058–1067.

202 Poeppelmeier, K.R., Leonowicz, M.E., and Longo, J.M. (1982) $CaMnO_{2.5}$ and $Ca_2MnO_{3.5}$: new oxygen-defect perovskite-type oxides. *J. Solid State Chem.*, **44** (1), 89–98.

203 Elshof, J.E., Bouwmeester, H.J.M., and Verweij, H. (1995) Oxygen transport through $La_{1-x}Sr_xFeO_{3-\delta}$ membranes. I. Permeation in air/He gradients. *Solid State Ionics*, **81** (1–2), 97–109.

204 Elshof, J.E., Bouwmeester, H.J.M., and Verweij, H. (1996) Oxygen transport through $La_{1-x}Sr_xFeO_{3-\delta}$ membranes. II. Permeation in air/CO, CO_2 gradients. *Solid State Ionics*, **89** (1–2), 81–92.

205 Hombo, J. and Nishimura, K. (1987) Conductivity response of the complex oxides $La_{1-x}Ca_xFeO_{3-\delta}$ to oxygen gas

concentration. *Denki Kagaku*, **55** (4), 307–310.

206 van Hassel, B.A., Elshof, J.E., and Bouwmeester, H.J.M. (1994) Oxygen permeation flux through $La_{1-y}Sr_yFeO_3$ limited by carbon monoxide oxidation rate. *Appl. Catal. A*, **119** (2), 279–291.

207 Kleveland, K., Einarsrud, M.-A., Schmidt, C.R., Shamsili, S., Faaland, S., Wiik, K., and Grande, T. (1999) Reactions between strontium-substituted lanthanum manganite and yttria-stabilized zirconia. II. Diffusion couples. *J. Am. Ceram. Soc.*, **82** (3), 729–734.

208 Wiik, K., Schmidt, C.R., Faaland, S., Shamsili, S., Einarsrud, M.-A., and Grande, T. (1999) Reactions between strontium-substituted lanthanum manganite and yttria-stabilized zirconia. I. Powder samples. *J. Am. Ceram. Soc.*, **82** (3), 721–728.

209 Bakken, E., Norby, T., and Stølen, S. (2002) Redox energies of perovskite-related oxides. *J. Mater. Chem.*, **12**, 317–323.

210 Rørmark, L., Stølen, S., Wiik, K., and Grande, T. (2002) Enthalpies of formation of $La_{1-x}A_xMnO_{3\pm\delta}$ (A=Ca and Sr) measured by high-temperature solution calorimetry. *J. Solid State Chem.*, **163** (1), 186–193.

211 Ralph, J.M., Kilner, J.A., and Steele, B.C.H. (2000) *Improving Gd-Doped Ceria Electrolytes for Low Temperature Solid Oxide Fuel Cells*, Materials Research Society, Warrendale, PA.

212 Zhang, T.S., Ma, J., Kong, L.B., Hing, P., Leng, Y.J., Chan, S.H., and Kilner, J.A. (2003) Sinterability and ionic conductivity of coprecipitated $Ce_{0.8}Gd_{0.2}O_{2-\delta}$ powders treated via a high-energy ball-milling process. *J. Power Sources*, **124** (1), 26–33.

213 Petrov, A.N., Cherepanov, V.A., and Zuev, A. (2006) Thermodynamics, defect structure, and charge transfer in doped lanthanum cobaltites: an overview. *J. Solid State Electrochem.*, **10**, 517–537.

214 Scholten, M.J., Schoonman, J., van Miltenburg, J.C., and Oonk, H.A.J. (1993) Synthesis of strontium and barium cerate and their reaction with carbon dioxide. *Solid State Ionics*, **61** (1–3), 83–91.

215 Mather, G.C. and Jurado, J.R. (2003) Nonstoichiometry and stability in water of undoped $SrCeO_3$. *Bol. Soc. Esp. Ceram. V*, **42** (5), 311–316.

216 Yokokawa, H., Kawada, T., and Dokiya, M. (1989) Thermodynamic regularities in perovskite and K_2NiF_4 compounds. *J. Am. Ceram. Soc.*, **72** (1), 152–153.

217 Stern, K.H. (1969) The effect of cations on the thermal decomposition of salts with oxyanions. *J. Chem. Educ.*, **46** (10), 645–649.

218 Shannon, R.D. (1976) Revised effective ionic radii and systematic studies of interatomic distances in halides and chalcogenides. *Acta Crystallogr.*, **A32**, 751.

219 Slater, J.C. (1930) Atomic shielding constants. *Phys. Rev.*, **36**, 57.

220 Shick, H.L. (1960) A thermodynamic analysis of the high temperature vaporization properties of silica. *Chem. Rev.*, **60**, 331.

221 Faaland, S., Einarsrud, M.-A., and Grande, T. (2001) Reactions between calcium- and strontium-substituted lanthanum cobaltite ceramic membranes and calcium silicate sealing materials. *Chem. Mater.*, **13** (3), 723–732.

222 Jacobson, N., Opila, E., Myers, D., and Copland, E. (2005) Thermodynamics of gas phase species in the Si–O–H system. *J. Thermodyn.*, **37**, 1130–1137.

223 Kaus, I., Wiik, K., Dahle, M., Brustad, M., and Aasland, S. (2007) Stability of $SrFeO_3$ based materials in the presence of SiO_x species in humid atmosphere at high temperatures and pressures. *J. Eur. Ceram. Soc.*, **27**, 4509–4514.

224 Jacobson, N., Myers, D., Opila, E., and Copland, E. (2005) Interactions of water vapor with oxides at elevated temperatures. *J. Phys. Chem. Solids*, **66**, 471–478.

225 Fergus, J.W. (2007) Effect of cathode and electrolyte transport properties on chromium poisoning in solid oxide fuel cells. *Int. J. Hydrogen Energy*, **32** (16), 3664–3671.

226 Panas, I., Svensson, J.-E., Asteman, H., Johnson, T.J.R., and Johansson, L.G. (2004) Chromic acid evaporation upon exposure of $Cr_2O_3(s)$ to $H_2O(g)$ and $O_2(g)$: mechanism from first principles. *Chem. Phys. Lett.*, **383** (5–6), 549–554.

227 Pradyot, P. (2003) *Handbook of Inorganic Chemicals*, McGraw-Hill, New York.
228 McColm, I.J. and Clark, N.J. (1988) *Forming, Shaping and Working of High-Performance Ceramics*, Blackie, Glasgow.
229 Wolfenstine, J., Armstrong, T.R., Weber, W.J., Boling-Risser, M.A., Goretta, K.C., and Routbort, J.L. (1996) Elevated temperature deformation of fine-grained $La_{0.9}Sr_{0.1}MnO_3$. *J. Mater. Res.*, **11** (3), 657–662.
230 Kleveland, K., Wereszczak, A., Kirkland, T.P., Einarsrud, M.-A., and Grande, T. (2001) Compressive creep performance of $SrFeO_3$. *J. Am. Ceram. Soc.*, **84** (8), 1822–1826.
231 Majkic, G., Wheeler, L.T., and Salama, K. (2000) Characterization of creep behaviour of $SrCo_{0.8}Fe_{0.2}O_{3-x}$. *Mater. Res. Soc. Symp. Proc.*, **575**, 349–354.
232 Lakki, A., Herzog, R., Weller, M., Schubert, H., Reetz, C., Görke, O., Kilo, M., and Borchardt, G. (2000) Mechanical loss, creep, diffusion and ionic conductivity of ZrO_2–8 mol% Y_2O_3 polycrystals. *J. Eur. Ceram. Soc.*, **20**, 285–296.
233 Goudiakas, J., Haire, R.G., and Fuger, J. (1990) Thermodynamics of lanthanide and actinide perovskite-type oxides. IV. Molar enthalpies of formation of $MM'O_3$ (M=Ba or Sr, M'=Ce, Tb, or Am) compounds. *J. Chem. Thermodyn.*, **22** (6), 577–587.
234 Morss, L.R. (1983) Thermochemical regularities among lanthanide and actinide oxides. *J. Less Common Met.*, **93** (2), 301–321.
235 Kanke, Y. and Navrotsky, A. (1998) A calorimetric study of the lanthanide aluminum oxides and the lanthanide gallium oxides: stability of the perovskites and the garnets. *J. Solid State Chem.*, **141** (2), 424–436.
236 Bale, C.W., Pelton, A.D., Thompson, W.T., Eriksson, G., Hack, K., Chartrand, P., Decterov, S., Jung, I.-H., Melancon, J., and Petersen, S. (1976) 2009 Factsage™, copyright Thermfact and GTT-Technologies, C.R.C.T., Ecole Polytechnique de Montreal.
237 Neumann, A. and Walter, D. (2006) The thermal transformation from lanthanum hydroxide to lanthanum hydroxide oxide. *Thermochim. Acta*, **445**, 200–204.
238 Shirsat, A.N., Ali, M., Kaimal, K.N.G., Bharadwaj, S.R., and Das, D. (2003) Thermochemistry of $La_2O_2CO_3$. *Thermochim. Acta*, **399**, 167–170.
239 Kishimoto, H., sakai, N., Horita, T., Yamaji, K., Brito, M.E., and Yokokawa, H. (2007) Cation transport behaviour in SOFC cathode materials of $La_{0.8}Sr_{0.2}FeO_3$ with perovskite structure. *Solid State Ionics*, **178**, 1317–1325.
240 Horita, T., Ishikawa, M., Yamaji, K., Sakai, N., Yokokawa, H., and Dokiya, M. (1998) Cation diffusion in $(La,Ca)CrO_3$ perovskite by SIMS. *Solid State Ionics*, **108**, 383–390.

10
Oxygen- and Hydrogen-Permeable Dense Ceramic Membranes
Jay Kniep and Jerry Y.S. Lin

Significant research over the past 30 years has focused on dense mixed ionic–electronic conducting oxide membranes, which can selectively transport oxygen ions or protons. This chapter provides a concise review of the different structures, synthesis and experimental methods, transport properties, doping strategies to improve chemical and mechanical stability, gas permeation under different conditions, and the use of dense ceramic membranes for gas purification or in a catalyzed membrane reactor system. Gas permeation models for various cases and the effect of a reducing gas in the downstream gas on the permeation rate through membranes are emphasized. Current challenges and future directions for dense mixed conducting membrane research are briefly discussed.

10.1
Introduction

Dense ceramic membranes with mixed ionic–electronic conducting characteristics that can selectively transport oxygen or hydrogen continue to attract significant interest [1–3]. These materials transport oxygen ions or protons as well as electrons and electron holes while maintaining electroneutrality (see Chapters 3 and 12 of the first volume). The driving force for ionic transport through the membranes is an oxygen or hydrogen partial pressure difference on either side of the membrane that introduces a chemical potential gradient across the membrane. As these materials transport oxygen or hydrogen in ionic form, the theoretical selectivity is 100%. There is no need for external electrodes and circuits due to the ability of electrons to transport through the materials. Ceramic materials are stable at elevated temperatures (600–1000 °C), so ceramic membranes can be used in environments that are detrimental to organic, silica, or palladium membranes.

Industrial applications of mixed ionic–electronic conducting membranes include gas separation, sensor materials, fuel cell materials, and catalyzed membrane reactors with chemical reactions. For industrial use, the membrane materials must meet a number of stringent requirements. The materials need to have economical

Solid State Electrochemistry II: Electrodes, Interfaces and Ceramic Membranes.
Edited by Vladislav V. Kharton.
© 2011 Wiley-VCH Verlag GmbH & Co. KGaA. Published 2011 by Wiley-VCH Verlag GmbH & Co. KGaA.

oxygen or hydrogen permeation values. Mechanical strength is required due to high pressures or vacuum used in industry to increase the driving force for ionic transport (see Chapter 9). Chemical stability under a variety of oxidizing and reducing environments resulting in reliable performance for long-term operation is crucial. For use in industry, the material will be produced on a large scale, so the cost of the materials should also be considered [1–3].

This chapter will provide a concise review of the progress made in mixed ionic and electronic conducting membranes over the past 30 years. The structure of different mixed conducting materials will be introduced. Doping strategies to improve the electrical and ionic transport properties, mechanical strength, and chemical stability of the membranes are summarized. The effect of various sweep gases on the permeation properties of the membranes will be discussed. The chapter concludes by giving an overview of applications of membranes, including gas separation and membrane reactors.

10.2
Structure of Membrane Materials

All membranes discussed in this chapter are either pure ionic conductors or mixed ionic (oxygen or protonic) and electronic conductors. In order for ions to transport through the dense membrane, a chemical potential gradient (via gas partial pressure environments) across the membrane must be present. As oxygen (as O^{2-}) or hydrogen (as H^+) species transport through the membrane in their ionic form, electrons are also transported through the membrane in order to maintain the electroneutrality of the membrane. The use of a second ceramic or metal phase with excellent electron conduction properties has been combined with pure ionic conductors in order to produce dual-phase membranes that are mixed ionic–electronic conductors, without the need for external circuits [1–3]. Figure 10.1 schematically shows the mechanism of mixed ionic–electronic conducting and dual-phase membranes for both oxygen and hydrogen separation.

The ability of any ceramic membrane to selectively transport oxygen or hydrogen is directly related to the material properties of the membrane. The properties of the materials, in turn, depend not only on the chemical composition but also on the material structure. General electrochemical properties of some common structures discussed in this section are shown in Table 10.1.

10.2.1
Fluorite Structure

The ideal fluorite structure, based on the mineral CaF_2, is a cubic structure with the general formula AO_2. The unit cell has cations in a simple face-centered cubic packing containing a cube of anions. The cations have a coordination number of 8 and the anions have a coordination number of 4. Detailed discussion on the fluorite structure type can be found in Chapter 2 of the first volume.

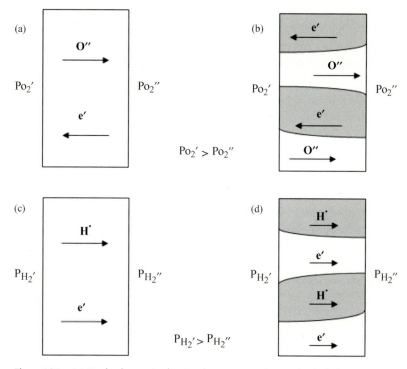

Figure 10.1 (a) Single-phase mixed ionic–electronic membrane; (b) dual-phase mixed ionic–electronic membrane; (c) single-phase protonic–electronic membrane; (d) dual-phase protonic–electronic membrane.

10.2.2
Perovskite Structure

The perovskite structure, based on the mineral $CaTiO_3$, is ideally a cubic structure having the general formula ABO_3 with A having the larger cation radii of the two metals (see Chapter 2 of the first volume). The unit cell can be described as A-site cations at all eight corners, O^{2-} ions located at the center of each of the cell face, and a single B-site cation at the center of the cell. Therefore, the A-site cation has a

Table 10.1 List of mixed ionic–electronic structures and their typical properties.

Structure	Ideal crystal structure	Electronic conductivity range (S cm^{-1})	Ionic conductivity range (S cm^{-1})	Ionic transference number range
Perovskite	Cubic	10–1000	0.01–0.1	10^{-4}–0.01
Fluorite	Cubic	10^{-6}–10^{-4}	10^{-4}–1	~1
Brownmillerite	Orthorhombic	10^{-5}–10^{-3}	10^{-4}–10^{-3}	0.1–1

coordination number of 12 and the B-site cation has a coordination number of 6. However, since the perovskite structure can have various representative compositional combinations such as $A^{1+}B^{5+}O_3$, $A^{2+}B^{4+}O_3$, or $A^{3+}B^{3+}O_3$, the number of different cations and their respective atomic radii that can be incorporated into the structure is numerous. With the wide range of cation combinations available for the perovskite structure, the actual unit cells are orthorhombic instead of cubic. The symmetry of the structures can be described by the tolerance factor (t), which quantifies the extent to which the perovskite structure varies from ideal cubic structure (where $t = 1$) and is defined as

$$t = \frac{(R_A + R_O)}{\sqrt{2}(R_B + R_O)}, \quad (10.1)$$

where R_A, R_B, and R_O are the respective ionic radii for ABO_3. The perovskite structure is stable if the tolerance factor is between 0.75 and 1 [4].

10.2.3
$Sr_4Fe_6O_{13\pm\delta}$ and Its Derivatives

A high permeability of $SrFeCo_{0.5}O_x$ was reported first by Balachandran et al. [5]. It was first described as a compound with a layered structure containing perovskite-like layers formed from FeO_6 octahedra sharing four equatorial vertices and rock salt-like layers formed from polyhedral with vertex and/or edge sharing, as for $Sr_4Fe_6O_{13\pm\delta}$ [6]. It was later found that $SrFeCo_{0.5}O_x$ is a phase mixture containing an intergrowth $Sr_4Fe_{6-x}Co_xO_{13+\delta}$, perovskite $SrFe_{1-x}Co_xO_{3-\delta}$, and spinel $Co_{3-x}Fe_xO_4$ or rock salt CoO phases [7, 8]. $Sr_4Fe_{6-x}Co_xO_{13+\delta}$ is a single-phase intergrowth phase at low levels of Co doping, comprises the intergrowth, perovskite, and spinel phases with intermediate Co doping, and consists of only a perovskite and spinel phase at high Co doping levels. The amount of the three phases present in the mixed phase compositions depends on the powder synthesis method, sintering temperature, and thermal history of the membrane [9, 10].

10.2.4
Brownmillerite and Other Perovskite-Related Structures

The ideal brownmillerite composition is $A_2B_2O_5$ and the orthorhombic structure is based on the mineral of the same name (Ca_2AlFeO_5); see also Chapter 2 of the first volume. The brownmillerite structure can be thought of as a series of highly defective perovskite oxides, where one-sixth of the oxide ions have been removed ($ABO_{2.5}$). The oxygen vacancies are ordered along the [101] direction resulting in alternating layers of BO_6 octahedra and BO_4 tetrahedra perpendicular to the b-axis [11, 12].

Another general type of structure is the Ruddlesden–Popper series ($A_{n+1}B_nO_{3n+1}$) of perovskite-related intergrowth oxides [13–15]. Ruddlesden–Popper-type materials consist of a layered structure in which AO–BO_2–AO blocks (perovskite blocks) alternate with either $B_2O_{2.5}$ blocks or AO rock salt blocks.

The alternating block depends on the value of n, which is the ratio of the number of perovskite blocks to that of the rock salt blocks.

While the perovskite structure ($n = \infty$) is one extreme of the Ruddlesden–Popper series, the other extreme ($n = 1$) is known as K_2NiF_4-type structures. The K_2NiF_4 (or A_2BO_4) structure consists of a series combination of perovskite layers alternating with AO rock salt layers [16, 17]. Similar to other intergrowth and mixed phase structures, oxygen ionic conduction in K_2NiF_4 structures can occur via vacancy hopping in the perovskite layers and via diffusion of interstitial oxygen in the rock salt layers.

10.2.5
Dual-Phase Membranes

While some materials are excellent ionic conductors and others have extremely high electronic conductivity, it is hard to identify a single-phase material that exhibits high ionic and electronic conductivity. Therefore, the idea of combining an ionic conducting phase with an electronic conducting phase in a single composite membrane has been studied extensively. Each phase would have to be continuous throughout the membrane in order for the ions or electrons to be transported throughout the membrane. For oxygen-conducting dual-phase membranes, both ceramic–metallic [18–20] and ceramic–ceramic [21–24] composite membranes have been studied. For proton-conducting membranes, where electronic conduction is limiting, various ceramic–metallic dual-phase membranes have been reported [25–27]. A review on the relevant microstructural aspects and other factors governing performance of the composite membranes is found in Chapter 11.

10.3
Synthesis and Permeation Experimental Methods

Various synthesis methods can be used to obtain mixed oxide ceramic powders. Some of the more common methods include sol-gel [28], coprecipitation [29–31], hydrothermal [32], and spray pyrolysis [33]. All these methods can result in the production of ceramic powders with good homogeneity and uniform, nanosized particles. However, the synthesis methods most extensively used are the solid-state reaction [8, 34, 35] and EDTA/citrate complex methods [36]; see also Chapter 9.

The solid-state reaction method consists of mechanically mixing and grinding the metal precursors. Suitable precursors include, but are not limited to, oxides, acetates, carbonates, chlorides, or nitrates of the desired metal ions. The precursors are usually mixed with a solvent during the grinding process. After grinding, the slurry is dried, ground again, and then calcined at a temperature high enough to thermally decompose the precursors and form the desired structure. As this process requires calcination at high temperatures for extended periods of time, it could be considered energy intensive. The combination of ball milling and calcining for long periods of time can lead to low powder purity and variable particle sizes.

The EDTA/citrate complex method is a wet method in which the desired cations are placed in a solution with a chelating agent, such as citric acid, ethylenediaminetetraacetic acid (EDTA), or glycine. During an isothermal reflux period, the metal ions are chelated with the chelating agent (such as citric acid) through the −OH and −COOH groups and act as bridges between the citric acid molecules to form polymeric molecules [37]. This is followed by evaporation of the water solvent, which leads to −COH and HOC− condensation and formation of a cross-linked, gel-like substance. The resulting gel is dried in air, self-ignited, grinded, and then calcined. The final powder product has high chemical homogeneity and is of very high purity.

Powders from the above-mentioned methods are then used to produce dense membranes of various geometries. Symmetrical disk membranes are produced by placing powder in a metal dye, pressed to a high pressure (150 MPa or more), and then sintered at a high temperature (1100–1700 °C). Symmetrical tubular or hollow fiber membranes are produced via extrusion by mixing the powder with a solvent and plasticizer and then sintering the extruded tube or fiber at an elevated temperature. Asymmetric disk membranes (thin dense layer supported by a porous support) can be made by a variety of methods including slip or tape casting, spin coating, and/or a dry pressing method [38, 39]. More recently, thin-walled hollow tube fibers have been produced using a phase inversion spinning technique [40, 41].

In order to measure the gas permeation flux through a dense membrane in a nongalvanic mode, a partial pressure gradient of the gas must be present over the cross section of the membrane. In order for the partial pressure gradient to be present, a good seal must be maintained. Therefore, for high-temperature gas permeation experiments, the sealing of the membrane is the most crucial step. The schematic of a generic high-temperature (700–900 °C) gas permeation setup can be found in Figure 10.2. In this figure, a gas-tight membrane and metal seal were

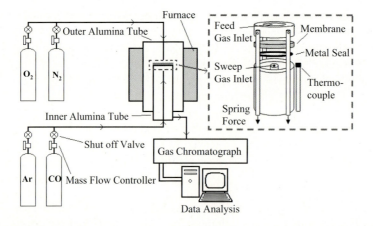

Figure 10.2 Experimental high-temperature oxygen permeation setup.

mounted on the inner alumina tube and held in place by spring pressure applied by the alumina spacer on top of the membrane. When a metal (usually silver or gold) seal is used as the sealing material, the system is heated up to the softening temperature of the metal. Then, flowing N_2 and Ar are introduced on either side of the membrane, respectively. As the metal softens, the dense membrane is pressed into the seal due to the spring force. The amount of N_2 in the Ar stream (and, therefore, the leakage rate through the seal) can be determined by running gas samples through a gas chromatographer with a packed column and a TCD detector. Once the N_2 content in the sweep stream was minimized, the system could be ramped down to experimental conditions for gas permeation experiments. Metal seals have been shown to be able to maintain a gastight seal under both oxidizing and reducing environments for extended periods of time [42, 43].

Another popular sealing method for high-temperature gas permeation experiments is the use of glass or glass–ceramic seals. Lin and coworkers [44] developed a ceramic–glass composite mixture that can be tailored to individual membrane properties. The properties include thermal expansion, wettability to the ceramic support and membrane, and chemical inertness. Generally, the ceramic–glass composite mixture contains 40–50 wt% membrane material powder, 20–50 wt% Pyrex glass, and 5–20 wt% additive, such as sodium aluminate or boron oxide. In order to seal with the ceramic–glass composite mixture, the powder is mixed with water to form a paste. The paste is then placed on the supporting alumina tube and the dense membrane is pressed into the paste. More paste is applied to plug the gaps and then the system is dried at room temperature for 2 h. Finally, the system is heated to the softening temperature of Pyrex, held there for 30 min, and cooled down to experimental conditions.

10.4
Gas Permeation Models

Oxygen or hydrogen permeation through a dense mixed conducting ceramic membrane is determined by several mass and electronic transport steps in series: the surface reaction step on the upstream membrane surface including several substeps of molecule adsorption and dissociation and charge transfer reaction, diffusion of at least two charged species (oxygen anion, vacancy or proton, electron and/or electron hole) in the membrane bulk, and the surface reaction on the downstream membrane surface including several substeps similar, but in a reversed direction, to that in the upstream membrane surface [45]. For transport of the two charged species through the bulk phase, the molar flux of the given species (oxygen anions, oxygen vacancies, or protons) can be described by [45–47]

$$J_1 = -\frac{\sigma_1 \sigma_2}{(\sigma_1 + \sigma_2)z_1^2 F^2}\left[\nabla\mu_1 - \frac{z_1}{z_2}\nabla\mu_2\right], \qquad (10.2)$$

where F is Faraday constant, and σ_i, z_i, and $\nabla\mu_i$ are the partial conductivity, charge number, and chemical potential gradient of species i, respectively (here 1 stands for oxygen vacancy or proton and 2 for electron or electron hole; $z_1 = 1$ for proton or -2 for oxygen ion; and $z_2 = -1$ for electron or 1 for electron hole). A brief analysis of the relevant formalism can be found in Chapter 3 of the first volume. According to the Wagner theory, a local equilibrium is assumed between the oxygen vacancy or proton (charged species 1), electron or hole (charged species 2), and neutral molecular oxygen or hydrogen. Thus,

$$\frac{1}{2}\nabla\mu_{X_2} = \nabla\mu_1 - \frac{z_1}{z_2}\nabla\mu_2 \tag{10.3}$$

with X being either O (oxygen) or H (hydrogen). Inserting Equation 10.3 into Equation 10.2 gives the molecular flux of hydrogen (H_2) or oxygen (O_2) through the membranes:

$$J_{X_2} = -\frac{\sigma_1\sigma_2}{4z_1^2(\sigma_1 + \sigma_2)F^2}\nabla\mu_{X_2} \tag{10.4}$$

At steady state and assuming that the bulk diffusion is the rate-limiting step, the oxygen or hydrogen permeation flux can be written in the integration form

$$J_{X_2} = \frac{RT}{4z_1^2 F^2 L}\int_{P''_{X_2}}^{P'_{X_2}} \frac{\sigma_1\sigma_2}{(\sigma_1 + \sigma_2)}\,d\ln P_{X_2}, \tag{10.5}$$

where L is the membrane thickness and P_{X_2} is the oxygen or hydrogen partial pressure with ′ and ″ indicating upstream and downstream pressures, respectively. Equation 10.5 can be simplified if the electrical conductivity for one charged species is much larger than that for the other. For example, for many perovskite-type mixed conducting membranes, the electronic conductivity (σ_2) is much larger than the oxygen ionic conductivity (σ_1) (see Table 10.1), and, in this case, the oxygen permeation flux equation (10.5) is simplified to

$$J_{O_2} = \frac{RT}{4z_1^2 F^2 L}\int_{P''_{O_2}}^{P'_{O_2}} \sigma_1\,d\ln P_{O_2}. \tag{10.6}$$

The partial conductivity is related to the concentration C_i and diffusivity D_i (or mobility) of the charged species by the Nernst–Einstein equation. In particular,

$$\sigma_i = \frac{z_i^2 F^2}{RT}C_i D_i. \tag{10.7}$$

In the literature, the diffusivity is often assumed to be constant and the major efforts were devoted to obtaining the relationship between the concentrations of the given defects (such as the oxygen vacancy) by means of the defect equilibrium thermodynamics [48]. Inserting the equilibrium relationship of the defect concentration versus oxygen or hydrogen partial pressure into Equation 10.6 with the help of Equation 10.7 and integrating results in the final permeation flux equation, which corrects the oxygen or hydrogen permeation flux to upstream and downstream pressures, membrane thickness, conductivity at a reference state (usually 1 atm), and the parameters for the defect equilibrium.

The major deficiency of the above approach is that it is difficult to incorporate the bulk flux equation (Equation 10.5) with the surface reaction equations for membranes for which bulk diffusion is not rate limiting. This deficiency was addressed by Lin et al. [45] who, following the ambipolar diffusion approach, obtained the following flux equation for oxygen permeation through mixed conducting membranes:

$$J_{O_2} = \frac{1}{2L}\int_{C_1'}^{C_1''} \frac{(C_2+4C_1)D_2 D_1}{C_2 D_2+4C_1 D_1} dC_1, \qquad (10.8)$$

where the indices 1 and 2 correspond to two types of the migrating species. For membranes with oxygen ionic conductivity much smaller than the electronic conductivity, such as perovskite-type membranes, Equation 10.8 is integrated into the solution of the first Fick's law:

$$J_{O_2} = \frac{D_1}{2L}(C_1''-C_1'). \qquad (10.9)$$

Equation 10.9 correlates the oxygen permeation flux with the oxygen vacancy concentration in the membrane on both the upstream and downstream membrane surfaces. If the bulk diffusion is rate limiting, the surface defect concentration can be correlated with the oxygen partial pressure in the gas phase via the gas–solid defect equilibrium. For example, it is often useful to use the following power law to correlate the oxygen vacancy (denoted by the subscript V here for clarity) with the oxygen partial pressure in the gas phase for the perovskite-type membranes:

$$C_V = C_V^o P_{O_2}^{-n}, \qquad (10.10)$$

where C_V^o is the oxygen vacancy concentration at $P_{O_2} = 1$ atm and n is the power coefficient indicating the creation speed of oxygen vacancy with the change of ambient oxygen partial pressure. Inserting Equation 10.10 into Equation 10.9 and using Equation 10.7 yields

$$J_{O_2} = \frac{RT\sigma_i^o}{8F^2 L}(P_{O_2}^{''-n} - P_{O_2}^{'-n}). \qquad (10.11)$$

Equation 10.8 can be easily used to derive flux equations for membrane with the bulk diffusion not being rate limiting. Rate equations for the surface reactions can be developed on the basis of a proposed surface reaction mechanism and these equations can be combined with Equation 10.8 to give explicit or implicit flux equation correlating oxygen permeation flux with transport properties of the membrane phase and surface reactions. Lin et al. [45] demonstrated this approach and obtained oxygen permeation flux equation for thin fluorite-structured membranes for which both bulk diffusion and surface reaction are important. Xu and Thomson [49], following the same approach, obtained the following oxygen permeation flux equation for perovskite-type membranes considering surface reaction resistance:

$$J_{O_2} = \frac{(k_r/k_f)(1/(p''_{O_2})^{0.5} - 1/(p'_{O_2})^{0.5})}{1/k_f(p'_{O_2})^{0.5} + (2L/D_V) + 1/k_f(p''_{O_2})^{0.5}}, \qquad (10.12)$$

where k_r and k_f are the rate parameters for the surface reaction. The above flux equations, (10.11) and (10.12), have often been used to describe experimental permeation data through perovskite-type membranes and, in many cases, they agree well with the experimental data. They clearly show how the oxygen permeation flux depends on the membrane thickness, transport properties, and operation conditions.

In summary, the theoretical framework for gas permeation through mixed conducting ceramic membranes is now available and many flux equations for oxygen permeation through mixed conducting membranes have been derived and reported in the literature. These equations were obtained for the cases with the partial conductivity for one charged species much larger than the other and, even for these special cases, the flux equations can be fairly complex and highly nonlinear with respect to the driving force due to the complex correlations for the defect equilibrium and surface reactions. In comparison, very limited work has been reported for flux equations for oxygen permeation through mixed conducting membranes with comparable conductivities for both of the charged species and for hydrogen permeation through mixed conducting membranes for which protonic, electronic, and electron hole conductivities are all important. In all of these more complex cases, understanding the oxygen or hydrogen permeation relies entirely on experimental studies.

10.5
Characteristics of Oxygen-Permeable Membranes

10.5.1
Electrical and Ionic Transport Properties

Generally for mixed oxygen ion electronic conducting materials, the charge carriers are oxygen ions (oxygen vacancies), electrons, and electron holes. Oxygen reacts on the surface of these materials and the oxygen ions are absorbed into and transferred through the membrane via oxygen vacancies and vacancy hopping [50]. As oxygen vacancies are crucial to oxygen ion transport, various doping strategies to introduce defects have been developed for different structures. However, for mixed phase membranes and layered structures, interstitial oxygen ions can also diffuse through the rock salt-like layers [16, 51]. Common techniques to measure the total or partial conductivity of a material include the four-point DC method [52] and AC impedance spectroscopy [53, 54]. Table 10.2 lists the total conductivity of various mixed oxygen ionic and electronic conductors at 900 °C in an air atmosphere.

For the perovskite structure ABO_3, materials with high total conductivity are, generally, heavily doped and take the form $A_{1-x}A'_xB_{1-y}B'_yO_{3-\delta}$, where $0 \leq x \leq 1$ and $0 \leq y \leq 1$. The parameter δ signifies the oxygen nonstoichiometry of the material or

10.5 Characteristics of Oxygen-Permeable Membranes

Table 10.2 Total conductivity of various mixed ionic–electronic conductors at 900 °C in air.

Composition	Structure	Total conductivity (S cm^{-1})	Reference
$La_{0.6}Sr_{0.4}Co_{0.2}Fe_{0.8}O_{3-\delta}$	Perovskite	252	[186]
$La_{0.2}Sr_{0.8}Co_{0.8}Fe_{0.2}O_{3-\delta}$	Perovskite	310	[186]
$La_{0.6}Ba_{0.4}Co_{0.2}Fe_{0.8}O_{3-\delta}$	Perovskite	123	[186]
$La_{0.2}Ba_{0.8}Co_{0.8}Fe_{0.2}O_{3-\delta}$	Perovskite	19	[186]
$BaCe_{0.15}Fe_{0.85}O_{3-\delta}$	Perovskite	32	[187]
$La_{0.3}Sr_{0.7}Fe_{0.8}Ga_{0.2}O_{3-\delta}$	Perovskite	40	[188]
$(La_{0.75}Sr_{0.25})_{0.95}Cr_{0.5}Mn_{0.5}O_{3-\delta}$	Perovskite	17	[189]
$SrCo_{0.8}Fe_{0.2}O_{3-\delta}$	Perovskite	158	[130]
$SrCo_{0.8}Fe_{0.1}Sn_{0.1}O_{3-\delta}$	Perovskite	81	[130]
$Ce_{0.6}Gd_{0.2}Pr_{0.2}O_{2-\delta}$	Fluorite	0.158	[30]
7.8 mol% Sc_2O_3–ZrO_2	Fluorite	0.15	[57]
8 mol% Y_2O_3–ZrO_2	Fluorite	0.135	[57]
3.4 wt% MgO–ZrO_2	Fluorite	0.0048	[57]
$[(ZrO_2)_{0.6}(CeO_2)_{0.4}]_{0.9}(CaO)_{0.1}$	Fluorite	1.6×10^{-6}	[68]
$SrFeCo_{0.5}O_x$	Mixed phase	23	[72]
$SrCoFeO_x$	Mixed phase	16.6	[42]
$Sr_4Fe_{4.8}Co_{1.2}O_{13\pm\delta}$	Mixed phase	9.5	[8]
$Sr_4Fe_{4.2}Co_{1.8}O_{13\pm\delta}$	Mixed phase	17	[92]
$Sr_4Fe_6O_{13\pm\delta}$	Intergrowth	0.65	[92]
$Sr_{1.6}La_{0.4}Fe_2O_{5.2}$	Brownmillerite	67.4	[82]
$Sr_{1.7}La_{0.3}GaFeO_{5.5}$	Brownmillerite	0.22	[82]
$Sr_{1.8}La_{0.2}Ga_{0.2}Fe_{1.8}O_{5.1}$	Brownmillerite	13.5	[82]

the amount of oxygen vacancies present. The oxygen nonstoichiometry of a material depends on the temperature and the oxygen partial pressure of the environment that the material is in. Lanthanides (most commonly lanthanum) or alkaline earth metals (such as barium, calcium, or strontium) typically occupy the A-site. The choice of the metal for the A-site has some effect on the conductivity properties of the material, but changing the ratio of the lanthanide and alkaline earth metal at the A-site has a much larger effect on the oxygen vacancy concentration of the material. This is due to the fact that the substitution of a +2 metal ion (such as barium, calcium, or strontium) for a +3 metal ion (lanthanum) results in a charge imbalance for the material. As these materials maintain electroneutrality, the difference in the charge between the metal cations and oxygen anions is equilibrated through a combination of the creation of oxygen vacancies and a change in the valence of some of the transition metals occupying the B-site [55]; see also Chapters 3 and 9 of the first volume.

While appropriate doping at the A-site increases the ionic conduction of the perovskite materials due to the introduction of oxygen vacancies, doping at the B-site may also increase their electronic conductivity. Electronic conduction occurs in the transition metal-containing perovskites via electron transfer between B^{n+}–O–$B^{(n+1)}$ pairs; therefore, transition metal cations that can have different valence numbers are attractive [55]. Iron is a good example of a transition metal that can be present in the +2, +3, or +4 oxidation states. Other transition metals that are commonly doped into perovskites include cobalt, aluminum, copper, nickel, titanium, magnesium, chromium, manganese, gallium, and zirconium. The choice of the transition metal can also have a dramatic effect on other material properties, such as thermal expansion coefficient, mechanical strength, and chemical stability.

While the total conductivity of fluorite-based structures is much lower than that of transition metal-containing perovskites, the oxygen ionic conductivity of various fluorite materials is comparable to, or in some cases higher than, perovskites (see Chapter 9 of the first volume). Zirconia and bismuth oxides are two fluorite materials that have attracted considerable attention due to their high ionic conductivity. Depending on the temperature, zirconia can be cubic, tetragonal, or monoclinic in structure. The cubic phase, which is present at high temperatures, has the highest oxygen ionic conductivity and is of most interest for oxygen transport purposes [56]. The cubic structure for zirconia can be stabilized at room temperature by doping with CaO, MgO, Y_2O_3, or Sc_2O_3 [57]. Bismuth oxide in its cubic (δ) form also has very high oxygen ionic conductivity at high temperatures due to high oxygen vacancy concentration in the material [58]. However, bismuth oxide has a relatively low melting point of 830 °C and a dramatic phase transition at 730 °C, which leads to a drastic volume change and mechanical strains [59]. The δ phase of bismuth oxide can be stabilized at room temperature by doping with Y_2O_3 or Er_2O_3 [59–61]. Bismuth oxide doped with SrO, CaO, or BaO may form a stabilized rhombohedral structure that also has high oxygen ionic conductivity [62–64].

While the stabilized zirconias and bismuth oxides still have high oxygen ionic conductivity, the materials are still primary ionic conductors and have minimal electronic conductivity. In order to make them mixed oxygen ionic–electronic conductors, additional doping must be added to the systems. Doping zirconia with TbO_2, TiO_2, and CeO_2 leads to the formation of variable valence redox couples and to higher electronic transport [65–68]. Doping of Sm_2O_3 into yttria-stabilized bismuth oxide has the same effect and increases the electronic conductivity [69–71]. Besides higher electronic conduction, doping with Sm_2O_3 leads to a material with good catalytic properties and a higher melting point than bismuth oxide.

While $SrFeCo_{0.05}O_x$ has a total conductivity of $\sim$23 S cm^{-1} at 900 °C in air, the oxygen ionic conductivity is $\sim$9 S cm^{-1}, which leads to an ionic transference number of 0.4 [72, 73]. The perovskite layer in $Sr_4Fe_6O_{13}$-type lattice serves as the primary conducting path for the electronic charge carriers while the rock salt layers may provide the main paths for the oxygen ions. Nonetheless, the intergrowth structure $Sr_4Fe_6O_{13+\delta}$ has rather low conductivity properties [74]. Isostructural $Sr_4Fe_{6-x}Co_xO_{13\pm\delta}$ materials have higher conductivity values than the undoped intergrowth composition, yet the conductivity values are still roughly an order of magnitude lower than mixed phase materials that contain an intergrowth, perovskite, and rock salt phases [7, 8]. A good list of conductivity and oxygen chemical diffusion coefficient values can be found in Ref. [75].

Brownmillerite-based structures tend to have oxygen ionic conductivity comparable to perovskites, but their total conductivity values are lower [76–79]. A good review of oxygen ionic conducting brownmillerite materials can be found elsewhere [80]. Although the concentration of oxygen vacancies in the brownmillerite structure is formally high, oxygen mobility is reduced as the vacancies are locked in the tetrahedral layers [81]. Small deviations from the ideal oxygen stoichiometry of the brownmillerite structure can lead to mobile vacancies in the octahedral layers [11]. The concentration of the mobile vacancies does not exceed, however, 0.02 per formula unit, which is an order of magnitude lower than the oxygen nonstoichiometry in perovskites [81]. However, a number of patents [82–84] contain oxygen ionic and electronic conductivity data for some brownmillerite compositions that are comparable to or higher than those of perovskite materials. The reason for the high transport properties of these brownmillerite materials is still unclear.

10.5.2
Oxygen Permeation under Air/Inert Gas Gradients

The pioneering work for oxygen permeation through dense perovskite membranes was performed by Teraoka et al. in the mid-1980s [85–87]. The oxygen permeation was measured through various 1.0 mm thick $La_{1-x}Sr_xCo_{1-y}Fe_yO_{3-\delta}$ membranes, with the highest flux being 3.2 ml cm^{-2} min^{-1} for $SrCo_{0.8}Fe_{0.2}O_{3-\delta}$ at 850 °C. The onset temperature of oxygen permeation lowered and the rate of oxygen permeation at a fixed temperature increased with an increase in the value of x. The substitution of Co^{3+} for Fe^{3+} also increased the oxygen permeation because Co^{3+} has a smaller ionic radius and lower bonding energy to oxide ions than Fe^{3+} [86]. With those trends, the higher oxygen permeation of $SrCo_{0.8}Fe_{0.2}O_{3-\delta}$ than $SrCoO_{3-\delta}$ at 850 °C emphasizes the importance of the perovskite structure with minimum distortions and ordering phenomena. While $SrCo_{0.8}Fe_{0.2}O_{3-\delta}$ is a perovskite phase, $SrCoO_{3-\delta}$ forms brownmillerite phase at 850 °C and does not transform into perovskite until ~900 °C [88]. Other cation substitution experiments showed that the oxygen permeation increased for B-site substitution as Cu > Ni > Co > Fe > Cr > Mn, A-site substitution as Ba > Ca > Sr > Na, and the A-site rare earth cation as Gd > Sm > Nd > Pr > La [87]. On the basis of these results, the optimal material was reported to be $Gd_{0.2}Ba_{0.8}Co_{0.7}Fe_{0.1}Cu_{0.2}O_{3-\delta}$, with an oxygen permeation flux of 5.5 ml cm^{-2} min^{-1} through a 1.5 mm thick membrane at 870 °C.

Although the work by Teraoka's group introduced a series of materials with good oxygen permeation properties, some experimental results have not been reproduced by other research groups. Even though various experimental parameters as well as synthesis and sintering of the material can cause some discrepancy in the oxygen permeations reported, there have been an order of magnitude differences in oxygen permeation values between different research groups [88]. Therefore, the focal point of research over the past 20 years has been to find a combination of dopants that yield a perovskite material that has high oxygen permeation and thermal and chemical stability and is mechanically strong under experimental conditions. The oxygen permeation for various perovskites and other compositions under an air and inert gas oxygen partial pressure difference can be found in Table 10.3.

Ma et al. [8] have reported oxygen permeation through a 2.9 mm thick $SrFeCo_{0.5}O_x$ membrane to be 0.75 ml cm^{-2} min^{-1} at 900 °C. Again, this value has not been reproduced in the literature and other research groups report oxygen permeation data

Table 10.3 Oxygen permeation flux through mixed ionic–electronic conducting membranes at 900 °C with an air and inert gas oxygen partial pressure gradient.

Composition	Structure	Geometry	Thickness (mm)	Oxygen flux (ml cm^{-2} min^{-1})	Reference
$Ba_{0.5}Sr_{0.5}Zn_{0.2}Fe_{0.8}O_{3-\delta}$	Perovskite	Disk	1.45	0.28	[126]
$La_{0.6}Sr_{0.4}Co_{0.8}Fe_{0.2}O_{3-\delta}$	Perovskite	Disk	0.98	0.147	[55]
$BaCe_{0.15}Fe_{0.85}O_{3-\delta}$	Perovskite	Disk	1	0.418	[122]
$La_{0.4}Ba_{0.6}Co_{0.2}Fe_{0.8}O_{3-\delta}$	Perovskite	Disk	2	0.48	[186]
$La_{0.2}Sr_{0.8}Co_{0.8}Fe_{0.2}O_{3-\delta}$	Perovskite	Disk	2	0.80	[186]
$Ba_{0.5}Sr_{0.5}Co_{0.8}Fe_{0.2}O_{3-\delta}$	Perovskite	Disk	1.5	1.4	[113]
$SrCo_{0.8}Fe_{0.2}O_{3-\delta}$	Perovskite	Disk	1	0.26	[88]
$SrCo_{0.8}Fe_{0.2}O_{3-\delta}$	Perovskite	Disk	1	1.1	[131]
$La_{0.8}Sr_{0.2}Ga_{0.7}Fe_{0.3}O_{3-\delta}$	Perovskite	Disk	0.5	1.05	[190]
$SrCo_{0.4}Fe_{0.5}Zr_{0.1}O_{3-\delta}$	Perovskite	Disk	0.8	0.93	[134]
$La_{0.2}Ba_{0.8}Co_{0.8}Fe_{0.15}Zr_{0.05}O_{3-\delta}$	Perovskite	Disk	1	1.83	[191]
$La_{0.85}Ce_{0.1}Ga_{0.3}Fe_{0.65}Al_{0.05}O_{3-\delta}$	Perovskite	Disk	1	0.21	[124]
$La_{0.2}Ba_{0.8}Co_{0.8}Fe_{0.2}O_{3-\delta}$	Perovskite	Disk	2	1.4[a]	[192]
$La_{0.2}Ba_{0.8}Co_{0.8}Fe_{0.2}O_{3-\delta}$	Perovskite	Disk	0.7	5[b]	[164]
$Bi_{1.5}Y_{0.5}O_3$	Fluorite	Disk	1.4	0.021	[193]
$Bi_{1.5}Y_{0.3}Sm_{0.2}O_3$	Fluorite	Disk	1.28	0.029	[20]
$GdCe_{0.8}Pr_{0.2}O_{2-\delta}$	Fluorite	Disk	1	0.033	[30]
$SrFeCo_{0.5}O_x$	Mixed phase	Disk	2.9	0.75	[8]
$SrFeCo_{0.5}O_x$	Mixed phase	Disk	0.97	0.073	[90]
$SrFeCo_{0.5}O_x$	Mixed phase	Disk	1.2	0.006	[42]
$Sr_4Fe_4Co_2O_{13\pm\delta}$	Mixed phase	Disk	1.8	0.0088	[9]
$SrFe_{1.125}Co_{0.375}O_x$	Mixed phase	Disk	1.2	0.0062	[194]
$Sr_4Fe_6O_{13\pm\delta}$	Intergrowth	Disk	1.85	0.00058	[92]
$Sr_4Fe_{3.4}Co_{2.6}O_{13\pm\delta}$	Mixed phase	Disk	1.5	0.029	[93]
$SrCoFeO_x$	Mixed phase	Disk	1.2	0.018	[42]
$Sr_{1.7}La_{0.3}Ga_{0.6}Fe_{1.4}O_{5+\delta}$	Brownmillerite	Tube	1	0.751	[83]
$Sr_{1.6}La_{0.4}Ga_{0.6}Fe_{1.3}Co_{0.1}O_{5+\delta}$	Brownmillerite	Tube	1	0.231	[83]
$Sr_{1.7}La_{0.3}Ga_{0.6}Fe_{1.1}Co_{0.3}O_{5+\delta}$	Brownmillerite	Tube	1	1.239	[83]
$Sr_{1.7}La_{0.3}Ga_{0.6}Fe_{1.2}Mg_{0.2}O_{5+\delta}$	Brownmillerite	Tube	1	0.9	[84]
$Sr_{1.6}La_{0.4}Ga_{0.4}Fe_{0.2}Mg_{1.4}O_{5+\delta}$	Brownmillerite	Tube	1	1.18	[84]
$La_{0.15}Sr_{0.85}Ga_{0.3}Fe_{0.7}O_{3-\delta}$ $-Ba_{0.5}Sr_{0.5}Fe_{0.2}Co_{0.8}O_{3-\delta}$	Dual phase	Disk	1.64	0.6	[101]
$Zr_{0.8}Y_{0.2}O_{0.9}-La_{0.8}Sr_{0.2}CrO_{3-\delta}$	Dual phase	Tube	1.23	0.0073	[102]
$Zr_{0.8}Y_{0.2}O_{0.9}-La_{0.8}Sr_{0.2}CrO_{3-\delta}$	Dual phase	Tube	1.23	0.0037	[102]
YSZ–Pd (40%)	Dual phase	Disk	2	0.018	[2]
YSZ–Pd (40%)	Dual phase	Disk	1.72	0.044	[19]
$(SrFeO_{3-\delta})_{0.7}(SrAl_2O_4)_{0.3}$	Dual phase	Disk	1	0.062	[24]

The difference between oxygen permeation flux values for the same composition can be due to a difference in driving force, geometry, and thickness, membrane synthesis, and sintering methods.
a) Measured at 834 °C.
b) Measured at 850 °C.

that are one to two orders of magnitude lower [9, 42, 89–91]. The difference in the flux values could be due to a difference in the amount of each phase (intergrowth, perovskite, and rock salt) present. $Sr_4Fe_6O_{13\pm\delta}$ has extremely low oxygen permeation, and Co doping into the intergrowth structure can increase the permeation properties [92]. However, the oxygen permeation of a single-phase intergrowth $Sr_4Fe_4Co_2O_{13\pm\delta}$ was still an order of magnitude lower than multiphase $Sr_4Fe_4Co_2O_{13\pm\delta}$ ($SrFeCo_{0.5}O_x$). Therefore, the perovskite phase is the main phase responsible for oxygen permeation through mixed phase materials. In addition, the oxygen permeation through the mixed phase $SrFeCo_{0.5}O_x$ was an order of magnitude lower than the permeation values through a single perovskite phase $SrFe_{0.75}Co_{0.25}O_{3-\delta}$ [9]. This reinforces the importance of the perovskite phase for oxygen permeation purposes as $SrFeCo_{0.5}O_x$ contains ~25 wt% perovskite phase [7, 51]. Besides the precursors used, synthesis method, and sintering temperature, the thermal history of the mixed phase is also important as the perovskite phase has been shown to slowly convert to the intergrowth phase at high temperatures over extended periods of time [9, 93].

Eltron Research [82–84] has reported oxygen permeation values for some brownmillerite materials that are equivalent to or exceed the permeation through perovskite materials. Brownmillerite-structured materials tend to have significantly lower oxygen permeation characteristics than perovskite-structured materials due to their low electronic conductivity and limited oxygen ionic conductivity [11, 88]. The reason for the high oxygen permeation through the various brownmillerite compositions (as a result of higher electrical conductivities) remains unclear.

Dual-phase membranes that combine a fluorite-structured phase for oxygen ionic conductivity and a metal phase for electronic conductivity have produced promising results [94–98]. In some cases, the improvement in the dual-phase membrane is orders of magnitude higher than the oxygen permeation through a pure fluorite ceramic membrane. Improvement in the oxygen permeation has also been reported for ceramic–ceramic dual-phase membranes [95, 99–102]. Despite the improvement in oxygen permeation, these materials still have lower oxygen permeation properties than perovskite membranes and the high material costs may inhibit dual-phase membranes from industrial or practical applications. A more detailed review on composite membranes is found in Chapter 11.

Experimental results on the dependence of oxygen permeation on membrane thickness and electrical properties as well as upstream oxygen partial pressure can be well explained by the oxygen permeation equations presented in Section 10.4. What has been neglected by most research groups is the effect of downstream conditions on the downstream oxygen partial pressure (P''_{O_2} in Section 10.4) and, hence, on oxygen permeation. For an inert gas, the downstream oxygen partial pressure depends on the oxygen permeation flow rate, inert gas flow rate, and mixing pattern of gases in the downstream chamber of the oxygen permeation device. Its values are typically in the range of about 10^{-3} atm (determined by the oxygen impurity level in the inert gas) and one order of magnitude lower than the upstream oxygen partial pressure [103, 104]. These variations due to the difference in the oxygen permeation setup and operation conditions may cause some difference in the oxygen permeation fluxes measured, even for the membrane of the same material under apparently identical conditions. For example, increasing temperature increases the electrical conductivity of the membrane and hence oxygen permeability. However, this also increases the downstream oxygen partial pressure and thus reduces the driving force. The true temperature effect on oxygen permeation should be considered under the same driving force conditions [42].

10.5.3
Oxygen Permeation in Reducing Gases

The oxygen permeation through various ceramic membranes increases anywhere from several fold to orders of magnitude when a reducing gas is in the sweep gas rather than in an inert gas, such as He or Ar. The oxygen permeation for various perovskites and other compositions under air/reducing gas gradients without catalysts can be found in Table 10.4. Under these conditions, the downstream oxygen partial pressure (P''_{O_2}) may vary from 10^{-25} atm (determined by the thermodynamic equilibrium for the chemical reaction between oxygen and reducing gas) to about one order of magnitude lower than the upstream oxygen partial pressure. The actual downstream oxygen partial pressure is determined by many factors including temperature, sweep gas flow rate, configuration of the permeation cell, and reaction kinetics between oxygen and reducing gas. Assuming a very fast reaction rate yielding instant equilibrium for the reaction involving the reducing gas, the downstream

Table 10.4 Oxygen permeation flux through mixed ionic–electronic conducting membranes at 900 °C with an air and reducing gas gradient (no catalyst).

Composition	Structure	Reducing gas	Thickness (mm)	Oxygen flux (ml cm^{-2} min^{-1})	Reference
$(LaCa)(CoFe)O_{3-\delta}$	Perovskite	CH_4	0.8	0.5[a]	[105]
$(LaCa)(CoFe)O_{3-\delta}$	Perovskite	90% H_2/N_2	0.8	7.7[a]	[105]
$(LaCa)(CoFe)O_{3-\delta}$	Perovskite	CO	0.8	9.1[a]	[105]
$Ba_{0.5}Sr_{0.5}Zn_{0.2}Fe_{0.8}O_{3-\delta}$	Perovskite	30% CH_4/He	1.25	2.25	[126]
$La_{0.7}Sr_{0.3}FeO_{3-\delta}$	Perovskite	CO/CO_2	1	1.47	[109]
$(La_{0.85}Ca_{0.15})_{1.01}FeO_{3-\delta}$	Perovskite	Natural gas mixture	0.95	2[b]	[125]
$Bi_{1.5}Y_{0.3}Sm_{0.2}O_3$	Fluorite	10% CH_4/He	1.2	0.058	[70]
$Bi_{1.5}Y_{0.3}Sm_{0.2}O_3$	Fluorite	10% CH_4/He	1.3	0.63	[70]
$Bi_{1.5}Y_{0.3}Sm_{0.2}O_3$	Fluorite	10% C_2H_6/He	1.2	0.924[c]	[195]
$SrFe_{1.125}Co_{0.375}O_x$	Mixed phase	15.9% CO/He	1.2	1.9	[194]
$SrFeCo_{0.5}O_x$	Mixed phase	CO	0.8	3.11	[42]
$SrCoFeO_x$	Mixed phase	CO	0.8	4.81	[42]

a) Measured at 950 °C.
b) Measured at 810 °C.
c) Measured at 875 °C.

oxygen partial pressure is determined by the thermodynamic equilibrium of the relevant redox reaction [103, 104].

Akin and Lin [103] studied the effect of oxygen reaction kinetics on the oxygen permeation through $Bi_{1.5}Y_{0.3}Sm_{0.2}O_{3-\delta}$ fluorite membranes using two different reactions (methane or ethane oxidation) on the sweep side. Although the thermodynamic equilibrium oxygen partial pressure for oxidative coupling of methane and selective oxidation of ethane are both extremely low, the oxygen permeation through $Bi_{1.5}Y_{0.3}Sm_{0.2}O_{3-\delta}$ was an order of magnitude higher with ethane than with methane. The difference in the oxygen permeation flux is due to the significantly faster oxidation reaction with ethane than with methane. Zhang et al. [105] have also shown an order of magnitude difference in the oxygen permeation through a $(LaCa)(CoFe)O_{3-\delta}$ perovskite membrane when methane, hydrogen, or carbon monoxide is present as the reducing gas.

As reported by Rui et al. [104], for a mixed conducting membrane the downstream oxygen partial pressure decreases from the value of 10^{-3} atm (determined by the oxygen impurity in the reducing gas) to the equilibrium oxygen partial pressure (e.g., 10^{-20} atm) and the oxygen permeation flux increases as the kinetic rate for the reaction between oxygen and reducing component increases. The detailed effects of the downstream conditions on oxygen permeation through mixed conducting membranes with a reducing gas having a finite reaction kinetic rate [104] can be used for the design of the membrane reactors, and can also explain the discrepancy in oxygen permeation fluxes through similar membranes reported by different research groups.

10.5.4
Membrane Stability and Mechanical Properties

In industrial settings, oxygen-conducting ceramic membranes are exposed to various gases, such as CO_2, SO_2, H_2S, and H_2O, that can have a detrimental effect on the oxygen permeation properties. Sr- and Co-containing perovskites have been shown to react with CO_2 at high temperature forming electronic insulating carbonate layers on the surface of the membrane [106–109]. The detailed carbonation kinetics have been studied by Lin and coworkers [110, 111]. Perovskites exposed to sulfur compounds at high temperatures show the formation of sulfates and other oxide phases on the surfaces of the poisoned membranes [108, 112]. Water vapor might also have a poisoning effect on perovskites under certain conditions. The oxygen permeation through $Ba_{0.5}Sr_{0.5}Co_{0.8}Fe_{0.2}O_{3-\delta}$ was lower when H_2O was present in the air feed and a metal hydroxide was formed on the surface of $LaNi_{1-x}Co_xO_{3-\delta}$ after steam and CO_2 reforming of methane reactions [113, 114].

Various strategies have been proposed to improve the chemical stability of ceramic membranes. Doping the B-site of perovskites with Zr has been shown to improve the chemical stability of membranes with respect to CO_2 and H_2O [115–118]. Russo et al. [119, 120] reported that adding MgO to $LaMn_{1-x}Mg_xO_{3-\delta}$ improved its resistance against sulfur poisoning and that Cr in $LaCr_{0.5-x}Mn_xMg_{0.5}O_{3-\delta}$ improved the resistance to sulfur poisoning but reduced the catalytic activity of the materials. Another strategy has been to synthesize Co-free materials to improve the chemical stability of materials. Although Co-based materials have been shown to have high oxygen permeation, cobaltites react readily with CO_2 as Co is easily reduced [121]. Co-free materials with good chemical stability and appreciable oxygen permeation include $BaCe_xFe_{1-x}O_{3-\delta}$, $La_{0.85}Ce_{0.1}Ga_{0.3}Fe_{0.65}Al_{0.05}O_{3-\delta}$, $(La_{1-x}Ca_x)_{1.01}FeO_{3-\delta}$, $Ba_{0.5}Sr_{0.5}Zn_{0.2}Fe_{0.8}O_{3-\delta}$, and $Sr_{2-x}La_xGa_{2-y}Fe_yO_{5+\delta}$ [82, 122–127].

Some oxygen permeating materials have mechanical stability issues as well. Balachandran et al. [5] reported that $La_{0.2}Sr_{0.8}Co_{0.8}Fe_{0.2}O_{3-\delta}$ tubular membranes broke within minutes when they were exposed to an air and 80% methane/argon environment at 850 °C. The large amount of oxygen vacancies on the reducing side resulted in lattice expansion that caused a mechanical strain across the membrane that led to fracture. $SrCo_{0.8}Fe_{0.2}O_{3-\delta}$ tubular membranes were also found to break due to lattice expansion under the same experimental conditions [128]. In addition, $SrCo_{1-x}Fe_xO_{3-\delta}$ exhibits a high-temperature phase change from brownmillerite to perovskite phase that raises mechanical issues due to a volume change [88]. $Ba_{0.5}Sr_{0.5}Co_{0.8}Fe_{0.2}O_{3-\delta}$ does not have a phase change, but the high oxygen non-stoichiometry of the material leads to a low creep resistivity resulting in a loss of mechanical integrity at high temperatures [11].

The mechanical strength of materials can be improved by different approaches, primarily by doping and microstructural engineering (see also Chapter 9). For example, the mixed phase material $SrFeCo_{0.5}O_x$ has been shown to be mechanically stable for over 1000 h under the same conditions that caused some perovskite materials to fracture [5, 128]. Partial doping of La [129], Sn [130], Ti [131, 132], or Zr [117, 131, 133, 134] has been shown to improve the mechanical stability of various perovskite materials. Although the chemical and mechanical stability of most compositions can be improved, the more stable materials almost always have lower oxygen permeation properties.

10.6
Characteristics of Hydrogen-Permeable Membranes

10.6.1
Electrical and Ionic Transport Properties

For hydrogen transport through dense, proton-conducting perovskite-structured ceramic membranes, the proton-conducting oxides absorb protons from water vapor or H_2 molecules in the feed gas. Unlike most oxygen ionic conductors, protons migrate through the membranes mainly via the interstitial mechanism (see also Chapter 7 of the first volume). In most important cases, the protons do not transport as free interstitial ions, but rather associate strongly with neighbor oxygen ions and migrate by hopping from one O^{2-} to another in the nearest-neighboring position and OH^- reorientation [135]; analysis of the relevant mechanisms can be found in Chapter 7 of the first volume. The equilibrium between the membrane and either wet or dry hydrogen-containing environments can be illustrated by the following equations:

$$H_2O + V_O^{\cdot\cdot} + O_O^X \leftrightarrow 2\,OH_O^{\cdot} \tag{10.13}$$

$$O_O^X + \tfrac{1}{2}H_2 \leftrightarrow OH_O^{\cdot} + e' \tag{10.14}$$

The equations follow the Kröger–Vink notation [136], such that $V_O^{\cdot\cdot}$ is a double-ionized oxygen vacancy, O_O^X is neutral lattice oxygen, and $OH_O^{\cdot}$ is a hydroxyl ion, which represents an interstitial proton associated with a lattice oxygen. The protons migrate through the membrane by hopping between adjacent lattice oxygen with the driving force for migration being a hydrogen partial pressure gradient across the membrane.

The charge carriers present in mixed protonic–electronic conducting membranes are protons, oxygen ions (oxygen vacancies), hydroxyl ions, electrons, and electron holes. Oxygen vacancies formed by doping aliovalent cations play an essential role in proton conduction [137]. In a nongalvanic mode (chemical diffusion), electrons are transported through the membrane based on the n-type or p-type conduction mechanism depending on the downstream oxygen partial pressure. The hydrogen permeation of these materials is determined by the protonic and electronic conductivity throughout the whole membrane. Both pure $SrCeO_3$ and $BaCeO_3$ exhibit low electronic conductivity; therefore, doping is crucial to enhancing proton and electron transport through the membrane. Strategies to improve the electronic conductivity of these materials include doping aliovalent ions, such as Tb [138], in Ce^{4+} site or by adding a metal phase (10–40 vol%) [26, 139–141].

The total conductivity of various mixed protonic–electronic conducting membranes in various atmospheres can be found in Tables 10.5 and 10.6. While $BaCe_{0.9}Nd_{0.1}O_{3-\delta}$ and $BaCe_{0.95}Y_{0.05}O_{3-\delta}$ exhibit rather high total conductivity values, oxygen ionic conduction can become dominant at high temperatures (above 800 °C) [142, 143]. Qi and Lin [144] studied the electronic conductivity of $SrCeO_{3-\delta}$-based materials doped with various metal ions and found that electronic conductivity is inversely related to the ionization potential of the dopants. While the total conductivity of various Zr-doped materials can be orders of magnitudes lower than that for other proton-conducting materials, these materials possess good chemical stability and mechanical strength. Dual-phase membranes, such as $Ni–SrCe_{0.9}Yb_{0.1}O_{3-\delta}$ or $Ni–BaY_xCe_{1-x}O_{3-\delta}$, with an electronic conducting metal phase (10–40 vol%) and a protonic conducting ceramic phase, have been shown to have appreciably high hydrogen permeability [141, 145].

10.6.2
Hydrogen Permeation under Oxidizing Conditions

The hydrogen permeation through mixed protonic–electronic conducting materials is usually measured without external load using a high-temperature gas permeation system (see, for example, Figure 10.2), when the driving force for hydrogen migration is a hydrogen partial pressure difference on either side of the dense membrane. Table 10.7 lists the hydrogen permeation through $SrCeO_{3-\delta}$-based membranes with

Table 10.5 Total conductivity of various mixed protonic–electronic conductors in wet or dry H_2 at 900 °C.

Composition	Total conductivity (S cm^{-1})	Reference
$BaCe_{0.9}Y_{0.1}O_{3-\delta}$	0.028	[140]
$BaCe_{0.9}Nd_{0.1}O_{3-\delta}$	0.025	[140]
$SrCe_{0.95}Tm_{0.05}O_{3-\delta}$	0.0081[a]	[47]
$SrCe_{0.75}Zr_{0.20}Tm_{0.05}O_{3-\delta}$	0.0073[a]	[47]
$SrCe_{0.95}Yb_{0.05}O_{3-\delta}$	0.007	[139]
$SrCe_{0.9}Y_{0.1}O_{3-\delta}$	0.0055	[139]
$CaZr_{0.9}Sc_{0.1}O_{3-\delta}$	0.0012	[196]
$SrZr_{0.95}Yb_{0.05}O_{3-\delta}$	0.003	[153]
$SrZrO_{3-\delta}$	0.0001	[153]
$BaCe_{0.9}Nd_{0.1}O_{3-\delta}$	0.033	[142]
$SrCe_{0.95}Yb_{0.05}O_{3-\delta}$	0.012	[142]
$SrCe_{0.95}Tb_{0.05}O_{3-\delta}$	0.0078[b]	[138]
$BaCe_{0.9}Y_{0.1}O_{3-\delta}$	0.055	[157]
$BaZr_{0.9}Y_{0.1}O_{3-\delta}$	0.0045	[157]
$BaCe_{0.8}Er_{0.2}O_{3-\delta}$	0.0065	[197]

a) Measured in 10% H_2/He.
b) Measured in 5% H_2/He.

Table 10.6 Total conductivity of various mixed protonic–electronic conductors in wet or dry air at 900 °C.

Composition	Total conductivity (S cm^{-1})	Reference
$BaCeO_{3-\delta}$	0.00015	[140]
$BaCe_{0.9}Nd_{0.1}O_{3-\delta}$	0.05	[140]
$SrCe_{0.75}Zr_{0.20}Tm_{0.05}O_{3-\delta}$	0.016	[47]
$SrCe_{0.95}Yb_{0.05}O_{3-\delta}$	0.013	[139]
$CaZr_{0.9}Sc_{0.1}O_{3-\delta}$	0.0025	[196]
$SrZr_{0.95}Yb_{0.05}O_{3-\delta}$	0.009	[153]
$BaCe_{0.85}Gd_{0.15}O_{3-\delta}$	0.04[a]	[198]
$SrCe_{09.5}Tb_{0.05}O_{3-\delta}$	7×10^{-5}	[138]
$SrCe_{09.5}Tm_{0.05}O_{3-\delta}$	0.02	[144]
$BaCe_{0.9}Y_{0.1}O_{3-\delta}$	0.09	[157]
$BaZr_{0.9}Y_{0.1}O_{3-\delta}$	0.05	[157]
$BaZr_{0.95}Rh_{0.05}O_{3-\delta}$	0.038	[199]
$BaCe_{0.8}Er_{0.2}O_{3-\delta}$	0.006	[197]
$BaCe_{0.9}Y_{0.1}O_{3-\delta}$	0.075	[54]
$BaCe_{0.3}Zr_{0.6}Y_{0.1}O_{3-\delta}$	0.011	[54]
$BaCe_{0.45}Zr_{0.45}Sc_{0.1}O_{3-\delta}$	0.0009	[158]
$BaCe_{0.4}Zr_{0.4}Sc_{0.2}O_{3-\delta}$	0.0022	[158]

a) Measured at 850 °C.

10.6 Characteristics of Hydrogen-Permeable Membranes

Table 10.7 Hydrogen permeation flux through mixed protonic–electronic conducting disk membranes with an oxidizing sweep gas at 900 °C.

Composition	Upstream	Downstream	Thickness (mm)	Hydrogen flux (ml cm^{-2} min^{-1})	Reference
SrCe$_{0.95}$Tm$_{0.05}$O$_{3-\delta}$	20% H$_2$/He	20% O$_2$/Ar	1	0.058	[47]
SrCe$_{0.75}$Zr$_{0.20}$Tm$_{0.05}$O$_{3-\delta}$	20% H$_2$/He	20% O$_2$/Ar	1	0.03	[47]
SrCe$_{09.5}$Tm$_{0.05}$O$_{3-\delta}$	10% H$_2$/He	20% O$_2$/Ar	3	0.031	[144]
SrCe$_{09.5}$Tm$_{0.05}$O$_{3-\delta}$	10% H$_2$/He	20% O$_2$/Ar	1.6	0.045	[144]
SrCe$_{09.5}$Tm$_{0.05}$O$_{3-\delta}$	10% H$_2$/He	20% O$_2$/Ar	1.2	0.048	[144]
SrCe$_{09.5}$Tm$_{0.05}$O$_{3-\delta}$	20% H$_2$/He	20% O$_2$/N$_2$	0.8	0.08	[39]
SrCe$_{09.5}$Tm$_{0.05}$O$_{3-\delta}$	20% H$_2$/He	20% O$_2$/N$_2$	0.15	0.205	[39]
SrCe$_{09.5}$Tm$_{0.05}$O$_{3-\delta}$	10% H$_2$/He	O$_2$	0.8	0.075	[39]
SrCe$_{09.5}$Tm$_{0.05}$O$_{3-\delta}$	10% H$_2$/He	O$_2$	0.15	0.165	[39]

an oxidizing sweep gas. Under these conditions, the permeating hydrogen reacts to form water on the sweep side, thus preserving a large hydrogen partial pressure gradient across the membrane. Cheng et al. [39] studied the effect of the feed and sweep gas concentrations on hydrogen permeation through SrCe$_{0.95}$Tm$_{0.05}$O$_{3-\delta}$ membranes. The hydrogen permeation flux increased with increasing upstream hydrogen partial pressure due to the net driving force for hydrogen permeation becoming greater; the protonic and electronic conductivity in the membrane increased due to higher concentrations of the charge carriers on the feed side of the membrane. Increasing the oxygen partial pressure on the sweep side also increased the hydrogen permeation as the electronic conductivity is improved and the driving force for hydrogen permeation becomes higher due to a lower hydrogen thermodynamic equilibrium partial pressure. Similar results have been reported for SrCe$_{0.95}$Yb$_{0.05}$O$_{3-\delta}$ membranes [146].

10.6.3
Hydrogen Permeation in Inert Sweep Gases

With an inert sweep gas, the hydrogen permeation through dense protonic conducting membranes (see Table 10.8) is lower due to a reduced hydrogen partial pressure gradient across the membrane and low electronic conductivity in the membrane near the sweep side. The hydrogen permeation through metal–ceramic dual-phase membranes is relatively higher than that for pure ceramic membranes due to the presence of the metal phase [26, 27, 145, 147, 148]. Song et al. [147] have reported that hydrogen can diffuse through both the metal and ceramic phases, and the fraction of hydrogen diffusing through the metal phase of a Ni–BaCe$_{0.8}$Y$_{0.2}$O$_{3-\delta}$ composite may reach 25% of the total hydrogen permeation flux under certain conditions. Wei et al. [43] investigated hydrogen permeation through SrCe$_{1-x}$Tb$_x$O$_{3-\delta}$ membranes. Although SrCe$_{0.95}$Tb$_{0.05}$O$_{3-\delta}$ possesses higher protonic conductivity than SrCe$_{0.95}$Yb$_{0.05}$O$_{3-\delta}$ under the same hydrogen partial pressure, hydrogen permeation through SrCe$_{0.95}$Tb$_{0.05}$O$_{3-\delta}$ membranes was not observed when nitrogen or air was used as

Table 10.8 Hydrogen permeation flux through mixed protonic–electronic conducting disk membranes with an inert sweep gas at 900 °C.

Composition	Upstream	Downstream	Thickness (mm)	Hydrogen flux (ml cm^{-2} min^{-1})	Reference
BaCe$_{0.95}$Eu$_{0.05}$O$_{3-\delta}$	H$_2$	He	1.72	0.0047[a]	[200]
Ni–BaCe$_{0.8}$Y$_{0.2}$O$_{3-\delta}$	Wet 4% H$_2$/He	N$_2$ (100 ppm H$_2$)	1	0.054	[27]
Ni–BaZr$_{0.2}$Ce$_{0.6}$Y$_{0.2}$O$_{3-\delta}$	Wet 4% H$_2$/He	N$_2$ (100 ppm H$_2$)	1	0.053	[27]
Ni–BaZr$_{0.4}$Ce$_{0.4}$Y$_{0.2}$O$_{3-\delta}$	Wet 4% H$_2$/He	N$_2$ (100 ppm H$_2$)	1	0.041	[27]
50 vol% Pd–YSZ	80% H$_2$/He	N$_2$ (100 ppm H$_2$)	0.21	2	[26]
Ni–BaZr$_{0.1}$Ce$_{0.7}$Y$_{0.2}$O$_{3-\delta}$	Wet 4% H$_2$/He	N$_2$ (100 ppm H$_2$)	0.978	0.056	[148]
Ni–BaZr$_{0.1}$Ce$_{0.7}$Y$_{0.2}$O$_{3-\delta}$	Wet 4% H$_2$/He	N$_2$ (100 ppm H$_2$)	0.266	0.12	[148]
Ni–BaZr$_{0.1}$Ce$_{0.7}$Y$_{0.2}$O$_{3-\delta}$	H$_2$	N$_2$ (100 ppm H$_2$)	0.266	0.81	[148]
SrCe$_{0.95}$Tb$_{0.05}$O$_{3-\delta}$	20% H$_2$/He	1% CO/Ar	1	0.0155	[43]
SrCe$_{0.95}$Tb$_{0.05}$O$_{3-\delta}$	80% H$_2$/He	1% CO/Ar	1	0.027	[43]

a) Measured at 850 °C.

the sweep gas [138] due to lower electronic conductivity in the permeate side layers. $SrCe_{1-x}Tb_xO_{3-\delta}$ materials are n-type electronic conductors [52, 138]; therefore, exposing the permeate side of the membrane to a reducing gas, such as carbon monoxide or hydrogen, can induce a high electronic conductivity across the entire membrane and appreciable hydrogen permeation through the membrane [43].

10.6.4
Membrane Stability

Although $SrCeO_3$- and $BaCeO_3$-based mixed protonic–electronic conducting membranes can offer extremely high hydrogen selectivity, thermal stability, and mechanical strength at high temperatures, the chemical stability of these materials under various environments must be improved in order to be used in industrial applications. In hydrogen separation applications, the metal oxide proton-conducting membrane is exposed to a reducing gas in both up and down gas streams. If the whole system is thermodynamically unstable, under kinetically favorable conditions (such as high temperature) the membrane material may be fully or partially reduced causing formation of new phases. Exposure to a CO_2- or H_2O-containing atmosphere at high temperatures may promote degradation in the material's performance. In CO_2-containing atmospheres, the materials form carbonates and metal oxides [149, 150]. This has a detrimental effect on the membrane's mechanical strength and hydrogen flux through the membrane as the carbonates and metal oxides inhibit the surface reactions. Cerium-containing perovskite-type ceramic membranes have also been found to be reactive in water vapor at elevated temperatures forming hydroxides, metal oxides, and other compounds on the membrane surface [151, 152].

Doped $SrZrO_3$ and $BaZrO_3$ have been shown to be mixed protonic–electronic conductors and have a relatively good chemical stability in both CO_2- and H_2O-containing atmospheres at high temperatures [151, 153, 154]. However, these materials have worse transport properties compared to $SrCeO_3$- or $BaCeO_3$-based materials. In order to produce a material with both high protonic conductivity and chemical stability, various researchers have investigated solid solutions containing alkaline earth metal cerates and zirconates [54, 155–159]. The resulting solid solutions exhibited enhanced chemical stability and a decrease in electrical conductivity with increasing Zr content. However, the Zr content needed for a compromise between chemical stability and electrical conductivity for industrial application varies depending on the alkaline earth metal and trivalent dopant present in the material. Zr-doped metal–ceramic dual-phase or pure ceramic membranes have been shown to have higher steady-state hydrogen permeation values than undoped materials in CO_2-containing atmospheres [27, 47, 148].

10.6.5
Comparison with Oxygen-Permeable Membranes

While mixed ionic–electronic conducting membranes have been found to be infinitely selective to either oxygen or hydrogen, oxygen-transporting membranes are about to enter the commercial phase due to the high flux values. In order for membranes to be feasible for industrial applications, the flux through the membrane needs to be 5–10 ml (STP) cm^{-2} min^{-1} or higher [121]. While many oxygen-conducting materials have reached that economic benchmark in terms of oxygen permeation fluxes with a reducing gas or chemical reaction on the sweep side, ceramic mixed conducting hydrogen-selective membranes have not achieved those oxygen permeation values. For membranes of comparable thickness and under similar conditions, the proton-conducting ceramic membranes offer hydrogen permeation fluxes about one to two orders of magnitude lower than the oxygen fluxes.

However, hydrogen-selective membranes have an advantage over oxygen-selective membranes in that hydrogen-selective membranes have been shown to be limited by bulk diffusion at a membrane thickness down to 2 μm [160]. On the other hand, oxygen permeation through mixed conducting membranes is often controlled by bulk diffusion and surface exchange kinetics as the membranes become thinner. Also, the critical thickness, or the thickness where oxygen permeation is completely controlled by surface exchange kinetics, depends on the composition of the membrane and the temperature, and oxygen partial pressure gradient of the system; the critical thickness of a typical oxygen-conducting material is 80 μm [161]. Therefore, if a thin-membrane fabrication technique is developed that is simple, highly reproducible, and can be scaled up for mass production, hydrogen-selective membranes could be used in an industrial setting for gas separation. Even with limited hydrogen permeation values, hydrogen-selective materials have potential use as fuel cell electrolytes or as a sensing material for high-temperature hydrogen sensors [162, 163].

Both oxygen- and hydrogen-selective membranes have chemical stability issues when they are exposed to gases containing carbon and sulfur. Many oxygen-selective mixed conducting metal oxide membranes are thermodynamically stable under the conditions of oxygen separation (both up- and downstream gas mixtures contain oxygen). However, hydrogen-selective mixed conducting metal oxide membranes are not in the thermodynamically stable conditions under the conditions of hydrogen separation. When either of the membranes is used for membrane reactor applications, the membrane is exposed, respectively, to a reducing and oxidizing gas and it is possible to control the conditions such that the whole system is thermodynamically stable. However, under such conditions, the membrane is exposed to an extremely large gradient of oxygen (or hydrogen) partial pressure that may cause segregation of the phases as a result of ion relocation within the membrane. This stability problem is common for both oxygen- and hydrogen-selective membranes.

10.7
Applications of Membranes

10.7.1
Gas Separation and Purification

Gas separation using dense ceramic mixed ionic–electronic conducting membranes is an environment-friendly, efficient, and economical separation process. Gas separation processes with these membranes are infinitely selective to either hydrogen or oxygen, which results in a high-purity product. Air Products and Chemicals [125, 164–168], Eltron Research [82, 83, 169, 170], and various other groups [171–173] have obtained numerous patents for fabrication methods and materials for oxygen separation. Mundschau et al. [174] have developed systems that integrate both oxygen- and hydrogen-selective membranes with coal gasifiers, natural gas syngas reactors, and water–gas shift reactors for the production of high-purity hydrogen and the sequestration of carbon dioxide.

Air Products and Chemicals is developing commercially applicable technology utilizing mixed oxygen ionic–electronic conducting materials [175]. The ITM (ion transport membrane) oxygen process uses mixed oxygen ionic and electronic conducting membranes to separate high-purity oxygen from air using high-pressure air to create an oxygen partial pressure difference. The ITM syngas process combines air separation and high-temperature syngas generation processes into a single compact ceramic membrane reactor, which has the potential for reducing the capital investment for gas to liquid plants and hydrogen distribution. On a smaller scale, the SEOS™ Oxygen Generator is an electrically driven oxygen generation and removal technology using ionic conducting membranes to produce high-purity oxygen that can be used both for medical applications and for industrial cutting and welding. The membranes may find applications to remove trace oxygen or hydrogen from gas streams by using a reducing gas on the sweep side to provide the driving force.

10.7.2
Membrane Reactors

Mixed ionic–electronic conducting membranes have been used for membrane reactors for a variety of reactions. Reactions that take advantage of using a membrane reactor with oxygen-selective materials include oxidative coupling of methane to ethylene or ethane [70, 71], selective oxidation of ethane to ethylene [103], and partial oxidation of heptanes to produce hydrogen [176]. Hydrogen-selective materials have been used in membrane reactors for the nonoxidative coupling of methane for C_2 [177]. Reactions to produce hydrogen, such as stream reforming of methane (SRM) or partial oxidation of methane (POM), have received significant attention recently due to increased interest in hydrogen as a green energy source and therefore are discussed in more detail next.

At present, the bulk of the hydrogen produced in the United States is produced using steam reforming of methane. In this process, steam and methane are fed into a high-temperature reactor (700–1000 °C) containing a reforming catalyst (usually supported nickel due to its wide availability and low cost). The main endothermic reaction that occurs during the process is

$$CH_4 + H_2O \leftrightarrow CO + 3H_2 \qquad \Delta H^0_{298} = 206 \text{ kJ mol}^{-1} \qquad (10.15)$$

Carbon monoxide also reacts with steam via the water–gas shift reaction to produce hydrogen and carbon dioxide.

$$CO + H_2O \leftrightarrow CO_2 + H_2 \qquad \Delta H^0_{298} = -41 \text{ kJ mol}^{-1} \qquad (10.16)$$

Although the steam reforming process is the main industrial process to produce hydrogen, it has some shortcomings. In particular, the main reaction is highly energy intensive, the process has low CO selectivity, and the high H_2 to CO production ratio is beyond the range needed for Fischer–Tropsch synthesis [178].

POM is another method that can be used to convert methane to syngas. Ideally, methane reacts with a less than stoichiometric amount of oxygen in a reactor in the presence of a catalyst to form hydrogen and carbon monoxide.

$$CH_4 + \tfrac{1}{2}O_2 \leftrightarrow CO + 2H_2 \qquad \Delta H^0_{298} = -36 \text{ kJ mol}^{-1} \qquad (10.17)$$

However, since the desired products are more reactive than methane, this mechanism has been reported only for extremely small reactor residence times [128, 179]. In general, the partial oxidation of methane to syngas occurs via the combustion and reforming reaction (CRR) mechanism [179, 180]. In this mechanism, the oxygen fed into the reactor completely combusts a portion of methane by the following reaction:

$$CH_4 + 2O_2 \leftrightarrow CO_2 + 2H_2O \qquad \Delta H^0_{298} = -802 \text{ kJ mol}^{-1} \qquad (10.18)$$

The remaining methane is then reformed by steam (reaction C) or by CO_2:

$$CH_4 + CO_2 \leftrightarrow 2CO + 2H_2 \qquad \Delta H^0_{298} = 247 \text{ kJ mol}^{-1} \qquad (10.19)$$

Although partial oxidation of methane has received significant attention recently as a viable alternative method to produce syngas from methane, there are still some barriers preventing commercial-size POM plants. The reactor must be carefully designed as the high exothermic value of reaction (10.18) could cause local hot spot problems and possible reactor runaway [181]. The main drawback of POM is that pure oxygen is required as downstream processing of syngas cannot tolerate nitrogen [128, 178, 180, 181]. Therefore, the cost of a POM plant is significantly higher due to the need of a cryogenic oxygen separation plant.

Membrane reactors can overcome many of the shortcomings of conventional packed bed reactors being used in industry for SRM or POM reactions. Selective removal of a product by a membrane can enhance the yield of a thermodynamically limited reaction [181–184]. Both the reaction stoichiometry and the interaction of the reactants can be controlled, which can eliminate the hot spot problem associated with POM [178, 181, 182]. With respect to a steam reforming membrane reactor, a hydrogen-selective membrane can combine the reactor and the downstream separation process into one unit, thus reducing costs [182]. An oxygen-selective membrane reactor can eliminate the need for a cryogenic oxygen separation plant for POM [178, 180, 181].

Figure 10.3 shows a schematic of a steam reforming of methane membrane reactor. Deckman et al. [185] have devised a membrane reactor system with a mixed

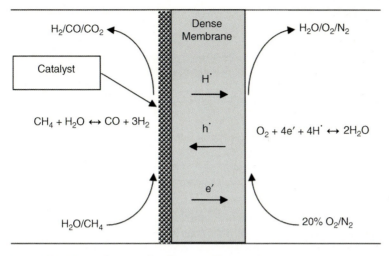

Figure 10.3 Steam reforming of methane membrane reactor.

protonic–electronic conducting membrane that employs steam reforming and water–gas shift reactions on one side of the membrane and hydrogen combustion on the other side. The heat of combustion is exchanged through the membrane both to heat the hydrocarbon fuel and to provide heat to drive the endothermic reforming reaction. Under certain conditions and high hydrogen permeation values, this membrane reactor system could produce enough heat to be energy independent. For a partial oxidation of methane membrane reactor (shown in Figure 10.4), methane is fed to one side of the membrane while 20% O_2/Ar (or air) is fed to the other side. Partial oxidation of methane membrane reactors has been studied extensively with promising methane

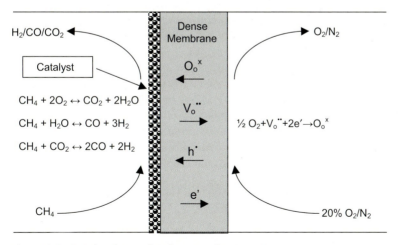

Figure 10.4 Partial oxidation of methane membrane reactor.

Table 10.9 Examples of partial oxidation of methane in the membrane reactors utilizing mixed ionic–electronic conducting membranes.

Composition	Catalyst	CH_4 conversion (%)	Oxygen flux (ml cm^{-2} min^{-1})	Duration (h)	Reference
$BaCo_{0.4}Fe_{0.4}Zr_{0.2}O_{3-\delta}$	LiLaNiO/γ-Al$_2$O$_3$	98	5.4–5.8	2200	[116]
$Ba_{0.5}Sr_{0.5}Co_{0.8}Fe_{0.2}O_{3-\delta}$	LiLaNiO/γ-Al$_2$O$_3$	97	11.5	500	[180]
$Sm_{0.4}Ba_{0.6}Fe_{0.8}Co_{0.2}O_{3-\delta}$	1 wt% Rh/MgO	90	0.59	—	[179]
$SrFeCo_{0.5}O_x$	Rh	98	2–4	500	[128]
$BaCe_{0.15}Fe_{0.85}O_{3-\delta}$	LiLaNiO/γ-Al$_2$O$_3$	96	4.2	140	[201]

conversion rates and high oxygen permeation values for extended periods of time. Table 10.9 summarizes various compositions that have been tested in partial oxidation of methane membrane reactor systems.

10.8
Summary and Conclusions

This chapter summarized the development of mixed ionic–electronic conducting materials and membranes that selectively transport oxygen or hydrogen. These materials have the ability to transport oxygen ions or protons as well as electrons and electronic holes while maintaining electroneutrality. Strategies for improving the ionic or electronic conductivity, mechanical, or chemical stability have evolved over the past 30 years. The improved ceramic and dual-phase membranes that have been developed show promise for use in gas separation areas, such as high-purity oxygen from air or in a reforming membrane reactor for hydrogen and syngas production. While numerous flux equations have been derived for the cases where the conductivity of one charged species is much larger than another, special cases such as mixed conducting membranes with comparable conductivities for all charged species still require significant attention. The effect of various sweep gases on the permeation properties of the materials was discussed in detail. Oxygen permeation through mixed conducting ceramic membranes with a reducing gas in the downstream is strongly affected by the downstream conditions including reactivity and reaction kinetic rate for oxidation reaction between oxygen and reducing gas. In order for ceramic membranes to be realized in an industrial setting, future research should focus on improving membrane fabrication methods and the mechanical or chemical stability of the membranes under various reducing environments.

Acknowledgments

The authors are grateful to the National Science Foundation and the Department of Energy for their support on work in mixed conducting ceramic membranes.

References

1. Lin, Y.S. (2001) *Sep. Purif. Technol.*, **25** (1–3), 39–55.
2. Liu, Y.Y., Tan, X.Y., and Li, K. (2006) *Catal. Rev.*, **48** (2), 145–198.
3. Leo, A., Liu, S.M., and da Costa, J.C.D. (2009) *Int. J. Greenhouse Gas Control*, **3** (4), 357–367.
4. Cook, R.L. and Sammells, A.F. (1991) *Solid State Ionics*, **45**, 311–321.
5. Balachandran, U., Dusek, J.T., Mieville, R.L., Poeppel, R.B., Kleefisch, M.S., Pei, S., Kobylinski, T.P., Udovich, C.A., and Bose, A.C. (1995) *Appl. Catal. A*, **133**, 19–29.
6. Ma, B. and Balachandran, U. (1997) *Solid State Ionics*, **100**, 53–62.
7. Guggilla, S. and Manthiram, A. (1997) *J. Electrochem. Soc.*, **144** (5), L120–L122.
8. Ma, B., Hodges, J.P., Jorgensen, J.D., Miller, D.J., Richardson, J.W., Jr., and Balachandran, U. (1998) *J. Solid State Chem.*, **141**, 576–586.
9. Armstrong, T., Prado, F., Xia, Y., and Manthiram, A. (2000) *J. Electrochem. Soc.*, **147** (2), 435–438.
10. Xia, Y., Armstrong, T., Prado, F., and Manthiram, A. (2000) *Solid State Ionics*, **130**, 81–90.
11. Vente, J.F., McIntosh, S., Haije, W.G., and Bouwmeester, H.J.M. (2006) *J. Solid State Electrochem.*, **10**, 581–588.
12. Kakinuma, K., Yamamura, H., Haneda, H., and Atake, T. (2001) *Solid State Ionics*, **140**, 301–306.
13. Guggilla, S., Armstrong, T., and Manthiram, A. (1999) *J. Solid State Chem.*, **145**, 260–266.
14. Manthiram, A., Prado, F., and Armstrong, T. (2002) *Solid State Ionics*, **152–153**, 647–655.
15. Armstrong, T., Prado, F., and Manthiram, A. (2001) *Solid State Ionics*, **140**, 89–96.
16. Al Daroukh, M., Vashook, V.V., Ullmann, H., Tietz, F., and Arual Raj, I. (2003) *Solid State Ionics*, **158**, 141–150.
17. Kharton, V.V., Viskup, A.P., Kovalesvsky, A.V., Naumovich, E.N., and Marques, F.M.B. (2001) *Solid State Ionics*, **143**, 337–353.
18. Chen, C.S., Boukamp, B.A., Bouwmeester, H.J.M., Cao, G.Z., Kruidhof, H., Winnubst, A.J.A., and Burggraaf, A.J. (1995) *Solid State Ionics*, **76**, 23–28.
19. Chen, C.S., Kruidhof, H., Bouwmeester, H.J.M., Verweij, H., and Burggraaf, A.J. (1996) *Solid State Ionics*, **86–88**, 569–572.
20. Kim, J. and Lin, Y.S. (2000) *J. Membr. Sci.*, **167**, 123–133.
21. Kharton, V.V., Kovalevsky, A.V., Viskup, A.P., Figueiredo, F.M., Yaremchenko, A.A., Naumovich, E.N., and Marques, F.M.B. (2001) *J. Eur. Ceram. Soc.*, **21**, 1763–1767.
22. Wang, H., Yang, W.S., Cong, Y., Zhu, X., and Lin, Y.S. (2003) *J. Membr. Sci.*, **224**, 107–115.
23. Wang, B., Zhan, M.C., Zhu, D.C., Liu, W., and Chen, C.S. (2006) *J. Solid State Electrochem.*, **10**, 625–628.
24. Kovalevsky, A.V., Kharton, V.V., Snijkers, F.M.M., Cooymans, J.F.C., Luyten, J.J., and Frade, J.R. (2008) *Solid State Ionics*, **179**, 61–65.
25. Zhang, G., Dorris, S.E., Balachandran, U., and Liu, M. (2003) *Solid State Ionics*, **159**, 121–134.
26. Balachandran, U., Lee, T.H., Chen, L., Song, S.J., Picciolo, J.J., and Dorris, S.E. (2006) *Fuel*, **85**, 150–155.
27. Zuo, C., Dorris, S.E., Balachandran, U., and Liu, M. (2006) *Chem. Mater.*, **18**, 4647–4650.
28. Selvaraj, U., Prasadarao, A.V., Komarneni, S., and Roy, R. (1991) *Mater. Lett.*, **12**, 311–315.
29. Bouwmeester, H.J.M., Kruidhof, H., Burggraaf, A.J., and Gellings, P.J. (1992) *Solid State Ionics*, **53–55**, 460–468.
30. Fagg, D.P., Marozau, I.P., Shaula, A.L., Kharton, V.V., and Frade, J.R. (2006) *J. Solid State Chem.*, **179**, 3347–3356.
31. Fan, W. and Niinisto, L. (1994) *Mater. Res. Bull.*, **29**, 451–458.
32. Zheng, W., Liu, C., Yue, Y., and Pang, W. (1997) *Mater. Lett.*, **30**, 93–97.
33. Qi, X., Lin, Y.S., and Swartz, S.L. (2000) *Ind. Eng. Chem. Res.*, **39**, 646–653.

34 Nigara, Y., Mizusaki, J., and Ishigame, M. (1995) *Solid State Ionics*, **79**, 208–211.

35 Qui, L., Lee, T.H., Liu, L.M., Yang, Y.L., and Jacobson, A.J. (1995) *Solid State Ionics*, **76**, 321–329.

36 Pechini, M.P. (1967) US Patent 3,330,697.

37 Zeng, Y. and Lin, Y.S. (2001) *J. Mater. Sci.*, **36**, 1271–1276.

38 Anderson, H.U., Nasrallah, M.M., and Chen, C.C. (27 1996) US Patent 5,494,700.

39 Cheng, S., Gupta, V.K., and Lin, J.Y.S. (2005) *Solid State Ionics*, **176**, 2653–2662.

40 Wang, H., Schiestel, T., Tablet, C., Schroeder, M., and Caro, J. (2006) *Solid State Ionics*, **177**, 2255–2259.

41 Tan, X., Li, K., Thursfield, A., and Metcalfe, I.S. (2008) *Catal. Today*, **131**, 292–304.

42 Kniep, J., Yin, Q., Kumakiri, I., and Lin, Y.S. (2009) *Solid State Ionics*, **180**, 1633–1639.

43 Wei, X., Kniep, J., and Lin, Y.S. (2009) *J. Membr. Sci.*, **345**, 201–206.

44 Qi, X., Akin, F.T., and Lin, Y.S. (2001) *J. Membr. Sci.*, **193**, 185–193.

45 Lin, Y.S., Wang, W., and Han, J. (1994) *AIChE J.*, **40**, 786–798.

46 Qi, X., Lin, Y.S., and Swartz, S.L. (2000) *Ind. Eng. Chem. Res.*, **39**, 646–653.

47 Kniep, J. and Lin, Y.S. (2010) *Ind. Eng. Chem. Res.*, **49**, 2768–2774.

48 Sunarso, J., Liu, S., Diniz da Costa, J.C., Meulenberg, W.A., Baumann, S., Serra, J.M., and Lin, Y.S. (2008) *J. Membr. Sci.*, **320** (1–2), 13–41.

49 Xu, S.J. and Thomson, W.J. (1999) *Chem. Eng. Sci.*, **54**, 3839–3850.

50 Zeng, Y. and Lin, Y.S. (1998) *Solid State Ionics*, **110**, 209.

51 Balachandran, U. and Ma, B. (2006) *J. Solid State Electrochem.*, **10**, 617–624.

52 Wei, X. and Lin, Y.S. (2008) *Solid State Ionics*, **178**, 1804–1811.

53 Haile, S.M., Staneff, G., and Ryu, K.H. (2001) *J. Mater. Sci.*, **36**, 1149–1160.

54 Zhong, Z. (2007) *Solid State Ionics*, **178**, 213–220.

55 ten Elshof, J.E., Bouwmeester, H.J.M., and Verweij, H. (1995) *Appl. Catal. A*, **130**, 195–212.

56 Dou, S., Masson, C.R., and Pacey, P.D. (1985) *J. Electrochem. Soc.*, **132** (8), 1843–1849.

57 Badwal, S.P.S. (1992) *Solid State Ionics*, **52**, 23–32.

58 Han, J., Zeng, Y., and Lin, Y.S. (1997) *J. Membr. Sci.*, **132**, 235–243.

59 Capoen, E., Steil, M.C., Nowogrocki, G., Malys, M., Pirovano, C., Löfberg, A., Bordes-Richard, E., Boivin, J.C., Mairesse, G., and Vannier, R.N. (2006) *Solid State Ionics*, **177**, 483–488.

60 Verkerk, M.J. and Burggraaf, A.J. (1981) *J. Electrochem. Soc.*, **128**, 75–82.

61 Bouwmeester, H.J.M., Kruidhof, H., Burggraaf, A.J., and Gellings, P.J. (1992) *Solid State Ionics*, **53–56**, 460–468.

62 Boivin, J.C. and Thomas, D.J. (1981) *Solid State Ionics*, **5**, 523–526.

63 Conflant, P., Boivin, J.C., and Thomas, D. (1976) *J. Solid State Chem.*, **18**, 133–140.

64 Conflant, P., Boivin, J.C., and Thomas, D. (1980) *J. Solid State Chem.*, **35**, 192–199.

65 Han, J., Zeng, Y., Xomeritakis, G., and Lin, Y.S. (1997) *Solid State Ionics*, **98**, 63–72.

66 Han, P. and Worrell, W.L. (1995) *J. Electrochem. Soc.*, **142** (12), 4235–4246.

67 Arashi, H. and Naito, H. (1992) *Solid State Ionics*, **53–56**, 431–435.

68 Nigara, Y., Mizusaki, J., and Ishigame, M. (1995) *Solid State Ionics*, **79**, 208–211.

69 Zeng, Y., Akin, F.T., and Lin, Y.S. (2001) *Appl. Catal. A*, **213**, 33–45.

70 Akin, F.T. and Lin, Y.S. (2002) *AIChE J.*, **48** (10), 2298–2306.

71 Zeng, Y. and Lin, Y.S. (2001) *AIChE J.*, **47** (2), 436–444.

72 Ma, B., Balachandran, U., Park, J.H., and Segre, C.U. (1996) *J. Electrochem. Soc.*, **143** (5), 1736–1744.

73 Ma, B. and Balachandran, U. (1997) *Solid State Ionics*, **100**, 53–62.

74 Bredesen, R., Norby, T., Bardal, A., and Lynum, V. (2000) *Solid State Ionics*, **135**, 687–697.

75 Bredesen, R. and Norby, T. (2000) *Solid State Ionics*, **129**, 285–297.

76 Kontoulis, I., Ftikos, C.P., and Steele, B.C.H. (1994) *Mater. Sci. Eng.*, **B22**, 313–316.

77 Takeda, Y., Imanishi, N., Kanno, R., Mizuno, T., Higuchi, H., and Yamamoto, O. (1992) *Solid State Ionics*, **53–56**, 748–753.

78 Kontoulis, I. and Steele, B.C.H. (1992) *J. Eur. Ceram. Soc.*, **9**, 459–462.

79 Goodenough, J.B., Ruiz-Diaz, J.E., and Zhen, Y.S. (1990) *Solid State Ionics*, **44**, 21–31.

80 Kendall, K.R., Navas, C., Thomas, J.K., and zur Loye, H.C. (1995) *Solid State Ionics*, **82**, 215–223.

81 McIntosh, S., Vente, J.F., Haije, W.G., Blank, D.H.A., and Bouwmeester, H.J.M. (2006) *Solid State Ionics*, **177**, 833–842.

82 Schwartz, M., White, J.H., and Sammels, A.F. (2000) US Patent 6,033,632.

83 Mackay, R., Schwartz, M., and Sammells, A.F. (2000) US Patent 6,165,431.

84 Mackay, R., Schwartz, M., and Sammells, A.F. (2003) US Patent 6,592,782.

85 Teraoka, Y., Zhang, H.M., and Yamazoe, N. (1985) *Chem. Lett.*, 1367–1370.

86 Teraoka, Y., Zhang, H.M., Furukawa, S., and Yamazoe, N. (1985) *Chem. Lett.*, 1743–1746.

87 Teraoka, Y., Nobunaga, T., and Yamazoe, N. (1988) *Chem. Lett.*, 503–506.

88 Kruidhof, H., Bouwmeester, H.J.M., Doorn, R.H.E.v., and Burggraaf, A.J. (1993) *Solid State Ionics*, **63–65**, 816–822.

89 Murphy, S.M., Slade, D.A., Nordheden, K.J., and Stagg-Williams, S.M. (2006) *J. Membr. Sci.*, **277**, 94–98.

90 Ikeguchi, M., Ishii, K., Sekine, Y., Kikuchi, E., and Matsukata, M. (2005) *Mater. Lett.*, **59**, 1356–1360.

91 Kim, S., Yang, Y.L., Christoffersen, R., and Jacobson, A.J. (1998) *Solid State Ionics*, **109**, 187–196.

92 Armstrong, T., Guggilla, S., and Manthiram, A. (1999) *Mater. Res. Bull.*, **34** (6), 837–844.

93 Xia, Y., Armstrong, T., Prado, F., and Manthiram, A. (2000) *Solid State Ionics*, **130**, 81–90.

94 Capoen, E., Steil, M.C., Nowogrocki, G., Malys, M., Pirovano, C., Löfberg, A., Bordes-Richard, E., Boivin, J.C., Mairesse, G., and Vannier, R.N. (2006) *Solid State Ionics*, **177**, 483–488.

95 Mazanec, T.J., Cable, T.L., and Frye, J.G., Jr. (1992) *Solid State Ionics*, **53–56**, 111–118.

96 Kim, J. and Lin, Y.S. (2000) *J. Membr. Sci.*, **167**, 123–133.

97 Chen, C.S., Boukamp, B.A., Bouwmeester, H.J.M., Cao, G.Z., Kruidhof, H., Winnubst, A.J.A., and Burggraaf, A.J. (1995) *Solid State Ionics*, **76**, 23–28.

98 Chen, C.S., Kruidhof, H., Bouwmeester, H.J.M., Verweij, H., and Burggraaf, A.J. (1996) *Solid State Ionics*, **86–88**, 569–572.

99 Kovalevsky, A.V., Kharton, V.V., Snijkers, F.M.M., Cooymans, J.F.C., Luyten, J.J., and Frade, J.R. (2008) *Solid State Ionics*, **179**, 61–65.

100 Kharton, V.V., Kovalevsky, A.V., Viskup, A.P., Figueiredo, F.M., Yaremchenko, A.A., Naumovich, E.N., and Marques, F.M.B. (2001) *J. Eur. Ceram. Soc.*, **21**, 1763–1767.

101 Wang, H., Yang, W.S., Cong, Y., Zhu, X., and Lin, Y.S. (2003) *J. Membr. Sci.*, **224**, 107–115.

102 Wang, B., Zhan, M.C., Zhu, D.C., Liu, W., and Chen, C.S. (2006) *J. Solid State Electrochem.*, **10**, 625–628.

103 Akin, F.T. and Lin, J.Y.S. (2004) *J. Membr. Sci*, **231**, 133–146.

104 Rui, Z., Li, Y., and Lin, Y.S. (2009) *Chem. Eng. Sci.*, **64**, 172–179.

105 Zhang, W., Smit, J., van Sint Annaland, M., and Kuipers, J.A.M. (2007) *J. Membr. Sci.*, **291**, 19–32.

106 Pei, S., Kleefisch, M.S., Kobylinski, T.P., Faber, C.A.J., Udovich, V., Zhang-McCoy, B.V., Dabrowdki, U.B., Mieville, R.L., and Poeppel, R.B. (1995) *Catal. Lett.*, **30**, 201–212.

107 Tong, J., Yang, W., Zhu, B., and Cai, R. (2002) *J. Membr. Sci.*, **203**, 175–189.

108 ten Elshof, J.E., Bouwmeester, H.J.M., and Verweij, H. (1995) *Appl. Catal. A*, **130**, 195–212.

109 ten Elshof, J.E., Bouwmeester, H.J.M., and Verweij, H. (1996) *Solid State Ionics*, **89**, 81–92.

110 Yang, Q. and Lin, Y.S. (2006) *Ind. Eng. Chem. Res.*, **45**, 6302–6310.

111 Lin, Y.S., Yang, Q., and Ida, J.-I. (2009) *J. Taiwan Inst. Chem. Eng.*, **40**, 276–280.

112 Li, S.G., Jin, W., Huang, P., Xu, N.P., Shi, J., and Lin, Y.S. (2000) *J. Membr. Sci.*, **166**, 51–61.

113 Shao, Z., Yang, W., Cong, Y., Dong, H., Tong, J., and Xiong, G. (2000) *J. Membr. Sci.*, **172**, 177–188.

114 Choudhary, V.R., Uphade, B.S., and Belhekar, A.A. (1996) *J. Catal.*, **163**, 312–318.

115 Tong, J., Yang, W., Zhu, B., and Cai, R. (2002) *J. Membr. Sci.*, **203**, 175–189.

116 Tong, J., Yang, W., Cai, R., Zhu, B., and Lin, L. (2002) *Catal. Lett.*, **78** (1–4), 129–137.

117 Caro, J., Wang, H.H., Tablet, C., Kleinert, A., Feldhoff, A., Schiestel, T., Kilgus, M., Kölsch, P., and Werth, S. (2006) *Catal. Today*, **118**, 128–135.

118 Taniguchi, N., Nishimura, C., and Kato, J. (2001) *Solid State Ionics*, **145**, 349–355.

119 Rosso, I., Garrone, E., Geobaldo, F., Onida, B., Saracco, G., and Specchia, V. (2001) *Appl. Catal. B*, **34**, 29–41.

120 Rosso, I., Saracco, G., Specchia, V., and Garrone, E. (2003) *Appl. Catal. B*, **40**, 195–205.

121 Bouwmeester, H.J.M. (2003) *Catal. Today*, **82**, 141–150.

122 Zhu, X., Wang, H., and Yang, W. (2004) *Chem. Commun.*, 1130–1131.

123 Zhu, X., Cong, Y., and Yang, W. (2006) *J. Membr. Sci.*, **283**, 38–44.

124 Dong, X., Zhang, G., Liu, Z., Zhong, Z., Jin, W., and Xu, N. (2009) *J. Membr. Sci.*, **340**, 141–147.

125 Dyer, P.N., Carolan, M.F., Butt, D., Van Doorn, R.H.E., and Culter, R.A. (2002) US Patent 6,492,290.

126 Wang, H., Tablet, C., Feldhoff, A., and Caro, J. (2005) *Adv. Mater.*, **17**, 1785–1788.

127 Schwartz, M., White, J.H., and Sammells, A.F.(10 2001) US Patent 6,214,757.

128 Balachandran, U., Dusek, J.T., Maiya, P.S., Ma, B., Mieville, R.L., Kleefisch, M.S., and Udovich, C.A. (1997) *Catal. Today*, **36**, 265–272.

129 Prado, F., Grunbaum, N., Caneiro, A., and Manthiram, A. (2004) *Solid State Ionics*, **167**, 147–154.

130 Fan, C., Liu, W., Deng, Z., Zuo, Y., Chen, C., and Bae, D. (2007) *J. Membr. Sci.*, **290**, 73–77.

131 Tong, J., Yang, W., Cai, R., Zhu, B., Xiong, G., and Lin, L. (2003) *Sep. Purif. Technol.*, **32**, 289–299.

132 Kharton, V.V., Shuangbao, L., Kovalevsky, A.V., and Naumovich, E.N. (1997) *Solid State Ionics*, **96**, 141–151.

133 Tong, J., Yang, W., Zhu, B., and Cai, R. (2002) *J. Membr. Sci.*, **203**, 175–189.

134 Wu, Z., Dong, X., Jin, W., Fan, Y., and Xu, N. (2007) *J. Membr. Sci.*, **291**, 172–179.

135 Flint, S.D. and Slade, R.C.T. (1997) *Solid State Ionics*, **97**, 457–464.

136 Kröger, F.A. (1964) *The Chemistry of Imperfect Crystals*, North-Holland Publishing Company, Amsterdam.

137 Uchida, H., Maeda, M., and Iwahara, H. (1983) *Solid State Ionics*, **11**, 117–124.

138 Qi, X. and Lin, Y.S. (1999) *Solid State Ionics*, **120**, 85–93.

139 Iwahara, H., Esaka, T., Uchida, H., and Maeda, N. (1981) *Solid State Ionics*, **3–4**, 359–363.

140 Iwahara, H., Uchida, H., Ono, K., and Ogaki, K. (1988) *J. Electrochem. Soc.*, **135** (2), 529–533.

141 Mather, G.C., Figueiredo, F.M., Fagg, D.P., Norby, T., Jurado, J.R., and Frade, J.R. (2003) *Solid State Ionics*, **158**, 333–342.

142 Iwahara, H. (1992) *Solid State Ionics*, **52**, 99–104.

143 Guan, J., Dorris, S.E., Balachandran, U., and Liu, M. (1997) *Solid State Ionics*, **100**, 45–52.

144 Qi, X. and Lin, Y.S. (2000) *Solid State Ionics*, **130**, 149–156.

145 Zhang, G., Dorris, S.E., Balachandran, U., and Liu, M. (2003) *Solid State Ionics*, **159**, 121–134.

146 Hamakawa, S., Hibino, T., and Iwahara, H. (1994) *J. Electrochem. Soc.*, **141** (7), 1720–1725.

147 Song, S.J., Moon, J.H., Lee, T.H., Dorris, S.E., and Balachandran, U. (2008) *Solid State Ionics*, **179**, 1854–1857.

148 Zuo, C., Lee, T.H., Dorris, S.E., Balachandran, U., and Liu, M. (2006) *J. Power Sources*, **159**, 1291–1295.

149 Tsuji, T., Kurono, H., and Yamamura, Y. (2000) *Solid State Ionics*, **136–137**, 313–317.

150 Kreuer, K.D. (1997) *Solid State Ionics*, **97**, 1–15.
151 Taniguchi, N., Nishimura, C., and Kato, J. (2001) *Solid State Ionics*, **145**, 349–355.
152 Tanner, C.W. and Virkar, A.V. (1996) *J. Electrochem. Soc.*, **143** (4), 1386–1389.
153 Yajima, T., Suzuki, H., Yogo, T., and Iwahara, H. (1992) *Solid State Ionics*, **51**, 101–107.
154 Iwahara, H., Yajima, T., Hibino, T., Ozaki, K., and Suzuki, H. (1993) *Solid State Ionics*, **61**, 65–69.
155 Ryu, K.H. and Haile, S.M. (1999) *Solid State Ionics*, **125**, 355–367.
156 Haile, S.M., Staneff, G., and Ryu, K.H. (2001) *J. Mater. Sci.*, **36**, 1149–1160.
157 Katahira, K., Kohchi, Y., Shimura, T., and Iwahara, H. (2000) *Solid State Ionics*, **138**, 91–98.
158 Azad, A.K. and Irvine, J.T.S. (2007) *Solid State Ionics*, **178**, 635–640.
159 Ahmed, I., Ericksson, S.G., Ahlberg, E., and Knee, C.S. (2008) *Solid State Ionics*, **179**, 1155–1160.
160 Hamakawa, S., Li, L., Li, A., and Iglesia, E. (2002) *Solid State Ionics*, **48**, 71–81.
161 Chen, C.H., Bouwmeester, H.J.M., van Doorn, R.H.E., Kruidhof, H., and Burggraaf, A.J. (1997) *Solid State Ionics*, **98**, 7–13.
162 Iwahara, H., Uchida, H., and Tanaka, S. (1983) *Solid State Ionics*, **9–10**, 1021–1025.
163 Iwahara, H. (1995) *Solid State Ionics*, **77**, 289–298.
164 Thorogood, R.M., Srinivasan, R., Yee, T.F., and Drake, M.P. (1993) US Patent 5,240,480.
165 Carolan., M.F. and Dyer, P.N. (1996) US Patent 5,534,471.
166 Carolan, M.F., Dyer, P.N., Motika, S.A., and Alba, P.B. (1998) US Patent 5,712,220.
167 Carolan, M.F., Dyer, P.N., and Motika, S.A. (1998) US Patent 5,817,597.
168 Carolan, M.F., Dyer, P.N., Dyer, K.B., Wilson, M.A., Ohm, T.R., Kneidel, K.E., Peterson, D., Chen, C.M., and Rackers, K.G. (2007) US Patent 7,279,027.
169 Mackay, R. and Sammells, A.F. (2000) US Patent 6,146,549.
170 Van Calcar, P., Mackay, R., and Sammells, A.F. (2002) US Patent 6,471,921.
171 Bauer, G., Krauss, H., and Kuntz, M. (1992) US Patent 5,108,465.
172 Burggraaf, A.J. and Lin, Y.S. (1992) US Patent 5,160,618.
173 Mazanec, T.J. and Velenyi, L.J. (1992) US Patent 5,160,713.
174 Mundschau, M.V., Xie, X., Evenson JIV, C.R., and Sammells, A.F. (2006) *Catal. Today*, **118**, 12–23.
175 Dyer, P.N., Richards, R.E., Russek, S.L., and Taylor, D.M. (2000) *Solid State Ionics*, **134**, 21–33.
176 Yang, W., Wang, H., Zhu, X., and Lin, L. (2005) *Top. Catal.*, **35** (1–2), 155–167.
177 Liu, Y., Tan, X., and Li, K. (2006) *Ind. Eng. Chem. Res.*, **45**, 3782–3790.
178 Wang, H., Cong, Y., and Yang, W. (2003) *Catal. Today*, **82**, 157–166.
179 Ikeguchi, M., Mimura, T., Sekine, Y., Kikuchi, E., and Mastsukata, M. (2005) *Appl. Catal. A*, **290**, 212–220.
180 Dong, H., Shao, Z., Xiong, G., Tong, J., Sheng, S., and Yang, W. (2001) *Catal. Today*, **67**, 3–13.
181 Tong, J., Yang, W., Cai, R., Zhu, B., and Lin, L. (2002) *Catal. Lett.*, **78** (1–4), 129–137.
182 Lu, G.Q., Diniz da Costa, J.C., Duke, M., Giessler, S., Socolow, R., Williams, R.H., and Kreutz, T. (2007) *J. Colloid Interface Sci.*, **314**, 589–603.
183 Kleinert, A., Grubert, G., Pan, X., Hamel, C., Seidel-Morgenstern, A., and Caro, J. (2005) *Catal. Today*, **104**, 267–273.
184 Chen, Y., Wang, Y., Xu, H., and Xiong, G. (2008) *Appl. Catal. B*, **80**, 283–294.
185 Deckman, H.W., Fulton, J.W., Grenda, J.M., and Hershkowitz, F. (2007) US Patent 7,217,304.
186 Stevenson, J.W., Armstrong, T.R., Carneim, R.D., Pederson, L.R., and Weber, W.J. (1996) *J. Electrochem. Soc.*, **143** (9), 2722–2729.
187 Zhu, X., Cong, Y., and Yang, W. (2006) *J. Membr. Sci.*, **283**, 158–163.
188 Kharton, V.V., Yaremchenko, A.A., Viskup, A.P., Patrakeev, M.V., Leonidov, I.A., Kozhevnikov, V.L., Figueiredo, F.M., Shaulo, A.L., Naumovich, E.N., and Marques, F.M.B. (2002) *J. Electrochem. Soc.*, **149** (4), E125–E135.

189 Tao, S. and Irvine, J.T.S. (2006) *Chem. Mater.*, **18**, 5453–5460.
190 Ishihara, T., Yamada, T., Arikawa, H., Nishiguchi, H., and Takita, Y. (2000) *Solid State Ionics*, **135**, 631–636.
191 Fan, C.G., Zuo, Y.B., Li, J.T., Lu, J.Q., Chen, C.S., and Bae, D.S. (2007) *Sep. Purif. Technol.*, **55**, 35–39.
192 Carolan, M.F., Dyer, P.N., LaBar, J.M., Sr., and Thorogood, R.M. (1993) US Patent 5,240,473.
193 Zeng, Y. and Lin, Y.S. (2000) *J. Catal.*, **193**, 58–64.
194 Ran, S., Zhang, X., Yang, P.H., Jiang, M., Peng, D.K., and Chen, C.S. (2000) *Solid State Ionics*, **135**, 681–685.
195 Akin, F.T. and Lin, Y.S. (2002) *J. Membr. Sci.*, **209**, 457–467.
196 Yajima, T., Kazeoka, H., Yogo, T., and Iwahara, H. (1991) *Solid State Ionics*, **47**, 271–275.
197 Qiu, L., Ma, G., and Wen, D. (2004) *Solid State Ionics*, **166**, 69–75.
198 Stevenson, D.A., Jiang, N., Buchanan, R.M., and Henn, F.E.G. (1993) *Solid State Ionics*, **62**, 279–285.
199 Shimura, T., Esaka, K., Matsumoto, H., and Iwahara, H. (2002) *Solid State Ionics*, **149**, 237–246.
200 Song, S.J., Wachsman, E.D., Rhodes, J., Dorris, S.E., and Balachandran, U. (2004) *Solid State Ionics*, **167**, 99–105.
201 Zhu, X., Wang, H.H., Cong, Y., and Yang, W. (2006) *Catal. Lett.*, **111** (3–4), 179–185.

11
Interfacial Phenomena in Mixed Conducting Membranes: Surface Oxygen Exchange- and Microstructure-Related Factors

Xuefeng Zhu and Weishen Yang

The transport behavior of ions, such as O^{2-} anions or protons, driven by chemical potential gradient across dense mixed conducting membranes is governed by many factors. In this chapter, the effects of surface exchange and ceramic microstructure are addressed. The surface-related aspects are discussed involving selected theoretical approaches and modeling. Relevant experimental methods are briefly described. The influence of microstructure on the gas permeation processes is analyzed for several types of mixed conducting membranes. The relationships between the membranes' stability and their microstructure and surface exchange kinetics are considered on the basis of existing experimental data.

11.1
Introduction

Solid-state nonmetallic conducting materials have been extensively developed and applied for various solid electrochemical devices during the last century. A large number of works reveal the great interest of researchers in energy storage and conversion devices, sensors, gas separation membranes, and other electrochemical applications of solid-state conducting materials [1–3]. The charge carriers may include electrons (or holes), protons, nonmetallic anions (such as O^{2-}, Cl^-, etc.), and metal cations (such as Li^+, Na^+, etc.). The major groups of ion-conducting materials are discussed in the first volume of this handbook. Mixed conductors are the materials that have more than one type of mobile charge carriers (see Chapter 3 of the first volume). The high-temperature electrochemical applications of mixed conductors are, in particular, electrodes for solid oxide fuel cells (SOFCs) and other electrochemical devices, and ceramic membranes for oxygen, hydrogen, and carbon dioxide separation. For these types of applications, the solid should have high values of both the ionic/protonic and electronic conductivities. The electrode applications have been reviewed by many researchers [4–7], including Chapter 12 of the first volume and Chapters 5, 6 and 9 of this book, and are thus outside the scope of this chapter focused on selected aspects of mixed conducting membranes for gas separation.

Solid State Electrochemistry II: Electrodes, Interfaces and Ceramic Membranes.
Edited by Vladislav V. Kharton.
© 2011 Wiley-VCH Verlag GmbH & Co. KGaA. Published 2011 by Wiley-VCH Verlag GmbH & Co. KGaA.

Up to now, there are three kinds of mixed conducting membranes, namely, oxygen-permeable membranes, hydrogen-permeable membranes, and carbon dioxide-permeable membranes (see also Chapter 10). Correspondingly, the charge carriers are electrons/holes and oxygen ions, electrons/holes and protons, and electrons/holes/oxygen ions and carbonate ions. Based on the transport mechanisms through these membranes, the theoretical permselectivity is 100% for oxygen and hydrogen membranes; however, it depends on the dominating carriers for CO_2-permeable membranes, which were thoroughly discussed in Refs [8–10]. The mixed conducting membranes for oxygen separation are especially discussed in this chapter, but the principles discussed for oxygen separation are similar to the ceramic membranes for hydrogen and CO_2 separation.

Dense ceramic oxygen-permeable membrane made from mixed oxygen ionic and electronic conductors can separate gaseous oxygen from air at elevated temperatures with infinite permeation selectivity (see Chapter 3 of the first volume). In such a separator, the membrane is exposed to an asymmetric oxygen partial pressure environment. The oxygen chemical potential gradation across the membrane makes the oxygen ions directionally transported from the high oxygen partial pressure side to the low oxygen partial pressure side, accompanied by the reverse transfer of electrons. These mixed conducting membranes are distinguished from the more conventional membranes that usually have pores as mass transporting path and are based on solution–diffusion, viscous diffusion, bulk diffusion, Knudsen diffusion, surface diffusion, activated diffusion, capillary condensation, molecular sieving, and other related mechanisms to realize selective separation. Cales and Baumard introduced the mixed conducting solid oxides concept for oxygen-permeable membranes about 30 years ago [11, 12], although the famous oxygen ions conductor, Y_2O_3-stabilized ZrO_2 solid solution, had been discovered by Nernst 110 years ago. Mixed ionic–electronic conducting is required for molecular oxygen transfer; however, fluorite oxides with low electronic conductivity exhibit poor permeability when they are applied as membranes. Perovskite-type membranes (see Chapter 9 of the first volume) have attracted attention since Teraoka and coworkers [13, 14] reported the high oxygen permeability of $LaCoO_3$-based perovskite oxides in 1985. High oxygen fluxes were later observed for dense ceramic membranes having perovskite-type structures at temperatures higher than 700 °C, especially for the compositional family $Ln_{1-x}(Ba,Sr,Ca)_xCo_{1-y}Fe_yO_{3-\delta}$ (Ln = lanthanide) [15–30]. The unique high-temperature oxygen permeation characteristics provide a possibility to integrate such membranes with catalytic oxidation reactions, such as partial oxidation of methane to syngas (POM) [26, 31–42], oxidative coupling of methane to C_2 (ethylene and/or ethane) (OCM) [43–49], and selective oxidation of ethane to ethylene (SOE) [50–53]. The investigations on mixed conducting membranes and membrane reactors for catalytic oxidation reactions have been extensively reviewed in the literature [31, 54–57].

When analyzing the most critical factors, one should note that oxygen permeation through mixed conducting membranes may be controlled by the dissociation and ionization of oxygen on the membrane surface at the high oxygen pressure side (feed side), by the transport of electrons/oxygen ions through membrane bulk, and/or by the recombination of oxygen ions to form oxygen molecules or other redox reactions at the low oxygen pressure side (permeate side). The first and third stages are usually called surface exchange limitation steps, and the second is called bulk diffusion limitation step. The bulk diffusion is governed by oxygen ionic or electronic conduction (or both). For the perovskite-related mixed conductors, the electronic conductivity is usually two to four orders higher than their ionic conductivity, so ionic diffusion is the rate-determining step (rds) (see Chapter 9 of the first volume); however, for fluorite-type electrolyte oxides, the ionic conductivity is high compared to their electronic conductivity, so electronic transport limits oxygen permeation. The permeation flux can be increased by reducing the thickness of the membrane when bulk diffusion is dominated, and oxygen flux follows well the classical Wagner equation (Chapter 3 of the first volume). Nonetheless, the reported fluxes of supported thin membranes are lower than the values expected on the diffusion limitation. The characteristic thickness L_c was introduced to simply analyze the permeation-determining factors [58, 59]. When the thickness of a dense membrane is lower than L_c, its oxygen flux can be only marginally improved by making the membrane thinner. L_c is defined via the ratio of oxygen self-diffusion and surface exchange coefficients. These two parameters depend not only on membrane material compositions but also on a number of other parameters, such as synthesis method of powders, sintering programs, impurities on surface and grain boundaries, and surface morphologies. That is to say, microstructure of membranes has great influence on oxygen permeation flux, which can explain most discrepancies in the literature data, such as inconsistent oxygen fluxes reported by different researchers on the same composition. This chapter discusses the microstructure and surface effects on oxygen permeation.

11.2 Surface Exchange

Figure 11.1 schematically shows the gradient of oxygen chemical potential (μ_{O_2}) across an oxygen-permeable membrane. The exchange process between gas-phase

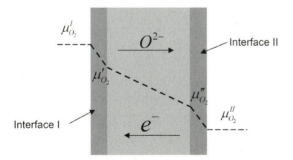

Figure 11.1 Oxygen chemical potential drop across a mixed conducting membrane.

oxygen molecules and membrane surfaces includes a series of reaction steps, and each of the steps may play an important role in oxygen permeation through membranes. The elemental steps such as adsorption, dissociation, charge transfer, surface diffusion, and oxygen ions incorporation into the lattice are involved on the interface zone of feed side. The same elemental steps occur on the interface zone of permeate side but in reverse direction, while the relative roles of the same steps may differ. Surface exchange reactions on both sides would become the rate-determining steps if the bulk transport, including both electronic and ionic, is fast enough or the thickness of membranes is thinner than L_c. Since the Wagner equation cannot well describe the overall permeation process through the membranes, proper equations need to be developed.

11.2.1
Theoretical Analysis of Surface Effects: One Relevant Approach

Appropriate assumptions are always necessary to allow the theoretical analysis, development of simple equations, and understanding the nature of interface effects. Following Virkar [60], who discussed the role of interfaces in transport through mixed conducting and ionic conducting membranes, a brief introduction to the interfacial analysis is presented here. The basic assumptions [60] are as follows: (i) the ionic and electronic transport properties in a given zone are constant regardless of position and oxygen chemical potentials; (ii) the diffusion resistance or concentration polarization associated with oxygen molecules transporting from surrounding atmosphere to the membrane interfaces on both sides can be totally ignored (caution that this hypothesis may be invalid for thin membranes on thick supports or membranes with porous layers); and (iii) all the steps (adsorption, dissociation, charge transfer, surface diffusion, and oxygen ions merging into the crystal lattices) are under isothermal conditions.

The best known diffusion equation describing a continuous flux (j_i, mol cm^{-2} s^{-1}) of ith species through a unit surface is the Fick's first law

$$j_i = -D_i \frac{dc_i}{dx}, \tag{11.1}$$

where D_i is the diffusion coefficient (cm^2 s^{-1}) and c_i is the concentration of ith species (cm^{-3}). The equation can be rewritten by introducing the electrochemical potential $\tilde{\mu}_i$ (see Chapters 1 and 3 of the first volume for definitions):

$$j_i = -\frac{D_i c_i}{k_B T} \frac{d\tilde{\mu}_i}{dx}, \tag{11.2}$$

where k_B is the Boltzmann constant. Due to internal short circuit in a mixed conducting membrane with high electronic conductivity, the electrochemical potential gradient is essentially equal to chemical potential gradient $d\mu_i/dx$.

The current density I_i (A cm^{-2}) corresponding to the directional transport of species i for a one-dimensional case is given by

$$I_i = z_i e j_i = -\frac{z_i e D_i c_i}{k_B T}\frac{d\mu_i}{dx} = -\frac{\sigma_i}{z_i e}\frac{d\mu_i}{dx}, \qquad (11.3)$$

where $\sigma_i = z_i^2 e^2 c_i D_i / k_B T$ is the partial conductivity of species i (S cm^{-1}), z_i is their charge (e.g., 2 for O^{2-}), and e is the absolute electron charge (C). It should be kept in mind that no net current flows through the membrane once the stationary state has been reached and equivalent amounts of electrons and oxygen ions are transferred in opposite directions. For a mixed electronic–ionic conductor, the key equilibrium of interest may be written as

$$\frac{1}{2}O_2 + 2e' \leftrightarrow O^{2-} \qquad (11.4)$$

At thermodynamic equilibrium,

$$\frac{1}{2}\mu_{O_2} + 2\mu_e = \mu_{O^{2-}}. \qquad (11.5)$$

For small deviations from the thermodynamic equilibrium, we can expand Equation 11.5:

$$\frac{1}{2}\delta\mu_{O_2} + 2\delta\mu_e = \delta\mu_{O^{2-}}, \qquad (11.6)$$

where δX denotes a small deviation in the thermodynamic potential X from the equilibrium state. Local equilibrium needs to be kept at any point of the membrane; this requires participation of at least two kinds of charge carriers in the reaction. For mixed conducting materials, fast transport of both electrons and oxygen ions makes it easy to maintain local equilibrium. Oxygen ionic and electronic current densities are given by

$$I_{O^{2-}} = \frac{\sigma_{O^{2-}}}{2e}\nabla\mu_{O^{2-}} = \frac{\sigma_{O^{2-}}}{2e}\frac{d\mu_{O^{2-}}}{dx} \qquad (11.7)$$

and

$$I_e = \frac{\sigma_e}{e}\nabla\mu_e = \frac{\sigma_e}{e}\frac{d\mu_e}{dx}. \qquad (11.8)$$

Substituting Equation 11.6 into Equation 11.7, we can get the following relation:

$$I_{O^{2-}} = \frac{\sigma_{O^{2-}}}{4e}\frac{d\mu_{O_2}}{dx} + \frac{\sigma_{O^{2-}}}{e}\frac{d\mu_e}{dx}. \qquad (11.9)$$

Figure 11.1 shows a schematic illustration of a mixed conducting membrane under an oxygen chemical potential gradient; the total differences of oxygen and electron chemical potentials are $(\mu_{O_2}^I - \mu_{O_2}^{II})$ and $(\mu_e^I - \mu_e^{II})$, respectively. Transport of oxygen ions and electrons occurs in the bulk and across the relatively thin interfacial zones, where the thickness may hardly be well defined by either experimental or theoretical methods. Virkar [60] assumed that the "interfacial zones" are of certain thickness, enabling to describe transport in terms of partial conductivities. Hereafter $\sigma'_{O^{2-}}$, σ'_e, $\sigma''_{O^{2-}}$, and σ''_e are the oxygen ionic and electronic conductivities in the interfacial layers I and II, respectively; l' and l'' are their thicknesses. Area-specific conductance (S cm^{-2}) and resistance (Ω cm^2) are introduced because it is impossible to separately determine partial conductivities and thickness of the interfacial zones, but the normalized conductance and resistance can be, in principle, measured experimentally. For example, for area-specific oxygen ionic conductance (κ) and resistance ($r_{O^{2-}}$) in the interfacial zone I,

$$\kappa'_{O^{2-}} = \frac{\sigma'_{O^{2-}}}{l'} = \frac{1}{r'_{O^{2-}}}. \tag{11.10}$$

This equation is based on the simplest assumption that the transport properties are constant across the interfacial zones, which is not the case for every membrane. Generally speaking, the transport properties at the membrane interfaces depend on the oxygen chemical potential exposed. Nonetheless, the simplification makes it possible to easily understand the nature of interfacial effects. The interfacial oxygen ionic or electronic conductivities are not equal to those in the membrane bulk; these properties are all greatly influenced by microstructure, surface composition, and other factors discussed below.

Assuming homogeneous transport across the interfacial zones, the oxygen ionic and electronic currents are given by (see Figure 11.1)

$$I'_{O^{2-}} = -\frac{\kappa'_{O^{2-}}}{4e}(\mu'_{O_2} - \mu^I_{O_2}) + \frac{\kappa'_{O^{2-}}}{e}(\mu'_e - \mu^I_e) = -\frac{\mu'_{O_2} - \mu^I_{O_2}}{4r'_{O^{2-}}e} + \frac{\mu'_e - \mu^I_e}{r'_{O^{2-}}e}, \tag{11.11}$$

$$I''_{O^{2-}} = -\frac{\kappa''_{O^{2-}}}{4e}(\mu''_{O_2} - \mu^{II}_{O_2}) + \frac{\kappa''_{O^{2-}}}{e}(\mu^{II}_e - \mu''_e) = \frac{\mu^{II}_{O_2} - \mu''_{O_2}}{4r''_{O^{2-}}e} + \frac{\mu^{II}_e - \mu''_e}{r''_{O^{2-}}e}, \tag{11.12}$$

$$I'_e = \frac{\kappa'_e}{e}(\mu'_e - \mu^I_e) = \frac{\mu'_e - \mu^I_e}{r'_e e}, \tag{11.13}$$

and

$$I''_e = \frac{\kappa''_e}{e}(\mu^{II}_e - \mu''_e) = \frac{\mu^{II}_e - \mu''_e}{r''_e e}, \tag{11.14}$$

where μ' and μ'' correspond to the chemical potentials at the interfaces I and II, respectively. In a steady state, all local current densities become equal to one another. Since no external voltage is applied, the net current through membranes is zero. That is, $I_{O^{2-}} + I_e = 0$; furthermore, $I_{O^{2-}} < 0$ and $I_e > 0$. By combining Equations 11.11 and 1.13, and Equations 11.12 and 11.14, respectively, and then rewriting the equations, one can get the relationships for oxygen chemical potentials:

$$\mu'_{O_2} = \mu^I_{O_2} - 4e(r'_{O^{2-}} + r'_e)|I_{O^{2-}}|, \tag{11.15}$$

$$\mu''_{O_2} = \mu^{II}_{O_2} + 4e(r''_{O^{2-}} + r''_e)|I_{O^{2-}}|. \tag{11.16}$$

That is, $\mu'_{O_2} < \mu^I_{O_2}$ and $\mu''_{O_2} > \mu^{II}_{O_2}$. Thus, for a mixed conducting membrane under a given oxygen chemical potential difference

$$\mu_{O_2}^{I} - \mu_{O_2}^{II} = \Delta\mu_{O_2}^{tot}, \tag{11.17}$$

the oxygen chemical potential decreases in the sequence $\mu_{O_2}^{I} > \mu_{O_2}' > \mu_{O_2}'' > \mu_{O_2}^{II}$, whatever the transport process is limited by interfacial exchange or bulk diffusion or both.

For the membrane bulk, the oxygen permeation flux can be written as

$$j_{O_2} = -\frac{1}{4^2 e^2 l_{bulk}} \frac{\sigma_{O^{2-}}^{bulk} \sigma_e^{bulk}}{\sigma_{O^{2-}}^{bulk} + \sigma_e^{bulk}} \Delta\mu_{O_2}^{bulk} = -\frac{1}{4^2 e^2} \frac{1}{r_{O^{2-}}^{bulk} + r_e^{bulk}} \Delta\mu_{O_2}^{bulk}, \tag{11.18}$$

where l_{bulk} is the thickness of the bulk diffusion zone, and the partial conductivities are considered independent of the oxygen chemical potential gradient. This assumption may be valid, in particular, when the gradient is small; otherwise, the averaged transfer coefficients should be considered (see Chapter 3 of the first volume). As $|I_{O^{2-}}| = 4ej_{O_2}$, one can easily obtain

$$\Delta\mu_{O_2}^{bulk} = \mu_{O_2}' - \mu_{O_2}'' = 4e|I_{O^{2-}}|(r_{O^{2-}}^{bulk} + r_e^{bulk}). \tag{11.19}$$

By combining Equations 11.15–11.17 and 11.19, one can derive the formula for total difference of the oxygen chemical potential across the membrane:

$$(\mu_{O_2}^{I} - \mu_{O_2}^{II}) = 4e|I_{O^{2-}}|(r_{O^{2-}}' + r_e') + 4e|I_{O^{2-}}|(r_{O^{2-}}^{bulk} + r_e^{bulk}) + 4e|I_{O^{2-}}|(r_{O^{2-}}'' + r_e''), \tag{11.20}$$

where the three terms on the right-hand side correspond to the reduced chemical potential at the interface I, bulk, and interface II, respectively. This equation may be reformulated as

$$\Delta\mu_{O_2}^{tot} = \Delta\mu_{O_2}' + \Delta\mu_{O_2}^{bulk} + \Delta\mu_{O_2}''. \tag{11.21}$$

All the three steps can be rate determining, depending on the specific resistances. Note again that the surface resistances are highly sensitive to the preparation processes, microstructures, morphologies, atmospheres, and other factors. That is why the membrane performance of the same compositions prepared by different groups often varies in a wide range.

Under conditions close to thermodynamic equilibrium, the permeation flux through an interface can be described as [58]

$$j_{O_2} = -j_{ex}^{i} \frac{\Delta\mu_{O_2}^{i}}{k_B T}, \tag{11.22}$$

where j_{ex}^{i} (m^{-2} s^{-1}) is the balance exchange rate in the absence of oxygen potential gradients at the interface. In the simplest cases, the latter quantity is directly related to the equilibrium surface exchange coefficient k_{ex} (m s^{-1}), which may be obtained from the experimental data of ^{18}O–^{16}O isotopic exchange [58]:

$$j_{ex}^{i} = \frac{1}{4} k_{ex} c_{O^{2-}}, \tag{11.23}$$

where $c_{O^{2-}}$ (m^{-3}) is the volume concentration of oxygen anions at equilibrium. Therefore,

$$k_{ex} = \frac{k_B T}{4e^2 c_{O^{2-}}} \frac{1}{r'_{O^{2-}} + r'_e} = \frac{k_B T}{4e^2 c_{O^{2-}}} \frac{\kappa^i_{O^{2-}} \kappa^i_e}{\kappa^i_{O^{2-}} + \kappa^i_e} = \frac{k_B T}{4e^2 c_{O^{2-}}} \kappa^i_{di}, \quad (11.24)$$

where the last quantity expresses interfacial ambipolar conduction, and all the parameters are defined on a per species (per ion, electron, atom, or molecule) basis. If defined on a per mole basis, Equation 11.24 should be rewritten as

$$k_{ex} = \frac{RT}{4F^2 C_{O^{2-}}} \kappa^i_{di}, \quad (11.25)$$

where R, T, and F have their usual meaning; $C_{O^{2-}}$ (mol m^{-3}) is the volume concentration of oxygen anions at equilibrium; κ^i_{di} (S m^{-2}) is the area-specific ambipolar conductance of the interface. Equations 11.24 and 11.25 explain why mixed conducting perovskite oxides such as $La_{1-x}Sr_xCoO_3$ have higher oxygen exchange coefficient than $La_{1-x}Sr_xMnO_3$ with very poor ionic conductivity. It should be noted that the interfacial conductance may not be considered equal to the bulk conductance as the interfacial transport mechanisms are much more complex. The surface diffusion and kinetic demixing lead to further complications. For example, enrichment of alkaline earth or lanthanoids is often found on membrane surface under oxygen permeation conditions [61], decreasing the interfacial transport.

The exchange coefficient described by Equation 11.25 is an equilibrium parameter. It may be related to the real parameter k^i as [62]

$$k^i = k_{ex}(2n_{ex} + 1/\Gamma), \quad (11.26)$$

where Γ is the thermodynamic factor given by

$$\Gamma = \frac{1}{RT} \frac{d\mu^i_{O_2}}{d \ln C_{O^{2-}}}. \quad (11.27)$$

For mixed conducting materials, $\Gamma \gg 1$ and n_{ex} coefficient usually lies between 0.5 and 1. This means that the real exchange coefficients are normally larger than the equilibrium coefficient.

Many mixed conductors, such as cobalt-containing perovskite oxides, exhibit a fast surface exchange, which is of great interest for other applications (e.g., fuel cell electrodes; see Chapter 13 of the first volume). However, for such materials, oxygen permeation is sometimes controlled by the exchange steps. Bouwmeester and Burggraaf [58, 59] introduced the characteristic thickness L_c to define the membrane thickness corresponding to transition from predominant bulk diffusion control to the state when permeation is governed by the interfacial exchange. For mixed conductors where the electronic conduction is dominant, the characteristic thickness is given by [58, 59]

$$L_c = \frac{D_s}{k_{ex}} = \frac{D^*}{k_{ex}} \quad (11.28)$$

and

$$D_s = \frac{\sigma_{O^{2-}}^{bulk} RT}{4F^2 C_{O^{2-}}},\qquad(11.29)$$

where D_s is the self-diffusion coefficient of oxygen ions derived from classical Nernst–Einstein relationship and D^* is the tracer diffusion coefficient equal to the self-diffusion coefficient if correlation effects can be neglected. The above equations are simplified and are only valid for small oxygen partial pressure gradients applied across membranes.

Substituting Equations 11.25 and 11.29 into Equation 11.28 gives

$$L_c = \frac{\sigma_{O^{2-}}^{bulk}}{\kappa_{di}^i} = \frac{\sigma_{O^{2-}}^{bulk}}{\sigma_{di}^i} l^i.\qquad(11.30)$$

The same equation can be obtained through Equations 11.20 and 11.21 if the sum of reduced oxygen chemical potential differences at the two membrane interfaces is equal to that in membrane bulk. From Equation 11.30 one can find that the characteristic thickness is a ratio between bulk oxygen ionic conductivity $\sigma_{O^{2-}}^{bulk}$ and interfacial ambipolar conductance κ_{di}^i or interface thickness multiplied by the ratio of bulk oxygen ionic conductivity to interfacial ambipolar conductivity σ_{di}^i. Here, interface-specific ambipolar conductance should be discussed, although electronic conduction is dominant for many perovskite-type mixed conductors. In particular, it is common for perovskite-type materials that elemental composition of membrane surfaces is very different from the bulk. This leads to the differences of the transport properties, so for some membranes their interfacial electronic conductance is even lower than the ionic conductance of the mixed conducting materials. Although the interface thickness may hardly be determined precisely, it is far thinner than the membrane thickness. The characteristic thickness of many membranes is in millimeter scale, so $\sigma_{O^{2-}}^{bulk} \gg \sigma_{di}^i$.

As a matter of fact, the characteristic thickness defined by Equation 11.28 is usually thinner than the real characteristic thickness under typical operation conditions. The exchange coefficient can be often approximated by a power function of the oxygen partial pressure with an exponent close to 0.5. Therefore, the exchange coefficient of permeate side, k_{ex}^{II}, is usually several times lower than that of feed side, k_{ex}^{I}. Considering that both coefficients should have contributions to the real characteristic thickness, a combined exchange coefficient can be defined as

$$k_{ex}^{I,II} = k_{ex}^{I} k_{ex}^{II}/(k_{ex}^{I} + k_{ex}^{II})\qquad(11.31)$$

and the real L_c values may be estimated as

$$L_c = \frac{D_s}{k_{ex}^{I,II}} = \frac{\sigma_{O^{2-}}^{bulk}}{\kappa_{di}^{I,II}} = \frac{\sigma_{O^{2-}}^{bulk}}{\sigma_{di}^{I,II}} l^{I,II},\qquad(11.32)$$

where $\kappa_{di}^{I,II}$, $\sigma_{di}^{I,II}$, and $l^{I,II}$ are the combined interface-specific conductance, interfacial conductivity, and interface thickness, respectively, all accounting for contributions of both sides of the membrane. Therefore, irrespective of the mechanisms, improvement of the interface conductivity is very important for membranes possessing slow surface exchange kinetics. It should also be specifically noted that the characteristic thickness is not an intrinsic material property, and largely depends on the roughness and pore structure of the membrane surface, similar to the interface conductance.

11.2.2
Modeling Formulas

Chemically driven oxygen permeation through mixed conducting membranes can be well described by Wagner equation if the permeation process is controlled by bulk diffusion. However, in many cases, the rate-determining step is not bulk diffusion but surface exchange or a joint control. A general transport equation derived from Equation 11.20 gives

$$j_{O_2} = -\frac{1}{4^2 F^2} \frac{1}{r' + r^{\text{bulk}} + r''} \Delta\mu_{O_2}^{\text{tot}} = -\frac{\Delta\mu_{O_2}^{\text{tot}}}{4^2 F^2} \frac{1}{r^{\text{tot}}}, \qquad (11.33)$$

where $-\Delta\mu_{O_2}^{\text{tot}}/4^2 F^2$ is the overall driving force and $r^{\text{tot}} = r' + r^{\text{bulk}} + r''$ is the total permeation resistance. Although Equation 11.33 can well describe the oxygen permeation process, it is inconvenient to obtain specific resistances at the interfaces. Empirical transport equations taking the oxygen partial pressure into account can be used to reveal that the permeation is controlled by bulk diffusion or surface exchange or both.

At a small oxygen deficiency variation and/or a limited oxygen pressure difference, acceptor dopants are among the majority defects for perovskite-type mixed conductive materials, in which the oxygen nonstoichiometry (δ) is a power function of oxygen pressure, $\delta \propto P_{O_2}^n$ [63]. If the oxygen vacancy ordering can be neglected or all the oxygen vacancies contribute to oxygen transport, an empirical equation can be suggested [58, 59]:

$$j_{O_2} = \frac{\beta_0}{L_{\text{th}}} \left(P_{O_2}^{\text{I }n_0} - P_{O_2}^{\text{II }n_0} \right), \qquad (11.34)$$

where $\beta_0 = D_v \delta_0 / 4 n_0 V_m$ in the case of dominant bulk limitations. Here, the parameters D_v, δ_0, V_m, and L_{th} are the diffusion coefficient of oxygen vacancies, nonstoichiometry at the reference oxygen pressure (1 atm), molar volume of the membrane material, and membrane thickness, respectively. The value of n_0 can be determined experimentally from the dependence of oxygen flux on oxygen pressure difference across the membrane. Actually, this equation can model the permeation flux dependence on oxygen partial pressure (whenever it is governed by bulk diffusion or surface exchange kinetics) by adopting different values of the exponent n_0, although the formula is theoretically derived from a bulk diffusion equation. Therefore, the slope coefficient β_0/L_{th} is directly related to the vacancy diffusion coefficient only if the permeation process is totally limited by the bulk diffusion; an exchange coefficient can be obtained using β_0/L_{th} as long as the exchange reactions prevail, as discussed below. In the latter case, although the physical meaning of n_0 in

Equation 11.34 becomes amphibolous, some researchers reported that the permeation is predominantly limited by surface exchange kinetics when n_0 is large and positive and is mainly controlled by the bulk processes when n_0 is small. For example, Huang and Goodenough [64] discussed the relationships between n_0 and oxygen permeation rate-determining steps in detail; the authors [64] argued that the permeation is mainly controlled by bulk diffusion when n_0 is negative, and is mainly restricted by exchange when n_0 is close to 0.5. Note that the values of n_0 in Equation 11.34 depend on thickness and temperature, so n_0 changes when the thickness is reduced, from diffusion-controlled to the interface-controlled regime. That is to say, n_0 is not an intrinsic parameter and varies for different membranes even if they are prepared via the same route and have the same composition; this parameter may hardly be used for extrapolation.

Jacobson et al. [62, 65–68] derived another general equation to describe the oxygen permeation, considering the dependence of net surface exchange flux governed by oxygen chemical potential in the gaseous phase and by the concentrations of ions and electronic charge carriers at the gas/solid interface. At thermal equilibrium and for a disk-type membrane,

$$\frac{4 n_P L_{th} j_{O_2}}{\langle c_i D_a \rangle} = \ln \frac{\sqrt{P_{O_2}^I/P_0} - 2 j_{O_2}/c_{i,1} k_{i0}}{\sqrt{P_{O_2}^{II}/P_0} + 2 j_{O_2}/c_{i,2} k_{i0}}, \tag{11.35}$$

where P_0 is the reference oxygen pressure, n_P is a constant, D_a is the ambipolar diffusion coefficient that is close to the ion diffusion coefficient ($D_{O^{2-}}$) when no limitations of the electronic transport is observed, and the denominator of the left term contains quantities averaged under a given gradient. In the surface exchange-controlled regime, the above equation reduces to

$$j_{O_2} = \frac{1}{2} \frac{c_{i,1} c_{i,2}}{c_{i,1} + c_{i,2}} k_{i0} \left[\left(P_{O_2}^I/P_0 \right)^{n_P} - \left(P_{O_2}^{II}/P_0 \right)^{n_P} \right]. \tag{11.36}$$

For the bulk diffusion control,

$$j_{O_2} = \frac{1}{4} \frac{\langle c_i D_a \rangle}{L_{th}} \ln \frac{P_{O_2}^I}{P_{O_2}^{II}}. \tag{11.37}$$

Here, $k_{i0} = k_{ex}(P_0/P_{O_2})^{-n_P}$, $D_a = t_e D_{O^{2-}}$, t_e is the electron transference number, and $c_{i,1}$ and $c_{i,2}$ are the oxygen ion concentrations at the interfaces, respectively. The oxygen pressure dependence of the permeation flux predicted by Equation 11.36 is similar to Equation 11.34, although these equations were derived from different assumptions. For example, the permeation properties of $SrCo_{0.8}Fe_{0.2}O_{3-\delta}$ membranes were analyzed by these two equations in Refs [59, 62], respectively. The predicted values of n_0 between -0.05 and -0.20 were obtained from the oxygen nonstoichiometry data; however, fitting of the oxygen flux data using Equation 11.34 results in 0.5 at 750 °C. Likewise, both Equations 11.36 and 11.37 cannot fit well with the permeation behavior of $SrCo_{0.8}Fe_{0.2}O_{3-\delta}$ membranes at 890 °C. In these cases, both interface exchange and bulk diffusion play an important role, so the above equations cannot clearly identify rate-determining steps; also, the exchange kinetics and ambipolar diffusion processes are more complex than those predicted by simplified theoretical models. Also note that Equation 11.36 dependence is essentially similar to Equation 11.34 and can be generalized as

$$j_{O_2} = \alpha_P \left(P_{O_2}^{I\ n_P} - P_{O_2}^{II\ n_P} \right), \tag{11.38}$$

where α_P is a constant comprising the oxygen vacancies diffusion coefficient and interface exchange coefficients.

Another important equation was developed by Xu and Thomson [69], who simplified Lin's model [70] analyzing the oxygen vacancies diffusion through membranes and the oxygen exchange reactions at interfaces; the main assumptions are as follows: (1) the electronic conductivity is much faster than ionic; (2) the vacancy diffusion coefficient D_v is independent of the oxygen partial pressure; (3) the forward and reverse interfacial reactions have the same rate constants (k_f and k_r, respectively); and (4) the mass action law can be used for the exchange reactions on both membrane surfaces. The resultant equation reads

$$j_{O_2} = \frac{(k_r/k_f)(P_{O_2}^{I-0.5} - P_{O_2}^{II-0.5})}{(1/k'_{ex}) + (2L_{th}/D_v) + (1k''_{ex})} = \frac{\Delta(P_{O_2})}{R'_{ex} + R^{bulk} + R''_{ex}} = \frac{\Delta(P_{O_2})}{R_t} \tag{11.39}$$

where $k'_{ex} = k_f P_{O_2}^{\prime 0.5}$, $k''_{ex} = k_f P_{O_2}^{\prime\prime 0.5}$, the numerator corresponds to the driving force for oxygen transport across the membrane, and the denominator is the total permeation resistance. A clear map showing the temperature and oxygen pressure effects on the distribution of separate contributions to the permeation resistance can be easily drawn using this model (Figure 11.2). Under a small oxygen pressure difference, the exchange coefficients at both interfaces are functions of the oxygen pressure, namely, $k'_{ex}/k''_{ex} = (P_{O_2}^I/P_{O_2}^{II})^{n_k}$, where n_k is close to 0.5. Since $P_{O_2}^I$ is usually much higher than $P_{O_2}^{II}$, the feed-side surface contribution to the total permeation resistance is often negligible compared to the bulk and permeate side. Moreover, the oxygen exchange activation energy at the permeate side is usually higher than that for vacancy diffusion, so the R''_{ex} increases quickly with decreasing temperature, as shown in Figure 11.2a. That is why the permeation fluxes exhibit normally two temperature ranges with different activation energies; the activation energy values are higher in the low-temperature regime, even if neither structural changes nor vacancy ordering occur. The inflection may just reveal that the rate-limiting step transfers from bulk to interfacial exchange at the permeate side. For most mixed conducting membranes, the exchange at the feed-side surface is not permeation determining, so the distribution of total permeation resistance remains essentially unaffected by increasing the feed-side oxygen pressure (Figure 11.2b). However, R''_{ex} tends to rise quickly when the permeate-side oxygen pressure decreases (Figure 11.2c). This can well explain experimental observations showing that in many cases the oxygen flux linearly increases with $\ln P_{O_2}^I/P_{O_2}^{II}$ at relatively small oxygen pressure gradients, but deviates from the linear dependence when the gradient increases. As discussed in Section 11.2.1, the oxygen permeation fluxes increase linearly with increasing reciprocal thickness in the bulk diffusion limitation regime until the interface exchange

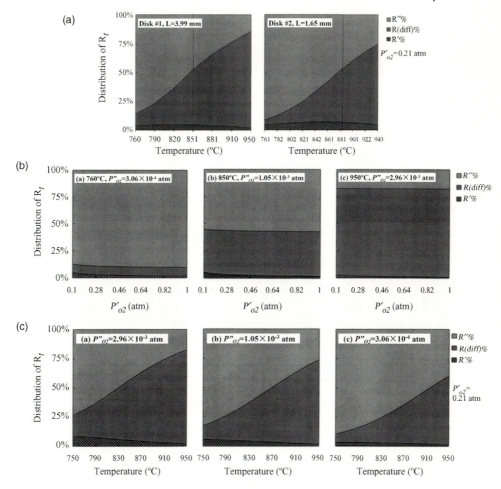

Figure 11.2 Relative roles of the permeation resistances on the steady-state oxygen permeation under different conditions: (a) effects of the membrane thickness and temperature; (b) effects of the oxygen partial pressure at the feed side; and (c) combined influence of the permeate-side oxygen pressure and temperature [69]. (Reproduced with permission from Ref. [69].)

resistance becomes significant. The characteristic thickness L_c is defined at the point where the exchange resistance $R'_{ex} + R''_{ex}$ is equal to bulk diffusion resistance. The effects of membrane thickness are displayed in Figure 11.2a. Hence, among the three model equations considered in this section, the last formalism is especially useful in identifying relative roles of the bulk diffusion and interfacial kinetics in the permeation process; that is why the last model is widely used in the literature [71–73].

In summary, all the model equations are not perfect, but they may provide useful information when used under proper conditions. In particular, Equation 11.38 is a general equation that can be used to model oxygen permeation process regardless of the rate-determining step; although the physical meaning of approximative n_P coefficient is not well defined, the equation provides a good basis for scaled-up engineering calculations.

11.2.3
Selected Experimental Methods

Considering the most common experimental techniques, information on diffusion and exchange coefficients can be obtained by the isotopic exchange [74–78], relaxation experiments [79–82], coulometric titration [83], impedance spectroscopy, and electrochemical polarization experiments [84, 85]. In this section, the most common approaches are briefly reviewed. Description of other methods mentioned in this section can be found in the references cited.

11.2.3.1 Isotopic Exchange

One of the direct methods to measure the oxygen self-diffusion coefficient and interfacial exchange coefficient of oxide materials is ^{18}O isotope exchange, followed by secondary ion mass spectrometry (SIMS) depth profiling and/or combined with the gaseous phase analysis. In the classical implementation, dense samples are to be evacuated at a given temperature and then to be equilibrated in an atmosphere with natural content of the oxygen isotopes (0.2% ^{18}O); once the samples had been cooled, the apparatus is evacuated again and ^{18}O labeled gas is introduced in the chamber followed by rapid heating and keeping at a target temperature for a certain annealing period. The analysis method depends on the ^{18}O penetration depth. For the depths smaller than 10 μm, the profiling method is usually used. For the profilometry mode, the primary ion beam is scanned across a region of the sample surface to produce a square profile, while the secondary ion intensities of ^{16}O and ^{18}O are recorded as a function of time. When the depths are relatively large, the annealed samples are cut in direction perpendicular to the exchanged surface, and then analyzed by laterally scanning the primary ion beam; the two isotopic secondary ion intensities are recorded. Mass transport during the isotope exchange across gas/solid interface is given by

$$k^*(C_g - C_s) = -D^* \left(\frac{\partial C(x)}{\partial x}\right)_{x=0}, \qquad (11.40)$$

where C_g and C_s are the normalized isotopic fractions of ^{18}O in the gas and solid, respectively, and k^* and D^* are the tracer surface exchange coefficient and tracer oxygen diffusion coefficient, respectively. Neglecting the correlation effects, the latter quantities are often considered essentially equivalent to the equilibrium exchange coefficient k_{ex} and self-diffusion coefficient, respectively. The analytical solution of the one-dimensional diffusion problem is given by Crank [86]:

$$C'(x) = \frac{C(x) - C_{bg}}{C_g - C_{bg}} = \mathrm{erfc}\left(\frac{x}{2\sqrt{D^*t}}\right) - \exp(h_i x + h_i^2 D^* t)\mathrm{erfc}\left(\frac{x}{2\sqrt{D^*t}} + h_i\sqrt{D^*t}\right), \qquad (11.41)$$

with

$$\text{erfc}(x) = \frac{2}{\sqrt{\pi}} \int_x^\infty \exp(-x^2)\,dx, \tag{11.42}$$

where $C(x)$ is the isotopic fraction obtained from the exchange experiment, C_{bg} is the natural background concentration of ^{18}O in the sample, and t is the time. The parameter $h_i = k^*/D^*$ corresponds to the factor describing the oxygen concentration ratio between the sample surface and the gas phase. Comparing with Equation 11.28, one may find that $L_c = 1/h_i$. Figure 11.3 shows the typical SIMS depth profile of $^{18}O/^{16}O$ exchanged ceramic samples. The two transport parameters can be directly obtained by fitting the experimental curves to Equation 11.41.

Recently, a pulse isotopic exchange technique was developed for rapid evaluation of the oxygen surface exchange rate under different operation conditions [87]. For this technique, the total oxygen surface exchange rate j_{ex}^{tot} is defined by the first-order rate equation under equilibrium conditions:

$$n_g \frac{\partial f_g^{18}}{\partial t} = -j_{ex}^{tot} S\left(f_g^{18} - f_s^{18}\right), \tag{11.43}$$

where n_g is the total number of oxygen atoms in the gaseous species, f_g^{18} and f_s^{18} are the atomic ^{18}O isotope fractions in the gas and solid phases, respectively, and S is the oxide surface area available for exchange. The exchange reaction involves a sequence of possible elementary steps and can be classified into two major steps, namely, dissociative adsorption of oxygen (or oxygen-containing species) and

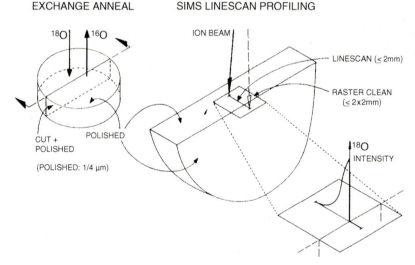

Figure 11.3 Schematic representation of the ^{18}O isotopic exchange technique with linescan SIMS profiling for ^{18}O isotopic concentration depth profile measurements in oxides of high diffusivity. (Reproduced with permission from Ref. [74].)

incorporation of oxygen into oxide lattice. The corresponding rates, designated as j_{ex}^{ad} and j_{ex}^{in}, respectively, have the following relation:

$$\frac{1}{j_{ex}^{tot}} = \frac{1}{j_{ex}^{ad}} + \frac{1}{j_{ex}^{in}}. \tag{11.44}$$

The oxygen exchange coefficients can be calculated according to Equation 11.24. As for other exchange techniques based on the analysis of $^{18}O/^{16}O$-containing species in the gaseous phase, powdered samples are employed in the experiment [87], so diffusion coefficients in the solid cannot be acquired. The exchange coefficients collected by the pulse method are similar to those obtained from electrical conductivity relaxation and isotopic exchange depth profiling methods [87]. For mixed conducting perovskites, the total exchange rate is dominated by the dissociative adsorption on the surface; the incorporation rate is two to three orders higher. For oxygen ionic conductors such as yttria-stabilized zirconia (YSZ), the total exchange rate is jointly controlled by dissociative adsorption and incorporation [87].

11.2.3.2 Electrical Conductivity Relaxation

For mixed conducting oxides, a sudden change in the oxygen partial pressure leads to changing intrinsic physical parameters until new equilibrium between the gas and solid phases is achieved. The equilibration kinetics, and thus the oxygen exchange rate, are hence reflected by the variations in physical parameters, such as electrical conductivity, mass, etc. The time dependence of these parameters relates to the rates of surface oxygen exchange and oxygen chemical diffusion in the bulk. Under relatively small alterations of the oxygen partial pressure, the mobility of change carriers can be considered constant; therefore, one may assume that the electrical conductivity linearly changes with the oxygen vacancy concentration. Again, if the surface reaction kinetics is essentially linear with respect to the oxygen concentration at the surface, the analytical solution for the plane geometry of width $2w_c$, thickness $2h_c$, and length $2l_c$ was given by Crank [86]:

$$g(t) = \frac{\sigma(t) - \sigma(0)}{\sigma(\infty) - \sigma(0)} = 1 - \sum_{i=1}^{\infty}\sum_{m=1}^{\infty}\sum_{n=1}^{\infty} \frac{2L_1^2 \exp(-\beta_i^2 D_{chem} t / h_c^2)}{\beta_i^2(\beta_i^2 + L_1^2 + L_1)}$$

$$\times \frac{2L_2^2 \exp(-\gamma_m^2 D_{chem} t / w_c^2)}{\gamma_m^2(\gamma_m^2 + L_2^2 + L_2)} \times \frac{2L_3^2 \exp(-\delta_n^2 D_{chem} t / l_c^2)}{\delta_n^2(\delta_n^2 + L_3^2 + L_3)}, \tag{11.45}$$

where $\sigma(0)$, $\sigma(\infty)$, and $\sigma(t)$ are the conductivity values in the initial state, after achieving a new equilibrium, and during the relaxation process, respectively. Here, D_{chem} is the chemical diffusion coefficient, t is the diffusion time, and L_1, L_2, and L_3 are dimensionless parameters determined by surface exchange and chemical diffusion coefficients: $L_1 = h_c k_{chem}/D_{chem}$, $L_2 = w_c k_{chem}/D_{chem}$, and $L_3 = l_c k_{chem}/D_{chem}$, respectively. β_i, γ_m, and δ_n are the nth roots of the equations: $\beta_i \tan \beta_i = L_1$, $\gamma_m \tan \gamma_m = L_2$, and $\delta_n \tan \delta_n = L_3$, respectively. If the sample is a bar where the ratio of the length to the width is large (for instance, >10), the length can be regarded as infinitely long, so the last term in Equation 11.45 can be neglected. Other simplified equations based on Equation 11.45 are also known in the literature, but it is not possible to give a full account of these models in this chapter. A four-probe method is usually used for conductivity measurements; the D_{chem} and k_{chem} values are derived by fitting of the relaxation data using Equation 11.45 or its derivatives as the regression model. The ratio of D_{chem} and k_{chem} is related to the characteristic thickness of mixed conducting membranes. In simple cases, these can also be converted to the self-diffusion coefficient and equilibrium oxygen exchange coefficient, respectively, via the use of thermodynamic factors (see Equation 11.27). It is necessary to note that the two parameters obtained by the relaxation method are averaged over the given oxygen partial pressure range.

11.2.4
Reduction and Elimination of Surface Exchange Limitations

Oxygen permeation flux is the most significant parameter determining application of the membrane technologies. If only the flux achieves a certain threshold, the technology can compete with other methods, such as the cryogenic air separation and pressure switch adsorption (PSA) in the case of oxygen generation. The thickness of mixed conducting membranes reported in the literature is normally of millimeter scale, which is much larger than that for polymer membranes and most other types of ceramic membranes. For instance, the thickness of zeolite membranes and palladium membranes is usually smaller than 10 μm. Therefore, reducing the thickness of mixed conducting membranes down to micron scale is among the most promising approaches to improve oxygen fluxes. If the permeation is controlled by bulk diffusion even for micron membranes (e.g., 10 μm), the fluxes may increase up to 50–200 ml cm^{-2} min^{-1}, which is even higher than those for microporous membranes and palladium membranes. However, surface exchange always presents a challenge, so it is necessary to know what factors affect the surface exchange rate and how the exchange kinetics can be enhanced.

Many researchers found that the same mixed conducting membranes have slightly higher permeation fluxes if the ceramics are polished after sintering at high temperatures. Even the simple use of sand paper for polishing has substantial effects on the oxygen flux [88], thus revealing important role of the membrane surface roughness. The surface area of ceramics increases with roughness; this provides additional active sites for the redox processes. Consequently, the flux can be significantly improved by coating a porous layer onto both sides of the membrane when the permeation is controlled by surface exchange reactions [30, 89–94]. The porous layers are usually made of cobalt-containing perovskite mixed conductors, such as $La_{0.6}Sr_{0.4}CoO_{3-\delta}$. Noble metals, such as Ag and Pt, are also used for oxygen activation. In addition to the intrinsic catalytic activity of the mixed conducting materials that has a great influence on the surface exchange kinetics, the porosity, pore size, and thickness of the porous layer also affect the exchange. Deng et al. [95, 96] calculated the oxygen flux through a very thin membrane where the oxygen chemical potential drop in the bulk can be neglected; that is, the oxygen permeation is limited by the oxygen exchange and transport in the porous layers. There is an optimum thickness for the active porous layer; too thick layers lead to increasing mass transport resistance in the pores and ionic transport resistance in the porous layer grains, while too thin layers have an insufficient activating effect. Similarly, the porosity of the layer has significant effects on oxygen exchange. A high porosity favors gas diffusion but limits ionic transport in the grains. Of course, a very important role is also played by the specific surface area of the porous layers. A simple model was developed to describe the porous coatings [95, 96]. Under a small oxygen pressure difference, the maximum improvement in the oxygen fluxes of surface-modified membranes compared to the membranes without modification (ξ) and the optimum thickness (L_p) are given by [58]

$$\xi = \sqrt{L_c S(1-\theta_p)/\tau_s} + \theta_p, \tag{11.46}$$

$$L_P = \sqrt{L_c(1-\theta_p)/S\tau_s}, \tag{11.47}$$

where S is the surface area in per unit volume and θ_p and τ_s are the porosity and tortuosity of the active layer, respectively. From Equations 11.46 and 11.47, one can see that improving the surface area is more effective than changing the layer porosity.

Numerous publications provide a clear evidence for the high efficiency of active porous layers to enhance oxygen exchange on the ceramic membrane surfaces. For thick membranes, the permeation is usually controlled by both bulk diffusion and surface exchange, so the enhancement in oxygen fluxes compared to nonmodified membranes usually lies in the range of 20–100%. For thin membranes, a porous support layer is needed to provide a sufficient mechanical strength. Although the dense side can also be modified with an active layer, the flux would not reach the values expected for the bulk diffusion limitations as the mass transport resistance through the support cannot be neglected. van Hassel [97] developed a model equation to predict the optimal thickness and pore sizes for activating layers to improve oxygen surface exchange on a 10 μm thick membrane, which is built on a multilevel pore-structured support. Figure 11.4 shows the calculated oxygen fluxes versus the porous layer thickness; a significant enhancement is predicted for smaller pore size of the active layer. The optimum thickness decreases with the decrease in pore size of the activating layer.

Sometimes the porous layer can be produced *in situ* by treating both sides of membranes with CO_2, CO, H_2, NH_3, and so on. These gases react with the perovskite-related oxides and produce a porous multiphase structure, which is converted into a porous perovskite layer after an air or pure oxygen flow is passed over the membrane. This effect is based on the reversible reactions involving ABO_3 perovskites, for example,

$$ABO_3 + CO_2 \leftrightarrow ACO_3 + BO_x + \frac{2-x}{2} O_2 \tag{11.48}$$

$$ABO_3 + (2-x)H_2 \rightarrow AO + BO_x + (2-x)H_2O$$

$$ABO_3 \leftarrow AO + BO_x + \left(1-\frac{x}{2}\right) O_2 \tag{11.49}$$

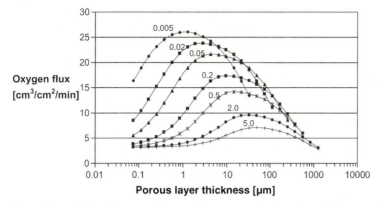

Figure 11.4 Calculated oxygen permeation fluxes versus the porous layer thickness on the air side of mixed conducting ceramic membrane. The marks indicate pore size (μm). (Reproduced with permission from Ref. [97].)

Note that most mixed conducting perovskites with high oxygen permeability contain alkaline earth metal cations in the A-sites and reducible transition metal ions in the B-sublattice. For example, a dense $La_{0.7}Sr_{0.3}Ga_{0.3}Fe_{0.7}O_{3-\delta}$ membrane tube was treated by exposing its internal surface to a H_2 stream and the shell side to the ambient air at 950 °C (Figure 11.5) [94]; the oxygen permeation measurement was performed by sweeping with helium to carry away the permeated oxygen. The H_2-treated membrane exhibited a large increase in oxygen permeation flux (by a factor of 4.5 at 950 °C and 10.5 at 800 °C) and a significant reduction in the apparent activation energy [94].

Figure 11.5 SEM images of $La_{0.7}Sr_{0.3}Ga_{0.3}Fe_{0.7}O_{3-\delta}$ membranes at the permeate side: (a) top and (b) fractured view of membrane without H_2 treatment; (c) top and (d) fractured view of membrane treated in H_2 for 16 h at 950 °C. (Reproduced with permission from Ref. [94].)

11.3
Microstructural Effects in Mixed Conducting Membranes

The definitions, fundamental approaches, and experimental methods relevant for microstructural analysis of materials can be found in numerous handbooks, such as Ref. [98]. In this chapter, the authors can only stress that microstructure of any ceramic material has a substantial influence on its physical properties, including electrical conductivity, thermal conductivity, strength, toughness, hardness, corrosion resistance, creep property, and so on. These properties may change a lot even for ceramic materials having the same composition. For example, the performance of the same materials reported by different researchers may often vary in a wide range. Although the conflicting results often reflect experimental difficulties (e.g., problems with sealing or controlling membrane operation conditions), the microstructural effects originating from the use of different precursors, synthesis method, and sintering programs cannot be ignored. The relevant factors include grain boundaries, phase composition and structure of ceramic surface, grain size, impurities or second phase, grain morphologies, extended defects, and so on. For mixed conducting materials used for air separation, the role of microstructural phenomena is very important as these affect both transport and mechanical properties at elevated temperatures necessary for the membrane operation (see also Chapter 9).

11.3.1
Selected Experimental Methods

For mixed conducting materials, the use of electron microscopy and local analysis techniques should always be coupled with the measurements of transport properties. Two methods are often used for this goal, namely, the isotopic exchange experiment followed by SIMS profiling analysis for oxygen diffusion coefficient along grain boundaries, and impedance spectroscopy to estimate grain boundary conductivity (see also Chapter 4 of the first volume). For isotopic exchange, Chung et al. proposed the following equation valid in case of diffusion along an isolated, planar, isotropic grain boundary [99–101]:

$$D_{gb}^* \delta_{gb} = D_b^{*3/2} t^{1/2} \left[10^{n_D} \left(-\frac{\partial \ln C(x,t)}{\partial \eta_D^{6/5}} \right)^{m_D} \right] \tag{11.50}$$

$$\eta_D = \frac{x}{\sqrt{D_b^* t}}, \tag{11.51}$$

where D_{gb}^* is the grain boundary diffusion coefficient, δ_{gb} is the grain boundary width, and n_D and m_D are the parameters determined by the tail slope of the plot of $\ln C(x,t)$ versus $\eta_D^{6/5}$. Only data within the specific range ($6 < \eta < 10$) can be used to estimate $D_{gb}^* \delta_{gb}$ according to Equation 11.50. A measure of the relative magnitudes of D_{gb}^* and D_b^* is given by the dimensionless parameter:

$$\beta_{gb} = \frac{D^*_{gb}\delta_{gb}}{D^{*3/2}_b t^{1/2}}. \tag{11.52}$$

For the impedance spectroscopy method, the grain boundary conductivity can be evaluated, in particular, using a brick layer model. The effective geometric factor for the grain boundaries, $(L/A)_{\text{eff}}$, may be defined as [102, 103]

$$(L/A)_{\text{eff}} = \frac{\delta_{gb}}{\bar{d}_g} \frac{L_{\text{sample}}}{A_{\text{sample}}}, \tag{11.53}$$

where $\bar{d}_g$ is the average grain size and L_{sample} and A_{sample} are the length and area of the sample, respectively. If the above relation is true and the grain boundary capacitance can be described as $C_{gb} = \varepsilon_0 \varepsilon_r / (L/A)_{\text{eff}}$, one may find

$$R_{gb} C_{gb} = \varepsilon_0 \varepsilon_r \varrho_{gb} = (2\pi f_{gb})^{-1} \tag{11.54}$$

or

$$\sigma_{gb} = (2\pi f_{gb}) \varepsilon_0 \varepsilon_r, \tag{11.55}$$

where R_{gb} and ϱ_{gb} are the grain boundary resistance and resistivity, respectively, and ε_0, ε_r, and f_{gb} are the vacuum dielectric constant, relative dielectric constant, and relaxation frequency, respectively. To estimate the grain boundary conductivity, the relative dielectric constant of the material bulk and relaxation frequency are then used. In addition to the isotopic exchange and impedance spectroscopy, the microstructural effects are often analyzed by measuring the oxygen permeation fluxes through ceramic membranes with different microstructures. If the oxygen flux increases with grain size, the grain boundaries are considered as blocking with respect to the oxygen transport, and vice versa.

11.3.2
Microstructural Phenomena in Perovskite-Type Membranes

In most cases, the effects of grain boundaries on ionic and electronic transport may hardly be predicted theoretically, although a significant progress is achieved in this area (see Chapters 1 and 4 of the first volume). For some materials where the concentration of charge carriers in the grain bulk is relatively low, the grain boundary regions with an excess free energy and a high population of structural defects may produce high-diffusivity paths. This type of behavior has been verified by the oxygen isotopic exchange experiments and SIMS profiling analysis for $La_{0.7}Ca_{0.3}CrO_{3-\delta}$ [104]. However, in other cases the grain boundaries lead to additional resistance to transport. For example, for solid electrolyte materials, the overall resistivity usually tends to decrease with increasing grain size when the grain boundary area decreases. There are many factors that can cause high boundary resistance, such as enrichment with resistive impurities, dopants, and second phase used as sintering aid, formation of an amorphous phase during sintering, or formation of defect clusters. Highly resistive impurities commonly segregate in the grain boundary regions, leading to transport limitations. In the solid electrolyte materials such as doped zirconia and ceria, silica and alumina are the most common impurities detected along with the grain boundaries; these components are often introduced during the preparation and calcination steps [105–109].

Great efforts for proper understanding of the grain boundary phenomena in the mixed conducting membranes were made during the recent years, but the relevant experimental information is still scarce and contradictive. The fast diffusion paths were identified along grain boundaries in lanthanum-deficient $La_{0.9}MnO_{3-\delta}$ but not in the stoichiometric manganite ceramics; in the former composition, the grain boundary ion diffusion coefficient is about three orders of magnitude higher than that in bulk [110]. An increase in the oxygen permeation fluxes with decreasing grain size was observed for mixed conducting $SrCo_{0.8}Fe_{0.2}O_{3-\delta}$ membranes, indicating that the boundaries provide a path for fast oxygen migration. Although the exact mechanism of this effect is not completely clear at this point, the increase of grain boundaries is hypothesized to affect both the surface exchange and diffusion processes [21]. However, other researchers argued that since the permeation experiments were performed under an air/helium gradient, the relatively low oxygen partial pressure at the membrane permeate side might lead to ordering of oxygen vacancies and to decreasing ionic conductivity [111], while small grains could contribute to stabilization of the disordered phase, as for bismuth vanadate-based materials [112]. For some materials such as $SrCo(Fe,Cu)O_{3-\delta}$, the grain boundaries were reported to decrease transfer of oxygen anions [113]; thus, the larger the grain size, the higher the oxygen permeation flux through such membranes. Similar phenomena were reported by Wang et al. [114] for $Ba_{0.5}Sr_{0.5}Co_{0.8}Fe_{0.2}O_{3-\delta}$ (BSCF). Arnold et al. [115] prepared dense membranes of BSCF and $Ba_{0.5}Sr_{0.5}Fe_{0.8}Zn_{0.2}O_{3-\delta}$ (BSFZ) via liquid-phase sintering with different grain sizes by adjusting amount of the BN sintering aid. It was found that the permeation fluxes through BSCF ceramics increase linearly with the average grain size area, whereas the permeability of BSFZ remains essentially unaffected by the grain size. The results of transmission electron microscopy (TEM) indicated that no amorphous components or interfacial phases were incorporated at the grain boundaries, so lattice misfit between the grains and associated strains near grain boundary regions could have negative impact on the oxygen ion transport [115]. At the same time, Diethelm et al. [116] found that an amorphous layer thinner than 1 nm (Figure 11.6) is present at the grain boundaries of $La_{0.5}Sr_{0.5}FeO_{3-\delta}$, and the apparent conductivity increases with increasing grain boundary area. Based on these observations, grain boundaries playing a significant role in oxygen transport and offering a more rapid path than bulk for oxygen ionic transport are suggested to explain the effects of grain size [116]. Analogously, the data on $La_{0.6}Sr_{0.4}Co_{0.2}Fe_{0.8}O_{3-\delta}$ show that both the oxygen flux and total electric conductivity monotonically increase with sintering temperature and grain size, showing a negative impact of grain boundaries [117]. However, these results are inconsistent with the observations that grain boundaries provide a fast oxygen transport path, made using the isotopic exchange experiment [111]. Figueiredo et al. [118] investi-

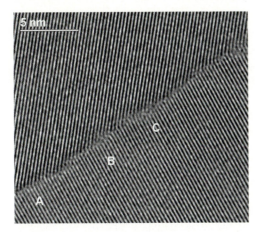

Figure 11.6 Example of the high-resolution TEM image of the interface between two grains in $La_{0.5}Sr_{0.5}FeO_{3-\delta}$ ceramics. The interface is wavy and interfacial edge dislocations are present where lattice planes join (A). In some areas the lattice plane contrast vanishes close to the interface (B and C) and some amorphous or foreign phase may be present in very thin layer (thickness <1 nm). (Reproduced with permission from Ref. [116].)

gated the effects of microstructure on oxygen permeation and conductivity of iron-substituted calcium titanate ceramic materials, prepared via the solid-state reactions method. A core–shell microstructure for smaller grains, consisting of a pure $CaTiO_3$ core and an iron-rich microdomain structure at the shell, was found responsible for the variations in the transport properties. The fast oxygen transport along the microdomain walls was suggested to explain the grain size dependence of oxygen fluxes [118]; simultaneously, for homogeneous $Ca(Ti,Fe)O_{3-\delta}$ ceramics, the ionic conductivity decreases with decreasing grain size and the grain boundaries thus represent an additional resistance. For the materials prepared via mechanical activation and sintering, the oxygen permeability increases linearly with the increase of grain size in the range from ~200 nm to ~10 μm, suggesting the grain boundaries have a negative impact on the high-temperature ionic transport [119]. It is worth noting that the iron-substituted calcium titanate membranes with the core–shell microstructure provide higher and more stable oxygen fluxes compared to the homogeneous ceramics. Another necessary comment is that the preparation methods of mixed conducting perovskite powders have a substantial influence on the microstructure and grain morphology of the ceramic membranes, and therefore have a large impact on oxygen transport. For instance, a $LaCoO_{3-\delta}$ ceramic membrane prepared by the traditional solid-state reaction method shows significantly higher oxygen permeation fluxes than those expected from the tracer diffusion data on single crystals, again demonstrating that grain boundaries may afford a fast diffusion path for oxygen transport [111, 120]. However, when the membranes with the same composition are synthesized using organic precursors, the oxygen permeability and ionic conductivity are much lower.

Herein, it can be concluded that contrary to simplified theoretical models, the realistic effects of grain boundaries on the oxygen transport in mixed conducting membranes are often complex and ambiguous. These effects depend on many factors, such as impurities, defects ordering, preparation method, membrane operation conditions, and so on; these make it difficult to expect the contribution of grain boundaries a priori. For selected membrane materials, these phenomena need to be systematically investigated with a very careful control of the experimental conditions.

11.3.3
Composite Membranes

The composite membranes discussed in this section include two major types: perovskite-type membrane materials where a second phase is purposively added or incidentally formed during processing, and dual-phase membranes that contain one phase for a dominant oxygen ionic transport and another phase mainly for electronic transport. In the composites, the influence of microstructure becomes even more complex as additional transport-effecting factors appear. These factors include, in particular, the overall composition, phase interaction and impurities formed at the interfaces, distribution of the constituent phases, and multiphase morphology of the membrane surfaces.

11.3.3.1 Perovskite Membranes with Second Phase or Impurities

The presence of impurities in mixed conducting perovskite-type membranes does not necessarily lead to degradation in the membrane performance. For example, the formation of $CaTiO_3$-based secondary phase by incomplete solid-state reactions during the synthesis of iron-doped $CaTiO_3$ materials results in a positive contribution to oxygen transport and stability [118]. The additions of MgO into $La_{0.6}Sr_{0.4}Fe_{0.9}Ga_{0.1}O_{3-\delta}$ ceramics can restrict the grain growth during the sintering process [121]. The grain size decreases with MgO amount as expected, but the oxygen flux and surface exchange coefficient all increase with magnesia additions [121]. Moreover, the phase impurities may have positive impact on the mechanical strength of the membrane materials, which is critically important for the practical applications. For a given composition of ceramics, the strength is closely related to its grain size. Therefore, a commonly used method to improve the mechanical strength of ceramic materials is based on the dispersion of second-phase particles to retard grain growth during sintering by pinning grain boundaries. Li *et al.* [122] reported that the three-point bending strength of $SrCo_{0.4}Fe_{0.6}O_{3-\delta}$ material can be markedly improved from 19.4 to 64.6 MPa by adding yttria-stabilized zirconia powders, although the oxygen permeability becomes lower. Similar results were obtained by Wu *et al.* [123] on the addition of Al_2O_3 into $SrCo_{0.8}Fe_{0.2}O_{3-\delta}$ membranes. It is worth noting that only a minor Al_2O_3 addition, 1 wt%, leads to losing more than 70% oxygen permeability [123]. Hence, the membrane powder preparation route should be carefully optimized to minimize impurities, such as Al_2O_3, which may come from glassware, mill balls, crucibles, and so on. Alumina and zirconia have a high reactivity with perovskite oxides and produce resistive phases, such as $SrZrO_3$, $La_2Zr_2O_3$, and $SrAl_2O_4$. These high-resistance components form at the grain boundaries, blocking the oxygen ionic and electronic transport. At the same time, compositional changes in the main perovskite phase may have a positive effect. Recently, Kharton and coworkers reported a series of alumina-doped $SrFeO_{3-\delta}$-based perovskite membranes with an improved stability [123–130]. While the solid solution formation range in $SrFe_{1-y}Al_yO_{3-\delta}$ system corresponds to $y = 0$–0.35, overstoichiometric additions of alumina lead to the segregation of insulating $SrAl_2O_4$ phase, but simultaneously increase sintering and, in some cases, enhance oxygen permeation (Figure 11.7) [124]. Similar tendencies were identified for $(SrFe)_{1-x}(SrAl_2)_xO_{3+x-\delta}$ composites, where moderate amounts of $SrAl_2O_4$ increase thermal shock stability and hardness without detrimental influence on the transport properties [128]. The increase in oxygen permeability is similar to that observed for strontium-deficient $Sr_{1-x}(Fe,Ti)O_{3-\delta}$ perovskites, where creation of the A-site vacancies promotes disorder in the oxygen sublattice and thus increases ionic conductivity [131].

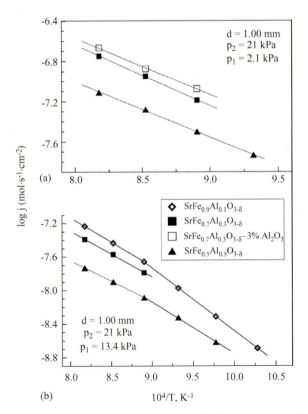

Figure 11.7 Temperature dependence of the oxygen permeation fluxes through $SrFe_{1-x}Al_xO_{3-\delta}$ membranes under fixed oxygen partial pressure gradients. (Reproduced with permission from Ref. [124].)

Another kind of composite systems is produced due to incomplete reaction of precursors or due to improper elemental ratio, including a cation nonstoichiometry in the A- or B-sites of ABO_3 perovskite phase and dopant amount exceeding the solid solution limits. As an important example, in the $BaCe_xFe_{1-x}O_{3-\delta}$ system, a mixture of two perovskite phases is obtained at $0.15 < x < 0.85$ [28]. The cerium-enriched phase, $BaCe_{0.85}Fe_{0.15}O_{3-\delta}$, shows lower oxygen permeability; therefore, the oxygen flux decreases by approximately 40% when the amount of cerium increases from 0.15 to 0.2, because of the segregation of Ce-rich phase [28]. Likewise, for Sn-doped $SrCo_{0.8}Fe_{0.2}O_{3-\delta}$ (SCF) composite membranes, the solid solubility of Sn in the perovskite is smaller than 2.5 mol%; at higher dopant concentrations, the membranes possess a composite structure comprising oxygen-permeable SCF phase and oxygen-impermeable $SrSnO_3$ [132]; the oxygen flux of the composite membrane is only 25% lower than that of SCF at 900 °C. Another attractive example is $SrFeCo_{0.5}O_z$ ceramics, reported by Balachandran et al. [40] as having a high oxygen permeability and good stability for methane conversion to syngas. Subsequent experiments clearly showed that the microstructure of these materials plays a very important role in the ionic transport processes. In particular, $SrFeCo_{0.5}O_z$ may comprise a layered phase, $Sr_4Fe_{6-x}Co_xO_{13+\delta}$ ($x < 1.8$), which shows low ionic conductivity; when the cobalt content exceeds the solution limit, a $Sr(Fe,Co)O_{3-\delta}$ perovskite phase with high ionic conductivity and spinel phase are produced [133–135]. For this material, relatively high oxygen fluxes were achieved for the membranes prepared using the standard solid-state reaction method, and low fluxes were reported for the membranes synthesized using the sol-gel process.

11.3.3.2 Dual-Phase Membranes

Dual-phase composite materials were suggested as possible candidates for the oxygen separation membranes since it is difficult to meet all the requirements (such as high permeability, stability, mechanical strength, etc.) in any single-phase material. In these membranes, one phase is usually ion conducting and another phase provides fast electronic transport. The oxygen ionic transport in the dual-phase membrane is strongly affected by the composite microstructure and reactivity of the components. The critical microstructural factors include grain size and shape, percolation, and distribution of the components in the bulk and on the surface. The dual-phase composites made of noble metal and oxide ionic conductors, such as $Ag–Bi_2O_3$ and Pd–YSZ, were first reported in the literature [136–138]; considering the costs of these materials, it is, however, difficult to expect wide use of these membranes in a large scale. The perovskite-type oxides were then proposed to replace noble metals for electronic transport. Ceria-based composite membranes containing perovskite-type $La_{0.7}Sr_{0.3}MnO_3$ (LSM) and fluorite-type $Ce_{0.8}Gd_{0.2}O_{1.9}$ (GDC) were reported to exhibit rather low oxygen permeation and a substantial degradation with time [139]. The degradation was attributed to the formation of highly resistive layers due to interdiffusion of the phase components. A similar interdiffusion during sintering and high-temperature operation was observed for $La_{0.8}Sr_{0.2}Fe_{0.8}Co_{0.2}O_3$–$(La_{0.9}Sr_{0.1})_{0.98}Ga_{0.8}Mg_{0.2}O_3$ composites where the two perovskite phases finally form a single-phase material containing all the elements [140]. From the above examples, one can find that the interaction between the two oxide phases plays a key role in oxygen transport and stability. Increasing the sintering temperature and time promotes phase interaction and components interdiffusion, especially when the two phases have similar crystal structures. If ceria- or zirconia-based oxygen ionic conductors are used in the dual-phase membranes, highly resistive phases (such as $(Ba,Sr)(Ce,Zr)O_3$ or $Ln_2Zr_2O_7$) are usually produced during sintering. Another problem is that the ionic transport between solid electrolyte grains is often blocked by electronically conducting phase if noble metals or perovskite-type electronic conductors, such as LSM, are used.

Recently, the authors' group developed a new series of ceramic dual-phase composite membranes with a high oxygen permeability and good stability for methane conversion to syngas [141–144]. The membranes usually have a doped ceria oxide for oxygen ionic transport and a Fe-based perovskite mixed conductor for both oxygen ionic and electronic transport. The same cations should be introduced as dopant both in ceria lattice and in perovskite lattice in order to minimize interdiffusion between the two phases; also, a proper amount of Sr^{2+} in the perovskite lattice is needed to keep a high ionic conductivity and to simultaneously lower reactivity with ceria. A variety of methods were used to prepare the composite powders of 75 wt% $Ce_{0.8}Gd_{0.2}O_{1.9}$ (GDC) and 25 wt% $Gd_{0.2}Sr_{0.8}FeO_{3-\delta}$ (GSF) where the composition corresponds to the volume ratio of approximately 71 : 29. These included (1) mixing of the two oxide powders, (2) mixing the GSF powder and GDC precursor solution, (3) mixing the GDC powder and GSF precursor solution, and (4) mixing of both precursor solutions together (the so-called one-pot method). In Figure 11.8, the derived dual-phase microstructures are labeled as GDC-GSF-1–GDC-GSF-4, respectively. No new phase was detected [143], including the one-pot method that did not produce any phase impurities such as $SrCeO_3$, $Gd_3Fe_5O_{12}$, and so on. The typical microstructures are illustrated by Figure 11.8, where the white grains are fluorite phase and the dark grains are the perovskite component. The composite preparation methods had a little effect on the grain sizes, but remarkably influenced the size distribution. The most uniform grain size was observed for GDC-GSF-4, prepared from the precursor solutions. The oxygen permeation fluxes (Figure 11.9) are controlled by the ambipolar conductivity. In the membranes with essentially inhomogeneous grain size distribution (e.g., GDC-GSF-2), some perovskite grains are surrounded by fluorite phase and are thus partially blocked. More homogeneous mixing of the two phases leads to higher oxygen permeability and lower activation energy.

For the dual-phase composite membranes comprising a solid oxide electrolyte and a pure electronic conductor, the interfacial exchange reactions are essentially localized in the vicinity of the triple-phase boundaries (TPBs), that is, the interface between gas, ionic conductor, and electronically conducting component (see Chapter 12 of the first volume). For the membranes comprising a mixed conducting component, however, the oxygen exchange can occur both at the surface of perovskite grains and near the TPBs. Consequently, the experimental results [141–144] showed a clear correlation between the composite homogeneity and oxygen permeability. Namely, the surface exchange reactions should play a more important role for less

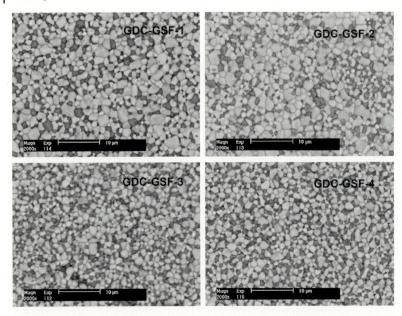

Figure 11.8 Backscattered electron images of the dual-phase GDC-GSF composite membranes prepared by different techniques. (Reproduced with permission from Ref. [143].)

homogeneous mixing of the phases, which is reflected by the behavior of GDC-GSF-2 (Figure 11.9).

For perovskite-type mixed conducting ceramics, grain boundary resistance plays an important role in ionic and electronic transport, as discussed in Section 11.3.2; the

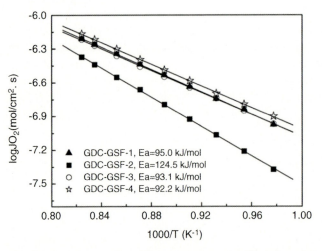

Figure 11.9 Arrhenius plots of the oxygen permeation fluxes through dual-phase composite membranes. Oxygen partial pressure: 21/0.5 kPa. (Reproduced with permission from Ref. [143].)

sintering temperature affects the grain size and morphology, thus influencing oxygen permeation. For the dual-phase composites, the behavior is, again, more complex. For example, there is an optimum sintering temperature (1425 °C) for the membrane consisting of 75 wt% $Ce_{0.85}Sm_{0.15}O_{1.925}$ and 25 wt% $Sm_{0.6}Sr_{0.4}FeO_{3-\delta}$ (SDC-SSF), prepared via the one-pot route [145]. Higher sintering temperature leads to an enrichment of the membrane surface with ceria-based phase having a low electronic conductivity, whereas the polished membranes display an important influence of grain growth. Increasing the grain size leads to less homogeneous distribution of the composite component grains and causes problems with the percolation of the mixed conducting phase [145].

Finally, unsteady permeation phenomena were repeatedly observed during the initial stage of dual-phase membrane operation [139, 141, 143, 146]. Usually these phenomena were discussed in terms of structural changes or reduction of initial oxygen desorption rates [147], although exact mechanisms require further investigations. In order to minimize influence of the oxygen surface exchange, $La_{0.6}Sr_{0.4}CoO_3$ (LSC) porous layers of $\sim$20 μm thickness were coated on one or both sides of the dual-phase membranes; for both surfaces coated with LSC, the membranes exhibited highest fluxes and minimum relaxation times, on the contrary to membranes with one porous layer or without surface layers [148]. This makes it possible to establish a correlation between the initial unsteady states and surface microstructure adjustment at elevated temperatures. For the dual-phase membranes where porous LSC is only deposited onto the feed-side surface, no microstructural alterations on the feed-side surface were detected after the permeation experiments [148]. However, the morphology of the surface without activating layer was changed remarkably, promoting the exchange.

11.3.4
Asymmetric Membranes

Since the first report on high oxygen permeability of cobalt-based perovskites in 1985 [13], the highest known fluxes still vary in the range from 2 to 4 $cm^3 min^{-1} cm^{-2}$, not sufficiently high for practical applications where an oxygen permeation flux more than 10 $cm^3 min^{-1} cm^{-2}$ is required with respect to the process efficiency and costs. One effective method to improve permeation flux is related to decreasing the membrane thickness from millimeter scale down to the micron scale. An increase of 10–1000 times can then be expected assuming that the oxygen transport is still controlled by bulk ambipolar diffusion. The potential problems are that the oxygen surface exchange becomes the rate-determining step, as discussed in Section 11.2.4, and that careful microstructural optimization of the supporting layer is necessary. The diffusion of oxygen molecules from gas phase to membrane surface may become rate determining for asymmetric membranes, where the pore structure, porosity, and support thickness all have critical impacts. Chang et al. [149] showed that the effects of porous support are operation mode dependent. Namely, the oxygen permeation fluxes flowing from the support to the thin dense layer are higher than those in the opposite direction; these effects also depend on the thickness [149].

These experimental results reveal that the asymmetric membrane performance is governed by both gas transport resistance and surface exchange. Note that the diffusion resistances in the supports are quite typical for zeolite and palladium membranes. For example, high permeation fluxes of water through a hollow fiber-supported membranes of Linde Type-A (LTA) zeolite depend mainly on the support [150]. In the case of asymmetric mixed conducting membranes, identification of optimum support microstructures and geometric parameters needs detailed experimental studies and modeling work.

11.4
Thermodynamic and Kinetic Stability

In this chapter, it is impossible to analyze all aspects related to the stability of mixed conducting membranes under their operation conditions. However, one should recall that the chemical (thermodynamic) stability is usually considered as a function of the lattice energies (or metal–oxygen bond energies) of the membrane materials. A common strategy to improve the stability of perovskite membranes is therefore based on doping with less reducible metal ions, such as Ga^{3+}, Al^{3+}, Ce^{4+}, Zr^{4+}, and Ti^{4+}, which still enable fast ionic transport (see Chapter 9 of the first volume). At the same time, metal–oxygen bond energies are not the only relevant factor determining the operation stability under an oxygen chemical potential gradient. Likewise, the thermal stability and mechanical stability of the mixed conducting materials are not directly related to their chemical stability. Whatever be the microscopic mechanisms, these properties can be significantly improved by optimization of ceramic microstructure and, often, surface exchange of the membranes.

11.4.1
Surface Limitations

In Section 11.2.1, it is explained that the oxygen chemical potential across a membrane decreases in the sequence $\mu_{O_2}^I > \mu_{O_2}' > \mu_{O_2}'' > \mu_{O_2}^{II}$ (Figure 11.1), regardless of the particular rate-determining steps. On introducing the thermodynamic stability limit of a given membrane material in reducing atmospheres ($\mu_{O_2}^{decomp}$), the membrane remains stable even if $\mu_{O_2}^{II} < \mu_{O_2}^{decomp}$ but $\mu_{O_2}'' > \mu_{O_2}^{decomp}$. In other words, the slow surface exchange kinetics at interface II can protect membranes from reduction at the permeate side. That is why some mixed conducting oxides are not thermodynamically stable under reducing environments, but the ceramic membranes can provide a stable operation for hundreds or thousands of hours in the regime of syngas generation. For example, the $(SrFe)_{0.7}(SrAl_2)_{0.3}O_{3.3-\delta}$ composite membranes can keep a constant oxygen flux and stable surface microstructure even after 1600 h operation under an air/H_2–H_2O–N_2 gradient, but both the oxygen flux and the membrane surface change a lot when an active porous layer made of GDC/Pt is coated on the permeate side [130]. BSCF membrane material is easily reduced by 5% H_2–Ar at 800 °C [30]; the oxygen permeation through this material is mainly controlled by bulk diffusion even when the ceramic membrane thickness is decreased down to 0.5 mm, but the stable operation of the membrane reactors for syngas production was successfully achieved for more than 500 h at 850 °C [26]. One main reason is that the BSCF reduced surface forms a porous structure with poor oxygen exchange kinetics, thus making $\mu''_{O_2} > \mu^{decomp}_{O_2}$ and protecting the membrane bulk against reduction by syngas. In addition, the formed porous structure of BSCF with good gas diffusion properties does not block oxygen transport to a significant extent. The authors' group has found that decreasing thickness of the BSCF membranes could not improve oxygen fluxes for syngas generation in the membrane reactors, which reveals that surface exchange limitations are prevailing under CH_4 oxidation conditions. Similar phenomena occur very often in the perovskite-type membrane reactors. Therefore, a material with relatively slow oxygen exchange can be used to coat membranes with insufficient chemical stability under operation conditions in order to prevent bulk degradation.

11.4.2
Microstructures and Kinetic Stability

Kinetic demixing (Chapter 9) under an oxygen chemical potential gradient is common for most membrane materials. One typical example is $La_{0.3}Sr_{0.7}CoO_{3-\delta}$, where phase demixing occurs at the oxygen lean side of the membrane, at 900 °C, when the membrane is exposed to streams of air and inert gas at opposite sides of the membrane [61]; no demixing occurs for the powder of the same composition annealed in either flowing air or nitrogen for several days. This makes it possible to unambiguously attribute phase changes to the action of oxygen partial pressure gradient across the membranes. Metal cations, like oxygen anions, can directionally diffuse in the membrane bulk, driven by oxygen partial pressure gradient. Their diffusion is, however, far slower than that of oxygen anions, so normally the cation contribution to the steady-state transport process is neglected when compared with oxygen and electrons. At the same time, the cation diffusion causes kinetic demixing and is hence critically important for the membrane stability. There is usually a large difference in the diffusivities of A- and B-site cations in the perovskite lattice, which results in compositional gradient and, then, in phase separation. From the theoretical point of view, it can be expected that the membrane materials with low cation diffusion coefficients or a small difference between the metal cation diffusion coefficients would provide a better kinetic stability. The second phase or phase impurities can suppress cation transport, thus facilitating a stable long-term operation of the membranes. Moreover, the dual-phase composite structure can hinder the metal cation diffusion between grains, favoring the kinetic stability. For example, after more than 1100 h operation at 950 °C under syngas generation conditions, the dual-phase 75 wt% $Ce_{0.85}Sm_{0.15}O_{1.925}$ and 25 wt% $Sm_{0.6}Sr_{0.4}Al_{0.3}Fe_{0.7}O_3$ membranes showed no significant microstructural alterations and no cation enrichment phenomena at the surfaces [142].

Acknowledgments

The authors gratefully acknowledge financial support of the National Science Fund (project 20801053), National Science Fund for Distinguished Young Scholars of China (20725313), and the Ministry of Science and Technology of China (Grant No. 2005CB221404).

References

1 Gellings, P.J. and Bouwmeester, H.J.M. (1997) *The CRC Handbook of Solid State Electrochemistry*, CRC Press.
2 Maier, J. (2004) *Physical Chemistry of Ionic Materials: Ions and Electrons in Solids*, John Wiley & Sons, Ltd.
3 Kharton, V.V. (2009) *Solid State Electrochemistry I: Fundamentals, Materials and Their Applications*, John Wiley & Sons, Ltd.
4 The American Ceramic Society (2006) *Progress in Solid Oxide Fuel Cells*, John Wiley & Sons, Ltd.
5 Jacobson, A.J. (2010) Materials for solid oxide fuel cells. *Chem. Mater.*, **22**, 660–674.
6 Tsipis, E.V. and Kharton, V.V. (2008) Electrode materials and reaction mechanisms in solid oxide fuel cells: a brief review. *J. Solid State Electrochem.*, **12**, 1367–1391.
7 Yano, M., Tomita, A., Sano, M., and Hibino, T. (2007) Recent advances in single-chamber solid oxide fuel cells: a review. *Solid State Ionics*, **177**, 3351–3359.
8 Chung, S.J., Park, J.H., Li, D., Ida, J.I., Kumakiri, I., and Lin, J.Y.S. (2005) Dual-phase metal-carbonate membrane for high-temperature carbon dioxide separation. *Ind. Eng. Chem. Res.*, **44**, 7999–8006.
9 Wade, J.L., Lackner, K.S., and West, A.C. (2007) Transport model for a high temperature, mixed conducting CO_2 separation membrane. *Solid State Ionics*, **178**, 1530–1540.
10 Rui, Z.B., Anderson, M., Lin, Y.S., and Li, Y.D. (2009) Modeling and analysis of carbon dioxide permeation through ceramic–carbonate dual-phase membranes. *J. Membr. Sci.*, **345**, 110–118.
11 Cales, B. and Baumard, J.F. (1982) Oxygen semipermeability and electronic conductivity in calcia-stabilized zirconia. *J. Mater. Sci.*, **17**, 3243–3248.
12 Cales, B. and Baumard, J.F. (1984) Mixed conduction and defects structure of ZrO_2–Y_2O_3–CeO_2 solid solution. *J. Electrochem. Soc.*, **131**, 2407–2413.
13 Teraoka, Y., Zhang, H.M., Furukawa, S., and Yamazoe, N. (1985) Oxygen permeation through perovskite-type oxides. *Chem. Lett.*, **11**, 1743–1746.
14 Teraoka, Y., Nobunaga, T., and Yamazoe, N. (1988) Effect of cation substitution on the oxygen semipermeability of perovskite-type oxides. *Chem. Lett.*, **3**, 503–506.
15 Qiu, L., Lee, T.H., Liu, L.M., Yang, Y.L., and Jacobson, A.J. (1995) Oxygen permeation studies of $SrCo_{0.8}Fe_{0.2}O_{3-\delta}$. *Solid State Ionics*, **76**, 321–329.
16 Stevenson, J.W., Armstrong, T.R., Carneim, R.D., Pederson, L.R., and Weber, L.R. (1996) Electrochemical properties of mixed conducting perovskite $La_{1-x}M_xCo_{1-y}Fe_yO_{3-\delta}$ (M=Sr, Ba, Ca). *J. Electrochem. Soc.*, **143**, 2722–2729.
17 Kharton, V.V., Li, S.B., Kovalevsky, A.V., and Naumovich, E.N. (1997) Oxygen permeability of perovskites in the system $SrCoO_{3-\delta}$–$SrTiO_3$. *Solid State Ionics*, **96**, 141–151.
18 Kharton, V.V., Kovalevsky, A.V., Tikhonovich, V.N., Naumovich, E.N., and Viskup, A.P. (1998) Mixed electronic and ionic conductivity of $LaCo(M)O_3$ (M=Ga, Cr, Fe or Ni) II. Oxygen permeation through Cr- and Ni-substituted $LaCoO_3$. *Solid State Ionics*, **110**, 53–60.

19 Kharton, V.V., Viskup, A.P., Bochkov, D.M., Naumovich, E.N., and Reut, O.P. (1998) Mixed electronic and ionic conductivity of LaCo(M)O$_3$ (M=Ga, Cr, Fe or Ni). III. Diffusion of oxygen through LaCo$_{1-x-y}$Fe$_x$Ni$_y$O$_{3\pm\delta}$ ceramics. *Solid State Ionics*, **110**, 61–68.

20 Tsai, C.Y., Dixon, A.G., Ma, Y.H., Moser, W.R., and Pascucci, M.R. (1998) Dense perovskite, La$_{1-x}$A$_x$Fe$_{1-y}$Co$_y$O$_{3-\delta}$ (A=Ba, Sr, Ca), membrane synthesis, applications, and characterization. *J. Am. Ceram. Soc.*, **81**, 1437–1444.

21 Zhang, K., Yang, Y.L., Ponnusamy, D., Jacobson, A.J., and Salama, K. (1999) Effect of microstructure on oxygen permeation in SrCo$_{0.8}$Fe$_{0.2}$O$_{3-\delta}$. *J. Mater. Sci.*, **34**, 1367–1372.

22 Kharton, V.V., Yaremchenko, A.A., Kovalevsky, A.V., Viskup, A.P., Naumovich, E.N., and Kerko, P.F. (1999) Perovskite-type oxides for high-temperature oxygen separation membranes. *J. Membr. Sci.*, **163**, 307–317.

23 Shao, Z.P., Yang, W.S., Cong, Y., Dong, H., Tong, J.H., and Xiong, G.X. (2000) Investigation of the permeation behavior and stability of a Ba$_{0.5}$Sr$_{0.5}$Co$_{0.8}$Fe$_{0.2}$O$_{3-\delta}$ oxygen membrane. *J. Membr. Sci.*, **172**, 177–188.

24 Chou, Y.S., Stevenson, J.W., Armstrong, T.R., and Pederson, L.R. (2000) Mechanical properties of La$_{1-x}$Sr$_x$Co$_{0.2}$Fe$_{0.8}$O$_3$ mixed-conducting perovskites made by the combustion synthesis technique. *J. Am. Ceram. Soc.*, **83** (6), 1457–1464.

25 Tong, J.H., Yang, W.S., Zhu, B.C., and Cai, R. (2002) Investigation of ideal zirconium-doped perovskite-type ceramic membrane materials for oxygen separation. *J. Membr. Sci.*, **203**, 175–189.

26 Shao, Z.P., Dong, H., Xiong, G.X., Cong, Y., and Yang, W.S. (2001) Performance of a mixed-conducting ceramic membrane reactor with high oxygen permeability for methane conversion. *J. Membr. Sci.*, **183**, 181–192.

27 Yang, L., Tan, L., Gu, X.H., Jin, W.Q., Zhang, L.X., and Xu, N.P. (2003) A new series of Sr(Co,Fe,Zr)O$_{3-\delta}$ perovskite-type membrane materials for oxygen permeation. *Ind. Eng. Chem. Res.*, **42**, 2299–2305.

28 Zhu, X.F., Wang, H.H., and Yang, W.S. (2004) Novel cobalt-free oxygen permeable membrane. *Chem. Commun.*, **9**, 1130–1131.

29 Wang, H.H., Tablet, C., Feldhoff, A., and Caro, J. (2005) A cobalt-free oxygen-permeable membrane based on the perovskite-type oxide Ba$_{0.5}$Sr$_{0.5}$Zn$_{0.2}$Fe$_{0.8}$O$_{3-\delta}$. *Adv. Mater.*, **17**, 1785–1788.

30 Zhu, X.F., Cong, Y., and Yang, W.S. (2006) Oxygen permeability and structural stability of BaCe$_{0.15}$Fe$_{0.85}$O$_{3-\delta}$ membranes. *J. Membr. Sci.*, **283**, 38–44.

31 Yang, W.S., Wang, H.H., Zhu, X.F., and Lin, L.W. (2005) Development and application of oxygen permeable membrane in selective oxidation of light alkanes. *Top. Catal.*, **35**, 155–167.

32 Lu, H., Tong, J.H., Cong, Y., and Yang, W.S. (2005) Partial oxidation of methane in Ba$_{0.5}$Sr$_{0.5}$Co$_{0.8}$Fe$_{0.2}$O$_{3-\delta}$ membrane reactor at high pressures. *Catal. Today*, **104**, 154–159.

33 Diethelm, S., Sfeir, J., Clemens, F., van Herle, J., and Favrat, D. (2004) Planar and tubular perovskite-type membrane reactors for the partial oxidation of methane to syngas. *J. Solid State Electrochem.*, **9**, 611–617.

34 Chen, C.S., Feng, S.J., Ran, S., Zhu, D.C., Liu, W., and Bouwmeester, H.J.M. (2003) Conversion of methane to syngas by a membrane-based oxidation–reforming process. *Angew. Chem., Int. Ed.*, **42**, 5196–5198.

35 Tong, J.H., Yang, W.S., Cai, R., Zhu, B.C., and Lin, L.W. (2002) Novel and ideal zirconium-based dense membrane reactors for partial oxidation of methane to syngas. *Catal. Lett.*, **78**, 129–137.

36 Ritchie, J.T., Richardson, J.T., and Luss, D. (2001) Ceramic membrane reactor for synthesis gas production. *AIChE J.*, **47**, 2092–2101.

37 Jin, W., Gu, X., Li, S., Huang, P., Xu, N., and Shi, J. (2000) Experimental and simulation study on a catalyst packed tubular dense membrane reactor for partial oxidation of methane to syngas. *Chem. Eng. Sci.*, **55**, 2617–2625.

38 Tsai, C.Y., Dixon, A.G., Ma, Y.H., Moser, W.R., and Pascucci, M.R. (1998) Dense perovskite, $La_{(1-x)}A'_xFe_{1-y}Co_yO_{3-\delta}$ (A'=Ba, Sr, Ca), membrane synthesis, applications, and characterization. *J. Am. Ceram. Soc.*, **81**, 1437–1444.

39 Tsai, C.Y., Dixon, A.G., Moser, W.R., and Ma, Y.H. (1997) Dense perovskite membrane reactors for partial oxidation of methane to syngas. *AIChE J.*, **43**, 2741–2750.

40 Balachandran, U., Ma, B., Maiya, P.S., Mieville, R.L., Dusek, J.T., Picciolo, J.J., Guan, J., Dorris, S.E., and Liu, M. (1998) Development of mixed-conducting oxides for gas separation. *Solid State Ionics*, **108**, 363–370.

41 Zhu, X.F., Wang, H.H., Cong, Y., and Yang, W.S. (2006) Partial oxidation of methane to syngas in $BaCe_{0.15}Fe_{0.85}O_{3-\delta}$ membrane reactors. *Catal. Lett.*, **111**, 179–185.

42 Wang, H.H., Tablet, C., Schiestel, T., Werth, S., and Caro, J. (2006) Partial oxidation of methane to syngas in a perovskite hollow fiber membrane reactor. *Catal. Commun.*, **7**, 907–912.

43 Wang, H.H., Cong, Y., and Yang, W.S. (2005) Oxidative coupling of methane in $Ba_{0.5}Sr_{0.5}Co_{0.8}Fe_{0.2}O_{3-\delta}$ tubular membrane reactors. *Catal. Today*, **104**, 160–167.

44 Zeng, Y., Lin, Y.S., and Swartz, S.L. (1998) Perovskite-type ceramic membrane: synthesis, oxygen permeation and membrane reactor performance for oxidative coupling of methane. *J. Membr. Sci.*, **150**, 87–98.

45 Akin, F.T. and Lin, J.Y.S. (2004) Oxygen permeation through oxygen ionic or mixed-conducting ceramic membranes with chemical reactions. *J. Membr. Sci.*, **231**, 133–146.

46 Akin, F.T. and Lin, Y.S. (2002) Oxidative coupling of methane in dense ceramic membrane reactor with high yields. *AIChE J.*, **48**, 2298–2306.

47 Tan, X., Pang, Z., and Liu, S. (2007) Catalytic perovskite hollow fibre membrane reactors for methane oxidative coupling. *J. Membr. Sci.*, **302**, 109–114.

48 Haag, S., van Veen, A.C., and Mirodatos, C. (2007) Influence of oxygen supply rates on performances of catalytic membrane reactors: application to the oxidative coupling of methane. *Catal. Today*, **127**, 157–164.

49 Lu, Y.P., Dixon, A.G., Moser, W.R., Ma, Y.H., and Balachandran, U. (2000) Oxygen-permeable dense membrane reactor for the oxidative coupling of methane. *J. Membr. Sci.*, **170**, 27–34.

50 Wang, H.H., Tablet, C., Schiestel, T., and Caro, J. (2006) Hollow fiber membrane reactors for the oxidative activation of ethane. *Catal. Today*, **118**, 98–103.

51 Akin, F.T. and Lin, Y.S. (2002) Selective oxidation of ethane to ethylene in a dense tubular membrane reactor. *J. Membr. Sci.*, **209**, 457–467.

52 Rebeilleau, D.M., Rosini, S., van Veen, A.C., Farrusseng, D., and Mirodatos, C. (2005) Oxidative activation of ethane on catalytic modified dense ionic oxygen conducting membranes. *Catal. Today*, **104**, 131–137.

53 Wang, H.H., Cong, Y., and Yang, W.S. (2002) High selectivity of oxidative dehydrogenation of ethane to ethylene in an oxygen permeable membrane reactor. *Chem. Commun.*, **14**, 1468–1469.

54 Zhu, X.F. and Yang, W.S. (2009) Mixed conductor oxygen permeable membrane reactors. *Chin. J. Catal.*, **30**, 801–816.

55 Sunarso, J., Baumann, S., Serra, J.M., Meulenberg, W.A., Liu, S., Lin, Y.S., and da Costa, J.C.D. (2008) Mixed ionic–electronic conducting (MIEC) ceramic-based membranes for oxygen separation. *J. Membr. Sci.*, **320**, 13–41.

56 Liu, Y.Y., Tan, X.Y., and Li, K. (2006) Mixed conducting ceramics for catalytic membrane processing. *Catal. Rev. Sci. Eng.*, **48**, 145–198.

57 Bouwmeester, H.J.M. (2003) Dense ceramic membranes for methane conversion. *Catal. Today*, **82**, 141–150.

58 Bouwmeester, H.J.M. and Burggraaf, A.J. (1997) Dense ceramic membranes for oxygen separation, in *The CRC Handbook of Solid State Electrochemistry* (eds P.J. Gellings and H.J.M. Bouwmeester), CRC Press.

59 Bouwmeester, H.J.M. and Burggraaf, A.J. (1994) Importance of the surface exchange kinetics as rate limiting step in oxygen permeation through mixed-conducting oxides. *Solid State Ionics*, **72**, 185–194.

60 Virkar, A.V. (2005) Theoretical analysis of the role of interfaces in transport through oxygen ion and electron conducting membranes. *J. Power Sources*, **147**, 8–31.

61 van Doorn, R.H.E., Bouwmeester, H.J.M., and Burggraaf, A.J. (1998) Kinetic decomposition of $La_{0.3}Sr_{0.7}CoO_{3-\delta}$ perovskite membranes during oxygen permeation. *Solid State Ionics*, **111**, 263–272.

62 Kim, S., Yang, Y.L., Jacobson, A.J., and Abeles, B. (1999) Oxygen surface exchange in mixed ionic electronic conductor membranes. *Solid State Ionics*, **121**, 31–36.

63 Mizusaki, J., Mima, Y., Yamauchi, S., Fueki, K., and Tagawa, H. (1989) Nonstoichiometry of the perovskite-type oxides $La_{1-x}Sr_xCoO_{3-\delta}$. *J. Solid State Chem.*, **80**, 102–111.

64 Huang, K., and Goodenough, J.B. (2001) Oxygen permeation through cobalt-containing perovskites surface oxygen exchange vs. lattice oxygen diffusion. *J. Electrochem. Soc.*, **148**, E203–E214.

65 Lee, T.H., Yang, Y.L., Jacobson, A.J., Abeles, B., and Milner, S. (1997) Oxygen permeation in $SrCo_{0.8}Fe_{0.2}O_{3-\delta}$ membranes with porous electrodes. *Solid State Ionics*, **100**, 87–94.

66 Lee, T.H., Yang, Y.L., Jacobson, A.J., Abeles, B., and Zhou, M. (1997) Oxygen permeation in dense $SrCo_{0.8}Fe_{0.2}O_{3-\delta}$ membranes: surface exchange kinetics versus bulk diffusion. *Solid State Ionics*, **100**, 77–85.

67 Kim, S., Yang, Y.L., Jacobson, A.J., and Abeles, B. (1999) Diffusion and surface exchange coefficients in mixed ionic electronic conducting oxides from the pressure dependence of oxygen permeation. *Solid State Ionics*, **106**, 189–195.

68 Qiu, L., Lee, T.H., Liu, L.M., Yang, Y.L., and Jacobson, A.J. (1995) Oxygen permeation studies of $SrCo_{0.8}Fe_{0.2}O_{3-\delta}$. *Solid State Ionics*, **76**, 321–329.

69 Xu, S.J. and Thomson, W.J. (1999) Oxygen permeation rates through ion-conducting perovskite membranes. *Chem. Eng. Sci.*, **54**, 3839–3850.

70 Lin, Y.S., Wang, W.J., and Han, J. (1994) Oxygen permeation through thin mixed-conducting solid oxide membranes. *AIChE J.*, **40**, 786–798.

71 Tan, X., Liu, S., Li, K., and Hughes, R. (2000) Theoretical analysis of ion permeation through mixed conducting membranes and its application to dehydrogenation reactions. *Solid State Ionics*, **138**, 149–159.

72 Tan, X., Liu, Y., and Li, K. (2005) Mixed conducting ceramic hollow-fiber membranes for air separation. *AIChE J.*, **51**, 1991–2000.

73 Tan, X. and Li, K. (2007) Oxygen production using dense ceramic hollow fiber membrane modules with different operating modes. *AIChE J.*, **53**, 838–845.

74 Chater, R.J., Carter, S., Kilner, J.A., and Steele, B.C.H. (1992) Development of a novel SIMS technique for oxygen self-diffusion and surface exchange coefficient measurements in oxides of high diffusivity. *Solid State Ionics*, **53–56**, 859–867.

75 Manning, P.S., Sirman, J.D., and Kilner, J.A. (1997) Oxygen self-diffusion and surface exchange studies of oxide electrolytes having the fluorite structure. *Solid State Ionics*, **93**, 125–132.

76 Routbort, J.L., Doshi, R., and Krumpelt, M. (1996) Oxygen tracer diffusion in $La_{1-x}Sr_xCoO_3$. *Solid State Ionics*, **90**, 21–27.

77 Ramos, T. and Atkinson, A. (2004) Oxygen diffusion and surface exchange in $La_{1-x}Sr_xFe_{0.8}Cr_{0.2}O_{3-\delta}$ ($x=0.2$, 0.4 and 0.6). *Solid State Ionics*, **170**, 275–286.

78 Fielitz, P. and Borchardt, G. (2001) On the accurate measurement of oxygen self-diffusivities and surface exchange coefficients in oxides via SIMS depth profiling. *Solid State Ionics*, **144**, 71–80.

79 ten Elshof, J.E., Lankhorst, M.H.R., and Bouwmeester, H.J.M. (1997) Oxygen exchange and diffusion coefficients of strontium-doped lanthanum ferrites by

electrical conductivity relaxation. *J. Electrochem. Soc.*, **144**, 1060–1067.

80. Preis, W., Bucher, E., and Sitte, W. (2002) Oxygen exchange measurements on perovskites as cathode materials for solid oxide fuel cells. *J. Power Sources*, **106**, 116–121.

81. Sahibzada, M., Morton, W., Hartley, A., Mantzavinos, D., and Metcalfe, I.S. (2000) A simple method for the determination of surface exchange and ionic transport kinetics in oxides. *Solid State Ionics*, **136–137**, 991–996.

82. Preis, W., Bucher, E., and Sitte, W. (2004) Oxygen exchange kinetics of $La_{0.4}Sr_{0.6}FeO_{3-\delta}$ by simultaneous application of conductivity relaxation and carrier gas coulometry. *Solid State Ionics*, **175**, 393–397.

83. Belzner, A., Gür, T.M., and Huggins, R.A. (1992) Oxygen chemical diffusion in strontium doped lanthanum manganites. *Solid State Ionics*, **57**, 327–337.

84. Diethelm, S., Closset, A., van Herle, J., and Nisancioglu, K. (2002) Determination of chemical diffusion and surface exchange coefficients of oxygen by electrochemical impedance spectroscopy. *J. Electrochem. Soc.*, **149**, E424–E432.

85. Bucher, E., Benisek, A., and Sitte, W. (2003) Electrochemical polarization measurements on mixed conducting oxides. *Solid State Ionics*, **157**, 39–44.

86. Crank, J. (1975) *The Mathematics of Diffusion*, Oxford University Press, Oxford.

87. Bouwmeester, H.J.M., Song, C., Zhu, J., Yi, J., Annaland, V.M., and Boukamp, B.A. (2009) A novel pulse isotopic exchange technique for rapid determination of the oxygen surface exchange rate of oxide ion conductors. *Phys. Chem. Chem. Phys.*, **11**, 9640–9643.

88. Kusaba, H., Shibata, Y., Sasaki, K., and Teraoka, Y. (2006) Surface effect on oxygen permeation through dense membrane of mixed-conductive LSCF perovskite-type oxide. *Solid State Ionics*, **177**, 2249–2253.

89. Teraoka, Y., Honbe, Y., Ishii, J., Furukawa, H., and Moriguchi, I. (2002) Catalytic effects in oxygen permeation through mixed-conductive LSCF perovskite membranes. *Solid State Ionics*, **152**, 681–687.

90. Etchegoyen, G., Chartier, T., and Del-Gallo, P. (2006) Oxygen permeation in $La_{0.6}Sr_{0.4}Fe_{0.9}Ga_{0.1}O_{3-\delta}$ dense membrane: effects of surface microstructure. *J. Solid State Electrochem.*, **10**, 597–603.

91. Figueiredo, F.M., Kharton, V.V., Viskup, A.P., and Frade, J.R. (2004) Surface enhanced oxygen permeation in $CaTi_{1-x}Fe_xO_{3-\delta}$ ceramic membranes. *J. Membr. Sci.*, **236**, 73–80.

92. Kharton, V.V., Kovalevsky, A.V., Yaremchenko, A.A., Figueiredo, F.M., Naumovich, E.N., Shaulo, A.L., and Marques, F.M.B. (2002) Surface modification of $La_{0.3}Sr_{0.7}CoO_{3-\delta}$ ceramic membranes. *J. Membr. Sci.*, **195**, 277–287.

93. Kovalevsky, A.V., Kharton, V.V., Snijkers, F.M.M., Cooymans, J.F.C., Luyten, J.J., and Marques, F.M.B. (2007) Oxygen transport and stability of asymmetric $SrFe(Al)O_{3-\delta}$–$SrAl_2O_4$ composite membranes. *J. Membr. Sci.*, **301**, 238–244.

94. Zhan, M., Ren, H., Tian, T., Wang, W., and Chen, C.S. (2008) Effects of hydrogen treatment on the surface microstructure and oxygen permeability of $La_{0.7}Sr_{0.3}Ga_{0.3}Fe_{0.7}O_{3-\delta}$ membrane. *Solid State Ionics*, **179**, 1382–1386.

95. Deng, H., Zhou, M., and Abeles, B. (1994) Diffusion reaction in porous mixed ionic–electronic solid oxide membranes. *Solid State Ionics*, **74**, 75–84.

96. Deng, H., Zhou, M., and Abeles, B. (1995) Transport in solid oxide porous electrodes: effect of gas diffusion. *Solid State Ionics*, **80**, 213–222.

97. van Hassel, B.A. (2004) Oxygen transfer across composite oxygen transport membranes. *Solid State Ionics*, **174**, 253–260.

98. ASM (1985) *Metallography and Microstructures*, ASM Handbook, vol. **9**, American Society for Metals, Metals Park, OH.

99. Chung, Y.C. and Wuensch, B.J. (1996) An improved method, based on

Whipple's exact solution, for obtaining accurate grain-boundary diffusion coefficients from shallow solute concentration gradients. *J. Appl. Phys.*, 79, 8323–8329.

100 Chung, Y.C. and Wuensch, B.J. (1996) Assessment of the accuracy of Le Claire's equation for determination of grain boundary diffusion coefficients from solute concentration gradients. *Mater. Lett.*, 28, 47–54.

101 Chung, Y.C., Kim, C.K., and Wuensch, B.J. (1996) Calculation of the contribution to grain boundary diffusion in ionic systems that arises from enhanced defect concentrations adjacent to the boundary. *J. Appl. Phys.*, 87, 2747–2752.

102 Abrantes, J.C.C., Labrincha, J.A., and Frade, J.R. (2000) Applicability of the brick layer model to describe the grain boundary properties of strontium titanate ceramics. *J. Eur. Ceram. Soc.*, 20, 1603–1609.

103 Chinarro, E., Jurado, J.R., Figueiredo, F.M., and Frade, J.R. (2003) Bulk and grain boundary conductivity of $Ca_{0.97}Ti_{1-x}Fe_xO_{3-\delta}$ materials. *Solid State Ionics*, 160, 161–168.

104 Kawada, T., Horita, T., Sakai, N., Yokokawa, H., and Dokiya, M. (1995) Experimental determination of oxygen permeation flux through bulk and grain boundary of $La_{0.7}Ca_{0.3}CrO_3$. *Solid State Ionics*, 79, 201–207.

105 Steele, B.C.H. (2000) Appraisal of $Ce_{1-y}Gd_yO_{2-y/2}$ electrolytes for IT-SOFC operation at 500 °C. *Solid State Ionics*, 129, 95–110.

106 Zhang, T.S., Ma, J., Chen, Y.Z., Luo, L.H., Kong, L.B., and Chan, S.H. (2006) Different conduction behaviors of grain boundaries in SiO_2-containing 8YSZ and CGO20 electrolytes. *Solid State Ionics*, 177, 1227–1235.

107 Martin, M.C. and Mecartneym, M.L. (2003) Grain boundary ionic conductivity of yttrium stabilized zirconia as a function of silica content and grain size. *Solid State Ionics*, 161, 67–79.

108 Guo, X., Tang, C., and Yuan, R. (1995) Grain boundary ionic conduction in zirconia-based solid electrolyte with alumina addition. *J. Eur. Ceram. Soc.*, 15, 25–32.

109 Kharton, V.V., Marques, F.M.B., Tsipis, E.V., Viskup, A.P., Vyshatko, N.P., Patrakeev, M.V., Naumovich, E.N., and Frade, J.R. (2004) Interfacial effects in electrochemical cells for oxygen ionic conduction measurements. III. Transference numbers vs. grain-boundary resistivity. *Solid State Ionics*, 168, 137–151.

110 Berenov, A.V., MacManus-Driscoll, J.L., and Kilner, J.A. (1999) Oxygen tracer diffusion in undoped lanthanum manganites. *Solid State Ionics*, 122, 41–49.

111 Kharton, V.V. and Marques, F.M.B. (2002) Mixed ionic–electronic conductors: effects of ceramic microstructure on transport properties. *Curr. Opin. Solid State Mater. Sci.*, 6, 261–269.

112 Steil, M.C., Fouletier, J., Kleitz, M., and Labrune, P. (1999) BICOVOX: sintering and grain size dependence of the electrical properties. *J. Eur. Ceram. Soc.*, 19, 815–818.

113 Kharton, V.V., Tikhonovich, V.N., Shuangbao, L., Naumovich, E.N., Kovalevsky, A.V., Viskup, A.P., Bashmakov, I.A., and Yaremchenko, A.A. (1998) Ceramic microstructure and oxygen permeability of $SrCo(Fe,M)O_3$ (M=Cu or Cr) perovskite membranes. *J. Electrochem. Soc.*, 145, 1363–1374.

114 Wang, H.H., Tablet, C., Feldhoff, A., and Caro, J. (2005) Investigation of phase structure, sintering, and permeability of perovskite-type $Ba_{0.5}Sr_{0.5}Co_{0.8}Fe_{0.2}O_{3-\delta}$ membranes. *J. Membr. Sci.*, 262, 20–26.

115 Arnold, M., Martynczuk, J., Efimov, K., Wang, H.H., and Feldhoff, A. (2008) Grain boundaries as barrier for oxygen transport in perovskite-type membranes. *J. Membr. Sci.*, 316, 137–144.

116 Diethelm, S., van Herle, J., Sfeir, J., and Buffat, P. (2005) Correlation between oxygen transport properties and microstructure in $La_{0.5}Sr_{0.5}FeO_{3-\delta}$. *J. Eur. Ceram. Soc.*, 25, 2191–2196.

117 Zeng, P., Ran, R., Chen, Z., Gu, H., Shao, Z.P., Diniz da Costa, J.C., and Liu, S. (2007) Significant effects of sintering temperature on the

118 Figueiredo, F.M., Kharton, V.V., Waerenborgh, J.C., Viskup, A.P., Naumovich, E.N., and Frade, J.R. (2004) Influence of microstructure on the electrical properties of iron-substituted calcium titanate ceramics. *J. Am. Ceram. Soc.*, **87**, 2252–2261.

performance of $La_{0.6}Sr_{0.4}Co_{0.2}Fe_{0.8}O_{3-\delta}$ oxygen selective membranes. *J. Membr. Sci.*, **302**, 171–179.

119 Chinarro, E., Jurado, J.R., Figueiredo, F.M., and Frade, J.R. (2003) Bulk and grain boundary conductivity of $Ca_{0.97}Ti_{1-x}Fe_xO_{3-\delta}$ materials. *Solid State Ionics*, **160**, 161–168.

120 Kharton, V.V., Naumovich, E.N., Kovalevsky, A.V., Viskup, A.P., Figueiredo, F.M., Bashmakov, I.A., and Marques, F.M.B. (2000) Mixed electronic and ionic conductivity of $LaCo(M)O_3$ (M=Ga, Cr, Fe or Ni). IV. Effect of the preparation method on oxygen transport in $LaCoO_3$. *Solid State Ionics*, **138**, 135–148.

121 Etchegoyen, G., Chartier, T., Julian, A., and Del-Gallo, P. (2006) Microstructure and oxygen permeability of a $La_{0.6}Sr_{0.4}Fe_{0.9}Ga_{0.1}O_{3-\delta}$ membrane containing magnesia as dispersed second phase particles. *J. Membr. Sci.*, **268**, 86–95.

122 Li, S., Jin, W., Xu, N., and Shi, J. (2001) Mechanical strength, and oxygen and electronic transport properties of $SrCo_{0.4}Fe_{0.6}O_{3-\delta}$–YSZ membranes. *J. Membr. Sci.*, **186**, 195–204.

123 Wu, Z., Jin, W., and Xu, N. (2006) Oxygen permeability and stability of Al_2O_3-doped $SrCo_{0.8}Fe_{0.2}O_{3-\delta}$ mixed conducting oxides. *J. Membr. Sci.*, **279**, 320–327.

124 Kharton, V.V., Shaula, A.L., Snijkers, F.M.M., Cooymans, J.F.C., Luyten, J.J., Yaremchenko, A.A., Valente, A.A., Tsipis, E.V., Frade, J.R., Marques, F.M.B., and Rocha, J. (2005) Processing, stability and oxygen permeability of $Sr(Fe, Al)O_3$-based ceramic membranes. *J. Membr. Sci.*, **252**, 215–225.

125 Kharton, V.V., Shaula, A.L., Snijkers, F.M.M., Cooymans, J.F.C., Luyten, J.J., Marozau, I.P., Viskup, A.P., Marques, F.M.B., and Frade, J.R. (2006) Oxygen transport in ferrite-based ceramic membranes: effects of alumina sintering aid. *J. Eur. Ceram. Soc.*, **26**, 3695–3704.

126 Shaula, A.L., Kharton, V.V., Vyshatko, N.P., Tsipis, E.V., Patrakeev, M.V., Marques, F.M.B., and Frade, J.R. (2005) Oxygen ionic transport in $SrFe_{1-y}Al_yO_{3-\delta}$ and $Sr_{1-x}Ca_xFe_{0.5}Al_{0.5}O_{3-\delta}$ ceramics. *J. Eur. Ceram. Soc.*, **25**, 489–499.

127 Kovalevsky, A.V., Kharton, V.V., Maxim, F., Shaula, A.L., and Frade, J.R. (2006) Processing and characterization of $La_{0.5}Sr_{0.5}FeO_3$-supported $Sr_{1-x}Fe(Al)O_3$–$SrAl_2O_4$ composite membranes. *J. Membr. Sci.*, **278**, 162–172.

128 Yaremchenko, A.A., Kharton, V.V., Valente, A.A., Shaula, A.L., Marques, F.M.B., and Rocha, J. (2006) Mixed conductivity and electrocatalytic performance of $SrFeO_{3-\delta}$–$SrAl_2O_4$ composite membranes. *Solid State Ionics*, **177**, 2285–2289.

129 Kovalevsky, A.V., Kharton, V.V., Snijkers, F.M.M., Cooymans, J.F.C., Luyten, J.J., and Marques, F.M.B. (2007) Oxygen transport and stability of asymmetric $SrFe(Al)O_{3-\delta}$–$SrAl_2O_4$ composite membranes. *J. Membr. Sci.*, **301**, 238–244.

130 Kharton, V.V., Kovalevsky, A.V., Yaremchenko, A.A., Snijkers, F.M.M., Cooymans, J.F.C., Luyten, J.J., Markov, A.A., Frade, J.R., and Marques, F.M.B. (2006) Oxygen transport and thermomechanical properties of $SrFe(Al)O_{3-\delta}$–$SrAl_2O_4$ composites: microstructural effects. *J. Solid State Electrochem.*, **10**, 663–673.

131 Kharton, V.V., Kovalevsky, A.V., Tsipis, E.V., Viskup, A.P., Naumovich, E.N., Jurado, J.R., and Frade, J.R. (2002) Mixed conductivity and stability of A-site-deficient $Sr(Fe,Ti)O_{3-\delta}$ perovskites. *J. Solid State Electrochem.*, **7**, 30–36.

132 Fan, C.G., Deng, Z.Q., Zuo, Y.B., Liu, W., and Chen, C.S. (2004) Preparation and characterization of $SrCo_{0.8}Fe_{0.2}O_{3-\delta}$–$SrSnO_3$ oxygen-permeable composite membrane. *Solid State Ionics*, **166**, 339–342.

133 Xia, Y., Armstrong, T., Prado, F., and Manthiram, A. (2000) Sol–gel synthesis, phase relationships, and oxygen permeation properties of

$Sr_4Fe_{6-x}Co_xO_{13+\delta}$ ($0 \leq x \leq 3$). *Solid State Ionics*, **130**, 81–90.

134 Fossdal, A., Sagdahl, L.T., Einarsrud, M.-A., Wiik, K., Grande, T., Larsen, P.H., and Poulsen, F.W. (2001) Phase equilibria and microstructure in $Sr_4Fe_{6-x}Co_xO_{13+\delta}$ ($0 \leq x \leq 4$) mixed conductors. *Solid State Ionics*, **143**, 367–377.

135 Bredesen, R. and Norby, T. (2000) On phase relations, transport properties and defect structure in mixed conducting $SrFe_{1.5-x}Co_xO_z$. *Solid State Ionics*, **129**, 285–297.

136 Lee, T.H., Yang, Y.L., and Jacobson, A.J. (2000) Electrical conductivity and oxygen permeation of $Ag/BaBi_8O_{13}$ composites. *Solid State Ionics*, **134**, 331–339.

137 Chen, C.S., Kruidhof, H., and Bouwmeester, H.J.M. (1997) Thickness dependence of oxygen permeation through stabilized bismuth oxide–silver composite. *Solid State Ionics*, **99**, 215–219.

138 Chen, C.S., Boukamp, B.A., Bouwmeester, H.J.M., Cao, G.Z., Kruidhof, H., Winnubst, A.J.A., and Burggraaf, A.J. (1995) Microstructural development, electrical properties and oxygen permeation of zirconia–palladium composites. *Solid State Ionics*, **76**, 23–28.

139 Kharton, V.V., Kovalevsky, A.V., Viskup, A.P., Yaremchenko, A.A., Naumovich, E.N., and Marques, F.M.B. (2000) Oxygen permeability of $Ce_{0.8}Gd_{0.2}O_{2-\delta}$–$La_{0.7}Sr_{0.3}MnO_{3-\delta}$ composite membranes. *J. Electrochem. Soc.*, **147**, 2814–2821.

140 Shaula, A.L., Kharton, V.V., and Marques, F.M.B. (2004) Phase interaction and oxygen transport in $La_{0.8}Sr_{0.2}Fe_{0.8}Co_{0.2}O_3$–$(La_{0.9}Sr_{0.1})_{0.98}Ga_{0.8}Mg_{0.2}O_3$ composites. *J. Eur. Ceram. Soc.*, **24**, 2631–2639.

141 Zhu, X.F. and Yang, W.S. (2008) Composite membrane based on ionic conductor and mixed conductor for oxygen permeation. *AIChE J.*, **54**, 665–672.

142 Zhu, X.F., Li, Q.M., Cong, Y., and Yang, W.S. (2008) Syngas generation in a membrane reactor with a highly stable ceramic composite membrane. *Catal. Commun.*, **10**, 309–312.

143 Zhu, X.F., Wang, H.H., and Yang, W.S. (2008) Relationship between homogeneity and oxygen permeability of composite membranes. *J. Membr. Sci.*, **309**, 120–127.

144 Li, Q.M., Zhu, X.F., and Yang, W.S. (2008) Single-step fabrication of asymmetric dual-phase composite membranes for oxygen separation. *J. Membr. Sci.*, **325**, 11–15.

145 Li, Q.M. (2009) Novel mixed conducting oxygen permeable membranes and preparation of dual-phase asymmetric membranes. Ph.D. thesis, Dalian Institute of Chemical Physics, Dalian.

146 Yi, J., Zuo, Y., Liu, W., Winnubst, L., and Chen, C. (2006) Oxygen permeation through a $Ce_{0.8}Sm_{0.2}O_{2-\delta}$–$La_{0.8}Sr_{0.2}CrO_{3-\delta}$ dual-phase composite membrane. *J. Membr. Sci.*, **280**, 849–855.

147 Xu, S.J. and Thomson, W.J. (1998) Stability of $La_{0.6}Sr_{0.4}Co_{0.2}Fe_{0.8}O_{3-\delta}$ perovskite membranes in reducing and nonreducing environments. *Ind. Eng. Chem. Res.*, **37**, 1290–1299.

148 Zhu, X.F. (2006) Investigation of novel high stability oxygen permeable ceramic membranes. Ph.D. thesis, Dalian Institute of Chemical Physics, Dalian.

149 Chang, X., Zhang, C., Dong, X., Yang, C., Jin, W., and Xu, N. (2008) Experimental and modeling study of oxygen permeation modes for asymmetric mixed-conducting membranes. *J. Membr. Sci.*, **322**, 429–435.

150 Wang, Z.B., Ge, Q.Q., Shao, J., and Yan, Y.S. (2009) High performance zeolite LTA pervaporation membranes on ceramic hollow fibers by dipcoating–wiping seed deposition. *J. Am. Chem. Soc.*, **20**, 6910–6911.

Index

a

AAEM *see* alkaline anion exchange membrane
acceptor-doped cobaltite as cathode
 material 224
activation overpotential 365
adhesion energy 31, 40, 62, 65
adsorption isotherm 75
– classification of 75
– for polyions 99
adsorption of a fluid on a solid 75
AFC *see* alkaline fuel cell
Ag_2S 8 f
AgGeSe 9
alkaline anion exchange membrane (AAEM)
 materials 190 ff
alkaline fuel cell (AFC) 183 f, 187 ff
alkaline membrane fuel cell (AMFC) 190 f
alkaline-doped polybenzimidazole (PBI)
 membrane 190
AMFC *see* alkaline membrane fuel cell
amorphous silica
– in composite electrolytes 60
analysis of cation diffusion
– creep measurement 430, 433
– diffusion-controlled solid-state reaction
 between two binary oxides 430, 432
– interdiffusion experiment 430 ff
– tracer or impurity experiment 430 f
analysis of electrochemical instabilities 128
analysis of oscillating reactions
– digital signal processing 139
analysis of pattern formation 142
analysis of spatial dynamics of electrochemical
 processes
– by two-dimensional imaging of electrode
 potential 137
– ellipsometry 137

– spatially resolved in situ infrared
 spectroscopy 137
– using a multielectrode array
 configuration 138
– using optical microscopy 137
analysis of temporal dynamics of
 electrochemical processes
– using impedance spectroscopy 136
– using linear sweep voltammetry 136
– using potentiostatic (galvanostatic)
 experiment 136
analysis of time series data of electrochemical
 systems 140
anion migration 7 f
anion-based resistive switching
 mechanism 6 ff
anode for SOFCs
– based on bimetallic nickel alloy 289 f
– based on double-perovskite materials 300 f
– based on perovskite materials 290 f
– based on perovskite-like materials 293
– cermet 286 ff
– composition and properties of materials
 for 294 f, 296 ff
– conductivity of materials for 294 f
– linear thermal expansion coefficients of
 material for 294 f
– micro-SOFC anode 307
– nickel-ceria-based cermets 267, 287 ff
– nickel-free cermet 290
– oxidic 286 ff
– SC-SOFC anode 305
– surface modification of 296 ff
anode of lithium-ion batteries
– intercalation nanomaterials for 393 f
– intermetallic compounds for 397
– titanium oxides for 396 f
anode poisoning 220 ff

Solid State Electrochemistry II: Electrodes, Interfaces and Ceramic Membranes.
Edited by Vladislav V. Kharton.
© 2011 Wiley-VCH Verlag GmbH & Co. KGaA. Published 2011 by Wiley-VCH Verlag GmbH & Co. KGaA.

anodic overpotential 287 f
anodic polarization resistance 291 f, 296 f, 300, 313
apatite-type solid electrolyte 285 f
application
– of ceramic membrane reactor 491 ff, 502
– of ionic memory 24
– of mixed conducting oxide materials 501 f
area-specific electrode conductivity 296 f, 314
Arnold tongue 161
array-level characterization 12
association of oppositely charged ions 95
asymmetric target pattern 159
atomic layer deposition 419
atomic switch see cation-based resistive switching memory
autocatalysis 156

b

back-end-of-line (BEOL) process 10, 20, 23
$(Ba,Sr)(Co,Fe)O_{3-\delta}$ (BSCF) 416, 445, 522
$BaZrO_3$ 425 f
bifurcation 126, 131
– cascade of period-doubling 134
– diagram 136, 140, 143, 161
– homoclinic 167
– Hopf (H) 133, 147 f
– saddle–loop 146
– saddle–node (SN) 133, 143
– secondary Hopf (torus) 134
biofuel cell 235
biomass-renewable fuels in fuel cells 185 f
bipolar plate for PEMFCs 201, 359
bipolar resistive switching phenomen 1, 3
bismuth-vanadate (BIMEVOX) solid electrolyte 286
bistability 126 f, 133, 143 f, 157
Born–Green–Yvon hierarchy 87
brick wall model 45
brownmillerite 470, 478
Bruggeman correlation 357
bulk defect 429, 445
– energy diagram of 37 f
– in oxygen permeation 503 f, 511
bursting oscillation in electrochemical systems 151 f
by-products of fuel cells 183, 185 f, 202, 250

c

calcia-stabilized zirconia (CSZ) 447
capillary evaporation 98
carbon dioxide permeable membrane 502
carbon metal fluoride nanocomposite (CMFNC) 392
carbon nanotube (CNT) 188, 198, 236, 393, 395
carbonaceous material in SOFCs 310
Carnahan–Starling equation of state 90
catalyst coated membrane (CCM) 336
catalyst layer (CL) 333 ff, 340 f, 345 ff, 351, 354, 361, 376 ff
– coating onto GDL 334
catalyst-sprayed membrane under irradiation (CSMUI) 339
cathode 384 ff
– electrochemical performance of $LiFePO_4$ 388
– higher voltage 386
– nanostructured layered lithium metal dioxides for 385
– olivine phosphates for 387
– silicates for 390 ff
– spinel compounds for 386 f
– surface coated material for 385, 387, 390
– transition metal fluorides for 392 f
– 4 V 384 ff
cathode for SOFCs 267 f
– $Ba_2[(Co,Mo,Nb)O_3]_{2-\delta}$ 268
– $BaCo(Mo,Nb)O_{5+\delta}$ 275
– $(Ba,Sr)(Co,Fe)O_{3-\delta}$ (BSCF) 268, 273, 275, 302
– $(Ba,Sr)_2CoO_{4+\delta}$ 268, 277
– $Ca_3Co_4O_{9-\delta}$ 268, 278
– cobaltite-containing 274 f, 277
– effect of doping on performance of 272 f, 275 ff
– ferrite-based 268 f, 275, 278
– $(Gd,Ca)(Mn,Ni)O_{3-\delta}$ 271
– $La_2NiO_{4+\delta}$-type 270, 273, 283 f
– $La_3Ni_2O_{7-\delta}$-type 270, 283 f
– $La_4Ni_3O_{10-\delta}$-type 270, 283 f
– $LaSrFeO_{4+\delta}$ 269, 279
– $(La,Sr)_{0.95}FeO_{3-\delta}$-type 269, 278, 283
– $(La,Sr)_4Ni_2CoO_{10-\delta}$ 270, 283 f
– manganese-containing material for 270 f
– manganite-based 284
– material for intermediated temperature (IT) applications 277
– micro-SOFC cathode 307
– nickelate-based 284
– nickel-containing material for 270 f, 279
– perovskite-based 268 f, 275 ff
– $Pr_2(Ni,Cu)O_{4+\delta}$ 270, 283 f
– $REBaCo_3Zn_4O_{7+\delta}$ 268, 273, 276, 278
– $RE(Ba,Sr)Co_2O_{5+\delta}$ 268, 275, 277
– $(RE,Sr)CoO_{3-\delta}$-type 267 f, 273, 278, 280
– SC-SOFC cathode 302
– $(Sr,Ce)(Mn,Cr)O_{3-\delta}$ 271

Index

- Sr(Co/Mn,Nb/Sb)O$_{3-\delta}$-type 268, 271 ff
- Sr$_2$[(Co,Fe)NbO$_3$]$_{2-\delta}$ 268, 272
- (Sr,K)FeO$_{3-\delta}$ 269, 279
- with double-perovskite structure 268, 272, 275
- YBaCo$_4$O$_{7+\delta}$-type 268, 276 ff
- zinc-doped cobaltite-containing 278

cathodic overpotential 299
- of half-cells with porous mixed conducting electrodes 275
- of mixed rare-earth oxides 272

cathodic polarization resistance 266 ff, 273 f, 276 f, 284, 313
cation diffusion 426 ff
- determination of 430 f
- in fluorite oxides 433 f
- in perovskite material 433, 435

cation migration 6
cation mobility 433
- effects on ceramic membrane material due to 435 ff

cation-based resistive switching mechanism 4 ff
CCM *see* catalyst coated membrane
ceramic membrane reactor 491 ff
ceramic oxide membrane 415
- based on BaZrO$_3$ powder 425
- fluorite structure 426 f, 430
- perovskite structure 424 ff, 430

ceramic powder of complex oxide 417
- densification of 420
- forming of 419
- preparation of 418

cermet (ceramic–metal composite) 64
cermet anode 219
chaos development 134, 150
chaos supression 161
chaotic attractor 141, 150
chaotic oscillations of dynamical systems 126, 134 f, 141 f, 150 ff
- effect of electric coupling on 155 f

chaotic phase synchronization 161
characterization of PEMFCs
- analysis of polarization curves 364 f
- atomic force microscopy (AFM) 353
- by current interrupt method 365 f
- CO stripping voltammetry 369
- composition analysis 361
- conductivity of PEMFCs 360
- current distribution mapping 374
- cyclic voltammetry (CV) 369
- differential scanning calorimetry (DSC) 362
- differential thermal analysis (DTA) 362

- diffusivity of PEMFCs 357
- electrochemical impedance spectroscopy (EIS) 366 ff
- energy dispersive spectrometry (EDS) 361
- gas composition analysis 371
- high-angle annular dark field scanning transmission electron microscopy (HAADF-STEM) 355
- high-resolution transmission microscopy (HR-TEM) 355
- hydrophilicity of PEMFCs 358
- Kelvin probe microscopy (KPM) 354
- linear sweep voltammetry (LSV) 370
- magnetic resonance imaging (MRI) 372
- measurment of external contact angle 359
- microstructure analysis 354 f
- neutron imaging 372
- permeability of PEMFCs 357
- porosimetry 356
- pressure drop measurement 371
- sample preparation 355
- scanning probe microscopy (SPM) 352 f
- scanning tunneling microscopy (STM) 353
- temperature distribution mapping 373
- thermal gravimetric analysis (TGA) 362 f
- using optical microscopy 349
- using scanning electron microscopy (SEM) 349 ff
- wetting properties of PEMFCs 358

charge distribution
- at solid-electrolyte solution interfaces 83 f
- in SCL 35 ff

charge inversion 83
charge transfer process 128, 130
- of the cathode CL 367 f

charge transport process in PEMFCs 367 ff
charge trapping 3, 19
chemical adsorption of ions 33
chemical expansion 267, 272, 278, 293, 439
- of oxide materials containing variable valence cations 440

chemical-induced stress 440
CL *see* catalyst layer
clustering 159
CMFNC *see* carbon metal fluoride nanocomposite
CMOL architecture *see* CMOS/hybrid ionic memory architecture 21
CMOS *see* complementary metal-oxide-semiconductor
CMOS/hybrid ionic memory architecture 21
CMOS-integrated ionic memory architecture 20
CNT *see* carbon nanotube

coarsening process   420
Coble creep   445
$CoFe_2O_4$ 404
colloidal particle   80, 105
– adsorption of ions on spherical   108 f
colloidal stability   80, 83, 108
combustion and reforming reaction (CRR)   492
complementary metao-oxide-semiconductor (CMOS)   1
composite interface   37
composite solid electrolyte   31 ff
– control of conductivity by variation of morphology   61 f
– defect equilibria in   32 ff
– design of   58 ff
– interface interactions in   32 f
– ionic "salt–glass" type   60
– layered composite of crystalline fluoride conductors   58, 62
– of the ionic salt–oxide type (MX–A)   31
– operating at elevated temperature   64
– oxide structure-based effects on conductivity of   58 f
– with inert matrix phase   60
– with inert oxide additive   61 f
composition analysis of PEMFCs   361
computer simulation of solid-electrolyte solution interfaces
– lattice summation method   102
– of ion–dipole mixtures   110
– reaction field method   103
– simulation of long-range forces   101 ff
conditioning of a metal oxide   6
conductive bridging RAM see cation-based resistive switching memory
conductivity measurement   360, 517
contact angle measurement   359 f
control   162
– of chaotic systems   163 f
– of critical behaviour   159 f
– of dynamics in electrochemical systems   159 f
conversion electrode   392, 401
– based on binary transition metal oxides   401 ff
– based on multinary oxideds   404
– based on transition metal oxysalts   404 f
– material for   392
conversion reaction   338, 392, 401, 403 ff
copper/YSZ cermet anode   221
corrosion
– in ACFs   189
$CoSn_2$ 398 f

Cr-doped $SrTiO_3$
– resistive switching characteristics of   6 f
creep parameter
– of fluorite oxides   445
– of perovskite oxides   445 f
crossbar array   1
crossbar array ionic memory architecture   21
CRR see combustion and reforming reaction
$CsCl–Al_2O_3$ interface   57
CSMUI see catalyst-sprayed membrane under irradiation
CSZ see calcia-stabilized zirconia
Cu/Cu_2S 16
Cu_2S 8
current generation via fuel cell   182
current measurement in fuel cells
– partial MEA method   374 f
– segmented cell method   374 f
– segmented plates method   375
– subcell method   374 f
– using printed circuit board (PCB) technology   376
current-induced Joule heating   4
cycling endurance   15, 18, 22

d

DCFC see direct carbon fuel cell
Debye–Hückel theory of electrolyte solution   79
DEFC see direct ethanol fuel cell
defect equilibria   32 ff
defect thermodynamics
– in an ionic crystal with Schottky defects   33 ff
– in composite electrolytes   37 f
degradation
– due to volatization of oxide components   453, 489
– of AFCs due to corrosion   189
– of electrocatalysts   203 ff, 213
– of electrodes   24, 219, 221 ff, 250, 263, 288 f, 300 f, 305 f, 315, 385
– of electrolytes   215, 227
– of fuel cells due to reaction with gaseous species   450 f
– of membranes   203 ff, 355, 362, 435 f, 432 f, 526, 531
– of PEMFCs   239, 249 f, 349 ff
– of SOECs   318 f
– of SOFCs   230 f, 265, 288 f, 433, 437
delamination   437
dense membrane system   416
densification process   420
density functional (DF) theory   88 ff

– comparison with computer simulation 96 f
density profile
– for ion–dipole mixtures confined between walls 111
– for the solvent primitive model system 110
– ionic 91, 96 f, 107
– of an electrolyte solution 79 ff
density profile equation 90
depletion of counterion profile 96
DET see direct electron transfer
DF theory see density functional theory
DFAFC see direct formic acid fuel cell
DHE see dynamic hydrogen electrode
diffusion in single crystals 429 f
dimethyl ether in liquid fuel cells 185 f
diode 21
dip coating 419
direct carbon fuel cell (DCFC) 183 f, 246
direct electron transfer (DET) by enzymes 235
direct ethanol fuel cell (DEFC) 210 f
direct formic acid fuel cell (DFAFC) 183 f, 208 ff
direct methanol fuel cell (DMFC) 183 f, 205 ff, 243
– passive 207 f
double-layer charging 128
DRAM see dynamic random access memory
dual-electrode setup 153
dynamic hydrogen electrode (DHE) 369
dynamic instability in electrochemical systems
– classification based on essential species 127
– classification based on nonlinear dynamics 131
– effect of noise on 168 f
– physicochemical origin of 126
dynamic random access memory (DRAM) 1 f, 19
dynamically patterned oscillatory system 135
dynamics of coupled electrodes 153
– formation of stationary patterns 156 ff
– global coupling 154
dynamics of electrochemical processes 136 ff

e
ECA see electrochemical surface area
EDTA/citrate complex method 472
EIS see electrochemical impedance spectroscopy
electric coupling
– of electrodes 153
– of migration currents 158
electric double-layer
– capacitance of 79, 96, 106
– draw back of double-layer theory 96
– effect of electrostatic image charges on the structure of 108
– effect of surface polarization on the structure of 107
– inverse capacitance of water–metal interphases 88
– model for solvent effects of water on 112 f
– simple model for solvent effects on 109 ff
– simulation of simple models for 104 f
– structure of 100
– theoretical description of 81 f, 89
– thermodynamics of 100
electric power-induced Joule heating 3
electric properties of interfaces 37 ff
electrical characteristics of ionic memories 12
electrical conductivity relaxation 516
electrical interaction in electrochemical systems 153
electrocatalysis in alkaline medium 192 ff, 196
electrocatalyst
– bimetallic alloy 196 f, 208, 210
– degradation in PEMFCs 203
– metal composite 196
– non-platinum-based 198, 209
– platinum-based 193, 197, 208
– transition-metal-based 193 f, 212
electrochemical cell 125
– dynamic instabilities in 125 ff
electrochemical dynamical system effect of noise on 167
electrochemical impedance spectroscopy (EIS) 366 ff, 521
electrochemical interface 74
– charge inversion at 83
– differential capicatance of 78
electrochemical surface area (ECA) 369
electrode degradation 24, 26
electrode diameter 12 f
electrode for high-temperature electrochemical cells 265 ff
electrode for PEMFCs
– components of 333
– performance of 334
electrode material 384 ff
electrode polarization 266, 269, 299, 317
electrodeposition 145, 159, 170
electrodissolution 135, 137, 139, 141 ff, 150 f, 154, 161, 168

electroforming cycle 15
electrolysis of carbon dioxide 318
electrolyte carbonation 189
electrolyte for fuel cells 184
electrolyzer *see* solid oxide electrolysis cell (SOEC) 317 ff
electron mediator 233
electron transfer mechanism in biofuel cells 236
electroneutrality 467
electrophoresis 419
electrophoretic deposition (EPD) 340 f
– of Pt/C nanocatalysts and Nafion solution 343
ellipsometry 169
embedded memory 25
energy conversion technology 180, 250
energy storage 384
enzyme catalyst for fuel cells 235
EPD *see* electrophoretic deposition
equation of state for hard-sphere fluid mixtures 91
erasing process *see* LRS-to-HRS switching
essential variable 127
evaporation–condensation 420
Ewald summation 102 f
excitability 127
extended time delay autosynchronization (ETDAS) 163
external field 75
extrusion 419

f

fabrication
– of ceramic membrane systems using wet ceramic techniques 419
– of dense ceramic 421
– of dense membrane 472
– of gas diffusion media 347 ff
– of hydrophobic PFTE-bonded gas diffusion electrode 334
– of low platinum loading electrode 340 ff
– of MEA using EPD 342 f
– of microporous layer (MPL) 347 ff
– of planar membranes 419
– of thin-film hydrophilic catalyst coated membrane (CCM) 336 f
– of tubular membranes 419
– of dense material 416 f
– of gas diffusion layer (GDL) 347
– of porous ceramic electrodes 417, 421
fabrication method for high-performance electrodes 333
– brushing 334

– direct spray coating 339
– direct spraying 334
– electrophoretic deposition (EPD) 340 ff
– electrospraying 345
– hot pressing 336
– inkjet printing 335, 338
– modified decal method 338
– Nafion impregnation 335
– plasma sputtering 345
– pulse electrodeposition 342
– screen printing 334, 337
– sol-gel method 346, 389
fast Fourier transform (FFT) method 139
fast spiking behavior 151
FC *see* fuel cell
feedback loop 127 ff, 166
FeRAM *see* ferroelectric random access memory
ferroelectric polarization reversal 3
ferroelectric random access memory (FeRAM) 1 f
$FeSn_2$ 398 f
FFT method *see* fast Fourier transform method
field programmable rectification 7
filament 7 f
– formation of 16
– growth between two electrodes 23
Flash memory-based solid-state drive (SSD) 25
flow field design 375 f
fluid–fluid interaction energy 75
fluid–solid interaction energy 75
FMT *see* fundamental measure theory
formic acid in liquid fuel cells 185 f
forming process of solid-state electrolyte 8
fracture of ceramic materials 442 f
fracture strength of ceramic materials 441 f
fracture toughness of ceramic materials 441 f
free energy of electrolyte
– electrostatic contribution to 92 ff
free surface of an ionic crystal 33 ff
Frumkin effect 130
Frenkel disorder 428
fuel cell (FC)
– application of 247 ff
– classification of 182
– current–potential and power versus current dependence in 182
– economic aspects 250
– electrolytes for 184
– emerging 232
– for electrochemical cells 183 f
– high-temperature 183, 185, 213, 228, 238, 312 f

– liquid 183 ff
– low-temperature 183 f
– thermodynamic efficiency of 179, 183
– thermodynamic principles of 180 f
fullerene 393 f
fundamental measure theory (FMT) 91

g

gas diffusion cathode 188, 212
gas diffusion electrode (GDE) 342 f
– CV analysis of 369
– hydrophobic PTFE-bonded 334 ff
gas diffusion layer (GDL) 199, 333 ff, 341, 345, 347 f, 351, 356 ff, 360, 365 ff, 371 ff
– fabrication of 347
gas permeation 473
– flux equation for mixed conducting membranes 475
gas purification 490 f
gas separation 490 f, 501
GDE see gas diffusion electrode
GDL see gas diffusion layer
Ge-based chalcogenide solid electrolyte 9
gel electrolyte 59
GeS 8
GeSe 8
Ginzburg–Landau equation (CGLE) 135
Gouy–Chapman formula see space charge layer 33
grain boundary conductivity 521 f
grain boundary diffusion 420, 429, 433 f, 445, 520
grain boundary resistance 528
grain growth in solid-state sintering 421 ff
grain size effect
– in nanocomposite solid electrolyte 42 ff
grand canonical thermodynamic potential 88 ff, 94
graphene 393

h

Hart's equation 429
heterogeneous doping 32
hidden N-NDR (HNNDR) system 130, 145
high resistance state (HRS) 3
high-temperature creep 444 f
high-temperature gas permeation experiment 472 f
high-temperature proton exchange membrane fuel cell (HT-PEMFC) 238
– based on ceramic PEM 240
– based on phosphoric acid-doped PBI composite membranes 238 f
– based on porous inorganic PEM 239 f

high-temperature solid-state electrochemistry 415
high-temperature steam electrolysis (HTSE) 317
HNC see hypernetted chain density profile approximation
homoclinic orbit 146
Hopf bifurcation point 148 f
hot pressing (HP) 422
hot spot 6
HP see hot pressing
HRR see hydrogen reduction reaction
HRS resistance/LRS resistance ratio 12 f, 17
HRS see high resistance state
HRS see high resistance state
HRS-to-LRS switching 3
HT-PEMFC see high-temperature proton exchange membrane fuel cell
HTSE see high-temperature steam electrolysis
hydrocarbons in liquid fuel cells 185 f, 216, 220, 232
hydrocarbons in SOFCs 309 f
hydrogen in fuel cells
– materials for storage of 183
hydrogen permeable membrane 502
– comparison with oxygen permeable membrane 489 f
– electrical and ionic transport properties of 484 ff
– stability of 489
hydrogen permeation see also hydrogen transport
– flux data 487 f
– in inert sweep gases 487 f
– under oxidizing conditions 485 f
hydrogen production 319, 492
hydrogen reduction reaction (HRR) 367
hydrogen transport 467, 473 f, 485 f
hydrogenperoxide reduction on platinum 152
hypernetted chain (HNC) density profile approximation 81 f
hysteresis 133
hysteresis loop 143

i

icon memory technology 1
impedance behavior of CLs 368
impedance spectroscopy 160, 521
individual device-level characterization 12
inert electrode material 8
inhomogeneous electrolyte 76 f, 92
interaction potential

– of a homogeneous system with electrostatic interactions 76
– of hard-sphere/hard-wall system 81
– of inhomogeneous electrolyte solutions 77 f
interdiffusion 430 ff, 443
interface energy 40
interface exchange coefficient 512, 514
interface interaction
– between ionic salt and oxide phase 37 ff
– between YSM and LSM 444
– chemical compatibility of contacting materials 443
– contribution of interatomic interactions between ions or atoms of adjacent phases 39
– contribution of the elastic energy of mechanical strains 39
– in composite electrolytes 32 ff
– in SOFCs 443
– interdiffusion 443
interface phase 31, 65
interfacial charge storage mechanism 403
interlayer for SOFCs $(CeGd)O_{2-\delta}$ (CGO) 267
interlayer for SOFCs $(Ce,La)O_{2-\delta}$ (CLO) 280, 291
interlayer for SOFCs $CeSmO_{2-\delta}$ (CSO) 280
International Technology Roadmap of Semiconductor (ITRS) 1 f
ion migration 1
ionic conductivity
– dependence on oxide grain size 50
– effect of heterogeneous doping on 32
– modelling of 51 ff
– of composites 32
– of nanocomposite solid electrolytes 45, 47, 63
– temperature dependence of 47 ff, 64
– of metal halide–Al_2O_3 composite 55, 61 f
ionic conductor with high diffusity 31
ionic liquid 232 f
ionic memory
– Ag/Ag_2S 9, 15
– Ag/Ag_2S-doped $AgPO_3$ (ASP) 11
– Ag/As_2S_3 11
– Ag/GeS 10, 23
– $Ag/GeSe$ 15, 18, 23
– Ag/glassy $(AgI)_{0.5}(AgPO_3)_{0.5}$ 11
– Ag/α-Si 11, 21
– Ag/Si_3N_4 11
– Ag/spin-on glass methyl silsesquioxane (MSQ) 11
– Ag/TiO_2 10
– $Ag/Zn_xCd_{1-x}S$ 11
– application of 24
– architecture for 20
– challenges of 22
– comparison with other memories 19
– Cu/Cu_2S 9, 26
– Cu/Cu-doped amorphous carbon (CuC) 11, 16
– Cu/GeS 10
– Cu/MoO_x 11
– Cu/SiO_x 10, 15 f, 24
– Cu/Ta_2O 10, 15, 18
– Cu/WO_3 10
– electrical characteristics of
– properties of 22
– random diffusion in 22 f
– switching speed of 19, 24
– thermal stability of 23
ionic memory array characteristics 18 f
ionic memory device 1, 12 ff
ionic memory switching mechanism 4
ionic salt-oxide composite
– ionic transport in 50 ff
– mechanical properties of 55 f
– physical origin of surface interactions in 59
– physical state of the ionic salt 59
ionic transport 467, 471, 476 ff, 484 ff
– in ionic salt-oxide composite 50 ff
– in solid electrolyte for SOFCs 299
ionic–electronic conducting membranes 467
ion-ion interaction 32 f
IR compensation 160
isostatic pressing 419
isotopic exchange technique 514 f, 520
I–V characteristic 7, 15
– asymmetric 21
– rectifying 11

j

Jahn–Teller effect 386
jellium model for metal electrodes 86

k

Karhunen–Loève (KL) decomposition 142
kinetic decomposition of ceramic membrane material 436
kinetic demixing of ceramic membrane material 435 f, 531
K_2NiF_4-type oxide 470
Kolmogorov entropy 141
Kuramoto transition 155

l

La$_{0.7}$Ca$_{0.3}$MnO$_3$ (LCMO) in resistive switching device 7
LaCoO$_{3-\delta}$ 416
LaGaO$_3$ 416
Landau–Lifshitz equation 52
Langmuir adsorption isotherm 33
La$_2$NiO$_{4+\delta}$ 416
lanthanum–strontium magnesium gallate (LSGM) 227
lanthanum–strontium manganite (LSM) cathode 218, 222 f
lanthanum-strontium-gallium-magnesium-oxided (LSGM) solid electrolyte 268 ff, 279
(La,Sr)(Co,Fe)O$_{3-\delta}$ (LSCF) 416, 424, 435, 445, 529
La$_{1-x}$Sr$_x$CoO$_{3-\delta}$ membrane 436
(La,Sr)MnO$_{3\pm\delta}$ (LSM) 416, 424, 445
lattice diffusion 420
layered metal dioxide 384 ff
Lichtenecker equation 52
LiCoO$_2$ 384 f
LiCoPO$_4$/C nanocomposite 388
Li$_2$CoSiO$_4$ 391
LiFePO$_4$ 387
Li$_2$FeSiO$_4$ 390 f
LiMn$_2$O$_4$ 386
LiMnPO$_4$ 389
Li$_2$MnSiO$_4$ 391
linear chemical expansion 440
linear stability analysis 131
linear thermal expansion coefficient (TEC) 268 ff, 294
– effect of cathode composition on 273
– of mixed conducting oxide materials 438
– of solid electrolytes 438
LiNiO$_2$ 385
liquid electrolyte solution 73 ff
liquid–vapor phase diagram 97 f
LISICON see lithium superionic conductor 390
lithium intercalation 393 f
lithium ion battery 383
– anode materials for 393 ff
– cathode materials for 384 ff
– charge/discharge cycle of 402
lithium solid-state battery 60
lithium superionic conductor (LISICON) 390
lithium-silicon intermetallic phase 400
lithium-tin intermetallic phase 397
Li$_2$TiSiO$_5$ 390
Li$_2$VOSiO$_4$ 392
long-range Coulomb force 101
low resistance state (LRS) 3
LRS-to-HRS switching 3
Lyapunov exponent 141

m

magnetic random access memory (MRAM) 1 f
Mansoori–Carnahan–Starling–Leland–Boublik equation of state 91, 94
mass transport mechanism during solid-state sintering 421
Maxwell–Garnett model 50
MC see Monte Carlo computer simulation
MCFC see molten carbonate fuel cell 214
MD see molecular dynamics
MEA see membrane electrode assembly
mean spherical approximation (MSA) 81, 94
mechanical deformation of ceramic materials 444
mediated electron transfer (MET) 233 ff
membrane electrode assembly (MEA) 194, 334, 336, 342
– performance of 339, 344
– SEM images of 352
memory
– classification of 2 f
– flash 1 f, 15
– historical development of 2
– density of 25
MEMS see microelectromechanical system
mesoporous oxide matrix 62
MET see mediated electron transfer
metacomposite solid electrolyte 62, 64
metal dissolution/passivation reaction 137
metal sulfide solid electrolyte 9
metal-insulator-metal structure (MIM structure) 3 f, 20
methane in liquid fuel cells 185 f
methanol in liquid fuel cells 185 f
MFC see microbial fuel cell
microbial fuel cell (MFC) 232 f
microelectromechanical system (MEMS) 243, 248
microfabrication technique 243 ff
microfluidic flow cell 153
microfluidic fuel cell 237 f
microorganism as catalyst for fuel cells 232 f
microporous layer (MPL) 333 f, 347 f, 358
micro-solid oxide fuel cell 242
microstructural effect
– effect of impurities in mixed conducting membranes on 524
– in asymmetric membranes 529 f
– in dual phase mixed conducting membranes 526

- in perovskite composite membranes with second phase 524 ff
- in perovskite-type mixed conducting membranes 521 ff, 530 f

microstructure analysis of PEMFCs 354 f
MIEC see mixed electronic-ionic conductor
migration of point defect 33
MIM structure see metal-insulator-metal structure
mixed conducting composite membrane 524
mixed conducting membrane
- characteristic thickness of 509, 513
- dual phase composite 526
- effects of impurities in solid electrolyte on 522
- grain boundary effects in 521 ff
- microstructural effects in 520 ff
- porosity effect on 517 ff
mixed conducting oxide material
- application of 501
mixed electronic–ionic conductor (MIEC) 266
mixed ionic–electronic conducting membrane 467
- based on brownmillerite-type oxide 469 ff, 478 ff
- based on fluorite-type oxide 468, 478 ff
- based on perovskite-type oxide 469 f, 474, 476 ff, 510 ff, 516
- dual-phase 468 f, 471, 481
- gas permeation through a 473 ff
- industrial application of 467, 490
- mechanism of 469
- single-phase 469
- stability of 483 f
- total conductivity of 477 f
mixed ionic–electronic conductor (MIEC) 10
mixed-mode oscillation (MMO) 134
modeling
- of electrolyte solutions 76 ff
- of solvent effects of water on electric double-layers 112 f
- of solvent effects on electric double-layers 109 ff
- of surface effects on mixed conducting membranes 510 ff
- of the conductivity of ionic salt-oxide composites 51 ff
- of the electrode–electrolyte solution interface 86 f
- of the interface between an ionic salt and an oxide 56
molecular dynamics (MD) simulation 100
- of ionic salt-oxide interfacial processes 56 f

- of the structure of sodium chloride aqueous solution at a goethite surface 113 ff
molten carbonate fuel cell (MCFC) 183 f, 213 ff, 310
- cathode for 214 f
- direct internal reforming (DIR-MCFC) 216 f
- stability of cell components 215 f
Monte Carlo (MC) computer simulation 100 f, 105
Moore's law 2
morphological instability of ceramic membrane material 437
Mott meta-I-insulator transition 3
Mott–Schottky model 35
MPL see microporous layer
MRAM see magnetic random access memory
MSA see mean spherical approximation
multilayer membrane structure 206
multi-walled carbon nanotube (MWCNT) 393, 395
MWCNT see multi-walled carbon nanotube

n

Nabarro-Herring creep 445
Nafion membrane 195, 205, 208, 232
NAND memory 1 f, 19
nanoBridge see cation-based resistive switching memory
nanocomposite solid electrolyte 62
- amorphous ionic salt phase in 45
- conductivity of 63
- Gibbs free energy of 42 f
- grain size effects in 42 ff
- phase transition in 43 f
nanodispersion 384, 401
nanoionic 430
nanostructured electrode 383
NDR system see negative differential resistance system
negative differential resistance (NDR) system 128 f, 169
negative electrode of lithium-ion batteries see anode
Nernst equation 181
Nexelion electrode 398
nickel oxide cathode for MCFCs 214
nickel/YSZ cermet anode 218 ff, 231, 241
NiO_x
- resistive switching characteristics of 7
N-NDR system 129, 143, 145 f, 160
noise-induced dynamics in electrochemical systems 167 ff
non-linear behaviour

– in applications of electrochemistry 167 f
– of homogeneous systems 127 ff
– of patterned systems 134 f
NOR memory 1 f, 19

o

ODE *see* system of ordinary differential equations
one-particle distribution function 77
on/off ratio *see* HRS resistance/LRS resistance ratio
operation voltage 15, 19
Ornstein–Zernike (OZ) equation 80 ff
ORR *see* oxygen reduction reaction
oscillation 126, 144 ff
– characterizing parameters of 148
– effect of electrode coupling on 154
– interpretation model of electrochemical 150
– timescale affecting the frequency of electrochemical 148
oscillator phase-locked 166
oscillatory electrochemical system 127
Ott–Grebogi–Yorke (OGY) feedback-type control method 162
overerasing 22
overprogramming 22
oxide solid electrolyte 10
oxygen chemical potential across a membrane 503, 507
oxygen content of ceramic materials 424, 440 ff
oxygen electrode 266, 317 f
– material for 266
oxygen exchange activation energy 512
oxygen exchange coefficient 508, 517
– measurement of 514
oxygen gas separation 502
oxygen ion conductor 449
oxygen ion migration 6
oxygen ionic conductivity 474
oxygen mobility 433
oxygen permeable membrane 436, 476, 502
– comparison with hydrogen permeable membrane 489 f
– electrical and ionic transport properties of 476 ff
– improving properties of 517 f
– improving the chemical stability of 483 f
– improving the mechanical strength of 484
oxygen permeation *see also* oxygen transport
– flux equation 475 f, 507
– influencing factors 502
– mechanism of 502
– membrane surface effects influencing 504 ff
– resistance 513
– transport equation of 510 f
– flux data 480, 482
– in reducing gases 482 f
– under air/inert gas gradients 479 ff
oxygen reduction reaction (ORR) 181, 188, 192 ff, 204, 218, 223 f, 283, 307, 364, 367
oxygen self-diffusion coefficient 514, 517
oxygen transport 467, 471, 473 f
– along grain boundaries 521 ff
oxygen vacancy 6, 447 f, 474
– diffusion coefficient 512

p

PAFC *see* phosphoric acid fuel cell
pair distribution function 76
partial oxidation of methane (POM) 491 ff
particle–particle and particle–mesh (PPPM or P3M) algorithm 104
pattern formation
– in electrochemical systems 158
– in oscillatory systems 159
– via synchronization 166
PCM *see* phase change memory
PEM *see* proton exchange membrane
PEMFC *see* polymer electrolyte membrane fuel cell
percolation threshold 50 ff
Percus–Yevick (PY) approximation 81
perfluorinated membrane 195
periodic forcing 160
permselectivity 502
perovskite-related cobaltite 267, 517
perovskite-related intergrowth oxide 470, 478
perovskite-type oxide material
– cation diffusion in 433
– creep parameter of 445 f
– fracture of 442 f
– microstructural phenomena in membranes comprising 521
– oxygen permeation through 509
– reaction with gaseous species 450f
– reaction with silica-rich sealing material 452
– thermodynamic stability of 447 ff
– volatilization of 453
perturbation of electrochemical systems 159 ff
phase change memory (PCM) 1 f
phase diagramm
– non-traditional 146, 149
phase transition

– in nanocomposite solid electrolytes   43 f
phosphoric acid fuel cell (PAFC)   183 f, 211 ff
pitting corrosion   168 f
planar multicell array (PMA)   243
PMA *see* planar multicell array
Poincaré map   140, 162
Poisson–Boltzmann (PB) approach to electrolyte solution   79
polarization curve   364
– N-shaped   129
– of a HT-PEMFC   240
– of a single hydrogen/air fuel cell   365
– of anode material for SOCFs   290 ff
– of electrocatalytic reactions in PEMFCs   199
– of mixed metal-oxide cathodes for SOFCs   276, 285
– S-shaped   130, 143
– Z-shaped   143
polyelectrolyte-functionalized multiwalled carbon nanotube (MWCNT)   198
polymer electrolyte   59
polymer electrolyte membrane
– degradation of   203
– layer-by-layer (LbL) self-assembly of   206
polymer electrolyte membrane fuel cell (PEMFC)   194 f
– by-products of   202
– comprising a stack of bipolar plates   200
– design of   200 ff
– electrode structure optimization   199
– fueled with liquid fuels   204 f
– micro-   243
– performance of   203
polymer-based memory   1 f
polymerized protonated polyaniline (PANI)/Nafion composite membrane   206
polytetrafluoroethylene (PTFE)-bonded electrode   188
POM *see* partial oxidation of methane
porosimetry of PEMFCs   356 f
positive electrode *see* cathode
potential IR drop   128
potential of zero charge (PZC)   78
potentiostatic electrochemical experiment   129
potentiostatic system   128
powder processing   416 f
power density of SC-SOFC   303 f
$Pr_{0.7}Ca_{0.3}MnO_3$ (PCMO) in resistive switching device   7
preparation
– corrugated thin-film SOFC   245
– of ceramic powder of complex oxides   417 f

– of electrochemical *in situ* nanodispersion   401 ff
– of lithium metal dioxide nanoparticles   385
– of lithium metal silicates   390 f
– of micro-SOFCs   243 ff
– of mixed oxide ceramic powders   471 f
– of nanostructured olivine phosphates   388 f
– of nanostructured quaternary lithium-mangan spinell   386 f
– of oxide solid electrolyte film   10 f
– of polymer electrolyte membranes   206 f
principal mode   142
programmable metallization cell *see* cation-based resistive switching memory
programming current   17 f
programming process *see* HRS-to-LRS switching
programming voltage   7
proportional feedback (SPF) algorithm   162
protic ionic liquid (PIL) electrolyte fuel cell   232
proton conduction mechanism   354, 360 f, 485
proton conductor   60, 416
proton exchange membrane (PEM)   184, 205, 238 ff
proton exchange membrane fuel cell *see also* polymer electrolyte membrane fuel cell (PEMFC)   183 f, 312
– analysis of temperature distribution in   373
– analysis of water management in   372
– architecture of   334
– characterization of   348 ff
– impedance spectra of   367
– IR transparent   373
– ohmic losses in   366
– porosity of   356
– properties of   331
– transport losses in   367 ff
– with embedded sensor   373
protonic-electronic conducting membrane
– dual-phase   469, 471
– electrical and ionic transport properties of   484 ff
– single-phase   469
– stability of   489
– total conductivity of   486
pseudocapacitance   145, 403
PTFE *see* polytetrafluoroethylene
Pyragas method   163
PZC *see* potential of zero charge

q

quasiperiodicity (irregular periodicity) 134

r

random diffusion in ionic memories 22 f
redox reaction 6
reference fluid density (RFD) approach to electrostatic free energy 92 ff
reprogrammable logic device 24
resetting process see LRS-to-HRS switching
resistive random access memory (RRAM) 1 f, 16
resistive switching memory see resistive random access memory
resistive switching process
– electronic effect 3, 19
– ionic effect 3, 19
– thermal effect 3, 19
restricted primitive model (RPM) of electrolyte solution 79 f, 89 f
retention 12 f
RFD see reference fluid density approach
RPM see restricted primitive model
RRAM see resistive random access memory
Ruddlesden–Popper oxide 470, 478

s

saddle node of periodic (SNP) orbits 133
scandia-stabilized zirconia (SSZ) electrolyte 272, 278
Schottky barrier 7
Schottky defect 34 ff, 428
SCL see space charge layer
screen printing 419
sealing for high-temperature gas permeation experiments 473
secondary ion mass spectrometry (SIMS) profiling analysis 514 f, 520
segmented flow-field plate 374 f
self-humidifying membrane 195, 199
self-inhibition 129
semiconductor 170
setting process see HRS-to-LRS switching
Si CMOS transistor 2
silicon-carbon-based anode 400
single-chamber solid oxide fuel cells (SC-SOFCs) 241
sintering
– cycle 421
– effect on morphology of heterogeneous composites 41 f

– of ceramic perovskite membrane material 424 f, 433
– of oxide electrolyte 424, 433
SiO_x 8
S-NDR system 130, 143, 145, 157
SNP see saddle node of periodic orbits
SOEC see solid oxide electrolysis cell
solid acid for HT-PEMFC 240 f
solid electrolyte 8 f
– apatite-type 285 f
– bismuth-vanadate (BIMEVOX) 286
– composite 31 ff
– for DCFCs 246
– for high-temperature steam electrolysis 317 f
– Ge-based chalcogenide 9
– grain size effect in nanocomposite 42 ff
– ionic conductivity of nanocomposite 45, 47, 63
– ionic transport in 299
– lanthanum-strontium-gallium-magnesium-oxided (LSGM) 268 ff, 279
– linear thermal expansion coefficient (TEC) of 438
– metacomposite 62, 64
– metal sulfide 9
– nanocomposite 62 ff
– phase transition in 43 f
– preparation of thin film of 10 f
– switching parameters of 13
solid electrolyte for SOFCs 218
– $(Ce,Gd)O_{2-\delta}$ (CGO) 268 ff, 272, 276 ff, 428, 445, 527
– $(CeNd)O_{2-\delta}$ (CNO) 268, 272
– $CeSmO_{2-\delta}$ (CSO) 267, 269, 272, 275, 278, 288
– $(La,Sr)(Ga,Mg)O_{3-\delta}$ (LSGM) 268 ff, 272, 276, 279 f, 288, 445
– LSAO 270 f
– micro-SOFC electrolyte 307
– scandia-stabilized zirconia (SSZ) 278, 281
– SC-SOFC electrolyte 305
– thermal expansion of 437
– yttria-stabilized zirconia (YSZ) 267 ff, 272, 280 ff, 288, 428, 434, 445
solid-electrolyte solution interface see also solid-fluid interface 74 ff
– integral equations describing the 80 ff
solid-fluid interface 74 ff
solid-fluid interface interaction theoretical approach to 74 ff

solid oxide electrolyte 10
– effect of temperature on conductivity of 64
– thermodynamic stability of 447 f
solid oxide electrolyzer (SOEC) 265
solid oxide fuel cell (SOFC) 64, 169, 183 f, 217 ff, 415
– anode materials for 218 ff
– anode-depending power densitiy of various cell compositions 280 ff
– anode-supported 230 f
– application of 247 ff, 319 f
– cathode materials for 222 ff
– cathode-supported 230
– chemical compatibility of materials for 443
– decomposition due to component volatilization 453
– decomposition due to reaction with gaseous species 450 f
– decomposition due to reaction with sealant 452
– degradation of 442 f, 450 f
– delamination of 437
– direct carbon (DC) 310 f
– direct hydrocarbon 309 f
– electrochemical reactors based on 320
– electrolyte materials for 225 ff
– electrolyte-supported 230
– interconnect-supported 230
– merged 310
– micro- 243, 306 ff
– oxide interconnect materials for high-temperature 228 f
– porous substrate-supported 230
– proton-conducting (PC) 312 ff
– reaction of cell components with chromium 453
– sealant material for 229
– single-chamber (SC) 241 f, 267, 301 ff
– stack design of 231
– symmetrical 308 f
solid proton-conducting (PC) electrolyte 312 f
solid–solid interaction energy 75
solid-state electrolyte memory see cation-based resistive switching memory
solid-state electrolyte see solid electrolyte
solid-state sensor 171
solid-state sintering 420 f
solvent primitive model (SPM) 110
space charge layer (SCL) 33
space charge model 32
spark plasma sintering (SPS) 422
spatiotemporal self-organization 136
SPF see proportional feedback algorithm
spin coating 419

SPM see solvent primitive model
spontaneous amorphization in nanocomposites 45 ff
spray coating 419
SPS see spark plasma sintering
SRAM see static random access memory
$Sr_4Fe_6O_{13\pm\delta}$ 470, 478
SRM see steam reforming of methane 491 ff
stand-alone memory 25
standard three-electrode electrochemical cell 136
standing wave 159
state of a dynamical system 126, 134
static random access memory (SRAM) 1 f, 19, 26
steady-state in two-variable dynamical systems 132
steam reforming of methane (SRM) 7, 483, 491 f
Stern model 33 ff
storage class memory 25
subcell 374
superionic metal sulfide 8
superionic oxide 34 f
superionic phase 47
surface charge of an ionic crystal 33
surface charge reversal 83
surface defect
– energy diagram of 37 f
surface diffusion 420
surface exchange limitation 503
surface fracture of ceramic materials 442
surface morphology characteristics of PEMFCs 348 ff
surface oxygen exchange 501 ff
– effect of membrane porosity on 518 f
– improvement of membranes for 517 f
– rate of 515 f
– reduction of surface exchange limitations 517 f
– theoretical analysis of influencing factors 504
surface potential
– of a doped ionic crystal 36
– of heterogeneous ionic crystal composites 38
– of a pure ionic crystal 35
surface spreading 39 f
sweep gas 487 f
switching current 15
switching cycle 15 f
switching energy 19
switching kinetics 17
switching parameter 12 f, 19

– of solid electrolytes   13 f
switching speed of ionic memories   19, 24
switching time   12 f
switching voltage   15
synchronization
– of chaotic oscillations   156
– of oscillators   153
– engineering   164
– via delayed feedback method   163
syngas production   320
system of ordinary differential equations (ODE)   131

t

Ta_2O_5 8
tape casting   419
testing of PEMFCs in operation
– chemical methods   371 f
– physical methods   371 f
– unsing electrochemical techniques   364 f
thermal expansion of oxide electrolyte   437
thermal stability
– of ionic memories   23
thermal stress   442
thermocouple   373
thermodynamic characteristics of interfaces   37 ff
thermodynamic stability   267, 279, 283, 287
– of an oxygen ion conductor   449
– of fluorite-type solid-oxide electrolyte   450
– of perovskite-type solid-oxide electrolyte   447 ff
thermodynamic stability criteria   39 f
thermodynamics of nanocomposite formation   31 ff, 39 ff, 64 f
thermomechanical stress   439
thickness of the solid-state electrolyte   12 f
three-dimensional stackable memory   1
timescale of non-linear behaviour   127, 139

tin composite oxide anode   398
TiO_2
– resistive switching characteristics of   7
transistor density   2
1-transistor–1-resistor structure (1T1R)   19
transition metal oxalate   404 ff
transmission line model of CLs   368
transport equation   510 f
tristability   143
turbulence   159
Turing mechanism   134
Turing pattern   159
two-terminal selection device   21
two-terminal structure   1
two-variable approximation of a multi-variable dynamical system   131 f

u

unstable periodic orbit (UPO)   162
UPO see unstable periodic orbit

v

vacancy migration   6
vacuum slip casting   419
vapor–liquid phase transition   92 ff, 97
voltammetry   369 f
volume thermal expansion   437

w

warm pressing   419
wetting effect   40, 358
writing process see HRS-to-LRS switching

y

Youngs's modulus of ceramic materials   439, 441 f
yttria-stabilized zirconia (YSZ)   218 ff, 226, 241, 516